Specialty Polymers

Specialty Polymers

Materials and Applications

Editor

Faiz Mohammad
D.Phil. (Sussex)

First published in the UK by

ANSHAN LTD
In 2007

6 Newlands Road,
Tunbridge Wells,
Kent.
TN4 9AT. UK

Tel: +44 (0) 1892 557767
Fax: +44 (0) 1892 530358

e-mail: info@anshan.co.uk
Web Site: www.anshan.co.uk

ISBN – 978 1904798 958

© 2007, I.K. International Publishing House Pvt. Ltd.

British Library Cataloguing in Publication Data
A catalogue record for this book is available from the British Library

Printed in India

Professor Peter J.S. Foot
Materials Research Group, SPC,
Kingston University London, Penrhyn Road,
Kingston upon Thames KT1 2EE (UK).
Tel. : (+44) 2085477485
Fax : (+44) 2085477562
Email : p.foot@kingston.ac.uk

Peter J.S. Foot

Foreword

During the initial period, the synthetic polymer industry was regarded with disapproval in many quarters. However, by the 1930s it became clear that without it, the material needs of mankind would soon outstrip the natural polymer supplies for clothing fibers, non-metallic structural materials and elastomers. During the World War II, several countries were forced to develop synthetic plastics and rubbers to replace unavailable natural commodities and this process turned into an apparently unstoppable expansion in the polymer industry.

Such an explosive growth has now ceased, but still the industry continues to expand steadily. Most of the excitement, in terms of scientific discovery and technical innovation, comes from Specialty Polymers, such as those discussed in the present book. The developments are driven by factors such as the global need for more efficient energy generation and storage, major biomedical and pharmaceutical breakthroughs, safe and clean new transportation systems and breathtaking advances in mass telecommunications and information technology. In all the cases, the technical requirements are increasingly stringent and the keynote is intelligent materials design, based on an unprecedented, sustained interdisciplinary cooperation. There have also been important developments in microprocessing and microstructural control, enabling full advantage to be taken of the improved molecular features.

Various aspects of the polymer-based advances are very well represented by the choice of topics in "**Specialty Polymers: Materials and Applications**". The international team of academic and industrial authors includes experts representing the major interacting disciplines of organic synthesis, polymer science, physics, engineering and medicine. The role of novel polymers and polymer composites in current applications is reviewed in detail, and the creation and structure-properties relationships of materials arising from on-going fundamental research are discussed. The pace of developments in this broad field being very rapid, it will be timely and helpful to have such a book, which will be of use to academic scholars, interested industrialists and postgraduate students as well as to advanced undergraduate students considering a career in advanced materials.

Peter J.S. Foot

Preface

Hermann Staudinger, a German chemist, was the first to propound that the macromolecular (polymeric) chemical structure for natural rubber is made up of many isoprene units and thus, macromolecules are composed of more than 10,000 atoms. In 1953, he received the Nobel Prize for this research, which countered the prevailing understanding that macromolecules are collections of small molecules. Later on, synthetic polymeric materials began replacing natural polymeric materials owing to their light weight, mechanical strength and environmental stability.

Moreover, chemists, material scientists, engineers and technologists, all joined together in the research and development work on polymers. Their systematic research on the preparation, characterization and utilization of polymers resulted in new, better and often a set of several desirable properties in a single polymer and the polymers started finding place in various engineering applications. The last few decades have witnessed a closer partnership between scientists, engineers, technologists and industrialists, as polymers have become materials of common interest. Nowadays, polymers have found practical applications in almost every walk of life including automotives, packaging, sanitary wares, detergents, cosmetics, pharmaceuticals, semiconductor devices, information technology, wireless and telecommunication etc.

Although the bulk of polymer production comprises of simple commodity polymers, the proportion of specially designed or tailor-made polymers for specific and sophisticated applications is increasing rapidly. The use of specialty polymers in specific and sophisticated applications holds the key to continuous scientific growth and technological advances in the new millennium. **Specialty Polymers: Materials and Applications** has been compiled to meet the need of the time.

"A good book is the purest essence of a human soul"   **—Thomas Carlyle**

The present book is the result of the contributions of 35 experts from international scientific community. This book thoroughly covers the latest topics on specialty polymers and their applications in the most sophisticated and specialized areas. It gives the latest and in-depth coverage to the chemistry, physics, material science, and technology and device applications of specialty polymers. Based on more than 2500 references, 250 figures, 50 tables, 140 equations, 15 schemes and 45 chemical structures, this book will certainly prove thought-provoking to the researchers working in the fields of chemistry, biochemistry, biotechnology, medicine, polymer chemistry, semiconductor physics, material science, electrochemistry, biology, electronics, photonics, solid state physics, nanotechnology, electrical and electronics engineering, optical engineering, device engineering, data storage etc. Based on thematic topics, it contains the following 16 chapters:

Chapter 1: *Liquid Crystal Conducting Polymers* contains an overview of research on the effects of imparting liquid crystalline properties to electronically conducting polymers.

Chapter 2: *Polyanilines: Materials and Applications* describes the synthesis, characterization and most promising applications of polyanilines, such as in microelectronics, lithography, resists, coatings, ultra-thin membranes, secondary batteries, multi-color electrochromic devices, conductive textiles and so on.

Chapter 3: *Polymer Nanofibers: Fabrication, Applications and Characterization* provides a review on the fabrication techniques with special emphasis on electrospinning, applications of polymer nanofibers and characterization techniques pertaining to single nanofibers and nanofibrous mats or foams.

Chapter 4: *Magnetic Polymer Microspheres* summarizes the structure, preparation methods and applications of magnetic polymer microspheres. Magnetic polymer microspheres are finding applications in many areas such as in biomedical field, organic synthesis, waste water treatment and so on, because of their two very important properties. First that they can be rapidly and easily separated by application of magnetic field and second that their surface can be modified to different functional groups.

Chapter 5: ***Intrinsically Conducting Polymers for Metallic Corrosion Protection*** reviews the metallic corrosion protection properties and mechanisms of intrinsically conducting polymers (ICPs) as well as their composites with special emphasis on polyaniline (PANI) and polypyrrole (PPY), which can be synthesized chemically or electrochemically.

Chapter 6: ***Inorganic Polymers for Advanced Applications*** covers the main group metalloid-containing inorganic polymers, which are made up of boron (group 13: polyborazines), silicon/germanium/tin (group 14: polysilanes, polygermanes, and polystannanes) and phosphorous (group 15: polyphosphazenes) atoms as advanced materials, with special emphasis on silicon-containing polymers with unusual properties.

Chapter 7: ***Light Weight Polymer Composite Materials for Automotive Industry*** reviews the important issues in automotive composites such as design, manufacturing, material selection, the use of computers in automotive composite development, concurrent engineering, natural fiber composites in automobiles, recycling of automotive composites and future trends. A brief history and definition of automotive polymer composites and the benefits and disadvantages of polymer composites in automotive industry are also covered.

Chapter 8: ***Conducting Polymer-Based Sensors*** illustrates the possibilities offered by conducting polymers in the area of chemical sensors by using various examples.

Chapter 9: ***Photo-Converters Based on Dye-Doped Polymers*** considers the main approaches to molecular design of dye-doped polymers matrices. It also discusses the prospects for the applications of dye-doped polymer materials as passive Q-switches of solid-state lasers, as active laser media, as luminescent solar converters, as photovoltaic cells and as electroluminescent emitters. The advantages of such matrices over polymers and dyes are also analyzed in it.

Chapter 10: ***Polymers in Medicine*** reviews the use of polymers to fabricate biomedical materials and devices for tissue engineering, gene delivery, wound treatment, heart valves, angioplasty balloons and stents, nerve regeneration, drug delivery, breast implants, dental devices, orthopedic devices, ophthalmic devices, gynecological devices etc.

Chapter 11: ***Luminescent Polymers*** covers the synthesis of substituted conjugated polymers to regulate the emission color and improve the EL efficiency and processibility of the luminescent polymers.

Chapter 12: ***Polymers in Electronics*** explains the role of conducting polymers in various electronic devices such as LEDs, photovoltaic cells and FETs.

Chapter 13: *Ion Conducting Polymers* describes the preparation and characterization of polymer gel electrolyte systems viz., PVDF:(PC+DEC):LiClO$_4$ and P(VDF-HFP):(PC+DEC):LiClO$_4$. It also illustrates the results of two polymer gel electrolyte systems with regard to ionic conductivity, X-ray diffraction, fourier transform, infra-red spectroscopy and scanning electron microscopy. This is done keeping in view their applications in electrochemical energy storage and conversion devices such as advanced rechargeable lithium batteries and fuel cells, power source for wireless telecommunication devices and portable information technologies, electric and hybrid vehicles, electrochemical sensors and so on.

Chapter 14: *Specialty Coatings and Adhesives* discusses the specific areas in solvent borne coatings, waterborne coatings, UV radiation curable coatings, which are ripe for future innovations and further development. These are discussed with special emphasis on the recently developed, sustainable resource based coating materials. The structure and reactivity associated with the alkyd, polyester, acrylic, amino, epoxy and polyurethane resins, commonly used in coating formulations, are described with preparation, performance and cure reactions in the first part. The second part covers a brief history of development of various primitive and advanced adhesives such as epoxy, phenol formaldehyde and urea formaldehyde, with special emphasis on conductive adhesives, waterborne adhesives and sustainable resource-based adhesives as well as their characteristic properties, with a view on their significance in new technology.

Chapter 15: *Recombinant Polypeptides in Therapeutics* provides the latest general information on various aspects of therapeutic applications of selected rDNA products, especially hormones of therapeutic interest, haemopoietic growth factors, human blood coagulation products, thrombolytic agents, anticoagulants, human interferons, human interleukins and therapeutic enzymes, in view of the increasing importance of rDNA products for human healthcare.

Chapter 16: *Electrically Conducting Polymer Composites* details the different methods for producing electrically conducting polymer composite materials, their electrical properties and a variety of applications.

This book has indeed been the result of remarkable cooperation of many distinguished experts, who came together to contribute comprehensive, in-depth and up-to-date review chapters. I am thankful to all the contributing authors and their co-authors for their valued contributions to the book. I would also like to express my gratitude to all the publishers, authors and others for granting us the copyright permissions to use their illustrations, especially, AAAS, Abdul Hadi Abdullah, American Chemical Society, American Institute of Physics, BCC Research, Cambridge University Press, Elsevier Science Ltd., Edisyams

Zainudin, Tuan Haji Nordin Osman, IEEE, IOP Publishing Ltd., IUPAC, John Wiley & Sons Inc., Kluwer Academic Publishers, Kvantovaya Elektronika and Turpion Ltd., Lee Ho Boon, MAIK "Nauka/Interperiodika", Nature, Naukova Dumka, Kiev, Nobel Foundation, Oficyna Wydawnicza Politechniki Wroclawskie, Old City Publishing, Saad Assi Mutasher, Saifulla Azizi, SciFinder Scholar, Sharifa Imihezri, Society of Chemical Industry, Springer-Verlag, St.Jude Medical, Inc., Sudan Engineering Society, Taylor & Francis Group LLC, The Chinese Society for Corrosion and Protection, The Electrochemical Society, Inc., The Society of Polymer Science, Japan and Wiley-VCH Verlag Gmbh & Co. KG. Although sincere efforts were made to obtain the copyright permissions from the respective owners and to include the citation with the reproduced material, I would still like to offer my sincere apologies to any copyright holder whose rights may have been unknowingly infringed.

I would also like to take this opportunity to express my gratitude to my mentors, Mr. Dinesh Chandra Jain (my Science teacher), Mr. Noor Mohammad (my elder brother and Math's teacher), Professor (Dr.) Wasi-ur-Rahman (Former Vice-Chancellor of Aligarh Muslim University), Dr. M.C. Nigam (Former Scientist of CIMAP, Lucknow), Dr. S.P. Manik (Former Executive Director of RDSO, Lucknow), Professor (Dr.) N.C. Billingham (University of Sussex) and Mr. Mohammad Khaleel Siddiqui (my local guardian during my stay at Sussex University). My special thanks also to my departmental colleagues Dr. Ali Mohammad, Dr. M.A. Quraishi, Dr. Asif Ali Khan, Dr. R.A.K. Rao, Dr. M. Mobin, Dr. M.Z.A. Rafiqui and Dr. Rais Ahmad, without whose continuous encouragement this book would have not been brought to its final form. Thanks are also due to my research students, Mr. Mohammad Khalid, Ms. Atika Khatoon, Mr. Shahid Parvez and Mr. Mohd. Omaish for their assistance from time to time. I would also like to acknowledge the sincere efforts of Ms. Sonia Mamgain, Mr. Arvind Mishra (from I.K. International Publishing House Pvt. Ltd.) in evolving this book into its final shape. Last but not the least, I am highly appreciative of the support of my brothers, sisters and the love of my family members, Shahina, Asif, Fatima, Tahir, Haleema, Zaib and Talib, during this exciting experience.

My sincere gratitude is also due to Prof. P.J.S. Foot for his insightful foreword.

Faiz Mohammad
Editor

Contributors

Ahmad, Ayesha
E-9, Johri Farm, Okhla, New Delhi—110025, India.

Ahmad, Mohammad
A-13, Medical Colony, Aligarh Muslim University, Aligarh—202002, India.
Email: mahmad202@yahoo.com

Ahmad, Sharif
Materials Research Laboratory, Department of Chemistry, Jamia Millia Islamia, New Delhi—110025, India.
Email: sharifahmad_jmi@yahoo.co.in

Al-Ahmed, Amir
Sensor Research Laboratory, Department of Chemistry, University of the Western Cape, Private Bag X17, Bellville—7535, Republic of South Africa.
Email: amiral@rediffmail.com

Bobacka, Johan
Åbo Akademi University, Process Chemistry Centre, Laboratory of Analytical Chemistry, Biskopsgatan 8, FIN-20500 Turku-Åbo, Finland.
Email: jbobacka@abo.fi

Brown, John W.
Materials Research Group, School of Chemical and Pharmaceutical Sciences, Kingston University London, Penrhyn Road, Kingston upon Thames KT1 2EE (UK).

Ding, Xiaobin
Chengdu Institute of Organic Chemistry, Chinese Academy of Sciences, Chengdu, 610041, People's Republic of China.
Email: xbding@cioc.ac.cn

Foot, Peter J.S.
Materials Research Group, School of Chemical and Pharmaceutical Sciences, Kingston University London, Penrhyn Road, Kingston upon Thames KT1 2EE (UK).
Email: p.foot@kingston.ac.uk

Goyal, Saurabh
Department of Electrical Engineering, IIT Bombay, Mumbai—400076, India.

Hou, Jianhui
CAS Key Laboratory of Organic Solids, Institute of Chemistry, Chinese Academy of Sciences, Beijing—100080, China.

Inamuddin
Analytical and Polymer Research Laboratory, Department of Applied Chemistry, Faculty of Engineering and Technology, Aligarh Muslim University, Aligarh—202 002, India.

Ishchenko, Alexander
Institute of Organic Chemistry, National Academy of Sciences of Ukraine, 5, Murmanskaya St., Kiev-94, 02094, Ukraine.
Email: alexish@i.com.ua

Ivaska, Ari
Åbo Akademi University, Process Chemistry Centre, Laboratory of Analytical Chemistry, Biskopsgatan 8, FIN-20500 Turku-Åbo, Finland.

Jing, Xinli
Department of Chemical Engineering, School of Energy and Power Engineering, Xi'an Jiaotong University, Xi'an, 710049, People's Republic of China.
Email: xljing@mail.xjtu.edu.cn

Khan, Rizwan H.
Interdisciplinary Biotechnology Unit, Aligarh Muslim University, Aligarh—202 002, India.
Email: rizwanhkhan@hotmail.com

Kim, Bo-Hye
Department of Korean Medicinal Supply, Dongshin University, Naju—520 714, Korea.

Kumar, Ashok
Department of Physics, Tezpur University, Tezpur—784028, Assam, India.
Email: ask@tezu.ernet.in

Lambe, Gillian J.
Materials Research Group, School of Chemical and Pharmaceutical Sciences, Kingston University London, Penrhyn Road, Kingston upon Thames KT1 2EE (UK).

Li, Yongfang
CAS Key Laboratory of Organic Solids, Institute of Chemistry, Chinese Academy of Sciences, Beijing—100080, China.
Email: liyf@infoc3.icas.ac.cn

Lim, Chwee Teck
Division of Bioengineering, National University of Singapore, Singapore—117576.
Department of Mechanical Engineering, National University of Singapore, Singapore—117576.
Nanoscience and Nanotechnology Initiative, National University of Singapore, Singapore—117576.
Email: ctlim@nus.edu.sg

Lindfors, Tom
Åbo Akademi University, Process Chemistry Centre, Laboratory of Analytical Chemistry, Biskopsgatan 8, FIN-20500 Turku-Åbo, Finland.

Mohammad, Faiz
Department of Applied Chemistry, Aligarh Muslim University, Aligarh—202002, India.

Nayeem, M.S.
Department of Chemistry, Aligarh Muslim University, Aligarh—202 002, India.

Ramakrishna, S.
Division of Bioengineering, National University of Singapore, Singapore—117576.
Department of Mechanical Engineering, National University of Singapore, Singapore—117576.
Nanoscience and Nanotechnology Initiative, National University of Singapore, Singapore—117576.

Rao, V. Ramgopal
Department of Electrical Engineering, IIT Bombay, Mumbai-400076, India.
Email: rrao@ee.iitb.ac.in

Riaz, Ufana
Materials Research Laboratory, Department of Chemistry, Jamia Millia Islamia, New Delhi—110025, India.

Saikia, D.
Department of Physics, Tezpur University, Tezpur—784028, Assam, India.

Sapuan, S.M.
Department of Mechanical and Manufacturing Engineering, University Putra Malaysia, 43400 Serdang, Selangor, Malaysia.
Email: sapuan@eng.upm.edu.my

Syed, Akheel Ahmed
Department of Studies in Chemistry, University of Mysore, Manasagangotri, Mysore—570006, India.
Email: syedakheelahmed@hotmail.com

Tan, E.P.S.
Division of Bioengineering, National University of Singapore, Singapore—117576.

Taqui, Syed Raihan
Department of Polymer Science and Technology, S.J. College of Engineering, Manasagangotri, Mysore—570 006, India.

Wang, Yangyong
Department of Chemical Engineering, School of Energy and Power Engineering, Xi'an Jiaotong University, Xi'an, 710049, People's Republic of China.

Woo, Hee-Gweon
Nanotechnology Research Center and Institute of Basic Sciences, Department of Chemistry, Chonnam National University, Gwangju—500 757, Korea.
Email: hgwoo@chonnam.ac.kr

Zhang, Junhua
State Key Laboratory of Polymer Material and Engineering, Sichuan University, Chengdu, 610065, People's Republic of China.

Zhang, Y.Z.
Division of Bioengineering, National University of Singapore, Singapore—117576.
Department of Mechanical Engineering, National University of Singapore, Singapore—117576.

Contents

1

Liquid Crystal Conducting Polymers

Gillian J. Lambe, John W. Brown and Peter J.S. Foot[*]

Materials Research Group, School of Chemical and Pharmaceutical Sciences,
Kingston University London, Penrhyn Road, Kingston upon Thames KT1 2EE (UK).
[*]Corresponding author's email: p.foot@kingston.ac.uk

Contents

Summary: In this chapter, research on the effects of imparting liquid crystalline properties on electronically conducting polymers is reviewed. The polymers generally have conjugated backbones such as those of polypyrrole, polythiophene or polyacetylene, and the liquid crystallinity is typically provided by a side-chain comprising a flexible spacer group culminating in a polarisable mesogen such as cyanobiphenyl. The aim of this research has been to take advantage of the self-organising properties of liquid crystals and their response to magnetic, electric or mechanical stress fields, in order to control or enhance the electrical properties of the conducting polymers backbone.

1. INTRODUCTION

Polymers with significant electronic conductivity have developed from laboratory curiosities to important commercial materials in the span of last 25 years [1-4]. By the early 1980s a wide range of conducting polymers had been synthesised, primarily those based on polyacetylene, poly(p-phenylene), polythiophene and polypyrrole. These materials had attracted wide attention in terms of basic science, and were already showing promise for applications extending from light-weight components in batteries and fuel cells, to antistatic coatings and materials for non-linear optics. Considerable effort was directed towards the production of materials which combined the inherent advantages of a conjugated backbone with processibility [5]. The most common and most successful approach to this objective was to attach pendent side groups to the aromatic system. Thus, for example, polymers based on acetophenone [6] and β-decyl pyrrole had been prepared and shown, in their undoped states at least, to be soluble. An extension of this idea was to incorporate side groups that in addition to improving the processibility of the conducting polymer systems, could also improve the materials' overall properties and offer the potential for mesophase formation.

2. AN OVERVIEW OF RESEARCH ON LCCPs

In 1989, le Moigne *et al.* [7] reported on the synthesis and polymerisation of new acetylenic and diacetylenic monomers incorporating mesogenic groups based on cholesteryl and methoxybiphenyl. Their structural and thermal behaviour was described and their mesomorphic properties were found to depend largely on the length and flexibility of the spacer. From the results outlined in this paper, it is clear that the role of the aliphatic chain is important in this series of monoacetylenic derivatives containing the cholesteric moiety. When there is no aliphatic chain, no stable mesophase is observed. When the aliphatic chain is short, only a cholesteric phase appears, and when it is long, the compound exhibits a stable smectic A phase. It was also interesting to note the absence of a liquid crystalline phase when the cholesteric moiety was replaced by methoxybiphenyl, the length of the aliphatic chain being kept the same. Microscopic orientation had been observed and obtained through the alignment of molecules in a liquid-crystalline state. Also the concept of "freezing-in" such a mesomorphic state by the polymerisation of a liquid crystal monomer had already been applied successfully in several cases [8].

1990 saw considerable attention being devoted to the production of materials which combined the inherent advantages of a conjugated backbone with processibility. The most common and most successful approach to this objective was to attach pendent side groups to the aromatic system [9,10].

In 1993, Langley *et al.* [11] attempted to develop synthetic routes to polypyrroles with pendent mesogenic groups (**Fig. 1.1**). The monomers were found to exhibit a monotropic liquid crystal phase and the ultimate aim was to convert liquid crystal pyrroles to conducting polymers. It was intended

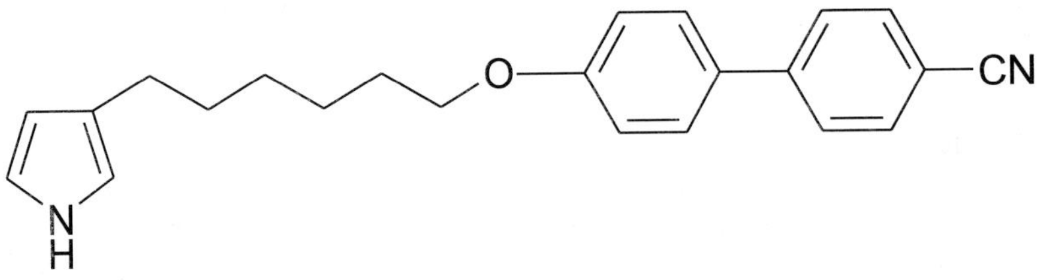

Fig. 1.1. Pyrrole with pendent mesogenic group attached.

that the presence of a mesophase could be utilised to improve the properties of the pyrrole-based polymers, by, for example, increasing the degree of order, which was known to improve conductivity.

However, the formation of the monotropic liquid crystal phase presented potential difficulties in the utilisation of the mesophase for forming conducting polymers. In this respect, the monomer described here is inferior to analogous thiophene derivatives. The low stability of the phase in the pyrrole case was presumed to be due to the availability of the N-H unit for intermolecular H-bonding, whereas in the case of thiophene derivatives such interactions are not possible [12,13]. Langley *et al.* [14] and Jego *et al.* [15] succeeded in synthesising β-substituted pyrroles with pendent mesogenic units; however the stability of the mesophase was apparently reduced by the presence of hydrogen bonding. This makes the use of a liquid crystalline phase to organize pyrrole systems rather more difficult than with polythiophenes [16]. However, relatively stable liquid crystal phases were achieved via mixing the polymer with small quantities of a low molar mass material with a stable liquid crystal phase.

In 1994, work published by Vicentini, Mauzac and Laversanne [17] reporting on the synthesis of new polyacetylene and polydiacetylene monomers substituted by mesogenic group. The monomers prepared were shown to exhibit thermotropic mesomorphic behaviour over a temperature range allowing polymerisation in an oriented phase. The following acetylenic series A_3C_m (**Fig. 1.2**) was studied. Some of the polymers were found to be liquid crystalline, showing a smectic phase for m = 1, 4 and 8. In the case of diacetylenic polymers, no mesomorphic behaviour was detected by optical observation using a polarising microscope. The electrical conductivity of the polymer A_3C_8 was studied under different conditions. The DC conductivity was 10^{-11} Scm^{-1} for the undoped sample but after doping with iodine vapour, it increased to 5×10^{-8} Scm^{-1}. The undoped liquid crystal polymer was also oriented in a magnetic field of 5T and heated into the isotropic state; after freezing in the liquid crystalline state a conductivity of 10^{-9} Scm^{-1} was achieved. These preliminary results showed for the first time without doubt that orientation of a side chain liquid crystal conjugated polymer could enhance its conductivity.

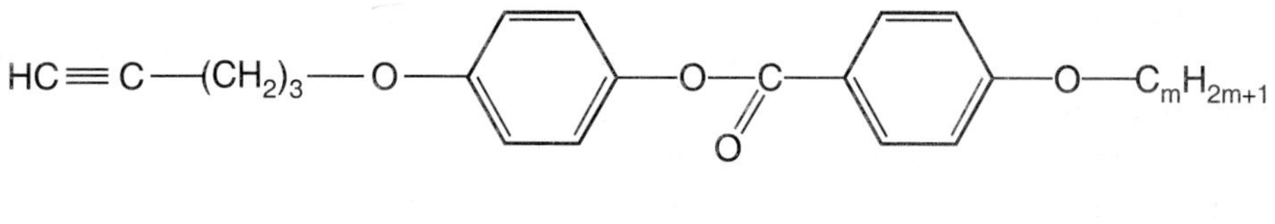

Fig. 1.2. Acetylenic series (A_3C_m) synthesised by Vicentini *et al.* [17].

By 1995, many insulating polymer liquid crystals, with flexible main chains and side chains containing mesogenic units, had been studied. However, only a few papers had described the synthesis and characterisation of polymeric liquid crystals with rigid main chains [18]. Yoshino *et al.* [19] highlighted the unique electrical and optical properties of the liquid crystalline phase of polyacetylene derivatives substituted with long side chains containing the mesogenic unit poly[5-(p-(*trans*-4'-n-octylcyclohexyl)phenoxy)-1-pentyne] (PPCH803A).

Much attention was devoted by Akagi *et al.* [20] to the synthesis and properties of liquid crystalline polyacetylene derivatives. They synthesised mono-substituted acetylenes having a liquid crystalline functionality, and then polymerised them with Ziegler-Natta and metathesis catalysis. The liquid crystalline group in the side chain of the co-polymer was composed of a cyclohexylphenoxy

(PCH) or biphenoxy (BP) moiety as a mesogenic core, a methylene chain $[-(CH_2)_3-]$ as a spacer, and an alkyl chain as a terminal group **(Fig. 1.3)**. All polymers exhibited a fan-shaped texture under the polarising optical microscope (characteristic of a smectic liquid crystalline phase), at a temperature that was consistent with DSC measurements. Electrical conductivities were evaluated following iodine doping, before and after magnetically induced alignment.

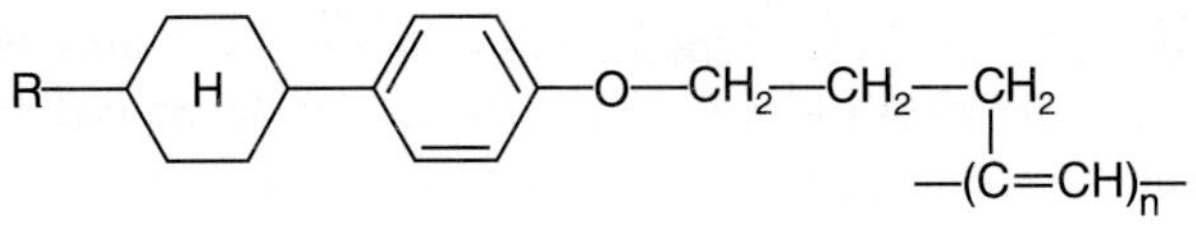

PPCHn03A $R = C_2, C_3, C_5-C_8$

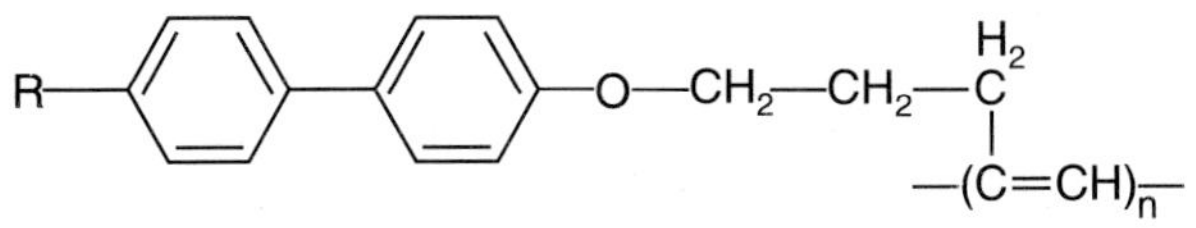

PBP503A $R = C_5$

Fig. 1.3. Polyacetylene derivatives. PPCHn03A: PCH stands for a phenylcyclohexyl moiety as a mesogenic core, also included is a methylene chain linked with an ether type oxygen atom, $-(CH_2)_3O$, as a spacer and an alkyl chain $-C_nH_{2n+1}$, where n = 2, 3 or 5-8 as a terminal group. PBP503A: BP stands for a biphenyl (BP) mesogen, methylene chain linked with oxygen as a spacer, and an n-pentyl group as a terminal moiety.

PCHn03A monomer showed schlieren textures characteristic of nematic phases in the polarising microscope. However, these mesophases were only observed in the cooling process, indicating monotropic nature. This is of particular interest because the PCH monomers, except those with shorter length of terminal alkyl group, show nematic phases while all polymers show smectic phases, which have a higher order than the nematic phase. This can be rationalised with a polymerisation effect i.e. a higher order molecular arrangement and/or packing was generated by the polymerisation of liquid crystalline molecules.

Orientation of the molecules was examined by shear stress and by magnetic field. The polymers showed not only an enantiotropic smectic A phase, by virtue of a side chain composed of phenylcyclohexyl or biphenyl mesogenic moiety, but also electrical conductivity upon iodine doping. Electrical conductivities of the polymers were respectively 10^{-8} and 10^{-7} Scm^{-1} upon iodine doping of the cast films. The alignment of the polymer chain accomplished by the side-chain orientation using an external magnetic field of 0.7 Tesla increased the electrical conductivity up to 10^{-6} Scm^{-1} and gave rise to a significant electrical anisotropy.

In the same year, Vicentini *et al.* [21] reported a new class of side chain liquid crystalline polymers containing a polypyrrole main chain **(Fig. 1.4)**. These materials were shown to be promising because they combined orientational behaviour and anisotropic electrical properties.

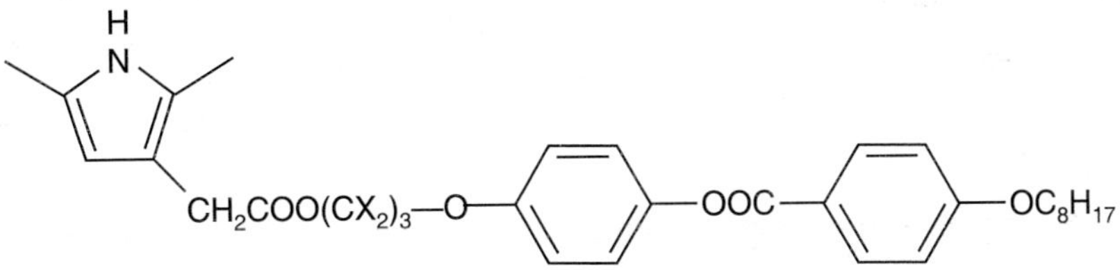

$$CH_2COO(CX_2)_3 - O - \langle C_6H_4 \rangle - OOC - \langle C_6H_4 \rangle - OC_8H_{17}$$

Where X = H for the hydrogenated polymer

(PPyCH) and X = D for the deuterated species (PPy(D))

Fig. 1.4. Substituted polypyrroles.

The substituted polypyrroles appeared to be good candidates for liquid crystal conducting polymers because of the known stability of the doped state in polypyrroles and their resistance to oxidation [22]. After polymerisation in the presence of oxidising agents, side-chain liquid crystal doped conductive polypyrroles were obtained [23]. All the polypyrroles obtained exhibited smectic A mesophases, and X-ray measurements confirmed their presence. In addition, the side-chain liquid crystal polypyrrole was aligned using a strong magnetic field [24]. Polymerisation by iodine vapour was performed in bulk, allowing the reaction to be effected on the oriented mesophase of the monomer. Room temperature conductivities of the polypyrrole synthesised by either process reached about 10^{-2}-10^{-4} Scm^{-1}. Conductivity was favoured in the plane perpendicular to the mesogen orientation axis.

The above authors' aim was to determine the conformation of the polypyrrole main chain in the oriented S_A phase in order to know if the mesogen/main chain anisotropy was responsible for the conduction anisotropy. Small-angle neutron scattering clearly established that the anisotropy of the planes containing the polypyrrole main chain was effectively excluded from the layers formed by the side-chains. In parallel, conductivity measurements revealed a large anisotropy of the conductivity at low frequencies [24]: $\sigma(\perp)/\sigma(\parallel) = 6$. The conductivity σ, in the direction perpendicular to the orientation axis of the mesogens (and therefore in the smectic planes) is higher than the conductivity σ_{11} in the direction of the mesogens. Substitution of the polypyrroles with mesogens proved to be an effective way to create 2-dimensional anisotropic conductivity in situ (via magnetic-field alignment).

In 1995, Ibison, Foot and Brown [25] reported the synthesis and polymerisation of the first N-substituted liquid crystal polypyrrole. The substituent was an organic spacer chain culminating in a mesogenic cyanobiphenyl group **(Fig. 1.5)**. The side chain (4-cyano-4'-alkoxybiphenyl) was bonded to the pyrrole nitrogen atom so as to enhance the orientation of the polymer backbone. Above a transition temperature the side-chains conferred liquid crystalline behaviour, and were susceptible to further alignment in the presence of magnetic fields. Similar studies had been performed using 3-substituted thiophenes [26], but there had been no previous reports of processible liquid crystalline pyrroles. The polymer resulting from chemical oxidation was a semi-conductor exhibiting a liquid crystalline phase slightly above room temperature, and was highly processible. The monomer molecule proved to be symmetrical, and its polymerisation produced head-to-tail coupling, resulting in increased chemical perfection of the polymer backbone.

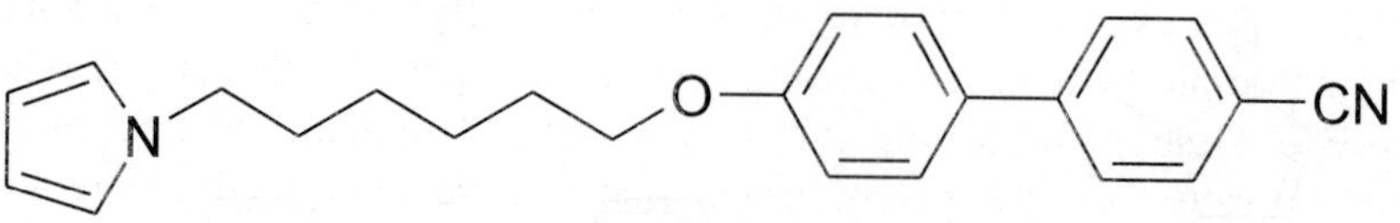

Fig. 1.5. Pyrrole monomer-1-(N-pyrrolo)-6-(4-cyano-4'-biphenyloxy)-hexane.

Gabaston, Foot, Brown and Teissier [27] reported the first synthesis of liquid crystalline aniline monomers and semi-conducting polyanilines with mesogenic side-chains. Electronic properties of polyaniline's are governed by two parameters, the oxidation level and protonation level; it possesses high conductivity in its emeraldine salt (50% oxidised and protonated) form [28]. The first objective of Gabaston *et al.* [27] was to synthesise and study aniline monomers with mesogenic side-chains. As aniline molecules are easily oxidised, it was decided to synthesise ortho-substituted nitrobenzenes, which were then hydrogenated to form 2-substituted anilines and subsequently polymerised. A systematic investigation was carried out in order to study the influence of the number of methylene units (n) in the bridging group (n = 2 to 10) on the liquid crystal transition temperatures of the precursors. Examples of precursors prior to reduction are given **(Fig. 1.6)**.

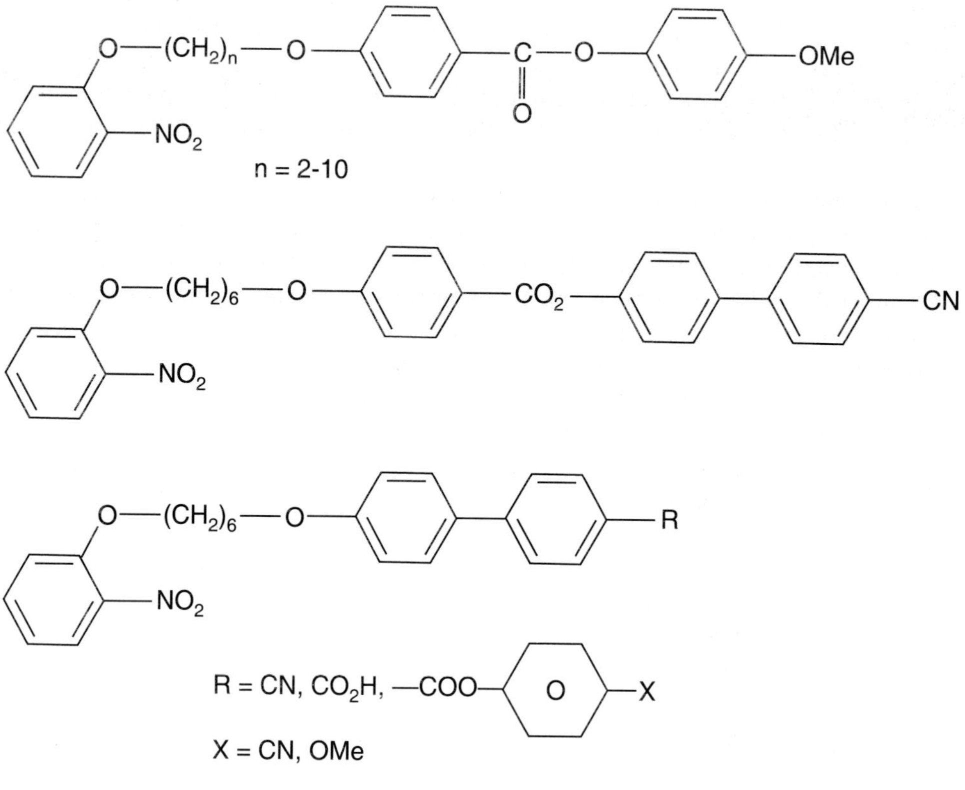

Fig. 1.6. Precursors prior to reduction.

The molecules with even spacer lengths were liquid crystalline above ambient temperature (schlieren and homeotropic textures of a monotropic nematic phase), whereas the odd numbers were not. Hydrogenation of these molecules formed ortho-substituted anilines. The anilines exhibited similar liquid crystalline behaviour, but after polymerisation, no mesogenic properties were initially observed. The large ortho-substituents were simply found to lower the polymer backbone glass transition temperature (T_g) by between 50 and 80°C. The synthesis of molecules with higher liquid crystal transition temperatures and with fixed spacer groups was then carried out, in order to form polymers with liquid crystalline properties above the polymer backbone T_g. This proved an important development as until then it had been difficult to obtain both the properties of liquid crystallinity and conductivity in a LCCP.

Around the same time, Ibison *et al.* [29] published research on the use of a focused laser beam to write patterns of local alignment directly onto films of liquid crystalline polymers based on pyrrole, thiophene and aniline (**Fig. 1.7**). Tracks of alignment were produced by slow movement (typically ~1 cm/s) of polymer films relative to the laser beam. The track widths were in keeping with the diameter of the focused laser beam. The irradiated region was viewed microscopically between crossed polarisers and was seen to be birefringent, implying that the optically anisotropic mesogenic side groups were aligned in a plane perpendicular to the incident laser beam. Each of the laser-aligned samples showed anisotropy in the apparent value of bulk conductivity. This technique is of potential interest as a direct method for lithographic conductivity processing of polymer films on the micro- or even nano-metre scale. It appears that the flow-induced alignment of the liquid crystalline side-chains produces an increased planarity of the polymer backbones, thereby enabling more effective transport of charge carriers.

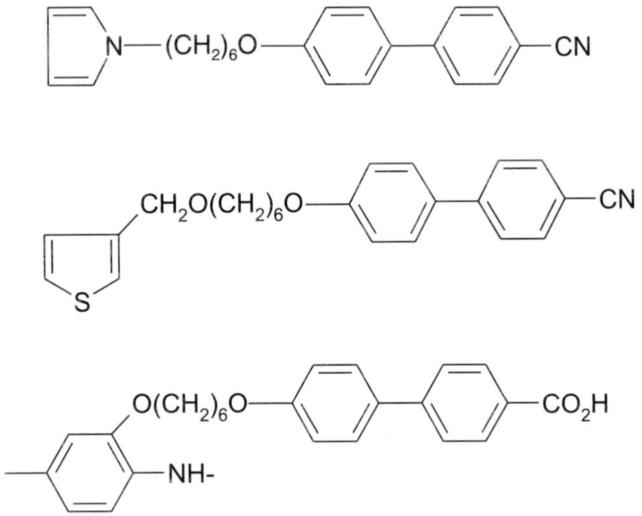

Fig. 1.7. Compounds investigated for the use of a laser to write patterns of local alignment directly onto films of liquid crystalline polymers.

In 1997, Sinigersky *et al.* [30] investigated doping a non-conjugated liquid crystal polymer in its different states with iodine. It was established that the polymer **(Fig. 1.8)** could not be doped in solution, but it was doped to different, but rather higher levels in the solid state and in frozen liquid crystalline and isotropic states. The products obtained exhibited intense optical absorption in the visible region, higher electrical conductivity and paramagnetism. The successful doping of this non-conjugated polymer in the solid state was explained by a collective interaction of several conjugated parts of the macromolecules (acting as extended conjugation) with iodine [30].

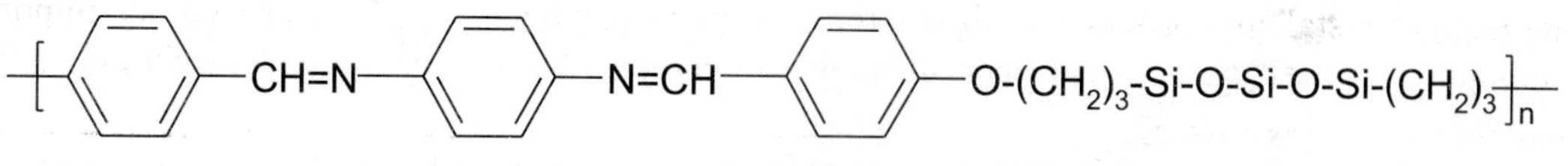

Fig. 1.8. Compound with short conjugated blocks and soft spacers.

The main reason, Sinigersky *et al.* [30] chose this polymer, was that the siloxane spacers were very soft, thus enabling easier spatial arrangement of the conjugated parts. These findings showed that a long, uninterrupted conjugation in the chain is not a prerequisite for the redox reaction of doping to proceed. The successful doping of the polymer in **Fig. 1.8**, whose structure and properties are typical for main-chain liquid crystalline polymers, opened wide possibilities for doping of other representatives of this class thus creating novel materials with interesting electrical and magnetic properties. Obviously the supramolecular arrangement plays a more important role in the realisation of the electrical conductivity rather than the long conjugation in this work.

Akagi, Goto and Shirakawa [31] synthesised chiroptical liquid crystalline polyacetylene derivatives to develop the first ferroelectric LC conducting polymers. The LC group consists of a chiral alkoxy moiety as a terminal group and a biphenyl moiety as a mesogenic core with a trimethylene or decamethylene segment linked with an ether-type atom as a spacer **(Fig. 1.9)**.

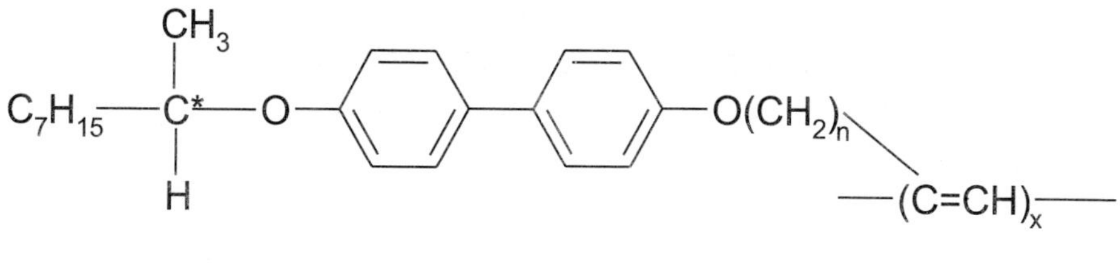

PBP9*00nA   (n=3, 10) : (S)-(-)- polymer

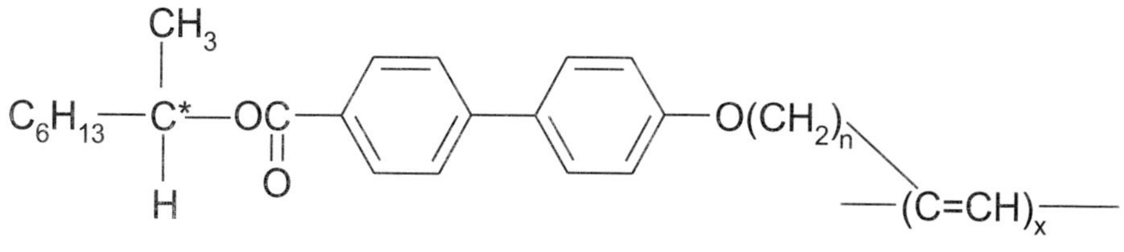

PBP8*C00nA   (n=10) : (S)-(-)- polymer

Fig. 1.9. Chiroptical liquid crystalline polyacetylene derivatives.

PBP9*003A (Mo) showed S_A and $S_C{}^*$ phases. The latter indicates the formation of ferroelectric liquid crystallinity (FLC). This polymer was the first conducting polymer to exhibit the ferroelectric liquid crystal phase. PBP8*C0010A (Rh) showed only a S_A phase, in spite of introduction of the carbonyl moiety in the chiral terminal alkoxy group designed to enhance a dipole moment of the LC side-chain. This is partly due to a shorter polyene segment (smaller degree of polymerisation) produced by the Rh- based catalyst compared to oxidation polymerisation with $FeCl_3$, although the catalyst is free from a side reaction with the carbonyl moiety. Also the optical configuration of each monomer is maintained during the polymerisation, irrespective of the catalyst employed.

Akagi *et al.* [32] also presented synthetic routes and properties of thermotropic liquid crystalline conducting polymers with small-bandgaps **(Fig. 1.10)**. According to a theoretical prediction based on band calculations, polyarylenemethines could be candidates for small bandgap conducting polymers. Akagi *et al.* [32] deduced that to synthesise small-bandgap conducting polymers, methine-bridged polymers with alternating aromatic and quinoid units might be promising.

Actually, Jenekhe and Chen [33] had already synthesised a periodic block polymer consisting of aromatic and quinoid segments and proposed a superlattice model for this system. The ultimate target that they envisaged was to synthesise novel small-bandgap polymers with liquid crystalline groups as a substituent to develop self-alignable conducting polymers.

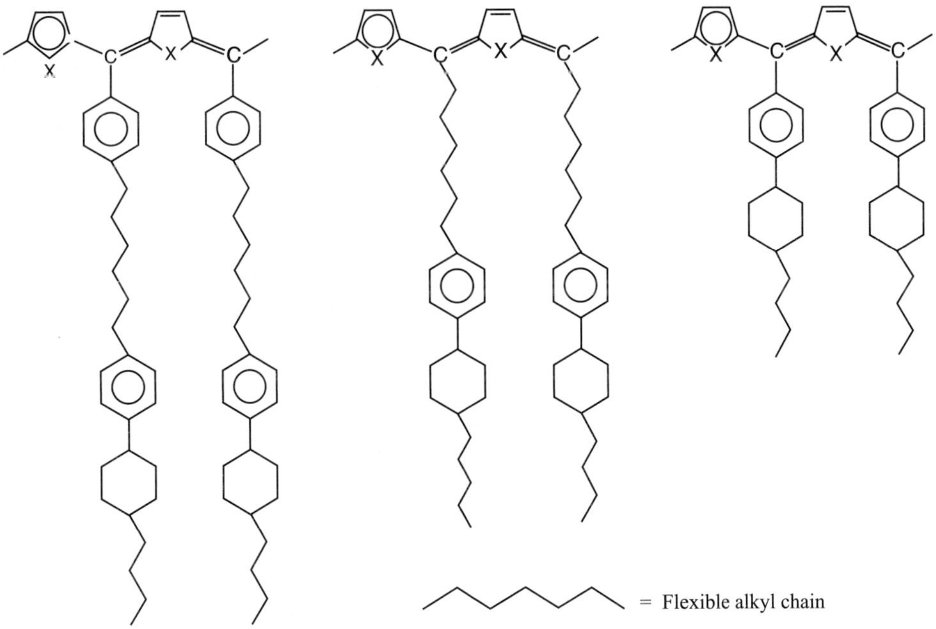

Fig. 1.10. Molecular structures of small-bandgap conjugated polymers with liquid crystalline groups where X = S.

They adopted a thienylene or pyrrolylene ring as a heteroarylene unit and a phenylcyclohexyl moiety as a mesogenic core in the liquid crystalline group. The thienylene-based polymers exhibited smectic E or smectic A phases. The band gaps evaluated for these polymers were 1.45-1.54 eV and 2.48 eV (semi-conductors), depending on an intramolecular p-electron resonance interaction between the main chain and the liquid crystalline side chain. Upon doping the conductivities increased from 10^{-6} Scm^{-1} to 10^{-5} S cm^{-1}-10^{-4} Scm^{-1}. The unexpected low values of the conductivities come from short effective conjugation lengths and are represented by degrees of polymerisation based on number average molecular weights. By improving polymerisation techniques, increasing conjugation length and by achieving macroscopic alignment of the polymer under liquid crystalline phase, conductivity values could be increased.

Few reports existed in 1997 about liquid crystalline polythiophenes, although soluble and fusible poly(3-substituted thiophenes) are easily obtained by conventional polymerisation of a thiophene monomer. Kijima *et al.* [34] investigated the synthesis of some novel 3-substituted thiophenes with a liquid crystalline substituent. Properties of these monomers, and the polymerisation method were investigated in order to obtain a soluble, fusible, and liquid crystalline polythiophene **(Fig. 1.11)**.

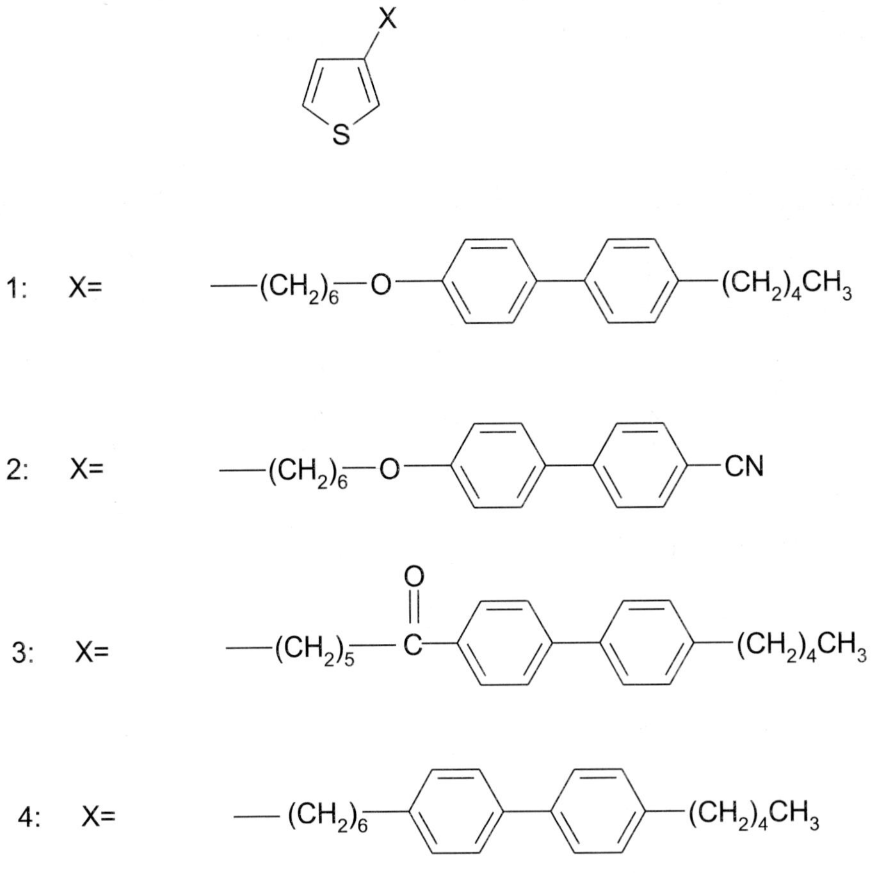

Fig. 1.11. Some novel 3-substituted thiophenes.

Kijima *et al.* [34] successfully synthesised, characterised and polymerised the four thiophenes with a mesogenic substituent at the 3-position. However, none of the polymers showed any mesophases. Three reasons were suggested for this:

1. The introduction of the terminal thiophene moiety reduces the rod-like shape of molecules and decreases the stability of a liquid crystalline state.
2. Mesogenic moieties in the monomers are oxidised in preference to oxidation of thiophene moieties during attempted oxidative polymerisation.
3. Irregular oxidative couplings among thiophene moieties and mesogenic moieties cause disordered polymers.

Up until 1997, Goto *et al.* [35] and Akagi *et al.* [36] had extensively reported on the synthesis and properties of polyacetylene derivatives with LC substituents in the side-chain. Next came the introduction of an azobenzene moiety as the mesogenic core of the liquid crystalline side chain to develop a novel liquid crystalline conjugated polymer with light switching function **(Figs. 1.12 & 1.13)** [37].

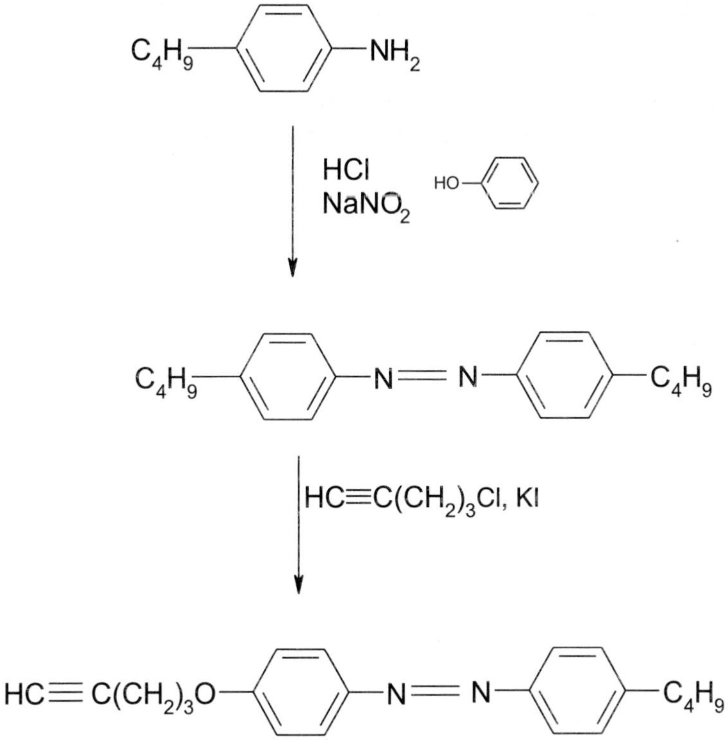

Fig. 1.12. A synthetic route for monomer AB403A.

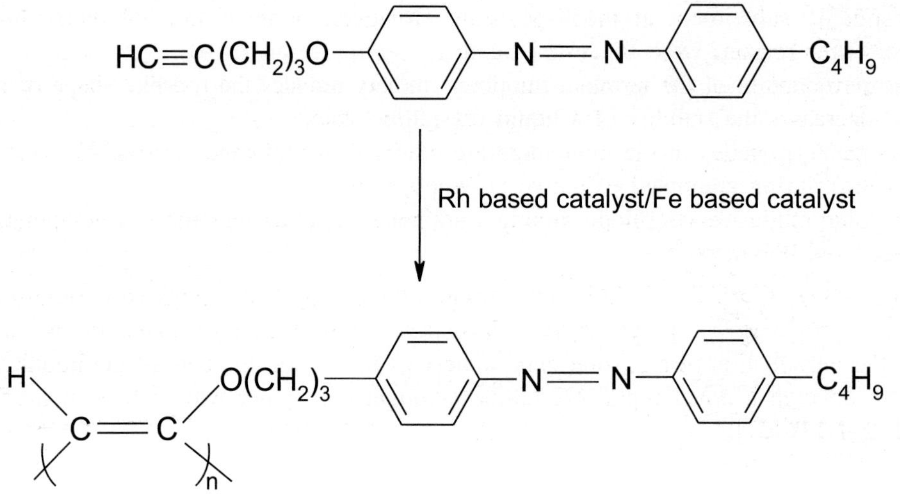

Fig. 1.13. A mono-substituted acetylene polymer with LC group.

Polarising microscope studies demonstrated that the monomer exhibited a droplet texture characteristic of a nematic phase and the polymer a polygonal texture assignable to a smectic phase. Photo-induced reversible isomerisation of the polymer backbone from *cis* to *trans* form, was demonstrated **(Fig. 1.14)**.

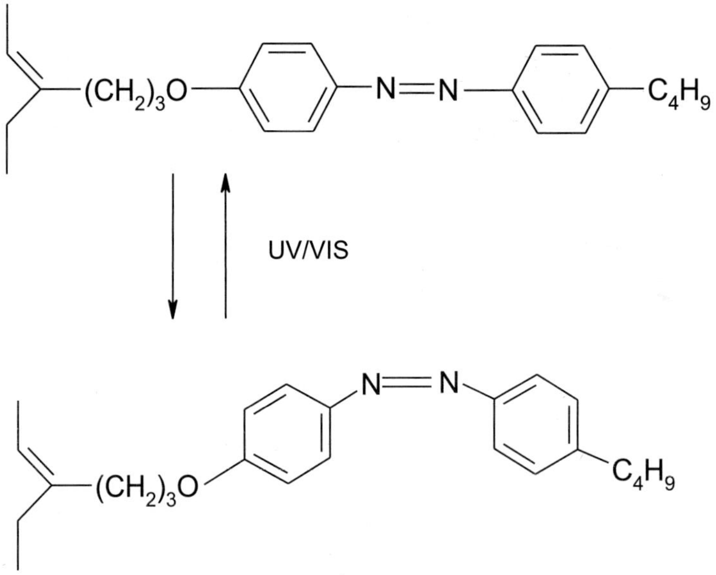

Fig. 1.14. Photo induced reversible isomerisation of the azobenzene moiety in liquid crystalline side chains.

Kijima *et al.* [38] reported again in 1998 on the synthesis and properties of liquid crystalline N-substituted pyrroles and their polymers. Contrary to the studies of the polyacetylene derivatives, liquid crystalline polymers with a conjugated hetero-aromatic ring system had scarcely succeeded with respect to thermal stability and reproducibility of properties, which might be accounted for by irregularity of their polymer structures. The structural irregularity arises from irregular couplings of a 3-substituted five-membered heteroaromatic ring (X = S or NH) (**Fig. 1.15**). The structural irregularity disturbs aggregation of the rod-like side chains, which suggests that stability of the liquid crystalline phase observed in monomers will be decreased by polymerisation. In fact, some polythiophenes and polypyrroles synthesised from liquid crystalline monomers did not show any mesophases [39]. In successful cases of polythiophenes [40] and polypyrroles [41], both have rather longer mesogenic substituents at the 3-position than those of liquid crystalline monosubstituted polyacetylenes [42]. Oxidative polymerisation methods can produce selective coupling at the pyrrole moiety of the monomers [43], giving a thermotropic liquid crystalline polymer, but cannot perfectly restrict the coupling to the 2,5-positions of the pyrrole moiety, which might account for the low workability of some of these materials. Structural analyses of the mesophase suggests that the unit layer consists of a linear polypyrrole main chain with perpendicular rod-shaped mesogenic side chains alternately oriented in opposite directions.

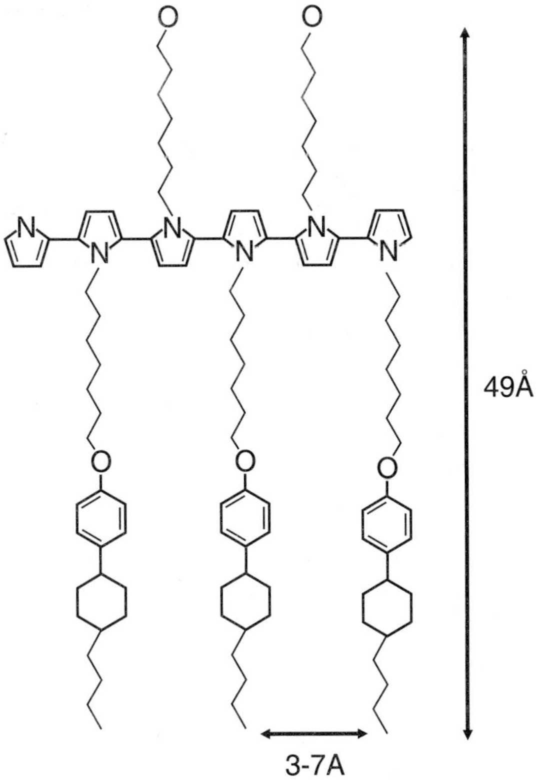

Fig. 1.15. Molecular structure of LC small-bandgap conjugated polymers.

Goto and Akagi [44] published a paper in 1999 on the synthesis and properties of thiophene and pyrrole based liquid crystalline small-bandgap conjugated polymers. The polymers consist of benzenoid and quinonoid structures of polythiophene or polypyrrole backbone with liquid crystalline substituent, where phenylcyclohexyl moiety was adopted as a mesogenic core (**Fig. 1.16**). Both kinds of polymers exhibited smectic liquid crystalline phases. The main idea was to develop self-alignable conducting polymers.

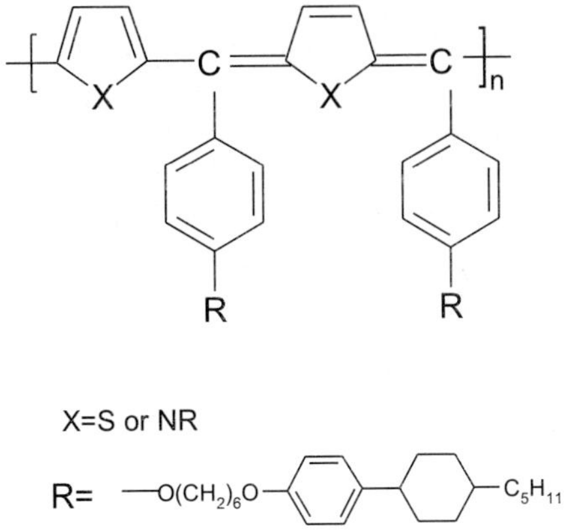

X=S or NR

R= —$O(CH_2)_6O$—⬡—⬡—C_5H_{11}

Fig. 1.16. Benzenoid and quinonoid structures of polythiophene or polypyrrole where phenylcyclohexyl moiety was adopted as a mesogenic core.

Photoluminescence (PL) spectra of the thiophene-based polymer with liquid crystallinity were obtained. The authors observed a substantial difference in intensity between spectra, measured parallel and perpendicular to the rubbing direction, giving a definite anisotropy.

The studies involving LCCP developed into the area of polarised fluorescence and orientational order parameters. Photoluminescence and electroluminescence of liquid crystalline aromatic conjugated polymers was reported in 2001 by Park *et al.* [45]. The liquid crystalline aromatic conjugated polymers were aligned by heating the substrates and then gradually cooling to LC temperature under an external magnetic field. They concluded that for aligned LC-poly(para-phenylene) [LC-PPP] sample, the polarised photoluminescence (PL) peak of for light polarised in the perpendicular direction to the magnetic field is 30% more intense than that of the parallel direction. Polarised optical microscopy shows the domains of aligned LC-PPP films. The PL spectra of LC-PPP at low temperature did not show any broadening or peak shift. The decrease in PL intensity as the temperature becomes high indicates the shortening of exciton life-time. Unlike LC-PPP, the LC-poly(para-phenylene vinylene) [LC-PPV] showed electroluminescent characteristics in a sandwich-structure device, ITO/LC-PPV/AL. The turn-on voltage of the LC-PPV device was 12V and a power law dependence in the I-V characteristic was observed.

Concurrently Abe *et al.* [46] continued to study the effects of mesogenic cores and length of spacers on the liquid crystallinity of N-substituted polypyrrole derivatives. They compared the properties of eight polypyrroles represented in **Fig. 1.17**.

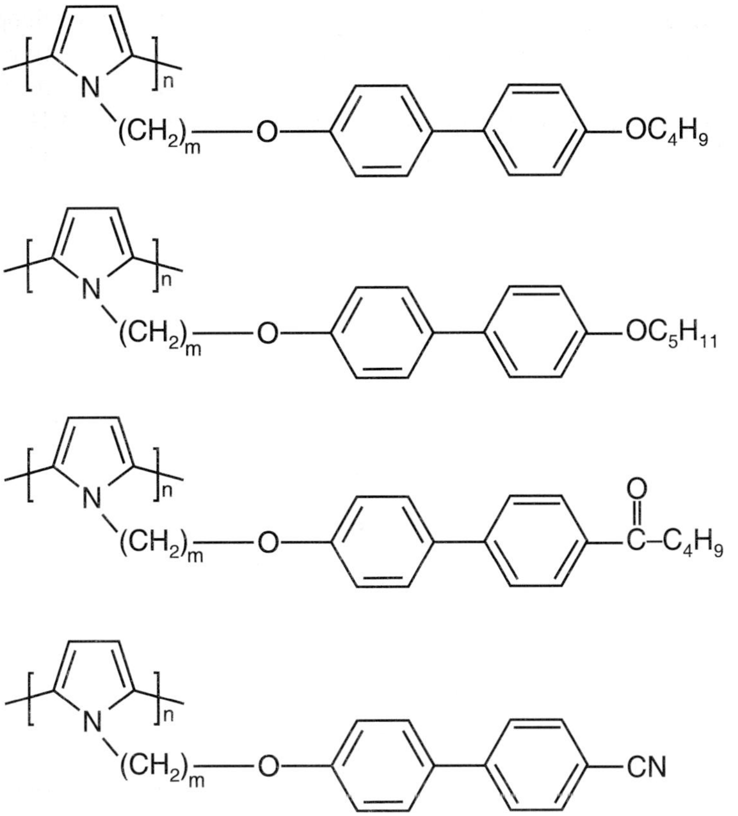

EB6, EB10

PCH6, PCH10

OB6, OB10

CB6, CB10

m = 6, 10

Fig. 1.17. Polypyrroles investigated by Abe *et al.* [46].

It was concluded that polar mesogenic side chains should be chosen to form large domains in the LC states. Selection of mesogenic cores was considered an important factor for the polarity of side-chains. In addition, since LC conjugated polymers often show smectic phases with high phase transition temperatures because of their rigid backbone, lowering the phase transition temperatures is indispensable for material applications. Introduction of a side-chain with intrinsic nematic properties into conjugated polymers was thought to be one of the choices for lowering the temperatures. Furthermore, introduction of long methylene spacers into LC side chain enhances the mobility of the side chain, which contributes growth of domains and/or appearance and stabilisation of highly-ordered smectic phases.

Conjugated polymers with liquid crystalline (LC) groups continue to draw current interest from the viewpoints of multifunctional electrical and optical materials [47-51]. Particularly, LC polythiophene derivatives are most intriguing polymers because its electrical and optical properties are expected to be controlled using molecular orientation of the liquid crystal side-chain. It is known that ferroelectric LC (FLC) molecules have the ability to respond more quickly to the electric field used as an external force than ordinary LCs, and that the ferroelectric liquid crystallinity is dependent on spontaneous polarisation generated in chiral smectic C phase [52,53].

Dai *et al.* [54] synthesised FLC polythiophene derivatives by introducing a chiral fluorine-containing LC group into the 3-position of the thiophene ring to improve the spontaneous polarisation of the liquid crystal moieties (**Fig. 1.18**).

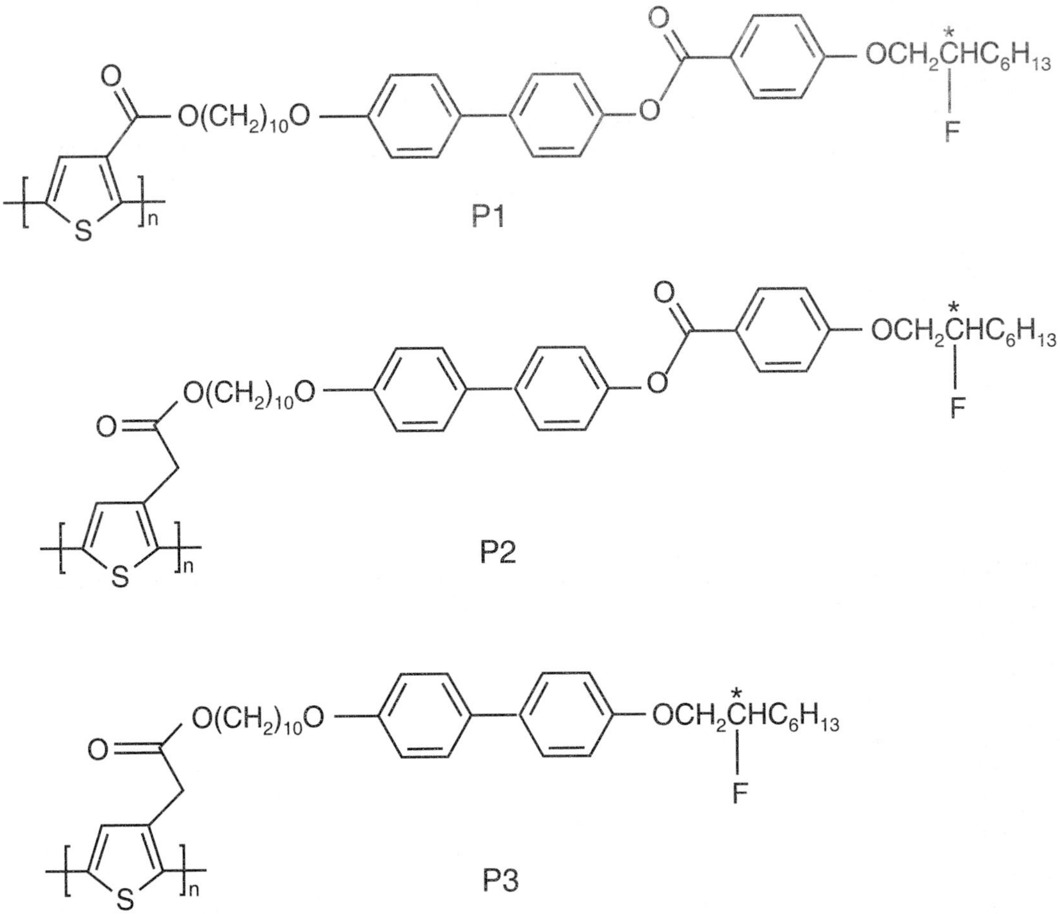

Fig. 1.18. Structures of chiral polythiophene derivatives.

One of the polymers (P2) showed a Smectic C phase with enantiotropic behaviour, whose temperature region was as wide as 20 degrees. It was the first time that the chiral Smectic C phase was observed in polythiophene derivatives. The ferroelectric behaviour was also confirmed by examining the temperature dependence of the dielectric constant of the polymer.

Imrie and Ryder [55] in 2001 described examples of a soluble, fusible liquid crystalline polypyrrole. This was an extension of work published by Ibison, Foot and Brown in 1996 [25]. Imrie and Ryder [55] detailed the synthesis and characterisation of two pyrrole-based monomers substituted at the N-position by a mesogenic group 4-cyanobiphenyl, separated from the pyrrole moiety by an alkyl chain of either 3 or 11 methylene units length (**Fig. 1.19**).

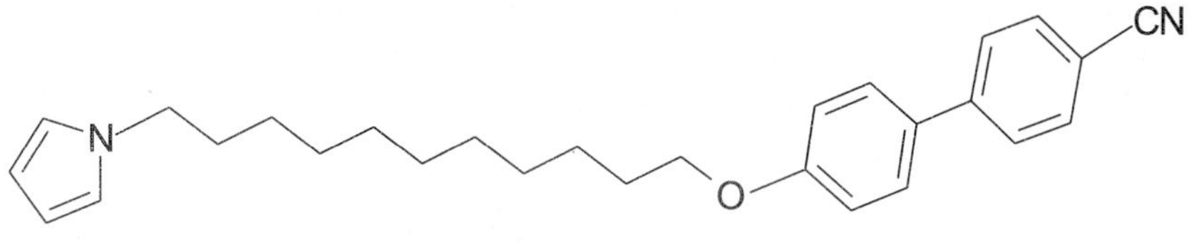

Py-11-CBip

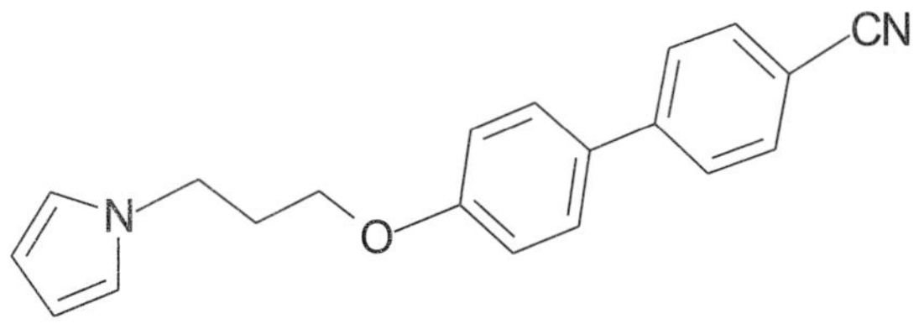

Py-3-CBip

Fig. 1.19. Structures and acronyms of the pyrrole-based monomers and polymers.

They investigated the thermal, spectroscopic and electrochemical properties of these polymers. This showed that the electropolymerisation of these compounds leads to insoluble and infusible materials that show no liquid crystalline phases. However, poly(Py-11-CBip) produced using homogeneous chemical oxidants produced a soluble and fusible polymer with a stable mesophase whose optical texture was consistent with a smectic A structure. The poor conductivity of all the materials discussed here is probably a consequence of the N-substitution and of the steric bulk of the substituents. Nevertheless the latter clearly has an important role in determining the overall stability and order of the

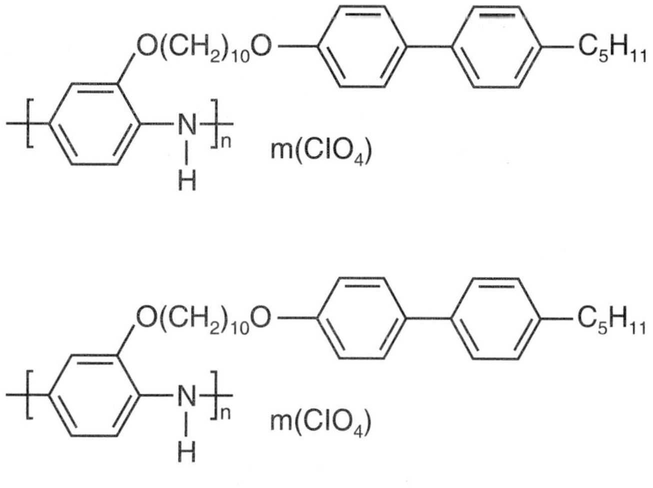

Fig. 1.20. Liquid crystalline polyaniline derivatives synthesised by Goto *et al.* [56].

mesophase. The soluble and fusible nature of poly(Py-11-Cbip) and the clearing temperature indicate that this is a linear polymer with relatively low defect population or cross-linking. For a given combination of alkyl spacer and mesogenic group, increasing backbone flexibility generally increases the clearing temperature while decreasing the glass transition temperature.

Like Gabaston *et al.* [27], Goto, Akagi and Itoh [56] synthesised liquid crystalline polyaniline derivatives by introducing a LC group into the ortho-position of the phenylene ring in the main chain **(Fig. 1.20)**. The polymers exhibited the mesophase, in both oxidised and reduced forms, showing deeply purple coloured fan-shaped textures under a polarising optical microscope, by which the liquid crystal phase was recognised as a smectic one. Subsequently, the polymer was macroscopically aligned in a LC phase by using a magnetic field of 10T. The formation of a mono-domain LC structure was confirmed through the observation of perfectly aligned optical texture.

Macroscopic alignment of the polymers in **Fig. 1.21** under magnetic force was conducted, the polymer being melted by heating on the quartz substrate and gradually cooled to LC temperature whilst applying an external magnetic field of 10T.

Akagi *et al.* [57] have reported on the synthesis of liquid crystalline polythiophene derivatives employing cyanobiphenyl and phenylcyclohexyl as mesogenic cores. After rubbing with a teflon rod, these materials showed 5-15 times improved conductivities in the direction of the rubbing relative to the conductivity of the random orientation. The teflon rod served to align the polymers by acting on the side-chains [58,59].

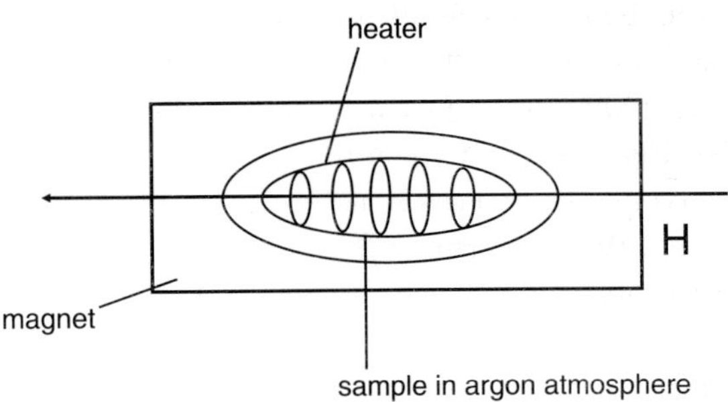

Fig. 1.21. Apparatus used to align LC polymer sample with external magnetic field.

It had been noticed that the great majority of liquid crystalline conducting polymers are side-chain polymer liquid crystals, with the mesogenic group terminally attached to the flexible spacer unit. Lambe *et al.* [60] investigated copolymers such as that in **Fig. 1.22**, in which the mesogen was laterally attached to the spacer, capable of orientating its molecular dipole essentially parallel to the conducting polymer backbone. The simple benzoic acid unit shown in **Fig. 1.22** would be expected to have very weak mesogenic properties, but hydrogen bonding between the acid groups appears to potentiate its ability to form a liquid crystal phase. Measurements of electronic conductivity versus temperature showed a smaller activation energy for conduction above the liquid crystal transition temperature than the one observed below it. This phenomenon was attributed to better conjugation due to the alignment of backbone units by the laterally-attached mesogens in the nematic liquid crystal phase.

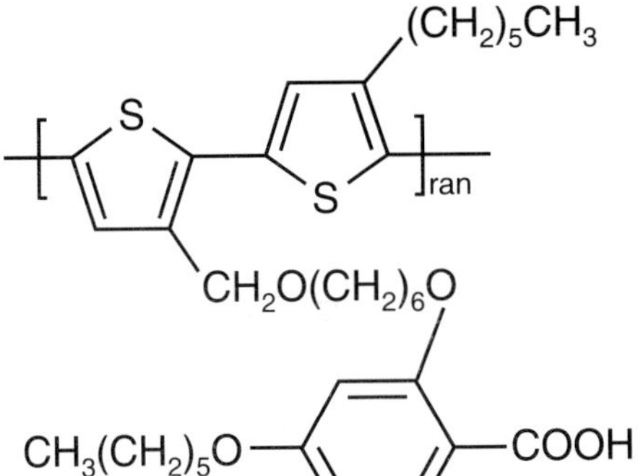

Fig. 1.22. Conducting polythiophene copolymer with laterally-attached mesogen.

Some promising recent developments in light-emitting polymer devices have involved the use of conjugated polyfluorenes and their copolymers. Poly(9,9-dioctylfluorene) exhibits both liquid crystalline self-organisation [61] and efficient blue fluorescence. The alternating copolymer of 9,9-dioctylfluorene with 2,5-bis(benzoyloxy)-1,4-phenylene [62] has blue fluorescence and a smectic mesophase above 200°C, while that with benzothiadiazole [63] **(Fig. 1.23)** is a green light-emitter with a relatively high electron mobility of around 10^{-3} cm^2/Vs. The latter copolymer also exhibits a very high optical gain [64], which may be of use for applications in laser devices or optical amplifiers.

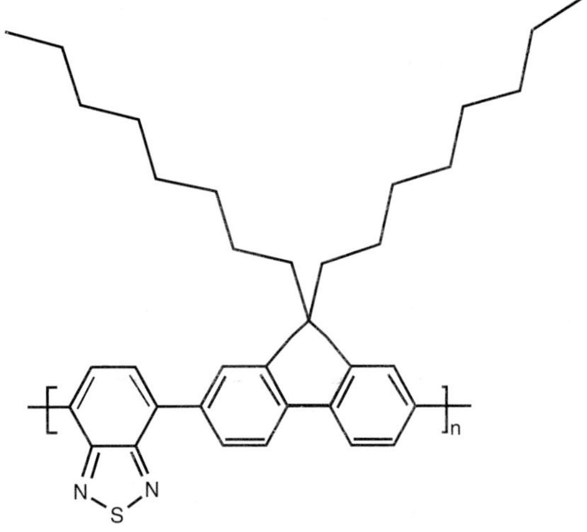

Fig. 1.23. LC polyfluorene copolymer with high electron mobility and green electroluminescence.

3. CONCLUSIONS AND FUTURE PROSPECTS

In conclusion, it seems fair to say that the ability of liquid crystalline functionality to assist the processing of conducting polymers and to allow the control of their electrical properties have been amply demonstrated in the past 15 years or so. The technique of laser microlithography based on these materials is likely to find applications in the production of polymer-based displays or sensor arrays, while the usefulness of polyfluorenes and other liquid crystal polymers in photonics is the subject of keen academic and industrial interest.

REFERENCES

1. M. Goosey; Polymers with Electrical and Electronic Properties, in Speciality Polymers, (Ed. R.W. Dyson), 84 (1987).
2. Handbook of Conducting Polymers, 2nd Edition (Eds. T.A. Skotheim *et al.*), Marcel Dekker, New York (1998).
3. Conducting Polymers and Plastics (Ed. J.M. Margolis), Chapman and Hall, New York (1989).
4. P.J.S. Foot and A.B. Kaiser; Conducting Polymers; in Kirk-Othmer Encyclopedia of Chemical Technology. J. Wiley and Sons Inc., New York (2004).
5. G. Tourillon and F. Garnier; J. Polymer Sci., Polymer Phys., 22, 33 (1984).
6. W.J. Feast; Chemistry and Industry, 263 (1985).
7. J. le Moigne, A. Soldera, D. Guillon and A. Skoulios; Liquid Crystals, 6, 627 (1989).
8. J. Hertz, F. Reiss-Husson, P. Rempp and V. Luzzati; J. Polym.Sci., C, 4, 1275 (1963).
9. H. Ringsdorf and A. Schneller; Br. Polym. J., 13, 43 (1981).
10. H. Finkelmann, H. Ringsdorf, J.H. Wendorff; Makromol. Chem., 179, 273 (1978).
11. P.J. Langley, F.J. Davis and G.R. Mitchell; Mol. Cryst.Liq. Cryst, 236, 225 (1993).
12. M.R. Bryce, A.D. Chissel, J. Copal, P. Kathirgamanathan and D. Parker; Synth.Met., 39, 397 (1991).
13. C. Thobie-Gautier, Y. Bouligand, A. Gorgues, M. Jubault and J. Roncali; Adv. Mater., 6, 138 (1994).
14. P.J. Langley, F.J. Davis and G.R. Mitchell; J. Chem. Soc., Perkin Trans. 2, 2229 (1997).

15. C. Jego, B. Agricole, F. Vicentini, J. Barrouillet, M. Mauzac and C.M ingotaud; J. Phys. Chem., 98, 13408 (1994).

16. H. Awagi, J. Takada and M. Koshioka; (Jpn. Kokai Tokyo Koho), Jpn. Patent No. 05 320 322 [93 320 322] (1993).

17. F. Vicentini, M. Mauzac and R. Laversanne; Liquid Crystals, 16, 721 (1994).

18. S.H. Jin, S.H. Kim, H.N. Cho, and S.K. Cho; Macromolecules, 24, 6050 (1991).

19. K. Yoshino, K. Kobayashi, K. Myojin, T. Kawai, H. Moritake and M. Ozaki; Mol. Cryst. Liq. Cryst., 261, 637 (1995).

20. K. Akagi, H. Goto, Y. Kadokura, H. Shirakawa, S.Y. Oh, and K. Araya; Synth.Met., 69, 13 (1995).

21. F. Vicentini, L. Noirez, G. Pepy and M. Mauzac; Europhys. Lett., 32, 657 (1995).

22. (a) D. Delabouglise, J. Roncali, M. Lemaire and F. Garnier; J. Chem. Soc., Chem. Commun., 475 (1989); (b) J. Ruhe, T.A. Ezquerra and G. Wegner; Synth.Met., 28, 177 (1989).

23. F. Vicentini, J. Barrouillet, R. Laversanne, M. Mauzac, F. Bibonne and J.P. Parneix; Liq. Cryst., 19, 235 (1995).

24. F. Vicentini, J. Barrouillet, R. Laversanne, M. Mauzac, L. Noirez, F. Bibonne and J.P. Parneix; J. Chim. Physique, 92, 795 (1995).

25. P. Ibison, P.J.S. Foot and J.W. Brown; Proc.E-MRS Spring Meeting, Symp. G, Strasbourg (1995); idem., Synth.Met., 76, 297 (1996).

26. C. Thobie-Gautier, Y. Bouligand, A. Gorgues, M. Jubault, and J. Roncali; Adv. Mater., 6, 138 (1994); J.W. Brown, P.J.S. Foot, L.I. Gabaston, P. Ibison and A. Prevost; Macromol. Chem. Phys., 205, 1823 (2004).

27. L. Gabaston, P.J.S. Foot, J.W. Brown and V. Teissier; Abst. Int. Conf. on Liquid Crystal Polymers, Beijing (Sept.1994); L.I. Gabaston, P.J.S. Foot and J.W. Brown; Chem. Commun., 429 (1996).

28. W.S. Huang, B.D. Humphrey and A.G. MacDiarmid; J Chem Soc., Faraday Trans.I, 82, 2385 (1986).

29. P. Ibison, P.J.S. Foot, J.W. Brown, L. Gabaston and R. Simon; Euromat Int. Conf. on Advanced Materials, Maastricht, Netherlands 1997 Proceedings, Vol. 2, p. 35 (Feder. of Eur. Materials Socs., 1997), [ISBN 90-803513-2-6].

30. V. Sinigersky, P-J. Medac, E. Marechal, I. Schopov; Macromol. Chem. Phys., 198, 919 (1997).

31. K. Akagi, H. Goto and H. Shirakawa; Synth.Met., 84, 313 (1997).

32. K. Akagi, H. Goto and H. Shirakawa; Synth.Met., 84, 385 (1997).

33. S.A. Jenekhe and X.L. Chen; Macromol., 28, 465 (1995).

34. M. Kijima, K. Akagi and H. Shirakawa; Synth.Met., 84, 237 (1997).

35. H. Goto, K. Akagi and H. Shirakawa; Synth.Met., 84, 373 (1997).

36. K. Akagi, H. Goto, Y. Kadokura, H. Shirakawa, S.-Y. Oh and K. Araya; Synth.Met., 69, 13 (1995).

37. M. Kijima, H. Hasegawa and H. Shirakawa; J. Polym. Sci., Part A, Polym. Chem., 36, 2691 (1998).

38. M. Kijima, K. Akagi and H. Shirakawa; Synth.Met., 84, 237 (1997).

39. R. Toyoshima, M. Narita, K. Akagi and H. Shirakawa; Synth.Met., 69, 289 (1995).

40. P.J. Langley, F.J. Davis and G.R. Mitchell; Mol. Cryst. Liq. Cryst., 236, 225 (1993); H. Hasegawa, M. Kijima and H. Shirakawa; Synth.Met., 84, 177 (1997).

41. (a) S.-Y. Oh, R. Ezaki, K. Akagi and H. Shirakawa; J. Polym. Sci., Part A, Polym. Chem., 31, 2977 (1993); (b) K. Akagi, H. Goto, Y. Kadokura, H. Shirakawa, S-Y- Oh and K. Karaya; Synth.Met., 69, 13 (1995).

42. P. Ibison, P.J.S. Foot and J.W. Brown; Synth.Met., 76, 297 (1996).

43. N. Koide and H. Iida; Mol. Cryst. Liq. Cryst., 261, 427 (1995).

44. H.Goto and K.Akagi; Synth.Met., 102, 1292 (1999).

45. J.H. Park, C.H. Lee, K. Akagi, H. Shirakawa and Y.W. Park; Synth.Met., 119, 633 (2001).

46. S. Abe, M. Kijima and H. Shirakawa; Synth.Met., 119, 421 (2001).

47. S-Y. Oh, K. Akagi, H. Shirakawa and K. Araya; Macromolecules, 26, 6203 (1993).

48. K. Akagi, H. Goto, Y. Kadokura, H. Shirakawa, S.Y. Oh, K. Araya; Synth.Met., 69, 13 (1995).

49. K. Akagi and H. Shirakawa; Macromol.Symp., 104, 137 (1996).
50. K. Akagi and H. Shirakawa; The Polymeric Materials Encyclopedia: Synthesis, Properties and Applications, CRC Press, 5, 3669 (1996).
51. K. Akagi and H. Shirakawa; in Electrical and Optical Polymer Systems: Fundamentals Methods and Applications (Eds. D.L. Wise *et al.*) Marcel Dekker, 28, 983 (1998).
52. X.M. Dai, H. Goto, K. Akagi and H. Shirakawa; Synth.Met., 102, 1289 (1999).
53. K. Akagi, H. Goto and H. Shirakawa; Synth.Met., 84, 313 (1997).
54. X.M. Dai, H. Narihiro, H. Goto, K. Akagi and H. Yokoyama; Synth.Met., 119, 397 (2001).
55. C.T. Imrie and K.S. Ryder; J. Mater.Chem., 11, 990 (2001).
56. H. Goto, K Akagi and K. Itoh; Synth.Met., 117, 91 (2001).
57. K. Akagi, J. Oguma and H. Shirakawa; J. Photopolym. Sci.Tech., 11, 249 (1998).
58. J. Oguma, K. Akagi and H. Shirakawa; Synth.Met., 101, 86 (1999).
59. K. Akagi, J. Oguma, S. Shibata, R. Toyoshima, I. Osaka and H. Shirakawa; Synth.Met., 102, 1287 (1999).
60. G.J. Lambe, J.W. Brown, J.A. Clipson and P.J.S. Foot; Macromol. Rapid Commun., 25, 1000 (2003).
61. M. Grell, D.D.C. Bradley, M. Inbasekaran and E. Woo; Adv. Mater., 9, 798 (1997).
62. S.J. Zheng, H. Goto and K. Akagi; Synth.Met., 135-136, 125 (2003).
63. A.J. Campbell, D.D.C. Bradley and H. Antoniadis; Applied Phys. Lett., 79, 2133 (2001).
64. R. Xia, M. Campoy-Quiles, G. Heliotis, P. Stavrinou and K.S. Whitehead; Synth.Met., 155, 274 (2005).

2

Polyanilines: Materials and Applications

Akheel Ahmed Syed[1]* and Syed Raihan Taqui[2]

[1]*Department of Studies in Chemistry, University of Mysore, Manasagangotri, Mysore—570006, India.*
[2]*Department of Polymer Science and Technology, S.J. College of Engineering,
Manasagangotri, Mysore—570006, India.*
**Corresponding author's email: syedakheelahmed@hotmail.com*

Contents

Summary: Polyanilines are attractive conductive materials because of their relative case of synthesis and good environmental, chemical and thermal stability as well as simple acid/base doping chemistry. Polyanilines have broad range of applications in microelectronics, in lithography, resists, coatings, ultra-thin membranes, secondary batteries, multi-color electrochromic devices, conductive textiles and so on. Therefore, the synthesis, characterization and most promising applications of polyanilines are discussed in this chapter.

1. INTRODUCTION

Polyanilines are a novel class of conducting polymers [1-3] which have been investigated extensively, because first, they belong to the class of organic polymers which have the conductivity as well as electronic and magnetic properties [4-17] of metals, combined with the mechanical properties [18] and processability [19,20] of conventional polymers; second they possess redox [21-24], ion-exchange [25,26], acid/base [27], and chromatic properties[28-29]; third, they have good environmental stability [1-3]; and fourth they can be relatively easily synthesized [1-3] and hence, finding applications in a wide variety of devices **(Fig. 2.1)**.

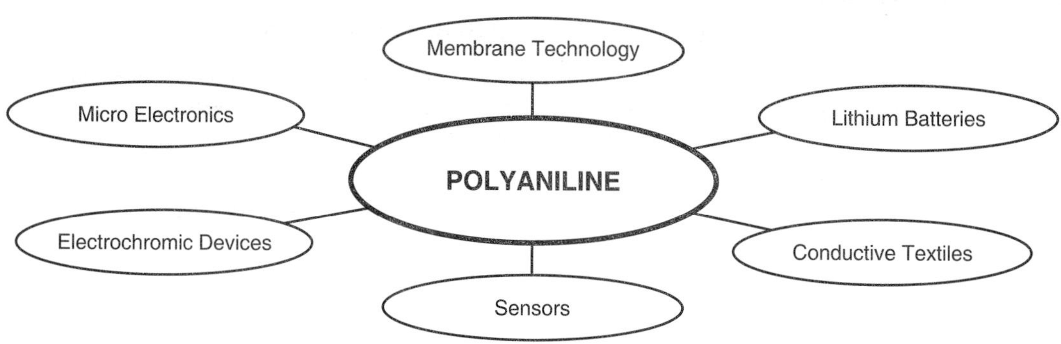

Fig. 2.1. An overview of some of the potential applications of polyanilines.

Polyanilines are the preferred conducting polymer system from an industrial point of view as in many applications they offer a number of advantages over other conducting polymers. They are generally soluble in some organic solvents [20,30-32] and environmentally stable and can be prepared by one-step synthesis using raw materials which are readily available [33]. Chemically they are more versatile, which allows them to be tuned to meet the needs of any given application. Indeed, many polyaniline derivatives exist today as a result of chemical modification of polymer backbone [1-3,34-39], dopant [40-45], and oxidation state [1,3,26]. For a better appreciation of the polyanilines an overview of the genesis, synthesis and technological applications of conducting polymers is desirable.

2. CONDUCTING POLYMERS: AN OVERVIEW

The emergence of polymeric materials with conjugated π-electron backbones and possessing unique electrical and electronic properties as against the conventional polymers, has drawn considerable attention [46]. These polymers, which constitute a new class of potentially useful electrical, electronic and optoelectronic materials, have come to be known as "Conducting Polymers" [47,48].

The term 'Conducting Polymer' was earlier used for insulating polymers mixed with a definite proportion of conducting material, such as, metal fibers to provide the necessary electronic conduction. The polymer matrix, acted as a solid adhesive and kept the conducting components together, providing the mechanical strength, without having any role in electrical conduction. Such materials are now known as conducting composites. Conducting polymers are unique by themselves as they exhibit electrical, electronic, magnetic and optical properties which are inherent to metals or semiconductors, at the same time retaining those mechanical properties which are common to conventional polymers. These properties open up new avenues for technological applications. As for example, the property of color change induced by the electrochemical doping-undoping makes their use in the manufacture of multichromic displays or electrochromic windows. The hole-injecting properties of these polymers under an applied potential or current have been sucessfully used in developing flexible light emitting diodes (LED) and light emitting photoelectrochemical cells. The photovoltaic effect of a conducting polymer/electrolyte interface, observed by irradiating it with energy higher than the polymer bandgap, has drawn much attention due to the possibility of producing photoelectrochemical cells.

Polyacetylene

Polyphenylene

Poly (phenylene-vinylene)

Polypyrrole

Polyaniline

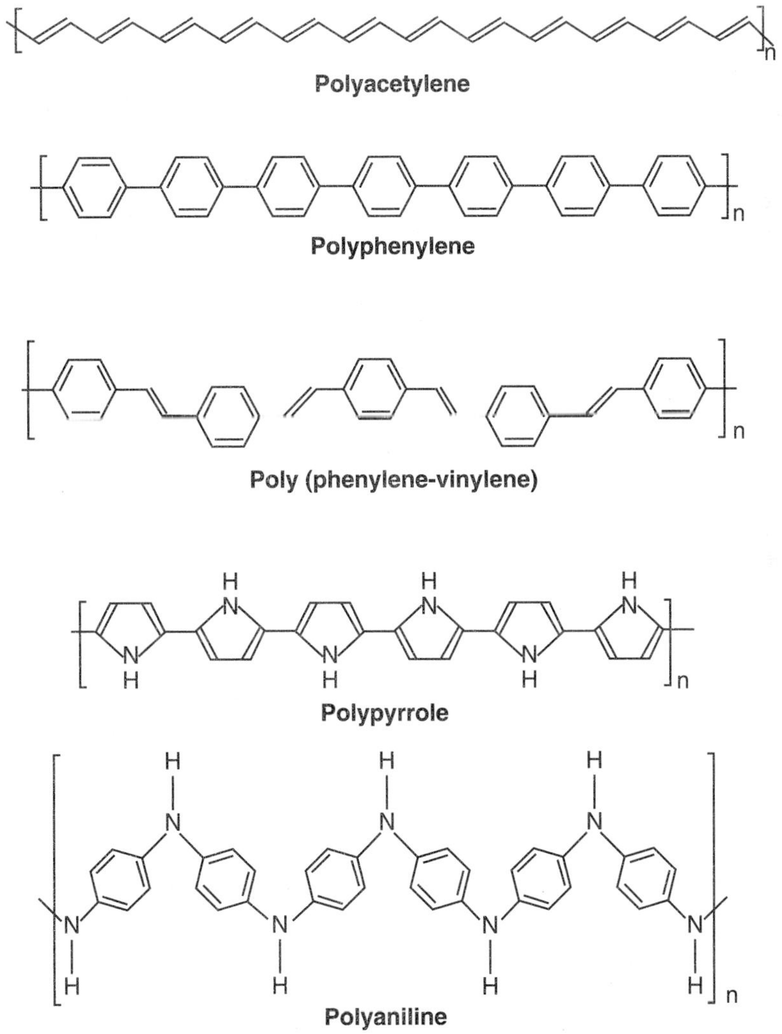

Fig. 2.2. Conjugation in conducting polymers.

The varied properties of conducting polymers can also be applied as active components in capacitors, biosensors, drug-releasing agents, gas separation membranes, ionexchange resins and chemical indicators. These unique properties are however different from those that originate from a physical mixture of insulating polymers with conducting fillers. In the case of intrinsic conducting polymers, the conductivity is related to the delocalisation of π bonded electrons over the polymeric backbone, coexisting with unusual electronic properties, such as low energy optical transitions, low ionization potentials, and high electron affinities. This phenomenon is as a consequence of the presence of conjugated C = C double bonds in the polymer backbone, as evident in the molecular structures of the repeating units of conducting polymers **(Fig. 2.2)**.

2.1. Genesis

A new epoch in the evolution of conducting polymers (CP) began with the discovery made in 1977 by MacDiarmid and co-workers, when they demonstrated the intrinsic semi-conducting and metallic properties of the linear organic polymer, 'polyacetylene' [47]. Thereafter, there was a surge of activity directed towards the synthesis and applications of the members of this new class of materials [49-62].

The phenomenal growth in the applications of CP was due to their inherent properties, such as low ionization potential, and low-energy optical transitions. Other features which are responsible for their popularity are: first, one-step chemical synthesis (in most of the cases), and second, attractive electropolymerization with electrochemical stoichiometry. The prospect of applying the skills of chemists to tailor electroactive polymers and improve their processability is the most important challenge behind the world-wide efforts.

2.2. Synthesis

There are two basic strategies for preparing macromolecules. The first involves condensation reactions, that is, reactions in which new bonds linking structural subunits are formed by the loss of small molecules eliminated from the starting compounds (monomers). The second is addition reactions in which all the atoms present in the monomers are incorporated into the polymer. Synthesis of polyaniline is as per the first type of reactions. However, in both theory and practice, a polymerization reaction using a monomer of the appropriate functionality can produce a linear polymer with a regularly repeated structural subunit. Even when this ideal situation occurs, it is important that the product is very unlikely to be a molecularly monodisperse sample, and it is expected to get a distribution of different-sized linear macromolecules.

2.2.1. Chemical Polymerization

Conjugated organic polymers can be synthesized as presented in **Fig. 2.3**. The syntheses can be broadly classified into two types; first, a one-step synthesis and second, a two-step or multi-step synthesis. The first approach represents the classical method in which an appropriate monomer is converted directly into a conjugated polymer, *via* addition or condensation process. The main disadvantage of this direct approach stems from the experimental observation that some of the structurally simplest conjugated polymers, such as, polyacetylene and polyphenylene, are essentially insoluble and intractable materials. This brings out several practical difficulties in its course. Thus, most of the molecular weight and molecular weight distribution determinations depend on solution phase measurements. Inherently insoluble polymers get precipitated from the solution during synthesis, restricting the molecular weight attainable. Also, conventional purification by 'reprecipitation' is ruled out in such materials. Insolubility and intractability also make both fabrication into a useful physical form and morphological modification difficult. Despite these obvious impediments, many successful syntheses have been accomplished by direct results.

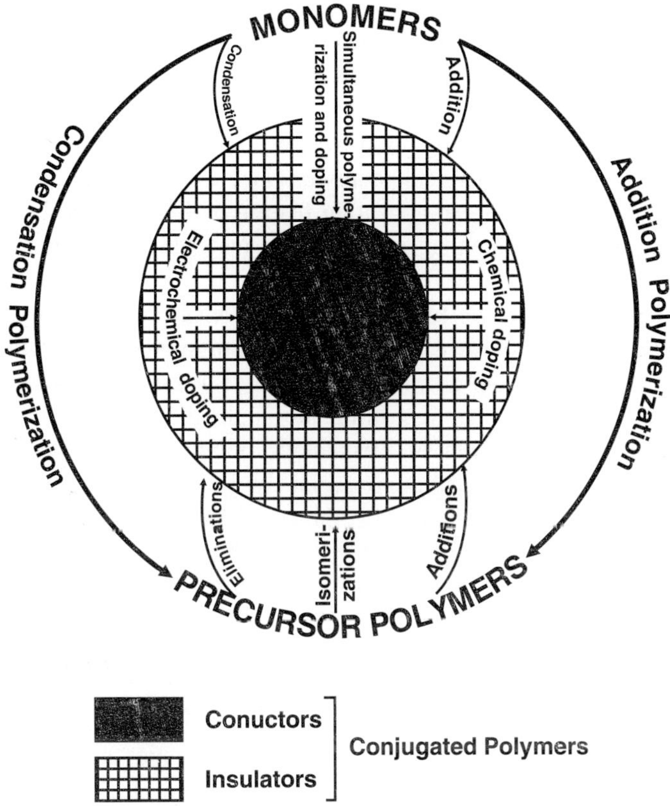

Fig. 2.3. Synthesis of conjugated organic polymers by different routes.

The main advantage of the two-step, or the multi-step route is an increased flexibility in the design of synthesis. Thus, in the first stage the precursor polymers may be formed by addition or condensation polymerization procedures (which is also the case for the direct synthesis). The second stage may be accomplished by a variety of reactions as for example, first, chemical synthesis of a conjugated polymer as a first step, followed by doping with electron acceptor or electron donor dopant as a second step; and second, chemical synthesis of a conjugated polymer followed by electrochemical doping. The feasibility of generating a conjugated conducting polymer by one-step chemical synthesis has given an impetus to chemists to look for alternate routes. Electrochemical polymerization has been proved an effective alternative in some instances.

2.2.2. Electrochemical Polymerization

About four decades ago, the synthesis of polymers was entirely monopolized by chemical methods. It was only during the last thirty-five years the investigations were contemplated in understanding the chemical phenomena connected with the formation of polymers in an electrolytic cell. From this point of view, electrochemical polymerization (ECP) is a new branch of electrochemistry that deals with phenomena distinctive of polymerization reactions. It is interesting to note that the production of macromolecules in an electrolytic cell, with a remarkable control over initiation, propagation and chain transfer processes, could be achieved by a simple galvanostat or potentiostat. Despite the versatility of the chemical methods of polymerization, electrochemical syntheses of polymers prove effective alternatives in some instances. This is because electrochemical reactions are considered to be much cleaner (with respect to possible pollutants) than chemical reactions. Moreover, electrons— regarded as a reagent—are inherently pollution-free, at least at the point of use. Thus, there are a number of examples where electrochemical routes produce unique products, and do so by reaction pathways, which in them are unique. An appealing feature of ECP is that they can be controlled so well (by means of the electrode potential and electrode material) that they can be made to give a high selectivity of desired products.

At present, there is at least a two-fold interest in the electrochemical synthesis of conducting polymers. First, ECP reactions provide a new method of polymerization, which have a fine control over the initiation and termination steps. Thus, they constitute a research tool offering new possibilities for conducting fundamental studies by permitting the application of fast transient, electrochemical methods that enable to study the rate of elementary polymerization steps. Secondly, ECP has considerable technological potential. If commercial applications of electrode reactions are any indication, the outlook for the development of commercial ECP processes is promising. Preliminary considerations suggest that several existing ECP technologies could be competitive with modern commercial polymerization processes; and future studies may uncover ECP processes for the synthesis of specialized polymers, which otherwise cannot be obtained economically by any other methods. However, ECP is sometimes fraught with artifacts due to the competing redox reactions solvent, of the electrolyte salt, and trace impurities.

Three electrochemical methods are generally employed for ECP and they include: first, constant current or galvanostatic; second, constant potential or potentiostatic; and third, potential scanning or sweeping method. The first one essentially consists of a two-electrode assembly dipped in an electrolyte solution containing the monomer. Passing a current of desired strength leads to either deposition of a polymer film on the electrode surface or formation of the insoluble polymer material in the solution. This takes place only in case of one-step synthesis. In constrast potentiostatic and potential supplying or scanning methods employ three-electrode assembly.

Most of the conducting polymers are produced by condensation polymerization. During electrochemical condensation polymerization, the reacting monomeric units split off, and are removed from the polymerizing system in the form of water, CO_2, or protons. Such polymerizations require at least one electron transfer for each monomer molecule, rather than one electron transfer for each polymer molecule, as can be seen in the case of some addition polymerizations. Most of the heterocyclic and aromatic monomers such as pyrrole, thiophene, aniline, etc., undergo polymerization *via* a condensation process, yielding conducting polymers. Protons are liberated during ECP and this decreases the pH of the medium [63].

Polyaniline is the simplest of the conducting polymers prepared by electrochemical techniques based on Letheby's original anodic oxidation of aniline [64]. The most convenient technique to demonstrate the preparation of polyaniline is, perhaps, by using an acidic solution of aniline in a one-compartment cell, equipped with two conducting electrodes. When the electric current flows through the system, the electrode reaction takes place, followed by the polymerization of aniline. Onc of the important features observed in the course of electropolymerization of aniline, pyrrole etc., is that it proceeds with electrochemical stoichiometry [65]. This allows the characterization of film-forming reactions.

The prospects of electrochemical techniques for the synthesis of polymers are bright. However, they are always associated with some undesirable phenomena occurring at the electrode during ECP. Also, the physical and chemical nature of electrodes including their surface, electrode overpotential, half cell reaction overpotential and current density are expected to influence the structures and properties of the resulting conducting polymers.

2.3. Quest for Higher Conductivities

Conductivity is the result of two important factors: the number of carrier-electrons and holes and carrier mobility, that is the ease with which an electron moves through a material. The electrical conductivity of most of the conductive polymers is in the same range as that of inorganic semiconductors. Semiconductors have a very low number of carriers, typically 10^{16} to 10^{18} per cc, but very high mobilities, typically 10^2 to 10^5 cm^2 per volt-second, which results from a high degree of crystallinity and purity and absence of any defects in these materials. However, unlike with inorganic semiconductors most of the conductive polymers are either amorphous or slightly crystalline. On the other hand, the number of carriers in conductive polymers, typically 10^{21} to 10^{23} per cc, is four to five orders of magnitude higher than in inorganic semiconductors. It is roughly equal to the number of cations per cubic centimeter, and is therefore as high as possible. But the mobility of conducting polymers is limited to 10^{-4} to 10^{-5} cm^2 per volt-second. For achieving higher conductivities higher mobilities are needed and these will come with more crystalline, better-oriented, defect-free materials.

Researchers tried several methods to align the chains more perfectly in conducting polymers to achieve higher mobilities. One of them is to polymerize the monomers in liquid crystal solvents. Another way is to align polymers by stretching polymer films or fibers. The utility of conducting polymers has been demonstrated in various tasks where they are effectively employed. These include photovoltaic, non-linear optics, microlithography, EMI shielding, anti-static fabrics, corrosion prevention in metals, rechargeable batteries and gas separation membranes. One of the latest applications is in the medical field concerened with nerve regeneration. The focus is now on improved processability of conducting polymers, especially polyaniline, polypyrrole and their derivatives, for application to anti-static fabrics and corrosion prevention in metals.

2.4. Applications of Conducting Polymers

The discovery of conducting polymers initiated a lot of fundamental investigations, and gave rise to the concept of a new class of materials, which combine typical properties of plastics with the electrical conductivity of metals. Electrically conducting polymers, which can be processed thermoplastically and possess the properties of mechanical behavior and corrosion stability of plastics, would open up entirely new fields of application. Electricity conducting cables, wallpaper which allow electrical heating, new materials for antistatic equipment, and electromagnetic interference shielding are a few areas in which electrically conducting polymers would find a big market.

Other interesting applications came up when it was found that conducting polymers could be reversibly oxidized and reduced electrochemically. A new type of lightweight battery using polymer electrodes has come up on the technical horizon. These batteries have a long life, a usable voltage of about 3 volts, and energy density several times higher than the currently available commercial nickel/cadmium and lead acid batteries. These batteries proved excellent for applications where long life, low-power, and reliable operation are required and these include backup sources for static random access memory, intelligent telephone, timer for video-cassette recorders and batteries for hand-held calculators, fax machines, television remote control and wrist watches. However, batteries that deliver high power at high rates are not yet developed. This is still a great challenge, which, if achieved, will open the way to long-range, light-weight electric vehicles.

Conducting polymers like polyaniline show a wide range of colors as a result of their many protonation and oxidation forms. The electrochromic properties of such polymers can be easily exploited to produce different electrochromic devices, including displays and thermal "smart windows". Unlike liquid crystal displays, electrochromic devices made of conducting polymers can be fabricated covering large areas and unlimited visual angles. As conducting polymers do not change colour after the applied voltage has been removed, electrochromic devices also exhibit memory function (bistability) and therefore they can be used to store information.

The response time of an electroactive polymer is determined by the time taken by the dopant ions to diffuse in and out of the polymer. Although the response times of most of the electroactive polymers are typically around 100 milliseconds and hence are not extremely fast, still they are fast enough for many practical applications such as "Smart windows", which, do not require very fast response times. During hot summer, smart windows which use conducting polymer to darken by absorbing a portion of the sunlight, save air-conditioning costs.

Polyisothianaphthene has shorter response times (about 10 milliseconds) than most of the conducting polymers and has sharper colour contrast. It is being tested for use in an electrochromic thermal window that would work in winter. In its transparent conducting state, it does not absorb much of the radiations in the infrared region of the spectrum. During the day, polyisothianaphthene windows permit visible light to enter and infared radiation to heat the room. At night, the polymer gets converted to its coloured insulating form, which absorbs extensively in the infrared region at the same time not allowing the room heat to escape. Conducting polymers to succeed in the "Smart window" market, will have to outperform tungsten trioxide, an inorganic electrochromic material with a life of 10^6 cycles. Conductive polymer-based windows at present have a life of about 10^5 cycles.

Electrically conductive fabrics and textiles impregnated with polypyrrole and polyaniline have the same strength and tactility as their commercially available counterparts, and have been considered for use in composites. Thus the conductive fabrics could be incorporated into the plastic composites used in airframes for charge dissipation during lightning strikes preventing damage to the airplane [66,67]. Such fabrics also distort radar signals and as such they can be used in military aircraft to mislead enemy radar. Several groups have developed latex compositions containing polypyrrole, polythiophene, or polyaniline colloid particles. These can be sprayed onto surfaces to give conductive coatings. Such coatings provide electromagnetic shielding and act as a weapon against electronic eavesdropping.

In summary, conducting polymers intrinsically exhibit unique combination of properties which make them appropriate alternatives for certain materials currently being used in microelectronics. The conductivity of these materials can be tuned to our requirement by chemical manipulation of the polymer backbone, by changing the nature of the dopant, by extant of doping, and by blending with other polymers. The additional advantages include their light weight, processability and flexibility.

Even though much progress has been made, challenges to the development of useful conductive polymers still remain. Materials that which have higher carrier mobilities and higher current densities are needed. Detailed structural information which is lacking will help develop theoretical models that will shed more light on conductivity mechanisms.

Breakthroughs can be achieved not only by improving existing materials, but also from the discovery of novel materials through creative chemical syntheses. Some of them may have unique properties and some may even be superconductors. The ease with which chemical modification and tuning can be achieved in conductive polymers is indicative of great prospects for future applications, which is comparable perhaps to the prospects envisioned several decades ago, when conventional plastics were first developed.

3. POLYANILINES

The polyanilines are a class of conducting polymers which have the general formula containing Y reduced and I-Y oxidized groups (**Fig. 2.4**).

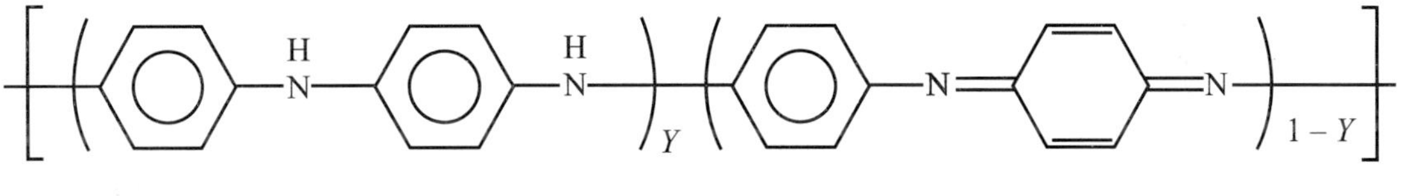

Fig. 2.4. General formula of polyaniline.

This polymer exists, at molecular level, only in discrete oxidation states with $Y = 0$, 0.5 or 1.0 [2]. These oxidation states are interconvertible by redox process. The emeraldine oxidation state ($Y = 0.5$) consists of alternating reduced and oxidized groups. It can be protonated, or doped by aqueous acids, with a concomitant increase in conductivity; the extent of protonation being dependent on the pH of the acid solution. The 50% protonated emeraldine oxidation form exhibits highest conductivity [68], and this is the significant point in the present discussion. All conducting forms can be converted to the corresponding insulator forms of the same oxidation state by treatment with an alkali such as NH_3 or NaOH. The redox and acid/base reactions of different oxidation states/ forms of polyanilines.

3.1. Historical Background

Polyaniline was first known in 1835 as "aniline black", a term used for any product obtained by the oxidation of aniline. A few years later, Fritzche [69,70] carried out the tentative analysis of the products obtained by the chemical oxidation of this aromatic amine. Thereafter, Letheby [64] discovered that the end product of anodic oxidation of aniline carried out at platinum electrode in aqueous sulphuric acid solution, is a dark brown precipitate. Subsequently [71-74] these results were verified and similar observations were made during the oxidation of aqueous hydrochloric acid solutions of aniline [75]. Bucherer [76,77] proposed a phenazene type structure for polyaniline which is more complex and is contrary to the present-day understanding. Green and co-workers [78-83] proposed a linear octameric structure of the quinone-imine type in the *para*-position for the product obtained by chemical oxidation of aniline. The base of the octamer is leucoemeraldine, which exhibits four distinct states of oxidation. They are: protoemeraldine, emeraldine, nigraniline and pernigraniline **(Fig. 2.5)**. Yasui [84] in 1935, suggested a reaction scheme for the anodic oxidation

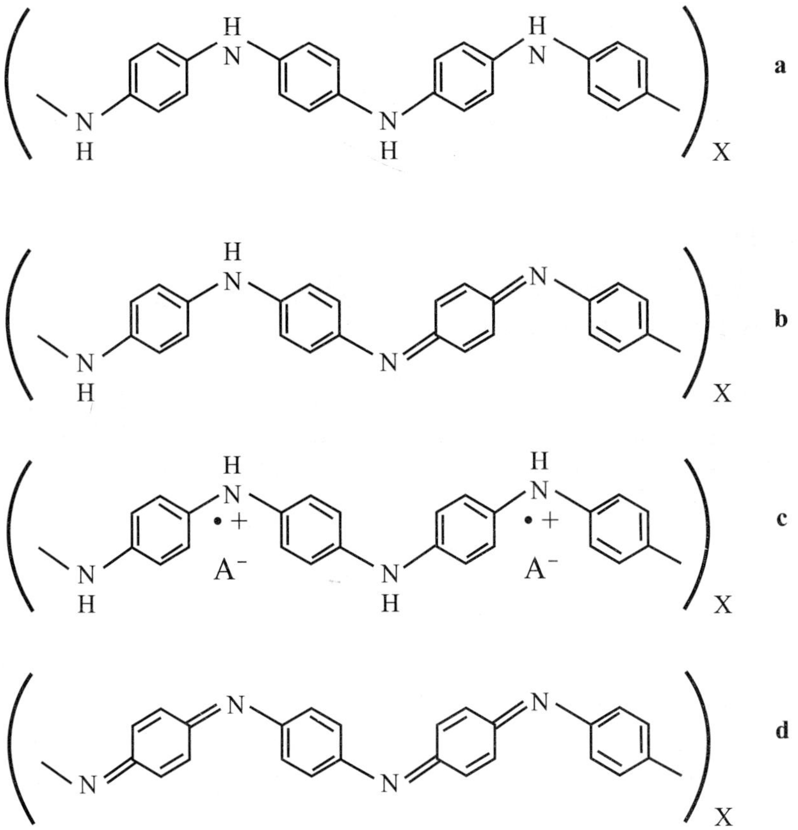

Fig. 2.5. Various states of oxidation and protonation of polyaniline: (a) leucoemeraldine base (b) emeraldine base, (c) emeraldine salt (d) pernigraniline base.

of aniline at a carbon electrode. No further progress was made until the middle of this century when Khomutov and Gorbachev [85,86] examined the results of Yasui, and verified Letheby's original observation about the green precipitate. In addition, on the basis of current-time curves, two different types of mechanisms were proposed for the electrode reaction. It was felt that further pursuit of the earlier research on oxidative condensation of aniline will help to have a better understanding of aniline polymerization and polyaniline structure which are of current interest.

3.2. Synthesis

Polyaniline can be synthesized either by the direct oxidation of aniline by chemical oxidants, or by anodic oxidation on an inert electrode (the abbreviation PANIC will be used for chemically synthesized polyaniline, and PANIE for electrochemically polymerized polyanilines).

3.2.1. Chemical Synthesis

PANIC is a precipitated product of aqueous solution containing ammonium peroxydisulphate (persulphate), acids like hydrochloric, sulphuric, nitric or perchloric and aniline. This is the classical approach to polyaniline synthesis in which aniline, the monomer, is converted directly to a conjugated polymer by condensation process. One of the disadvantages of this direct approach stems from the experimental observation that an excess of the oxidant and higher ionic strength of the medium produce material [4,87,88] that is essentially intractable.

Ortho- and *meta*-substituted aniline derivatives cannot be polymerized due to the steric and weak inductance effects of the substituents. The *para*-position should be free for radical coupling. During oxidative condensation of aniline, the solution progressively becomes coloured and turns into a black precipitate. The coloration of the solvent is probably due to the soluble oligomers. The intensity of the color depends on the nature of the medium and the concentration of the oxidant.

Many variations in the synthesis of PANIC have appeared in the literature. Major parameters which affect the course the reaction and the nature of the final product include nature of the medium, concentration of the oxidant, duration of the reaction and temperature of the medium. In addtion, we have studied the combined effect of concentration of ammonium persulphate (oxidant) and duration of polymerization reaction on the yield of insoluble precipitate (polyaniline) which will be discussed later.

Medium: While choosing a medium for the synthesis of PANIC the following are the important factors to be considered for getting desirable results: first, low ionic strength, second, volatility and third, noncorrosive nature of the medium. There is no medium that satisfies all of these requirements. Sulphuric acid due to its low volatility, a thin film of acid is left on the polyaniline powder after drying (to remove water) under dynamic vacuum. Hydrochloric acid and a novel anhydrous medium [89]–a eutectic mixture of HF and NH_4F with an average formula of NH_4F–2.3 HF (henceforth referred to as NF) are more volatile than sulphuric acid, but both are corrosive. One advantage of the NF medium is that the percentage yield of PANIC is maximum, compared to any other medium [89]. Thus, the oligomers formed in the reaction are at a minimum, and thereby an ordered material can be expected. We have tried [89] sulphuric acid containing $0.5M$ sodium sulphate (pH = 1) as a medium for the synthesis of PANIC. Sulphuric acid with low concentration avoids undesirable cations [4,87,88] while the presence of appreciable amounts of sulphate ions enhances the growth rate of the polymer chain [5]. It is usually admitted that the faster growth rate of nucleation yields better results. The thin film of sulphuric acid present on the particles of the powder could be eliminated by equilibration of the powder with 1M HCl [5,45]. This treatment replaces the sulphate ions by chloride ions on the polymer. This has been demonstrated in those polyanilines prepared under various conditions and this is a general feature of all polyanilines. Elimination of sulphate ions is of great importance if PANI is intended for use in a battery material in nonaqueous solvents such as propylene carbonate-1M lithium perchlorate medium, because the presence of sulphate ions in the polymer matrix precipitates lithium sulphate [45]. Use of salts, such as sodium sulphate or sodium chloride, in sulphuric or hydrochloric acid, may act as buffer and enhance the conductivity of the medium. Thus, the net effect will be augmentation of percentage yield of the insoluble precipitate and improvement in the 'quality' of the polymer obtained.

Effect of persulphate/aniline molar ratio (Z) and duration of reaction: The superiority of ammonium persulphate as an oxidizing agent for the condensation of aniline has already been established [90]. The present study shows that oxidant/monomer (persulphate/aniline) molar ratio is an important factor affecting the yield, conductivity and ion-exchange capacity of the resulting polyanilines [25-27]. Armes and Miller [91] have reported the optimum reaction conditions for the polymerization of aniline in a hydrochloric acid medium by ammonium persulphate. They observed that the yield of polymer was maximum when the persulphate/aniline initial mole ratio(r) was 1.0 to 1.5, whereas, yields were lower when r < 1 or r > 1.5.

Results of the syntheses carried out in the present study, at room temperature, by varying persulphate/aniline molar ratio and duration of reaction are shown in **Tab. 2.1**. By choosing a particular duration of reaction, (for example, 1 hour), we observed that the variation of yield with respect to 'Z' value, and found a region, *viz.* 1 < Z < 1.5, where the yield remained stable. The results are in full agreement with the data published by Armes and Miller [91] for oxidative condensation of aniline employing persulphate in hydrocholoric acid medium at 20°C.

For Z-values < 1 or > 1.5, the yields were very much lower. For example, when the 'Z' value was 0.25, a very low yield of ~20% was obtained; and at higher 'Z' value of 5.00 the yield of PANI was ~46%. The decrease in yield of PANI at 'Z' values >1.5 may be due to the degradation of

Tab. 2.1. Effect of persulphate/aniline initial mole ratio (Z) and duration of reaction on the yield of polyaniline (room temperature).

Reaction time (hours)	*Yield (%) of insoluble polymer** *						
$z^{**} \rightarrow$ *0.25*	*0.50*	*1.00*	*1.50*	*2.00*	*3.00*	*4.00*	*5.00*
0.25	23.4	46.5	91.7	87.8	77.6	67.0	54.0	47.2
0.50	21.5	45.1	94.1	88.4	77.3	56.4	54.8	46.2
1.00	20.2	45.2	90.2	89.2	78.4	60.5	50.2	48.5
2.00	20.5	46.0	88.3	89.5	86.6	55.7	54.4	46.8
4.00	20.8	46.4	89.5	89.9	72.3	60.2	58.5	45.7
6.00	21.2	45.8	89.9	90.2	77.2	61.3	55.7	45.8
8.00	22.8	45.6	90.1	91.8	76.7	61.8	54.5	44.9

*The anion (monovalent) content of 0.5 per ring is taken into consideration. The extraneous water present in the powder is not taken into account. Mere drying in dynamic vacuum is not sufficient to eliminate water completely.
**Different persulphate/aniline initial mole ratios (Z).

polymer molecules, resulting in aqueous soluble fractions (evidenced from the red colour of the filtrate-preliminary GPC analysis of which indicated a molecular weight of, 100). The lower yields at 'z' values <1.0 may be due to use of the lesser amount of persulphate than what is required by stoichiometry.

Syntheses carried out at each 'Z' value, for reaction times ranging from 15 minutes to 8 hours clearly indicate that the formation of polyaniline is almost complete in about 30 minutes. Therefore, it is concluded that reaction time is not an important factor affecting the yield, at least when the reaction time is >30 minutes (using persulphate as the oxidant). As against this Armes and Miller, [91] observed complete polymerization of aniline within 2-3 hours, when persulphate was an oxidant in a hydrochloric acid medium, at 20°C.

Effect of temperature: The process of chemical oxidation of aniline consists of a slow endothermic reaction, which is dependent on pH, temperature and the concentration of the reactants in which the dissolved oxygen in the solvent has no influence (in contrast to oxidation-reduction reagents which reduce the induction time) and a fast exothermic reaction which is temperature dependent and varies with the concentration of the oxidant. Furthermore, the polymerization reaction rate varies with temperature in the range 0-80°, whereas, the total enthalpy of the reaction (H = 372 kJ/mole) remains almost constant [88].

Similarly the syntheses were carried out at low temperature ~4°C under similar conditions as those at room temperature, except the aniline solution was cooled to ~4°C before the addition of ammonium persulphate and the reaction mixture was kept below 4°C by keeping the reaction vessel in a freezing mixture, or ice-bath. The syntheses were carried out at persulphate/aniline molar ratios of Z = 0.25, 0.50, 1.0, 1.5, 2.0, 3.0, 4.0 and 5.0. For each 'Z' value the syntheses were also carried out for reaction times of 1.0, 2.0, 3.0, 4.0, 5.0, 6.0 and 8.0 hours. The results are given in **Tab. 2.2**.

Tab. 2.2. Effect of persulphate/aniline initial mole ratio (Z) and duration of reaction on the yield of polyaniline (~4°C).

Reaction time (hours)	Yield (%) of insoluble polymer*							
$z^{**} \rightarrow$	0.25	0.50	1.00	1.50	2.00	3.00	4.00	5.00
1.00	20.4	44.9	90.2	85.6	51.6	35.8	35.5	33.2
2.00	20.8	45.3	90.2	84.0	46.4	38.2	36.7	35.1
3.00	21.2	45.4	89.1	84.3	47.7	28.2	28.8	31.6
4.00	20.9	46.3	90.5	92.5	43.6	34.6	33.3	31.5
5.00	20.7	45.0	95.0	90.9	48.3	31.8	31.1	32.1
6.00	21.6	46.6	92.0	89.9	47.2	34.1	34.1	32.1
8.00	20.5	45.8	90.9	90.2	46.9	33.3	34.8	33.2

*The anion (monovalent) content of 0.5 per ring is taken into consideration. The extraneous water present in the powder is not taken into account. Mere drying in dynamic vacuum is not sufficient to eliminate water completely.

**Different persulphate/aniline initial mole ratios (Z).

Representation of the results of chemical syntheses in 2-D and 3-D diagrams: It is customary to present the experimental data through two-dimensional graphs. When there are two parameters determine the dependent variable, then there are two options to represent the data: to plot different lines which may be tagged with the respective variable (for example, concentration of persulphate or persulphate/aniline molar ratio, in the present case). However, it is rather difficult to get a clear view of how the dependent variable (yield) varies with respect to the determining parameters (persulphate/aniline molar ratio (Z)) and duration of reaction and to draw 3-D graphs representing the variation of yield with the other two variables.

Generally, it is difficult to draw smoothened free-hand 3-D graphs and if attempted, it may not be free from certain kinks. The reason is that: without knowing the nature and extent of such variations, it is very difficult to do justice to 3-D representations. The alternate approach is that the kinetics, could be modelled to build a mathematical relationship (if possible) between dependent variable (yield) and independent variable (persulphate/aniline molar ratio, duration of reaction and temperature). Several options are available for accomplishing such a task; first: to empirically describe the observed behaviour with the associated risks that the resultant model may be applicable to the data set (or experiment) at hand only, if several of the influential factors have not been controlled or have been missed due to some reasons; second, to correctly visualize what may be the relationship between the end product (yield) and the reactants, where equilibrium constants, Arrhenius equations and Nernst equations could be effectivly used. Such models are possible, if exact stoichiometry of the reactions or reactions involved are known.

The task will be rather difficult in certain cases where such a relationship has not been established, or the exact stoichiometry has not been proved (e.g., reaction between phenols and Gibb's reagent, and reactions involving clathrate compounds [92]). Under such circumstances the only option is empirical modelling. The empirical method has two approaches: one is fitting appropriate polynomials (or multinomials) to get statistically satisfactory fitts. This is a very general and one of the versatile methods. However, it is difficult to know the precise nature of functional relationships. The other approach is to use known functions to depict an observed behavior or trend with varying degrees of complexities. With the latter, the nature of variations of dependent variables with each of the independent variable (reactant, and other conditions) can be understood better and it may provide, further insight into the possible mechanisms by better understanding of the nature of the reaction. Such information may be used to define the stoichiometry as well as other favorable conditions, which lead to an all-round much better control over the dependent variable. This approach has been adopted in the present case.

$$\text{yield} = y_0 + a_0 \left(\frac{t_2}{1.0 - a_1\sqrt{t} + a_2 t + a_2 t^2} \right) \text{X}\ldots\ldots$$

$$\ldots\ldots \left(\frac{r^2}{1 - a_4\sqrt{r} + a_5 r} \right)^3 \text{X}\ldots\ldots$$

$$\ldots\ldots \left(\frac{T^2}{1 + a_6\sqrt{T} + a_7 T + a_8 T^2} \right); \tag{2.1}$$

The yield data collected (%, weight basis) with time using different ratios of oxidant/aniline and temperature were fitted to the function [93] using the flexible polyhedron search method of Nelder and Mead [94].

$$\text{Yield (\%)} = 17.47 + 13732.30 \left[\frac{t^2}{1.00-4.68\ \sqrt{t}+4.75t+227.60t^2} \right] X.......$$

$$....... \left[\frac{r}{1.00-1.92\ \sqrt{r}+1.40r} \right]^3 X.......$$

$$....... \left[\frac{T^2}{1.00-51361.5\ \sqrt{T}+1536.9T-8.39T^2} \right] \tag{2.2}$$

In the given function: t is the time (h), r is the molar ratio (persulphate/aniline), T is temperature (K) and y_0, a_0, a_1, ..., a_8 are constants to be determined. These individual functions were selected based on the initial plots except for temperature.

The standard deviations of the data were minimized using Nelder and Mead's flexible polyhedron method [94]. The method is reported to be quite efficient, and gives consistent results. The fitted equation is given below, with residual mean of sum of squares of 2.91 and multiple correlation coefficient of 0.94 explaining more than 87% of the variation.

The 3-dimensional and 2-dimensional diagrams illustrating the variation in yield with respect to persulphate/aniline molar ratio and time of reaction for syntheses at room temperature are displayed in **Fig. 2.6** and those for syntheses at low temperature (~4°C) in **Fig. 2.7**.

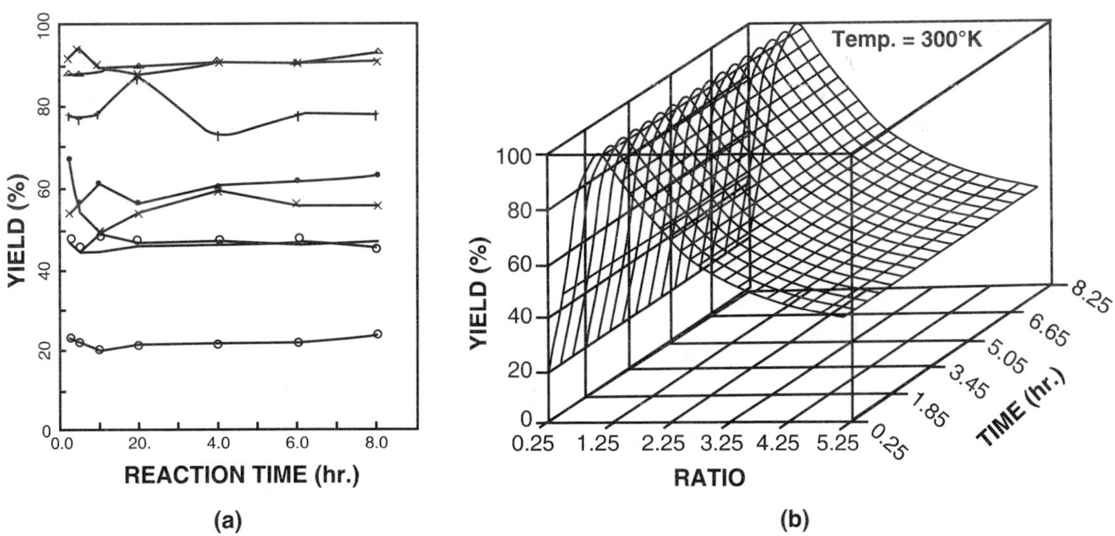

Fig. 2.6. (a) **2-D diagram representing the variation of yield (at room temperature) with respect to persulphate/aniline, molar ratio (Z) and time of reaction z = 0.25 (—O—O—O—) 0.50 (—⊗—⊗—⊗—), 1.00 (—× — × — ×—), 1.5 (— Δ — Δ — Δ —), 2.0 (— † — † — † —), 3.00 (—·—·—·—), 4.00 (—✦—✦—✦—) and (—☐ — ☐ — ☐ —) and (b) 3-D diagram representing the same results effected with the help of computer.**

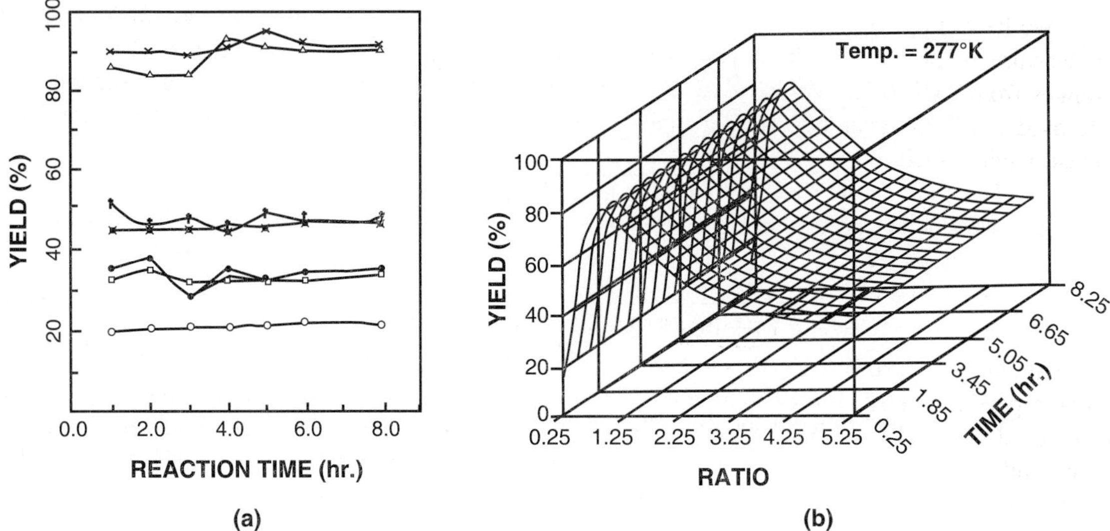

Fig. 2.7. (a) 2-D diagram representing the variation of yield (at room temperature) with respect to persulphate/aniline, molar ratio (Z) and time of reaction z = 0.25 (—O—O—O—) 0.50 (— ⊗ — ⊗ — ⊗ —), 1.00 (—× — × — ×—), 1.5 (— Δ — Δ — Δ —), 2.0 (— † — † — † —), 3.00 (—·—·—·—), 4.00 (—✦—✦—✦—) and 5.00 (—□—□—□—) and (b) 3-D diagram representing the same results effected with the help of computer.

Syntheses carried out at low temperature revealed that the temperature of the medium has a limited influence on the yields at (< 1.5) 'Z'-values. At higher values the yield of PANI was less than that obtained at room temperature.

Effect of oxidant: Ammonium persulphate, potassium dichromate, ceric sulphate, sodium vanadate, potassium ferricyanide, potassium iodate and hydrogen peroxide (copper and iron salts have also been employed which will be discussed later) have been used as oxidants for the polymerization of aniline in acidic media. Of these ammonium persulphate is most extensively used and the yield, elemental composition, conductivity and extant of oxidation of the resulting polymer are essentially independent of the value of the initial aniline / persulphate mole ratio (r), when $r < 1.15$ [91]. However, when $r > 1.15$, it results in over-oxidation of the polyaniline, with a concomitant decrease in the conductivity [91] and the yield of the polymer [90]. A marked change in morphology is also observed [90]. At higher concentrations ($r > 1.15$) of the oxidants, such as cerium(IV) sulphate, potassium dichromate, sodium vanadate and potassium ferricyanide, a complexation reaction probably occurs which results in products containing a higher percentage of the metal [90] that is indicated by elemental composition data. This can be further confirmed by igniting the sample and analyzing the residue, which indicates the presence of the metal [90]. As regards the duration of the reaction *vis-à-vis* the percentage yield of the polyaniline, it was found that yield percentage increases with time (0-2 hr), when $r \leq 1.15$. Thereafter, the duration of the reaction has only a limited influence on the yield, conductivity and elemental composition of the polymer [90,91].

Effect of standard reduction potential: Apart from the major parameters discussed above, we have studied the effect of the standard reduction potential (SRP) of the chemical oxidant on the nature and yield of the final product, the polyaniline. The results of our studies along with the conclusions drawn by Pron *et al.* [95] are summarized below:

We have employed [25-27] five oxidants, namely: ammonium persulphate, cerium(IV) sulphate, potassium dichromate, sodium vanadate and potassium ferricyanide. The SRP of these oxidants ranges from 2.01 to 0.36 V *vs.* NHE. Pron *et al.* [95] have employed potassium iodate and hydrogen peroxide, as well as ammonium persulphate and potassium dichromate. Our major conclusions from these studies include:

1. The SRP of the oxidant greatly affects the yield of polyaniline.
2. Ammonium persulphate is a better oxidant for polymerization of aniline, because of:
   - High SRP
   - Non-metallic oxidizing species
   - Non-interference of the reduction product, namely, the sulphate-ion.

In constrast to this according to Pron *et al.* (1) the process of polymerization of aniline is much less dependent on the redox potential than on the degradation process; (2) degradation is significantly slowed down in the case of chemical preparation; (3) hydrogen peroxide is ruled out as a reagent in the synthesis of PANIC; and (4) potassium iodate is the most convenient, since it gives good quality samples in a wide range of synthesis parameters. However, it is necessary to note that the latter employed different temperatures and reaction times for different oxidizing agents (**Tab. 2.3**). Hence, the conclusion that potassium iodate is a better oxidant for the synthesis of PANIC appears to be arbitrary. Nevertheless, another communication reports [96] the use of potassium iodate as the oxidant in the preparation of colloidal polyaniline.

Tab. 2.3. Chemical synthesis of polyaniline (PANI) using different oxidants under varied reaction conditions.

Oxidant	Nature of species	Reaction time (h)	Temp. °C	Redox couple	SRP	Reaction medium	Ratio**	Yield %
*KIO_3	Non-metallic	15.0	Zero	IO_3^-/I_2	1.195	2M HCl	0.61	100.0
*KIO_3	Non-metallic	3.0	Zero	IO_3^-/I_2	1.195	2M HCl	0.61	48.5
*$(NH_4)_2S_2O_8$	Non-metallic	1.5	Zero	$S_2O_8^{2-}/SO_4^{2-}$	2.010	2M HCl	0.61	82.4
*H_2O_2	Non-metallic	15.0	RT	H_2O_2/H_2O	1.775	4M HCl	0.61	60.0
$(NH_4)_2S_2O_8$	Non-metallic	1.0	RT	$S_2O_8^{2-}/SO_4^{2-}$	2.010	1M H_2SO_4	1.00	99.0
$Ce(SO_4)_2$	Metallic	1.0	RT	Ce^{4+}/Ce^{3+}	1.440	1M H_2SO_4	1.00	75.4
$K_2Cr_2O_7$	Metallic	1.0	RT	$Cr_2O_7^{2-}/Cr^{3+}$	1.330	1M H_2SO_4	1.00	96.0
Na VO_3	Metallic	1.0	RT	VO_2/VO^{2+}	1.000	1M H_2SO_4	1.00	94.4
$K_3[Fe(CN)_6]$	Metallic	1.0	RT	$Fe(CN)_6^{4-}/Fe(CN)_6^{3-}$	0.360	1M H_2SO_4	1.00	2.1

*part of the data from Pron *et al.* [95]; RT = Room Temperature; SRP = Standard Reduction Potential vs NHE.
**Normalized aniline/oxidant ratio.

The redox couples of ammonium persulphate ($S_2O_8^{2-}/SO_4^{2-}$), cerium(IV) sulphate (Ce^{4+}/Ce^{3+}), potassium dichromate (Cr^{6+}/Cr^{3+}), sodium vanadate (VO^{2+}/VO^{2+}) and potassium ferricyanide ($Fe(CN)_6^{4-}/Fe(CN)_6^{3-}$) have standard reduction potential values of 2.01, 1.44, 1.33, 1.00 and 0.36 V respectively *versus* NHE. Of the four oxidants used in our study, ammonium persulphate has the following characteristic features: first, it has a non-metallic oxidizing species, namely, persulphate ion, $S_2O^{2-}_8$, second, non-interference of the reduction product, namely, sulphate ion (because the synthesis is carried out in sulphuric acid medium) and third, it has the highest standard reduction potential.

The higher reduction potential value of $S_2O^{2-}_8/SO^{2-}_4$ redox couple is interesting, as it may initiate nucleation faster. A large number of nuclei may help to build up a longer chain. This is supported by the highest yield of PANI **(Tab. 2.3)**. (It is presumed that the larger the chain length, the lower is the solubility.) This opens up a wide scope for study of molecular weight of polyanilines synthesized with different oxidants in the same medium. Until the molecular weight data become available, it will be difficult to arrive at a definite conclusion about structure-activity relationship.

The use of oxidizing species containing metals for the polymerization of aniline leaves behind certain residual impurities as for example the polyaniline obtained by the oxidative action of potassium dichromate (or sodium vanadate) on aniline contains a residue of 0.93% chromium (7.3% vanadium). This is evident when known amount of polyaniline sample was ignited (in a silica crucible heated in an electric bunsen) and the residue was analyzed by gravimetric method. We have also observed that the presence of residual impurities affects the magnetic properties of the polymer [6]. However, the polyaniline obtained by the oxidation of cerium(IV) sulphate on aniline did not contain any metallic residue. This can be attributed to the diamagnetic properties of cerium, which do not participate in the formation of complexes with the same facility as chromium or vanadium. However, low yield, low conductivity, and poor performance as an anion-exchange resin [25] and high cost of cerium(IV) sulphate are the limitations for the synthesis of polyaniline.

In brief, polyaniline obtained as an insoluble residue during the oxidative condensation of aniline is separated by filtration, washed with a copious amount of acidic solution of desired ionic strength to remove oligomers, and dried under dynamic vacuum for *ca.* 48 h. The resulting dark powder is generally transferred to a soxhlet apparatus to remove low-molecular-weight species and extracted with CH_3CN, C_2H_5OH or CH_3OH to get a colorless extract. The drak powder is dried under dynamic vacuum.

3.2.2. Chemical Synthesis of Emeraldine Hydrochloride and Emeraldine Base

Synthesis of emeraldine hydrochloride can be summarized as follows [8]. An aqueous solution of ammonium persulphate is added slowly to a solution of aniline in 1.0 mole/dm³ aqueous hydrochloric acid (both solutions, precooled to 1°C). The reaction mixture is stirred for about an hour (~5°C). The precipitate is removed by filtration and washed repeatedly with 1.0*M* hydrochloric acid followed by drying under dynamic vacuum for 48 h. The find material is emeraldine hydro-chloride [1].

The moist emeraldine hydrochloride precipitate cake is suspended in ammonium hydroxide solution for ~15 h. The separated powder is washed and dried under dynamic vacuum. The powder obtained is emeraldine base [68]. In conclusion, the chemical route for preparation of polyaniline is good method, but at present, electrochemical methods also provide a satisfactory product.

3.2.3. Electrochemical Polymerization

There is two-fold interest in the electrochemical polymerization (ECP) process. First, ECP reactions provide a new method of polymerization with a fine control of the initiation and termination steps. Second, ECP has technological potential for applications. Besides, one of the important features of

ECP reactions of conducting polymers is that they proceed with electrochemical stoichiometry. This is because electrochemical reactions are often devoid of any possible pollutants, compared to chemical reactions. Moreover, electrons, as a reagent, are inherently pollution-free, at least, at the point of use.

Electrochemical methods generally employed for the polymerization of aniline include: first, constant current or galvanostatic; second, constant potential or potentiostatic; and third, potential scanning/cycling or sweeping methods. The first method essentially consists of a two-electrode assembly dipped in an electrolyte solution containing the monomer. Passing a current density of *ca.* 1 mA/cm^2 will result in the deposition of a PANIE film on the surface of a platinum foil electrode. Polymerization of aniline at constant potential produces a powder which adheres poorly to the electrode [97]. On the other hand, electrooxidation of aniline by continuous cycling between the predetermined potentials produces an even polymeric film which adheres strongly to the electrode surface [21-23,97]. This thin film can be cycled between conducting oxidized and reduced states. Thicker films can be produced and can be peeled off from the electrode surface which have free-standing and electrically conducting properties. As these films are in the oxidized state, they represent polyaniline cations, and their overall charge balance can be achieved by incorporation of counter anions from the electrolyte of the solution. Such counter anions include F$^-$ [21-23], Cl$^-$ [1], CIO$_4^-$ [5], SO$_4^{2-}$ [5,21-23,45] and BF$_4^-$ [1,99].

The anodic oxidation of aniline is generally carried out on an inert electrode. Though the usual anode material would be platinum, or conducting glass, many metals such as Fe [100-102], Cu [103] and Au [7] have been employed. Graphite [104], vitreous carbon [105,106] stainless steel [90] and n-type silicon [107] can also be used. Metals such as Ag or Al, which get oxidized more readily than aniline monomer, would obviously be not a good choice for the anode. Until recently, it was believed that PANIE deposited from aprotic solutions such as CH$_3$CN is electro-inactive, and, therefore, it was thought that electroactive polyaniline could be formed only from acidic aqueous solutions. However, it has been proved that electroactive polyaniline can be synthesized in an aprotic solvent, such as propylene carbonate solution containing an organic acid (CF$_3$COOH) and an electrolyte, lithium perchlorate [108]. The ECP of aniline is described as a bimolecular reaction involving a radical cation intermediate. The reaction has a ΔH equal to 121 kJ/mole^{-3}, αn equal to 1.2-1.3 [109], where $n = 2$ [110]. Here, n is the number of electrons consumed during the electrochemical polymerization and α is the transfer coefficient. It is reported that the number of electrons (n) consumed during ECP of aniline are 2.60-2.70 [22,23], 2.16 [97] and 2.25-2.29 [5].

In summary, electrochemical methods have definite advantages over chemical methods in the synthesis of polyaniline, because of the reliability of the techniques. Moreover, the stoichiometric electropolymerization reaction can be a general procedure for the preparation of organic polymer films, with electroactive properties and good electrode behavior. Polyaniline is one such example.

4. POLYANILINE IN MICROELECTRONICS

Materials representing the entire spectrum of conductivity are at the heart of the microelectronics industry [111-116]. Conducting materials are used extensively in interconnection applications for electrostatic discharge, protection of integrated circuits for electromagnetic interference shielding and electronic equipment such as mobile electronic "gadgets", desktop personal computers and large super computers.

Conducting polymers have a wide range of applications in microelectronics. In the area of lithography, they have been shown to provide an effective, simple, spin-apply process for charge dissipation, in particular for e-beam lithography and for SEM metrology. However, application of these polymers in this area will be guided by their commercial availability and cost. The lithographic performance of these systems has not currently reached a stage where they can compete with conventional resists. They require further improvements for the sucessful applications.

Conducting polymers have been applied for metallization of plated through holes for printed circuit board technology and this process is being currently used by a number of companies worldwide. Conducting polymer-based coatings have been developed and they offer excellent ESD protection and process numerous advantages over currently used materials. A number of coating formulations are either already commercially available or in the process of being manufactured. Protection of metals such as silver and copper from corrosion using conducting polymers has been shown to be quite promising. Conducting polymers have today certainly entered into the marketplace. As the materials become commercially available and more and more properties are evolved, greater application potential may be realized in the future with this class of polymers.

Polyanilines during the last two decades have shown remarkable potential as candidates in the field of microelectronics. The purpose of this section is to provide an overview of the present status of understanding, knowledge and the progress made by polyanilines in the field of lithography, metallization and their use for the protection of metals and electronic devices.

4.1. Lithography

Lithography is essentially based on radiation-sensitive polymers called resists. These materials when irradiated undergo chain scissioning, cross-linking, molecular rearrangement or other changes that create a solubility difference between the irradiated or exposed areas and the nonirradiated or unexposed areas of the polymer [117-120]. Subsequently when developing is done, the more soluble regions are selectively removed. In positive-acting resists the exposed regions become soluble, whereas

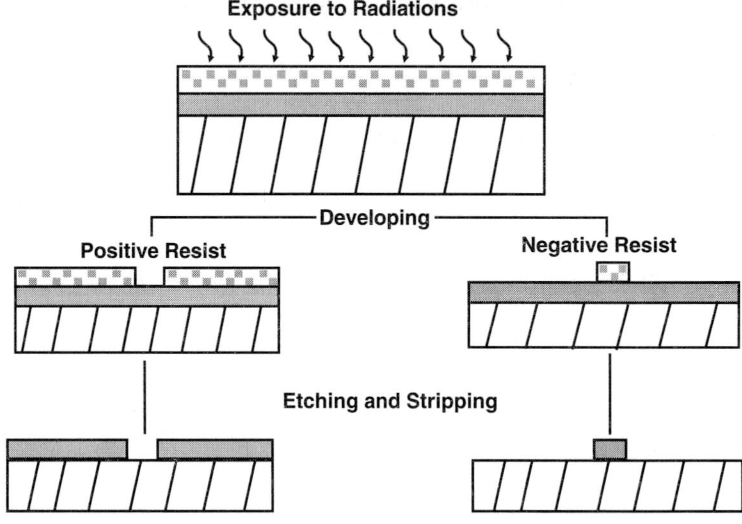

Fig. 2.8. Schematic diagram of delineation by lithography.

in negative-acting resists the exposed regions become insoluble. In either case a pattern gets generated in the resist as shown in **Fig. 2.8**. This pattern is subsequently transferred to the underlying substrate for example, (silicon dioxide or metal) by etching processes followed by removal of the resist [119-121].

4.1.1. Resists

Resists are patterned with photons, e.g., ultra-violet radiation, electron beams, X-rays, and ion beams. Photolithography has been the dominant technology in the industry to date. In this case, the resist is exposed to ultraviolet radiation through a quartz/chrome mask containing the pattern to be transferred. Electron beam (e-beam) technology is also currently used but on a much smaller scale. It is used to fabricate high resolution, low volume speciality chips or prototype chips and also to fabricate the masks for photolithography. Ion beam and X-ray as well as e-beam lithography are currently under investigation as possible future alternatives to optical lithography [119,120,122,123].

4.1.2. Electron Beam Lithography

Electron beam lithography is a direct-write method in which a focused beam of electrons is directly scanned over the resist [119,120]. No masking is required as the pattern is computer-generated. It is a technology capable of providing high resolution as the beam of electrons can be focused to tens of nanometers [124]. It is capable of giving excellent level-to-level pattern overlay. However, currently it is considered to be a low throughput technology and is limited to fabrication of high resolution speciality/prototype chips and for fabrication of masks [119].

During the e-beam patterning process, charging of the resist is a major problem [125-127]. Thick insulating resist materials can trap charge and delay bleed-off through the underlying silicon. The trapped charge as well as any surface charge can deflect the path of the e-beam and result in image distortion as well as level-to-level registration errors. To circumvent this problem, conducting materials that function as discharge layers are incorporated into resist systems as coatings above or below the imaging resist.

In the initial stage [128] when polyaniline was evaluated as an e-beam in lithography the polymer was incorporated into a multilayer resist system (**Fig. 2.9**) as a conducting interlayer between the imaging resist and the planarizing underlayer. The multilayer structure was designed to test the

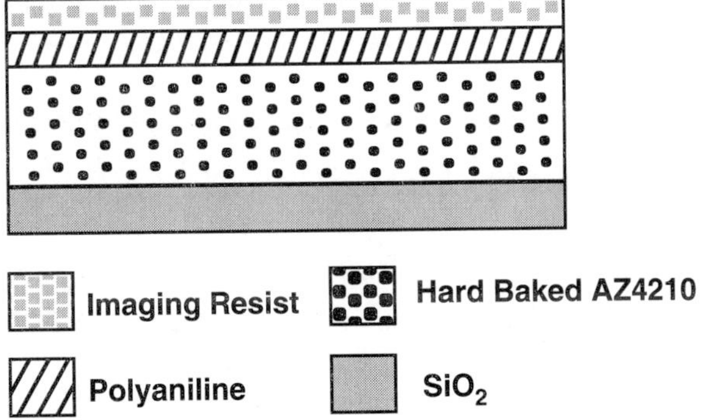

Fig. 2.9. Polyaniline as a conducting interlayer in a multilayer resist.

efficiency of the conducting polyaniline as charge dissipator. In this case relatively thick layers of insulators were used so as to enhance charging. The underlayer consisted of a 2.8 μm film of a hard-baked or cross-linked AZ4210 material and 5000 Å of silicon dioxide. Polyaniline in the base form (2000 Å) was spin-applied onto the AZ material. The base polymer was doped by dipping in dilute aqueous hydrochloric acid solution. As this conducting salt is insoluble, the resist (a typical diazonaphthoquinone-novolac formulation) [119] could be directly coated on top of the polyaniline without any interfacial problems.

The multilayer structure was subjected to a discharge test that evaluates resist charging during the e-beam writing process. Initially, a 20 × 20 matrix of 2 μm squares was written across a 5 min chip. Because of its low density, charging during the writing of this pattern is negligible. The chip was then completely overwritten except for a 10 × 10 μm area centered on each inner square. Following exposure, the resist was developed and the images were inspected to determine the degree of pattern displacement [128].

A pattern displacement greater than 5 μm across a 5 mm chip was observed in the case where a conducting discharge layer was not incorporated into the resist structure and the use of polyaniline was found to give zero displacement [128]. These results demonstrate that polyaniline is very effective at eliminating resist charging even though in this particular configuration it is used as a thin interlayer between thick insulating layers. In addition, in this configuration the polyaniline is not grounded. The substrate (i.e., silicon wafer) is grounded, but the connections are made on the top and bottom surfaces of the wafer. The polyaniline functions by bleeding off charges from the resist and preventing charge build-up in the resist layer that would deflect the e-bearm and create placement errors. Further, in this study [128], the conductivity of polyaniline was varied to determine the required level of conductivity in order to eliminate resist charging. The conductivity was tuned by altering the pH of the aqueous acid dopant solution [8,20]. It was found that a conductivity greater than 10^{-4} S/cm was required to prevent pattern displacement. However this conductivity requirement was for the interlayer structure. The level of conductivity needed for charge dissipation was found to depend on the resist structure configuration.

The polymer processing was done in two steps while valuating polyaniline as a charge dissipator in resist systems. The base form of the polymer, which is generally in the more soluble form of polyaniline was first spin-applied to a surface. The sample was further converted to conducting form by dipping in an aqueous acid solution. The doping reaction takes several hours to complete the diffusion of the dopant into the bulk of the film. This two-step process, involving soaking of the sample in aqueous acid solution is not desirable. Besides increasing the number of processing steps, the prolonged exposure of substrates to acid solutions poses problems like contamination, reliability and corrosion.

In a subsequent study [129], a simplified, one-step process was developed to apply the polyaniline to resist systems. A method involving inducing the doping *in situ* in the polymer was developed eliminating the need for external acid solutions. This was accomplished by incorporating salts into the polymer systems that would decompose upon irradiation or thermal treatment to generate the active dopant species, i.e. a protonic acid, *in situ* in the material [129-131]. The incorporated salts are of two types—onium and amine triflate. Onium salts, such as, triarylsulfonium and diaryliodonium decompose upon exposure to ultraviolet radiation or to an e-beam to generate a protonic acid [132,133]. When a polyaniline base film containing an onium salt was irradiated, conductivity of 0.1 S/cm was obained [130].

Thermally induced doping is a simpler method compared to radiation-induced doping as it involves only application of the film and baking. Baking is normally done on any film to remove solvent. Thus, no additional steps are needed. Polyanilines doped by this method were evaluated for charge dissipation in the same way as described above for the aqueous acid-doped materials and were found to be equally effective [129].

In the studies described so far, the polyaniline was incorporated below the imaging resist. In these systems, once the resist was exposed and developed, the polyaniline continues to remain in the open areas as shown in **Fig. 2.10(a)**. In order to transfer the pattern to the underlying substrate, the exposed polyaniline must be removed. This is achieved by oxygen reactive ion etching (RIE) [119,121,128] after the imaging resist is made etch-resistant by an appropriate silylation process [134].

Although the method of using polyaniline below the resist works quite well, an easier and appropriate approach will be to apply the conducting layer on top of the resist and remove the conductor simultaneously with the development of the resist. The relevant configuration and process are depicted in **Fig. 2.10(b)**. As the conductor is applied directly on the resist, it must fulfill certain requirements. First, the solvent used to coat the conductor must not dissolve the resist or induce any interfacial problems. Thus, polar solvents such as N-mythylpyrrolidone (NMP), which are commonly used to process polyaniline, are not suitable as they dissolve the commonly used resists. The conductor must not degrade the lithographic performance of the resist. It should not introduce any contamination. Second, it should be completely removed, if possible during the development of the resist. Most resists currently used in the industry are developed in aqueous solutions, and thus, the conductor must get to dissolved in these aqueous systems.

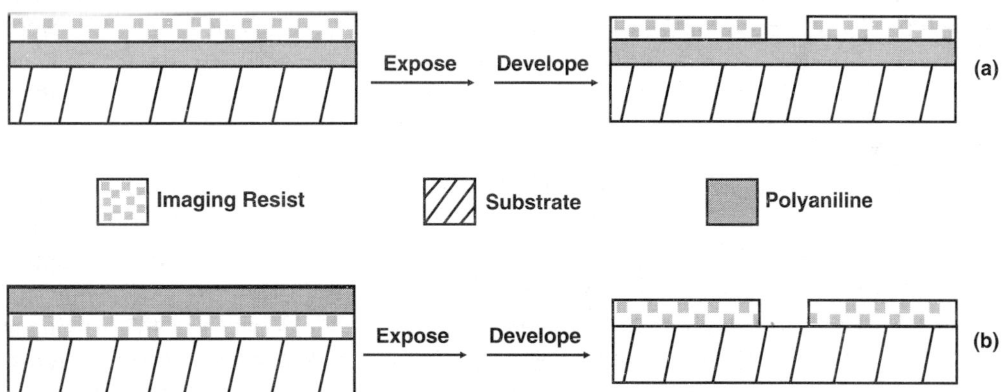

Fig. 2.10. (a) Polyaniline as an interlayer in a multilayer resist structure and (b) Polyaniline as a topcoat discharge layer in a multilayer resist structure.

In the last few years a number of polyaniline derivatives have been developed that are soluble in conducting form [39,40,43]. Most of them are organic solvent based and although some of these could in principle be used as topcoat discharge layers, but water-soluble polyanilines are more preferred. Such polymers eliminate environmental concerns of organic solvents. Reduction of solvents is an issue of increasing importance especially in the electronics industry. Hence, water-based systems would be more preferred with current resists.

A number of water-soluble polyaniline derivatives have been developed in recent years [34-36,44,135-139]. Among these one method used to introduce water solubility has been to incorporate sulfonate groups onto the polymer backbone. There are several routes to accomplish this. One process used sulfonation of polyaniline base by treating the polymer with fuming sulphuric acid [34,136]. This resulted in a sulfonic acid ring-substituted derivative that was alkaline-soluble but only upon conversion to the nonconducting sulfonate salt form. Another method of introducing sulfonate groups was by deprotonating polyaniline base and reacting with a sultone, such as 1,3-propanesultone [137,138]. This gives rise to an water-soluble *N*-substituted polyaniline derivative. Another route involved is the polymerization of sulfonated aniline monomers such as sodium salt of diphenylaminesulfonic acid [34,35]. IBM introduced a series of water-soluble polyanilines [44,139] referred to as PanAquas (trademark of IBM Corporation). These polymers are highly soluble in neutral water in the conducting form. They are prepared by one-step straightforward synthesis involving a template-guided polymerization. In this process, the aniline monomer is first complexed to a polymeric acid. Once the complex is formed, the aniline is polymerized in a controlled fashion to allow the polyaniline chain as it grows to wrap around the polyacid chain. By this the formation of an interpenetrating network is prevented. As a result, the polyaniline/polyacid blend that is isolated directly in the conducting form is watersoluble. A number of derivatives can be made by this method by varying the nature of the aniline monomer (variations in R) and also the nature of the polyacid. These polymers have conductivity of the order of 10^{-2}-10^{-1} S/cm.

The PanAquas provide a simple and effective discharge solution for e-beam lithography [139]. In particular, the unsubstituted derivative where, R = H was spin-applied onto the surface of a number of common resists such as novolacs, acrylates, and chemically amplified systems used in the industry [119] and was found to be compatible. The performance of the resist was in no way altered. A 2000 Å layer was found to be quite effective in eliminating charging of the resist. The PanAqua can be completely removed during the alkaline development of the resist [139].

4.2. Metallization

In microelectronics, "metallization" generally refers to the deposition of a patterned film of conducting material on a substrate to establish interconnections between electronic components [112]. Over the last few years, polyanilines have been used to provide a new route to metallization, in particular in printed circuit board (PCB) technology [140-144].

PCB [145-147] are of very complex in nature ranging from single-sided boards where circuitry is found only on one side to double-sided boards, to boards consisting of multilayers of circuitry. The extent of complex depends on the specific interconnection needs of a given product. Connections between the two sides of a board and layer-to-layer connections are made with copper-plated through holes (PTH) [145-147]. PTH allow greater circuit density as they provide crossover capability. One circuit can cross over another by entering a PTH, continuing on the other side and so on. Through holes are drilled into the laminate substrate and are then copper plated.

Two common metallization schemes for PCB are in operation [145-147]. In one of the schemes conducting strike layer, generally a thin layer of copper is deposited by electroless plating. This copper layer renders the surface sufficiently conducting to allow a thicker copper layer to be electrolytically deposited in selected regions that are defined by a photoresist process. In an another scheme an all electroless process is used.

Electroless deposition uses noble metal salts, like $PdCl_4^{2-}$, as seeds. The salts are applied to the PCB surface followed by reduction of the noble metal to the zero-valent state. The zero-valent noble metal particles are the active sites for heterogeneous copper reduction in electroless plating. The precious metal seeds are expensive. Electroless baths are generally unstable and require close monitoring. The baths can be either too stable, which results in PTH voids, or too active, which results in homogeneous decomposition of the bath. Futher, the most commonly used reducing agent in electroless baths formaldehyde, is toxic and poses environmental concerns.

An alternative to the presently used methods is the use of polyaniline as an electrode for direct electrolytic metallization of copper [140]. Because of its solubility, the polymer can be directly deposited onto a PCB by a dip-coating process. In this study [140], PCB was coated with a thick (~1 µm) layer of polyaniline applied from an aqueous acetic acid solution. Today, many soluble polyaniline derivatives have been developed, and a variety of solvents including water can be used to apply the polymer. The polyaniline coated board is in fact electrolytically copper plated.

4.3. Polyaniline for Protection of Metal Corrosion

Metals such as copper (Cu) and silver (Ag) are widely used in microelectronics for wiring, EMI shielding, and other purposes. These metals readily corrode in a variety of ambients. In oxygen saturated water, Ag and Cu dissolve with a corrosion rate of about 1×10^{-7} and 1×10^{-5} A/cm^2, respectively, or 0.002 and 0.2 nm/min, as can be measured in the potentiodynamic polarization curve [148-150]. With increased anodic potential, metal dissolution rapidly increases to exceed 10^{-1} A/cm^2, corresponding to catastrophic metal removal at a rate of at least 35 nm/s [148-150]. In the presence of an applied potential and humidity, not an uncommon situation for devices in operation, these metals dissolve from the more positive metallic part and plate at the more negative part as dendrites [148-151] which can cause short circuits. In addition, with increasing line density and decreasing dimensions, ion accumulation alone without accompanying dendrite formation can destroy the designed electrical performance of the product. Inhibitors such as benzotriazole (BTA) provide excellent corrosion protection for both the metals [148-151]. But, these azole-type inhibitors do not provide sufficient protection at elevated temperatures as in soldering application nor do they protect against an applied potential [150,151]. Therefore, new materials are needed to protect Ag and Cu against corrosion and dissolution particularly that can provide protection at high temperature and in the presence of an applied potential.

The use of polyaniline for corrosion protection of metals, particularly stainless steel, has been investigated over a decade. In the earliest study [155] polyaniline was electrochemically deposited on ferritic stainless steel to provide anodic protection that significantly reduces corrosion rates in acid solutions. Since then numerous studies have confirmed the corrosion-preventing properties of polyanilines. Further, non-electrochemical methods of applying polyaniline have been demonstrated [52-54,156-161]. In a recent study [159], dispersions based on doped polyaniline, Versicon (trademark of Allied Signal Corp.) were found to passivate mild steel, stainless steel and copper.

The use of polyaniline to protect Cu and Ag, particularly, at high temperature and under an applied potential is of interest to the microelectronics industry [152-154]. A number of soluble polyanilines were evaluated in both doped and nondoped forms. These materials were tested by two methods that were designed to closely resemble conditions under which metals are exposed during application in any of the electronic products, as for example in a PCB. In most of the cases, the corrosive environment varies depending on the relative humidity, or the amount of adsorbed water on the surface. One method uses a three-electrode electrochemical cell with a water droplet as an electrolyte as described earlier [162]. This setup basically consists of a sample working electrode, which is the coated metal masked with plating tape to expose 0.32 cm^2 area, a platinum mesh counter electrode, and a mercurous sulfate reference electrode. A filter paper disk separates the electrodes. A droplet of water is introduced as an electrolyte. The corrosion potential is monitored for ≈15 min, and the polarization resistance is measured by scanning the potential ± 20 mV from the corrosion potential. The potentiodynamic polarization curve is measured from 0.25 V cathodic of the corrosion potential. The corrosion rate is evaluated by extrapolation of the cathodic and anodic currents to the corrosion potential. In this study [153], the dissolution rates at high anodic potentials are closely monitored. In a second method, patterned metal lines are tested for ease of dendrite formation by placing a water droplet across adjacent metal lines with an applied potential between the lines.

4.4. Polyaniline for Protection of Electronic Components

Electrostatic charge (ESC) and electrostatic discharge (ESD) constitute a serious and expensive problem for many industries, particularly for microelectronics [163-166]. It has been estimated that $15 billion a year are attributed to ESD damage alone by the U.S. electronics industry [163-164]. Electrostatic charge can accumulate to thousands of volts. The static charge can attract airborne particles, as is often observed on cathode ray tubes (CRT). This becomes a significant concern of contamination if particles are attracted on critical surfaces such as device wafers. The accumulated charge will eventually get discharged in the form of a "lightning bolt," which can destroy devices on IC [163-165].

Conducting materials are extensively used in clean rooms during their manufacture to protect devices against ESC and ESD. In addition, conductors are incorporated into plastic packages that are used to transport sensitive electronic components such as chips, modules and PCB. The materials currently used include ionic conductors, carbon- or metal-filled resins, and in certain cases metal coatings [167]. These materials do not offer the ideal solution to protect ESD. The use of ionic conductors is an inexpensive approach and has significant drawbacks. These materials exhibit very low surface conductivity (10^{-9} to 10^{-11} Ohms$^-$/□) and thus are not dissipative. The conductivity is humidity dependent, as an electrolyte water is necessary for ionic conduction. In addition, ionic conductors can readily be removed by water, which precludes any washing of structural parts of the container. The use of ionic conductors is not a reliable method for ESD protection. Electrical conductors, on the other hand, are stable systems and the conductivity is not humidity dependent. But, these are more expensive materials. Carbon filled systems are subjected to contamination due to sloughing of the carbon particles. Besides, relatively high loading levels are required to attain a given level of conductivity. Such high loading not only degrades the mechanical/physical properties of the host polymer but also recycling of the plastic carriers becomes more difficult.

Polyanilines offer an alternative to ESD protection with numerous advantages over currently used materials. The conductivity can be tuned to the requirement and can easily meet the high end of the dissipative range. The conductivity is stable compared to ionic conductors. Besides, they offer a high degree of transparency. These polymers have been used as fillers in a number of host resins. In addition, coating formulations that can be applied directly onto plastic surfaces have also been developed.

Polyaniline in the form of dispersible powder (Versicon) can be blended with a number of thermoplastic and thermoset resins to achieve excellent ESD properties [165-170]. Soluble polyanilines have also been blended with low loading and appropriate polymers to reach the required level of conductivity [171,172].

Of particular interest for ESD protection of electronic component packages are coating formulations. The coatings can be applied directly by spray technique on already fabricated packages or they can be applied onto plastic sheets that are subsequently thermoformed into packages. A number of coatings based on polyanilines have been developed and are commercially available. One such system [44,173] is a curable waterbased coating reported by IBM. Formulations based on soluble polyanilines have the advantage that they do not pose contamination due to particles. This is particularly important for microelectronics where actual devices are in contact with the conducting carriers. Any particle sloughing would contaminate the devices. Some of the polyanilines coating formulations described earlier are also commercially used for ESD protection.

4.5. Prospects in Microelectronic Industries

There are many more potential applications for polyanilines in microelectronics beside those discussed already. Polyanilines can also be considered candidate for interconnection technology. The use of polyanilines for wiring has been widely speculated since the emergence of these systems in the late 1970s. Such an application, need copper like conductivity. Polyacetylene is the only conducting polymer that has such conductivity, but its environmental instability and lack of processability come in the way of its application. Enhanced conductivity in some of the processable and environmentally stable polymers is necessary before they can be considered viable conductors for interconnection technology.

There is great scope for the use of polyanilines in various devices which may lead to development of new technologies in IC fabrication and flat panel displays. Its unique application potential may give rise to considerable academic interest and industrial developments. However, much work is needed to be done in material properties prior to these applications in new technologies. Higher mobilities are required from the semi-conducting conjugated polymers before they can compete with silicon based devices. Enhancement of the long-term stability of the polymers is very essential before polymer-based light-emitting diodes become viable. Since early eighties remarkable progress has been made in understanding and developing the various properties of polyanilines. As the knowledge regarding the material properties continue to improve, some of what are currently viewed as "far out" applications may become enabling technologies in the near future.

5. POLYANILINES FOR MEMBRANE TECHNOLOGY

Membrane based purification system functions as a simple and direct method of purifying commercially important and often difficult to separate mixtures. Membranes have a wide range of applications as separating oxygen and nitrogen from air, desalinating seawater, removing organics from waste streams, and separating pure ethanol from its azeotropic mixture with water. Membrane purification systems, particularly polymer membranes, have been developed over the past several decades into energy-efficient alternatives to standard industrial purification techniques [174-181].

Polymers as membranes have distinct advantages in the purification of many liquids and gases since they can be easily synthesized and processed into flexible thin sheets or high surface area hollow fibers to allow large fluxes of gases or liquids through the membranes. Polymer membranes are basically tested for their permselectivity. If a given polymer has a low selectivity or insufficient permeability, an alternate polymer is selected. When a promising candidate appears, derivatives with different side chains or copolymers or blends are synthesized in an attempt to enhance the physical properties of the parent polymer.

Polyaniline after synthesis can be modified through doping to induce structural changes to achieve enhanced separating abilities. Realization of the large differences in the permeability of gases through doped and undoped polyaniline films led to detailed investigations of polyaniline for the separation of gases [182-186]. For the full realization of the potential of polyaniline as separation membranes, a thorough understanding of fundamental structure-property relationship of the polyaniline and dopants must be explored, which is beyond the scope of this article.

5.1. Preparation of Free-Standing Polyaniline Films

Robust free-standing films of polyaniline are prepared from the emeraldine base polymer powder. The brown powder is dispersed in N-methyl-2-pyrrolidinone (NMP) (5-8% *w/v*) with continuous grinding. The deep blue viscous polymer solution is uniformly spread onto glass plates and dried in a convection oven at 110°C. The resulting strong free-standing films are removed from the glass plates by soaking in distilled water followed by drying. These polymer films can be doped using acids to achieve the desired properties. The dopant counterion (X^-) is electrostatically associated with the polymer backbone to maintain charge neutrality. By this close proximity in the preformed film local rearrangements take place in the polymer structure to accommodate the extra space required for the counterions [187]. The size of the counterion decides the amount of rearrangement and also the amount of "void" space or free volume opened up on its removal by undoping. Increase in free volume can be realized after removing dopants by measuring the increased permeation of gases through the membrane. Correlation between counterion size and permeability has been observed for a range of gases, where larger the dopant counterion, higher will be the permeability of gases through the undoped membrane [186]. A more detailed description of membrane separation will be helpful in understanding the potential utility of polyaniline as a membrane.

5.2. Separation Process

The separation process involves three distinct parts: first, sorption of the constituent feed mixture onto the membrane (adsorption); second, movement through the membrane (diffusion); and third, loss out the other side of the membrane (desorption). The term "permeation" indicates the overall process of mass transport across the membrane. In liquid separations this is termed as "prevaporation" to emphasize the fact that the permeant undergoes a phase change from liquid to vapor while passing through the membrane.

Physical features of the separation process such as surface area and thickness and physical and chemical characteristics of the membrane itself, such as crystallinity and substituent groups, all together yield an effective separation membrane. Manipulation of these factors and study of the consequences of these changes to flux and separation can further enhance our knowledge and understanding of the structure-property relationships of polymer membranes.

The selection of a membrane for any particular type of separation is determined by its average pore size; e.g., 10-100 μm pore size is useful for conventional filtration, 0.1-10 μm is for microfiltration, 50-1000 Å is required for ultrafiltration and less than 50 Å is needed for reverse osmosis, gas separation and pervaporation. The latter are also called nonporous membranes and the separation depends on molecular interactions between the permeant and the membrane itself to effect separation.

5.2.1. Polyaniline for Gas Separation

Polyaniline films (as-cast) selectively permeant the gases. In general, larger the size of the gas molecule the lower is its permeability through as-cast polyaniline film. The only one exception for this is oxygen. Permeability of oxygen is actually higher than that of argon even though oxygen (kinetic diameter 3.46 Å) is slightly larger than argon (kinetic diameter 3.40 Å). This may be due to the relatively high solubility for oxygen in polyaniline [188-191].

Doping of polyaniline lowers its permeability to oxygen and nitrogen. Here the permeability of polyaniline to oxygen and nitrogen is compared in four different states of doping: as-cast (untreated), doped (exposed to acid), undoped (doped film exposed to base), and redoped (undoped film exposed to acid). The as-cast polyaniline shows higher permeability to oxygen (0. 174 barrer) than nitrogen (0.019 barrer) with a selectivity of about 9.1. Fully doping the polyaniline film with aqueous hydrochloric acid lowers the permeability to oxygen and nitrogen as dopants that easily penetrate the film to dope each imine nitrogen and fill void space of polymer. Undoping with ammonium hydroxide completely removes the chloride dopants as determined by elemental analysis [186]. The undoped polyaniline film is now more permeable to all gases compared to the as-cast film even though they have chemically equivalent compositions. The doping/undoping process also removes residual NMP used as the casting solvent, accounting for a small increase in free volume [192]. The undoped film has permeability of 0.546 barrer for oxygen, 0.063 barrer for nitrogen and a selectivity of 8.7. Redoping can be applied to reduce the permeability of larger nitrogen molecules than that of the smaller oxygen resulting in high selectivity.

The pores represent microscopic pathways through which gas molecules penetrate as segments of aniline rings flip due to thermal vibrations [193,194]. When fully doped with HCl, the protons find each imine nitrogen carrying along with them the large solvated chloride counterions to maintain electroneutrality. The dopants and counterions fill void spaces in the polymer and lower the permeability to all gases. Undoping completely removes the dopants and counterions. This increases pore connectivity resulting in higher permeability to all gases. There is a large driving force on doping for dopants to associate with imine nitrogens, which changes the morphology; undoping, on the other hand, removes dopants as there is no driving force for the polyaniline structure to return to its as-cast form. Redoping will then begin to fill in void spaces left by dopants and affects the diffusion of gases and their solubility by changing gas-polymer interactions. The permselectivity can be tuned by the quantity of dopants put back in to polyaniline films.

The mechanism behind high selectivities in partially doped polyaniline, likely involves increased solubility for oxygen and enhanced control over diffusion. The increased oxygen solubility could be due to the interaction of paramagnetic molecular oxygen with impaired electrons created in the conducting polymer backbone through doping process [190,191,195]. Interactions of oxygen with free spins have been observed in other types of materials [196,197].

The oxygen/nitrogen selectivity of as-cast polyaniline (~9) is comparable to commercial polymers such as polysulfone (a = 5.6 [189]) and other much-studied polymers like polyimides (4 < a < 8 [189]). However, the O_2/N_2 selectivity of redoped polyaniline is the highest reported for any polymer [186]. The commercial polymers generally have higher oxygen permeabilities (1.4 barrers for polysulfone) than polyaniline (–0.2 barrer). A major challenge for polyaniline is to develop pinhole-free ultrathin membranes (such as asymmetric hollow fibers) that have high flux to compensate the relatively lower permeabilities.

In this connection, thin films (1.5-4 μm) of polyaniline grown on alumina (Anopore support membranes) were tested for gas transport properties [198]. Films of 1.5 μm thickness gave, O_2/N_2 selectivities of < 1 indicating Knudsen-type diffusion, which may be due to microporous defects. Selectivity coefficients of slightly higher than 9 were obtained for thickness of 3 μm or greater. This is comparable to the work of Anderson *et al.* [186] on free-standing films who demonstrated that supported films of 3 Am can be prepared free of defects. Free-standing films of below 10 Am thickness were microporous and showed Knudsen diffusion [198]. Increasing the thickness of free-standing films led to higher selectivity values reaching >9 for thicker than 20 μm.

The effect of doping level was studied for both alumina-supported and free-standing polyaniline films [198]. Exposing them to 1.0 M aqueous HCl, corresponding to 13, 22, 31 and 38% doping levels, did doping of thin films. Permeability of all gases decreased with increasing doping level. Permeability for oxygen was about 0.5-0.15 barrer, while for nitrogen, it was about 0.06-0.01 barrer. Oxygen/nitrogen selectivities for the supported first membranes increased from about 9.5 (undoped) to 15.2 (38% doped) and then fell slightly at full (50%) doping. Free-standing films also showed similar trends, with O_2/N_2 selectivities increasing from about 9.5 (undoped) to 14.9 (36% doped) to 15.2 (50% doped). Kuwabata and Martin [198] suggested that repeated doping and undoping of polyaniline films could potentially enhance selectivities, but they noted no discernible effect on selectivities after five repeated dopings and undopings.

Other work [199] on gas permeability through polyaniline membranes confirmed the general effects of doping, undoping, and partial redoping and reported O_2/N_2 selectivity of 14.2. The difference in reported O_2/N_2 selectivities could be due to uncertainty in the low nitrogen permeability values, membrane preparation methods, and lack of homogeneity in the doping process.

The use of other dopants also led to relatively high selectivities [198], as for example, a 36% sulfuric acid-doped sample had an O_2/N_2 selectivity of 13.4, while a 36% *p*-toluenesulfonate-doped sample had an O_2/N_2 selectivity of 14.9. A thin film composite doped with 38% nitric acid showed O_2/N_2 selectivity of 14.8.

The halogen acid series HF, HCl, HBr, and HI have been studied to find the effect of dopant size [186]. Polyaniline films were fully doped with each halogen acid and followed by completely undoping using ammonium hydroxide, a progression decreased in the order HF, HCl, HBr and HI. This is consistent with the sizes of the solvated halogen acids in which solvated HF is the largest of the series (in contrast to halides in a crystalline lattice) [200,201]. Further, it demonstrates that in general smaller gases permeat polyaniline more easily than larger gases. One notable deviation is the increased permeation of oxygen through all polyaniline membranes relative to that of argon. This is most likely due to the high solubility of oxygen in polyaniline [188,189].

5.3. Polyaniline Blends

Blending is physically mixing two components to form a composite. In this case, solutions of polyamic acid (the precursor to polyimide) and polyaniline were prepared separately using the solvent N-methylpyrrolidone (NMP) and then mixed in 50/50 weight ratio, cast into films at 100°C, and cured at 200°C for 1 h and at 300°C for 15 min. The bluish color of the blend, formed from a golden yellow polyamic acid solution and brown polyaniline emeraldine base powder dispersed in NMP suggests that polyaniline is in its doped acid form is typically blue.

The 50/50 mixing of polyaniline and polyamic acid forms a highly miscible blend with molecular level (acid-base) binding of the two polymers, as indicated by FTIR spectroscopy. The blend exhibits all absorption characteristics (1485, 1504, 1567, 1599, 1660, and 1732 cm^{-1}) of polyamic acid (1660 cm^{-1}), polyimide (1730 cm^{-1}) and acid (1486 and 1568 cm^{-1}) and base (1505 and 1595 cm^{-1}) forms of polyaniline. This indicates that chemical doping of polyaniline with the polyamic acid precursor occurs. The amide nitrogens from the polyamic acid remain even though full curing temperatures for imidization were used, likely due to the strong interaction with the imine nitrogens of the polyaniline backbone preventing ring closure.

In contrast to the small molecule ionically bound protonic acid dopants discussed so far, large molecule permanently bound dopants are of interest to further elucidate the effects of dopant state of polyaniline on gas permeability. Combining the controlled dopability of polyaniline with the best features of a polymer studied extensively for gas separations, polyaniline-polyimide blends were developed [202-204].

Gas permeability tests performed with the blended films indicated a definite increase in permeability for all gases compared to polyimide and the base form of polyaniline [204]. Usually, there exists an inverse relationship between permeability of two gases and the separation factor for those gases [205-208]. But, this is not true with the present blend. The separation factors for the blend are comparable to those of polyaniline (as-cast) for H_2/N_2 ($\alpha = 200$) and O_2/N_2 ($\alpha = 9$) are closer to polyimide for CO_2/CH_4 ($\alpha = 58$). Thus, this blend appears to have resulted in improved properties compared to the parent polymers with enhanced permeability and good selectivity.

From these studies it can be concluded that polyanilines form an interesting candidate of membrane materials with tailorable properties. Doping and undoping of polyaniline results in increased permeabilities compared to as-cast films owing to the localized morphological changes by the addition and removal of dopant counterions. Redoping with small amounts of dopants can result in very high selectivities for some important gas pairs such as O_2/N_2, H_2/N_2 and CO_2/CH_4 [186,198]. In fact, independent research teams found that polyaniline has the highest current O_2/N_2 selectivity among any polymer membranes, falling above a permselectivity threshold for state-of-the-art membrane technology [186,198,199,209]. Changing the properties of polyaniline through the use of protonic acids and different polymer blends will provide substantial scope for future research.

5.4. Polyaniline Membranes for Pervaporation

Pervaporation is a process that can separate liquid mixtures by selectively allowing the passage of a single component. Important among these include the separation of azeotropes, dehydration of organics and removal of organic compounds from water [190]. Since most liquids contain molecules in the

2-10 Å size regime [210], any separation by pervaporation requires fully dense pinhole-free membranes of the type used for gas separation. With the ability to separate gases on this size scale, polyaniline films have the potential to separate liquids [211].

The transport of water through supported polyaniline membranes grown electrochemically was investigated [212]. A 25-30% increase in water permeation was observed with sulfuric acid-doped polyaniline compared to the undoped polymer. The doped polymer will have a more open structure, thereby accounting for the enhanced water permeation. But, a simpler explanation could be the increased hydrophilicity of doped polyaniline. Analogous increases were also observed in methanol transport through both polyaniline and polypyrrole [213,214].

Pervaporation of methanol and methyl *tert*-butyl ether (MTBE) using hollow fibers coated with polyaniline was studied [215]. A microporous hollow-fiber ultrafiltration membrane made of cellulose was coated with a conjugated polymer. When compare this an uncoated hollow fiber had only a slight preference for transporting methanol ($\alpha_{methanol/MTBE}$ = 1.06), and this was attributed to the more polar methanol being more soluble in the somewhat polar cellulose-based film. Polyaniline-coated fibers showed a selectivity of 4.9. The study demonstrated that polyaniline could be used selectively to transport liquids. However, many of these studies have been limited by the stability of the membranes and low permeation rates.

5.4.1. Free-standing Membranes

Chemically synthesized free-standing polyaniline membranes are very robust and quite stable in aqueous solutions. Hence, interesting to explore their liquid permeation properties and identify potentially useful applications.

A free-standing polyaniline membrane (40-100 μm thick) was placed on a porous stainless steel support and sealed with an O-ring. About 200 ml of feed solution at room temperature was put in the feed reservoir, and the downstream side of the pervaporator was evacuated. The permeant collector was cooled to ~65°C to maintain a low vapor pressure on the downstream side. Due to the pressure difference, first the liquid adsorbs onto permeants, and desorbs from the membrane in the vapor phase and condenses in a graduated collector. Liquid permeation through the membrane is allowed to proceed for approximately 24 h to attain a steady-state of flow condition. The volume of liquid permeanting through the membrane was monitored with time and recorded from the graduated collection vessel. Flux was then calculated.

Pervaporation was next carried out to test the selectivity of polyaniline membranes toward acetic acid-water mixtures. Different feed ratios of acetic acid to water were pervaporated through both undoped and doped polyaniline, where the feed water content *versus* the permeant water content, was plotted. For comparison, the vapor-liquid equilibrium curve for acetic acid-water [216] was plotted just above the line of no separation. It was observed that undoped polyaniline has a marginal preference for permeating water over acetic acid. However, this small preference at an average water permeability of about 0.5 g-mm/(m²-h) is too low to have any impact. More interesting is fully HCl-doped polyaniline, which permeants water over acetic acid in a much more selective fashion. In fact, even with a mixture of 85% acetic acid, 15% water, at least 93 wt% of the permeant was water. It should be remembered that when undoped membranes are used in the presence of acids, the polyaniline will partially get doped by the acid, the extent however depends on the pH of the acid used. But, when a doped membrane is used, as long as the dopant has a stronger interaction with polyaniline than the feed acid, very little acid-dopant exchange will occur. The total permeability of the fully HCl-doped polyaniline membrane essentially parallels the water permeability.

To test the effect of different molecular size permeants on HCl-doped polyaniline, formic, acetic, and propionic acids were used. The feed content is plotted against the permeant content as determined by ^{1}H NMR. The smallest permeant, formic acid (HCOOH), has a molecular size of ~3.5 Å in aqueous solution [210]. Although preferentially held back by doped polyaniline, formic acid is still transported with the water, leading to relatively poor separation. Acetic acid (CH_3COOH) with a size of ~4.5 Å [210] is readily held back by doped polyaniline, while water preferentially permeants. The propionic acid (CH_3CH_2COOH) with the largest molecular size of (~5.5 Å [210]), is completely rejected by an HCl-doped polyaniline membrane. Water permeants preferentially at all composition, and at >50% water in the feed essentially no propionic acid is detected. Even at the lowest (26 wt%) value tested at least 98 wt% water was transported through the doped polyaniline membrane.

To test the ability of doped polyaniline to its selective permiability by mixtures, approximately equal amount of formic, acetic, and propionic acids along with water were put in the feed reservoir. The feed composition was 27.7 wt% water, 21.8% formic acid, acetic acid, and 24.0% propionic acid. The results of pervaporation through doped polyaniline suggest that water has preferentially gone from 27.7% in the feed to 94.5%. While, formic acid reduced to 4.3%, acetic acid to 1.1% and propionic acid to 0.2%. In brief, HCl-doped polyaniline can separate a mixture of formic, acetic and propionic acid as follows; the smaller formic acid molecules permeante to some extent, acetic acid permeation is considerably less, and propionic acid is essentially rejected.

These tests clearly indicate that HCl-doped polyaniline prevents the permeation of molecules larger than about ~4.5Å but allows smaller molecules (especially water) to pass through. This clearly indicates that it is useful for effecting separations ranging from the concentration of fruit juices to the separation of azetotropes. However, low flux limitations will have to be compensated by asymmetric membranes, which could provide higher fluxes. As polyaniline has very high thermal stability, increased flux could be obtained by larger temperature differences. As selectivity gets adversely affected by higher temperatures, testing is clearly warranted.

In summary, polyaniline is the most promising candidate for membrane separations because of its high chemical and thermal stability and its simple acid/base doping chemistry. The doping process can be used to control hydrophilicity for liquid separations and gas permselectivities for gas separations. A doping-undoping and partial redoping processes provide us with the most selective membrane known so far for separation of oxygen-nitrogen.

6. LITHIUM-POLYANILINE BATTERIES

Lithium, because of its low equivalent weight and high standard electrode potential, is an appropriate an anode material for high-energy-density battery system. For the past two decades, a great deal of effort has been made to develop high-energy and/or high-power-density lithium batteries for space and defense programs, for electric-vehicle propulsion and for certain consumer markets such as power sources for human implantable devices and various electronic devices [218].

Lithium batteries are a promising high-energy-density power source useful over a wide temperature range. With proper choice of cathode material and electrolyte solution, these batteries can be designed to operate around 3.0 or 1.5 V. Therefore, some of these systems have become attractive alternative power sources for a number of applications where conventional aqueous alkaline-based batteries are being used at present. During the last three decades, a number of primary lithium batteries suitable for ambient-temperature have been produced and tried for commercial and military applications. Several systems have reached the advanced stage of development ready for commercial production [219].

The outstanding technical success of primary lithium batteries spurred interest in ambient-temperature secondary lithium batteries. During the last decade, significant advances have been made in research and development with the use of organic electrolyte solutions for rechargeable lithium cells. Lithium batteries that show a reasonable number of discharge cycles have been built. Recently there have been research and development activity on rechargeable lithium cells by use of organic conducting polymers.

Ever since, it was reported that electrochemical doping and undoping of conductive polymers could be performed on a reversible basis [220] active research has been conducted on two types of secondary batteries. First, conducting polymer batteries with high performance features such as, light weight, suitability for fabricating electrodes from porous film, etc. and second, on clean batteries with no toxic metals. Besides, conducting polymers secondary batteries have the following features:

- Doping concentration can be controlled with the current and voltage conditions.
- Doping and undoping can be performed on a reversible basis.
- The doped dopant can be maintained with comparative stability.
- p-type (anion) doping.
- n-type (cation) doping (not possible in case of polyaniline material).

6.1. Principles of Batteries using Conductive Polymers

The first polymer material proposed for use as a secondary battery electrode was polyacetylene $[(CH)_x]$, and the configuration of the battery was $(CH)_x/LiClO_4$ in PC/Li (PC = propylene carbonate) [221]. The reaction of the battery is expressed as follows:

Positive Electrode:

$$(CH)_x + xyCl_4^- - xye^- \underset{\text{discharge}}{\overset{\text{charge}}{\rightleftharpoons}} [CH^{+Y}(ClO_4)y]_x \qquad (2.3)$$

Negative Electrode:

$$xyLi^+ + xye^- \underset{\text{discharge}}{\overset{\text{charge}}{\rightleftharpoons}} xyLi \qquad (2.4)$$

Total Reaction:

$$(CH)_x + xyClO_4^- \underset{\text{discharge}}{\overset{\text{charge}}{\rightleftharpoons}} [CH^{+Y}(ClO_4)y]_x \qquad (2.5)$$

There are six possibilities for the use of conducting polymers (CP) as electrode material for secondary batteries. They are:

Type I	:	CP [p-doped] / Li
Type II	:	CP [p-doped] / CP [n-doped]
Type III	:	CP [p-(high)-doped] / CP [p-(low)-doped]
Type IV	:	CP [n-(low)-doped] / CP [n-(high)-doped]
Type V	:	Inorganic materials / CP [n-doped]
Type VI	:	CP [undoped] / Li

Of these, only in the case of Type VI does the doping reaction of conductive polymer correspond to the discharging reaction of the battery. For all of the others, the undoping reaction is equivalent to the discharging reaction. Types III and IV cannot use more than a part of the reversible doping level of the conductive polymer for the charging and discharging reaction. As a result, their discharge capacity is limited and their cell voltage is also low. For Type VI, both cell voltage and discharge capacity are lower than Type I in which p-doped polymer is used as the positive electrode. However, of these cell configurations, the concentration of the dopant in the electrolyte solution for Types I and II increases during the cell discharging reaction, and decreases as the charging reaction progresses. Consequently, the cell capacity is limited by the amount of dopant in the electrolyte solution. In contrast, for other types (III, IV, V and VI) the concentration of the dopant in the electrolyte solution does not change throughout the charging and discharging reactions.

Polyaniline as the electrode material of a secondary battery belongs to Type I cell configuration. Hence, the electrochemical doping-undoping reaction is equivalent to the charging-discharging reaction. As a result, the following items are particularly important as conditions for a superior conductive polymer for use as a secondary battery electrode:

- High doping amount per unit weight and unit volume.
- Good reversibility of doping and undoping.
- Fast diffusion of dopant in polymer molecules.
- High potential for p-type doping; low potential for n-type doping.
- Chemical, electrochemical and thermal stability.
- Ease of handling.
- Stability with respect to electrolyte solution.
- High conductivity.

The secondary battery should satisfy as many of these requirements as possible in order to obtain good capacity, charging and discharging efficiency, charging and discharging cycle performance and storage characteristics.

6.2. Electrolytes

The properties required for the electrolyte solution in batteries that utilize polyaniline as the electrode are as follows:

- No chemical reaction with the active material of the positive and negative electrode.
- Chemical and electrochemical stability.
- High ionic conductivity over a wide temperature range.
- No obstruction of the progress of the electrode reaction.
- Safe and non-toxic properties.

In addition, in the cell configuration of Type I the concentration of the dopant in the electrolyte solution decreases during charging and increases during discharging. In such cases, the cell capacity is limited by the amount of dopant in the electrolyte solution. In order to create a high-capacity battery, a suitably large quantity of dopant must be dissolved in the electrolyte solution. Alternatively solid electrolytes [222,223] can be used provided they exhibit similar properties discussed above.

6.3. Synthesis of Electrode Material

The synthesis of polyaniline is divided in two categories—electrochemical synthesis and chemical synthesis.

6.3.1. Electrochemically Synthesized Polyaniline

As mentioned earlier, the properties of electrochemically synthesized polyaniline are greatly influenced by the electrolysis conditions. For example, surface morphologies of polyaniline were fibrous in the case where 2.0 M $HClO_4$ was used, but granular for 0.5 M $HClO_4$. When 1.0 M $HClO_4$ was used the morphology was short fibrils, an intermediate structure between those obtained using 0.5 M and 2.0 M $HClO_4$. Cyclic voltammograms and charge-discharge characteristics of $Li/LiClO_4$ in propylene carbonate (PC)/polyaniline cells obtained by using 2.0 M $HClO_4$, a coulombic efficiency of 100% was obtained at a 67 A h kg^{-1} (polymer + initial dopant weight basis) charging, which led to a cell voltage of 3.6 V [224].

Taguchi *et al.* [225] obtained a maximum doping level of 164 A h kg^{-1} (polymer weight basis) for fibrous polyaniline. At a charge capacity of 83 A h kg^{-1}, a $Li/LiClO_4$ in PC/polyaniline cells showed a 100% coulombic efficiency over 500 charge-discharge cycles. The self-discharge rate was less than 2% per day at room temperature. In another communication 100% coulombic efficiency is reported at a charge capacity of 120 A h kg^{-1}. The $Li/LiClO_4$ in PC/polyaniline cells retained 93% of its original capacity after 30 days of storage [226].

Genies *et al.* [227] electrochemically synthesized polyaniline from eutectic $NH_4F.2.53$ HF mixture containing 0.6 M aniline. A discharge capacity of 130 A h kg^{-1} was obtained in $Li-Al/LiClO_4$ in a PC/polyaniline cell. A button cell and sub-C-type cell using this polyaniline showed discharge capacities of 11.2 mA h and 68 mA h, respectively.

Takei *et al.* studied the influence of electrolysis current densities [228]. The amount of polyaniline obtained was almost the same under current densities between 0.2-1.0 mA cm^{-1}, but fibril diameter was smaller and the specific surface area was larger for those obtained at higher current densities.

Polyaniline can also be synthesized from nonaqueous solutions [229,230] and the performance is influenced by the solution composition. Osaka *et al.* [230] obtained polyaniline from a PC solution containing 0.5 M aniline, 0.5 M $LiClO_4$ and various concentrations of CF_3COOH. The electroactivity was higher for polyaniline synthesized with higher concentrations of CF_3COOH [230]. An energy density of 379 W h kg^{-1} (polyaniline weight basis) was obtained for $Li/LiClO_4$ in a PC/polyaniline cell. Protons from strong acids are considered necessary for electrochemical synthesis in nonaqueous solutions. Besides, the surface morphology of polyaniline was affected remarkably by the addition of polymer electrolyte such as poly(styrenesulphonate), poly(2-acrylamido-2-methyl-1-propane-sulphonate) or poly(vinylsulphonate), becoming granular [231].

6.3.2. Chemically Synthesized Polyaniline

In chemical synthesis, monomers are polymerized using appropriate catalysts or reagents. The morphological, physical and electrochemical properties are influenced by reaction conditions such as the species and concentrations of catalyst, reagent and solvent, the monomer concentration, the reaction temperature, and so on. Chemical synthesis is considered to be suitable for producing large amounts of homogeneous polymer. Polyaniline is chemically synthesized by the polymerization of aniline using various oxidizing agents. Genies *et al.* [232] compared some oxidizing agents such as permanganate, dichromate, iodate and persulphate for polymerization of aniline in NH_4F-2.3HF, and found that a high yield of polyaniline was obtained using $(NH_4)_2S_2O_8$. In Li-Al / $LiClO_4$ in PC / polyaniline cells, 113 A h kg^{-1} was obtained by charge-discharge cycling between 1.7 V and 3.9 V, at a current density of 0.5 mA cm^{-1}. A doping level of 30 mol% (about 89 A h kg^{-1}) was also obtained for polyaniline synthesized from aqueous solution using $(NH_4)S_2O_8$ [233].

For the synthesis of polyaniline designed for battery electrodes, $Fe(ClO_4)_3$ and $Cu(BF_4)_2$ are preferable as oxidizing agents, because the products contain ClO_4^- or BF_4^- anions, which are commonly used in lithium secondary batteries. Polyaniline synthesized by $Cu(BF_4)_2$ has a fibrous morphology [234].

Charge-discharge testing of the Li-Al / $LiClO_4$ / polyaniline cells was conducted using electrolytes consisting of various two-ingredient solvents [234-236]. The charge-discharge characteristics of the battery were greatly influenced by the electrolyte. In the cases where PC-DME, EC-DME or γ-BL-DME were used, a discharge capacity of about 110 A h kg^{-1} was obtained. Even at –20°C, a discharge capacity of about 90 A h kg^{-1} was obtained when PC-DME or γ-BL-DME was used as the solvent. The conductivity of the electrolyte solution and the diffusion coefficient of dopants in polymer electrodes are considered to have a large influence on the charge-discharge characteristics of the cells.

The theoretical energy density depends on the concentration of the dopant in the electrolyte solution, as well as the specific capacities of the positive and negative electrodes. To develop batteries with a higher energy density, it is necessary to develop electrolyte solutions which have both a higher concentration of dopant and a higher conductivity over a wide temperature range.

One of the advantages of polymer electrodes is their ease of processing. This feature is thought to be particularly important in batteries with special shapes, such as the ultra-thin or flexible types. The demand for special secondary batteries like these is expected to grow as battery applications diversify in the future. Lithium batteries with conducting polymer electrodes in general and polyaniline in particular are likely to be used to meet such demands.

7. ELECTROCHROMIC DEVICES

Electrochromic device is essentially a rechargeable battery in which the electrochromic electrode is separated by a suitable solid or liquid electrolyte from a charge-balancing counter-electrode and the colour changes occur by charging and discharging the electrochemical cell with applied potential of a few volts. A conductive, transparent glass coated with electrochromic material constitutes the electrochromic electrode of these devices, which can work either in the reflective or transmissive mode. The counterelectrode can be of any material that provides a reversible electrochemical reaction in devices operating in the reflective mode (like electrochromic displays); by contrast, in variable light transmission electrochromic devices (like electrochromic windows) it has to be either colourless in both oxidized and reduced states or electrochromic in a complementary mode to the primary electrochromic material since the entire system is in the optical path.

Electrochromic devices are today considered suitable for technologically important goals such as optical information displays, architectural glazing windows for energy control, anti-glare rear-view mirrors and sun roofs for automobiles and consumer sunglasses. The interest in electrochromic devices lies in the fact that they have a number of specific advantages, such as high optical contrast with continuous variation of the transmittance and no dependence on viewing angle, optical memory, UV stability, and wide operation temperature ranges. These favorable characteristics may ultimately overcome the well-known deficiencies of liquid crystal displays and thus place electrochromic devices in a prominent position for the production of large visual angle panels.

Surprisingly, despite these many attractive features, almost four decades of research and development by a large number of academic and industrial groups have not been sufficient to promote large-scale development and to introduce electrochrornic devices in the marketplace. There are a variety of reasons to account for this failure to break through, including advances in the competing liquid crystal device technology and, particularly, the lack of reliable and efficient electrochromic materials which resulted in the production of devices with limited lifetime, slow response times and restricted addressability. However, consistent progress has been achieved in the most recent years, especially with the characterization of various new electrochromic materials and of advanced assembly structures. Conjugated polymers with aromatic-like structure behave more promisingly as electrochromic materials when the p-doping-undoping process is involved because n-doping is greatly affected by the nature of the cation in the charge balancing process; *n*-doped forms are also less stable than *p*-doped once.

The aim of this section is to review the progress and development in the electrochromic technology with particular attention to devices having a polyaniline-based, laminated configuration, which finds wide applications in the automobile and building industries.

7.1. Basic Concepts

Conjugated polymers have an electronic band structure. The energy gap (E_g between the highest occupied π electron band (valence band) and the lowest unoccupied one (conduction band) determines the intrinsic optical properties of the polymers. The color changes elicited by doping are due to the modification of the polymer's band electronic structure. The one-dimensional character of the polymer system energetically favors the localization on the chain of the charge created by doping and the relaxation of the lattice around this charge. This confinement of the charge then creates the defects that produce in the gap new electronic states, which cause the color changes. In polymers, with a nondegenerate ground state and an aromatic-like structure the lattice relaxation is towards the quinoid structure and the defects, polarons (single charge of spin 1) and bipolarons (double charge, spinless), are related to the amount of stored charge in the polymer, usually indicated as the doping level. As the doping level increases, the bipolaron states overlap in bipolaron bands, and the bipolaron bands may merge with valence and conduction bands, leading to a metallic-like state.

Electrochromism is the phenomenon related to changes in colour induced by reversible electrochemical processes in selected materials. In fact, the phenomenon of electrochromism, namely, of light modulation controlled by an external electric pulse, can be exploited for the realization of devices of diversified interest. In fact, an electrochromic material changes colour in a persistently but, in a reversible manner by an electrochemical reaction. Accordingly, conjugated polymers that can be repeatedly driven from insulating to conductive states electrochemically with high contrast in colour are promising materials for electrochromic device technology.

Spectroelectrochemical studies have been of great importance in gaining insight into basics of conjugated polymer doping. However, few reports have been so far devoted to ascertaining the performance of these materials for use in electrochromic devices. High electrochromic efficiency, short electrochromic response time, long switching life and optical memory are the most important requisites for electrochromic materials in device technology.

Electrochromic efficiency, expressed as C^{-1} cm^2, is the ratio of the variation of optical density to the injected charge as a function of unit area, and it is a wavelength-dependent quantity. The electrical power needed to operate an electrochromatic device is related to electrochromic efficiency an important parameter to discriminate among electrochromic materials for large-area devices.

Electrochromic response time, is the time the polymer takes in response to the potential pulse to go from fully coloured state 1 to fully coloured state 2. As conducting polymers are ion-insertion electrochromic materials, very fast electrochromic response times cannot be expected when polymer films are made thick enough to achieve sufficiently high differences in optical density between doped and undoped forms.

Long switching life is essentially related to the coulombic efficiency of the doping-undoping process, *i.e.* the ratio of the reduction to the oxidation charges involved in the electrochromic process. Long switching life can be assured only at 100% coulombic efficiency values. However, different technological applications of electrochromic devices can also require different electrochromic response times and switching lives. A time range from several to a few hundred milliseconds and a life of up to 10^6-10^7 switches are demanded for display applications, while a few seconds and 10^5 switches can be satisfactory for variable light transmission devices (like electrochromic windows). The optical memory is the persistence of the coloured state even when the driving voltage is removed. For polymer-based electrochromic materials it is related to stability in the electrolytic medium of the doped and undoped forms of the polymer.

The optical memory is the persistence of the colored state even when the driving voltage is removed. For polyaniline-based electrochromic material it is related to stability in the electrolytic medium of the doped and undoped forms of the polymer.

7.2. Electrochromic Performance

Electrochromic performance of polyaniline is usually rated on the basis of the following:
1. Cyclic voltammetry curves, to identify the potential range of the doping-undoping process.
2. Electronic spectra in the visible and near IR regions at different injected charge recorded *in situ* at different potentials, to select the switching potentials and calculate electrochromic efficiency.
3. Optical and electrical data over time following the application of repeated square wave potential switches, to evaluate electrochromic response time and injected charge over time (optical data are usually evaluated by detecting the transmitted light; the source is monochromatic light at the maximum absorption wavelength of the undoped form and/or polychromatic light.

4. Cyclic voltammetry curves as well as spectra of doped and undoped forms, after different numbers of switches to investigate polymer switching-life.

5. Light transmitted by doped and undoped polymer recorded *in situ* over time in open circuit conditions, to estimate optical memory. The information regarding film preparation method and thickness is also important in assessing polymer electrochromic performance.

Most of the studies on the electrochromic performance of conducting polymers are carried out in liquid electrolyte [237-250]. However, the data reported in the early papers are not exhaustive enough to evaluate the parameters necessary for the use of the electrochromic material in device technology. These data are generally of electrical response time only (*i.e.* the time the polymer takes to store charge) and may not be representative of electrochromic response time.

Polyaniline is a peculiar conducting polymer because it is the only one, along with its derivatives, whose electrical properties not only depend on its oxidation state but also on its protonation state and, hence, on the acidity of the electrolytic medium involved. Polyaniline shows two electrochemical redox processes with color changes from transparent yellow to greenish-blue and then to dark blue. The electrochemical processes have been extensively investigated in aqueous acid solutions and in organic electrolytic media and several redox mechanisms involving protonation/deprotonation and/or anion insertion/extraction have been proposed. Genies *et al.* [251] have also demonstrated that the polaron-bipolaron model is viable for two electrochemical processes of polyaniline in acid aqueous solution.

The multicolor changes of polyaniline, depending on the redox process involved at different potentials, are attractive for device applications and a few studies of electrochromic phenomena both in acid aqueous solutions and in organic electrolyte have been carried out. Kobayashi *et al.* [241,242] indicated that to improve significantly polyanilines switching-life in acid aqueous solution, the potential pulse has to be narrowed to the first redox process, from –0.2 to 0.6 V *vs* SCE, where the color changes from transparent yellow to green.

Two kinds of redox reaction are suggested for the electrochromic process in this potential range. One, involving protonation/deprotonation, occurs at potentials that are less positive than 0.3 V *vs* SCE, and gives the cyclic voltammetry curve with color change from yellow to yellowish-green; the other, mainly occurring at potentials more positive than 0.3 V and leading to green color, producing not a voltammetric wave but a large capacitative current with anion insertion to compensate for the produced charge. Very fast electrochromic response time, 100 milliseconds, and long switching-life, up to 10^5 switches for polyaniline film of *ca.* 0.05 µm thickness switched between - 0.2 and 0.45 V in 2 M HCl, are reported. Chronocoulometric response to a potential pulse from -0.1 to 0.6V *vs* SCE over a timescale of 100 ms was also reported in 1987 by LaCroix *et al.* [252] for a polyaniline film (0.05-1.5 µm thickness range) in 1-2 M sulphuric acid, with a response that is initially linear in time and at the end changes to $t^{1/2}$ dependence. Decreasing solution acidity changed the kinetics of the reaction: in 0.1 M sulphuric acid the switching time was about three times longer and the time dependence of the process was linear with $t^{1/2}$ indicating that ion diffusion was now important in the kinetics of the process. The electrochromic response time of polyaniline in 1-2 M acid solution, which is faster than that of the other heterocyclic polymers, might be due to the fact that the electrochromic process of polyaniline mainly involves a protonation/deprotonation redox reaction, and proton motion is faster than anion motion.

Polyaniline electrochromic phenomena have been investigated in organic liquid electrolytes such as PC-LiClO$_4$, and it has been reported that switching potential range can be extended up to the second redox process, with color change from transparent colorless to dark blue. In fact, unlike in aqueous solutions, repeated cyclic voltammetry curves in PC-LiClO$_4$ between 2.5 and 4.5 V *vs* Li up to 100 cycles do not give evidence of degradation [253]. The darker color contrast in organic liquid media is an advantage in practical applications, where test results of a polyaniline-based electrochromic device operating in PC-LiClO$_4$ are reported. LaCroix *et al.* [254] reported ultrafast electrical response data on a timescale of 100 µs for polyaniline film (*ca.* 0.2 gm thickness) in 2 M sulphuric acid using an ultramicroelectrode (*ca.* 0.2 mm^2 area) with appropriate correction for uncompensated ohmic drop. However, the optical response data are reported in arbitrary units and thus they are difficult to evaluate with a view to practical uses. In general, it would seem that 100 µs are not a viable reference timescale for electrochromic response, at least for conventional electrochromic devices.

7.3. Polyaniline-Based Electrochromic Devices

Several variable-light transmission electrochromic devices based on polyaniline with a counter-electrode, that is, electrochromic in the complementary mode have been developed. The device using polyaniline and WO$_3$ operates in PC-LiClO$_4$ [253]. WO$_3$, which changes from transparent-colorless to pale blue by a cathodic lithium ion insertion process, is the most investigated and promising electrochromic material for device applications among the ion-insertion inorganic compounds [255]. The device, 5 cm × 5 cm, was assembled by facing an ITO glass electrode coated with polyaniline film (0.6 µm thickness by electrochemical synthesis) and an ITO glass electrode coated with WO$_3$ (0.6 µm thickness by vacuum deposition); the 0.1 mm clearance between the two electrodes was filled with the electrolyte. The device's driving potential pulse was extended up to second electrochemical process. Polyaniline colors anodically to dark blue and WO$_3$ cathodically to pale blue; the device's overall process is shown below, with color change from transparent-colorless to dark blue:

$$(\text{polyaniline})_n + n\text{WO}_3 + ny\text{LiClO}_4 = n\text{Li}_y\text{WO}_3 + [(\text{polyaniline}) + {}^y(\text{ClO}_4^-)_y]_n \qquad (2.6)$$
colorless   dark blue

The test results of this device indicated:
(i) Transmittance change from 80% to 30% in is from 80% to 4% in 1 min.
(ii) Transmittance change between 80% and 30% up to 10^6 repeated cycles is between 70% and 7% at more than 10^5 cycles.

These findings seem to indicate that the polyaniline-WO$_3$ system satisfies the requisites for practical use in device technology.

In brief, electrochromism touches upon diversity of applications. For example, the electrochromic performance of polyanilines meets the most important requisites for their being use in electro-optic devices. Yet, when compared with liquid crystals, the materials currently employed in this technology, polyanilines, do not represent an improvement as far as switching response time is concerned; however they do have the important advantages of an unlimited visual angle and an open-circuit optical memory. Besides, a variety of color contrast can be achieved with polyanilines, which is the most significant characteristic of ion-insertion organic electrochromic materials as opposed to ion-insertion inorganic electrochromic materials. This feature is related to the conformational 'flexibility' of conjugated polymers, which implies that polyanilines can become virtually irreplaceable for the fabrication of multi-color electrochromic devices.

8. POLYANILINE FOR CONDUCTIVE TEXTILES

Textiles are among the oldest materials known to humankind. In addition to their initial use as apparel, they were used in several early structural materials. Composite structures made from cotton fabric and phenolic resins were developed in the early 1900s. Since then, the importance of textile-reinforced materials has grown, and textiles now play a crucial role in many engineering materials including polymers, ceramics, and metals.

A textile is a material made from fibers or other extended linear materials such as thread or yarn. Classes of textiles include woven, crocheted, knitted, knotted (as in macrame) or tufted cloth, and non-woven fabrics such as felt. Materials made from fibers such as fiberglass, carbon fiber, and ceramic fibers which are infiltrated by a matrix of another material are considered fiber-reinforced composite materials.

With the rapid development of the electrical and particularly the electronics industry, a need arose for flexible conducting and semiconducting materials. Flexible, highly conducting materials have been prepared by weaving thin wires of various metals such as brass and aluminum. Semiconductive textiles, including yarns and woven, nonwoven, and knitted fabrics, have been produced by impregnating textile substrates with conductive carbon or metal powders. The inclusion of appropriate binders allows a sufficient amount of these powders to be incorporated to impart the desired properties. These products have found wide application in the fields of electromagnetic interference (EMI) shielding and static dissipation. The development of filament-size fibers of stainless steel allowed the manufacture of conductive yarns, and subsequently fabrics, by blending various amounts of these fibers with synthetic or natural fibers. Carbon fibers and whiskers have also been incorporated by blending into textiles [256]. In addition, 100% carbon fibers, yarns, and fabrics are widely used in composite structures, particularly in the aerospace industry.

Applications of conductive textiles may be divided into static dissipation, EMI shielding, heating elements, composite structures, and military applications. The advantages of using conductive textiles lie in the strength and flexibility of the substrate, the wide range of substrates available, and the wide range of surface conductance available in these fabrics. In addition, anisotropic materials can be constructed with almost any ratio. Amongst the application polyaniline composite structures have shown promise for military applications.

8.1. Routes

One of the most cost-effective routes for producing conductive plastics is the incorporation of conductive fillers, particularly carbon [257]. However, the incorporation of a sufficient amount of carbon, up to 40%, to allow percolation causes a significant deterioration of mechanical properties in the polymer/filler blend, which results in considerable processing problems in the production of textile fibers. Commercial products based on nylon and polyester have been developed using highly filled polymers, either in the core or as a sheath of the fiber, that retain at least some of the strength of the unfilled polymer. Products of this nature are available from most commercial fiber producers.

Yet another approach to conductive textiles involves the coating of textiles with metals. Silver-, copper-, and nickel-coated textiles using various synthetic fibers are now commercially available. A wide variety of processes can be used, including vapor deposition, sputtering, reduction of complexed copper salts, and electrodeless plating using noble metal catalysts [258,259]. Conductive copper sulfide can be deposited on synthetic fibers to yield moderately conductive fibers that are widely used in static-dissipating carpets due to their relatively light color [260,261].

Conducting polymers offer an interesting alternative to coated or filled plastics and textiles. However, these materials have several severe processing limitations. Although solution-spun fibers of polyaniline have been prepared and thin films of many conjugated polymers can be produced electrochemically. These fibers and films are brittle, expensive to produce, and difficult to manufacture on a large scale. Considering these limitations, textiles of various kinds represent a reasonable choice as a substrate for thin coatings of such polymers.

8.2. Deposition of Polyaniline on Textiles

Widespread industrial production of polyaniline conductive textiles is still lagging behind numerous earlier optimistic forecasts. This may be due to concerns that the highly toxic benzidine moiety may be formed during the oxidative polymerization of aniline under acidic conditions and may be incorporated in the polymeric structure. The first *in situ* polymerization of aniline on the surface of textiles was reported by Milliken Research Corporation [262] and has since been used by other investigators [263-265]. To obtain polymerization on the surface of the textile substrate only, the polymerization must be carefully controlled. Usually the polymerization is initiated by using a persulfate catalyst, such as potassium or ammonium persulfate, as has been described for the polymerization of aniline in aqueous media under acidic conditions [266]. This reaction exhibits a considerable induction period that varies from several minutes in concentrations of 0.5 mol/L to rather long periods in the desirable concentration of 0.03 mol/ L. If sodium or ammonium vanadate is used for the oxidation, this initiation period is eliminated and the polymerization on the surface of textiles can easily be controlled [267]. It is believed that the aniline forms a complex with the vanadium compound to allow a faster reaction to the polymerizing species. It is actually sufficient to use catalytic amounts of vanadium and conduct the bulk of the oxidation with persulfate. No fiber-fiber bonding occurs under these conditions, and the strength and flexibility of the substrate are preserved. Tzou and Gregory [268] studied the oxidative chemical polymerization kinetics of aniline with ammonium persulfate in aqueous hydrochloric acid solutions with and without textile substrates and noted an increase in reaction rates in the presence of a textile substrate. However, special precautions must be taken with polyaniline-coated textiles during the final washing and rinsing operation. It is therefore necessary to wash with water containing sufficient amounts of the counterion as demonstrated for polyaniline powders [266]. Because of the solubility and processibility of polyaniline doped with camphorsulfonic or dodecylbenzenesulfonic acid, polyaniline can be processed into fibers that may subsequently be incorporated into textiles.

Several variations on the production of polyaniline coated textiles have been reported. Researchers at the Huazhong University in China studied the *in situ* polymerization of aniline on poly(vinyl alcohol) fibers using ammonium persulfate as the oxidizing agent. Conductive fibers with good mechanical properties and stability were obtained [269]. In addition, researchers at the Atomic Energy Commission in Monts, France developed a continuous process to treat textiles with polyaniline. The textile substrate is first impregnated with a relatively concentrated hydrochloric acid solution of aniline(1-20%) and dried. Subsequently, the fabric is exposed to a solution of potassium dichromate in hydrochloric acid. The polyaniline-coated fabric is washed in 2 N hydrochloric acid solution and dried. This process can be repeated up to three times, resulting in a uniform, coherent layer of chloride-doped polyaniline with a conductivity of up to 0.5 ohm/□. A single pass takes 10-20 min, and, depending on the number of passes, surface resistances as low as 10 *D/D* may be reached. The intrinsic conductivity of the polyaniline deposited by this procedure reaches about 20 S cm^{-1} [270,271]. The fiber to be treated, preferably poly(p-phenylene terephthalamide) is first soaked with aniline hydrochloride and then exposed to the oxidizing agent [272]. Fabrics treated in the fashions exhibit surface resistances of several hundred ohm/□.

8.3. Properties of Polyaniline Conductive Textiles

8.3.1. Mechanical Properties

Unlike other conducting polymers, polyaniline exhibits good mechanical properties. It can be prepared as oriented crystalline fibers of excellent strength. The use of a textile substrate represents a convenient method of introducing mechanical strength, flexibility, and processibility to polyaniline for practical applications. Fabrics coated with a thin layer of polyaniline have essentially the same mechanical properties as the textile substrate, with minor variations depending on processing conditions.

8.3.2. Electrical Properties

The electrical properties of conducting textiles depend on the mass of the substrate, the diameter of the individual textile fibers, the thickness of the adsorbed layer, and the intrinsic volume conductivity of the conducting polymer. Volume conductivities for polyaniline-coated textiles as high as 20 Scm^{-1} have been reported. By adjusting the concentration of the reaction media, the surface resistance can be tailored to any desired value, with a lower limit somewhat dependent on the substrate weight.

8.3.3. Microwave Properties

The range of surface resistance obtainable with conductive textiles based on polyaniline (10-10,000 ohm/□) allows these materials to be used for a number of applications. The use of such fabrics as reinforcements to polymeric resins results in composites that have interesting and well-controlled electrical properties. These unusual electrical properties are quite evident in the microwave region of the electromagnetic spectrum. The literature on microwave properties of such composites is not extensive, but some results have been reported [271,273-276].

8.4. Stability of Polyaniline Conducting Textiles

In many applications, the stability of conducting materials is an important concern. Metal-coated fabrics, fabrics coated with inorganic conducting particles, and carbon-filled materials generally demonstrate stabilities that are equivalent to those of the parent conducting materials when monitored in a noncorrosive environment. Corrosion, or the degradation of the substrate, which causes interruptions in the percolating network, is normally the limiting factor in the ultimate lifetime of these materials. Organic materials, including conducting polymers, are much less stable in air owing to their relatively high reactivity with a variety of atmospheric chemicals, most notably oxygen. The stability of conducting polymers and conducting polymer-coated textiles has been shown to be improved dramatically when the material is isolated from the environment with a protective coating or laminate [277].

The stability of polyaniline-coated textiles has not been extensively studied. Several studies have been published on the thermal degradation of polyaniline films and powders [278-282]. The stability has been found to be highly dependent on the form of the polymer (film, pressed pellet or powder), its thermal and chemical history, and the dopant employed. The dependence on sample form indicates that the degradation is a diffusion controlled process [278]. Among conducting polymers, the stability of polyaniline is somewhat better than that of the poly(3-alkylthiophenes) but not as good as that of electrochemically generated arylsulfonate-doped polypyrrole. Polyaniline films that employ methanesulfonic acid as the dopant are more stable in moist air than in the nitrogen atmosphere indicating that oxidation by oxygen is not the primary degradation mechanisms in these films [282].

8.5. Composite Structures

Glass fabrics are widely used for the reinforcement of resins such as polyester, epoxy, imide, and others. The ability to combine the reinforcing properties of such a fabric with the electrical properties of polyaniline coating offers intriguing possibilities for a wide variety of industrial applications. The use of textiles based on carbon or graphite fibers offers similar properties, but the range of conductance can be better controlled with coated fabrics. In addition, the wide variety of substrates made from Kevlar, ceramic fibers, glass or quartz, etc. allows applications that cannot be met with carbon fibers. The use of such material has been suggested for airplanes to facilitate deicing [283].

Polyanilines have been suggested for use in the welding of plastics. A strip of conducting polymer-coated material can be used to convert microwave radiation into thermal energy [284]. Similarly, composite structures can be manufactured using conducting polymer-coated fabrics as reinforcement materials for thermoset resins. The advantage lies in the highly efficient microwave absorbance characteristics of conducting polymers. The use of resins in conjunction with polyaniline-coated fabrics yields composite structures that protect the polyanilines from oxygen and therefore imparts additional stability to these conductive fabrics. The range of conductance attainable with polyaniline-coated fabrics makes these textiles ideal for use as radar-absorbing materials. These materials show promise for use in a number of military applications. The microwave absorption characteristics of polyaniline-coated fabrics have been further exploited in multilayer products such as Salisbury screens and Jaumann absorbers [283].

In brief, the *in situ* polymerization of aniline in the presence of a substrate has been shown to offer an alternative to traditional processing techniques to overcome the inherently poor mechanical properties of these polymers. Textiles provide an ideal substrate because of their high surface area, wide range of mechanical properties, high flexibility, and large-scale availability. Besides, the use of non-reflective, radar-absorbing, lightweight materials for military camouflage remains a high potential application for conductive textiles. At present, either other markets are too small in yardage to support economical manufacturing or competitive materials (carbon fiber, metal foil, sputtered coatings) are better suited for the application. As stability concerns with respect to conducting polymers are addressed further and economies of scale to lower the cost of manufacture, conductive textiles based on polyanilines should become commercially important materials.

9. CONCLUSIONS AND FUTURE PROSPECTS

Polyanilines are attractive as conducting polymers because of their relative ease of synthesis and good environmental, chemical and thermal stability and its simple acid/base doping chemistry. In order to exploit these materials for commercial applications, it will be necessary to synthesize at low cost. The use of aqueous solutions and very cheap starting materials offer prospects of producing polyanilines in bulk, in the form of powder with approximately the same properties as the polyaniline films produced by anodic oxidation.

Polyanilines have a broad range of applications in microelectronics. In the area of lithography, they have been shown to provide an effective, simple, spin-apply process for charge dissipation, in particular for e-beam lithography and for SEM metrology. Implementation of these polymers in this area will depend on their commercial availability and cost. A number of resists based on polyaniline have been developed. However, the lithographic performance of these systems is not currently at a stage that can compete with conventional resists. They require further improvements for future use. Polyanilines have been shown to be applicable for metallization of plated through holes for printed circuit board technology. This process is currently in use in a number of companies worldwide. Polyaniline-based coatings have been developed that offer excellent ESD protection and offer numerous advantages over currently used materials. A number of coating formulations are either already commercially available or in the process of being made so. Corrosion protection of metals using polyanilines has been shown to be quite promising.

Polyaniline is the most promising candidate among the conducting polymers for membrane separations. The doping process can be used to control hydrophilicity for liquid separations and gas permselectivities for gas separations. A doping-undoping and partial redoping process leads to the most selective membrane known so far for oxygen-nitrogen separations. A future challenge is to develop pinhole-free ultra-thin membranes to achieve high fluxes.

One of the major problems facing commercialization of lithium battery technology is an unfocused approach to product development. For instance, the clear chemical advantage of applying technology to portable electronic devices has led to a large amount of work, being focused on reducing the operating temperature. In this direction lithium polyaniline batteries show a great promise.

The electrochromic performance of polyanilines meets the most important requisites for their use in electrochromic devices. Yet, when compared to liquid crystals, the polyanilines currently employed in this technology do not represent an improvement as far as switching response time is concerned; however, they have the important advantages of variety of colour contrast that is most significant characteristics. This property enables polyanilines as most promising candidates for the fabrication of multi-colour electrochromic devices.

The *in situ* polymerization of anilines in the presence of a substrate has been shown to offer an alternative to traditional processing techniques to overcome the inherently poor mechanical properties of these polymers. Textiles provide an ideal substrate because of their high surface area, wide range of mechanical properties, high flexibility and large-scale availability. This technique enables the production of large quantities of polyanilines composites from relatively inexpensive starting materials in aqueous solution. The use of nonreflective, radar-absorbing, light-weight materials for military camouflage remains a high potential application for conductive textiles. As stability concerns with respect to polyanilines are addressed further and economies of scale to lower the cost of manufacture, conductive textiles based on polyaniline should become commercially important materials.

In brief, polyanilines have certainly emerged into the market place. As the materials become commercially available and their properties continue to evolve, greater potential may be realized with these novel conducting polymer—the polyanilines.

REFERENCES

1. W.S. Huang, B.D. Humphrey and A.G. MacDiarmid; J. Chem. Soc., Faraday Trans.1, 82, 2385 (1986).
2. A.G. MacDiarmid and A.J. Epstein; Chem. Soc., 88, 317 (1989).
3. A.A. Syed and M.K. Dinesan; Talanta, 38, 815 (1991)
4. R. de Surville, M. Jozefowicz, L.T. Yu, J. Perichon and R. Buvet; Electrochim. Acta, 13, 1451 (1968).
5. A. Kitani, J. Izumi, J. Yano, Y. Hiromoto and K. Sasaki; Bull.Chem. Soc.Jpn., 57, 2254 (1984).
6. J.P. Travers, J. Chroboczek, F. Devreux, F. Genoud, M. Nechtschein, A.A. Syed, E.M. Genies and C. Tsintavis; Mol. Cryst. Liq. Cryst., 121, 195 (1985).
7. E.W. Paul, A.J. Ricco and M.S. Wrighton; J. Phys. Chem., 89, 1441 (1985).
8. J.C. Chiang and A.G. MacDiarmid; Synthetic Metals, 13, 193 (1986).
9. M. Nechtschein and C. Santier; J. Phys. (Paris), 47, 935 (1986).
10. S.H. Glarum and J.H. Marshall; J. Phys. Chem., 90, 6076 (1986).
11. J.P. Travers and M. Nechtschein; Synthetic Metals, 21, 135 (1987).
12. M. Nechtschein, C. Santier, J.F. Travers, J. Chroboczek, A. Alix and M. Ripert; Synthetic Metals, 18, 311 (1987).
13. W.S. Huang, A.G. MacDiarmid and A.J. Epstein; J. Chem. Soc., Chem. Commun., 1784 (1987).
14. A.J. Epstein, J.M. Ginder, F. Zuo, H.S. Woo, D.B. Tanner, A.F. Richter, M. Angelopoulos, W.S. Huang and A.G. MacDiarmid; Insulator-to-metal transition in polyaniline: Effect of protonation in emeraldine, 21, 63 (1987).
15. E.M. Genies and M. Lapkowski; J. Electroanal. Chem., 220, 67 (1987).
16. W.W. Focke and G.E. Wnek; J. Electroanal. Chem., 256, 343 (1988).
17. H.H.S. Javadi, R. Laversanne, A.J. Epstein, R.K. Kohli, E.M. Scherr and A.G. MacDiarmid; Synthetic Metals, 29, E 439 (1989).
18. A. Kitani, M. Kaya, S.I. Tsujioka and K. Sasaki; J. Polym. Sci., Part A: Polym. Chem., 26, 1531 (1988).
19. A. Andreatta, Y. Cao, J.C. Chiang, A.J. Heeger and P. Smith; Synthetic Metals, 26, 383 (1988).
20. M. Angelopoulos, A. Ray, A.G. MacDiarmid and A.J. Epstein; Synthetic Metals, 21, 21 (1987).
21. E.M. Genies, A.A. Syed and C. Tsintavis; Mol. Cryst. Liq. Cryst., 121, 181 (1985).
22. E.M. Genies and C. Tsintavis; J. Electroanal. Chem., 195, 109 (1985).
23. E.M. Genies and C. Tsintavis; J. Electroanal. Chem., 200, 127 (1986).
24. R. Jiang, S. Dong and S. Song; J. Chem. Soc., Faraday Trans.I, 85, 1575 (1989).
25. A.A. Syed and M.K. Dinesan, Reactive Polymers, 17, 145-157 (1992).
26. A.A. Syed and M.K. Dinesan, Analyst, 117, 61-62 (1992).
27. A.A. Syed and M.K. Dinesan, Synthetic Metals, 36, 209-215 (1990).
28. R. Jiang and S. Dong; Synthetic Metals, 24, 255 (1988).
29. R. Jiang and S. Dong; Synthetic Metals, 85, 1585 (1989).
30. M. Angelopoulos, G.E. Asturias, S.P. Ermer, A. Ray, E.M. Scherr and A.G. MacDiarmid; Mol. Cryst. Liq. Cryst., 160, 151 (1988).
31. L.W. Shacklette and C.C. Han; Mater. Res. Soc. Symp. Proc., 328, 157 (1994).
32. K.T. Tzou and R.V. Gregory; Synthetic Metals, 69, 109 (1995).
33. A.G. MacDiarmid, J.C. Chiang, A.F. Richter, N.L.D. Somasiri and A.J. Epstein; in Conducting Polymers (Ed. L. Alcacer), pp.105-120, Reidel, Dordrecht, (1985).
34. J. Yue and A.J. Epstein; J. Am. Chem. Soc., 112, 2800 (1990).
35. C. DeArmitt, S.P. Armes, J. Winter, F.A. Uribe, S. Gottesfeld and C. Mombourquette; Polymer, 34(I), 158 (1993).
36. M.T. Nguyen, P. Kasai, J.L. Miller and A.F. Diaz; Macromolecules, 27, 3625 (1994).
37. Y. Wei, R. Hariharan and S.A. Patel; Macromolecules, 23, 758 (1990).
38. M. Leclerc, J. Guay and L.H. Dao; Macromolecules, 22, 649 (1989).
39. Y.H. Liao, M. Angelopoulos and K. Levon; J. Polym. Sci., Polym. Chem. Ed., 33, 2725 (1995).

40. M. Angelopoulos, S.P. Ermer, S.K. Manohar and A.G. MacDiarmid; Mol. Cryst. Liq. Cryst., 160, 223 (1988).
41. Y. Cao, P. Smith and A.J. Heeger; Synthetic Metals, 48, 91 (1992).
42. M. Angelopoulos, N. Patel and R. Saraf; Synthetic Metals, 55, 1552 (1993).
43. K. Tzou and R.V. Gregory; Synthetic Metals, 53, 365 (1993).
44. M. Angelopoulos, N. Patel and J.M. Shaw; Mater. Res. Soc. Symp. Proc., 328, 173 (1994).
45. N. Mermilliod, J. Tanguy, M. Hoclet and A.A. Syed; Synthetic Metals, 18, 359-364 (1993).
46. J.E. Frommer and R.R. Chance; in Encyclopedia of Polymer Science and Engineering, Vol. 5, pp. 462-507, Wiley, New York, 2nd Edition, (1986).
47. H. Shirakawa, E.J. Louis, A.G. MacDiarmid, C.K. Chiang and A.J. Heeger; J. Chem. Soc. Chem. Commun., 578 (1977).
48. Hand Book of Conducting Polymers (Ed. T.A. Skotheim), Vols. I and II, Marcel Dekker, New York, (1986).
49. A.J. Heeger; Polym. J., 17, 201 (1985).
50. J.E. Frommer; Acc. Chem. Res., 19, 2 (1986).
51. R.J. Waltman and J. Bargon; Can. J. Chem., 64, 76 (1986).
52. B.D. Malhotra, N. Kumar and S. Chandra; Prog. Polym. Sci., 12, 179 (1986).
53. A.K. Bakhshi, J. Ladik and M. Steel; Phys. Rev. B, 35, 704 (1987).
54. Saxena and J.D. Gunton; Phys. Rev. B, 35, 3914 (1987).
55. A.J. Heeger, S. Kivelson, J.R. Schrieffer and W.P. Su; Rev. Nod. Phys., 60, 781 (1988).
56. A.O. Patil, A.J. Heeger and F. Wudl; Chem. Rev., 88, 183 (1988).
57. G.L. Baker; in Electronic and Photonic Applications of Polymers (Eds. M.J. Bowden and S.R. Turner) pp. 271-296, Advances in Chemistry Series, American Chemical Society, (1988).
58. Scrosati; Prog. Solid State Chem., 18, 1 (1988).
59. S. Maiti; J. Scientific and Industrial Research, 48, 577 (1989).
60. S. Stafstrom and J.L. Bredas; J. Molecular Structure (Theochem), 188, 393 (1989).
61. J. Heinze and M. Dietrich; Materials Science Forum, 42, 63 (1989).
62. E.M. Genies, A. Boyle, M. Lapkowski and C. Tsintavis; Synthetic Metals, 36, 139 (1990).
63. G.B. Street; in Hand Book of Conducting Polymers, (Ed. T.A. Skotheim) p. 270 Marcel Dekker, New York, (1986).
64. H. Letheby; J.Chem. Soc., 15, 161 (1862).
65. A.F. Diaz; Chem. Scr., 17, 145 (1981).
66. W.L. Öbel, Mater. Sci., 16(4), 73 (1990).
67. P.R. Newman and P.H. Cunningham, U.S. Patent 4, 942, 078 (1990).
68. J.C. Chiang; Chemical, electrochemical and infrared studies of polyaniline, Ph.D. thesis, University of Pennsylavania, (1987).
69. J. Fritsche; J. Prakt. Chem., 20, 453 (1840).
70. J. Fritsche; J. Prakt. Chem., 28, 198 (1843).
71. J.J. Coquillon; Compt. Rend., 81, 408 (1875).
72. J.J. Coquillon; Compt. Rend., 82, 228 (1876).
73. A. Rosenstiehl; Compt. Rend., 81, 1257 (1875).
74. F. Goppelsroeder; Compt. Rend., 82, 1392 (1876).
75. L. Gilchrist; J. Phys. Chem., 8, 539 (1904).
76. H.T. Bucherer; Ber., 40, 3412 (1907).
77. H.T. Bucherer; Ber., 42, 2931 (1909).
78. A.G. Green and A.E. Woodhead; J. Chem. Soc., 97, 2388 (1910).
79. A.G. Green and A.E. Woodhead; J. Chem. Soc., 101, 1117 (1912).
80. A.G. Green and W. Johnson; Ber., 46, 3769 (1913).
81. A.G. Green and S. Wolff; Ber., 44, 2570 (1911).
82. A.G. Green and S. Wolff; Ber., 46, 33 (1913).
83. A.G. Green and S. Wolff; J. Soc. Dyers Colorists, 29, 105 (1913).
84. T. Yasui; Bull.Chem. Soc. Jpn., 10, 305 (1935).
85. N.E. Khomutov and S.V. Gorbachev; Zh. Fiz. Khim., 24, 1101 (1950).
86. N.E. Khomutov and S.V. Gorbachev; Sovehsch. Eleetrokhim, 579 (1950).
87. M. Jozefowicz, J.H. Perichon, L.T. Yu and R. Buvet; Br. Patent, No. 1216569, (1970).
88. L.T. Yu, M.S. Borredon, M. Jozefowicz, G. Belorgey and R. Buvet; J. Polym. Sci., Part C, Polymer Symposia, 16, 2931 (1967).

89. A.A. Syed, M.K. Dinesan and E.M. Genies; Bull. Electrochem., 4, 737 (1988).
90. A.A. Syed and M.K. Dinesan; Department of Chemistry, University of Mysore, Unpublished results.
91. S.P. Armes and J.F. Miller; Synthetic Metals, 22, 385 (1988).
92. D. Svobodova, P. Krenek, M. Fresnke and J. Gasparie; Mirkochim. Acta, 11, 197 (1978).
93. V. Kafarov; Cybernetic Methods in Chemistry and Chemical Engineering, pp. 360-401, Mir Publishers, Moscow, (1976).
94. J.A. Nelder and R. Mead; The Computer Journal, 7, 308 (1965).
95. A. Pron, F. Genoud, C. Menardo and M. Nechtschein; Synthetic Metals, 24, 193 (1988).
96. S.P. Armes and M. Aldissi; J. Chem. Soc., Chem. Commun., 88 (1989).
97. A.F. Diaz and J.A. Logan; J. Electroanal. Chem., 111, 111 (1980).
98. A.A. Syed M.K. Dinesan, in Solid-State Ionic Devices, B.V.R. Chowdari and S. Radhakrishna (Eds.) pp. 481-487, World Scientific Pub.Co., Singapore (1988).
99. A.G. MacDiarmid, J.C. Chiang, M. Halpern, W.S. Huang, S.L. Mu, N.L.D. Somasiri, W. Wu and S.I. Yaniger; Mol. Cryst. Liq. Cryst., 121, 173 (1985).
100. G. Mengoli, M.T. Munari, P. Bianco and M.M. Musiani; J. Appl. Polym. Sci., 26, 4247 (1981).
101. M.M. Musiani, G. Mengoli and F. Furlanetto; J. Appl. Polym. Sci., 29, 4433 (1984).
102. D.W. DeBerry; J. Electrochem. Soc., 132, 1022 (1985).
103. G. Mengoli, M.T. Munari and C. Folonari; J. Electroanal. Chem., 124, 237 (1981).
104. N. Oyama, Y. Ohnuki, K. Chiba and T. Ohsaka; Chem. Lett., 1759 (1983).
105. R.L. Hand and R.F. Nelson; J. Am. Chem. Soc., 96, 850 (1974).
106. R.L. Hand and R.F. Nelson; J. Electrochem. Soc., 125, 1059 (1978).
107. R. Noufi, A.J. Nozik, J. White and L.F. Warren; J. Electrochem. Soc., 29, 2261 (1982).
108. T. Osaka, S. Ogano and K. Naoi; J. Electrochem. Soc., 135, 539 (1988).
109. D.M. Mohilner, R.N. Adams and W.J. Argersinger Jr.; J. Am. Chem. Soc., 84, 3618 (1962).
110. J. Bacon and R.N. Adams; J. Am. Chem. Soc., 90, 6596 (1968).
111. Polymers in Microelectronics, D.S. Soane and Z. Martynenko (Eds.), Elsevier, Amsterdam (1989).
112. Microelectronics Packaging Handbook, R.R. Tummala and E.J. Rymaszewski (Eds.), Van Nostrand Reinhold, New York (1989).
113. Handbook of Electronic Package Design, M. Pecht (Ed.), Marcel Dekker, New York, (1991).
114. Electronic Packaging and Interconnection Handbook, C.A. Harper (Ed.), McGraw-Hill, New York (1991).
115. R.L. Cadenhead, Materials and electronic phenomena, in Electronics Materials Handbook (C.A. Dostal Ed.) pp. 89-120, ASM International, Materials Park, OH (1989).
116. Polymer Materials for Electronic Applications (Eds. E.D. Feitz and C.W. Wilkins), ACS Symp. Ser., ACS, Washington, DC (1982).
117. Polymers in Microlithography, (Eds. E. Reichmanis, S.A. MacDonald and T. Iwayanagi), ACS Symp. Ser., ACS, Washington, DC (1989).
118. J.M. Shaw; in Imaging Processes and Materials (Eds. J. Sturge, V. Walworh and A. Shepp,), pp. 567-586, Van Nostrand Reinhold, New York (1989).
119. Introduction to Microlithography, (Eds. L.F. Thompson, C.G. Wilson and M.J. Bowden), ACS, Washington, DC (1994).
120. Semiconductor Lithography, (Ed. W.M. Moreau), Plenum, New York, (1988).
121. Glow Discharge Processes, (Ed. B. Chapman), Wiley, New York, (1980).
122. H.J. Smith; J. Vac. Sci. Technol. B, 13(6), 2323 (1995).
123. Proceedings of 39th International Conference on Electron, Ion and Photon Beam Technology and Nanofabrication, (Ed. D. Kern), American Vacuum Society, American Institute of Physics, New York (1995).
124. J.M. Ryan, A.C.F. Hoole and A.N. Broers; J. Vac. Sci. Technol. B, 13(6), 3035 (1995).
125. G.O. Langner; Proc. Microcirc. Eng. 1979, 261 (1979).
126. K.D. Cummings and M. Kiersh; J. Vac. Sci. Technol. B, 7(6), 1536 (1989).
127. H. Itoh, K. Nakamura and H. Hayakawa; J. Vac. Sci. Technol. B, 7(6), 1532 (1989).
128. M. Angelopoulos, J.M. Shaw, R.D. Kaplan and S. Perreault; J. Vac. Sci. Technol. B, 7(6), 1519 (1989).
129. M. Angelopoulos, J.M. Shaw, K.L. Lee, W.S. Huang, M.A. Lecorre and M. Tissier; J. Vac. Sci. Technol. B, 9(6), 3428 (1991).
130. M. Angelopoulos, J.M. Shaw, W.S. Huang and R.D. Kaplan; Mol. Cryst. Liq. Cryst., 189, 221 (1990).
131. M. Angelopoulos, J.M. Shaw and K.L. Leung; Mater. Res. Soc. Symp. Proc., 214, 137 (1991).
132. J.V. Crivello and J.H.W. Lam; Macromolecules, 10, 1307 (1977).

133. J.V. Crivello and J.H.W. Lam; J. Polym. Sci., Polym. Chem. Ed., 17, 977 (1979).

134. J.M. Shaw, M. Hatzakis, E. Babich, J.R. Paraszczak, D. Witman and K.J. Stewart; J. Vac.Sci. Technol. B, 7(6), 1709 (1989).

135. J.M. Liu, L. Sun, J.H. Hwang and S.C. Yang; Mater. Res. Soc. Symp. Proc., 247, 601 (1992).

136. J. Yue, Z.H. Wang, K.R. Cromack, A.J. Epstein and A.G. MacDiarmid; J. Am. Chem.Soc., 113(7), 2665 (1991).

137. S.A. Chen and G.W. Hwang; J. Am. Chem. Soc., 116(17), 7939 (1994).

138. S.A. Chen and G.W. Hwang; J. Am. Chem. Soc., 117(40), 10055 (1995).

139. M. Angelopoulos, N. Patel, J.M. Shaw, N.C. Labianca and S. Rishton; J. Vac. Sci. Technol.B, 11(6), 2794 (1993).

140. W.S. Huang, M. Angelopoulos, J.R. White and J.M. Park; Mol. Cryst. Liq. Cryst., 189, 227 (1991).

141. H. Stuckmann and J. Hupe; Blasberg-Mitteilungen 11, 3-27 (1991).

142. K. Beator, B. Hildesheim, B. Bressel and H.J. Grapentin; Metalloberflache, 46, 384 (1992).

143. J. Hupe; Blasberg-Mitteilungen 15, 14-18 (1995).

144. S. Gottesfeld, F.A. Uribe and S.P. Armes; J. Electrochem. Soc., 139(I), L14 (1992).

145. J. Fjelstad; in Printed wiring board technology: current capabilities and limitations, Electronics Materials Handbook (Ed. C.A. Dostal), pp. 507-512, ASM Int., Materials Park OH (1989).

146. L. Lynch and R.A. Nesbitt; in Rigid printed wiring board fabrication techniques, Electronics Materials Handbook (Ed. C.A. Dostal), pp. 538-555, ASM Int., Materials Park, OH (1989).

147. D.P. Seraphim, D.E. Barr, W.T. Chen, G.P. Schmitt and R.R. Tummala; in Printed circuit board packaging, Microelectronics Packaging Handbook (Eds. R.R. Tammala and E.J. Rymaszewski), pp. 853-922, Van Nostrand Reinhold, New York (1989).

148. J.J. Steppan, J.A. Roth, L.C. Hall, D.A. Jeannotte and S.P. Carbone; J. Electrochem. Soc., 134, 175 (1987).

149. R. Walker; J. Chem. Ed., 57, 789 (1980).

150. V. Brusic, M.A. Frisch, B.N. Eldridge, F.P. Novak, F.B. Kaufman, B.M. Rush and G.S. Frankel; J. Electrochem. Soc., 138, 2253 (1991).

151. V. Brusic, G.S. Frankel, J. Roldan and R. Saraf; J. Electrochem. Soc., 42 2591 (1995).

152. V. Brusic, M. Angelopoulos and T.Graham; Proc.Soc.Plastics Eng.53rd Annual Technical Conference (1995).

153. V. Brusic, M. Angelopoulos, T. Graham and P. Buchwalter; Proc.188th Electrochem. Soc., Chicago, IL (1995).

154. V. Brusic, M. Angelopoulos and T. Graham; J. Electrochem. Soc., 144(2) 436 (1997).

155. D.W. Berry; J. Electrodhem Soc., 132, 1022 (1985).

156. A.G. MacDiarmid; Disucssions, Int. Conf. Synthetic Metals, Kyoto, Japan (1986).

157. D.A. Wrobleski, B.C. Benicewicz, K.G. Thompson and C.J. Bryan; ACS Polym. Prep., 35, 265 (1994).

158. S. Sathiyanarayanan, S.K. Dhawan, D.C. Trivedi and K. Balakrishnan; Corros. Sci., 33, 1831 (1992).

159. B. Wessling; Adv. Mater. 6(3), 226 (1994).

160. W.K. Lu, R.L. Elsenbaumer and B. Wessling; Synthetic Metals, 71, 2163 (1995).

161. S. Jasty and A.J. Epstein; Polym. Mater. Sci. Eng., 72, 565 (1995).

162. V. Brusic, M. Russak, R. Schad, G. Frankel, A. Selius, D. DiMilia and D. Edmonson; J. Electrochem. Soc., 136, 42 (1989).

163. S.L. Law, S, Mucha and S. Banks; Electron.Packag. Prod., 31(5) 82 (1991).

164. L. Brown and D. Burns; Electron. Packag. Prod., 50 (1990).

165. N.F. Colaneri and L.W. Shacklette; IEEE Trans. Instrum. Meas., 41, 291 (1992).

166. L.W. Shacklette, C.C. Han and M.H. Luly; Synthetic Metals, 57, 3532 (1993).

167. Proceedings of Electrical Overstress/Electrostatic Discharge Symposium (EOS16), Las Vegas, NV, published by EOS/ESD Assoc. (1994).

168. V.G. Kulkarni, W.R. Matthew, J.C. Campbell, C.J. Dinkins and P.J. Durbin; Proc.Soc.Plast.Eng., 665, (1991).

169. V.G. Kulkarni, J.C. Campbell and W.R. Mathew; Synthetic Metals, 57, 3780 (1993).

170. V.G. Kulkarni; in Intrinsically Conducting Polymers: An Emerging Technology: (Ed.M. Aldissi), p. 45, Kluwer, Dordrecht (1993).

171. C.Y. Yang, Y. Cao, P. Smith and A.J. Heeger; Synthetic Metals, 53, 294 (1993).

172. J.E. Osterholm, J. Laakso, O. Ikkala, H. Ruohonen, K. Vakiparta, E. Virtanen, H. Jarvinen and T. Taka; Polym. Prepr., 35(I), 244 (1994).

173. M. Angelopoulos, J.D. Gelorme and J.M. Shaw; Proceedings of Electrical Overstress/Electrostatic Discharge Symposium: Las Vegas, NV (1994).

174. Polymers for Gas Separation, N. Toshima (Ed.), VCH, New York (1992).

175. R.E. Kesting and A.K. Fritzsche; Polymeric Gas Separation Membranes, Wiley, New York, (1993).

176. S.A. Stern; J. Membrane Sci., 94, 1 (1994).
177. W.J. Koros and G.K. Fleming; J. Membrane Sci., 83, 1 (1993).
178. W.J. Koros, M.R. Coleman and D.R.B. Walker; Anna. Rev. Mater.Sci., 22, 47 (1992).
179. Pervaporation Membrane Separation Processes, R.Y.M. Huang (Ed.), Elsevier, New York (1991).
180. S.M. Zhang and E. Drioli; Separ. Sci. Technol., 30, 1 (1995).
181. T.M. Aminabhari, R.S. Khinnavar, S.B. Harogoppad and U.S. Aithal; J. Macromol. Sci. Rev. Macromol. Chem. Phys., 34(2), 129 (1994).
182. M.G. Kanadtzidis; Chem. Eng. News, 68(49), 36 M (1990).
183. B.R. Mattes, M.R. Anderson, H. Reiss and R.B. Kaner; ACS Polym. Mater. Sci. Eng., 64, 336 (1991).
184. M.R. Anderson, B.R. Mattes, H. Reiss and R.B. Kaner; Synthetic Metals, 41-43, 1151 (1991).
185. R.B. Kaner, M.R. Anderson, B.R. Mattes and H. Reiss; U.S. Patents 5, 095, 586 (March 17, 1992) and 5,358,556 (Oct. 24, 1994).
186. M.R. Anderson, B.R. Mattes, H. Reiss and R.B. Kaner; Science, 252, 1412 (1991).
187. J. Maron, M.J. Winokur and B.R. Mattes; Macromolecules, 28, 4475 (1995).
188. J. Pellingrino, R. Radebaugh and B.R. Mattes; Macromolecules, 29, 4985 (1996).
189. M.R. Anderson, An investigation of conducting polymer materials as gas separation membranes, Ph.D. Thesis, Univ. Calif., Los Angeles, (1992).
190. L. Rebattet, M. Escoubes, E. Genies and M. Pined; J. Appl. Polym. Sci., 58, 923 (1995).
191. L. Rebattet, M. Escoubes, M. Pineri and E.M. Genies; Synthetic Metals, 71, 2133 (1995).
192. J.A. Conklin; Polyaniline films for membrane based gas permeability studies, Ph.D. Thesis, Univ. Calif., Los Angeles, (1994).
193. S. Kaplan, E.M. Conwell, A.F. Richter and A.G. MacDiarmid; Macromolecules, 22 1669 (1989).
194. S. Kaplan, E.M. Conwell, A.F. Richter and A.G. MacDiarmid; Synthetic Metals, 29, 235 (1989).
195. B.R. Mattes, M.R. Anderson, H. Reiss, and R.B. Kaner; in Intrinsically Conducting Polymers: An Emerging Technology (Ed. M. Aldissi), p. 61, Kluwer, Boston (1993).
196. M. Nechtsein, F. Devreux, F. Genoud, M. Guglielmi and K. Holczer; Phys. Rev. B, 27, 61 (1983).
197. F. Bensebaa and J.J. Andre; J. Phys. Chem., 96, 5739 (1992).
198. S. Kuwabata and C.R. Martin; J. Membrane Sci., 91, 1 (1994).
199. L. Rebattet, M. Escoubes, E. Genies and M. Pineri; J. Appl. Polym. Sci., 57, 1595 (1995).
200. E.S. Amis and J.F. Hinton; Solvent Effects on Chemical Phenoma, pp. 46-181, Academic Press, New York (1973).
201. B.E. Conway, Ionic Hydration in Chemistry and Biophysics, p. 59, Elsevier, Amsterdam (1981).
202. M. Angelopoulos, N. Patel and R. Saraf; Synthetic Metals, 55-57, 1552 (1993).
203. B.R. Mattes, M.R. Anderson, J.A. Conklin, H. Reiss and R.B. Kaner; Synthetic Metals, 55-57, 3655 (1993).
204. S.C. Huang, J.A. Conklin, T.M. Su, I.J. Ball, S.L. Nguyen, B.M. Lew and R.B. Kaner; ACS Polym. Mater. Sci. Eng. 72, 323 (1995).
205. T.H. Kim, W.J. Koros and G.R. Husk; Separ. Sci. Technol., 23, 1611 (1988).
206. S.A. Stern, Y. Mi, H. Yamamoto and A.K. St. Clair; J. Polym. Sci., Polym. Phys., 27, 1887 (1989).
207. T.A. Barbari, W.J. Koros and D.R. Paul; J. Membrane Sci., 42, 69 (1989).
208. L.M. Robeson, W.F. Burgoyne, M. Langsam, A.C. Savoca and C.F. Tien; Polymer, 35, 4970 (1994).
209. L.M. Robeson; J. Membrane Sci. 62, 165 (1991).
210. Lange's Handbook of Chemistry, (Ed. J. Dean), Table 5-2, McGraw-Hill, New York, (1992).
211. S.C. Huang, J.A. Conklin, T.M. Su, I.J. Ball, S.L. Nguyen and R.B. Kaner; ACS Polym. Mater. Sci. Eng., 72, 168 (1995).
212. V.M. Schmidt, D. Tegtmeyer and J. Heitbaum; J. Electroanal. Chem., 385, 149 (1995).
213. V.M. Schmidt, D. Tegtmeyer and J. Heitbaum; Adv. Mater., 4, 428 (1992).
214. V.M. Schmidt and J. Heitbaum; Synthetic Metals, 41, 425 (1991).
215. C.R. Martin, N. Liang, V. Menan, R. Parthasarathy and A. Parthasarathy; Synthetic Metals, 55-57, 3766 (1993).
216. J.C. Chu, S.L. Wang, S.L. Levy and R. Paul; Ju Chin Vapor-Liquid Equilibrium Data, J.W. Edwards, Ann Arbor, MI, (1956).
217. D.L. Feldheirn and C.M. Elliot; J. Membrane Sci, 70, 9 (1992).
218. Practical Lithium Batteries (Ed. Y. Matsuda), p. 129, JEC Press (1988).
219. K.M. Abraham and S.B. Brummer; In Lithium Batteries (Ed. J.P. Gabano), p. 371, Academic Press, London (1983).
220. P.J. Nigrey, A.G. MacDiarmid and A.J. Heeger; Electrochim. Acta, 595 (1979).

221. P.J. Nigrey, D. Jr., MacInnes, D.P. Nairns and A.J. MacDiarmid; J. Electrochem. Soc., 128, 1651 (1981).

222. T. Nagatomo, C. Ichikawa and O; J. Electrochem. Soc., 134, 305 (1987).

223. F. Beniere, D. Boils and J. Hanepa; J. Electrochem. Soc., 132, 2100 (1985).

224. N. Furukawa and Koji Nishio; in Applications of Electroaction polymer, (Ed.Bruno Scorosati), p. 169, Chapwall & Stall, London (1993).

225. S. Taguchi and T. Tanaka; J. Power Sources, 20, 249 (1987).

226. F. Goto, K. Abe and K. Okabayashi; J. Power Sources, 20, 243 (1987).

227. E. Genies, P. Hany and Ch. Santier; Synthetic Metals, 28, C647 (1989).

228. K. Takei, K. Isihihara, T. Iwahori and T. Tanaka; Denki Kagaku, 58, 934 (1990).

229. T. Osaka, S. Ogano, K. Naoi and N. Oyama; J.Electrochem. Soc., 136, 306 (1989).

230. T. Osaka, T. Nakajima, K. Naoi and B.B. Owens; J. Electrochem. Soc., 137, 2139 (1990).

231. K. Hyodo, M. Omae and Y. Kagami; Electrochim. Acta, 36, 87 (1991).

232. E.M. Genies, P. Hany and Ch. Santier; Synthetic Metals, 18, 751 (1988).

233. M. Mizumoto, M. Namba, S. Nishimura, H. Miyadera, M. Koseki and Y. Kobayashi; Synth.Met., 28,(1-2), 639 (1989).

234. K. Nishio, M. Fujimoto, N. Yoshinaga; *et al.* 40th ISE Meeting, Extended Abstracts, p. 553 (1989).

235. K. Nishio, M. Fujimoto, N. Yoshinaga, *et al.* 29th Battery Symposium in Japan, Abstracts, p. 227 (1988).

236. K. Nishio, M. Fujimoto, N. Yoshinaga and N. Furukawa; 30th Battery Symposium in Japan, Abstracts, p. 127 (1989).

237. A.M. Mastragostino, Marinangeli, A. Corradini and S. Giacobbe; Synthetic Metals, 28, C501-C506 (1989).

238. K. Yoshino, K. Kaneto and Y. Inuishi; Jpn. J. Appl. Phys., 22, L157-LI58 (1983).

239. F. Garnier, G. Tourillon, M. Gazard and J.C. Dubois; J. Electroanal. Chem., 148, 299-303 (1983).

240. S. Kuwabata, H. Yoneyama and H. Tamura; Bull. Chem. Soc. Jpn., 57, 2247-53 (1984).

241. T. Kobayashi, H. Yoneyama and H. Tamura; J. Electroanal. Chem., 161, 419-23 (1984).

242. T. Kobayashi, H. Yoneyama and H. Tamura; J. Electroanal. Chem., 177, 281-91 (1984).

243. A. Kitani, J. Yano and K. Sasaki; J. Electroanal. Chem., 209, 227-32 (1986).

244. E.M. Genies, M. Lapkowski, C. Santier and E. Vieil; Synthetic Metals, 18, 631-6 (1987).

245. M. Kaneko, H. Nakamura and T. Shimomura; Makromol. Chem, Rapid Commun., 8, 179-80 (1987).

246. H. Yashima, M. Kobayashi, K.B. Lee, D. Chung, A.J. Heeger and F. Wudl; J. Electrochem. Soc., 134, 46-52 (1987).

247. M. Akhtar, H.A. Weakliem, R.M. Paiste and K. Gaughan; Synthetic Metals, 26, 203-8 (1988).

248. A. Corradini, A.M. Marinangeti, M. Mastragostino and B. Scrosati; Solid State Ionics, 28-30, 1738-42 (1988).

249. M.A. De Paoli, S. Panero, S. Passerini and B. Scrosati; Adv. Mater., 2, 480-2 (1990).

250. M.A. Rodrigues, M.A. De Paoli and M. Mastragostino; Electrochimica Acta, 36, 2143-6 (1991).

251. E.M. Genies, A. Boyle, M. Lapkowski and C. Tsintavis; Synthetic Metals, 36, 139-82 (1990).

252. J.C. LaCroix and A.F. Diaz; Makromol. Chem., Macromol. Symp., 8, 17-37 (1987).

253. T. Asaoka, K. Okabayashi, T. Abe and T. Yoshida; 40th ISE Meeting (Kyoto), Ext. Abs., 1, 245-6 (1989).

254. J.C. Lacroix, K.K. Kanazawa and A. Diaz; J. Electrochem. Soc., 136, 1308-13 (1989).

255. C. Arbizzani, M. Mastragostino, S. Passerini, R. Pillegi, B. Scrosatt; Electrochim.Acta, 36(5/6) 837 (1991).

256. W. Lobel; Mater. Sci.16(4), 73 (1990).

257. E.K. Sichel (ed.), Carbon Black Polymer Composites Marcel Dekker, New York, (1982).

258. H. Ebneth; Melliand Textilber. 62, 297 (1981).

259. W.C. Smith; J. Coated Fabr. 17, 242 (1988).

260. M.Okoniewski; Melliand Textilber., 71, 94 (1990).

261. S. Tomibe, R. Gomibuchi and K. Takahashi; Electrically conductive fiber and method of making same, U.S. Patent 4, 410, 593 (1983).

262. R.V. Gregory, W.C. Kimbrell and H.H. Kuhn; Synthetic Metals, 28, C823 (1989).

263. E.M. Genies, C. Petrescu and L. Olmedo; Synthetic Metals, 41, 665 (1991).

264. D.C. Trivedi and S.K. Dhawan; Synthetic Metals, 59, 267 (1993).

265. M. Ikuo, Manufacture of electrically conductive polymer substrate coated with polyaniline, Jpn, Patent 03, 119, 612 (1991).

266. A.G. MacDiarmid, J.C. Chiang, A.F. Richter and N.L.D. Somasiri; Polyaniline: Synthesis and characterization of the emeraldine oxidation state by elemental analysis, in Conducting Polymers (L.Alcacer, ed.), Reidel, Dordrecht, (1987), p 105.

267. H.H. Kuhn and W.C. Kimbrell; U.S. Patent 4, 981, 718 (1991).

268. K. Tzou and R.V. Gregory; Synthetic Metals, 47, 267 (1992).
269. H. Liu, X. Li, X. Gou and H. Xie; Gaofenzi Cailiao Kexue YII Gongcheng, 10(6), 22 (1994).
270. J.L. Forveille and L. Olmedo; Synthetic Metals, 65, 5 (1994).
271. F. Jousse, L. Delnaud and L. Olmedo; Processing of conductive fabrics in composite applications, 40th Int. Sampe. Symp. Exhib., (1995), p. 360.
272. L.C. Sengupta and W.A. Spurgeon, Dielectric properties of polymer matrix composites prepared with conductive polymer treated fabrics, 6th Int. SAMPE Electron Conf., (1992), p. 46.
273. C.H. Hsu; Electrically conductive articles, Eur. Patent 0355 518 A2 (1990).
274. T.C.P. Wong, B. Chambers, A.P. Anderson and P.V. Wright; Electron.Lett., 28, 165 (1992).
275. T.C.P. Wong, B. Chambers, A.P. Anderson and P.V. Wright; Fabrication and evaluation of conducting polymer composites as radar absorbers, 8th Int. Conf. Antennas and Propag., 993, p. 934.
276. P.V. Wright, T.C.P. Wong, B. Chambers and A.P. Anderson; Adv. Mater. Opt. Electron., 4, 253 (1994).
277. T. Haigwara, M. Yamaura and K. Iwata; Synthetic Metals, 25, 243 (1988).
278. H.H. Kuhn, W.C. Kimbrell, G. Worrell and C.S. Chen; Tech. Pap.- Soc. Plast. Eng., 37, 760 (1991).
279. K.G. Neoh, E.T. Kang, S.H. Khor and K.L. Tan; Polym. Degrad. Stab., 27, 107 (1990).
280. Y. Wei and K.F. Hsueh, Polym. Sci., Part A: Polym. Chem., 27, 435 (1989).
281. V.G. Kulkarni, L.D. Campbell and W.R. Mathew; Synthetic Metals, 30, 32 (1988).
282. Y. Wang and M.F. Rubner; Synthetic Metals, 47, 255 (1992).
283. P.R. Newman and P.H. Cunningham, Electrically heated structural composite and method of its manufacture, U.S. Patent 4,942,078 (1990).
284. A.J. Epstein, J. Joo, C.-Y. Wu, A. Benator, C.F. Faisst, J. Zegarski, and A.G. MacDiarmid, Polyaniline: Recent advances in processing and applications to welding of plastics, in Intrinsically Condllcting Polymers (M. Aldissi, ed.), Kluwer, Dordrecht, (1993), p. 65.

3

Polymer Nanofibers: Fabrication, Applications and Characterization

E.P.S. Tan[a], Y.Z. Zhang[a,b], S. Ramakrishna[a,b,c] and Lim Chwee Teck[a,b,c,*]

[a]*Division of Bioengineering, National University of Singapore, Singapore—117576.*
[b]*Department of Mechanical Engineering, National University of Singapore, Singapore—117576.*
[c]*Nanoscience and Nanotechnology Initiative, National University of Singapore, Singapore—117576.*
Corresponding author's email: ctlim@nus.edu.sg

Contents

- ❏ **INTRODUCTION**
- ❏ **FABRICATION METHODS**
- ❏ **APPLICATIONS**
- ❏ **CHARACTERIZATION**
- ❏ **CONCLUSIONS AND FUTURE PROSPECTS**
- ❏ **REFERENCES**

Summary: Research on polymer nanofibers with diameter in the nanometer range has seen a rapid growth over the last ten years. In particular, electrospun polymeric nanofibers have been researched into most extensively due to their potential for mass production and the large variety of polymers that can be converted to nanofibers using this technique. This chapter presents a review of the fabrication techniques with emphasis placed on electrospinning, applications of polymer nanofibers and characterization techniques pertaining to single nanofibers and nanofibrous mats or foams. Challenges and future trends in processing and characterization of polymeric nanofibers are also discussed.

1. INTRODUCTION

Some of the first polymer fibers used by man were from natural sources such as silk and cotton. It was only until the last century that various forms of synthetic polymer fibers were produced for a wide range of applications such as modern apparel, home furnishings, medicine and industry [1]. As research fields such as nanotechnology and biomedical engineering began to take off in recent years, the vast advantages that nanofibers possess, i.e. superior mechanical properties and large surface area to volume ratio, become apparent. With the need for regeneration of tissues that can mimic natural collagen fibers, filtration systems with high efficiency, tougher composites and extremely sensitive biological sensors, research into large scale production of polymeric nanofibers which are viable in terms of chemical, physical and mechanical properties is required. As a result, polymeric nanofibers have attracted the attention of many researchers in recent years. The aim of this chapter is to present a review of the fabrication techniques, applications of polymeric nanofibers and characterization techniques for single nanofibers and nanofibers in the form of mats and foams. Challenges and future trends are also presented.

2. FABRICATION METHODS

Research on fabrication methods remains one of the most important topics for polymeric nanofibers. Several fabrication techniques such as electrospinning [2], melt blowing [3,4], phase separation [5,6], self-assembly [7-9], and template synthesis [10,11] have been employed to produce suitable polymer nanofibers for meeting different application needs. A general comparison of these fabrication methods is presented in **Tab. 3.1**. Among them, electrospinning, a term derived from "electrostatic spinning", has become the most popular and preferred technique in the last decade because this method is simple, cost-effective and able to produce continuous nanofibers of various materials from polymers to ceramics. In addition, the electrospinning seems to be the only method which can be further developed for large scale production of continuous nanofibers for industrial applications. As such, the electrospinning related fabrications of polymeric nanofibers will be the main focus here.

2.1. Principle of Electrospinning

Historically, the electrospinning process has been known for more than 70 years [12]. A very basic and controllable laboratory setup for electrospinning mainly consists of four components, viz., spinning dope feeding pump (e.g., a syringe pump), a spinneret system (e.g., a metallic needle connected to a syringe), a high-voltage power supply up to several tens of kilovolts and a grounded collecting device (e.g., aluminum foil) as shown in **Fig. 3.1**. In a typical electrospinning process, when a high voltage is applied from zero to several or even tens of kilovolts depending on the electrospinnability of the polymeric solutions or melts, an electrical field is simultaneously established between the spinneret and collecting device. The ball-shape drop pendent on the nozzle exit simultaneously deformed, as a consequence of the force interactions between the Coulombic force (exerted by the external electric field) and the surface tension of the polymer solution, into a conical shape (commonly termed as the Taylor's Cone). As the electric field strength increases to a threshold value, the electrostatic forces overcome the surface tension which results in an ejection of polymer liquid jet. This jet is then subjected to an extremely high ratio of stretching process and rapid evaporation of solvents. This finally leads to the formation of nanometer to micron sized fibers which are deposited on the collecting device.

   As described in the process, the mechanism of forming nanoscale polymeric fibers with electrospinning has recently been identified as a result of the bending instability [13] or whipping [14,15] of the charged jet, which was previously described phenomenally as splitting or splaying

Tab. 3.1. Comparison of various processing methods for producing polymer nanofiber or nanofibrous structure.

Processing methods	Descriptions	Fiber dimensions	Features	References
Electrospinning	To produce nanofibers by electrically charging a polymer solution or melt. The simplest setup consists of only a syringe or pipette to hold polymer solution, two electrodes and a DC high voltage power generator.	Diameter: 3 nm to several micrometers; Length: continuous	- a 'top-down' approach; - simple and cost effective - can be applied to many materials; - fibers produced are continuous and typically distributed randomly; - applicable for industrial production.	[2,18,171]
Melt-blown	Based on traditional melt blowing fiber spinning technology, but using smaller orifices (with diameter as small as few tens of micrometers) made by microfabrication techniques and in combination with high velocity streams of heated air gas applied on extruded molten polymers.	Diameter: 150 nm to 1000 nm; Length: continuous	- a 'top-down' approach; - fiber size is dependent on the orifice size of extrusion mould, hence, difficult to get fiber diameter smaller than 100 nm; - still under development.	[3,4]
Phase separation	Typically consisting of five steps, i.e., raw material dissolution, phase separation and gelation, solvent extraction, freezing, and freeze-drying.	Diameter: 50 nm to 500 nm; Length: few micrometers	- resulting in nanofibrous foam directly after the freeze-drying; - relatively long time to obtain a batch of nanofibrous foam; - applicable to certain special (able to gel) polymers such as PLLA and its blend.	[5,6]
Self-assembly	A process whereby atoms, molecules and molecular aggregates organize and arrange themselves through weak and non-covalent forces such as hydrogen bonding, electrostatic interactions, and hydrophobic forces into stable and structurally well-defined functional entities at the meso- and nanoscale dimensions.	Diameter: well below 100 nm; Length: up to few micrometers.	- a 'bottom-up' approach; - self-assembled materials through purposeful manipulation potentially offering novel property and functionality that cannot be achieved by conventional organic synthesis; - longer preparation time in certain circumstances.	[7,9-11,172]
Template synthesis	To use nanoporous membranes available commercially as template to synthesize or extrude nanoscale fibers.	Diameter: a few to hundreds nm; Length: micrometers	- nanoscale tubules and fibrils from polymers, metals, carbons, etc available; - resultant fiber diameters are monodispersed.	[10,11]

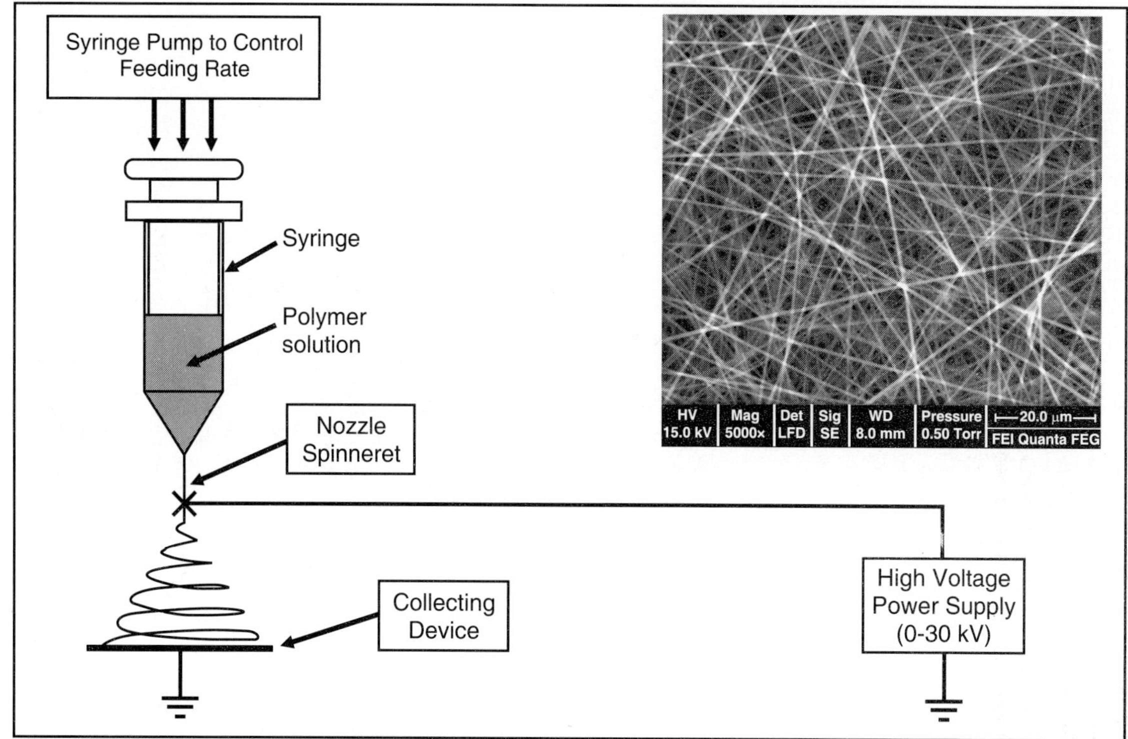

Fig. 3.1. Schematic diagram of a basic laboratory electrospinning setup and a representative SEM image showing nonwoven electrospun Gelatin nanofibers.

[2,16]. To date, more than 100 different types of materials have been electrospun into fibers with diameters ranging from a few nanometers to tens of microns [2,17,18].

2.2. Process Variables in Electrospinning

Electrospinning seems to be quite simple and straightforward for producing nanofibers. However, apart from its complicated nanofibers forming mechanism, there are many variables that may influence the control of the process, in particular, the resultant fiber morphology and diameter. These parameters can generally be classified into the following three types: (i) spinning fluid properties such as viscosity and/or concentrations, conductivity, surface tension and solvent properties; (ii) operation variables which include the applied electrical field strength, solution feeding rate, gap distance between spinneret orifice and the collection device and (iii) ambient conditions, for instance, temperature, humidity, electromagnetic field interferences and air flow. A comprehensive summary of the above variables is given in **Tab. 3.2**.

2.3. Aligned Nanofibers

As mentioned earlier, the basic setup shown in **Fig. 3.1** usually produces randomly arrayed fibrous membranes or nonwovens. By modifying certain components in the simple setup, especially the

Tab. 3.2. Variables affecting the electrospinning process.

Influence factors		Materials electrospun	General conclusions	Ref.
Fluid Properties	Viscosity (e.g., contributions from Mw, solution concentrations)	PE, PP melts Polyurethaneurea copolymer in DMF	- For a given voltage, the higher the temperature of the melt the smaller was the fiber diameter (as a result of reducing viscosity)	[173]
			- High solution temperature favored forming uniform fiber diameter, and increased deposition rate.	[174]
			- A 3rd power relationship was established between fiber diameter and concentration.	
			- A trimocal distribution of fibers in diameter was found in electrospinning high concentrations.	
			- Concentration also affected the fiber morphology from curly, wavy, and straight structures at high concentration to the beads on string at low concentration.	
		Polyetherimide (PEI) in 1,1,2-trichloroethane (TCE)	- fiber diameters were strongly influenced by solution concentration.	[175]
		Acrylic resin/DMF solution	- fiber diameters increased with 0.5 power of solution viscosity.	[176]
		PEO in water	- fiber diameters increased with solution concentration through a 0.5 power law relationship.	[177]
			- higher concentration of 8% above gave rise to a bimodal distribution of fiber diameters.	
		PDLA/DMF; PLLA in methlyene chloride and DMF at a ratio of 1.5:1	- higher concentration of solution favored the formation of uniform and bead-free nanofibers.	[178]
		PEO in mixture of water, ethanol, NaCl	- higher viscosity favored formation of beads free fibers.	[179]
		Silk-like protein polymer dissolved in formic acid	- With variation in polymer concentration from 0.8% to 12.1% at constant field strength 4 kV/cm and deposition distance of 2 cm, morphology was changec from beaded coating to filamentous coating.	[180]
		Elastin-mimetic peptide polymers in water	- At 5 wt% solution and regardless of flow rate, obtained fibers were short and fracmented with a triangle or spindle-shaped morphology; at above 10 wt%, long uniform fibers with diameters of 300 to 400 nm over all flow rates were generated; a new morphological pattern of flattened or ribbon-shaped fibers with average width of ca. 3 μm was observed at 15% and 20% of the concentrations.	[61]
	Conductivity	PEO in mixture of water, ethanol, NaCl	- higher net charge density helped formation of thinner fibers without beads.	[179]
		Polyurethaneurea copolymer in DMF	- Adding a small amount of triethylbenzylammonium chloride salt, conductivity of the studied polymer solution increased dramatically as a corresponding high flow rate was observed (flow rate is nearly linear to applied voltage).	[174]
		PDLA/DMF; PLLA in methlyene chloride and DMF at a ratio of 1.5:1	- Fiber diameter decreased with addition of salts. - The adcition of small amount of salts could also greatly change the morphology of electrospun fibers. - Different salts had different effects on the fiber morphologies	[178]

		Polyetherimide (PEI) in 1,1,2-trichloroethane (TCE)	- Addition of carbon black induced smaller size fibers under the same electrospinning conditions compared with pure PEI solution.	[175]
	Surface tension	PEO in mixture of water, ethanol, NaCl	- Reduced surface tension favored the formation of fibers without beads	[179]
	Solvent properties	Poly(vinyl chloride) (PVC) in THF, DMF and mixture of THF & DMF	- Solvent volume ratio was strongly correlated with the diameter of the electrospun PVC fibers with respect to other processing parameters. - For PVC in THF, the resultant fiber had a broad diameter distribution ranging from 500 nm to 6 mm; whereas with DMF only, a narrow distribution had the average diameter of 200 nm was obtained. Using a mixed solvent of THF and DMF with volume ratio ranging from 80/20 to 20/80, fiber diameter obtained was less than 1 µm.	[181]
		Cellulose acetate in solvents of acetone, acetic acid, and dimethylacetamide (DMAc)	- None of mono-solvent alone enables producing continuous fibers, however, mixing solvent of DMAc with either acetone or acetic acid was found suitable solvent system. - The mixture ratio 2:1 of acetone to DMAc was the most versatile solvent as cellulose acetate in the concentration of 12.5-20% can be electrospun into fibers with diameters from 100 nm to 1 µm.	[182]
Operation Variables	Flow / feeding rate	Acrylic resin/DMF solution	- At a specified concentration (12.5%) and constant voltage and gap, the effect of flow rate on fiber diameter was small over a tenfold range of flow rate.	[176]
		PDLA/DMF; PLLA in methlyene chloride and DMF at a ratio of 1.5:1	- Fiber diameter increased with flow rate.	[178]
	Electrical field strength or voltage	PE, PP melts	- Fiber diameter decreased as the applied voltage increased	[173]
		Polyetherimide (PEI) in 1,1,2-trichloroethane (TCE)	- Higher voltage produced narrower diameter fibers	[175]
		PDLA/DMF; PLLA in methlyene chloride and DMF at a ratio of 1.5:1	- Fiber diameter increases with voltages applied	[178]
		Acrylic resin/DMF solution	- Seems there was an optimum voltage exist to obtaining minimum diameter fibers	[176]
		PEO/Water solution	- There was an initiation voltage and stop voltage in-between electrospinning can be conducted.	[16]
		Elastin-mimetic peptide polymers in water	- Both the initiation voltage and stop voltage increased with solution concentrations	[61]
		Silk-like protein polymer dissolved in formic acid	- Increasing applied electric field strength rendered a decrease in fiber diameter.	[180]
		PEO in water	- For the studied 7% of PEO with a Mw 400K, it was noted the increasing electrospinning voltage from 5.5 kV to 9.0 kV, the morphology of electrospun fibers changed from straight and defect-free to prevalent bead-on-string structure with increased bead defect density.	[177]

Parameters		Materials	Observations	Ref.
		Polyurethaneurea copolymer in DMF PAN in DMF	- Jet diameter increases with applied voltage - Increasing voltage favored the formation of more jets from the droplet at tip - Applied electrical voltage had a double influence oppositely on the fiber diameters depending on the tip to collector (a rotating cylinder) distance: at 5 cm, the average fiber diameter increased with increasing voltage, while at 9 cm, the average fiber diameter decreased as voltage increasing. The double effects of voltage were associated to evaporation of DMF.	[174] [183]
	Collections (e.g., using stationary, or rotating devices, material types, etc.)	PE, PP melts	- Higher take-up speeds gave rise to lower fiber diameters due to increased tension in the spinline.	[173]
		Cellulose acetate in solvents of acetone, acetic acid, and dimethylacetamide (DMAc)	- Variation of collectors (e.g., copper mesh, Al foil, water, and paper) had effect on the obtained fiber morphology and packing states. - Fibers collected on paper had smooth surfaces, uniform sizes, and few defects. Fibers collected on water were more varied in size and more densely packed. Conductive Al foil and water favored tightly packed and thick membrane structure, whereas loosely packed fibrous structure can be formed on nonconductive paper. Accordingly, the resultant pore volume and porosity are different.	[182]
		PEO in mixture solvent of water and ethanol at the ratio of 60:40 PAN in DMF	- Collector is a rotating sharp wheel. - Other than common non-woven mats, aligned electrospun nanofibers with diameters ranging from 100-300 nm were produced - Fiber diameter decreased with increasing rotating speed, and increasing rotating speed is helpful in fiber orientation, obtaining narrow width of non-woven mat on the cylinder.	[21] [183]
	Collecting distance	Silk-like protein polymer dissolved in formic acid Polyetherimide (PEI) in 1,1,2-trichloroethane (TCE) PAN in DMF	- Decreasing deposition distances resulted in wet fibers or beads on the surface of collecting substance. - Insufficient collecting distance led to bead shaped by-products. - Fiber diameter decreased with increasing collecting distance.	[180] [175] [183]
	Collecting time	Silk-like protein polymer dissolved in formic acid	- Deposition time was the factor determining thickness of electrospun fibrous coating, at this point, higher electrical field strength will shorten the time to obtain same thickness of coating.	[180]
Ambient Conditions	Temperature		- Ambient temperature will affect the surface tension, viscosity of the spinning solutions.	
	Relative humidity	Acrylic resin/DMF solution	- Dry air (<5% RH) could dry out the jet and made the spinning carry out for only 1 or 2 min, while humid air (> 60% RH) got wet fibers tangled on the grounded screen.	[176]
	Gas surrounding	Acrylic resin/DMF solution	- Gases of normal air, helium and CCl_2F_2 were used. - Spinning could not be done in a helium atmosphere due to its low break-down voltage of 2500 V. In CCl_2F_2 gas ambient, bigger diameter of fibers could be made than that of spinning in air surrounding conditions.	[176]
	Air flow velocity Others (e.g., field interferences)		- Possibly reduce the yield of electrospun nanofibers; - Slightly affect the stability of Taylor cone. - Seriously affect the electrospinning process, for instance, during electrospinning those nanofibers hang anywhere in the chamber actually results in stop of the process.	

collecting device, aligned nanofibers can be obtained. This is one of the most significant achievements in the recent development of electrospinning technology. Huang *et al.* [18] have given a comprehensive review on various techniques developed for preparing aligned nanofibers. Here, only two of them, which have been widely recognized and used, will be mentioned. Both are based on a strategy of controlling the macroscopic electric field by employing different collectors to obtain deposition of parallel electrospun fibers.

2.3.1. Using a High Speed Rotating Thin Wheel

It has been known that the alignment of electrospun fibers can be achieved only to some extent by using a high speed rotating cylinder as the collecting device [19,20]. Significant improvement on aligning electrospun fibers was made by Theron *et al.* [21] who described a modified and enhanced way to deposit and align nanofibers on a tapered and grounded wheel-like bobbin as shown in **Fig. 3.2**. Both the tip-like edge, which can substantially concentrate the electrical field, and the high speed rotation, are responsible for the good alignment. With the thin wheel device, well aligned polymer nanofibers have been fabricated for engineering blood vessel tissue [22] and exploring the structure-property relationship of single nanofibers [23].

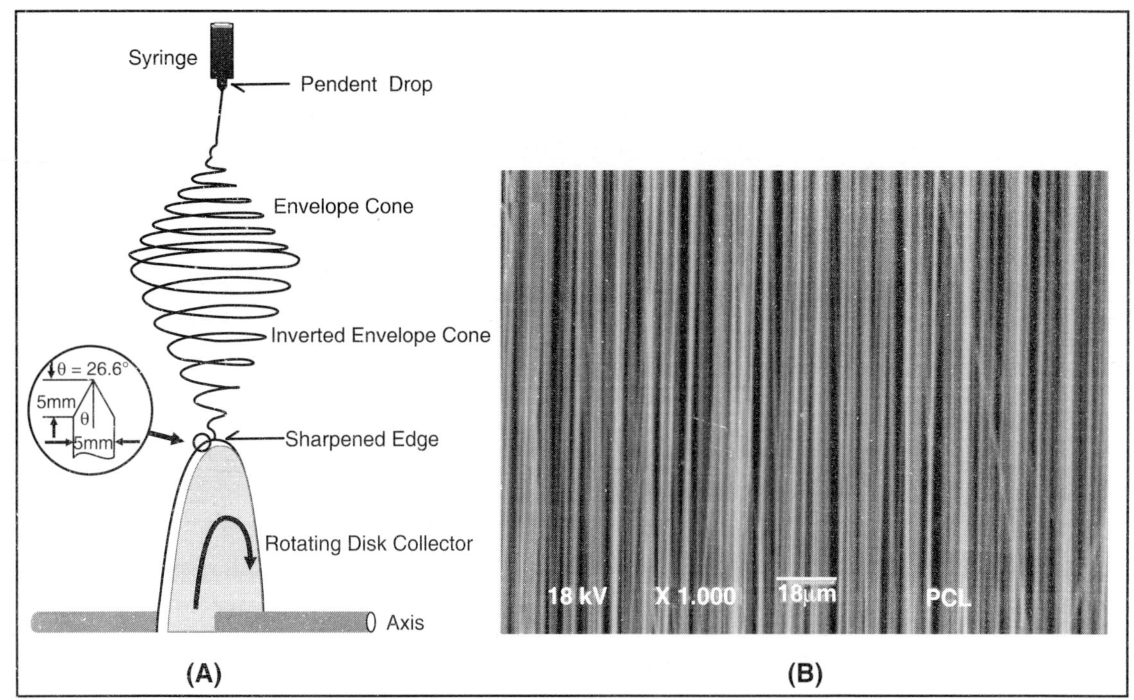

Fig. 3.2. (A) A setup used to collect uniaxial nanofibers. Reproduced with permission from A. Theron *et al.*, Nanotechnology, 12(3), 384-390 (2001). Copyright © 2001, IOP Publishing Limited. (B) thus obtained PCL fibers. Reproduced with permission from R. Inai *et al.*, Nanotechnology, 16, 208-213 (2005). Copyright © 2001, IOP Publishing Limited.

2.3.2. Using a Frame Collector

This technique has been recognized and reported by several research groups [18,24-28]. For preparing aligned fibers, their common "gadget" is to simply use a conductive frame or strips separated from micrometers to several centimeters as the counter electrode. Its capability of obtaining better alignment of electrospun nanofibers was particularly well demonstrated by Li *et al.* [26,27]. For example, apart from properly paralleled array, stacked layers with different alignment directions achieved by configuring the placement of strips, was also demonstrated **(Fig. 3.3)**. The achievement of alignment was attributed to the electrostatic charging effects as interpreted elsewhere [24,26,28]. Obviously, compared to the previous technique, this method can produce aligned nanofibrous membrane over a large area. Furthermore, it provides a convenient means to collect separated or single nanofibers for nanofiber characterization [24]. An even larger area of nanofibers can be collected by rotating a multi-frame cylindrical structure as reported by Ramakrishna and Lim's research group [18,28]. Very recently, using a framed cylinder to obtain a sheet of aligned nanofibers was also reported by Katta *et al.* [29].

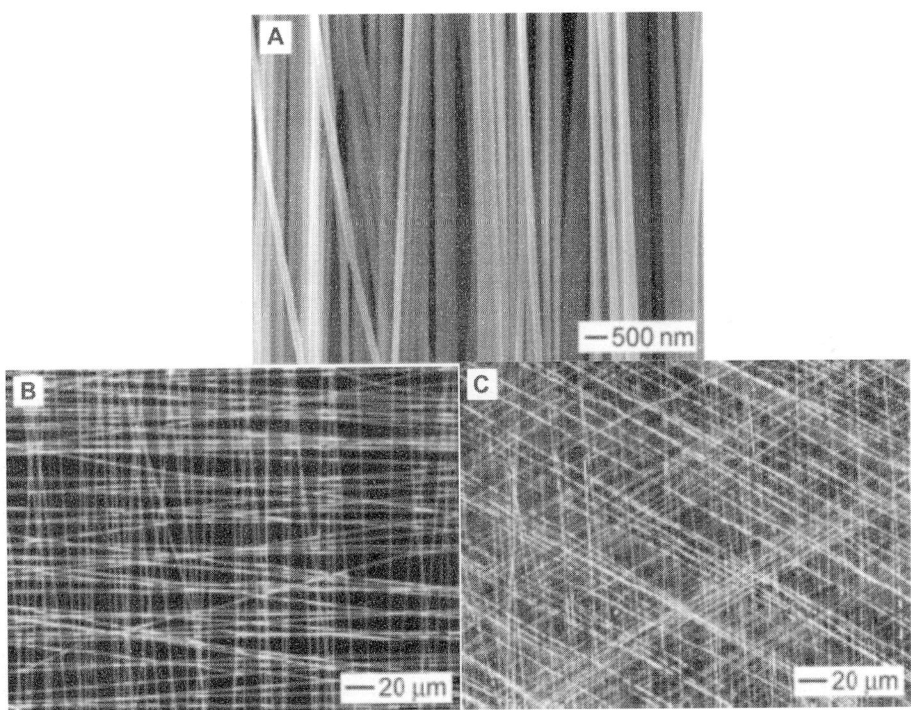

Fig. 3.3. (A) Paralleled array of carbon nanofibers. Reproduced with permission from D. Li *et al.*, Nano Letters, 3(8), 1167-1171 (2003). Copyright © 2003, American Chemical Society. (B and C) Stacked alignment structure from electrospun poly(vinyl pyrrolidone) (PVP). Reproduced with permission from D. Li *et al.*, Advanced Materials, 16(4), 361-366 (2004). Copyright © 2004 Wiley-VCH Verlag GmbH & Co. KG.

2.4. Functional and Uniquely Structured Nanofibers

Although polymeric nanofibers with certain functions can be directly prepared from electrospinning specialty polymers, for example, electrospinning of the conductive polyaniline and piezoelectric poly(vinylidene fluoride) (PVDF) [30,31], functionalizing nanofibers can be most conveniently achieved through electrospinning of blends that contain function-imparting components. For instance, nanofibers incorporated with drugs or bioactive molecules can be used for controlled releases and improving cell affinity [32-34], nanofibers containing magnetic nanoparticles showed supermagnetic behavior [35,36], and incorporation of carbon nanotubes (CNTs) gave rise to nanofibers with improved mechanical, electrical and thermal performance [37,38].

In practice, one may also modify the post-electrospun polymeric nanofibers to induce desired function for the nanofibrous membranes. In this regard, physical surface coating is the simplest approach to adopt. For example, surface coated polymer nanofibers can be used to develop highly sensitive bio- or chemical-sensors [39,40]. Alternatively, surface chemical modification techniques, e.g., to improve the hydrophilicity and biochemistry affinity of cell-scaffold interactions for tissue engineering applications, can be used [41,42]. It is worth noting that uniquely structured nanofibers can also be prepared based on those post-electrospun nanofibers. For example, foam-like 3-D porous nanofibers have been made through a simple leaching process in our lab as shown in **Fig. 3.4**. This technique will provide the highest possibility of creating extremely large specific surface area of ~100-1,000 m^2/g [43], which is one to two orders of magnitude larger than the current plain electrospun nanofibers. Such super large surface area will be most desired for cell adhesion, catalysis, filter, and fuel cell membrane applications.

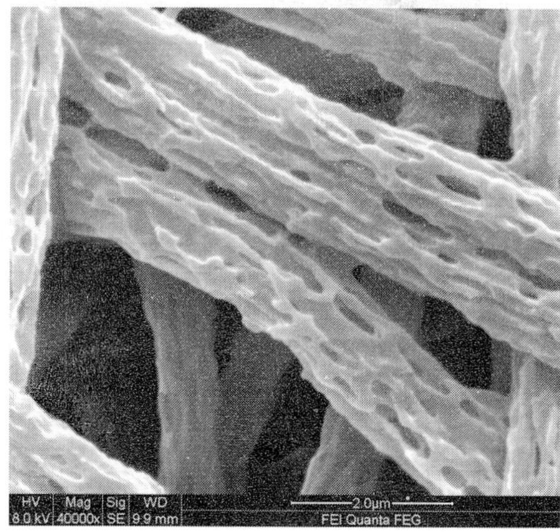

Fig. 3.4. A FESEM image of 3-D porous fibers of PCL developed by a leaching process from bioactive Gelatin/PCL composite fibers.

Apart from the above methods introduced, recently developed coaxial electrospinning provides a novel route to design and develop new and functional nanofiber structures [44-48]. Coaxial electrospinning is essentially a modification or extension of the ordinary electrospinning technique. The major difference lies in its compound spinneret which consists of a set of concentric inner and outer capillary tubes by which different fluids can be separately fed into and integrated into a

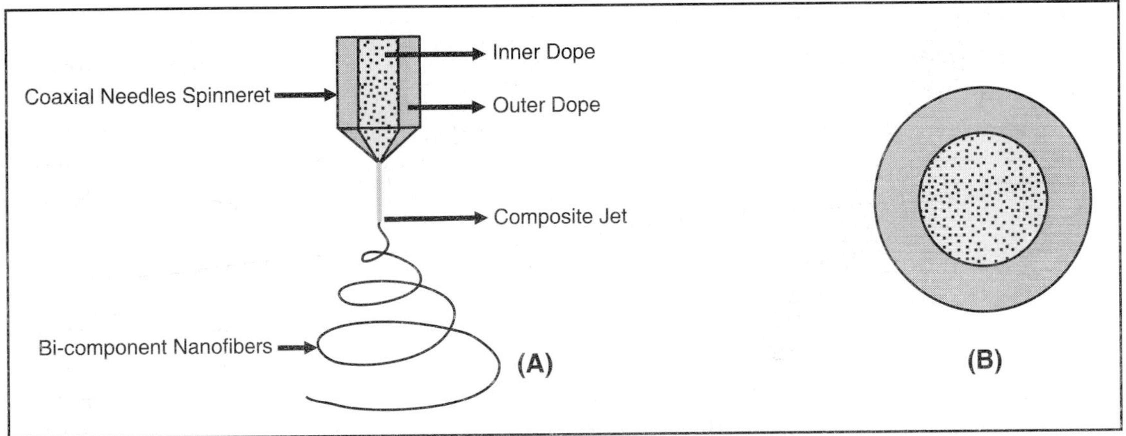

Fig. 3.5. (A) Coaxial electrospinning and (B) cross-sectional view of resultant bicomponent composite fiber. Reproduced with permission from Y.Z. Zhang *et al.*, Biomacromolecules, 6(5), 2583-2589 (2005). Copyright © 2005, American Chemical Society.

core-sheath structured composite fiber as they are charged and emitted from the spinneret [49] (**Fig. 3.5**). With coaxial electrospinning, at least four types of nanofibers with different microstructures can be envisioned and developed (**Fig. 3.6**).

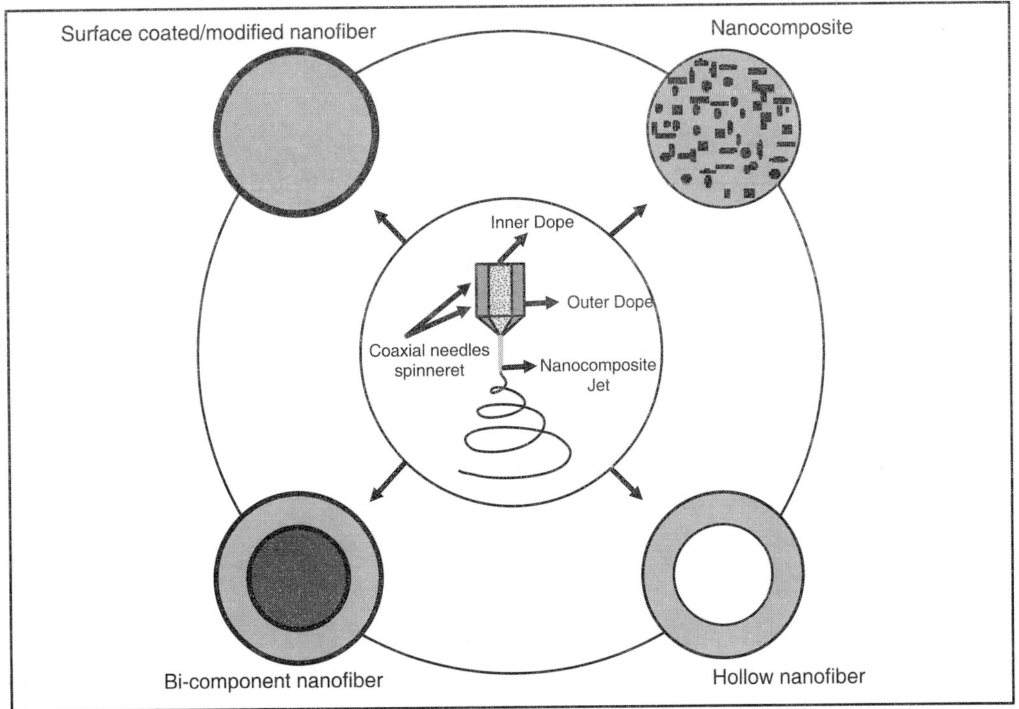

Fig. 3.6. Cross-sectional views of a variety of novel and functional polymeric nanofibers developed from coaxial electrospinning.

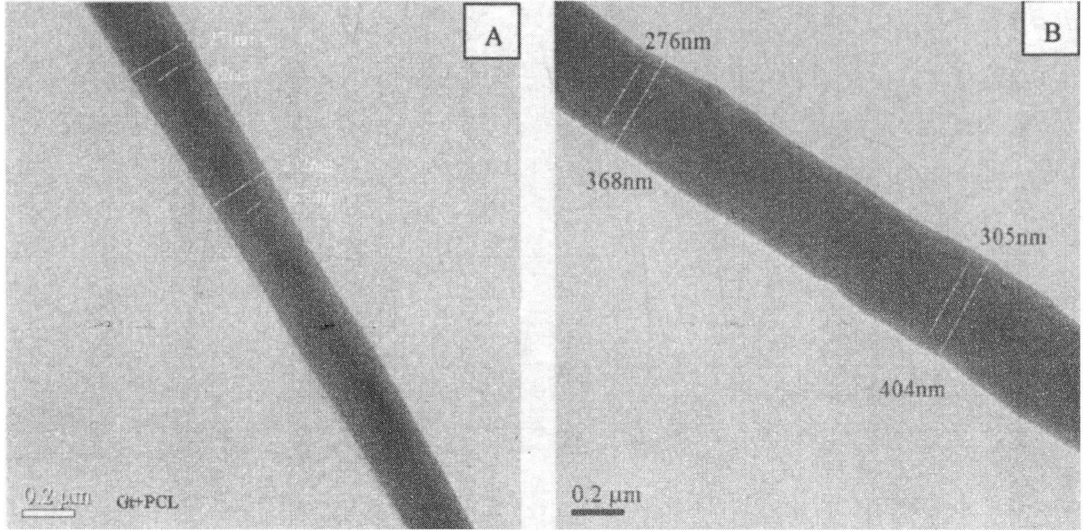

Fig. 3.7. TEM images of nanofibers from coaxially electrospinning of PCL/TFE (shell fluid) and Gelatin/TFE (core fluid). (A) bi-component and (B) surface coated. Reproduced with permission from Y. Zhang *et al.*, Chem.Mater, 16(18), 3406-3409 (2004). Copyright © 2004, American Chemical Society.

The basic fiber form from coaxial electrospinning is either bi-component or nanofiber with surface-coating which is dependent on the control of shell-thickness during processing **(Fig. 3.7)**. Furthermore, the most prominent advantage of coaxial electrospinning is its capability of converting those materials, which cannot be electrospun with the ordinary electrospinning process, into fiber form. For example, fiber form of poly(dodecylthiophene) (PDT), which cannot be electrospun due to its low molecular weight, has been prepared as the core component in the shell of poly(ethylene oxide) (PEO) in the work by Sun *et al.* [45]. Similarly, Yu *et al.* [48] successfully incorporated polyaniline (PAni) (core) in the poly(vinyl alcohol) (PVA) (shell) for making conductive nanowires. Furthermore, based on the core-shell nanofibers, one may also derive new material or structure. For instance, by performing a post-treatment, metal Pd nanowire in the poly(L-lactide) (PLA) was obtained after annealing the PLA-Pd(OAc)$_2$ for 2 h at 170°C [45]. By removing the shell material, even fine nanofibers (with average diameter less than 100 nm) can be conveniently prepared [48]. However, this is usually very difficult to achieve with the usual electrospinning process. Similarly, by removing the core constituent, hollow nanofibers can be produced, resulting in a novel two-stage method of fabricating nanotubes instead of the previously reported three-stage process [50-52]. Nanoparticles loaded composite nanofiber is another possible form with coaxial electrospinning. In this regard, self-assembled FePt magnetic nanoparticles (ca.4 nm) have been encapsulated in poly(ε-caprolactone) (PCL) nanofibers in our laboratory [53] **(Fig. 3.8)**. Likewise, if carbon nanotubes (CNTs) are used as core component, conductive nanocables with better mechanical properties can be developed. Compared with electrospinning of blends with nanoparticles poorly dispersed in the polymer solution, this approach might provide an effective route to load high volume and well dispersed nanoparticles in polymer matrix.

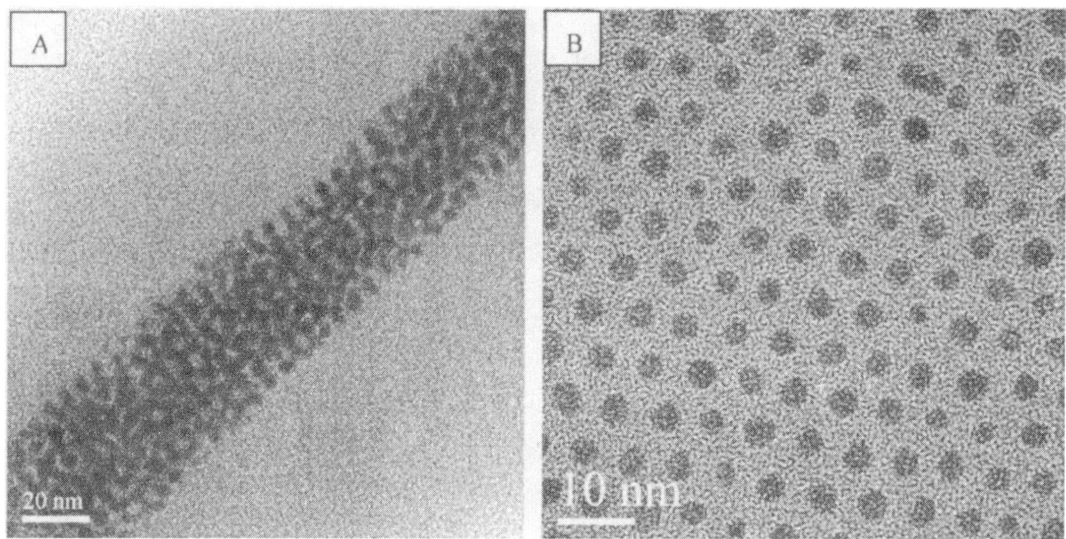

Fig. 3.8. TEM images of (A) coaxially electrospun FePt magnetic nanoparticles encapsulated in poly(ε-caprolactone) (PCL) and (B) self-assembly of FePt nanoparticles on a silica substrate before encapsulation.

Tubular shaped or hollow nanofibers have been useful in a number of applications in developing novel field-emitter displays and magnetic media, biocatalyst, purification and separation, sensor, controlled release, and so on. Other than conventional multi-step approaches, both Li *et al.* and Loscertales *et al.* have creatively demonstrated the feasibility of directly fabricating hollow nanofibers via combining the coaxial electrospinning and sol-gel chemistry in processing [47,54] as shown in **Fig. 3.9.**

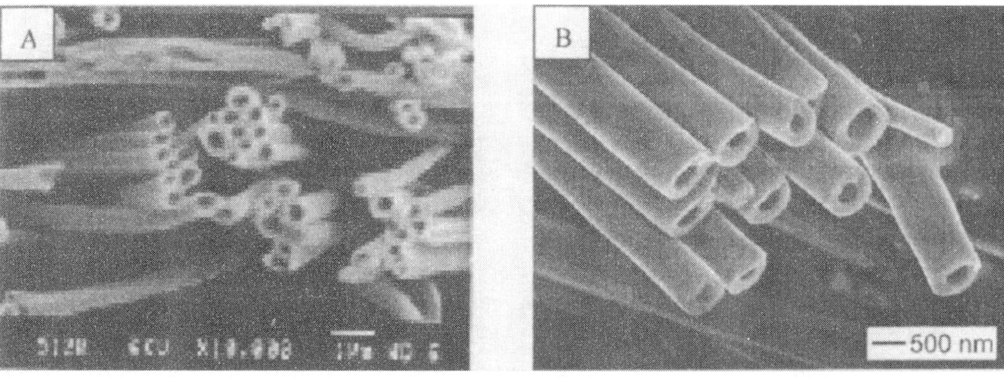

Fig. 3.9. Hollow nanofibers prepared from coaxial electrospinning of: (A) an aged sol of tetraethyl orthosilicate (TEOS, outer fluid) and glycerin (inner fluid). Reproduced with permission from I.G. Loscertales *et al.*, Journal of the American Chemical Society, 126(17), 5376-5377 (2004). Copyright © 2004, American Chemical Society. (B) a PVP solution (in ethanol, 0.03 g/mL) that contained 0.5 g/mL of Ti(OiPr)$_4$ (shell solution) and heavy mineral oil (core fluid), followed by extraction and calcinations treatment. Reproduced with permission from D. Li *et al.*, Nano Letters, 4(5), 933-938 (2004). Copyright © 2004, American Chemical Society.

Without doubt, coaxial electrospinning will propel the electrospinning technology into a revolutionary and more advanced stage in terms of developing a variety of functional nanofibers and for widening more potential applications. However, currently investigation on this technique is still quite limited and sometimes, the generated conclusions are contradicting between different researchers. Many questions are raised from processing mechanism to process control and they need to be addressed before realizing its actual values and applications. With regard to the formation of core component, while Yu *et al.* [48] stressed the role of low interfacial tension between the two fluids in allowing production of fiber form in the core, they also speculated the possible actions of electrostatic force as a consequence of charging effect, which is similar to that occurring in an ordinary electrospinning. On the other hand, Li *et al.* [47] suggested the rapid stretching of the sheath can cause strong viscous stress, which will be passed onto the oil core. Tangential to the jet line, the shear stress would stretch the core component and elongate it along with the sheath solution via mechanism such as viscous dragging and/or contact friction [47]. Another issue is under what kind of conditions can high yield of core-shell structured nanofibers be produced, and how those commonly appreciated processing variables such as applied electric field strength, solution viscosity and/or concentrations, and flow rate affect the control of sheath-thickness as well as the resultant fiber diameters. Presently, our work indicated that by altering the inner polymer solution concentrations, both the inner and overall diameter of such coaxially electrospun bi-component nanofibers can be consequently changed [46]. Li *et al.* [47] investigated the influences of varying flow rate and electrical strength. They found increasing the feeding rates led to larger inner diameter, and both the inner and outer diameters of the core-shell fibers decreased as the electrical field was enhanced.

3. APPLICATIONS

Compared to other forms of materials, electrospun polymeric nanofibers possess very high specific surface area and can easily be manufactured into 2-D and 3-D structures for numerous applications. This has been demonstrated recently by academic researchers and industrial engineers for exploring applications in biomedical engineering, filtration and separation, high performance nanocomposites, sensors, and other functional electrical, optical and catalytic technologies.

3.1. Biomedical Engineering Applications

One of the most exciting and promising direction for biomedical related applications is engineering biological tissues [55]. Tissue engineering is an interdisciplinary technology to repair and regenerate faulty human tissues by using knowledge of bioengineering, life sciences and clinical sciences so as to resolve the problems of donor shortage and permanent immunosuppressive medication encountered in traditional organ transplantation. It has been appreciated that nanoscale polymeric fibers would be extremely suitable for developing bioactive scaffolds in tissue engineering. This is because from a biological viewpoint, almost all of the human tissues and organs, such as bone, skin, tendon and cartilage, are in a nanofibrous form of collagen characterized by well organized hierarchical fibrous structures [56].Therefore, for successful tissue engineering to take place, one may employ the biomimetic methodology to develop synthetic nanofibrous extracellular matrix (ECM) which are able to biomimic or resemble not only the nanofibrous physical structures, but also biochemistry attributes. In line with this strategy, polymer nanofibers have been utilized for engineering tissues such as cartilages [57-59], bones [60], arterial blood vessels [22,61-63], heart [64] and nerves [5,65].

To prepare these scaffolds, electrospinning has now been the most extensively used fabrication method. Advantages of using electrospinning to prepare tissue scaffolds are as follows: (i) it is capable of producing ultra-thin fibers with diameters ranging from several micrometers down to a few nanometers, which are able to mimic the native structure of ECM; and (ii) it is versatile in a sense that various monopolymers, blends of polymers, and compositions of polymers with other materials or additives such as inorganics, growth factors, other cell regulatory biomolecules, and even living cells can be used to develop functionally active nanofibrous structures. The scaffolds produced provide a highly porous microstructure with interconnected pores and extremely large surface area to volume ratio which is conducive for tissue growth. Apart from tissue engineering scaffolds, polymeric nanofibers have also been used as dressings for protection of wound and promoting healing [66-68]. Another potential field for polymeric nanofibers in the biomedical context is in the area of controlled releases [33,69,70]. This can be coupled with the development of tissue engineering scaffolds (or nanofibrous dressings). Presently, such drug-loaded nanofibers are usually manufactured by electrospinning blends of polymer carriers and drugs. Subjected to the drug-polymer interactions, burst releases were usually observed to take place during releases. Recently developed coaxial electrospinning could provide an appropriate solution for this problem, especially for cases where it is necessary to protect bioactive proteinous drugs from denaturation because of sample preparation and harsh releasing environment.

3.2. Filtration and Separation

Electrospun nanofibers for industrial filtration applications have a history of over 20 years [71,72]. As the fiber is very fine, filtration efficiency in capturing submicron sized particles can be greatly improved compared to that using conventional microfibers. With the versatile electrospinning process, filtration efficiency can be further enhanced after incorporating some active substances into the nanofibers. Apart from its use as industrial filters, electrospun nanofibers have been exploited for military protective clothing. In this regard, Gibson *et al.* [73-75] systematically investigated the transport properties of electrospun nanofibrous membranes and found that light-weight, high efficiency in capturing aerosol particles and lower resistance to water vapor transportation are the desired merits in developing high performance protective clothing. Likewise, electrospun polymer nanofibers with functionalized surface can be extended for potential use in high efficient biomolecular or protein separation. The high surface area to mass ratio makes nanofibers an ideal substrate for molecular separation. The principle of separation is similar to that of traditional affinity chromatography. In order to fulfill the function of molecular separation, besides considering proper surface functionalization of the nanofibrous membrane, structural and material properties of the nanofibrous membranes need to be considered so that the membranes can withstand the imposed forces acting on them during the filtration process. At the present, polymer nanofibrous membrane for molecular separation is still a concept that needs to be realized in the near future.

3.3. Nanocomposites

Like other fiber form materials (e.g., carbon, glass, and Kevlar fibers) for developing light-weight and high-strength engineering composites, the extremely fine nanofibers are expected to give rise to superior reinforcement effect in nanocomposites. Kim and Reneker [76] investigated the reinforcing effect of using electrospun polybenzimidazole (PBI) nanofibers for an epoxy matrix and a rubber matrix. They found the bending Young's modulus and fracture toughness of the epoxy based nanocomposites were increased marginally, whereas the fracture energy increased significantly. For the rubber based nanocomposites, the Young's modulus was ten times and the tear strength was twice as high as that of the pristine rubber material. Bergshoef and Vancso [77] reported that electrospun Nylon-4,6 nanofiber (30-200 nm) reinforced epoxy nanocomposites exhibited a higher stiffness and strength and were fully optically transparent in the visible spectrum of 400-700 nm. Despite the fact that current research pertaining to the polymeric nanofibers reinforced nanocomposite is quite limited, high specific-surface-area polymer nanofibers will have great potential in addressing the poor interlaminar fracture toughness problem in conventional composites. Furthermore, by inducing secondary nanostructured components such carbon nanotubes (CNTs), nanocomposite nanofibers can be prepared for even better strength-reinforcement effect in nanocomposite development [37,38].

3.4. Sensors

Sensitivity is one of the most important concerns in the development of sensors for the detection of gases and biological substances at low concentration. Improvement of sensitivity will require using sensor films with larger surface area to unit mass ratio [78]. Polymeric nanofibers are the ideal candidates for this kind of application. Kwoun *et al.* [79,80] pioneered the work of developing chemical and biochemical sensors by making use of a poly(lactic acid co glycolic acid) (PLAGA) nanofiber film as a new sensing interface. Researchers at the University of Massachusetts Lowell [81,82] have demonstrated that sensors made from electrospun nanofiber (100-400 nm in diameter) membranes containing fluorescent poly(acrylic acid)-poly(pyrene methanol) (PAA-PM) for detecting metal ions (Fe^{3+} and Hg^{2+}) and 2,4-dinitrotulene (DNT) exhibited a sensitivity of almost three orders of magnitude higher than that of thin films of the same material. By assembling hydrolyzed poly (2-3-thienyl) ethanol butoxy carbonyl-methyl urethane) (H-PURET) onto the surface of nanofibrous cellulose acetate membrane [82], high sensitivity in detecting extremely low concentration (ppb) of methyl viologen (MV^{2+}) and cytochrome c(cyt c) was achieved. The high specific surface area of materials and efficient interaction between sensing material and detected substance are responsible for this significant improvement.

3.5. Other Functional Applications

Due to the well-known fact that the rate of electrochemical reactions is proportional to the surface area of the electrode, nanofibrous porous membranes are ideal candidates in developing high performance battery as demonstrated by Choi *et al.* [83]. Kim and Yang have evaluated the performance of carbon nanofibers as electrode material for making super capacitor with an observed maximum capacitance of 175 F/g [84]. In a different direction, Jia *et al.* demonstrated using polymer nanofibers for preserving and immobilizing a biologically active enzyme [85]. It was found that the nanofibrous enzyme possessed high activity in both aqueous and organic media and could be easily recovered with a longer half-life. Very recently, Demir *et al.* [86] investigated the catalytic activity of palladium-incorporated nanofibers by selective hydrogenation of dehydrolinalool and found that the catalytic activity was 4.5 times higher than that of the traditional Pd/Al_2O_3 catalyst.

4. CHARACTERIZATION

The methods of characterizing single nanofibers and nanofibrous mats and foams using suitable instruments or techniques are discussed in this section.

4.1. Imaging

Microscopy techniques such as atomic force microscopy (AFM), scanning electron microscopy (SEM), transmission electron microscopy (TEM) and optical microscopy have been used to observe the microstructure, surface morphology and deformation behavior of single nanofibers. The application of these four techniques for single polymeric nanofibers will be presented in this section. **Tab. 3.3** shows the summary of various properties that can be obtained using the different imaging techniques.

Tab. 3.3. Visualizing single polymeric nanofibers.

Properties	*Instruments*
Fiber diameter	AFM, SEM, TEM
Surface morphology	AFM, SEM
Internal morphology	TEM
Molecular orientation	AFM, optical birefringence
Surface roughness	AFM
Fiber deformation behavior	SEM. TEM

4.1.1. Atomic Force Microscopy

The AFM has been used to study the morphology of a wide range of polymers [87,88]. Due to its ability to resolve structures in the sub-nanometer range by the direct probing of a sharp tip on the sample surface, it is an attractive instrument to use for the study of polymeric nanofibers. As AFM can be applied to non-conductive samples, no coating is required for sample preparation and the polymeric nanofibers can be imaged as-fabricated.

The diameter of the nanofibers can be measured using the AFM [89-92]. In order to measure the diameter accurately, the nanofiber has to be deposited on a flat substrate [92]. The vertical distance between the highest point of the fiber cross-section and the substrate surface is taken to be the diameter measurement. Due to the AFM tip geometry, the horizontal dimension measurements are overestimated [90]. If this is not possible due to the limitation of fabrication techniques, Sirinivasan *et al.* [89] proposed a technique to measure diameters of nanofibers that are overlapping each other. In this method, two fibers crossing each other on the surface were chosen. The upper horizontal tangent of the lower fiber was taken as the reference. The vertical distance above this reference was considered to be the exact diameter of the upper nanofiber.

The morphology of the nanofiber surface or cross-section can be observed using various modes of AFM, such as height, friction and phase [88]. Height images correspond to the topography of the sample. Friction images map out the surface variation of friction between the AFM tip and sample. Phase images can be used to map variations in surface properties such as elasticity, adhesion, and friction. Longitudinal and transverse sections of electrospun silk/polyethylene oxide (PEO) fibers [93] were scanned in height mode to reveal the globular microstructure. The phase images revealed a skin-core structure of the silk fibroin/PEO fibers after methanol treatment. Fibers usually have structures oriented in the direction of fiber axis. This can be revealed using the AFM. Phase images of longitudinal sections of melt-spun poly(lactic acid) fibers revealed microfibrillar morphology with alternating lamellae of crystalline and amorphous regions existing within each microfibril [94]. Shish-kebab morphology of poly(L-lactic acid) (PLLA) fibers produced by the phase separation method can be observed by scanning the surface of the nanofiber using phase imaging [95] **(Fig. 3.10)**. When the scan area is reduced to a much smaller scale, the crystal orientation of the polymer within the nanofiber can be observed. Height and friction images of the surface of electrospun PEO nanofibers (approximately 10 × 10 nm) revealed the orientation of the PEO chains [91].

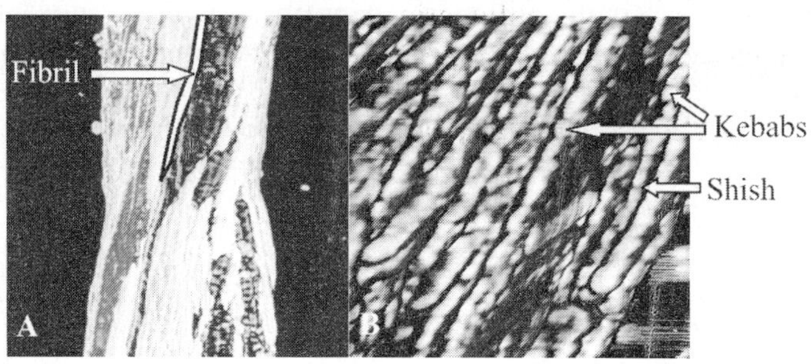

Fig. 3.10. (A) AFM phase image of a PLLA nanofiber revealing the fibrillar structure and (B) a close-up view of the surface of the nanofiber showing the shish-kebab morphology. Reproduced with permission from E.P.S. Tan *et al.*, Appl.Phys.Lett, 84(9), 1603-1605 (2004). Copyright © 2004, American Institute of Physics.

For large fibers with diameters in the range of several hundreds of nanometers to a few microns, the surface roughness of the fiber can be measured. Micro-roughness of nylon-6 and poly(ethylene terephthalate) melt spun fibers was measured using the AFM [96]. Although melt spun fibers have dimensions that are much larger than nanofibers, the surface roughness of nanofibers can still be obtained in a similar manner using a smaller area of analysis. Demir *et al.* [90] measured the surface roughness of electrospun polyurethane fibers to reveal that electrospun fibers are more than two times rougher than wet-spun fiber of the same polymer.

4.1.2. Scanning Electron Microscopy (SEM)

The SEM is widely used in the study of polymers as it is easy to operate and the images produced are easy to interpret [97]. Although charging occurs in the SEM when the sample is non-conductive, as is the case of most polymers, a thin layer of conductive coating such as gold or platinum can be deposited on the polymeric nanofibers before imaging.

The most common application of SEM for nanofibers is to observe the general appearance of the fibers and measure the dimensions of the fibers or features on the fibers. The change in morphology is observed as the processing conditions change during the fabrication of nanofibers by electrospinning [98-101]. For instance, the diameter of nanofibers was observed to increase with polymer concentration. At the same time, fewer beads were formed and the diameter of the fibers became more uniform along the length. At higher magnifications, surface characteristics of the nanofibers can be seen such as surface waviness [101] and pore structures [102,103]. In a study by

Wang *et al.* on melt-drawn nanofibers, the lamellar structures on the surface of the nanofibers can be observed using SEM [104]. When the nanofibers are freeze-fractured, internal structures can be observed using the SEM [93].

Besides microstructural examination, the SEM can be used to study the deformation behavior of single polymeric nanofibers. Elongated microcracks were observed on the surface of melt-drawn isotactic poly(1-butene) nanofibers, which were precursors to fiber splitting events during fiber drawing [104]. Multiple neck formation was observed on electrospun nanofibers that were stretched [92,105] **(Fig. 3.11A)**. At near to failure strains several fibrils can be seen connecting the two broken ends of a PEO nanofiber **(Fig. 3.11B)**. At failure, several fibrils can be seen sticking out from the nanofiber cross section as well as the core-shell morphology of the nanofiber [92] **(Fig. 3.11C)**.

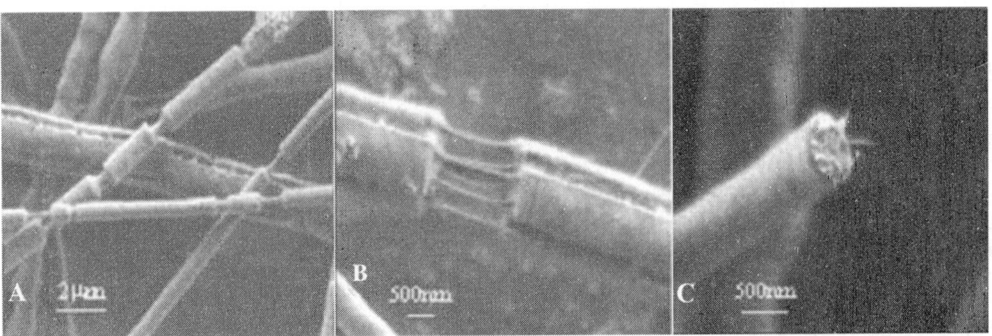

Fig. 3.11. (A) High resolution scanning electron micrographs of multiple neck formation in PEO electrospun nanofibers, (B) fibrillar structure in nanofiber and (C) failed nanofiber revealing core-shell structure. Reproduced with permission from E. Zussman *et al.*, Appl.Phys.Lett, 82(22), 3958-3960 (2003). Copyright © 2003, American Institute of Physics.

4.1.3. Transmission Electron Microscopy (TEM)

Like the SEM, TEM utilizes electron beam to image samples. In the TEM, the electrons are transmitted through the sample and recorded on an imaging plate in order to provide information about the ultrastructure of the sample. Polymeric nanofiber samples can be coated with carbon or gold to improve contrast of images. Polymeric samples may have a tendency to drift under the electron beam [106]. Thus low dose TEM techniques are often used to minimize the exposure of the sample to the electron beam [107].

Although TEM and SEM images are similar, TEM is able to image features that are beneath the fiber surface. This is useful in observing the distribution and alignment of carbon nanotubes (CNT) dispersed in polymer nanofibers [108,109]. The core-shell structure of polycaprolactone (PCL)-gelatin bi-component nanofibers can be seen using the TEM without the need for sectioning [46] **(Fig. 3.7)**. The fibrillar structure of electrospun silk-like polymer with fibronectin functionality can be seen using bright field TEM. Details of the crystallite arrangement can also be seen.

The TEM can be used to study the deformation behavior of nanofibers. Polyacrylonitrile (PAN) nanofibers with CNTs as the reinforcing material were found to deform by crazing [110]. The interaction between the CNTs and PAN during deformation was studied. The high magnification capability of TEM enabled this as fracture sites can be observed at a greater detail as compared to SEM.

4.1.4. Optical Microscopy

The presence of molecular chain orientation and/or crystallization can be deduced by using polarized optical microscopy [93]. Birefringence of electrospun polymeric nanofibers has been observed under cross-polarized optical microscopes [93,111,112]. When the electrospun nanofiber mat is turned by 45°, fibers that originally appeared bright will appear dark. Birefringence is defined as the difference in refractive indices parallel and perpendicular to the direction of orientation [113]. This birefringent behavior of electrospun polymeric nanofibers implied that the molecules are oriented parallel to the fiber axis.

4.2. Physical and Chemical Structures and Properties

The techniques outlined in this section are concerned with the arrangement of molecules as well as structural features in nanofibers. The method of obtaining thermal properties of the nanofibers will also be presented. **Tab. 3.4** shows the summary of various properties that can be obtained from the different characterization techniques.

Tab. 3.4. Physical and chemical structures and properties of nanofibers.

Properties	*Instruments and techniques*
Chemical composition	Raman and IR spectroscopy, NMR, XPS
Conformation	Raman and IR spectroscopy, NMR
Molecular orientation	WAXD, SAED
Crystal information	WAXD, SAED
Crystallinity	WAXD, DSC
Thermal properties	DSC

4.2.1. Spectroscopic Techniques

Spectroscopic techniques are generally used for molecular structural analysis [114]. Infrared and Raman spectroscopy are both vibrational spectroscopy techniques that are the most popular methods for identifying a polymer [114]. Thus these two spectroscopic techniques are useful for identifying or confirming the presence of components in composite nanofibers such as CNT-filled polymeric nanofibers [109,115]. Fourier transform infrared spectroscopy (FT-IR) was used to study whether blends of polymers result in chemical reaction between the constituent polymers [116]. Blend nanofibers were found to have both IR features of the components with no new peaks. This implied that there was no chemical reaction between the components. Both FT-IR and Raman spectroscopy were used to observe the conformational changes of electrospun silk during electrospinning process [105]. Besides identification of polymers, Raman spectroscopy provides information on the relative conjugation length [117]. A hypothesis concerning the reduction of defects in polymer chains with decreasing thickness was tested for polypyrrole nanotubes [118]. An increase in a specific conjugation length, normally correlated to a decrease in chain defects, was observed with the decrease in nanotube diameter, thus supporting the hypothesis.

Other spectroscopic techniques such as nuclear magnetic resonance (NMR) spectroscopy can be used as a complimentary tool to vibrational spectroscopy techniques. Solid-state ^{13}C-NMR spectroscopy was found to be more effective as an analytical tool for detecting conformational transition of silk fibroin (which existed in the form of blend with chitosan) than infrared spectroscopy [119]. X-ray photoelectron spectroscopy (XPS) can detect elemental information of up to a depth of approximately 100 Å from the surface of a sample [120]. This was found to be useful to show that the shell of bi-component nanofibers with a core-shell structure did not form blends or react chemically with the core [46] (**Fig. 3.12**).

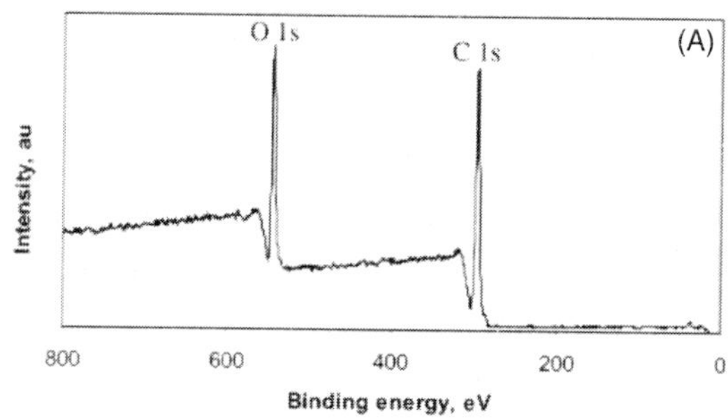

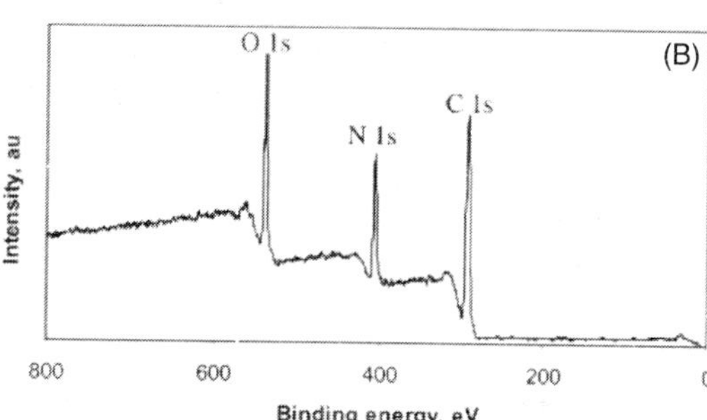

Fig. 3.12. XPS spectra of: (A) PCL-gelatin composite nanofibers and (B) pure gelatin nanofibers. The absence of N peak for the composite nanofibers indicated that traces of nitrogen were not found in the PCL shell since gelatin possesses nitrogen-containing amino groups. Reproduced with permission from Y. Zhang *et al.*, Chem.Mater, 16(18), 3406-3409 (2004). Copyright © 2004, American Chemical Society.

4.2.2. Diffraction Techniques

Polymer crystal orientation and degree of crystallinity of nanofibers can be measured using X-ray diffraction. On a smaller scale, crystal habit analysis can be performed under TEM using selected area electron diffraction (SAED). Diffraction studies are usually performed in combination with imaging (section 4.1) for a more complete picture of the arrangement of molecules and structural features within the nanofibers.

Generally, wide angle X-ray diffraction (WAXD) has been used to identify the type of crystal (e.g., monoclinic) [92] and type of crystalline phase in polymeric nanofibers in the form of polymer blends [93], multi-phase fibers [105] or composites [115]. This can be done by observing the peaks that correspond to crystallographic planes of specific crystals in the WAXD pattern. Rapid fiber formation of nanofibers from polymer solution during electrospinning process usually results in nanofibers with low crystallinity for polymers with glass transition temperatures above electrospinning temperature as the molecular chains are not given sufficient time to form crystallites. WAXD pattern of such as-spun nanofibers usually exhibit no crystalline peaks. When the nanofibers are annealed, the crystalline peaks appear in the X-ray pattern [23,100] (**Fig. 3.13**). The crystallinity of the nanofibers can be calculated quantitatively by analyzing the intensity of X-ray scattering [121].

Preferred orientation of the polymer chains within the nanofibers have been studied using WAXD and SAED. A high degree of molecular orientation along fiber axis has been found in various studies [92,100,105] using WAXD. More details regarding orientation can be provided by SAED. For instance, variation in orientation along the length of a single nanofiber was observed for electrospun PLA nanofibers [106]. Crystallite size and misorientation angle of the molecular chains within the nanofibers can also be deduced quantitatively using SAED [122].

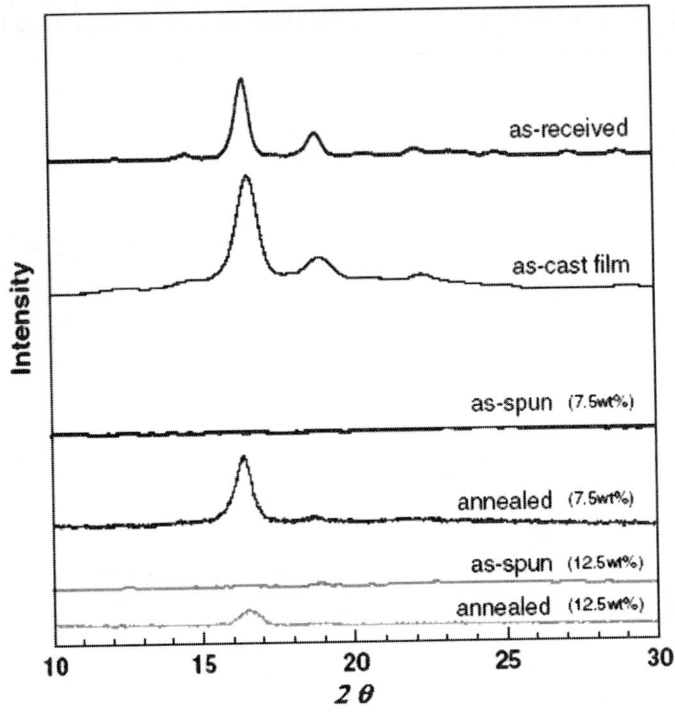

Fig. 3.13. XRD diagram of PLLA nanofibers electrospun from different polymer concentrations. Reproduced with permission from R. Inai *et al.*, Nanotechnology, 16, 208-213 (2005). Copyright © 2005, IOP Publishing Limited.

Small angle X-ray diffraction (SAXD) can be used to study the lamellar structure of semi-crystalline polymers [121]. However, this has not been widely used as the interpretation of the SAXD pattern is complicated. This is due to more than one structural features that contribute at the same time to the observed X-ray pattern [100], such as void morphology and surface scattering of nanofibers.

4.2.3. Thermal Analysis

Differential scanning calorimetry (DSC) has been used as the primary tool for obtaining thermal properties of polymeric nanofibers as the nanofiber sample is heated from room temperature or below glass transition temperature to a temperature of 10-40 K above melting temperature. Thermal properties such as glass transition temperature, crystal melting temperature and degree of crystallinity are usually obtained from the DSC thermograms [5,100,123]. DSC has also been useful in studying the effects of chemical treatment such as methanol treatment on the thermal properties of silk nanofibers [93].

In some studies, two heating runs of nanofibers are carried out to compare the characteristic features of nanofibers formed by the rapid process of electrospinning and the polymer from which the nanofiber was made [23,102]. In a study by Bognitzki *et al.* [102], the thermogram of electrospun poly(L-lactic acid) (PLLA) nanofibers for the first heating run was found to have a distinctive aging peak adjacent to the glass transition and an exothermic crystallization peak with a maximum at about 80°C **(Fig. 3.14)**. These features of the thermogram on heating the sample are typical of the rapid structure formation during electrospinning. The second heating run does not show an aging peak adjacent to the glass transition and the exothermic crystallization peak is shifted by 20 K to a higher temperature. This may be due to a certain alignment of the polymer chains in the as-spun fibers which allowed crystallization to occur at a lower temperature than in an isotropic sample.

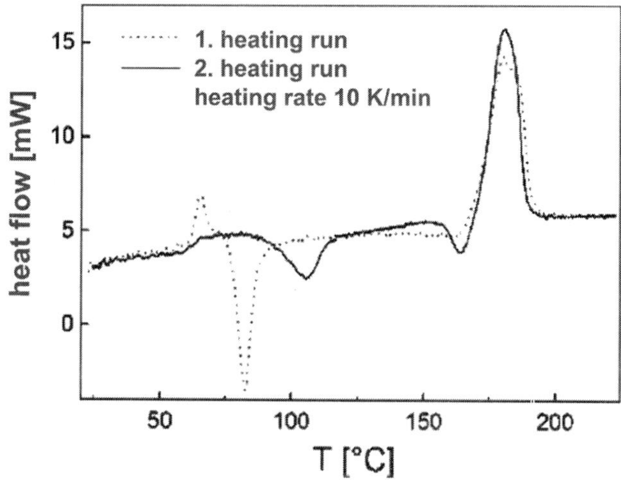

Fig. 3.14. DSC thermogram obtained for electrospun PLLA nanofibers (first heating run) and for a sample cooled down from melt with 10 K/min (second heating run. Reproduced with permission from M. Bognitzki *et al.*, Adv.Mater., 13, 70-72 (2001). Copyright © 2001, Wiley-VCH Verlag GmbH & Co KG.

Generally, the transformation of as-received polymer to electrospun nanofibers results in the reduction of glass transition temperature, melting temperature and degree of crystallinity [100]. This is mainly due to the rapid solidification process and high elongation rate of electrospinning, which results in less time for molecular chains to rearrange and form crystals.

4.3. Mechanical Properties of Individual Nanofibers

It is important that polymeric nanofibers are able to withstand the stresses and strains imposed by the environment that they are in. Therefore it is necessary to characterize the mechanical properties of individual nanofibers. However, it is a challenge to obtain the mechanical properties of single polymeric nanofiber due to the difficulties in: (i) handling of extremely small fibers, (ii) finding suitable mode of observing fiber deformation, (iii) sourcing for accurate and sensitive force transducer, (iv) sourcing for accurate and high resolution actuator and (v) preparing single-stranded nanofibers.

Although methods for testing nano-scale materials have been established for materials such as carbon nanotubes [124-127], these methods cannot be easily replicated for polymeric nanofibers as the tests have to be carried out in the SEM or TEM. Due to the non-conductive nature of most polymeric nanofibers, such methods are not suitable. With the availability of AFM and other instruments that can measure loads in the nano-Newton range, mechanical testing of polymeric nanofibers is now possible. The difficulty in handling the nanofibers still remains as the main challenge, but with novel techniques used in combination with existing equipment, this can be overcome. Tensile test, three-point bend test, nanoindentation and other techniques of testing polymeric nanofibers will be presented in this section.

4.3.1. Tensile Test

Of all the mechanical tests, tensile test may be the most challenging to perform because in most cases direct manipulation of the fiber is required. In a study by Tan *et al.* [128], a piezo-resistive AFM tip was used to perform tensile test of a single electrospun poly(ethylene oxide) (PEO) nanofiber. Aligned electrospun nanofibers were produced [26] to obtain several aligned fibers across two parallel strings mounted on a wooden frame. One end of a selected nanofiber was attached to a movable optical microscope stage and the other end to a piezo-resistive AFM cantilever tip (**Fig. 3.15**). The nanofiber was stretched by moving the microscope stage and the force was measured by the cantilever deflection. The stress-strain curve of the nanofiber was obtained from the force-extension data. The use of AFM cantilevers in tensile test is suitable for fibers ranging from tens of nanometers to several hundred nanometers in diameter. One disadvantage of this method is it is time consuming to manipulate and test single fibers.

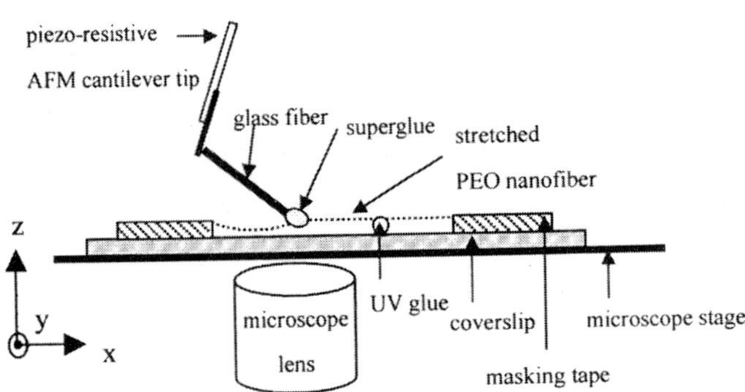

Fig. 3.15. Schematic diagram of the tensile test of the nanofiber using a piezo-resistive AFM tip. Reproduced with permission from E.P.S. Tan *et al.*, Appl.Phys.Lett., 86(6), 073115 (2005). Copyright © 2005, American Institute of Physics.

A commercial nano tensile testing system can be used to conduct the tensile test of continuous fibers of a few millimeters to several centimeters in length. In order to prepare single-stranded tensile test samples without manipulation, as-fabricated fibers need to be collected on a frame that can be mounted on the tensile tester. Electrospun polycaprolactone (PCL) [24] and poly (L-lactic acid) (PLLA) [23] nanofibers have been tested using this method. The tensile test of PCL fibers will be described briefly to demonstrate the tensile test of ultrafine fibers. Several strands of PCL fibers were collected on a cardboard frame with a 10 mm gap between a series of parallel strips. **Fig. 3.16** shows one such partition containing a single nanofiber being mounted on the nano tensile tester. The cardboard partitions were cut along the discontinuous lines before stretching the fiber. The mechanical properties were observed to vary with fiber diameter. Fibers with smaller diameter had higher strength but lower ductility due to the higher 'draw ratio' that was applied during the electrospinning process. Although this allows many samples to be tested to failure conveniently, it may not be suitable for fibers with diameters in the range of tens of nanometers. Observation of the nanofibers during the tensile test may be a challenge as this system does not provide a mode of observation.

Fig. 3.16. Tensile sample mounted on a nano tensile tester (NanoBionix, MTS, USA). Reproduced with permission from E.P.S. Tan *et al.*, Biomaterials, 26(13), 1453-1456 (2005). Copyright © 2005, Elsevier Science Ltd.

4.3.2. Three-Point Bend Test

AFM has been used to conduct three-point bend tests of polymeric nanofibers as forces in the nano-Newton range can be applied easily. Measurement of fiber deformation in the sub-nanometer range is also possible with the AFM. As the working principle of AFM is based on the interaction between atoms of the AFM tip and the sample upon contact or near contact [129], there is no requirement for the sample to be conductive. Generally, beam mechanics is used to analyze the results of the bend tests.

Most samples for three-point bend test are prepared by depositing the nanofibers in a random manner on substrates with holes or grooves and with a section of the fiber suspended over them. The adhesion between the sample and substrate is usually sufficient upon deposition for the bend test to be conducted without slippage at the fixed ends. Samples tested using this method of preparation include conductive polymer nanofibers [118,130], biological materials [131] and biodegradable polymer nanofibers [95].

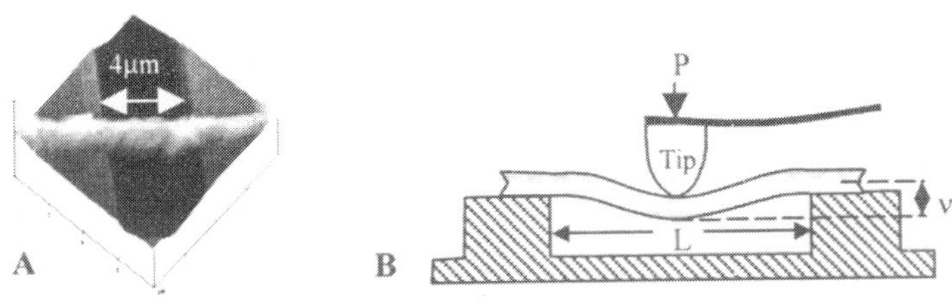

Fig. 3.17. (A) AFM contact mode image of a single nanofiber (400 nm diameter) suspended over an etched groove and (B) schematic diagram of a nanofiber with mid-span deflected by an AFM tips.

In a study by Tan *et al.* [95], poly(L-lactic acid) (PLLA) nanofibers produced by phase separation was used to demonstrate this test. The Young's modulus of PLLA nanofiber (E) was obtained by performing a nanoscale three-point bend test on a single nanofiber suspended over an etched groove in a silicon wafer **(Fig. 3.17A)** using AFM. An AFM tip was used to apply a small elastic deflection at the mid-span of the nanofiber along its suspended length **(Fig. 3.17B)**. E is found from beam bending theory [132] for a beam with two ends fixed as:

$$E = \frac{PL^3}{192vI} \tag{3.1}$$

where P is the maximum force applied, L is the suspended length, v is the deflection of the beam at midspan, and I is the second moment of area of the beam (where $I = \pi D^4/64$ and D is the beam diameter). The method of obtaining v from the force curves using AFM is shown in **Fig. 3.18**. E is found to be 1.0 ± 0.2 GPa for PLLA fibers that satisfy the assumptions of an elastic beam undergoing pure bending, i.e. fibers with sufficient length to diameter ratio of greater than 10.

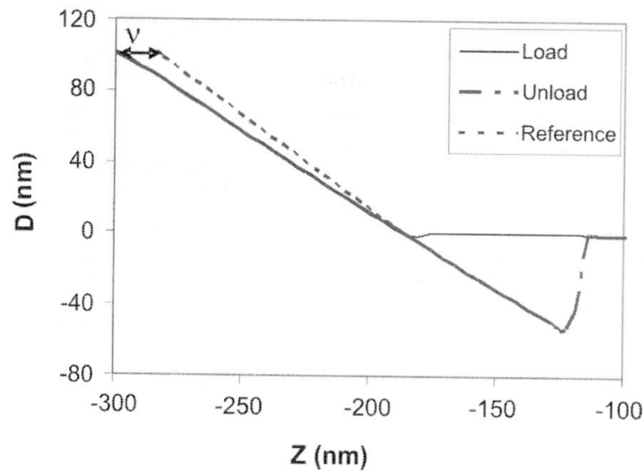

Fig. 3.18. Plot of cantilever deflection (*D*) vs. vertical displacement of the z-piezo (*Z*). A reference curve is obtained by measuring the cantilever deflection over the *Z* piezo displacement on silicon wafer. The loading and unloading curves are obtained by using the AFM tip to deflect the mid-span of the nanofiber. The deflection of the fiber, *n* is the difference between the loading and the reference curve. Reproduced with permission from E.P.S.Tan *et al.*, Appl.Phys.Lett, 84(9), 1603-1605 (2004). Copyright © 2004, American Institute of Physics.

There is a need to ensure that the possibility of an indentation being made on the nanofiber by the AFM tip during deflection is minimal or eliminated. This can be done ensuring that no indentation is made by applying the same maximum force to portions of nanofiber lying on the substrate.

4.3.3. Nanoindentation

The elastic and elastic-plastic behavior of the material can be obtained from nanoindentation data. Nanoindentation of nanofibers has not been widely studied due to the difficulty in probing the curved surface of the fiber. Nonetheless, the elastic modulus values obtained in the few reported studies agree with the bulk material properties or with other related works.

Elastic-plastic nanoindentation of materials has generally been performed using AFM-based nanoindentation systems. Such systems are unable to detect initial contact loads of less than 1μN [133] and thus not suitable for probing polymeric nanofibers which are softer than metals [134]. An AFM tip was used to perform nanoindentation of electrospun silk fibers instead of using AFM-based nanoindentation system in a study by Wang *et al.* [93]. As the diameter of the nanofiber is much larger than the contact area, elastic-plastic nanoindentation is applicable in this study.

The elastic modulus from elastic-plastic indentation is obtained from the slope of the initial portion of the unloading curve (S), which has been fitted by a power-law relation. This was first formulated by Oliver and Pharr [135] as shown in **eq. 3.2**:

$$S = 2\beta \sqrt{\frac{A}{\pi}} E_r \tag{3.2}$$

where β is a constant that depends on the geometry of the indenter, A is the projected area of the indent and E_r is the reduced elastic modulus, given by **eq. 3.3**:

$$\frac{1}{E_r} = \frac{1-v^2}{E} + \frac{1-v_i^2}{E_i} \tag{3.3}$$

where E and v are the elastic modulus and Poisson's ratio for the sample and E_i and v_i are the same quantities for the indenter, respectively.

Calculating the exact value of modulus from **eq. 2** may be difficult due to uncertainties of cantilever stiffness and AFM tip geometry. To overcome this problem, an indent was first made on epoxy with a known value of elastic modulus (E_1). The contact radius of the plastic indent (r_1) and slope of unloading curve at maximum load for epoxy (S_1) were then obtained. Using the same tip, an indent is made with the same loading force on the nanofiber. Contact radius of the plastic indent (r_2) and slope of unloading curve at maximum load for the nanofiber (S_2) were obtained. All these values are substituted into **eq. 3.4** to calculate the modulus of the nanofiber (E_2).

$$\frac{S_1}{S_2} = \frac{r_1 E_1}{r_2 E_2} \tag{3.4}$$

Creep can be significant when elastic-plastic nanoindentation is performed on polymer nanofibers. Viscoelastic creep was found to increase the initial slopes of the unloading curves and resulting in overestimation of modulus values of polymers [133]. In some cases, the slope was found to be negative [136,137] due to creep. In such cases, measurement of elastic modulus is not possible.

Elastic nanoindentation has been performed using the AFM as it is the most suitable device for probing mechanical properties of soft polymeric nanofibers. However, there are a number of factors that need to be considered:

(i) **Effect of underlying substrate:** The elastic modulus values can be overestimated if the diameter of the nanofiber is too small (<200 nm) [138].

(ii) **Ambiguous tip shape and cantilever spring constant:** Although actual values may deviate from values provided by the manufacturer, nominal tip radius and cantilever spring constants are often used.

(iii) **Fiber surface roughness:** All real surfaces have certain roughness due to numerous asperities. These asperities can act like a soft layer such that contact is extended over a larger area than it would be if the surfaces were perfectly smooth. Hence, the contact pressure for a given load will be reduced [139].

(iv) **Curvature of fiber surface:** Unlike most nanoindentation studies where the surface of the sample is flat, the curved surface of the fiber has to be taken into account when Hertz contact theory is used.

(v) **Non-perpendicular loading:** Although the AFM tip is not perpendicular to the sample surface and may cause slip and friction between the AFM tip and the sample surface during indentation [134], the non-perpendicular load factor is not significant [140]. AFM tip slippage can be detected when the loading and unloading force curves for elastic indents do not coincide.

(vi) **Adhesion force:** The effects of adhesion due to the capillary effect of water condensation between tip and sample can become significant [141] in a high humidity environment (>70% relative humidity). Such adhesion forces have to be accounted for as shown in a study by Park *et al.* [142].

Ko *et al.* [109,143] used the AFM to perform nanoindentation of electrospun polyacrylonitrile (PAN) nanofibers. The indentation depth was assumed to be of comparable size to the AFM tip and the tip radius (5 nm) was small compared to nanofiber diameter (50-500 nm), such that the fiber curvature can be ignored. The reduced modulus is given by **eq. 3.5**:

$$\frac{P}{\delta} = 2\sqrt{\frac{A}{\pi}}E_r \tag{3.5}$$

where P is the normal force, δ is the indentation depth and A is the contact area.

In a study by Park *et al.* [142] on nanoindentation of polypyrrole nanotubes, Hertz theory of contact mechanics was used with the sample radius, tip radius, and the tip-sample adhesion force considered for the calculation of elastic modulus. With these added considerations, the reduced modulus is given by **eq. 3.6**:

$$E_r = \sqrt{\frac{k_c^3}{6F_L R^*}} \tag{3.6}$$

where F_L is the external loading force on the system, $F_L = F_c + F_{ad}$. F_c is the cantilever force and F_{ad} is the adhesion force. k_c is the contact stiffness, which is taken from the slope of the force vs indentation depth curve. The force-indentation depth relationship obtained was linear, thus k_c is a constant. R^* is the effective radius of tip-sample system given by **eq. 3.7**:

$$\frac{1}{R^*} = \frac{1}{R_{tip}} + \frac{1}{R_{sample}} \tag{3.7}$$

where tip radius (R_{tip} = 20 nm) is of comparable size to the sample radius (R_{sample} = 60 nm).

In a study by Tan *et al.* [144], nanoindentation of PLLA nanofibers produced by the phase separation method was performed using the AFM to obtain the elastic modulus. Hertz theory of normal contact of elastic solids was used to analyze the data [145]. The nanofiber was modeled as a cylinder and the AFM tip was modeled as a sphere. The relative elastic modulus, E_r is given by:

$$E_r = \sqrt{\frac{9P^2}{16R_e \delta^3}} \qquad (3.8)$$

where P is the force applied, δ is the indentation depth and R_e is the equivalent radius for a spherical indenter in contact with an infinitely long cylinder. R_e is given by:

$$R_e = \sqrt{\frac{R_t^2 R_f}{R_t + R_f}} \qquad (3.9)$$

where R_t is the AFM tip radius and R_f is the radius of the nanofiber. The elastic modulus of the nanofiber, E_f is related to the relative elastic modulus E_r by the following relation:

$$E_f \approx E_r (1 - v_f^2) \qquad (3.10)$$

where v_f is the Poisson's ratio of the nanofiber. The absence of indentation marks after indentation indicated that deformation of the nanofiber by the tip was elastic. As the nanoindentation depth is of the same order of magnitude as the surface roughness, the variation in modulus values obtained for each fiber was larger than that of the three-point bend test [95]. Although the modulus value obtained (0.7 ± 0.2 GPa) was lower than that from the three-point bend test, the elastic modulus value after incorporating roughness correction was found to be approximately 1GPa [144]. This result after roughness correction agrees well with that of the three-point bend test [95].

4.3.4. Other Methods

There are also techniques that involve relating the elastic modulus of the nanofiber to its resonance behavior. Cuenot *et al.* [146,147] used a method based on the measurement of resonance frequencies of AFM cantilevers in contact with polypyrrole nanofibers to measure their mechanical properties. The nanofibers were suspended over the pores of a membrane. The resonance frequency was measured by modulating the cantilever deflection with an oscillating electric field applied between the sample holder and AFM head. Frequency behavior of the nanomechanical system consisting of the cantilever and sample was modeled as a cantilever with the tip in contact with two springs. The nanofiber stiffness was thus determined from the resonance frequency measurement using this model.

Boussaad *et al.* [148] measured the mechanical properties of a polymer wire bridged across the two prongs of a microfabricated tuning fork. They made use of the resonant frequency of the tuning fork with the polymer wire to obtain the Young's modulus of the nitrocellulose/toluene sulfonamide formaldehyde resin.

4.3.5. Current Limitations and Future Work

Since most polymers are non-conductive, it remains a challenge to find ways to observe deformation and failure of polymer nanofibers. Conventional methods of using SEM or TEM are not suitable as samples need to be conductive. A fair amount of micro-manipulation using sophisticated instruments is required in most of the experiments. Thus preparing and manipulating nanofibers still remains as the biggest challenge to date. Most of the mechanical tests performed to date do not take into account of the viscoelastic nature of polymeric nanofibers because actuators and load sensors that are available are unable to measure very small oscillatory movements on a nanoscale.

With the current limitations in mind, here are some of the areas that can be further explored for the mechanical characterization of nanofibers: (i) development of systems or methods for *in-situ* observation of the sample under test, (ii) determination of actual mechanism of deformation and failure in relation to its nano and molecular structures, (iii) performing of dynamic material characterization and (iv) development of accurate constitutive relations for the nanofiber by taking into consideration its size and any possible deviation from bulk property values.

4.4. Nanofiber Mats and Foams

Polymeric nanofibers are fabricated in many forms (as mentioned in section 2) such as membranes, mats or foams, depending on the fabrication methods used. Electrospun polymeric nanofibers are fabricated in the form of random [105,149] or aligned [23,106] nanofiber membranes or mats. Nanofibers produced by the phase separation method usually result in foams [5,123]. **Tab. 3.5** shows the summary of various properties that can be obtained from the different characterization techniques.

Tab. 3.5. Characterization of nanofiber mats and foams.

Properties	Details	Instruments and techniques
Porous structure	Pore morphology	SEM
	Pore size and distribution	Mercury porosimetry
	Porosity	Mercury porosimetry, liquid displacement methods, calculated based on known fiber density
Permeability	Air and vapor permeability	Gas permeation analyzer, dynamic moisture permeation cells
Mechanical	Tensile and compression properties, relationship between individual nanofibers and bulk mat or foam	Universal mechanical tester
Biological	Biodegradability	Mass loss, surface morphology change
	Cell proliferation	Optical microscopy, biological assays, confocal microscopy

4.4.1. Porosity

The porous structure of nanofiber mats or foams can be observed with the imaging techniques mentioned in section 4.1, namely AFM, SEM, TEM and birefringence. For convenience, all nanofiber mats and membranes will be referred as foams in this section. Relative density of foam is defined as the ratio of foam density (ρ^*) to the density of polymer (ρ_s) from which the nanofibers are made. The fraction of pore space in the foam is its porosity, which is given by ($1-\rho^*/\rho_s$). It is important to characterize the porosity of the nanofiber foams as it affects other properties such as air and vapor permeability, mechanical properties and cell penetration in biodegradable scaffolds.

The porosity and pore size distribution of a nanofiber foam can be found via mercury porosimetry [149]. Although this method allows quantitative measurements of physical parameters related to the porous structure, it has not been widely used for polymeric nanofiber mats or foams. This is because the nanofibrous samples have a tendency to collapse as they are penetrated with mercury under pressure. The pore size measured would then be inaccurate. Liquid displacement methods are more popular for obtaining the porosity of polymer foams [150,151]. Ethanol is usually used as the displacement liquid because it penetrates easily into the pores and does not induce shrinkage or swelling [150]. Yang *et al.,* made use of this method to obtain the porosity of nanofibrous PLLA foams produced by phase separation [5].

Porosity of nanofibrous foams can be obtained by non-invasive methods if the density of individual nanofibers (ρ_s) is known. The skeletal density of PLLA nanofibrous foam was calculated based on the crystallinity of the nanofibers [123,152], which is given by:

$$\rho_s = \frac{1}{\dfrac{1-X_c}{D_a} + \dfrac{X_c}{D_c}} \tag{3.11}$$

where X_c is the degree of crystallinity of the polymer, D_a is the density of amorphous form of the polymer and D_c is the density of 100% crystalline polymer. The density of the foam (ρ^*) is calculated from its weight and volume.

4.4.2. Air and Vapor Permeability

Air and vapor permeability of nanofiber mats or membranes are of interest in applications such as filtration, protective clothing and wound dressings. Gas permeation analyzers [153] or dynamic moisture permeation cells [154] have been used to measure the air and/or vapor permeability of nanofiber mats. In a permeation cell, gas flows of known temperature and water vapor concentration enter the test cell. The fluxes of gas and water vapor transported through the test sample are determined by measuring the temperature, water vapor concentration, and flow rates of the gas leaving the cell [154].

4.4.3. Mechanical Properties

The type of mechanical tests used to characterize nanofibers depends on whether the fibers are fabricated in the form of a membrane, mat or foam. Nanofiber membranes, mats and foams can be characterized by tensile test. Only nanofibrous foams can be characterized by compression test. The mechanical tests are carried out on strips or blocks of samples using conventional universal testing machines. Most of the mechanical tests reported in the literature are tensile tests of electrospun polymeric nanofiber mats [105,116,149]. Tensile tests were performed to observe the effects of changing processing parameters on mechanical properties. The processing parameters of interest include polymer concentration [155], polymer blend ratio [116] and velocity of drum collector [99]. **Fig. 3.19** shows the effect of varying polymer concentration for gelatin nanofibers [155]. Differences in mechanical properties were found for electrospun nanofibers collected on rotating drums in the directions parallel and perpendicular to the direction of fiber collection [99,156]. Nanofiber mats are generally found to be stronger and stiffer in the direction parallel to fiber collection due to the higher percentage of fibers oriented along that direction.

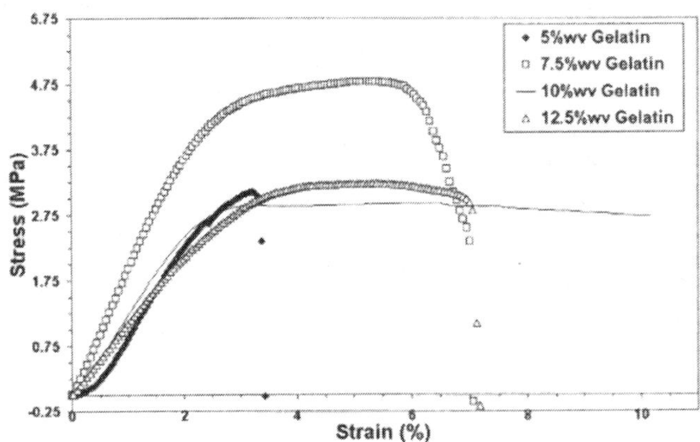

Fig. 3.19. Stress-strain curves of electrospun gelatin nanofiber mats at different polymer concentrations. Reproduced with permission Z.-M. Huang *et al.*, Polymer, 45(15), 5361-5368 (2004). Copyright © 2004, Elsevier Science Ltd.

Various highly porous polymeric foams have been characterized by compression test [151,157,158] and the mechanical properties were modeled using the theory of cellular solids [159]. Such models relate the foam structure and mechanical properties of the skeletal members (e.g. nanofibers) to the bulk foam. Nanofibrous foams of PLLA have been characterized by tensile [123] and compression [160] tests. In a study by Tan *et al.* [160], bulk properties of PLLA nanofibrous scaffolds were characterized from nanomechanical properties of single nanofibers [95]. The elastic modulus of a single nanofiber obtained previously from three-point bend tests performed using the AFM was used to model the elastic modulus and elastic collapse stress of the bulk scaffold under compression. Various open-cell solid models were examined for the modeling of bulk scaffold behavior.

The simplest and most commonly used model is one suggested by Gibson and Ashby [159] whereby the open-cell foams are modeled as a staggered cubic array of struts **(Fig. 3.20A)** [159]. This has been used to model structures such as bone [161] and cell cytoskeleton [162]. The primary mode of deformation is by bending of the cell edges when the foam is loaded in compression as shown in **Fig. 3.20B**. The relationship between bulk foam modulus (E^*), solid elastic modulus (E_s), density of foam (ρ) and density of the solid skeleton (ρ_s) for this model is given by **eq. 3.12** as:

$$\frac{E^*}{E_s} = C_1 \left(\frac{\rho^*}{\rho_s} \right)^n \tag{3.12}$$

where the constants C_1 and n depend on the microstructure of the solid material. Experimental evidence suggests that $n = 2$ for open-cell foams. The pre-factor $C_1 \approx 1$ as determined empirically by Gibson and Ashby [159]. As the elastic distortion increases, the axial load on the cell edge (F) increases as well. If F reaches the Euler load (F_{crit}) for the edge, it buckles **(Fig. 3.20c)**. The stress at which this occurs is the elastic collapse stress (σ_{el}^*), which is given by:

$$\frac{\sigma^*_{el}}{E_s} = C_2 \left(\frac{\rho^*}{\rho_s} \right)^2$$

(3.13)

where $C_2 \approx 0.05$ as determined empirically by Gibson and Ashby.

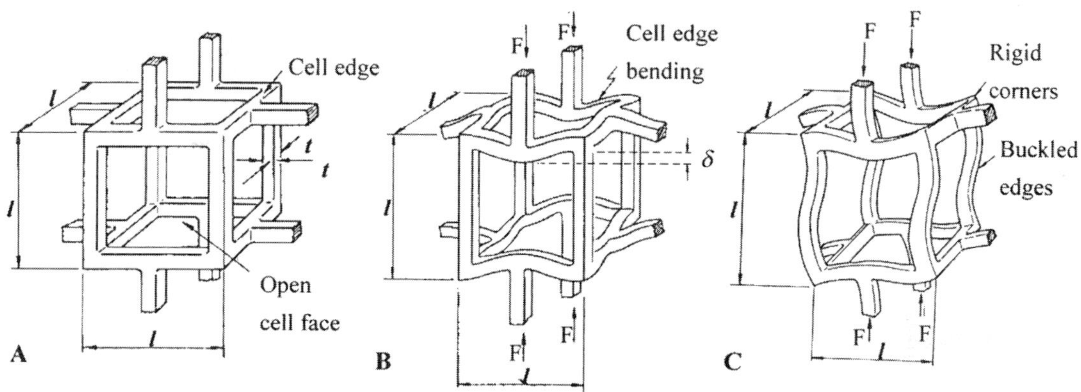

Fig. 3.20. (A) A cubic model for an open-cell foam showing the edge length, *l*, and the edge thickness, *t*, (B) Cell edge bending during linear elastic deformation and (C) Elastic buckling in the cell walls of an open-cell foam. Reproduced with permission from L.J. Gibson *et al.* in Cellular solids: Structure and properties. Copyright © 1997, Cambridge University Press.

The model suggested by Gibson and Ashby was found to overestimate the foam modulus and elastic collapse stress of PLLA foams [160]. This is due to the fact that real foams seldom have a staggered cubic structure. The presence of irregularities and defects also reduce the modulus and strength of foams [163]. Other models such as 'tetrakaidecahedral' [164] **(Fig. 3.21A)** and Voronoi foam model [165] **(Fig. 3.21B)** are shown to be more realistic models of nanofibrous foams [160].

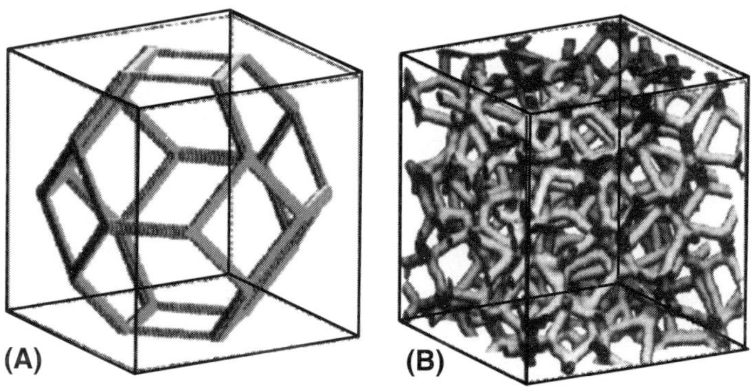

Fig. 3.21. (A) A unit cell of the tetrakaidecahedral model and (B) Open-cell Voronoi foam with $r^*/r_s = 0.15$. Reproduced with permission A.P. Roberts *et al.*, Journal of the Mechanics and Physics of Solids, 50(1), 33-55 (2002). Copyright © 2002, Elsevier Science Ltd.

4.4.4. Biological Properties

As mentioned earlier, one major application of polymeric nanofibers is in the field of tissue engineering. Therefore it is of interest to study biological properties of nanofiber mats and foams such as biodegradability and cell proliferation. Biodegradability studies have been performed on polylactide and polyglycolide nanofiber scaffolds [166,167]. The scaffolds were immersed in phosphate buffer solution at 37°C for a period of time during the degradation study. The change in structure, morphology and mass loss were evaluated at different stages of degradation. Biodegradation studies are of importance as the nanofiber mats or foams have to remain viable during the service lifetime. Cell proliferation studies have been performed on electrospun polymeric nanofiber mats [149,153,168,169] and foams [5]. The types of cells seeded on the scaffold include fibroblasts [149], smooth muscle cells [168], cardiomyocytes [169] and nerve stem cells [5] (**Fig. 3.22**). Optical microscopy, biological assays and laser scanning confocal microscopy are generally used to determine the number of living cells in the scaffold after a certain number of days of cell culture.

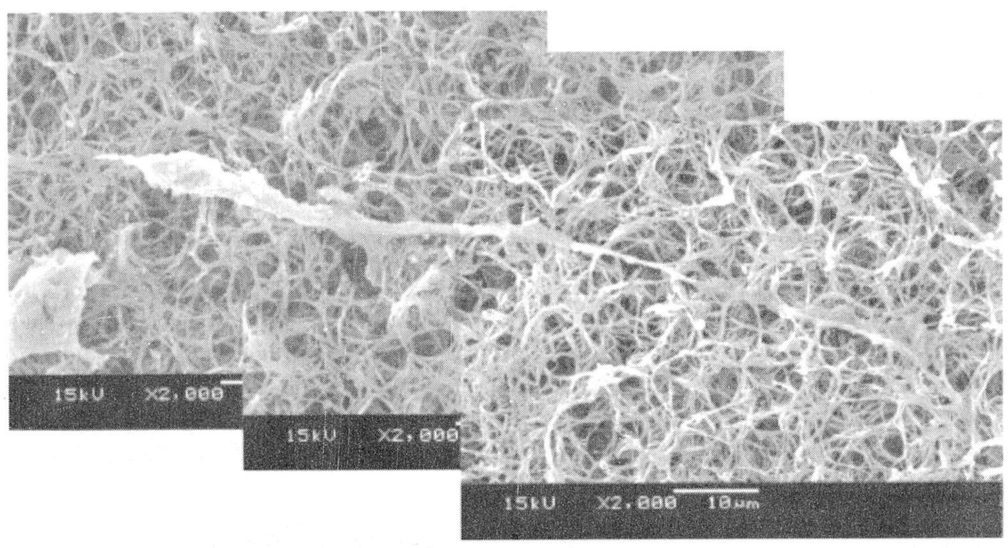

Fig. 3.22. SEM micrograph of a differentiated nerve stem cell with long neurite on PLLA nanofibrous scaffold. Reproduced with permission from F. Yang *et al.*, Biomaterials, 25(10), 1891-1900 (2004). Copyright © 2004, Elsevier Science Ltd.

5. CONCLUSIONS AND FUTURE PROSPECTS

The interest in polymeric nanofibers has been growing rapidly in recent years due to their vast potential in a wide range of applications, especially in the field of biomedical engineering. Other areas that receive attention include filtration and separation, sensor technology and nanocomposites due to the uniquely high surface area to volume ratio of polymeric nanofibers. The attractiveness of polymeric fibers is enhanced by the ease of fabrication, especially with the use of the electrospinning technique. This technique also allows aligned fibers and various forms of nanofibers such as coaxial, porous and composite nanofibers to be produced rapidly and cost-effectively.

Before the nanofibers can be used for various applications, characterization of the nanofiber properties has to be performed to ensure the viability of the product. The physical structure and morphology can be characterized via microscopy techniques. Various physical and chemical properties can be investigated by spectroscopic and diffraction techniques and thermal analysis. Mechanical characterizations of single nanofibers are performed mainly by tensile tests, three-point bend tests and nanoindentation. Characterization of nanofiber mats and foams include the measurement of porosity, air and vapor permeability, and mechanical and biological properties, depending on the application.

Although there has been much advancement in polymeric nanofiber fabrication techniques and characterization in recent years, more research is still required to actualize many potential products that are being developed at the laboratory level.

REFERENCES

1. A Short History of Manufactured Fibers; http://www.fibersource.com/f-tutor/history.htm
2. D. Reneker and I. Chun; Nanotechnology, 7(3), 216-223 (1996).
3. B. Gu, J. Badding and A. Sen; Polymer Preprints, 44(2), 142 (2003).
4. G. Ward; Filtration & Separation, 38(9), 42-43 (2001).
5. F. Yang, R. Murugan, S. Ramakrishna, X. Wang, Y.-X. Ma and S.Wang; Biomaterials, 25(10), 1891-1900 (2004).
6. P. Ma and R. Zhang; Journal of Biomedical Materials Research, 46(1), 60-72 (1999).
7. J.D. Hartgerink, E. Beniash and S.I. Stupp; Science, 294, 1684-1688 (2001).
8. Q.Cheng; Polymer Preprints, 44(2), 87 (2003).
9. M. Guler, B. Rabatic, R. Claussen and S. Stupp; Polymer Preprints, 44(2), 102 (2003).
10. L. Feng, S. Li, H. Li, J. Zhai, Y. Song, L. Jiang and D. Zhu, Angewandte Chemie (International Ed, In English), 41(7), 1221-1223 (2002).
11. C. Martin; Chem. Mater, 8, 1739-1746 (1996).
12. A. Formhals, *US patent 1, 975,504* 1934.
13. D.H. Reneker, A.L. Yarin, F. Hao and S. Koombhongse; Journal of Applied Physics, 87(9, Part 1-3), 4531-4547 (2000).
14. Y.M. Shin, M.M. Hohman, M.P. Brenner and G.C. Rutledge; Polymer, 42(25), 09955-09967 (2001).
15. Y.M. Shin, M.M. Hohman, M.P. Brenner and G.C. Rutledge; Applied Physics Letters, 78(8), 1149-1151 (2001).
16. J. Doshi and D.H. Reneker; Journal of Electrostatics, 35, 151-160 (1995).
17. A. Frenot and I.S. Chronakis; Current Opinion in Colloid & Interface Science, 8(1), 64-75 (2003).
18. Z.-M. Huang, Y.-Z. Zhang, M. Kotaki and S. Ramakrishna; Composites Science and Technology, 63(15), 2223-2253 (2003).
19. E.D. Boland, G.E. Wnek, D.G. Simpson, K.J. Pawlowski and G.L. Bowlin; J.Macromol. Sci. Pure Appl.Chem, A38(12), 1231-1243 (2001).
20. J.A. Matthews, G.E. Wnek, D.G. Simpson and G.L. Bowlin; Biomacromolecules, 3(2), 232-238 (2002).
21. A. Theron, E. Zussman and A.L. Yarin; Nanotechnology, 12(3), 384-390 (2001).
22. C.Y. Xu, R. Inai, M. Kotaki and S. Ramakrishna; Biomaterials, 25(5), 877-886 (2004).
23. R. Inai, M. Kotaki and S. Ramakrishna; Nanotechnology, 16, 208-213 (2005).
24. E.P.S. Tan, S.Y. Ng and C.T. Lim; Biomaterials, 26(13), 1453-1456 (2005).
25. R. Dersch, T. Liu, A.K. Schaper, A. Greiner and J.H. Wendorff; Journal of Polymer Science: Part A: Polymer Chemistry, 41, 545-553 (2003).
26. D. Li, Y. Wang and Y. Xia; Nano Letters, 3(8), 1167-1171 (2003).
27. D. Li, Y. Wang and Y. Xia; Advanced Materials, 16(4), 361-366 (2004).
28. S. Chia, C. Lim, Y. Zhang and S. Ramakrishna, Highly productive electrospinning apparatus with multiple spinneret system, 2003. p. US provisional application no.60/475921.

29. P. Katta, M. Alessandro, R.D. Ramsier and G.G. Chase; Nano Letters, 4(11), 2215-2218 (2004).

30. A.G. MacDiarmid, W.E. Jones Jr., I.D. Norris, J. Gao, A.T. Johnson Jr., N.J. Pinto, J. Hone, B. Han, F.K. Ko, H. Okuzaki and M. Llaguno; Synthetic Metals, 119(1-3), 27-30 (2001).

31. W.D. Bates, C.P. Barnes, Z. Ounaies and G.E. Wnek; Polymer Preprints, 44(2), 114 (2003).

32. Y.K. Luu, K. Kim, B.S. Hsiao, B. Chu and M. Hadjiargyrou; Journal of Controlled Release: Official Journal of the Controlled Release Society, 89(2), 341-353 (2003).

33. E.-R. Kenawy, G.L. Bowlin, K. Mansfield, J. Layman, D.G. Simpson, E.H. Sanders and G.E. Wnek; Journal of Controlled Release: Official Journal of the Controlled Release Society, 81(1-2), 57-64 (2002).

34. Y.Z. Zhang, H.W. Ouyang, C.T. Lim, S. Ramakrishna and Z.-M. Huang; Journal of Biomedical Materials Research, Part B: Applied Biomaterials, 72B(1), 156-165 (2005).

35. M. Wang, H. Singh, T.A. Hatton and G.C. Rutledge; Polymer, 45(16), 5505-5514 (2004).

36. D. Li, T. Herricks and Y. Xia; Applied Physics Letters, 83(22), 4586-4588 (2003).

37. F. Ko, Y. Gogotsi, A. Ali, N. Naguib, H. Ye, G. Yang, C. Li and P. Willis; Advanced Materials, 15(14), 1161-1165 (2003).

38. Y. Dror, W. Salalha, R.L. Khalfin, Y. Cohen, A.L. Yarin and E. Zussman; Langmuir, 19(17), 7012-7020 (2003).

39. X. Wang, Y.-G. Kim, C. Drew, B.-C. Ku, J. Kumar and L.A. Samuelson; Nano Letters, 4(2), 331-334 (2004).

40. C.L. Drew, X. Ziegler, D. Wang, X. Bruno, F.F. Whitten, J. Samuelson, L.A. Kumar; J. Nano Letters, 3(2), 143-147 (2003).

41. Z. Ma, M. Kotaki, T. Yong, W. He and S. Ramakrishna; Biomaterials, I26(15), 2527-2536 (2005).

42. K.-N. Chua, W.-S. Lim, P. Zhang, H. Lu, J. Wen, S. Ramakrishna, K.W. Leong and H.-Q. Mao; Biomaterials, 26(15), 2537-2547 (2005).

43. S. Megelski, J.S. Stephens, D.B. Chase and J.F. Rabolt; Macromolecules, 35(22), 8456-8466 (2002).

44. I.G. Loscertales, A. Barrero, I. Guerrero, R. Cortijo, M. Marquez and A.M. Ganan-Calvo; Science, 295(5560), 1695-1698 (2002).

45. Z. Sun, E. Zussman, A. Yarin, J. Wendorff and A. Greiner; Advanced Materials, 15(22), 1929-1932 (2003).

46. Y. Zhang, Z.-M. Huang, X. Xu, C.T. Lim and S. Ramakrishna; Chem.Mater, 16(18), 3406-3409 (2004).

47. D. Li and O. Xia; Nano Letters, 4(5), 933-938 (2004).

48. J.H. Yu, S.V. Fridrikh and G.C. Rutledge; Advanced Materials, 16(17), 1562-1566 (2004).

49. Y.Z. Zhang, J. Venugopal, Z.M. Huang, C.T. Lim and S. Ramakrishna; Biomacromolecules, 6(5), 2583-2589 (2005).

50. M. Bognitzki, H. Hou, M. Ishaque, T. Frese, M. Hellwig, C. Schwarte, A. Schaper, J.H. Wendorff and A. Greiner; Advanced Materials, 12(9), 637-640 (2000).

51. R.A. Caruso, J.H. Schattka and A. Greiner; Advanced Materials, 13(20), 1577-1579 (2001).

52. H. Hou, J. Zeng, A. Reuning, A. Schaper, J.H. Wendorff and A. Greiner; Macromolecules, 35, 2429-2431 (2002).

53. T. Song, Y.Z. Zhang, T.J. Zhou, C.T. Lim, S. Ramakrishna and B. Liu; Chemical Physics Letters, 415, 317-322 (2005).

54. I.G. Loscertales, A. Barrero, M. Marquez, R. Spretz, R. Velarde-Ortiz and G. Larsen; Journal of the American Chemical Society, 126(17), 5376 -5377 (2004).

55. Y.Z. Zhang, C.T. Lim, S. Ramakrishna and Z.-M. Huang; Journal of Materials Science: Materials in Medicine, 16(10), 933-946 (2005).

56. K.E. Kadler, D.F. Holmes, J. Trotter and J.A. Chapman; Biochemical Journal, 316(1-11) (1996).

57. W.-J. Li, C.T. Laurencin, E.J. Caterson, R.S. Tuan and F.K. Ko; Journal of Biomedical Materials Research, 60(4), 613-621 (2002).

58. W.-J. Li, K.G. Danielson, P.G. Alexander and R.S. Tuan; Journal of Biomedical Materials Research, 67A(4), 1105-1114 (2003).

59. A. Fertala, W.B. Han and F.K. Ko; Journal of Biomedical Materials Research, 57, 48-58 (2001).
60. H. Yoshimoto, Y.M. Shin, H. Terai and J.P. Vacanti; Biomaterials, 24(12), 2077-2082 (2003).
61. L. Huang, R.A. McMillan, R.P. Apkarian, B. Pourdeyhimi, V.P. Conticello and E.L. Chaikof; Macromolecules, 33(8), 2989-2997 (2000).
62. K. Nagapudi, W.T. Brinkman, J.E. Leisen, L. Huang, R.A. McMillan, R.P. Apkarian, V.P. Conticello and E.L. Chaikof; Macromolecules, 35, 1730-1737 (2002).
63. X.M. Mo, C.Y. Xu, M. Kotaki and S. Ramakrishna; Biomaterials, 25(10), 1883-1890 (2004).
64. X. Zong, H. Bien, C. Chung, L. Yin, K. Kim, D. Fang, B. Chu, B. Hsiao and E. Entcheva; Polymer Preprints, 44(2), 96-97 (2003).
65. G.A. Silva, C. Czeisler, K.L. Niece, E. Beniash, D.A. Harrington, J.A. Kessler and S.I. Stupp; Science, 303(5662), 1352-1355 (2004).
66. G. Wnek, M. Carr, D. Simpson and G. Bowlin; Nano Letters, 3(2), 213-216 (2003).
67. E.-R. Kenawy, J.M. Layman, J.R. Watkins, G.L. Bowlin, J.A. Matthews, D.G. Simpson and G.E. Wnek; Biomaterials, 24(6), 907-913 (2003).
68. M.-S. Khil, D.-I. Cha, H.-Y. Kim, I.-S. Kim and N. Bhattarai; Journal of Biomedical Materials Research, 67B(2), 675-679 (2003).
69. G. Verreck, I. Chun, J. Peeters, J. Rosenblatt and M.E. Brewster; Pharmaceutical Research, 20(5), 810-817 (2003).
70. J. Zeng, X. Xu, X. Chen, Q. Liang, X. Bian, L. Yang and X. Jing; Journal of Controlled Release: Official Journal of the Controlled Release Society, 92(3), 227-231 (2003).
71. T. Grafe; International Nonwovens Journal, 12(1), 51-55 (2003).
72. A. Suthat and G. Chase; Chemical Engineer (London), 26-28 (2001).
73. P. Gibson, H. Schreuder-Gibson and D. Rivin; Colloids and Surfaces A: Physicochemical and Engineering Aspects, 187-188, 469-481 (2001).
74. P.W. Gibson, H.L. Schreuder-Gibson and D. Rivin; AIChE Journal, 45(1), 190-195 (1999).
75. H. Schreuder-Gibson, P. Gibson, K. Senecal, M. Sennett, J. Walker, W. Yeomans, D. Ziegler and P.T. Tsai; Journal of Advanced Materials, 34(4), 44 55 (2002).
76. J.-S. Kim and D.H. Reneker; Polymer Composites, 20(1), 124-131 (1999).
77. M.M. Bergshoef and V.G. Julius; Advanced Materials, 11(6), 1362-1365 (1999).
78. Z. Xu, J. Vetelino, R.Lec and D.Parker; Journal of Vacuum Science & Technology A (Vacuum, Surfaces and Films), 8(4), 3634-3638 (1990).
79. S.J. Kwoun, R.M. Lec, B. Han and F.K. Ko; A novel polymer nanofiber interface for chemical sensor applications. Proc. 2000 IEEE/EIA International Frequency Control Symposium and Exhibition, June 7-9.2000, Kansas City, Missouri.
80. S.J. Kwoun, R.M. Lec, B. Han and F.K. Ko; Polymer nanofiber thin films for biosensor applications. Proc. IEEE 27th Annual Northeast Bioengineering Conference, March 31-April 1.2001.University of Connecticut, Storrs, Connecticut.
81. X. Wang, C. Drew, S.-H. Lee, K.J. Senecal, J. Kumar and L.A. Samuelson; Nano Letters, 2(11), 1273-1275 (2002).
82. X. Wang, Y.-G. Kim, C. Drew, B.-C. Ku, J. Kumar and L. Samuelson; Nano Letters, 4(2), 331-334 (2004).
83. S.W. Choi, S.M. Jo, W.S. Lee and Y.-R. Kim; Advanced Materials, 15(23), 2027-2032 (2003).
84. C. Kim and K.S. Yang; Applied Physics Letters, 83(6), 1216-1218 (2003).
85. H. Jia, G. Zhu, B. Vugrinovich, W. Kataphinan, D. Reneker and P. Wang; Biotechnology Progress, 18(5), 1027-1032 (2002).
86. M.M. Demir, M.A. Gulgun, Y.Z. Menceloglu, B. Erman, S.S. Abramchuk, E.E. Makhaeva, A.R. Khokhlov, V.G. Matveeva and M.G. Sulman; Macomolecules, 37(5), 1787-1792 (2004).
87. G.J. Vansco, H. Schonherr and D. Snetivy; in Scanning Probe Microscopy of Polymers (Ed. B.D. Ratner and V.V. Tsukruk), p. 67-93, American Chemical Society (1998).
88. V.V. Tsukruk; Rubber Chemistry and Technology, 70(3), 430-467 (1997).
89. G. Srinivasan and D.H. Reneker; Polym Int, 36(2), 195-201 (1995).

90. M.M. Demir, I. Yilgor, E. Yilgor and B. Erman; Polymer, 43, 3303-3309 (2002).
91. R. Jaeger, H. Schonherr and G.J. Vancso; Macromolecules, 29, 7634-7363 (1996).
92. E. Zussman, D. Rittel and A.L. Yarin; Appl.Phys.Lett, 82(22), 3958-3960 (2003).
93. M. Wang, H.-J. Jin, D.L. Kaplan and G.C. Rutledge; Macromolecules, 37, 6856-6864 (2004).
94. J.A. Cicero and J.R. Dorgan; J.Polym.Environ, 9(1), 1-10 (2001).
95. E.P.S. Tan and C.T. Lim; Appl.Phys.Lett, 84(9), 1603-1605 (2004).
96. Y.K. Kamath, S.B. Ruetsch, E. Petrovicova, L. Kintrup and H.J. Schwark; Journal of Applied Polymer Science.85(2), 394-414 (2002).
97. D. Campbell, R.A. Pethrick and J.R. White; in Polymer Characterization: physical techniques, p. 293-323, Stanley Thornes Publishers, United Kingdom (2000).
98. X. Yuan, Y. Zhang, C. Dong and J. Sheng; Polym Int.53, 1704-1710 (2004).
99. K.H. Lee, H.Y. Kim, M.S. Khil, Y.M. Ra and D.R. Lee; Polymer.44, 1287-1294 (2003).
100. X. Zong, K. Kim, D. Fang, S. Ran, B.S. Hsiao and B. Chu; Polymer.43, 4403-4412 (2002).
101. C.-M. Hsu and S. Shikumar; Journal of Materials Science.39, 3003-3013 (2004).
102. M. Bognitzki, W. Czado, T. Frese, A. Schaper, M. Hellwig, M. Steinhart, A. Greiner and J.H. Wendorff; Adv.Mater., 13, 70-72 (2001).
103. S. Megelski, J.S. Stephens, D.B. Chase and J.F. Rabolt; Macromolecules, 35, 8456-8466 (2002).
104. J. Wang, X. Kuang and S. Yan; Journal of Polymer Science: Part B: Polymer Physics, 42, 2703-2709 (2004).
105. J. Ayutsede, M. Ghandi, S. Sukigara, M. Micklus, H.-E. Chen and F. Ko; Polymer, 46, 1625-1634 (2005).
106. R. Dersch, T. Liu, A.K. Schaper, A. Greiner and J.H. Wendorf; Journal of Polymer Science Part A: Polymer Chemistry, 41, 545-553 (2003).
107. D.C. Martin and E.L. Thomas; Polymer, 36(9), 1743-1759 (1995).
108. W. Salalha, Y. Dror, R.L. Khalfin, Y. Cohen, A.L. Yarin and E. Zussman; Langmuir, 20, 9852-9855 (2004).
109. F. Ko, Y. Gogotsi, A. Ali, N. Naguib, H.H. Ye, G.L. Yang, C.Li and P. Willis; Advanced Materials, 15(14), 1161-1165 (2003).
110. H. Ye, H. Lam, N. Titchenal, Y. Gogotsi and F. Ko; Appl.Phys.Lett., 85(10), 1775-1777 (2004).
111. H. Fong and D.H. Reneker; Journal of Polymer Science Part B-Polymer Physics, 37, 3488-3493 (1999).
112. J.-S. Kim and D.H. Reneker; Polymer Engineering and Science, 39(5), 849-854 (1999).
113. D.W.V. Krevelen; p. 221, Elsevier Scientific Publishing Company, Amsterdam, Oxford, New York (1970).
114. D. Campbell, R.A. Pethrick and J.R. White; in Polymer Characterization: Physical Techniques, p. 1-14, Stanley Thornes Publishers, United Kingdom (2000).
115. H. Hou, J.J. Ge, J. Zeng, G. Li, D.H. Reneker, A. Greiner and S.Z.D. Cheng; Chem.Mater., 17, 967-973 (2005).
116. B. Ding, E. Kimura, T. Sato, S. Fujita and S. Shiratori; Polymer, 45, 1895-1902 (2004).
117. S. Demoustier-Champagne and P.-Y. Stavaux; Chem.Mater., 11, 829-834 (1999).
118. S. Cuenot, S. Demoustier-Champagne and B. Nysten; Physical Review Letters, 85(8), 1690-1693 (2000).
119. W.H. Park, L. Jeong, D.I. Yoo and S. Hudson; Polymer, 45, 7151-7157 (2004).
120. B.D. Ratner, A. Chilkoti and D.G. Castner; Clin.Mater., 11, 25-36 (1992).
121. D. Campbell, R.A. Pethrick and J.R. White; in Polymer Characterization: Physical Techniques, p. 194-235, Stanley Thornes Publishers, United Kingdom (2000).
122. C.J. Buchko, L.C. Chen, Y. Shen and D.C. Martin; Polymer, 40, 7397-7407 (1999).
123. P.X. Ma and R. Zhang; J Biomed Mater Res., 46, 60-72 (1999).
124. B.G. Demczyk, Y.M. Wang, J. Cumings, M. Hetman, W. Han, A. Zettl and R.O. Ritchie; Materials Science and Engineering A-Structural Materials Properties Microstructure and Processing, 334 (1-2), 173-178 (2002).

125. P. Poncharal, Z.L. Wang, D. Ugarte and W.A.D. Heer; Science, 283, 1513-1516 (1999).
126. M.F. Yu, M.J. Dyer, G.D. Skidmore, H.W. Rohrs, X.K. Lu, K.D. Ausman, J.R.V. Ehr and R.S. Ruoff; Nanotechnology, 10, 244-252 (1999).
127. M.F. Yu, O. Lourie, M.J. Dyer, K. Moloni, T.F. Kelly and R.S. Ruoff; Science, 287, 637-640 (2000).
128. E.P.S. Tan, C.N. Goh, C.H. Sow and C.T. Lim; Appl.Phys.Lett., 86(6), 073115 (2005).
129. B. Cappella and G. Dietler; Surface Science Reports, 34(1-3), 1-104 (1999).
130. J.L. Duvail, P. Retho, C. Godon, C. Marhic, G. Louarn, O. Chauvet, S. Cuenot, N.L.D.D. Pra and S. Demoustier-Champagne; Synthetic Metals, 135(1-3), 329-330 (2003).
131. W. Xu, P.J. Mulhern, B.L. Blackford, M.H. Jericho and I. Templeton; Scanning Microscopy, 8(3), 499-506 (1994).
132. A.C. Ugural; in Mechanics of Materials, p. 152-213, McGraw-Hill, Inc. (1993).
133. M.R. VanLandingham, J.S. Villarrubia, W.F. Guthrie and G.F. Meyers; Macromolecular Symposia, 167, 15-43 (2001).
134. X. Li, H. Gao, C.J. Murphy and K.K. Caswell; Nano Letters, 3(11), 1495-1498 (2003).
135. W.C. Oliver and G.M. Pharr; Journal of Materials Research, 7(6), 1564-1583 (1992).
136. B.J. Briscoe, L. Fiori and E. Pelillo; Journal of Physics D-Applied Physics, 31(19), 2395-2405 (1998).
137. M.O.- Tiesma, Y.A. Toivola and R.F. Cook; Load-Displacement Behavior During Sharp Indentation of Viscous-Elastic-Plastic Materials. Proc. Mat.Res.Soc.Symp. 2001: Materials Research Society.
138. J. Domke and M. Radmacher; Langmuir, 14(12), 3320-3325 (1998).
139. K.L. Johnson; in Contact Mechanics, p. 397-423, Cambridge University Press, Cambridge (1992).
140. M.S. Bischel, M.R. Vanlandingham, R.F. Eduljee, J.W. Gillespie and J.M. Schultz; Journal of Materials Science, 35(1), 221-228 (2000).
141. T. Sumomogi, K. Hieda, T. Endo and K. Kuwahara; Applied Physics A Materials Science & Processing, 66, 299-303 (1998).
142. J.G. Park, S.H. Lee, B. Kim and Y.W. Park; Applied Physics Letters, 81(24), 4625-4627 (2002).
143. J.J. Mack, L.M. Viculis, A. Ali, R. Luoh, G. Yang, H.T. Hahn, F.K. Ko and R.B. Kaner; Advanced Materials, 17(1), 77-80 (2005).
144. E.P.S. Tan and C.T. Lim; Appl.Phys.Lett., 87(12), 123106 (2005).
145. K.L. Johnson; in Contact Mechanics, p. 84-106, Cambridge University Press, Cambridge (1992).
146. S. Cuenot, C. Fretigny, S. Demoustier-Champagne and B. Nysten; Journal of Applied Physics, 93(9), 5650-5655 (2003).
147. S. Cuenot, C. Fretigny, S.D.- Champagne and B. Nysten; Phys.Rev. B. 69, 165410 (2004).
148. S. Boussaad and N.J. Tao; Nano Letters, 3(8), 1173-1176 (2003).
149. W.-J. Li, C.T. Laurencin, E.J. Caterson, R.S. Tuan and F.K. Ko; J. Biomed.Mater. Res., 60, 613-621 (2002).
150. J. Guan, K.L. Fujimoto, M.S. Sacks and W.R. Wagner; Biomaterials, 26, 3961-3971 (2005).
151. Y. Zhang and M.Q. Zhang; Journal of Non-Crystalline Solids, 282(2-3), 159-164 (2001).
152. R. Zhang and P.X. Ma; Journal of Biomedical Materials Research, 52, 430-438 (2000).
153. M.-S. Khil, D.-I. Cha, H.-Y. Kim, I.-S. Kim and N. Bhattarai; J Biomed Mater Res Part B: Appl Biomater., 67B, 675-679 (2003).
154. P. Gibson, H. Schreuder-Gibson and D. Rivin; Colloids and Surfaces A: Physicochemical and Engineering Aspects, 187-188, 469-481 (2001).
155. Z.-M. Huang, Y.Z. Zhang, S. Ramakrishna and C.T. Lim; Polymer, 45(15), 5361-5368 (2004).
156. J.A. Matthews, G.E. Wnek, D.G. Simpson and G.L. Bowlin; Biomacromolecules, 3, 232-238 (2003).
157. J.H. DeGroot, H.W. Kuijper and A.J. Pennings; Journal of Materials Science-Materials in Medicine, 8(11), 707-712 (1997).
158. Z. Liu, C.S.L. Chuah and M.G. Scanlon; Acta Materialia, 51, 365-371 (2003).
159. L.J. Gibson and M.F. Ashby; in Cellular Solids: Structure and Properties, p. 175-234, Cambridge University Press (1997).
160. E.P.S. Tan and C.T. Lim; J. Biomed. Mater. Res. A, 77A, 526 (2006).
161. L.J. Gibson and M.F. Ashby; in Cellular Solids: Structure and Properties, p. 429-452, Cambridge University Press (1997).

162. R.L. Satcher and C.F. Dewey; Biophysical Journal, 71, 109-118 (1996).

163. X.E. Guo and L.J. Gibson; International Journal of Mechanical Sciences, 41, 85-105 (1999).

164. H.X. Zhu, J.F. Knott and N.J. Mills; Journal of the Mechanics and Physics of Solids, 45(3), 319-343 (1997).

165. L.J. Gibson and M.F. Ashby; in Cellular Solids: Structure and Properties, p. 15-51, Cambridge University Press (1997).

166. K. Kim, M. Yu, X. Zong, J. Chiu, D. Fang, Y.S. Seo, B.S. Hsiao, B. Chu and M. Hadjiargyrou; Biomaterials, 24, 4977-4985 (2003).

167. Y. You, B.-M. Min, S.J. Lee, T.S. Lee and W.H. Park; Journal of Applied Polymer Science, 95, 193-200 (2005).

168. C.Y. Xu, R. Inai, M. Kotaki and S. Ramakrishna; Biomaterials, 25, 877-886 (2004).

169. M. Shin, O. Ishii, T. Sueda and J.P. Vacanti; Biomaterials, 25(17), 3717-3723 (2004).

170. A.P. Roberts and E.J. Garboczi; Journal of the Mechanics and Physics of Solids, 50(1), 33-55 (2002).

171. D. Li and Y. Xia; Advanced Materials, 16(14), 1151-1170 (2004).

172. X. Yan, G. Liu, F. Liu, B. Tang, H. Peng, A. Pakhomov and C. Wong; Angew.Chem.Int.Ed., 40(19), 3593-3596 (2001).

173. L. Larrondo and R.S. John Manley; Journal of Polymer Science: Polymer Physics Edition, 19, 909-929 (1981).

174. M.M. Demir, I. Yilgor, E. Yilgor and B. Erman; Polymer, 43(11), 3303-3309 (2002).

175. S.-G. Lee, S.-S. Choi and C.-W. Joo; Journal of the Korean Fiber Society, 39(1), 1-13 (2002).

176. P.K. Baumgarten; Journal of Colloid and Interface Science, 36(1), 71-79 (1971).

177. J.M. Deitzel, J. Kleinmeyer, D. Harris and N.C. Beck Tan; Polymer, 42(1), 261-272 (2001).

178. X. Zong, K. Kwangsok, F. Dufei, R. Shaofeng, B.S. Hsiao and B. Chu; Polymer, 43(16), 4403-4412 (2002).

179. H. Fong, I. Chun and D.H. Reneker; Polymer, 40(16), 4585-4592 (1999).

180. C.J. Buchko, L.C. Chen, Y. Shen and D.C. Martin; Polymer, 40(26), 7397-7407 (1999).

181. K.H. Lee, H.Y. Kim, Y.M. La, D.R. Lee and N.H. Sung; Journal of Polymer science: Part B: Polymer Physics, 40, 2259-2268 (2002).

182. H. Liu and Y.-L. Hsieh; Journal of Polymer Science: Part B: Polymer Physics, 40, 2119-2129 (2002).

183. Y.-S. Kang, H.-Y. Kim, Y.-J. Ryu, D.-R. Lee and S.-J. Park; Polymer (Korea), 26(3), 360-366 (2002).

Magnetic Polymer Microspheres

Junhua Zhang[1] and Xiaobin Ding[2*]

[1]*State Key Laboratory of Polymer Material and Engineering, Sichuan University, Chengdu, 610065, P.R. China.*
[2]*Chengdu Institute of Organic Chemistry, Chinese Academy of Sciences, Chengdu, 610041, P.R. China.*
[*]*Corresponding author's email: xbding@cioc.ac.cn*

Contents

- **INTRODUCTION**
- **STRUCTURE OF POLYMERIC MICROSPHERES**
- **SELECTION OF MAGNETIC MATERIALS AND POLYMER MATRICES**
- **PROPERTIES OF MAGNETIC POLYMER MICROSPHERES**
- **PREPARATION OF MAGNETITE**
- **PREPARATION OF MAGNETIC POLYMER MICROSPHERES**
- **FUNCTIONALIZATION OF MAGNETIC POLYMER MICROSPHERES**
- **APPLICATIONS OF MAGNETIC POLYMER MICROSPHERES**
- **CONCLUSIONS AND FUTURE PROSPECTS**
- **ACKNOWLEDGEMENTS**
- **REFERENCES**

Summary: Magnetic polymer microspheres have two important properties viz. they can be rapidly and easily separated by application of magnetic field and their surface can be modified to different functional groups. Thus, they are finding applications in many areas such as in the biomedical field, in organic synthesis, in wastewater treatment and so on. In this chapter, we have summarized the structure, the preparation methods and the applications of magnetic polymer microspheres.

1. INTRODUCTION

What are magnetic polymer microspheres? Magnetic polymer microspheres are the kind of nano- or microparticles produced by combining inorganic magnetic materials (e.g., Fe_3O_4, Fe_2O_3, nickel and cobalt etc.) and polymer together through physical or chemical methods. As a new kind of functional materials, magnetic microspheres have been widely investigated since 1970 and have been found suitable materials for many applications such as, industrial wastewater treatment, clinical detection, targeted drug release, bacterial separation, organic synthesis and so on. The attributes of magnetic microspheres are the magnetic response and magnetic separation properties. The surface of magnetic microspheres can be functionalized with different functional groups (such as hydroxyl, carboxyl, amino etc.) to meet the requirements of different applications.

2. STRUCTURE OF MAGNETIC POLYMER MICROSPHERES

Generally, magnetic polymer microspheres have three kinds of typical structures as shown in **Fig. 4.1**, **A**: the core is of polymer and the shell is of magnetic material [1-6], **B**: this type has a three layered structure, the inner and the outer layers are of polymer and the middle layer is of magnetic material [7] and **C**: the core is of magnetic material and the shell is of polymer [8-16]. **A** type has a disadvantage that is the magnetite is easier to lose leading to the deterioration of magnetic property of the microspheres during use. **B** and **C** types are better because the magnets are incorporated well inside the microspheres, however, the preparation of **B** type is more complicated than that of **C** type. **C** type of magnetic polymer microspheres is well investigated because of their potential applications in many areas.

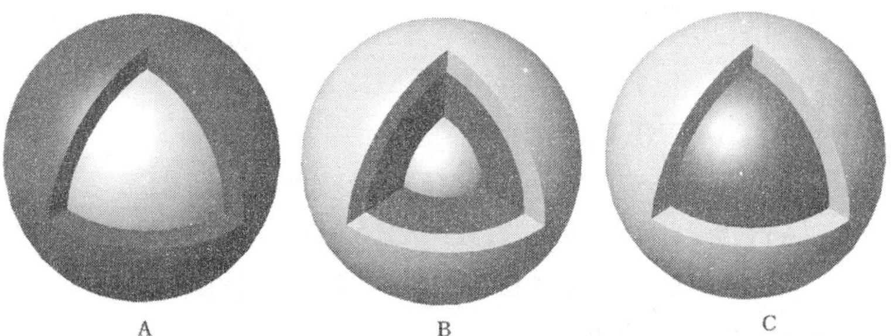

Fig. 4.1. Magnetic polymer microspheres. A: the core is of polymer and the shell is of magnetic material, B: the inner and the outer layers are of polymer and the middle layer is of magnetic material and C: the core is of magnetic material and the shell is of polymer.

3. SELECTION OF MAGNETIC MATERIALS AND POLYMER MATRIX

The magnetic materials used in magnetic polymer microspheres are usually iron, iron oxide, iron and cobalt alloy, nickel and cobalt etc. [17]. However, nickel and cobalt magnetic particles are dangerous for human body, so people seldom use them [18]. The polymer matrix can be made from several types of polymers, including natural and synthetic polymers. They have their advantages and disadvantages; hence, one must select the suitable polymer matrix according to the requirements of applications.

Magnetic polystyrene particles are known for their excellent size distribution and spherical shape but their surface is very hydrophobic and results in a high amount of unspecific binding of proteins on the particle surface. Magnetic silica particles efficiently absorb proteins, especially, DNA on their surface. They are highly stable, even in organic solvents. But up till now, the magnetic silica particles of narrow size distribution and ideal spherical shape are not available.

Magnetic polysaccharide particles are important for many *in vivo* applications, including magnetic resonance imaging, magnetic field assisted radionuclide therapy or hyperthermia. They have the advantages of biocompatibility and availability in a size range below 300 nm. However, their particle shape is irregular and particle matrix is soft that cause them to be sensitive to mechanical stress.

Magnetic poly(lactic acid) particles play an important role in magnetic drug targeting and controlled drug delivery and in the magnetic field assisted radionuclide therapy. They are biodegradable and their degradation time in the blood can be adjusted by controlling the molecular weight and composition of the basic poly(lactic-co-glycolic acid) or poly(lactic acid) matrix. Because of the hydrophobic surface, poly(lactic acid) particles stick to the plastic surfaces such as pipette tips, cuvettes or Eppendorf tubes. This results into the difficulties in particle handling as well as analytical errors [19-22]. PMMA is regarded as a suitable matrix for immobilization of enzyme due to its biocompatibility and adequate mechanical strength required for most of the biomedical and biotechnological applications [23].

Poly(vinylalcohol) (PVAL) is a polymer that is frequently used as matrix for the immobilization of various enzymes and cells because of its ease of availability, its low price, its hydrophilic character and its hydroxyl groups capable of chemical reaction. In addition, its chemical structure can cause protein stabilization by attachment to the polymer chain or due to its resemblance with polysaccharides, known stabilizing agents for proteins [24].

Chitosan can be used as a base material for magnetic carriers. It is a polyaminosaccharide with many significant biological (biodegradable, biocompatible and bioactive) and chemical properties (polycationic, hydrogel and contains reactive OH and NH_2 groups). Hence, it is used in many types of applications such as wastewater treatment, chromatographic support, enzyme immobilization and drug-delivery systems. Chitosan is a very cheap biopolymer because it can be produced by alkaline deacetylation of chitin, which is the second most abundant natural polymer after cellulose [25].

4. PROPERTIES OF MAGNETIC POLYMER MICORSPHERES

The magnetic polymer microspheres consist of two main components, the magnetic materials and the polymer matrix. The magnetic component gives magnetic responsiveness to the magnetic polymer microspheres i.e. the microspheres can move toward some direction under an additional magnetic field. The polymer component, on the other hand, gives the choice of selection among the wide variety of polymers with desired functional groups according to the requirements of different application areas. In the following lines, we will discuss the concept and characterization of magnetic properties and the surface modification of the microspheres.

Hirschein [26] proposed the relationship between the force F and the magnetic polymer microspheres as given by **eq. 4.1**.

$$F = (X_v - X'_v) \, V \, H \, (dH/dX) \qquad\qquad (4.1)$$

where F is the force of magnetic polymer microspheres, dH/dX is the strength of magnetic field, X_v is the susceptibility of magnetic polymer microspheres, X'_v is the susceptibility of the medium and V is the volume of the magnetic polymer microspheres. According to the **eq. 4.1**, F increases with increase in V and H. Bigger microspheres (diameter >10 µm) can be separated under weaker magnetic field but they tend to sediment. However, smaller micropsheres (diameter <0.03 μm) can be dispersed in a solvent without obvious sedimentation, but it requires higher magnetic strength for separation. In addition, F has relationship with X_v, which is determined by the materials of magnetite and the diameter of magnetite. For example, if the diameter of Fe_3O_4 < 300 Å, it shows superparamagnetism otherwise ferromagnetism.

5. PREPARATION OF MAGNETITE

Magnetic polymer microspheres can be prepared by physical or chemical methods. Different methods lead to different size, size distribution and magnetic material content of the microspheres, which strongly influence the magnetic properties of the microspheres [27,28]. As the magnetic polymer microspheres consist of two main parts, one is the magnetite and the other is polymer matrix. To prepare magnetite, usually Fe_3O_4 is the commonly used material because: (i) it is safe for human, (ii) the diameter of Fe_3O_4 magnetite is always less than 30 nm and (iii) the microspheres have superparamagnetic properties after being combined with polymer matrix i.e. the magnetic properties of microspheres disappear soon after the external magnetic filed disappears and remagnetization is not possible. Usually there are two main methods to prepare magnetite [29].

5.1. Coprecipitation

This method involves the precipitation of bivalent and trivalent metal ions by using some base such as NaOH and $NH_3 \cdot H_2O$ in suitable concentration. The metal ions may be of same metal or different metals.

$$Fe^{2+} + 2\ Fe^{3+} + 8\ OH^- \rightarrow Fe_3O_4 + 4\ H_2O \tag{4.2}$$

For example, 3.9 g of $FeCl_2$ (19 mmol) and 10.9 g of $FeCl_3$ (39 mmol) ($Fe^{3+}/Fe^{2+} = 2$) were mixed with 435 ml of water and precipitated by using 45 ml of ammonia solution (28%) at 80°C. The resulting black dispersion of magnetite was separated and oxidized with 75 ml of HNO_3 (2M)

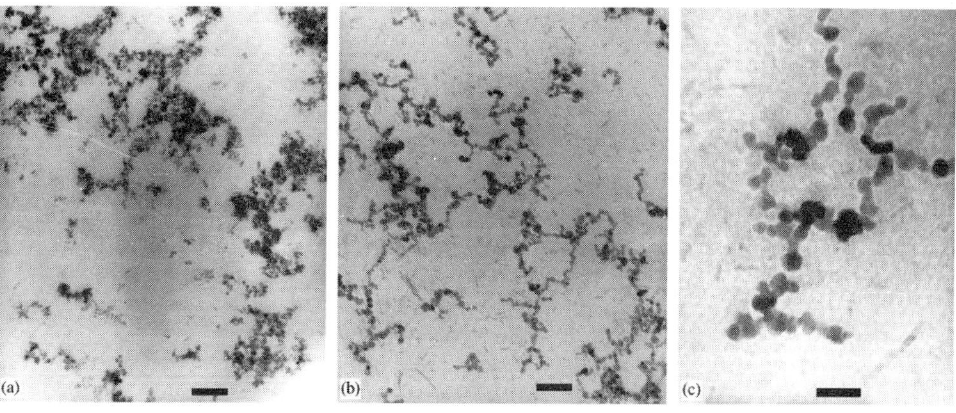

Fig. 4.2. TEM images of (a) Fe-dextran preparation: scale bar = 50 nm. (b) Fe-PVA preparation: scale bar = 50 nm, and (c) Fe-PVA preparation: scale bar = 20 nm. Reproduced with permission from H. Pardoe *et al.*; J.Magnetism and Magnetic Mater., 225, 41 (2001). Copyright © 2001, Elsevier Science Ltd.

for 10 min and then with 75 ml of a $Fe(NO_3)_3$ (0.33 M) stock solution at 100°C for 30 min. The brown dispersion was centrifuged, washed three times with acetone and peptized with 50 ml of HNO_3 (2M) for 15 min. The resulting red-brown paste was then dispersed into 60 ml of water and the traces of acetone were removed *in vacuo* at 60°C. Finally, the total volume was brought to 60 ml by adding water. The average particle size was found to be in the range of 10 ± 5 nm as determined by transmission electron microscopy (TEM) whereas dynamic light scattering indicated an average diameter of 50 ± 5 nm suggesting the slight aggregation of oxide particles [30,31].

Pardoe *et al.* [32] also reported a similar procedure with slight modification. They prepared magnetite nanoparticles in the presence of either dextran or polyvinyl alcohol yielding cluster- and necklace-like aggregates as shown in **Fig. 4.2**. Kim *et al.* [33] reported the synthesis and surfactant-coated superparamagnetic monodispersed iron oxide nanoparticles and investigated the influence of molar concentration of NaOH on particle size **(Tab. 4.1)**.

Tab. 4.1. The influence of NaOH on the paticles. Reproduced with permission from D.K. Kim *et al.*, J. Magnetism and Magnetic Mater., 225, 30 (2001). Copyright © 2001, Elsevier Science Ltd.

Sample	*pH*	*NaOH (M)*	D_{XRD} *(Å)*
S1	14	0.9	13
S2	14	1.0	17
S3	14	1.1	29
S4	14	1.5	30
S5	12.5	1.5	55
S6	11.54	1.5	60

Cobalt ferrite based ionic ferrofluids (FFs) can also be synthesized by co-precipitation method [34]. Stable cobalt ferrite based ($CoFe_2O_4$) ionic FFs were obtained after performing by a three-step procedure. Cobalt ferrite nanoparticles were synthesized by coprecipitating Co (II) and Fe (III) in alkaline medium. The starting solution, containing 6 ml of concentrated nitric acid, 250 ml of aqueous solution of cobalt nitrate (0.005 M) and 250 ml of aqueous solution of iron chloride (0.010 M), was heated to 95°C under vigorous stirring. Aqueous solution of sodium hydroxide (1 M) was then quickly added to the stirring solution until pH 12 was achieved. About 120 ml of the sodium hydroxide solution was required to bring the pH to about 12. Three different samples were prepared using different stirring conditions at 2700, 5400, and 8100 rpm. After removing the aqueous solution, nitric acid (2 M) was added to the precipitate until pH 3 was reached. The precipitate was left overnight on a magnet. In the second step, the passivation of the nanoparticle surface was achieved by Fe (III) enrichment by treatment with ferric nitrate. Samples were treated with 10 ml of hot ferric nitrate (0.5 M) under stirring for about 3 min. In the third step, the peptization of the nanoparticles was obtained by controlling the surface charge density with reducing the ionic strength of the aqueous medium. An aqueous triethylamine (0.1 M) solution was added drop wise to the acid precipitate until pH 6 was reached. The precipitate was then washed through several cycles with acetone and deionized water to reduce the ionic strength of the medium. Finally, perchloric acid was added to the precipitate until pH 2 was reached at which point peptization occurred.

The control on the size of nanoparticles was observed in the first step of synthesis by changing the stirring speed. It was found that higher the stirring speed, smaller was the average particle diameter. However, the standard deviation associated with the lognormal particle-size distribution did not change when the nanoparticle size increased by about 50% as discussed below. The understanding of the nanoparticle growth mechanism through chemical condensation reaction in solution and under vigorous stirring, has not been the focus of these investigations, despite the huge impact it may have from fundamental and applied viewpoints.

In addition, Santra *et al.* [35] proposed the preparation of silica-coated iron oxide nanoparticles by coprecipitation reaction of ferrous and ferric salts with inorganic bases in the presence of three different nonionic surfactants. The effect of the surfactants on the particle size, crystallinity, and magnetic properties was investigated.

5.2. Redox Precipitation

This method uses the redox reagent to oxidize or deoxidize metal ions to metal or metal oxide. Metal ions can be redoxed in solution, in polymer matrix porous structure or on the surface of polymer matrix as shown in **eq. 4.3** and **eq. 4.4**. The type, size and shape of the particles influence the magnetic response property. 70 g of polyethylene glycol (PEG, M_w = 4000) was dissolved in 80 ml

$$Fe^{2+} + 2\ OH^- \rightarrow Fe(OH)_2 \tag{4.3}$$

$$Fe(OH)_2 + O \rightarrow Fe_3O_4 + 3\ H_2O \tag{4.4}$$

of distilled water in a 250 ml round flask equipped with a condenser, a nitrogen inlet and a stirrer. 60 ml of ferrous sulphate (0.18 M) was added to the mixture. Nitrogen was bubbled into the solution for 30 min under continuous stirring and 20 ml of aqueous solution containing 0.6 ml of 30% H_2O_2 was added to the system. Reaction was allowed to proceed for 6 hours at 50°C at pH 13. The products were washed and separated by magnetic field several times to neutral and redispersed in distilled water [36].

Grasset *et al.* [37] reported that Yttrium aluminum iron garnet $Y_3Fe_{5-x}Al_xO_{12}$ compounds were produced by the citrate gel process by varying the aluminum content x from 0 to 2. Particles with an average size of about 100 nm and a specific surface area ranging from 6 to 28 m^2/g were obtained. Polycrystalline samples of the general formula $Y_3Fe_{5-x}Al_xO_{12}$ with $0 \leq x \leq 2$ were initially prepared from a solution of $Fe(NO_3)_3 \cdot 9H_2O$, $Al(NO_3)_3 \cdot 9H_2O$, $Y(NO_3)_3$ and citric acid in appropriate stoichiometric ratios at 65°C for 3 hours until gelation. Gels were dried at 110°C for an hour, cooled, and then heated up to 800°C or 1150°C in air at a heating rate of 5°C/min and heat treated at this temperature for 2 hours.

6. PREPARATION OF MAGNETIC POLYMER MICROSPHERES

6.1. Solvent Evaporation Method

Arica *et al.* [38] reported the preparation of magnetic poly(methylmethacrylate) microspheres (MPMMA) with a 6-carbon spacer-arm by solvent evaporation method. Initially, poly(methylmethacrylate) (PMMA) microspheres were synthesized by suspension polymerization. PMMA microspheres dispersion was stabilized in chloroform and then suitable magnetite was added with a strong stirring. The magnetite particles of diameter range from 100 to 200 μm, completely encapsulated in the PMMA microspheres, were produced after the solvent evaporation.

6.2. Sonochemical Method

Avivi *et al.* [39] reported a sonochemical method for preparing magnetic proteinaceous microspheres. Using high-ultrasound, the magnetic microspheres composed of iron oxide filled and coated globular bovine serum albumin (BSA) were prepared. The average diameter of microspheres was around 4.2 μm with a wide size distribution ranging from 2 to 8 μm and the magnetite content was in the range of 39-42%.

In this process decane and iron pentacarbonyl $Fe(CO)_5$ (7.43×10^{-4} M) were layered over a 5% w/v protein solution. The bottom of the high intensity ultrasonic horn was positioned at the water-organic interface. The mixture was irradiated for 3 min employing and acoustic power of ≈ 150 Wcm^{-2} at an initial temperature of 23°C in the reaction cell. The cell pH was adjusted to 7.0 by adding HCl. This procedure was performed again with a 7.66×10^{-3} M aqueous solution of iron acetate, $Fe(CH_3CO_2)_2$.

After the synthesis, the products were separated from the unreacted protein and from the residues of iron acetate or iron pentacarbonyl by centrifugation (1000 rpm for 5 min). The magnetic micropsheres were washed a few times with sufficient volumes of water to remove the residues of the precursors. The washings were continued until the solution contained no protein and no acetate ions. Residues of unreacted iron pentacarbonyl were removed by washing with pentane.

6.3. Double Emulsion Technique

Gómez-Lopera *et al.* [21] proposed the double emulsion method for preparing colloidal particles formed by a magnetite nucleus and a biodegradable poly(DL-lactide) polymer coating. The technique involves the formation of a w/o emulsion of an aqueous solution of, a pharmaceutical drug in most instances, in an organic solution of the polymer. This emulsion is used to prepare a new emulsion in an aqueous surfactant solution so as to finally obtain a w/o/w double emulsion. A 5% w/v solution of poly(DL-lactide) in dichloromethane was first prepared. 10 ml of 20% w/v aqueous suspension of magnetite particles was added to 40 ml of poly(DL-lactide) solution, the mixture was homogenized at 14,000 rpm for 5-10 s. A whitish, slightly transparent emulsion was thus obtained. This emulsion was added to 200 ml of a 1% w/v poly(vinyl alcohol) (PVAL) solution and homogenized as above to obtain the second emulsion of white color. Organic solvent was evaporated by stirring the solution with a magnetic stirrer set at 900 rpm at room temperature. The polymer suspension was then subjected to a cleaning procedure that included repeated cycles of centrifugation (12,000 rpm) and redispersed in Milli-Q water and finally serum replacement in a stirred filtration cell with a membrane of 0.2 μm pore size. In order to ensure that the suspension was sufficiently clean, both the conductivity and surface tension of the supernatant liquid were measured. The polymer particles were dried at room temperature and stored in dry polyethylene containers.

6.4. Dispersion Polymerization

Dispersion polymerization has been widely adapted to prepare magnetic polymer microspheres because one can get micro-sized particles by this method. [40-44]. In presence of magnetite, monomers are dispersed in a mix solvent (methanol/water [40-43,45] or toluene/2-methylpropan-1-ol [44]). Using dispersion polymerization, one can get microspheres of diameter ranging from 1 to 100 μm with magnetite contents <10% and the size distributions are always polydisperse. In this process, polyglycidylmethacrylate microspheres were prepared by dispersion polymerization of glycidylmethacrylate (GMA) using AIBN as an initiator in an alcohol/water medium in presence of iron oxide needles. Polymerization was carried out in a 100 ml reaction vessel equipped with an anchor stirrer. 2 g of iron oxide was added to 1.8 g of poly(N-vinylpyrrolidone) dissolved in 68 g of aqueous alcohol and the mixture was sonicated at 5°C for 6 min at 50% output. A solution of 0.24 g of AIBN in 12 g GMA was poured into the reaction vessel that was purged with nitrogen for 10 min. However, to obtain crosslinked microspheres, 0.96 g of ethylene dimethacrylate must be added 2 hours after the start of the polymerization. The reaction was allowed to proceed for 20 hours at 70°C (65°C when methanol was used as a solvent) under continuous stirring at 500 rpm. The resulting particles were purified by repeated magnetic separation from methanol and decantation and were finally dried [45].

6.5. Emulsion Polymerization

Emulsion polymerization [46,47], including emulsifier free emulsion polymerization [48], inverse microemulsion polymerization [49] and miniemulsion polymerization [50], are also widely used to get magnetic polymer microspheres. Generally, small sized particles with a narrow size distribution can be produced by this method. This is important for medical applications. The size distributions of these particles range from several to 100 of nanometers.

6.5.1. Preparation of Magnetic Polystyrene Latex

Emulsion polymerization of styrene (St) was carried out in the presence of a ferrofluid, using ammonium persulfate (APS) or potassium persulfate (KPS) as an initiator. Required amounts of St, ferrofluid, and additives were placed in a 120 ml glass pressure bottle. The total volume of the reaction mixture was adjusted to 60 ml by addition of water. After having been gently bubbled with a fine stream of nitrogen for about 5 min, the mixture was stirred for 1 hour with constant horizontal and elliptic shaking using a Taiyo incubator. The mixture was then heated to 70°C for 20 hours with continuous shaking. The coagulum was removed by filtration and dried at 80°C [46].

Miniemulsion polymerization is a recent and new type of polymerization process. In miniemulsion polymerization, both the particle nucleation and the subsequent propagation reactions occur primarily in submicrometer monomer droplets. Each droplet can be regarded as an individual reactor. Miniemulsion is typically obtained by shearing a system containing monomer, water, surfactant, costabilizer and initiator. The droplet diameter could be controlled by the type and amount of surfactant and costabilizer used, the volume fraction of the disperse phase and the homogenization process [50]. For example, the costabilizer (SA, stearyl alcohol) and the oil-soluble initiator (AIBN) were dissolved in the styrene monomer. The magnetite particles (Fe_3O_4, 57 wt %, based on the toluene phase) were then directly dispersed in the styrene. The resultant stable oil based dispersion was added to a solution of surfactant (SDS, sodium lauryl sulfate) in water. The mixture was stirred for 1 hour, followed by ultrasonication for 120 min to obtain a stable miniemulsion. During sonication, low temperature was maintained to avoid any undesirable polymerization. The recipe used to prepare the water based Fe_3O_4:styrene miniemulsion is shown in **Tab. 4.2**. To yield magnetic polymer latex, the above miniemulsion was polymerized at 65°C for 12 hours under mechanical stirring at 500 rpm. The latex was subsequently subjected to water vapor distillation to remove any unreacted monomer.

Tab. 4.2. The recipe used to prepare the water-based Fe_3O_4/styrene miniemulsion. Reproduced with permission from I. Csetneki *et al.*; J. Polym.Sci., Part A, Polym.Chem., 4802, 19 (2004). Copyright © 2004, John Wiley & Sons, Inc.

Component	Amount (g)
Styrene	18.2
Ferrofluid	0.735-7.35
AIBN	0.52
SA	0.72
SDS	0.14
Distilled water	120.0

6.5.2. Inverse Microemulsion Polymerization

Iron oxide dispersion was prepared using the method based on the coprecipitation of $FeCl_2$ and $FeCl_3$. The monomer (Acrylamide, Am) and cross-linker (N,N'-methylene-bis-acrylamide, MBA) were added to a suspension of iron oxide nanoparticles (6 wt%) in the desired amount as shown in **Tab. 4.3**. The aqueous phase is dispersed in an aerosol OT (AOT) toluene solution to form a w/o microemulsion under nitrogen. The microemulsion was then purged with nitrogen for 30 min. The polymerization of monomers was carried out with AIBN or 2,2'-azobis(2-amidinopropane) dihydrochloride (V_{50}) at 60°C. After polymerization, the magnetic polymeric particles were dispersed in water by two methods viz. (i) precipitation-redispersion in which the particles were recovered by precipitation with an excess of methanol and thoroughly washed and vacuum dried before redispersion in water using ultrasonic bath and (ii) phase inversion in which the polymerized w/o dispersion was inverted to o/w structure by adding an excess of water. The solvent, toluene, and the emulsifier, AOT, were removed by magnetic separation and washing [51].

Tab. 4.3. Influence of AOT/H_2O ratio, MBA amount and different treatment methods on the size of magnetic polymeric latex and particle at 25°C[a]. Reproduced with permission from Y. Deng *et al.*; J. Magnetism and Magnetic Mater., 257, 69 (2003). Copyright © 2002, Elsevier Science Ltd.

Run	AOT/H_2O[b]	MBA (10^{-6} mol)	D_H^c (nm)	D_H^d (nm)	D_H^e (nm)
1201A	0.25	12.72	74	170	144
1106A	0.25	25.97	73	145	124
1222A	0.25	38.69	71	216	106
1222B	0.25	51.41	75	134	114
1222C	0.25	64.66	75	224	86
1110B	0.21	25.97	83	223	166
1109B	0.16	25.97	91	192	174
1110A	0.12	25.97	126	285	186
1109C[f]	0.09	25.97	—	—	—
1128A[g]	0.25	12.72	76	—	260
1201A	025	12.72	74	170	1454

[a]Am: 1.41 mmol, toluene: 81.40 mmol, Fe_3O_4 dispersion 0.53 g (6% wt), [b]Molar ratio of AOT/H_2O, [c]Magnetic polymeric latex without further treatment, [d]Precipitation and redispersion method, [e]Phase inversion method, [f]Unstable dispersion >10 hours and [g]Sample with V_{50} as initiator.

6.6. Suspension Polymerization

To prepare particles of larger size, especially of several hundred micrometers, suspension polymerization is usually adopted [52-54]. However, it is difficult to get monodispersed diameter distribution by this method.

6.6.1. Preparation of Polystyrene Particles Covered with Magnetite

Polystyrene particles coated with $SrO\cdot Fe_2O_3$, γ-Fe_2O_3 or RuO_2 can be synthesized by suspension polymerization. For example, 2 g of γ-Fe_2O_3 was dispersed in 200 ml of ion exchange water. Thus obtained suspension was mixed with 30 g of styrene containing 0.7 g of lauroyl peroxide as an initiator. The ternary complex disperse system was then mixed vigorously in a tall beaker of 300 ml capacity with a homogenizer at 5000 rpm for 15 min. A final reaction mixture was obtained by subsequently adding 180 ml water. The complex disperse system thus obtained consists of small monomer droplets surrounded by small inorganic particles at the monomer/aqueous phase boundary. Polymerization was carried out by putting the reaction mixture into a 500 ml four-neck separable flask preheated at 80°C and stirred at 750 rpm for 8 hours. The polymerized product was obtained by filtering and drying under a reduced pressure at 25°C for 10 hours [4,5].

6.6.2 Preparation of the Magnetic Polystyrene Microspheres by Suspension Polymerization

20 g of magnetite (Fe_3O_4) was soaked with 260 ml of oleic acid in 500 ml of distilled water and the solution was shaken until all of the magnetite transferred to the oleic-acid phase. The oleic-acid-grafted magnetite aggregated and separated in the distilled water after 10 min in the grafting reaction. The two-phase solutions composed of distilled water, oleic acid and the oleic-acid-grafted magnetite were repeatedly washed with ethanol to remove the unreacted oleic acid and were then vacuumed at 80°C. Styrene (4.26 wt %), DVB (0.54 wt %) and AIBN (0.09 wt %) were mixed and decanted into a three-necked flask, in which the magnetite or the oleic-acid-grafted magnetite (0.48 wt %) was contained. The mixture was stirred for 1 hour at room temperature under N_2 atmosphere. Distilled water (94.54 wt %), in which polyvinylalcohol (0.09 wt %) was dissolved, was carefully added into the reactor with continuous stirring. The suspension polymerization for the magnetic polymer microspheres was performed for 16 hours at 70°C under N_2 atmosphere. After the completion of polymerization, the product was washed with excess of distilled water and sieved with a 60 μm strainer to remove the remaining polyvinylalcohol and nonencapsulated magnetite. Pure product was obtained by drying under vacuum at 60°C [55].

6.6.3. Preparation of Magnetic Poly(vinyl alcohol) (PVAL) Microspheres

Magnetic polyvinyl alcohol microspheres were prepared by applying a suspension cross-linking procedure. 4.3% (w/w) of Fe_3O_4 and 16.6% (v/v) of HCl (3 M) were added to a 5% (w/v) aqueous solution of PVAL (M_n 224000). The mixture was sonicated for 1 min in an ultrasonic bath at 20°C and then added to a three-fold volume (relative to PVAL phase) of a vegetable oil of viscosity of 130-190 mPa s. Suspension cross-linking was carried out at 700 rpm. Immediately after starting the suspension process, 4.7% (v/v relative to the PVAL phase) of a 25% glutardialdehyde solution was added. Cross-linking proceeded for a further 1 min. The microspheres were isolated from the oil phase by centrifugation for 1 min at 9000 rpm. The beads were then successively washed with n-hexane and 2-butanone. Intensive washing with several portions of water then followed for several hours [56].

6.7. Anionic Polymerization

Arias *et al.* [57] described anionic polymerization method to prepare core-shell magnetic colloidal nanoparticles. The core is of Fe_3O_4 magnetite and the shell is of a biodegradable polymer (poly(ethyl-2-cyanoacrylate)). In this method, the monomer (1%, w/v) was added dropwise to 50 ml of an 0.75% (w/v) magnetite aqueous suspension polymerization medium containing HCl (2×10^{-3} N) under continuous mechanical stirring at 1000 rpm. After 3 hours, neutralization of the medium was achieved by adding 1 ml of a KOH (10^{-1} M) solution and to end the polymerization. A whitish suspension was obtained which was then subjected to a cleaning procedure that included repeated cycles of centrifugation and redispersion in Milli-Q water and finally serum replacement in a stirred filtration cell with a membrane of pore size of 16 mm. To ensure that the suspension was sufficiently clean, the conductivity of the supernatant liquid was measured. The polymer particles were dried at 35°C in a vacuum oven and stored in dry polyethylene containers **(Tab. 4.4)**.

Tab. 4.4. Relative amount of polymer bound to magnetite (mass of polymer per unit mass of magnetite), coating thickness and redispersibility characteristics of the composite particles obtained for different relative initial amounts of monomer and magnetite. Reproduced with permission from J.L. Arias *et al.*; J. Controlled Release, 77, 309 (2001). Copyright © 2001, Elsevier Science Ltd.

Initial Monomer/ Magnetite Weight Ratio	Relative Weight of Coating Polymer	Coating Thickness (nm)	Redispersibility
1:4	0.028860 ± 0008	—	Very good
2:4	0.18360 ± 001	40 ± 20	Very good
3:4	0.19360 ± 001	50 ± 20	Very good
1:1	0.123660 ± 0009	90 ± 20	Very good
4:3	0.60260 ± 001	46 ± 20	Fair
4:2	1.41860 ± 003	62 ± 20	Difficult
4:1	1.48760 ± 005		None

7. FUNCTIONALIZATION OF MAGNETIC POLYMER MICROSPHERES

7.1. Surface Modifications

Different applications demand different surface characteristics of magnetic microspheres i.e. different functional groups on the surface. The simplest way to functionalize is to copolymerize with monomers, which have functional groups such as OH [58], NH_2 and COOH [59], CHO [60,61] etc. However, the functional groups on the surface prepared by copolymerization are very limited and not very easy to use, especially in solid organic synthesis [62-64]. The catalyst carriers [65] as spacers are necessary and very useful. Dr. Liu [66] synthesized a series of macromonomers with M_w = 1500, 800 and 400. Macromonomers were then copolymerized with styrene in the presence of magnetic fluid. Magnetic microspheres with hydroxyl groups on the spacers were obtained as shown in **Scheme 4.1**. Other modification on the magnetic polymer microsphere surface is the incorporation

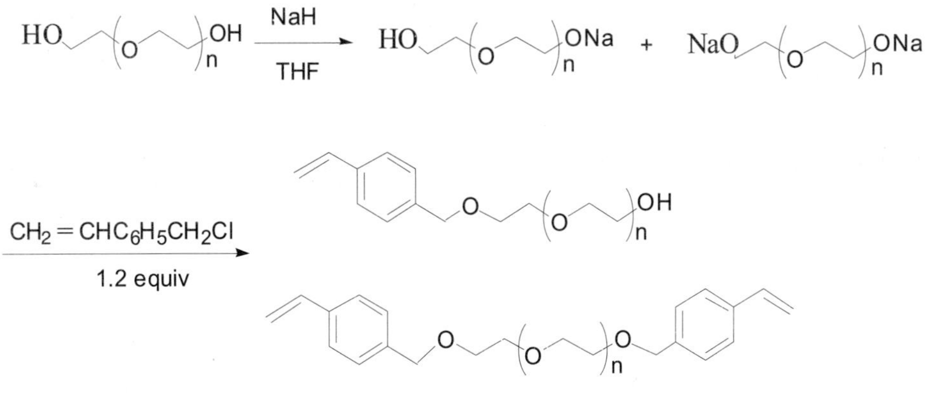

Scheme 4.1. Procedure used to prepare MPEO.

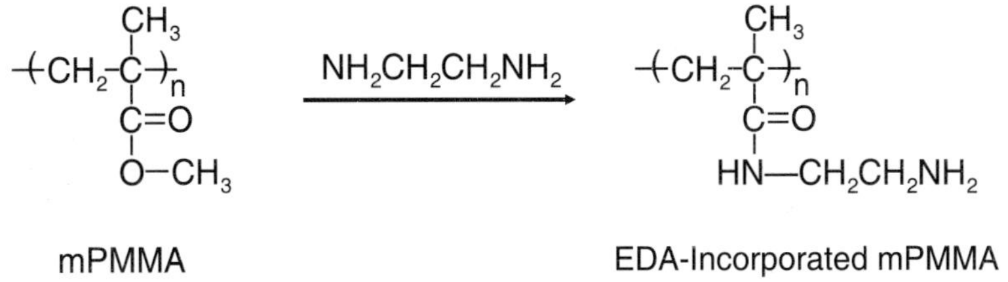

mPMMA   EDA-Incorporated mPMMA

Scheme 4.2. Chemical coupling reaction between the magnetic polymer and EDA. Reproduced with permission from Adil Denizli *et al.*; J. Appl.Polym.Sci., 78, 81 (2000). Copyright © 2000, John Wiley Sons, Inc.

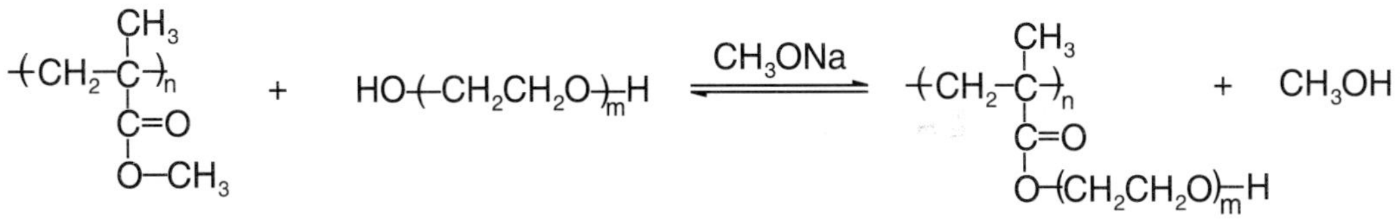

Scheme 4.3. The replacement of -OCH$_3$ groups on the polymer surface by PEG 400. Reproduced with permission from M. Odaasi *et al.*; J. Appl.Polym.Sci., 93, 2501 (2004). Copyright © 2004, John Wiley Sons, Inc.

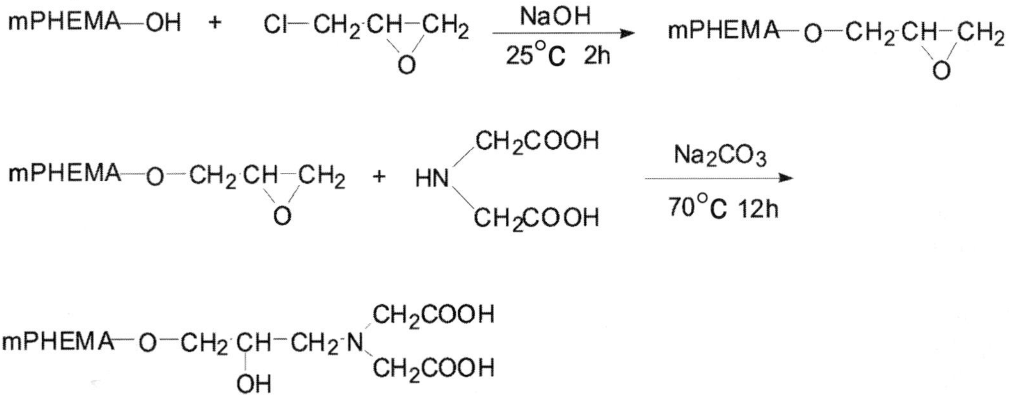

Scheme 4.4. Schematic diagram for the preparation of magnetic mPHEMA beads. Reproduced with permission from M. Odaasi *et al.*; J. Appl.Polym. Sci., 93, 2501 (2004). Copyright © 2004, John Wiley Sons, Inc.

(P)—OH + Cl-Cibacron Blue $\xrightarrow{\text{NaOH}}$ (P)—O—Cibacron

Scheme 4.5. CB was covalently coupled to the PEG-modified nanospheres via a nucleophilic reaction between the chloride of its triazine ring and the hydroxyl groups of the nanospheres. Reproduced with permission from X.Q. Liu *et al.*; Langmuir, 20, 10278 (2004). Copyright © 2004, American Chemical Society.

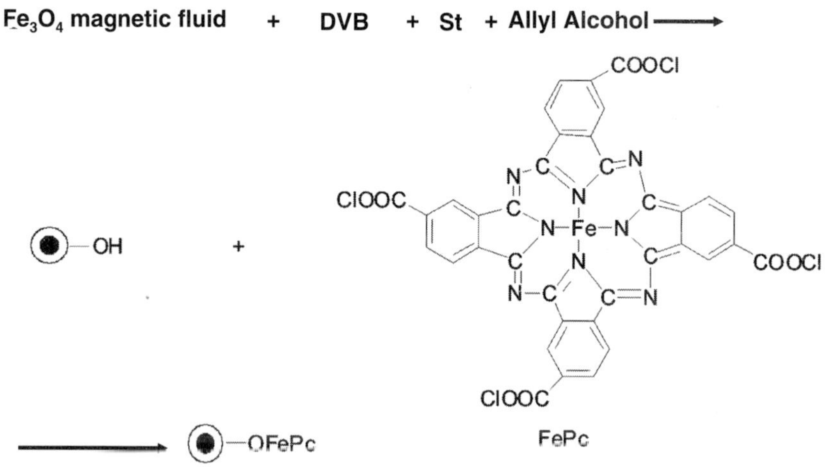

Fe_3O_4 **magnetic fluid** + **DVB** + **St** + **Allyl Alcohol** ⟶

(●)—OH + FePc

⟶ (●)—OFePc

Scheme 4.6. The synthetic route to PSA and PSFePc. Reproduced with permission from J.H. Zhang *et al.*; Polymer International, 51, 61 (2002). Copyright © 2002, Society of Chemical Industry.

of ethylene diamine onto polymethymethacrylate (PMMA) microbeads [67] as shown in **Scheme 4.2**. A suitable length of poly(ethylene glycol) (PEG) chain may also be bound onto PMMA beads [68] as given in **Scheme 4.3**. Odaasi *et al.* [68] first synthesized magnetic poly(2-hydroxyethylmeth-acrylate) (mPHEMA) beads. These beads were modified by iminodiacetic acid (IDA) in order to use these kinds of beads for separating metal ions. The reaction is shown in **Scheme 4.4**. Liu *et al.* [69] reported the joining of a PEG chain covalently with PMMA beads as well as with an affinity dye, Cibacron blue F3G-A (CB) covalently coupled to the PMMA-PEG beads via a nucleophilic reaction between the chloride of its triazine ring and the hydroxyl groups of the beads as shown below in **Scheme 4.5**. Zhang *et al.* [70-72] tried to covalently bind photoconducting materials onto magnetic polymer microspheres. These kinds of microspheres show photoconductivity and have considerable applications as a new diagnosis or detection reagent according to photoconductive signal in **Scheme 4.6**.

Grüttner and Teller [73] reported a type of silica-fortified magnetic nanopaticles of the diameter ranging from 200-400 nm with an iron oxide content of 75-80%. The polymer matrix is of polysaccharide as shown in **Fig. 4.3**. Zhang [74] *et al.* reported the modification of superparamagnetic magnetite nanoparticles with poly(ethylene glycol) (PEG) and folic acid to improve the intracellula uptake and the ability to target specific cells as shown in **Figs. 4.4 & 4.5**.

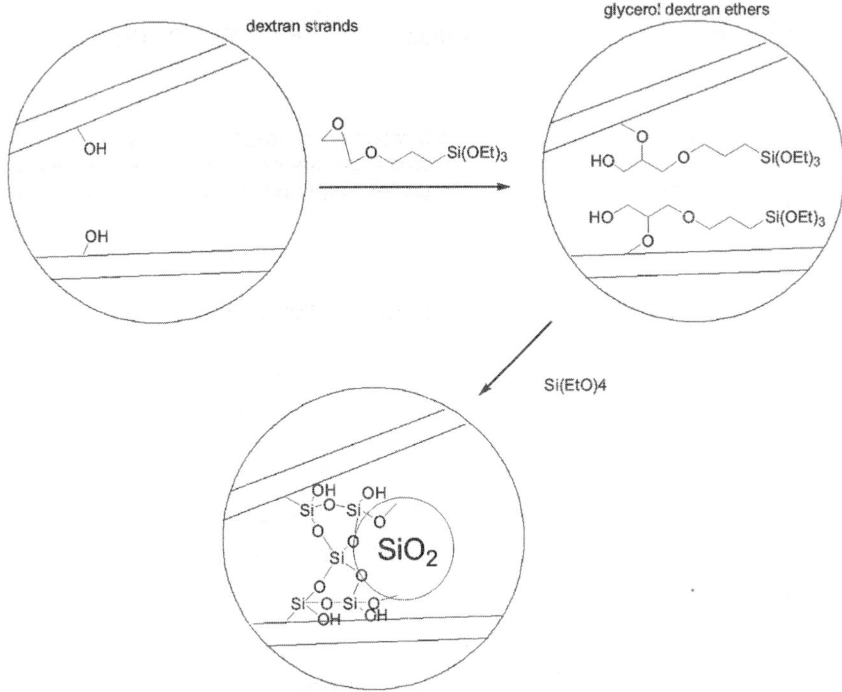

Fig. 4.3. Scheme for cross-linking dextran strands in the matrix of nanomagnetic particles by intercalation of nano-scaled silica. Reproduced with permission from C. Grüttner and J. Teller; J. Magnetism and Magnetic Mater., 194, 8 (1999). Copyright © 1999, Elsevier Science Ltd.

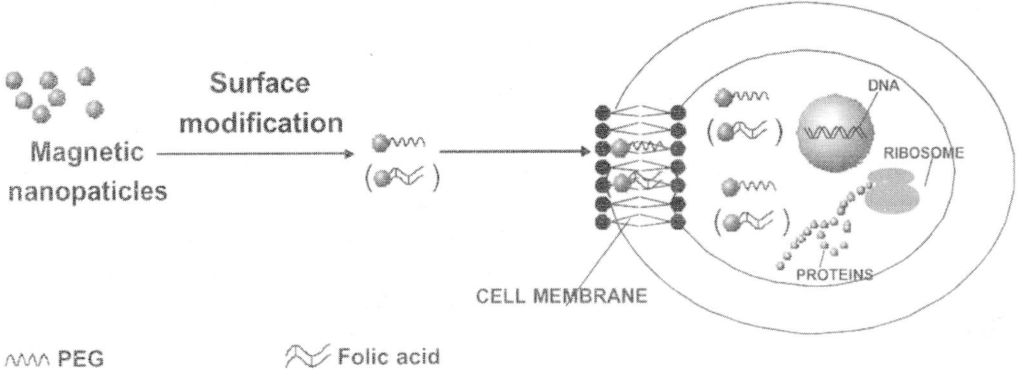

Fig. 4.4. Schematic diagram of uptake of surface modified magnetite nanoparticles into cells. Reproduced with permission from Y. Zhang *et al.*; Biomaterials, 23, 1553 (2000). Copyright © 2002, Elsevier Science Ltd.

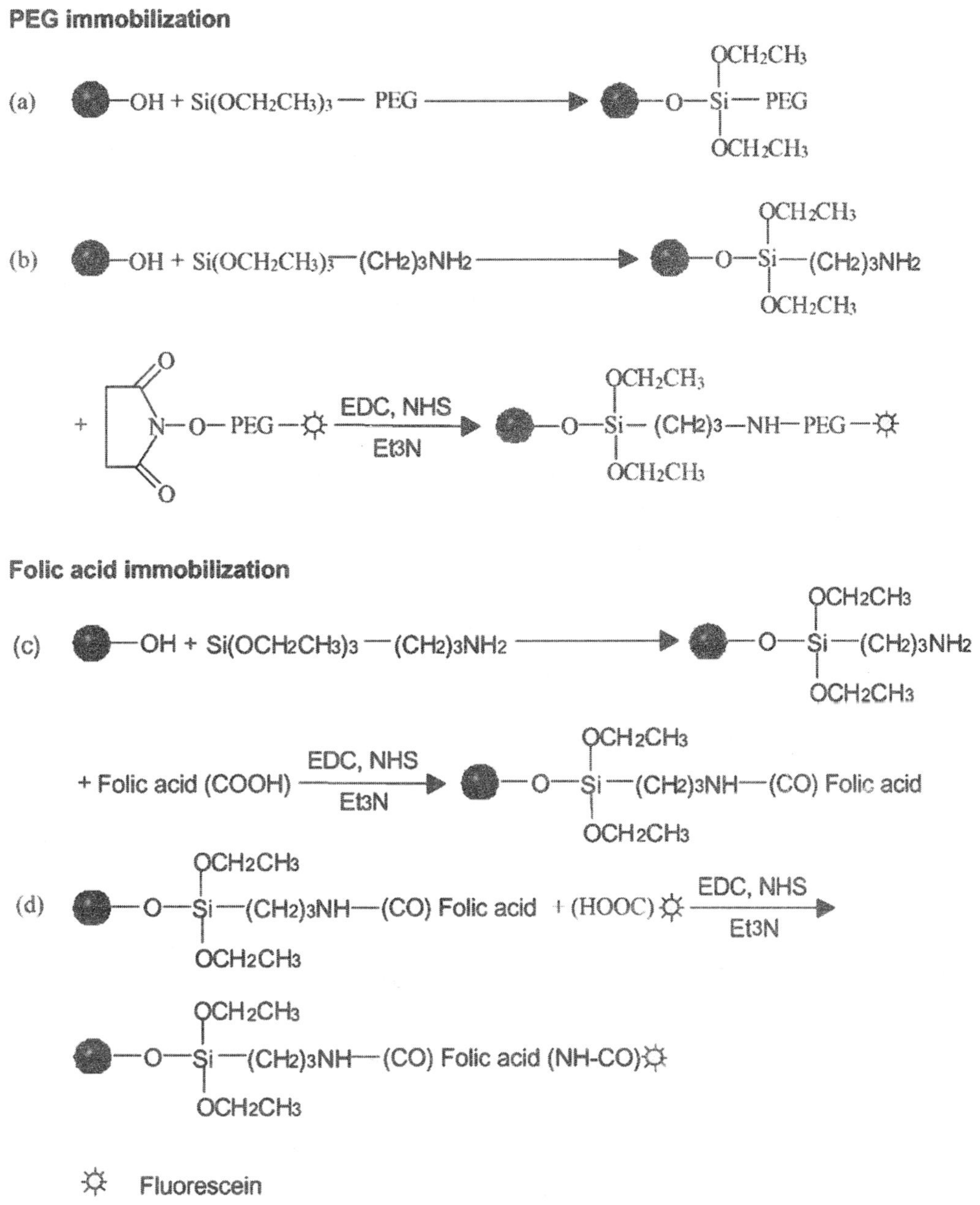

PEG immobilization

Folic acid immobilization

Fig. 4.5. Chemical reaction schemes for immobilizing (a) PEG, (b) PEG-fluorescein, (c) folic acid and (d) folic acid-fluorescein on the surface of magnetite nanoparticles. Reproduced with permission from Y. Zhang *et al.*; Biomaterials, 23, 1553 (2000). Copyright © 2002, Elsevier Science Ltd.

Modifying Magnetite Nanoparticles with PEG

The superparamagnetic magnetite nanoparticles with diameter about 10 nm as produced were washed twice in ethanol, centrifuged, dried at 110°C for 1 hour and vacuum-dried overnight to remove adsorbed water. 15 mg dried nanoparticles were dispersed in a solution of 3 mM of methoxy-PEG-silane in 5 ml of toluene. After vortexing and sonication, the mixture was incubated at 60°C for 4 hours. The suspension was centrifuged and the precipitate was sonicated in toluene for 10 min, washed with toluene and ethanol and then vacuum-dried.

Modifying Magnetite Nanoparticles with PEG-Fluorescein

15 mg dried nanoparticles were dispersed in a solution of 3 mmol of 3-aminopropyl-trimethoxysilane in 5 ml of toluene. The mixture was vortexed, sonicated and incubated at 60°C for 4 hours. The suspension was centrifuged and the precipitate was sonicated in toluene for 10 min and washed with toluene and ethanol. The precipitate was added to 10 ml of NHS-PEG-fluorescein (15 mM) solution in phosphate-buffered saline (PBS, pH 7.4) containing 1-ethyl-3-(3-dimethylamino-propyl) carbodiimide (EDC) (75 mM) and incubated at 37°C for 4 hours. After centrifugation and washing with deionized water, the precipitate was vacuum-dried overnight.

Modifying Magnetite Nanoparticles with Folic Acid

15 mg dried nanoparticles were dispersed in a solution of 3 mmol of 3-aminopropyl-trimethoxysilane in 5 ml of toluene. The mixture was vortexed, sonicated, and incubated at 60°C for 4 hours. The suspension was centrifuged, the precipitate was sonicated in toluene for 10 min and washed with toluene and ethanol. The precipitate was added to the mixture of 1 ml of folic acid (10 mM) solution in dimethylsulfoxide (DMSO), 1.5 ml of NHS (15 mM) and 1.5 ml of EDC (75 mM) solution in water while triethylamine was used as a catalyst. The pH was adjusted to 9 and after incubation at 37°C for 4 hours, the suspension was centrifuged and the precipitate was washed with deionized water and vacuum-dried overnight.

Modifying Magnetite Nanoparticles with Folic Acid-Fluorescein

The magnetite nanoparticles modified with folic acid obtained above were used for further attachment of fluorescein. The nanoparticles were added to the mixture of 4 ml of fluorescein (0.015 M) solution in methanol, 1.5 ml of NHS (15 mM) solution and 1.5 ml of EDC (75 mM) solution in water while triethylamine was used as a catalyst. The pH was adjusted to 9 and after incubation at 37°C for 4 hours, the mixture was centrifuged and the precipitate was washed by deionized water and vacuum-dried overnight.

6.2. Thermo-Sensitive Magnetic Particles

It is very well known that the thermo-sensitive polymers are a kind of very important intelligent polymers. This kind of polymers has lower critical solution temperature (CST). They are hydrophilic below CST and hydrophobic above CST. As the magnetic polymer particles can be separated rapidly from their mixture or solution, combination of magnetic property with thermal sensitivity to develop a new kind of thermo-sensitive magnetic polymer particles, has become a potential subject of investigation. So, it becomes necessary to introduce this kind of functional magnetic polymer microspheres. On one hand, this kind of particles has magnetic responsiveness and on the other hand, the target compound absorbed on the particle can be desorbed from it easily by changing the temperature of the system. Thus, the particles can be used as potential carriers to deliver the desired compound, as separation process is very simple and easily operable.

A novel class of multifunctional particles with thermo-sensitivity as well as magnetic property was synthesized by dispersion polymerization in the presence of magnetic fluid for adsorbing human serum albumin (HSA) [54]. The sorption and desorption behavior of the magnetic thermo-sensitive particles with respect to the change in the temperature were investigated [58]. The results showed that the sorption was dependent mainly on the properties of the surface of the particles. By increasing temperature, particles deswelled and became susceptible to adsorb large amounts of proteins, which could be desorbed at lower temperature.

Chang and Su [75] reported the preparation and characterization of thermo-sensitive magnetic particles. The authors hydroxylated the magnetic particles by hydrogen peroxide and then linked γ-methyacryloyl oxypropyl trimethoxysilane (KH-570) onto the hydroxylated particles and finally copolymerized the modified particles with N-isopropylacrylamide (NIPAAM) as given in **Scheme 4.7**.

Kondo and Fukuda [76] reported the thermo-sensitive magnetic microspheres prepared by two-step emulsifier-free emulsion polymerization. Magnetic particles coated with polystyrene or poly(styrene-co-divinylbenzene) were synthesized and N-isopropylacrylamide was then added to the system to continue polymerization for a suitable period. This kind of thermo-sensitive magnetic microspheres showed reversible transition between dispersion and flocculation by controlling temperature and salt concentration.

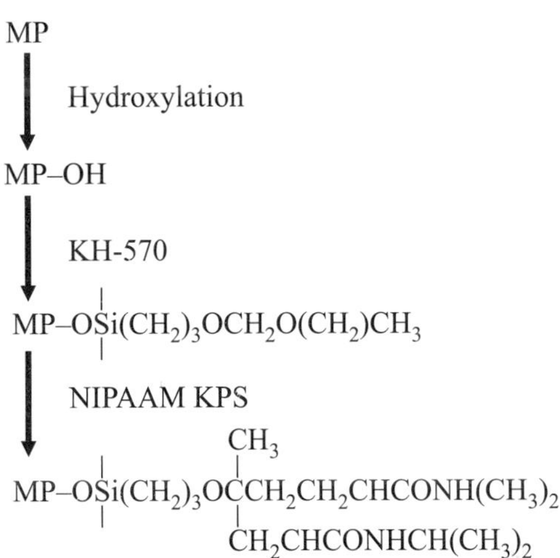

Scheme 4.7. Formation of modified magnetic particles and magnetic powders. Reproduced with permission from Y. Chang and Z.X. Su; Mater.Sci. Engg. A, 333, 155 (2002). Copyright © 2002, Elsevier Science Ltd.

8. APPLICATIONS OF MAGNETIC POLYMER MICROSPHERES

Magnetic polymer microspheres have been attracting much attention for their extensive applications in many fields such as in protein and enzyme immobilization, immunoassay, RNA and DNA purification, cell isolation, drug targeting, industrial wastewater treatment, ion-exchange chromatography, solid state organic synthesis, support for catalysts and so on.

8.1. Immobilization of Enzymes

The conventional supports for enzymes immobilization are restricted to soluble substrates [77]. Immobilization of enzymes onto insoluble polymeric matrices is a very effective way to stabilize them [78]. By immobilization, the storage properties and the pH stability of enzymes are often improved [79]. Enzymes immobilized on magnetic polymer microspheres have the advantages over soluble substrates such as: (i) magnetic polymer microspheres can easily release the enzymes as undesirable centrifugation, sample dilution and loss of carriers are inevitable if non-magnetic carriers are used [80], (ii) the immobilized enzymes can repeatedly be used lowering the cost of the process [81-83], (iii) it can improve biocompatibility, immunocompetence, hydrophilic or hydrophobic properties of enzymes [84] and (iv) several kinds of enzymes can be immobilized on magnetic polymer microspheres [85].

Usually, the enzymes can be immobilized onto magnetic polymer microspheres by the aid of reagent with double functional groups such as 1,1'-carbonyldiimidazole, glutaraldehyde etc. [24,86,87] as evident from **Scheme 4.8**.

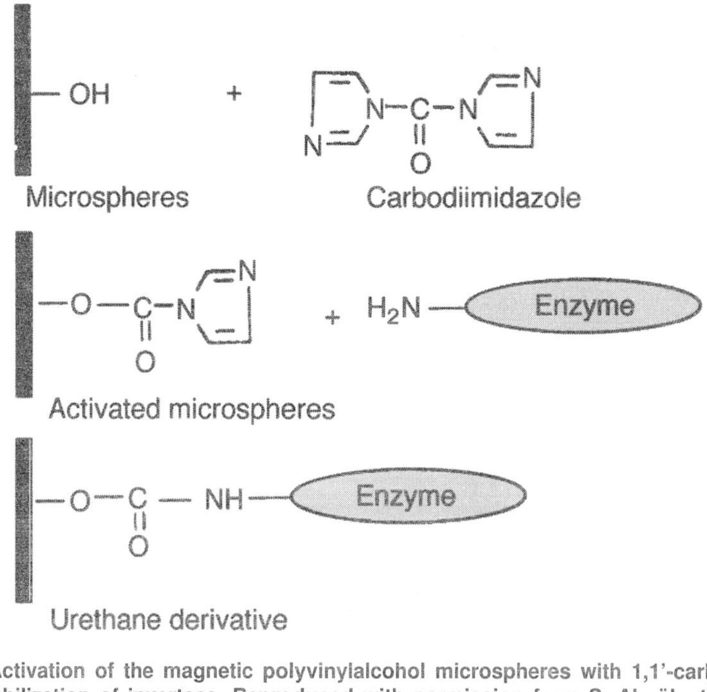

Scheme 4.8. Activation of the magnetic polyvinylalcohol microspheres with 1,1'-carbonyldiimidazole and covalent immobilization of invertase. Reproduced with permission from S. Akgöl *et al.*; Food Chemistry, **74, 281 (2001). Copyright © 2001, Elsevier Science Ltd.**

Activation of magnetic polyvinylalcohol (PVAL) microspheres

The magnetic PVAL microspheres were prepared by suspension polymerization as described before [39]. The activation of hydroxyl groups of magnetic PVAL microspheres for covalent immobilization of enzyme was achieved by reaction with 1,1'-carbonyldiimidazole (CDI). 5 g of magnetic PVAL microspheres were immersed in 10 ml of CDI solution in acetone (20 mg ml^{-1}) and stirred at 25°C for 24 hours. After filtration, the microspheres were washed twice with acetone, dried under reduced pressure at 25°C and kept at 4°C until use.

Immobilization of invertase onto magnetic PVAL microspheres

The activated magnetic PVAL microspheres were equilibrated in phosphate buffer (50 mM and pH 7.0) and immersed in the same fresh medium containing invertase (2 mg ml^{-1}). The immobilization was carried out at 4°C for 24 hours while continuously stirring the medium. Ionically bound enzyme was removed by washing with 10 ml of saline solution (0.5 M) and then with acetate buffer (50 mM and pH 5.5) and was stored at 4°C in fresh buffer until use.

Determination of immobilization efficiency

The amounts of protein in the enzyme solution and in the wash solutions were determined by using Coomassie Brilliant Blue. The amount of bound enzyme was calculated by using **eq. 4.5**.

$$q = [(C_i - C_f) \cdot V]/W \tag{4.5}$$

where q is the amount of bound enzyme onto magnetic PVAL microspheres (in mg g^{-1}), C_i and C_f are the initial and the final concentrations of the enzyme in the reaction medium (in mg ml^{-1}) respectively, V is the volume of the reaction medium (in ml), W is the weight of the microspheres (in g). All the data used in this formula are the averages of at least duplicated experiments.

Activity assays

The activities of both the free and the immobilized invertase preparations were determined by measuring the amount of glucose liberated from the invertase-catalyzed hydrolysis of sucrose per unit time. In the determination of the activity of the free enzyme, the reaction medium consisted of 2.5 ml of acetate buffer (50 mM and pH 5.5), 0.1 ml of sucrose (300 mM). Following a pre-incubation period of 5 min at 35°C, the assay was started by the addition of the 0.1 ml of enzyme solution (10 mg ml^{-1}) and the incubation was continued for 5 min. In order to terminate the enzymatic reaction, the reaction medium was then placed in a boiling water bath for 5 min. The same assay medium was used for the determination of the activity of the immobilized enzyme. The enzymatic reaction was started by the introduction of ten disks into the 10 ml of assay medium and was carried out at 35°C with shaking in a water bath. After 15 min, the reaction was terminated by removing the magnetic PVAL microspheres from the reaction mixture. Sucrose hydrolysis performance of the free and immobilized preparations was determined by measuring the glucose content of the media. Assay mixture containing 25 mg of GOD, 6 mg of POD and 13.2 mg of o-dianisidine in 100 ml of phosphate buffer (0.1 M, pH 7.0). A 2.5 ml aliquot of assay mixture and 0.1 ml of enzymatically hydrolyzed sample were mixed and then incubated at 35°C for 30 min in a water bath. After addition of 1.5 mml of sulphuric acid (30%), absorbance was measured in a UV/VIS spectrophotometer, at a fixed wavelength of 525 nm. The activities of the free and the immobilized invertase were expressed in mmol of glucose $min^{-1}mg^{-1}$ of enzyme and mmol of glucose $min^{-1}g^{-1}$ of enzyme-MPVAL microspheres respectively. To determine the pH and temperature profiles for the free and immobilized enzymes, activity assays were carried out over the pH range of 4.0-8.0 in the temperature range of 20-60°C. The results of dependence on pH, temperature, storage and repeated use are presented in a normalized form, considering the highest value of each set being the value of 100% activity.

8.2 Drug Targeting

In order to improve the usage level of a drug and to reduce its side effects, the drug targeting technology has attracted attention of academia and industrialists. Taking orally or by artery pipe, magnetic polymer microspheres can be used as drug carriers to deliver the desired drug to the special site in the human body under an external magnetic field [88-90]. Drug can be released at the site of the pathological changes and the magnetic polymer microspheres can be excreted out safely. Lübbe *et al.* [91,92] performed the first clinical cancer therapy trial using magnetic micropheres of about 100 nm diameter in the treatment of advanced solid cancer. The results showed the low toxicity of this method and the accumulation of the magnetic microspheres in the target area. Wu *et al.* [93] prepared a kind of isotope labeled magnetic gelatin microspheres, which were injected in rabbits at different intervals of magnetic field and intensity. The results showed that the radioactivity is 15 times more at the target site when magnet is used than that when magnet is not used. Chang [94,95] developed a kind of nanoparticles with the diameter ranging from 158-299 nm, Adriamycin and anti-human cancer monoclonal antibody were conjugated to the nanoparticles. *In vitro* and *in vivo* experiments showed that the nanoparticles bound the cancer cells and killed 80% of them *in vitro* and inferred that 60-65% of the nanoparticles could be targeted at the predetermined site. Häfeli *et al.* [96] proposed the radiolabeling of magnetic particles with rhenium-188 for cancer therapy. The magnetic carriers could effectively target to the solid tumors.

In general, a suitable magnetic carrier must have the properties such as: (i) small size of less than ≈ 1.4 μm to allow capillary distribution and uniform perfusion at the desired target site, (ii) appropriate magnetic responsiveness to technically achievable local fields and gradients at flow rates found in physiological systems, (iii) the ability to carry a wide variety of chemotherapeutic agents with sufficient drug space to allow the delivery of adequate amounts of biologically active drugs without excessive loading of the organism with magnetizable material, (iv) controllable or predictable release rates of the drug(s) from the carrier at the desired target site, (v) surface properties permitting maximum biocompatibility and minimal antigenicity, (PIBCA in rats: 242 mg/ kg, LD of polyhex-50) and (vi) biodegradability of either elimination or minimized toxicity of breakdown products [35].

8.3. Cell Separation

Cell separation is a very important technique in cytology and high efficiency cell separation is all-important and essential in clinical detection and treatment. For example, during the radioactive therapy, bone marrow should be drawn out and the cancer cells in bone marrow must be separated out. However, the conventional method is centrifugation, which is time-consuming and some cells are always destroyed during the process. Immunomagnetic microspheres can be specifically bind target cells and give them magnetic response. By binding a monoclonal antibody to polystyrene magnetic microspheres, after purging cancer cells from bone marrow, the cancer cells were effectively removed while most of the bone marrow cells were recovered [97]. Another example is the detection of disseminated cells in tumor patients. The conventional histopathological methods are not sensitive enough to detect the small number of circulating tumor cells in peripheral blood [98]. The magnetic microspheres have become increasingly important for rapid and sensitive isolation of living cells [99]. Sieben *et al.* [100] investigated the cell separation efficiencies of the size, matrix and surface material of different magnetic particles and established a more effective method for tumor cell separation from peripheral blood. Other reports [101,102] showed that the magnetic polymer microspheres are very efficient in blood purification as well as cell labeling and separation.

Haik *et al.* [103] proposed a magnetic device for cell separation shown in **Fig. 4.6**. The blood enters the system at the top left corner and mixes with the lectin-bound microspheres in mixing chamber 2 (MC2). For the first part of the system, the directional valve (DV1) is aligned to allow for the whole blood mix to flow to the separation chamber 1 (SC1). Once the mix is in SC1, the magnetic field is turned on and the red cells accumulate at the bottom. Subsequently, the remaining blood passes out of SC1 and to DV2. Since this blood is to be treated in the UVAR system, the

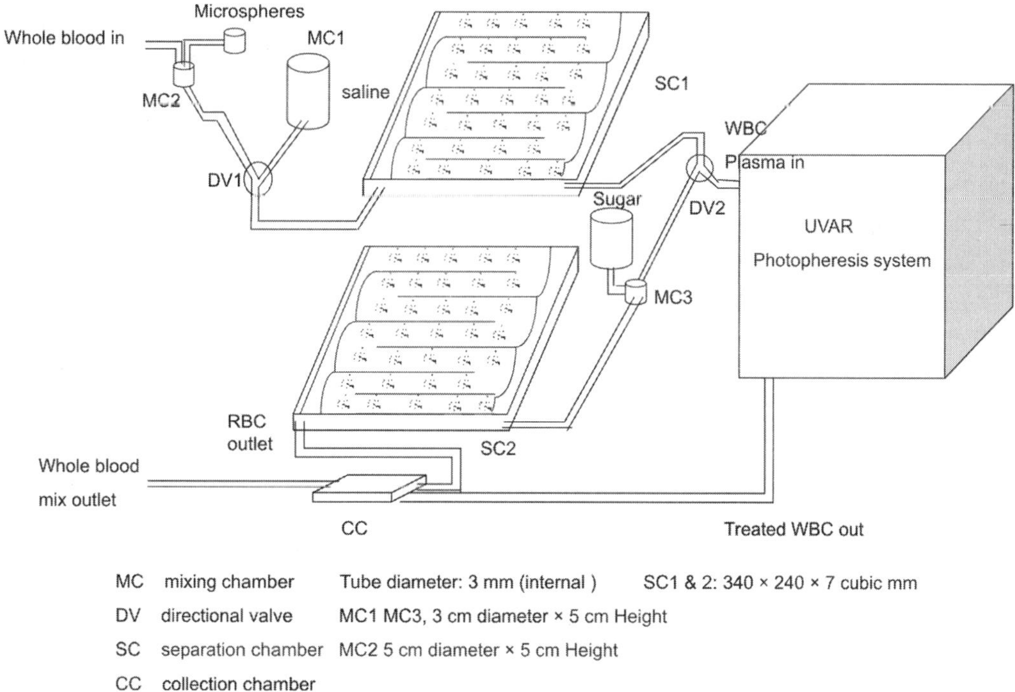

Fig. 4.6. Magnetic device used for continuous separation of red blood cells from blood. Reproduced with permission from Y. Haik *et al.*; J. Magnetism and Magnetic Mater., 194, 254 (1999). Copyright © 1999, Elsevier Science Ltd.

directional valve DV2 is set such that the blood goes directly to the UVAR unit. The red blood cells, which are left behind in the SC1 are then flushed out of SC1 in the next step. DV1 is switched to allow saline to flow into SC1. The RBCs are then washed out of SC1 along with the saline solution. The mix leaving SC1 then goes to DV2. DV2 is now switched such that the flow goes to MC3. In MC3, the specific sugar is added to the RBC and saline mix. This will cause the lectins to separate from the blood cells. The fluid then leaves MC3 and enters separation chamber 2 (SC2), where an arrangement of identical to that used for SC1 is utilized. The magnetic field is switched on and the lectin-bound magnetic microspheres are collected on the magnet surface while the RBCs are set free and flow out of the chamber. If needed, the lectin left in SC2 can be flushed out using saline solution and by switching off the magnetic field. The RBCs that exit SC2 are then directed to the collection chamber where they join the treated white blood cells and plasma.

8.4. Immunoassay

The aim of immunoassay is to confirm the concentration of immune active molecules such as antibodies or antigens in solvents. The requirements of immunoassay are the rapid, accurate and highly sensitive measurements. The conventional methods have radio-immuno-assay (RIA) and enzyme-linked-immuno-sorbent-assay (ELISA), however, the RIA method is seldom used now because it is very difficult to deal with the radioactive material used in RIA. ELISA is very time-consuming and the sensitivity is not sufficiently high. Chen *et al.* [104] prepared nanoparticles of diameter of 90-100 nm with carboxyl groups on the surface of the particles. The particles covalently bound with hepatitis B surface antibody (HBsAb) and were used to detect hepatitis B surface antigen (HBsAg). Under an additional magnetic field, the complex of antibody and antigen were separated from sample solution and the enzyme labeled at the complex catalyzed the reaction of 2-aminohydroxybenzene and hydrogen peroxide in the buffer solution. The electroactivity was detected using differential pulse voltammetry. The detection limit was found to be 0.06 ng ml^{-1}, obviously this sensitivity is much higher than that of ELISA technique. Mcneill *et al.* [105] developed a simple technique that utilizes magnetic polymer beads to detect cytokines. The results showed that the assay is rapid, simple and requires less sample handling in comparison to the standard ELISA technique. At the same time, this technique facilitates the timely and accurate quantification and the sensitivity.

8.5. Industrial Wastewater Treatment

Heavy metals such as lead, mercury, arsenic, copper and cadmium are dangerous for human body as well as their accumulation in soil and water is poisoning them and making them harmful for the environment. Magnetic polymer micropheres, as carrier in the removal of heavy metals, have attracted the interest worldwide [106-108]. Denizli *et al.* [107] reported the use of the magnetic polymethyl-methacrylate microbeads carrying ethylene diamine in the removal of heavy metal irons (copper, lead, cadmium and mercury) from aqueous solutions containing different amounts of these ions (5-700 mgL^{-1}) and at different pH values (2.0-8.0). The results showed that the magnetic polymer microspheres are very efficient to remove heavy metal ions from wastewater and the micropheres can be reused after treatment with 0.1 M HNO$_3$.

8.6. Magnetic Ion-Exchangers

Recently, the use of magnetic ion-exchange resins in environmental technology has become one of the most important applications of magnetic polymer microspheres because they can be separated and collected rapidly and easily by magnetic control [109]. Odaasi *et al.* [68] prepared porous magnetic chelator support for albumin as shown in **Scheme 4.9**.

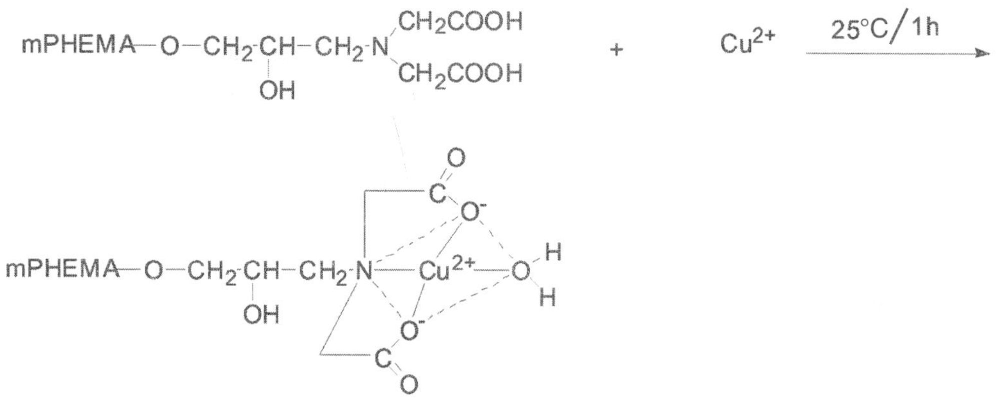

Scheme 4.9. Schematic diagram for the preparation of magnetic mPHEMA metal-chelate affinity beads. Reproduced with permission from M. Odaasi *et al.*, J. Appl.Polym. Sci. 93, 2501 (2004). Copyright © 2004, John Wiley & Sons, Inc.

8.7. Photonic Bandgap Applications

It is also very interesting that magnetic polymer microspheres are finding place in photonic bandgap (PBG) applications. Saado *et al.* [110] developed magnetic polystyrene spheres covered with gold nanoparticles. The magnetic spheres are self-assembling on a liquid surface in 2-D or 3-D structures, the symmetry and lattice parameters may be controlled by application of external magnetic field. In micro- and submicro-sized diameters, the band gap of PBG structures appears in the visible and infrared range. The similar work [111,112] demonstrates the possibility to design and construct tunable PBG devices in the microwave and mm-wave ranges also.

8.8. Bacterial Separation

Efficient separation of bacteria from natural samples may lead to more successful application of the detection techniques [113]. Magnetic separations generally use antisera for specific cell recognition and provide the bridge between target cells and the magnetic particles [114-115] or use polycolonal antibody to recognize mycolic acid containing bacteria [116].

A schematic diagram of indirect immunomagnetic separation (IMS) procedure is shown in **Fig. 4.7**. The primary antibody (anti-mycolic acid antibody) was previously bound to the target

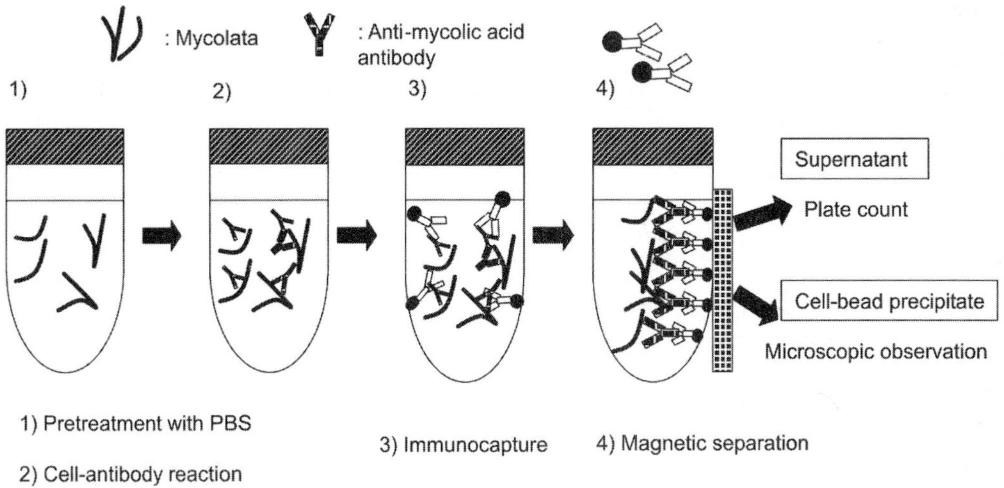

Fig. 4.7. IMS procedure used in this study. In step 1, G. amarae SC1 cells were suspended in PBS and incubated with anti-mycolic acid antibody; excess antibody was removed. In step 2, G. amarae SC1 cells were immunocaptured and concentrated by IMS with magnetic beads coated with sheep anti-rabbit IgG antibody. In step 3, the supernatant was carefully taken and subjected to plate counting. The cells bound to the beads were suspended in deionized water and subjected to phase contrast microscopy. Reproduced with permission from S. Morisada *et al.*; J. Microbiol. Methods, 51, 141 (2002). Copyright © 2002, Elsevier Science Ltd.

cells. The antibody-associated cells were captured by immunomagnetic beads (Dynabeads M-280, sheep anti-rabbit antibody-coated bead; Dynal, Oslo, Norway), which recognize the primary antibody on cells. A suitable dilution ratio of the antiserum was determined by fluorescent antibody staining. 1 ml of G.amarae SC1 culture containing about 1×10^8 CFU ml^{-1} was washed with 1 ml of phosphate buffered saline (PBS; 10 mM sodium phosphate and 0.15 M NaCl, pH 7.2) plus 0.05% (w/v) Tween 20 by centrifugation at 3000 g for 10 min at 4°C. Washed cells were resuspended in 1 ml of PBS and allowed to stand for 1 hour at room temperature with gentle shaking. After PBS was removed, cell concentration was adjusted to about 1×10^6 CFU ml^{-1} with a blocking solution of PBS containing 1% Block Ace. The antiserum that was diluted with the blocking solution was added to the cell suspension to give final dilution ratios of 1:50 to 1:1000 and the mixture was incubated for 1 hour at room temperature with gentle shaking. Unbound excess antibodies were removed by centrifugation with PBS plus Tween 20 and with PBS. The secondary antibody, fluorescein isothiocyanate (FITC)-labeled goat anti-rabbit IgG (H + L) which was diluted with the blocking solution, was added to a dilution ratio of 1:200 and the suspension was incubated and washed in the same manner as that of the primary antibody. FITC-labeled cells were suspended in 50 μl of deionized water and subjected to epifluorescent microscopy. The stained G.amarae SC1 cells were observed with an Olympus BX-FLA-2 (magnification ×1000) with an appropriate filter unit (470 to 490 nm). Under the experimental conditions, the sufficiently labeled cells displayed a bright green fluorescence [117].

8.9. Hyperthemia

Hyperthemia is a useful method in cancer therapy by heating certain organs or tissues to temperature between 41°C and 46°C [118]. The conventional methods include whole body hyperthermia [119], radiofrequency hyperthermia [120], microwave hyperthermia [121] and implantable needles [122]. However, the problem of these techniques is the difficulty in uniform heating of only the tumor region up to the required temperature without causing any damage to normal tissues [123]. Magnetic particles can be localized to a suitable site and can be heated under an alternating magnetic field by hysteresis loss [124-126]. So they have attracted the attention for their use in hyperthermia, however, clinical applications of magnetic polymer microspheres are seldom reported. In recent years, Hergt *et al.* [127] proposed maghemite nanoparticles covalently coated with polyethylene glycol for hyperthermia. Results showed that the nanoparticles exhibit excellent magnetic properties under the influence of AC-fields.

9. CONCLUSIONS AND FUTURE PROSPECTS

Magnetic polymer micro- and nanoparticles are receiving increasing attention for their application in the fields of medicine, biology, biotechnology, industrial wastewater treatment, ion-exchangers and so on because they have magnetic response properties under magnetic fields and their surface can be modified to desired functional group according to the requirements of the application while the size, size distribution of magnetic polymer particles can be controlled by different preparation methods. In recent years, many newer applications are explored, for example, in magnetic resonance imaging (MRI) contrast enhancement [128,129], markers [130], eye surgery [131] and so on. However, there are numerous challenges ahead. One area is the preparation of uniform-sized particles, especially for biomedical applications as the uniform size of the particles is very important in drug delivery to control the rate of release of drug. The second is to synthesize multifunctional groups on magnetic polymer microspheres wide variety of applications. We believe that magnetic polymer microspheres can be used in future in more and more areas of application due to their simple separation process and multifunctional groups.

ACKNOWLEDGEMENTS

The authors gratefully acknowledge the financial support from National Nature Science Foundation of China (No. 59073119, 59903007, 59573011 and 50273040).

REFERENCES

1. Y. Wang, L. Feng and C. Pan; J. Appl.Polym. Sci., 70, 2307 (1998).
2. A. Warshawsky and D.A. Upson; J. Polym. Sci, Polym.Chem., 27, 2963 (1989).
3. A. Warshawsky and D.A. Upson; J. Polym.Sci, Polym.Chem., 27, 2995 (1989).
4. K. Tokuoka, M. Senna, H.J. Kuno; J. Mater.Sci., 21, 493 (1986).
5. M. Tanaka, K. Hayashi; J. Mater.Sci., 25, 987 (1990).
6. J. Lee, M. Senna; Coll.Polym.Sci., 273, 76 (1995).
7. K. Furusawa, K. Nagashima and C. Anzai; Coll.Polym. Sci., 272, 1104 (1994).
8. X.B. Ding, W. Li, Z.H. Zheng and W.C. Zhang; J. Appl. Polym.Sci., 79, 1847 (2001).
9. J.G. Deng, C.L. He, Y.X. Peng, J.H. Wang, X.P. Long and P. Li, A.S.C. Chan; Synth.Metals, 139, 295 (2003).

10. J.G. Deng, Y.X. Peng, C.L. He, X.P. Long, P. Li and A.S.C. Chan; Polymer International, 52, 1182 (2003).
11. J.G. Deng, C.L. He, X.P. Long, Y.X. Peng, P. Li and A.S.C. Chan; Acta Polymerica Sinica, (3), 393 (2003).
12. X.Y. Liu, X.B. Ding, Z.H. Zheng, Y.X. Peng, X.P. Long, K. Chang, X.C. Wang and A.S.C. Chan; Acta Polymerica Sinica, (1), 104 (2003).
13. X.Y. Liu, X.B. Ding, Z.H. Zheng, Y.X. Peng, K. Chang, X.C. Wang, A.S.C. Chan and X.P. Long; Polymer International, 52, 235 (2003).
14. X.Y. Liu, X.B. Ding, Z.H. Zheng, Y.X. Peng, K. Chang, X.C. Wang, A.S.C. Chan and X.P. Long; Chem.Res. in Chinese Univ., 8, 458 (2002).
15. J.G. Deng, Y.X. Peng, X.B. Ding, J.H. Wang, X.P. Long, P. Li and A.S.C. Chan; Chinese J. Chem.Phys., 2, 149 (2002).
16. J.H. Zhang, X.B. Ding, Y.X. Peng and M. Wang; Acta Polymerica Sinica, (6), 730 (2001).
17. R. Matsuno, K. Yamamoto, H. Otsuka and A. Takahara; Macromolecules, 37, 2203 (2004).
18. L. Chang and L. Li; J. Jinan Univ. (Natural Sci.), 25, 375 (2004).
19. J. Mauduit, N. Bukh and M. Vert; J. Controlled Release, 23, 221 (1993).
20. I. Yamakawa, Y. Tsushima, R. Machida, S. Watanabe; J. Pharm.Sci., 81, 899 (1992).
21. S.A. Gómez-Lopera, R.C. Plaza and A.V. Delgado; J. Coll.Interface Sci., 240, 40 (2001).
22. J. Mauduit, N. Bukh and M. Vert; J. Controlled Release, 23, 209 (1993).
23. A. Denizli, G. Özkan and M.Y. Aarica; J. Appl.Polym.Sci., 78, 81 (2001).
24. S. Akgöl, Y. Kaçar, A. Denizli and M.Y. Arica; Food Chem., 74, 281 (2001).
25. E.B. Denkbas, E. Kiliçay, C. Birlikseven and E. Öztürk; React.Funct.Polym., 50, 225 (2002).
26. B.L. Hirschein; Chem.Techn., March, 170 (1982).
27. K. Zhang, Q. Fu, Y.H. Huang and D.H. Zhou; Space Mater.Tech., 6, 1 (2004).
28. Y.M. Wang and L.X. Feng; Polym.Mater.Engg., 14 (5), 6 (1998).
29. X. Yang, J. Qing and C.Y. Tao; Chemical World, 3, 185 (2005).
30. R. Massart; U.S. Patent 4, 329241 (1982).
31. C. Flesch, C. Delaite, P. Dumas, E. Bourgeat-lami, and E. Duguet; J. Polym.Sci., Part A, 42, 6011 (2004).
32. H. Pardoe, W. Chua-anusorn, T.G.S. Pierre and J. Dobson; J.Magnetism and Magnetic Mater., 225, 41 (2001).
33. D.K. Kim, Y. Zhang, W. Voit, K.V. Rao and M. Muhammed; J. Magnetism and Magnetic Mater., 225, 30 (2001).
34. P.C. Morais, V.K. Garg, A.C. Oliveria, L.P. Silva, R.B. Azevedo, A.M.L. Silva and E.C.D. Lima; J. Magnetism and Magnetic Mater., 225, 37 (2001).
35. S. Santra, R. Tapec, N. Thedoropoulou, J. Dobson, A. Hebard and W.H. Tan; Langmuir, 17, 2900 (2001).
36. X.B. Ding, Z.H. Sun, G.X. Wan and Y.Y. Jiang; React.Funct.Polym., 38, 11 (1998).
37. F. Grasset, S. Mornet, A. Demourgues, J. Portier, J. Bonnet, A. Vekris and E. Duguet; J. Magnetism and Magnetic Mater., 234, 409 (2001).
38. M.Y. Arica, H. Yavuz, S. Patir and A. Denizli; J. Molecular Catalysis B, Enzymatic, 11, 127 (2000).
39. S. Avivi, I. Felner, I. Novik and A. Gedanken; Biochimica et Biophysica Acta, 1527, 123 (2001).
40. X.B. Ding, Z.H. Sun, W.C. Zhang, Y.X. Peng, A.S.C. Chan and P. Li; Coll. Polym.Sci., 278, 459 (2000).
41. X.B. Ding, Z.H. Sun, W.C. Zhang, Y.X. Peng and G.X. Wan; J. Appl.Polym. Sci., 77, 2915 (2000).
42. X.Y. Liu, X.B. Ding, Z.H. Zheng and W.C. Zhang; Chem.Res.Chinese U., 18, 458 (2002).
43. X.B. Ding, Z.H. Sun and G.X. Wan; Acta Polymerica Sinica, 5, 628 (1998).
44. D. Horák, J. Bohácek and M. Subri; J. Polym.Sci., Part A, Polym.Chem., 38, 1161 (2000).
45. D. Horak; J. Polym.Sci.,Part A, Polym.Chem., 39, 3707 (2001).
46. H. Noguchi, N. Yanase, Y. Uchida and T. Suzuta; J. Appl.Polym.Sci., 48, 1539 (1993).
47. N. Yanase, H. Noguchi, H. Asakkura and T. Suzuta; J. Appl.Polym.Sci., 50, 765 (1993).

48. A. Kondo and H. Fukuda; Coll. Surf., 153, 435 (1999).
49. Y.H. Deng, L. Wang, W.L. Yang, S.K. Fu and E. Abdelhamid; Chem.J.Chinese Univ., 24, 920 (2003).
50. I. Csetneki, M.K. Faix, A. Szilágyi, A.L. Kovács, Z. Németh and M. Zrinyi; J. Polym.Sci., Part A, Polym.Chem., 42, 4802 (2004).
51. Y. Deng, L. Wang, W. Yang, S. Fu and A. Elaïssari; J. Magnetism and Magnetic Mater., 257, 69 (2003).
52. C.L. Yang, Y.P. Guan, J.M. Xing, J.G. Liu, Z.T. An and H.Z. Liu; Sci. China, Ser. B, Chem., 34, 265 (2004).
53. W. Li, X.B. Ding, Z.H. Zheng, W.C. Zhang, J.G. Deng, Y.X. Peng, A.S.C. Chan and C.W. Yip; J. Appl.Polym. Sci., 81, 333 (2001).
54. S. Mornet, F. Grasset, J. Portier and E. Duguet; Eur.Cell.Mater., 3, 110 (2002).
55. Y. Lee, J. Rho and B. Jung; J. Appl.Polym. Sci., 89, 2058 (2003).
56. D. Müller-Schulte and H. Brunner; J. Chromatography A, 711, 53 (1995).
57. J.L. Arias, V. Gallardo, S.A. Gömez-Lopera, R.C. Plaza and A.V. Delgado; J. Controlled Release, 77, 309 (2001).
58. J. Oster, J. Parker and L. Brassard; J. Magnetism and Magnetic Mater., 225, 145 (2001).
59. R.F.M. Dekker; Appl Biochem Biotech, 22, 289 (1989).
60. J.C. Kang, S. Buka, S.S. Xie and S.L. Wei; Acta Pharmaceutica Sinica, 33, 52 (1998).
61. X.H. Li and Z.H. Sun; J. Appl.Polym. Sci., 58, 1991 (1995).
62. J.J. Pastor, I. Fernández, F. Rabanal and E. Giralt; Org.Lett., 4, 3831 (2002).
63. K.S. Lam, T. Sroka, M.L. Chen, Y. Zhao, Q. Lou, J.Z. Wu and Z.G. Zhao; Life Science, 62, 1577 (1998).
64. K.S. Lam, M. Lebl and V. Krchòák; Chem. Rev., 97, 411 (1997).
65. D. Bozhinova, B. Galunsky, G. Yueping, M. Franzreb, R. Köster and V. Kasche; Biotechnol. Lett., 26, 343 (2004).
66. X.Y. Liu; "Study on the Synthesis and Applications of Amphiphilic and Magnetic Polymer Microspheres", 2001; M.Sc. Thesis, Chengdu Institute of Organic Chemistry, Chinese Academy of Sciences.
67. Adil Denizli, Güleren Özkan and M. Yakup Arica; J. Appl.Polym. Sci., 78, 81 (2000).
68. M. Odaasi, L. Uzun and Adil Denizli; J. Appl.Polym. Sci., 93, 2501 (2004).
69. X.Q. Liu, Y.P. Guan, Z.Y. Ma and H.Z. Liu; Langmuir, 20, 10278 (2004).
70. J.H. Zhang, X.B. Ding, Y.X. Peng and M. Mang; Polymer International, 51, 617 (2002).
71. J.H. Zhang, X.B. Ding, Y.X. Peng and M. Mang; Coll. Polym.Sci., 280, 674 (2002).
72. J.H. Zhang, X.B. Ding, Y.X. Peng and M. Mang; J. Appl.Polym.Sci., 3, 277 (2002).
73. C. Grüttner and J. Teller; J. Magnetism and Magnetic Mater., 194, 8 (1999).
74. Y. Zhang, N. Kohler and M.Q. Zhang; Biomaterials, 23, 1553 (2000).
75. Y. Chang and Z.X. Su; Materials Science and Engineering A, 333, 155 (2002).
76. A. Kondo and H. Fukuda; Coll. Surf., A, Physiochem. Engg. Aspects, 153, 435 (1999).
77. E.V. Leemputten and M. Horisberger, Biotech.Bioengg., XVI, 385 (1974).
78. X.H. Li and Z.H. Sun, J. Appl.Polym.Sci., 58, 1991 (1995).
79. Z. Bilkova, M. Slovakova, D. Horak, J. Lenfeld and J. Churacek; J. Chromatography, 770, 177 (2002).
80. J. Lenfeld; Angewante, 212, 147 (1993).
81. B. Pietes, G. Bardeletti and P.R. Coulet; Appl. Biochem.Biotechnol., 32, 37 (1991).
82. M.Y. Arica, H. Yavuz, S. Yatir, A. Denizli; J. Molecular Catalysis B, Enzymatic, 11, 127 (2000).
83. T. Bahar and S. Celebi; Enzyme and Microbial Technol., 26, 28 (2000).
84. J. Tukova, L. Petkov, J. Sajdok, J. Kas and M.J. Benes; J. Chromatography, 500, 585 (1990).
85. M.A. Burns, G.I. Kvesitadze and D.J. Graves; Biotechnol.Bioengg., XXVII, 137 (1985).
86. M. Koneracká, P. Kopèansky, M. Antalík, M. Timko, C.N. Ramchand, D. Lobo, R.V. Mehta and R.V. Upadhyay; J. Magnetism and Magnetic Mater., 201, 427 (1999).

87. M. Koneracká, P. Kopèansky, M. Timko, C.N. Ramchand, A.D. Sequeira and M. Trevan; J. Molecular Catalysis B, Enzymatic, 18, 13 (2002).

88. M. Sako, S. Yokogawa, K. Sakomoto, S. Adachi, S. Hirota, S. Okada, S. Murao Invest. Radiol., 17, 573 (1982).

89. M. Rettenmaier, J. Stratton, M. Berman, A. Senyei, K. Widder, D. White, P. Disaia, Gynecol.Oncol., 27, 34 (1987).

90. M. Sako, S. Ohtsuki, M. Arai, S. Hirota, H. Watanabe, S. Kimura. J. Jpn.Sco. Cancer Ther., 18, 92 (1983).

91. A. Lubbe, C. Bergemann, H. Riess, F. Schriever, P. Reichardt, K. Possinger, M. Matthias, B. Dorken, F. Herrmann, R. Gurtler, P. Hohenberger, N. Haas, R. Sohr, B. Sander, A. Lemke, D. Ohlendorf, W. Huhnt, D. Huhn, Cancer Research, 56, 4686 (1996).

92. A.S. Lübbe, C. Alexious and C. Bergemann; J. Surg.Res., 95, 200 (2001).

93. C.B. Wu, Y.L. Zhao, S.M. He and S.L. Wei; Acta Pharmaceutica Sinica, 28, 464 (1993).

94. J. Chang; Chinese J. Biomed.Engg., 15, 97 (1996).

95. J. Chang; Chinese J. Biomed.Engg., 15, 354 (1996).

96. U. Häfeli, G. Pauer, S. Failing and G. Tapolsky; J. Magnetism and Magnetic Mater., 225, 73 (2001).

97. J.C. Kang and S.L. Wei; Acta Pharmaceutica Sinica, 32, 536 (1997).

98. B. Wörmann, G.G. Griesinger and W. Hiddemann; Internist, 38, 1083 (1997).

99. G. Bepler, A. Koehler, P. Kiefer, K. Havemann, K. Beisenherz, G. Jaques, C. Gropp and M. Haeder; Differentiation, 37, 158 (1988).

100. S. Sieben, C. Bergemann, A. Lübbe, B. Brockmann and D. Rescheleit; J. Magnetism and Magnetic Mater., 225, 175 (2001).

101. T. Bjerke, S. Nielsen, J. Helgestad, B.W. Nielsen and P.O. Schiøtz; J. Immunol.Methods, 157, 49 (1993).

102. A. Rembaum, R.C.K. Yen, D.H. Kempner and J. Ugelstad; J. Immunol.Methods, 52, 341 (1982).

103. Y. Haik, V. Pai and C.J. Chen; J. Magnetism and Magnetic Mater., 194, 254 (1999).

104. Q. Chen, T.Z. Peng and A.L. Liu; Acta Chimica Sinica, 62, 2447 (2004).

105. A. Mcneill, J. Kastrup and B.T. Bäckström; Scandinavian J. Immunol., 60, 287 (2004).

106. Y. Terashima, H. Ozaki and M. Sekine; Water Res., 20, 537 (1986).

107. A. Denizli, D. Tanyolaç, B. Salih and A. Özdural; J. Chromatography A, 793, 47 (1998).

108. B.A. Bolto; J. Macromol.Sci., Chem., A14, 107 (1980).

109. D. Bursill; Water Sci Technol.Water Suppl., 1, 1 (2001).

110. Y. Saado, T. Ji, M. Golosovsky, D. Davidov, Y. Avni and A. Frenkel; Optical Mater., 17, 1 (2001).

111. K. Zahn, R. Lenke and G. Maret; Phys.Rev.Lett., 82, 2721 (1999).

112. R. Bubeck, C. Bechinger, S. Neser and P. Leiderer; Phys.Rev.Lett., 82, 3364 (1999).

113. J. Porter and R.W. Pickup; J. Microbiol.Methods, 33, 221 (1998).

114. E. Skjerve, L.M. Rorvik and O. Olsvik; Appl.Environ.Microbiol., 56, 3478 (1990).

115. A.R. Bennett, S. MacPhee and R.P. Betts; Lett.Appl.Microbiol., 22, 237 (1996).

116. L.S.L. Yu, J. Uknalis and S. Tu; J. Immunol.Methods, 256, 11 (2001).

117. S. Morisada, N. Miyata and K. Iwahori; J. Microbiol.Methods, 51, 141 (2002).

118. A. Jordan and R. Scholz; International Syposium on Magnetic Carriers-Biological and Clinical Applications, 1999, October 22-24, p. 7, Wuhan, China.

119. R. Cavaliere, E.C. Ciocatto, B.C. Giovanella, C. Heidelberger, R.O. Johnson and M. Margottini; Cancer, 20, 1351 (1967).

120. N. Ikeda, O. Hayashida, H. Kameda H. Ito and I. Matsuda; Int. J. Hyperthermia, 10, 553 (1994).

121. J.C. Lin and J. Wang; Int. J. Hyperthermia, 3, 37 (1987).

122. P.R. Stauffer, T.C. Cetas, A.M. Feltcher, D.W. Deyoung, M.W. Dewhirst, J.R. Oleson and R.B. Roemer; IEEE Trans.Biomed.Engg., BME, 31, 76 (1984).

123. M. Shikai, M. Yanase, M. Suzuki, H. Honda, T. Wakabayashi, J. Yoshida and T. Kobayashi; J. Magnetism and Magnetic Mater., 194, 176 (1999).

124. M. Mitsumori, M. Hiraoka, T. Shibata, Y. Okuno, S. Masunaga, M. Koishi, K. Okajima, Y. Nagata, Y. Nishimura, M. Abe, K. Ohura, M. Hasegawa, H. Nagae and Y. Ebisawa; Int. J. Hyperthermia, 10, 785 (1994).
125. R.E. Rosensweig; J. Magnetism and Magnetic Mater., 252, 370 (2002).
126. A. Jordan, R. Scholz, P. Wust, H. Fähling and R. Felix; J.Magnetism and Magnetic Mater., 201, 413 (1999).
127. R. Hergt, R. Hiergeist, I. Hilger, W.A. Kaiser, Y. Lapatnikov, S. Margel and U. Richter; J. Magnetism and Magnetic Mater., 270, 345 (2004).
128. D. Högemann, L. Josephson, R. Weissleder and J.P. Basilion; Bioconjugate Chem., 11, 941 (2000).
129. L. Babes, B. Denizot, G. Tanguy, J.J.L. Jeune and P. Jallet; J. Coll.Interface Sci., 212, 474 (1999).
130. H. Brückl, M. Panhorst, J. Schotter, P.B. Kamp and A. Becker; IEE Pro.-Nanobiotechnol., 152, 41 (2005).
131. J.P. Dailey, J.P. Philips, C. Li and J.S. Riffle; J. Magnetism and Magnetic Mater., 194, 140 (1999).

5

Intrinsically Conducting Polymers for Metallic Corrosion Protection

Xinli Jing[*] and Yangyong Wang

Department of Chemical Engineering, School of Energy and Power Engineering, Xi'an Jiaotong University, Xi'an, 710049, People's Republic of China.
Corresponding author's e-mail: xljing@mail.xjtu.edu.cn

Contents

Summary: This chapter reviews the metallic corrosion protection properties and mechanisms of intrinsically conducting polymers (ICPs) as well as their composites with emphasis on polyaniline (PANI) and polypyrrole (PPY), which can be synthesized chemically or electrochemically. The excellent corrosion prevention performances of ICP coatings and the ways to prepare the coatings are well documented. The solution cast approach for coating preparation is not recognized owing to the poor solubility of ICPs in common organic solvents and the poor mechanical and adhesional properties of the obtained coatings. The same problem exists for the electrodeposited ICP coatings due to the inconvenience in coating preparation, especially for those large sized infrastructures, such as bridges, ships and so on. The composite or blend approach, in which ICPs were solution mixed, or dispersion blended with conventional polymers, is an effective way to enhance the mechanical and adhesional properties of the ICP coatings, furthermore, to improve the corrosion prevention properties of the coatings a little or more. Kinds of mechanisms, such as passivation, electronic barrier, etc. have been put forward to date, which made it indispensable and imperative to elucidate the corrosion protection mechanisms of ICPs. Anyway, ICPs are good candidates for metallic corrosion protection and further works are still needed to optimize the performances of ICPs based anti-corrosive coatings.

1. INTRODUCTION

Corrosion, i.e. the destruction or deterioration of a material (especially metals) due to reaction with its environment, has long been an important problem, bringing tremendous economic losses in excess of 100 billion dollars annually worldwide [1]. Among the methods used for corrosion protection, the use of coatings, especially organic coatings, is the most popular method for corrosion protection of steel and other metals [2,3]. A typical coating system consists of three individual coating layers. The first layer, a conversion coating, is the product of substrate pretreatment and is usually a very thin (5-60 nm) inorganic layer that provides corrosion protection and improved adhesion between the substrate and the second layer, i.e. the primer, which is the principal provider of corrosion protection and is comprised of pigments or inhibitors and polymeric matrix. Finally, a topcoat that mainly serves as the main barrier against environmental influences such as extreme climates and ultra-violet rays and also for decoration.

Generally, organic coating systems protect metals against corrosion by four mechanisms, viz. the physicochemical (barrier), the electrochemical (inhibition or cathodic protection), ion exchange and the adhesional mechanisms [4,5]. Organic coatiang systems protect by physicochemical mechanism contain pigments of a lamellar, flaky, or plate-like shape, which greatly increase the length of the diffusion pathways for oxygen and water and decrease the permeability of the coating systems. Systems that protect by electrochemical mechanism consist of inhibitors which are dissolved by the electrolyte entering from the environment to retard corrosion due to cathodic process or both. Systems that protect by ion exchange mechanism consist of ion exchangers that hinder the transport of Cl^- and Fe^{2+} to the substrate.

Tremendous papers have been published on corrosion protection of various metals, such as Fe, Al, Al alloys, Cu, to mention just a few, with intrinsically conducting polymers (ICPs) since the earliest work by DeBerry [6] who observed the effective passivation of steel in sulfuric acid solutions by electrodeposited polyaniline (PANI). As candidates for metallic protection against corrosion, ICPs exhibit several advantages such as light weight, inexpensiveness, ease of synthesis, effective for corrosion protection e.g., PANI coating was said three to ten times more effective than zinc coating [7] and environmentally friendly as compared with those inhibitive metal oxides such as chromate or molybdate which pose potential environmental and health concerns [8,9] although are effective as corrosion inhibitors. ICPs for metallic corrosion protection can be synthesized chemically or electrochemically, used in different oxidation and doping states, processed in solution or dispersion, utilized alone or in composite (or blend) with other organic (or inorganic) materials and applied to various metals, ferrous or non-ferrous. The objective of this chapter is to review the past works concerning corrosion protection with ICPs, mainly PANI and polypyrrole (PPY), although there have been a few attempts with other ICPs like polythiophene (PTH) and its derivatives [10-15] for the use. It should be emphasized that the chapter is only intended to provide a brief summary of the literature and detailed, to some extent, with the approaches of coating preparation, testing media, and the results, which are not without controversy. Although 279 articles have been referenced,- omissions are inevitable due to the myriad of investigations carried out on this subject.

The organization of the rest of the chapter is as follows. In Section 2, we briefly describe the fundamentals of ICPs and especially the synthesis, structures, properties and processing of PANI and PPY. In Section 3, we discuss the corrosion protection of various metals with PANI, PPY and their composites according to the approaches by which the ICPs and ICPs containing films or coatings were formed. In Section 4, we present the generally accepted mechanisms proposed for corrosion protection with ICPs. Section 5 is the concluding remarks of the chapter.

2. SYNTHESIS, STRUCTURE AND PROCESSING OF POLYANILINE (PANI) AND POLYPYRROLE (PPY)

2.1. Fundamentals

ICPs are termed organic polymers that possess the electrical, electronic, magnetic, and optical properties of a metal while retaining the mechanical property, processibility, etc. commonly associated with a conventional polymer, and more commonly known as "synthetic metals"[16]. Since the first ICP, polyacetylene (PA), was successfully synthesized in 1977 by Shirakawa *et al.* [17], great interests have been aroused and a series of ICPs have been synthesized subsequently **(Tab. 5.1)** and researched extensively. Many potential applications such as electrochromic display [18,19], corrosion protection [1,6], electrostatic discharge [20], electromagnetic interference shielding [21], field-effect transistor [22], to mention the principle, have been suggested although the major drawback of intractability made these polymers difficult to process. With a process commonly referred to as doping, which can be accomplished in several ways, for instance, chemical doping by charge transfer, electrochemical doping, photodoping, doping by acid-base chemistry (only PANI undergoes this form of doping) and so on [23], a variety of interesting and important phenomena can be realized as summarized by Heeger [24] according to the chemical, electrochemical, photo and interfacial related properties of these polymers. And with the Nobel Prize in Chemistry for 2000 awarded to Alan J. Heeger, Alan G. MacDiarmid and Hideki Shirakawa for "the discovery and development of electrically conductive polymers" [25-27], the research of ICPs reached its zenith.

Tab. 5.1. Chemical structures of some intrinsically conducting polymers (ICPs).

Polyacetylene (PA)		Polyparaphenylenevinylene (PPV)	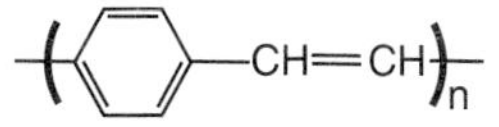
Polythiophene (PTH)	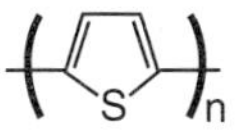	Polyparaphenylene (PPP)	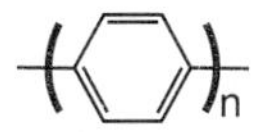
Polypyrrole (PPY)	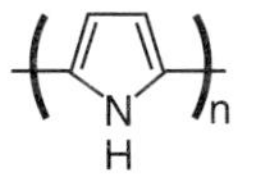	Polyparaphenylenesulphide (PPS)	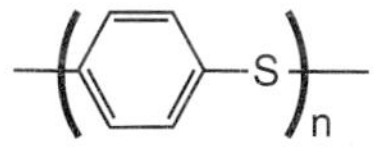
Polyaniline (PANI)	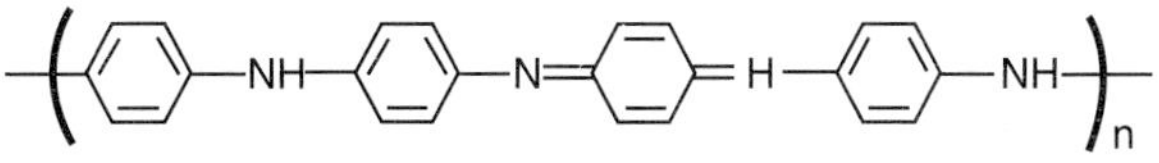		

Among the ICPs ever found, PANI and PPY are two important ones due to their low cost, high and tunable electrical conductivity, good environmental stability and unique chemical and electrochemical properties. They can be synthesized both chemically and electrochemically. Chemical oxidation method prefers for the mass production of ICPs while electrochemical method is suitable for preparation of uniform ICP films or coatings.

2.2. Polyaniline (PANI)

Although polyaniline (PANI) was first reported in 1862 [28] and known as "aniline black" from then on, it was not until the year 1985 when MacDiarmid *et al.* [29] synthesized PANI in an acidic system by chemical oxidation polymerization and firstly found its electric conductivity that PANI was known as a new synthetic metal. In the years followed, numerous papers were dedicated to the polymerization of PANI with different structures under various conditions, to the elucidation of polymerization mechanisms [30], to the optimization of the synthesizing and processing parameters to improve its processibility, and to the exploitation of potential applications of PANI.

2.2.1. Synthesis of Polyaniline (PANI)

Generally, PANI can be synthesized by direct oxidation of aniline with the use of appropriate oxidants or by the use of electrochemical oxidation on various electrodes. The chemical synthesis is relatively simple and cheap while the electrochemical one is clean. Commonly, the chemical synthesis is preferable over the electrochemical one, whereas, the electrochemical one is better suited for *in situ* spectroscopic investigations. The structures, morphologies, physico-chemical properties of the polymer depend strongly on the synthesis methods and conditions [31,32].

PANI in emeraldine salt (ES) form can be successfully prepared in aqueous acidic solutions with various oxidants such as ammonium persulfate or peroxydisulfate [29], hydrogen peroxide [33], potassium bichromate [34], to name the principle. The acids adopted include hydrochloric acid (HCl), sulfuric acid, etc. in different concentrations. Hydrochloric acid was adopted in common use and will not stain on the PANI prepared after completely dried for its volatility. The total reaction equation of aniline polymerized with peroxydisulfate in hydrochloric acid was illustrated in **Fig. 5.1** and the green PANI obtained in ES form can be dedoped with ammonia to prepare coffee-colored dedoped PANI, which is in emeraldine base (EB) form. As far as the mechanisms are concerned, the anilinium cation-radical mechanism is generally accepted for oxidative polymerization of aniline [35]. Accordingly, anilinium cation-radicals are formed at the first step, and then two

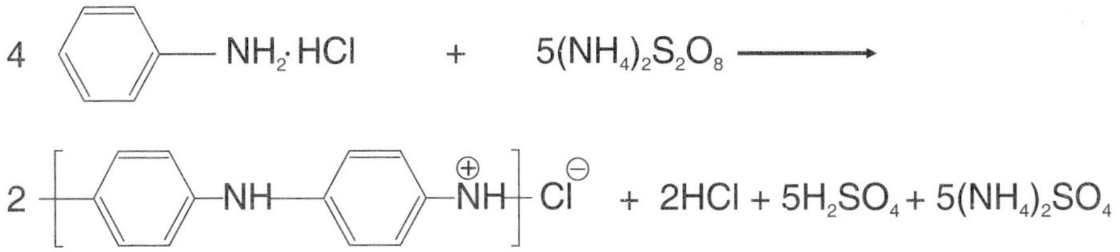

$$4 \; \text{C}_6\text{H}_5\text{—NH}_2\cdot\text{HCl} \; + \; 5(\text{NH}_4)_2\text{S}_2\text{O}_8 \longrightarrow$$

$$2 \left[\text{—C}_6\text{H}_4\text{—NH—C}_6\text{H}_4\text{—}\overset{\oplus}{\text{N}}\text{H—} \right] \text{Cl}^{\ominus} \; + \; 2\text{HCl} + 5\text{H}_2\text{SO}_4 + 5(\text{NH}_4)_2\text{SO}_4$$

Fig. 5.1. Total reaction equation for aniline polymerized in hydrochloric acid with ammonium persulphate as oxidant.

cation-radicals combine to form N-phenyl-1,4-phenylenediamine which will be readily oxidized to trimers, tetramers and so on until the high molecular weight polymer is formed.

Electrochemical synthesis is the anodic oxidation of aniline on inert electrodes (e.g., the most usually platinum) or recently on other materials such as iron, copper, zinc, aluminum [36] and son on. Several electrochemical modes including galvanostatic, potentiostatic and cyclic voltammetric ones were employed for electrochemical polymerization of aniline. The common reaction media for electrochemical synthesis are acidic whereas there are also reports of PANI synthesis in basic media under specific conditions. The electrochemically prepared PANI is purer than chemically prepared one and is beneficial for characterizations, while the electrochemical methods are unfavorable for mass production of PANI, for example, in the case of industrialized production.

Other methods such as plasma polymerization [37-39], electroless polymerization [40,41], solid-state polymerization [42-43], etc. were also used for polymerization of aniline and their derivatives.

2.2.2. Structures and Redox Properties of Polyaniline (PANI)

The first molecular structure of PANI was given by MacDiarmid *et al.* [44] in EB form and ES form **(Fig. 5.2)**. The average oxidation state, (1-y) can be varied from zero to give the completely reduced PANI, "leucoemeraldine", to 0.5 to give the "half-oxidized" PANI, "emeraldine", and to 1 to give the completely oxidized PANI, "pernigraniline" **(Fig. 5.3)**. In principle, the imine nitrogen atoms in PANI molecule can be protonated in whole or part to give the corresponding salts and the degree of protonation of the polymer depends on its oxidation state and on the pH of the aqueous acid [45]. Five different phases of PANI can be distinguished, although people are not sure hitherto whether the oxidation states of PANI at the molecular level are ranging from 0 to 1 continuously [46], as schematically shown in **Fig. 5.3**, which are related either by protonation-deprotonation along the horizontal axis, or by redox equilibrium along the vertical axis, or by a combination of both [47]. Among these phases, only the ES is electrically conducting.

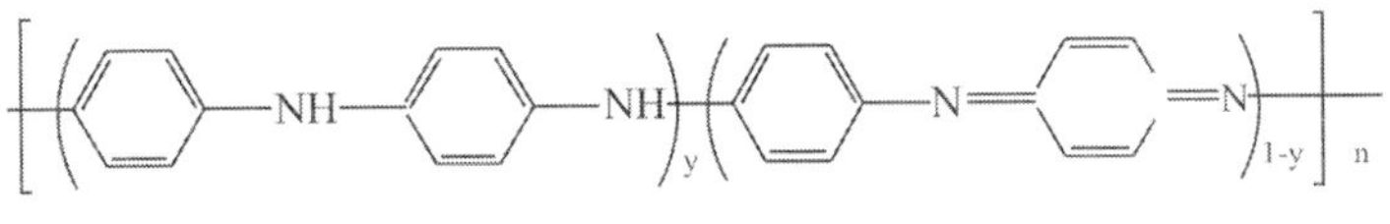

EB

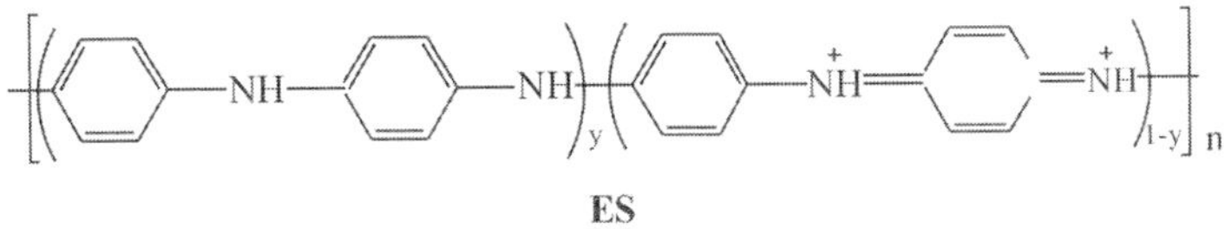

ES

Fig. 5.2. Molecular structures of polyaniline (PANI) in emeraldine base (EB) form and emeraldine salt (ES) form.

In addition, Wang [48] suggested EB to be a mixture of undoped emeraldine with doped emeraldine based on the ion exchange experiment and published electron spectroscopy for chemical analysis data. The proposed modified EB formula is shown in **Fig. 5.4** with the couterion of hydroxide ion, and thanks to these hydroxyl counterions, which will exchange with the aggressive anions and make the aggressive anions trapped by the coatings, excellent corrosion protection was achieved with EB as compared with ES [49].

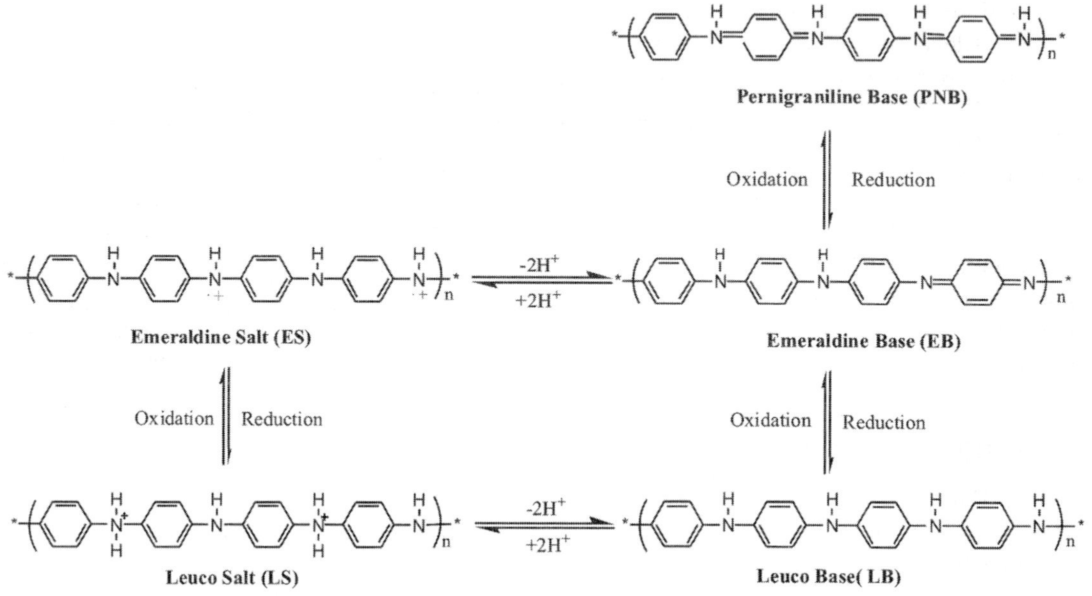

Pernigraniline Base (PNB)

Oxidation | Reduction

Emeraldine Salt (ES)

$-2H^+$ / $+2H^+$

Emeraldine Base (EB)

Oxidation | Reduction

Oxidation | Reduction

Leuco Salt (LS)

$-2H^+$ / $+2H^+$

Leuco Base(LB)

Fig. 5.3. Schematic phase diagram of polyaniline (PANI).

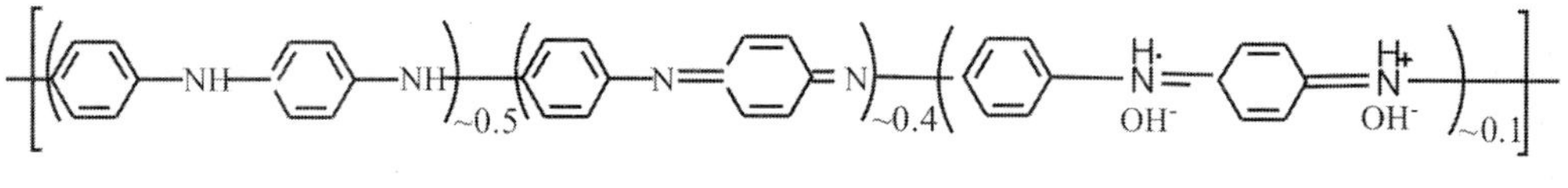

Fig. 5.4. Emeraldine base (EB) formula of polyaniline (PANI) as modified by Wang. Reproduced with permission from J.G. Wang; Synth.Met., 132(I), 49 (2002). Copyright © 2002, Elsevier Science Ltd.

2.2.3. Processing of Polyaniline (PANI)

As mentioned above, the reversible control of electrical properties by either changing the oxidation state of the PANI backbone or via protonation of the imine nitrogen atoms makes PANI a unique polymer among the ICPs ever found. However, PANI is intractable, i.e., neither could it be processed in the melt state, since it decomposes at the temperatures below a softening or melting point, nor be processed in solution, since it cannot be dissolved in conventional organic solvents except poorly soluble in few strong polar solvents like N-methylpyrrolidone (NMP), N, N-dimethylformamide (DMF) and dimethyl sulfoxide (DMSO). To date, significant efforts directed to improving the solubility and processibility of PANI have been undertaken and the works were summarized by Namazi *et al.* [50] as follows:

- Structural modification of PANI by ring or N-substitution.
- Electrochemical polymerization of aniline in polymer matrix.
- Doping PANI with functionalized protonic acids in place of inorganic acids such as HCl.
- Blending with other polymers using a cosolvent.

And, as we supplemented below:

- Blend or composite of particulate PANI with conventional polymers [50-52].
- *In-situ* polymerization of aniline with conventional polymers [53-56] or inorganic materials [57-60].
- Graft polymerization of aniline on other materials, e.g., on Si (100) surface [61].
- Enzymatic synthesis of molecular complexes of PANI with DNA [62], and so on [63].

No matter in which approach it is prepared, from solution cast or *in-situ* polymerization (chemically or electrochemically), the major drawback of pure PANI film or coating has been its poor mechanical properties and adhesion, and thereafter the PANI composite or blend seems to be more preferable for use in that it possesses the mechanical properties of the insulating host matrix and the electrical properties of the conducting PANI guest [64].

2.3. Polypyrrole (PPY)

Polypyrrole (PPY) was first reported by Angeli in 1916 and was known at the time as "pyrrole black" [23]. It is also one of the most investigated materials for its easy polymerization, good electrochemical activity at higher pH [65] and wide practical applications in gas sensors [66-69] and glucose sensors [70], electrochromic devices [71-72], batteries [73], actuators [74-76] or even bi-ionic actuator [77], electromagnetic interference shielding [21] and so on. The electrical conductivity of PPY can be switched between the conducting and the insulating states through doping and dedoping cycles. Moreover, the conductivity of oxidized PPY can be increased to a level of 10^3 Scm^{-1}, or higher, by doping PPY with smaller counteranions that are coplanar with the polymer chains [78]. Actually, PPY is a kind of mixed conductor via both electronic and ionic mechanisms. This property makes it a potential conductive polymer that can be used in the cathodic material of lithium batteries [73,79]. Under applied anodic potential, PPY film can be positively charged (**Fig. 5.5**). This process is followed by the formation of a cation radical in the pyrrole chain with corresponding formation of a salt with dopant ion to maintain the electroneutrality.

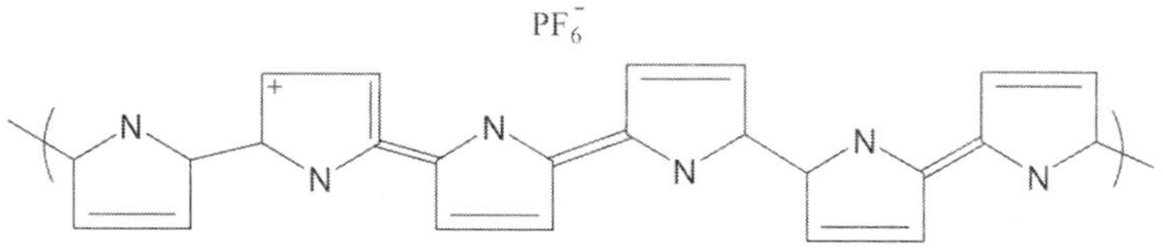

Fig. 5.5. Positively charged polypyrrole (PPY) molecule. Reproduced with permission from J.O. Iroh and K. Levine; J. Power Sources, 117(1-2), 267 (2003). Copyright © 2003, Elsevier Science Ltd.

Polypyrrole (PPY) films display distinct color either in reduced state (yellow) or in its oxidized state (black), and they absorb the light intensively in the longer wavelength bands [81]. According to Syritski *et al.* [82], the mechanisms for the reduction and oxidation of a PPY film can be grouped into two categories depending on the mobility of the dopant ion within the polymer matrix as illustrated in **Fig. 5.6** for a PPY doped with D$^-$ anion and immersed in CA electrolyte solution.

In the case of an immobile doping ion entrapped in the polymer, an excess of negative charges on its backbone originating under electrochemical reduction is compensated by cation incorporation from the solution (**Fig. 5.6A**). On subsequent re-oxidation, charge neutrality of the polymer can be supported by both cation expelling and anion incorporation from the solution (**Fig. 5.6B**). When the polymer was doped with an anion of good mobility, the dedoping process can then take place (**Fig. 5.6C**) and during re-oxidation, the injection of positive charges onto the polymer backbone is accompanied by capture of the anions from the solution (**Fig. 5.6D**). PPY can be easily synthesized by either chemical and electrochemical oxidative polymerization or other methods such as plasma polymerization [83].

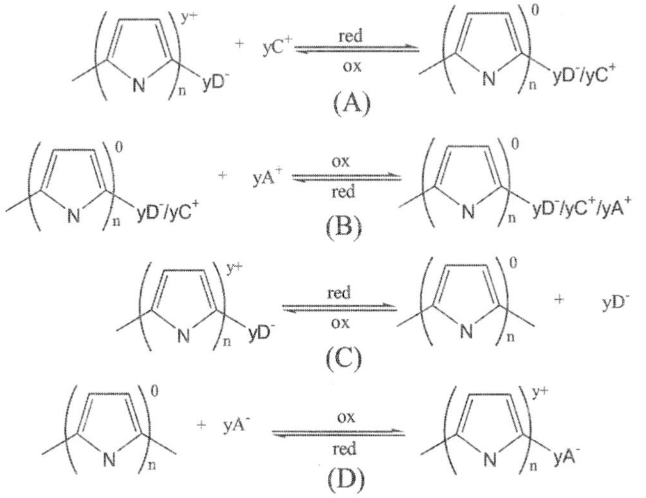

Fig. 5.6. Mechanisms for the reduction and oxidation of polypyrrole (PPY). Reproduced with permission from V. Syritski *et al.*; Electrochim.Acta, 48(10), 1409 (2003). Copyright © 2003, Elsevier Science Ltd.

2.3.1. Synthesis of Polypyrrole (PPY)

In chemical oxidative method, conducting and stable PPY is often synthesized by polymerization of pyrrole in aqueous solution containing an oxidant and a surfactant. The oxidants used generally are as follows: ferric sulfate, ammonium persulfate, hydrogen peroxide, and many kinds of salts containing transition metal ions, for example, Fe^{3+}, Cu^{2+}, Cr^{6+}, Ce^{4+}, Ru^{3+} and Mn^{7+}. The surfactants used cover the anionic surfactants such as sodium dodecylbenzenesulfonate, dodecylbenzenesulfonic acid (DBSA), sodium dodecyl sulfate, poly (ethylene oxide)-*n*-alkyl-3-sulfopropyl ether potassium salt; cationic surfactants such as tetradecyltrimethylammonium bromide; and non-ionic surfactants such as poly (ethylene oxide) etc. [84]. PPY prepared by chemical oxidative polymerization is often in powder form, whereas, some coral-like nanowires and nanowire networks of conducting PPY [85], nanotubes and nanofibrils of PPY can also be prepared through this way [86,87].

The electrochemical synthesis of PPY can be carried out in a variety of electrolytes, aqueous or nonaqueous, acidic or alkaline. The first electrochemically synthesized PPY film was reported in 1979 on a platinum electrode by the anodic oxidation of pyrrole in acetonitrile [88]. The mostly used electrolytes for electrochemical synthesis of PPY were acetonitrile and water [89], whereas, PPY films can also be synthesized from alcohol solutions as the one reported firstly by Xu *et al.* [90], in which a PPY film with good mechanical property, high conductivity ($10^2 Scm^{-1}$)

and nice environmental stability was obtained. Stable, adherent and conducting PPY films can be synthesized not only in acidic solutions as the commonly used, but also in alkaline solutions [91-93]; not only at inert electrode like platinum, but also at a variety of other electrodes such as iron [94], pure aluminum [95], aluminum alloys [96], titanium [97], nickel [98], copper [99,100], magnesium [93], zinc [101] and other materials, e.g., ITO [102].

The mechanism for electrochemical polymerization of pyrrole has not yet been fully elucidated and one of the mechanisms reviewed by Deronzier *et al.* [103] was shown in **Fig. 5.7**. The initial step involves an oxidation to a pyrrole radical cation that couples with another radical cation, or with pyrrole followed by a further charge transfers. Two protons are eliminated, forming a neutral dimer. The latter is immediately oxidized at the applied potential, since it is more oxidizable than the monomer, leading to a new radical cation. The continuing reaction is a lengthening of the oligomer chain, and thus leading to the growth of PPY on the electrode surface.

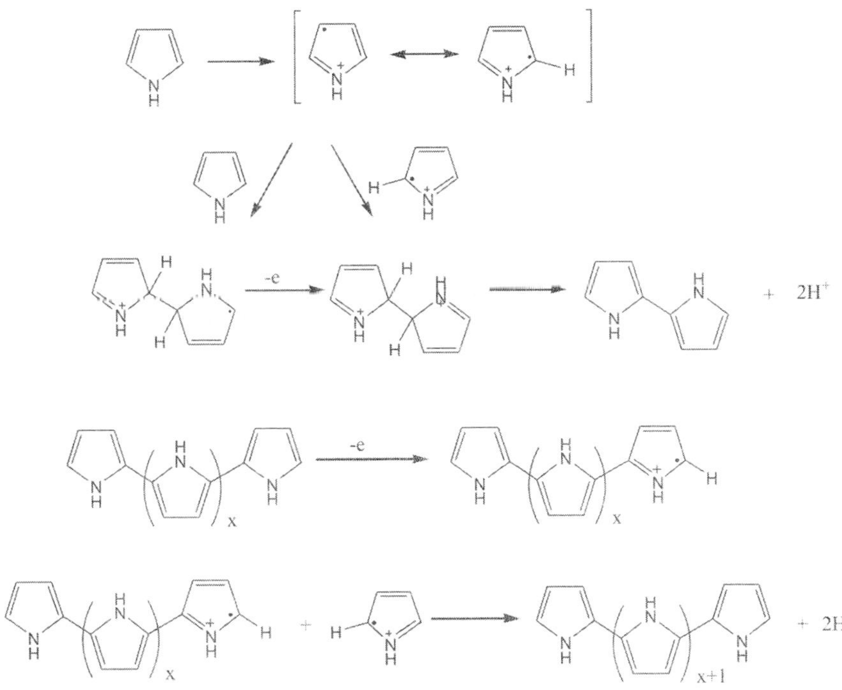

Fig. 5.7. Mechanism for electropolymerization of pyrrole. Reproduced with permission from A. Deronzier and J.C. Moutet; Coord.Chem.Rev., 147, 339 (1996). Copyright © 1996, Elsevier Science Ltd.

2.3.2. *Processing of Polypyrrole (PPY)*

Great efforts have been made to improve the processibility of PPY, i.e. the insolubility and infusibility as the same for PANI. The ways can be concluded as shown below:

- Synthesis of poly (3-alkylpyrrole) with an alkyl group equal or greater than a butyl group [104,105] and poly (*N*-substituted pyrroles) [106] to improve the solubility in organic solvents.
- Synthesis of block and graft copolymers containing conventional and conducting sequences, where the conventional sequences increase the solubility of the resultant block or graft copolymer [107,108].
- Doping with relatively large dopants such as DBSA, naphthalenesulfonic acid (NSA), or mixed acid containing DBSA and camphor sulfonic acid (CSA) [109] to increase the solubility in various organic solvents [110-112].
- Preparation of PPY composite [80,113,114] or blend [115] with conventional polymers, inorganic materials, e.g., organoclay [116], carbon nanotubes [70], textile fabrics [117], Fe_3O_4 [118], TiO_2 [119], etc., in core-shell structure [120,121] or nano dispersions [121] by *in-situ* polymerization, mixing of PPY solution with a second material in solution or dispersion [67,112,116].
- Synthesis of block copolymers of pyrrole and aniline [122], thiophene [123] etc. to improve the mechanical properties of the resultant polymers.
- Preparation of composite or blend of PPY with other ICPs, such as PANI [124-127], PTH [128] etc.

3. CORROSION PROTECTION WITH POLYANILINE (PANI) AND POLYPYRROLE (PPY)

3.1. Polyaniline (PANI)

Polyaniline (PANI) for metallic corrosion protection was adopted in various processing approaches including solution casting or spinning, electrochemical deposition, composite (or blend) with conventional polymers from solutions or dispersions, to form PANI or PANI containing coatings, and tested in diverse corrosion medias such as: aqueous HCl, aqueous NaCl solution, tap water and so on. We reviewed below the metallic corrosion protection properties and performances of various PANI coatings according to their preparing approaches, and detailed with the testing methods and the results.

3.1.1. Solution Prepared PANI Coatings

In this method, PANI coatings were commonly prepared by dissolving chemically synthesized PANI, in EB or ES form, in an organic solvent such as NMP, DMF, and then spray, cast or spin coated onto metal surfaces. Dried in air or an oven, the coupons were subjected to corrosion testing.

Wei *et al.* [129] used electrochemical measurements of corrosion potential, polarization resistance and corrosion current to test the effect of EB coating prepared on cold rolled steel (CRS) coupons in 5-wt% NaCl solution. The results showed that EB exhibited good corrosion protection in aqueous NaCl medium as evidenced by the increase of corrosion potential and polarization resistance and by the decrease in the corrosion current in comparison with the uncoated CRS, and the EB coating was better than the acid-doped PANI coating [130]. Using X-ray photoelectron spectroscopy (XPS) and potentiodynamic studies, Fahlman *et al.* [131] also discovered the noble shift of open circuit potential (OCP) of EB coated CRS coupons as compared with those non-coated ones. Wrobleski *et al.* [132] found solution prepared EB coating from NMP was effective as corrosion-resistant primer and this coating system performed better than the epoxy or polyurethane (PU) coating alone. Mirmohseni *et al.* [133] investigated EB coating prepared from NMP solution on iron samples in various corrosive environments such as NaCl (3.5-wt%) solution, tap water and acids. All results indicated EB was preferred over conventional polymers, with the noble shift of OCP and significant decrease of corrosion current owing to the passive layer formed on iron by the redox properties of PANI. Santos *et al.* [134] pointed out EB films spray on steel plates from NMP solution was capable to protect untreated steel against corrosion in aqueous NaCl solution saturated with air. In order to further investigate the corrosion protection property of EB, Fahlman *et al.* [135] cast EB solution on the back surface of the CRS coupons, not on the surface to be protected, and the coating was referred to as a protective undercoat. All samples were handled in ambient air for a minimum of 12 hours before being exposed to corrosive environment. Using XPS study, a thin top layer of Fe_2O_3 of ~15 Å in thickness followed by a thicker layer of Fe_3O_4 were found for all the EB undercoated samples. EB on stainless steel (SS) passivated SS by superficially oxidizing the SS surface. In the meantime, EB itself was reduced to leuco base (LB), and air oxidized it back to EB then. When pitting corrosion occurred, EB healed it in the same way as pointed by Ahmad *et al.* [1]. In another paper [136], SS electrodes coated with EB solution (in NMP) were tested in highly corrosive H_2SO_4 solution by a variety of experimental methods, including the OCP measurement, Auger depth profiling and the scanning reference electrode technique (SRET). In the OCP measurement, a dip and recovery of the OCP was firstly reported, in addition to the high positive OCP values for PANI coated iron and SS in acidic solutions reported earlier [6]. Dramatic dip and slower recovery was observed for the thinner film and more highly concentrated solutions, which was interpreted by the cracks in the PANI film shown by scanning electron microscopy (SEM) images and was supported by the work of Ahmad *et al.* [1] as will be detailed below. On the basis of the result that the fastest recovery of the OCP to more positive values was found for oxygen bubbled solution and the slowest for nitrogen bubbled one, and the fact of "self-healing" phenomenon of the intentionally cut pinholes, the authors confirmed that not all of the metal surface must be coated in order to obtain passivation.

In the work of Ahmad *et al.* [1], EB films were deposited from DMSO solution onto the surface of SS and the corrosion protection performance of the coated samples were evaluated in both H_2SO_4 and chloride-containing (0.2 M H_2SO_4 + 0.15 M Na_2SO_4 + 0.1 M NaCl) solutions. The EB deposited SS-304 was dipped first in H_2SO_4 of pH 4 for ca. 20 h to convert the EB to sulfate salt form and then dipped 2 cm in H_2SO_4 of pH 5 for OCP measurements. After 1 h, a 1 cm-long scratch was made on SS-304 to the metal surface. The OCP decreased but in a few minutes rose again, although fluctuated, it stayed at –0.15 V after 7 days. A similar experiment was repeated with SS-430 and tested in 0.4 M H_2SO_4. The initial OCP decreased with time and to 0.44 V after 2 h. When a deep scratch was made, OCP measurement showed a decrease of 0.025 V, but it increased again in 3 min. Half of the EB layer was removed by a second after 3h, but the OCP drop for ca. 2-3 min and rose again. All these results indicated EB on SS-430 had remarkable capacity to protect SS against corrosion from 0.4 M H_2SO_4. However, in chloride-containing solution, the OCP dropped quickly from the initial 0.4 V to about 0.1 V, increased a little and dropped to –0.56 V in ca. 17 h. Repeated experiments showed that the as deposited EB without pre-treatment of the steel could not protect the steel from pitting corrosion for a long time, which was probably due to the film porosity and the film adhesion to the metal surface. With pre-treatment of the steels by using of phosphoric acid, polyphosphoric acid, alizarin sulfonate or chromotropic acid, the then deposited EB films protect SS from corrosion in chloride-containing acid media.

Pud *et al.* [137] prepared coatings of pure PANI from NMP solution (Versicon™ from Monsanto) cast on mild steel substrates and subsequently doped them with *p*-toluenesulfonic acid (TSA) to form doped coatings. Prepared also were the sulfonic acids like CSA, DBSA and dodecylsulfonic acid (DSA) doped PANI coatings formed from the outset in the doped state through casting from xylene or chloroform solutions. All the coated panels were topcoated with a curable epoxy resin and tested in 3.5-wt% NaCl and 0.1 M HCl solutions at room temperature. Significant decrease of corrosion currents was confirmed for the pure undoped PANI coating and coating doped with TSA in neutral medium, but in acidic medium, the undoped PANI one acted much weaker than the TSA doped one. The action of the doped coatings from xylene or chloroform was much worse than that of both abovementioned kinds of coatings. In the neutral medium this set even increased the corrosion currents of steel. Moreover, no gray oxide layer was observed at the panel surface for the directly doped coatings, but formed for undoped PANI followed by doping which was in agreement with the observations of Fahlman *et al.* [135] who have shown by XPS that the coating of PANI in the EB form produced iron oxide layers both at steel and iron surfaces. In the authors' opinion, the difference in corrosion currents can be probably explained by the appearance or absence of iron oxide layers that passivate the steel surface. And the undoped PANI or possibly its complex with Fe may catalyze surface oxidation which the directly doped ones cannot do at least when they were applied by casting from xylene or chloroform solutions.

PANI coatings, both in EB and self-doped sulfonated form, were cast and spun from solutions onto Al and Al alloys (Al 3003 and Al 2024-T3) by Epstein *et al.* [138]. Potentiodynamic polarization studies of the coated samples in 0.1 M NaCl solution showed that there were modest indications of potential corrosion protection capability of PANI for Al and Al 3003 alloy, while for Al 2024-T3 alloy, the corrosion currents measured for both EB and sulfonated PANI coated coupons were tenfold reduced from that of the uncoated materials, indicating much lower corrosion rates. The corrosion potentials for EB and sulfonated PANI coated Al 2024-T3 alloys were −480 mV (All potentials in this chapter were *vs.* saturated calomel electrode (SCE) unless specified) and −880 mV as compared to −990 mV of the bare Al 2024-T3 alloy. XPS study showed a reduction of copper concentration at the surface of the coated coupons and it was suggested that the EB and sulfonated PANI coatings facilitated the extraction of copper from the surface of Al 2024-T3 thereby reduced the galvanic couple between aluminum and copper that usually leads to accelerated corrosion of Al 2024-T3 alloys.

The amino-terminated aniline oligomers in EB form were also solution cast on CRS and tested in 5-wt% NaCl solution [139]. Measurements of corrosion potential, polarization resistance and corrosion current all revealed oligomers had better anti-corrosive performance than the conventional EB film, which was attributed to the possibility that oligomers could have better "adhesion" (i.e. stronger interactions) with the metal surface. The oligomer-cured epoxy resin coatings also exhibited excellent corrosion protection effects. Some time later, Sein Jr *et al.* [140] calculated the electronic structure and thermodynamic properties of aniline trimers adsorbed onto cluster models of the Al (100) and Fe (100) surfaces by using the ZINDO/1 semi-empirical method and interpreted the effects of progressive oxidation and protonation of the trimers on corrosion inhibition in terms of these calculations. The calculations demonstrated that the interaction of aniline trimers with Fe (100) surface was far greater than with an Al (100) surface. It was also suggested that the ability of the emeraldine dihydrochloride to lower the HOMO of the Fe (100) cluster might be a significant factor in its ability to inhibit corrosion on iron (and steel) surfaces. More recently, the same authors presented a hybrid *ab initio*/density functional theory (DFT) calculation on the thermodynamic parameters for several competing reaction sequences occurring during the interaction of aniline trimers with the iron or aluminum surface and a model for corrosion inhibition by aniline trimers was advanced based on the analysis of both equilibrium (thermodynamics and potentials) and rate (kinetics) [141]. It was shown that the emeraldine dihydrochloride was the most effective form for corrosion inhibition, as compared with the emeraldine hydrochloride.

EB films spun on iron rods were tested in chloride or sulphate solutions at neutral or slightly acidic pH and it was indicated that ES was produced near the metal surface by *ex-situ* Raman spectroscopy. Combined with the impedance results, it was assumed that the production of Fe^{2+} and the precipitation under an oxide form would produce H^+ and contribute to the formation of ES, although it was not clear whether oxygen can be reduced and in which part of the film [142]. In another paper, neutralized PANI coatings were applied to mild steel surface as a solution and dried at room temperature in air. Doped PANI used was prepared by exposing the coated samples to an aqueous solution of TSA for 4 to 24 h, and then rinsed with distilled water and air-dried. Electrochemical measurements indicated PANI coatings provided significant corrosion protection to mild steels exposed to dilute HCl and NaCl and further, the corrosion protection extended to areas of bare steel that gaivanically coupled to the PANI coated steel. The level of corrosion protection provided by doped PANI in dilute HCl was more significant than in neutral saline conditions and the extended protection to bare area was surprising and of potential commercial application. But the coatings exhibited rather poor adhesion to steel substrates. It was evident that the levels of protection provided by PANI depend on the form of PANI and the nature of the corrosion environment [143]. PANI in both doped and undoped form were used in a typical steel phosphating process and the results compared with Cr (VI) post-treatment based on impedance measurements showed that PANI effectively stabilized coating systems in aggressive environment (1 M NaCl, pH = 4) [144].

Chemically synthesized PANI was doped with CSA and dissolved in *m*-cresol with sonicating overnight, and then, with filtration of the undissolved polymer particles, the resultant solution was cast onto glass substrates coated with an electrically conductive film of fluorine doped tin oxide, and onto surface of acid etched Al alloy coupons. Galvanic coupling of CSA doped PANI to Al alloy in 3.5-wt% aqueous NaCl solution resulted in rapid polarization of the conductive polymer to a mixed potential dominated by the alloy. The polarization resulted in the conversion of the conducting PANI to the nonconducting LB form which offered a barrier effect rather than an electrochemical process and led to the reduced corrosion on Al alloy. Zero resistance ammeter measurements of galvanic couples between CSA doped PANI and Al alloy revealed that PANI, particularly in the presence of oxygen, polarized the Al alloy slightly positive of its initial corrosion potential and increased the anodic corrosion current density at the exposed alloy [145]. A similar analysis was applied to DBSA doped PANI coated copper by Brusic *et al.* [146], who observed that the corrosion potential of PANI is about 200 mV more positive than that of copper, and thus, higher dissolution rates should be expected on copper in contact with PANI. However, the observation that PANI inhibited pitting corrosion of Al alloy in Cl^- containing media at defects in the PANI coatings suggested a corrosion inhibition mechanism other than anodic stabilization. A possible mechanism involving inhibition by residual solvent, aniline monomer and PANI oligomers remaining in the PANI coatings was also postulated, but seemed less likely. According to the work by Cook *et al.* [147], in which EB, *n*-dodecylphosphonic acid (DPA) doped PANI and a commercially obtained ES (Ormecon Chemie) were dispersed in *m*-cresol and dropped on the surfaces of mild steel with filtration of the undissolved residues for primer formation and subsequently topcoated with four layers of a polyacrylic resin, the significantly improved protection of mild steel from 0.1 M HCl and 0.1 NaCl solutions came from the inhibition effect of the dopants, which were released with reduction of ES, not from the anodic protection in the form of passivation. Besides, the EB coating only provided some degree of protection in acidic medium.

Brusic *et al.* [146] deposited various PANIs on copper and silver by spin-drying (with thickness of 95-510 nm), and the PANIs include unsubstituted PANI in base and salt (doped with either HCl or DBSA) form and substituted PANIs with either ethoxy or propyl groups in the *ortho* position. Corrosion tests were adopted which most closely resemble the conditions to which copper and silver might be exposed during the useful life of electronic product, and were thus conducted in water or a temperature/humidity/bias chamber. The results showed that the substituted PANI in base form exhibited the best corrosion protection, even superior to the commonly used corrosion inhibitor benzotriazol. The protection property of the base form was primarily due to the barrier effect and the superior performance of the ethoxy derivative was attributed to the better coverage and better adhesion of the polymer to the metal surface. The corrosion protection of doped unsubstituted PANI depended on the processing conditions but was worse than the base polymer in all cases. The DBSA doped PANI even had an adverse effect on corrosion protection that was most probably a result of the excess DBSA presented in the film.

Tallman *et al.* [148] cast solutions of doped PANI, which were obtained from Monsanto in xylene and butyl cellosolve, on steel panels with or without epoxy topcoats. Electrochemical impedance spectroscopy (EIS) and electrochemical noise method (ENM) were adopted to probe the behavior of these coated samples under immersion in aqueous 3-wt% NaCl solution. Upon immersion, the polymer and/or steel underwent electron-transfer reactions, as evidenced by the significant current flow and potential oscillations in ENM experiments. The immediate commence of these events indicated that water and ion ingress into the coatings rapidly. Oscillations in current and noise resistance persisted for about 5 days, which reflected a poor barrier property for PANI coating, and during which time the oxygen sparging experiment provided evidence for an interaction between PANI and oxygen, at least during the early stages of immersion. The pH value of the immersion electrolyte decreased during this same time, after which the pH remained rather constant. Tafel plots showed lower corrosion rates and more noble corrosion potentials for the PANI coated samples as compared to bare steel. All electrochemical data confirmed that significant interactions between PANI and steel occurred during the first 5 days of immersion, but that a continuing interaction occurred throughout the entire immersion period. Moreover, EIS carried out on an epoxy topcoated PANI coating indicated a superior performance of the system to that of bare steel coated with epoxy alone. In another paper by the same group [149], the same PANI and same preparing and testing procedures were adopted on Al alloy samples under immersion in a dilute Harrison solution (an aqueous solution containing 0.35-wt% ammonium sulfate and 0.05-wt% NaCl). For the PANI coated alloys without epoxy topcoat, although an strong metal-polymer interaction was characterized by EIS and ENM data and a decrease in immersion electrolyte pH during the first few days of immersion, the coatings on each alloy failed after ca. 20 to 25 days, the same time-to-failure as observed with epoxy coating alone. With an epoxy topcoat, the PANI coated alloy exhibited significantly improved corrosion protection compared to an epoxy topcoat alone.

Commercially obtained PANI, which was in the ES form doped with dinonylnapthalene sulfonic acid (DNNSA) and received as a liquid concentrate comprised of ~47 wt% PANI-DNNSA solids in a mixture of butyl cellosolve and xylene solvent, was cast on cold-rolled A06 grade plain carbon steel (PCS) after dilution with AR-grade xylene and dried at 80°C for 24 h to prepare the ES coating, i.e. PANI-DNNSA coating [150]. The EB coating was obtained by immersion of the PANI-DNNSA coated PCS in sodium hydroxide solution (1.0 M dissolved in 50/50 water/methanol solution) for 15-20 min. Corrosion tests of both the coatings and conventional epoxy coating were carried out in aerated 3.5-wt% NaCl (pH 5 to 6) solution at ambient conditions for up to 10 days. Visual inspection showed that the color of the ES coating changed from the initial green to transient pale green/yellow and finally to blue EB after 2 h to 4 h of immersion; with longer immersion period, the specimen produced small localized rust spots of less than a millimeter in diameter, which increased in size and number over the immersion period. Specimen blistering was detected between 48 h and 72 h. In the case of the EB, the coating retained its blue color during immersion. Localized rust spots with diameter less than 1 millimeter appeared after 24 h immersion and the spots increased in size and number with immersion time. Blistering was detected at 72 h and intensified with further immersion. Film breakdown was observed upon extended immersion. Quantitative testing by the total iron ion analysis of the immersion solution indicated that the corrosion of ES coating showed a three-stage behavior. The corrosion began immediately after immersion and continued at the same rate for the first 3 h, which was resulted from both the many pinhole defects of the ES coating and the electroactive nature of the coating, i.e. due to the acidic character of ES, galvanic coupling of the more noble ES to steel caused ennoblism but not passivation. Between 3 h to 10 h, the iron concentration was constant, which may be associated with the more effective passivating oxide provided by the ES transformed EB, as showed by visual inspection and UV-visible-NIR spectra. After ca. 10 h, the iron concentration again began to increase. The constant iron concentration between 3 and 10 h of immersion, which has no relation with the iron absorbing on the polymer coating as indicated by some other experiments, suggested that the corrosion ceased or slowed significantly during this time. The EB coating showed four distinct stages of iron dissolution with the last three stages similar to the behavior of ES coating. In the first 4 to 5 h of immersion, very little of iron dissolution was detected, though the pinhole defects of the coating, which was due to the formation of passivating oxide by EB. The Pourbaix diagram for iron in water suggested also that EB was likely to produce passivating ferric hydroxide surface oxides, while ES may shift the metal to active corrosion region of the diagram due to the pH difference of EB and ES. The increased corrosion rate between 5 and 10 h was caused by the conversion of the EB to ES by the iron ions, a process known as pseudo-protonic doping. ES was converted to LB simultaneously and re-oxidized back to EB by dissolved oxygen, and then the plateau region, i.e. the almost ceased corrosion between 10 and 27 h was observed. However, the cycle of reaction did not continue indefinitely. The corrosion rate abruptly increased after ca. 27 h of immersion, which was mostly associated with film breakdown. The reasons for the breakdown were not known, but according to the authors, the swelling/de-swelling of the coating that is associated with the conversion of EB to ES and back to EB may induce fatigue-like stresses that could be responsible for cracking and delamination. Though the corrosion protection properties of the PANI coatings were inferior to the epoxy coatings of similar thickness, the mechanisms of protection were clearly different. EB provided excellent corrosion protection because of the stable oxide layer formed on the steel surface, which was not likely for ES due to its acidity.

The same PANI-DNNSA coating and EB coating were prepared as those described in [150] by the same group [151]. The EB film was re-protonated by immersion in various acid solutions. The re-protonation acids include HCl, tartaric acid (TA), oxalic acid (OA), DBSA, polymethoxyaniline sulphonic acid (PMAS) and nitrilotri (methyl phosphoric acid) (NTPA). All the PANI films were topcoated with either epoxy or PU. The thickness of the topcoat was ~3.5 μm for epoxy and 10 μm for PU. Corrosion test of the topcoated PCS specimens was carried out in stagnant aerated solution of 3.5-wt% NaCl (with pH from 5 to 6) at room temperature. In the case of epoxy topcoated samples, the ES primers showed improved corrosion protection even though the ES had appeared to be converted to EB by the amine-cured epoxy. A number of corrosion sites were formed during the initial immersion of the ES/epoxy topcoat system coated PCS, but the sites did not grow into large blisters and distinct color changes were noted at each corrosion site. Typically a localized corrosion spot of ca. 0.25 mm diameter was dark-black Fe_2O_3, and the polymer around the sites was re-transformed back to ES form by the local high concentration of metal ions, which is known as "pseudo-protonic" doping [150]. Upon longer immersion, large blisters of several millimeters in diameter were developed on the coating systems, and it was deemed that the coatings were failed at this moment. The differences between the ES primers with different dopants in terms of the time before gross blistering were evident, and the estimated order of longevity was: ES (NTPA) ~2 years > ES (PMAS) 1-2 years > ES (DNNSA) 1-2 years > ES (OA) ~1 year > ES (DBSA) 3-6 months. Besides, the development of large blisters considered responsible for the coating failure had little actual relation with the substrate corrosion. Large area of shiny surface was observed on removal of the coating after 3 years immersion, and dark grey/black oxide surface beneath the large blisters, not like that red/brown corrosion product of epoxy-only system. In the case of the PU topcoat, all the ES/PU systems showed lower corrosion rates relative to the PU-only coated PCS, which rusted locally during the second day of immersion and deteriorated further with prolonged immersion due to its porosity; in addition, all the ES/PU systems showed the same color changes during immersion, i.e. from the initial olive green, to pale green, to yellow, to pale blue and finally to the dark blue EB, which indicated the galvanic coupling between the PANI and the metal. The speed of color change from green to blue appeared to correlate with the corrosion rates and the order from slowest to fastest was: ES (PMAS)> ES (HCl) > ES (TA) > ES (DNNSA) >ES (DBSA). Comparison of all the PU-topcoated results showed that the EB/PU system exhibited the lowest corrosion rate throughout the 11-day immersion period. The study clearly showed that the combination of PANI primers with a barrier topcoat provided superior corrosion protection to steel in saline media in comparison with the topcoat only; and that EB was superior to ES, in which different dopants also had a significant effect on the corrosion rates due to the different extent of galvanic activity of the corresponding ES and the PCS; and also that the type of topcoat affected the corrosion rates.

A recent work [152] at Rockwell Scientific considered the corrosion protection capability of PANI coatings for Al 2024-T3 in neutral NaCl solution. EB film on Al 2024-T3 panel (7.6×7.6 cm^2) was accomplished by coating EB solution (obtained from Aldrich dissolved in a small volume of NMP) using a No. 13 Meyer bar and air-cured then. A region of ca. 8 cm^2 on the surface area was exposed to a solution containing 0.02 M of the inhibiting anion dopant A for 24 h with the help of a gasketed clamp-on glass cell to form the doped region on the film. The remaining part, i.e. the as-received coating region was referred to as the undoped region. Similarly, a proprietary PANI coating, PANDA, was prepared in the same way, except that the doped region was exposed to an inhibitor B. The samples were then rinsed and dried before corrosion testing, a salt fog exposure. Though the EB from Aldrich resulted porous poorly protective coating, less corrosion occurred in the doped region and the coating away from the doped region was entirely reduced. Due to the porosity of the coating, it was not clear, according to the authors, whether the treatment of the doping inhibitor inserted a releasable inhibitor in the coating or whether it acted by adsorbing on the base of the pores. However, the fact that the scribed region sustained less damage in the doped region than the undoped region suggested that there was some extent of the coating in releasing the inhibitor. In the case of the PANDA coating, considerable white corrosion product appeared in the scribe in the undoped region, while a slight tarnish was observed of the scribe in the doped region. In addition, the entire film in the undoped region changed to blue reduced form, as it appeared. By combining the results from a PANI blend coating (Zipperling Corrpassiv 900226/19®), in which white corrosion product was observed in the scribe away from the undoped region and the region adjacent to the scribe in the undoped region exhibited considerable apparent reduction/alkaline hydrolysis. The authors pointed out that the anionic inhibitors residing in the doped portion of the PANI coating was caused to release by the scribe polarization and the release of the inhibitors slowed substantially than the corrosion rate of the Al in neutral chloride salt fog. As such, the PANI or PANI containing coatings become "smart" in that corrosion inhibitors were released on demand, such as being breached in this study. Later, the same group [153] evaluated the corrosion inhibiting performance of a commercially available Ligno-PANI pigment and a PANI coating, which were additionally treated with anionic inhibitor such as the ammonium of 1-pyrrolidine dithiocarbamate or the dipotassium salt of 2,5 dimercapto 1,3,4 thiadiazole, in 5-wt% NaCl solution and pointed out that the controlled release of the inhibitors via ion exchange and reduction [147] process rendered the PANI based corrosion inhibiting coatings "smart". Salt fog exposure demonstrated that the effectiveness of the oxygen reduction reaction inhibitors incorporated PANI coatings for Al 2024-T3.

More recently, corrosion protection of aluminum alloy (Al 3003) by both undoped PANI and PANI doped with TSA and DBSA in aqueous 3.5-wt% NaCl and 0.1 M HCl solutions were investigated by Ogurtsov *et al.* [154]. The EB coatings were obtained by casting 2-wt% solution of neutralized Versicon ^{TM} in NMP and subsequent drying at 60°C for 3 h; the PANI-TSA coatings were prepared by doping of the EB coatings in 1M TSA solution for 24 h and then drying at 60°C for 2 h; the PANI-CSA and the PANI-DBSA coatings were produced by casting chloroform and xylene solutions correspondingly and drying at room temperature for 15 min. All the coatings were in thickness of ca. 15 μm and topcoated with a curable epoxy resin. All types of PANI coatings showed considerable shift of the corrosion potential (up to 300 mV) in the anodic direction. However, as indicated in the case for protection of mild steel [137], the magnitude of the shift correlated poorly with the corrosion inhibition factor γ. γ values for all tests were cited from the paper and re-tabulated in **Tab. 5.2**. It was evident that the undoped PANI was the most effective one in both media. The slightly better performance of PANI-TSA, as compared with PANI-CSA and PANI-DBSA, was due to, on one hand, the distribution of the dopants in the coatings. In the case of PANI-TSA, TSA concentration in the coating decreased from the coating/solution interface to the coating/metal interface, resulting in the incompletely doped or undoped PANI near to the Al 3003 surface, differing from the homogeneous distribution of CSA and DBSA in the corresponding coatings. On the other hand, to the interaction of the PANI-TSA with the alloy surface, which could probably be partly formed in the stage of NMP solution casting and changed the electrochemical behavior of the whole system. Comparison of the work on aluminum with literature [137] on mild steel showed that the electrochemical behavior of PANI on these substrates was different, as in the case of mild steel [137], the best result (γ = 18.75) was PANI-TSA in 0.1 M HCl, not like the best γ achieved in neutral medium with undoped PANI in this work.

Tab. 5.2. Inhibition factors (γ) for all PANI coatings in both neutral and acidic solutions. Reproduced with permission from N.A. Ogurtsov *et al.*; Synth.Met., 143(1), 43 (2004). Copyright © 2004, Elsevier Science Ltd.

Testing medium	*Coating*	*γ*
	Undoped PANI	12
	PANI-TSA	5.0
3.5-wt% NaCl water solution	PANI-CSA	0.79
	PANI-DBSA	0.86
	Undoped PANI	4.4
0.1 M HCl water solution	PANI-TSA	1.4
	PANI-CSA	0.66
	PANI-DBSA	0.91

Although many interesting works concerning solution prepared PANI coatings for metallic corrosion protection were mentioned above, Araujo *et al.* [155] found that the solution cast prepared EB films did not present essential properties to be proposed as an anti-corrosive coating on mild and galvanized steel, even topcoated with a high-performance paint. Such bad behavior did not depend on the composition of the testing solution, however, it was resulted from the porosity and poor adhesion of the EB film on the substrates as pointed out.

3.1.2. Co-solution Prepared PANI Composite Coatings

Composite or blend of PANI possesses the mechanical properties of the host matrix and the electrical properties of PANI guest [64], and the advantage is equally true for PANI composite coatings in corrosion protection. PANI containing coatings through co-solution approach were developed by several groups of investigators.

EB was diluted into the coatings with the help of a surfactant, 4-dodecylphenol (DDPH), to form PANI containing coatings and the PANI was found to be oxidized by the hardener. The oxidized form of PANI provided improved corrosion protection than that of EB since the value of the standard electrode potential for the oxidized form of PANI was higher than that of EB. Additionally, the surfactant improved the wet adhesion property between the coating and metal surface [156]. The reasons of effective corrosion protection presented by PANI-DDPH were summarized [157] as follow. Firstly, PANI contributed the formation of a metal oxide layer between PANI coatings and metal surfaces; secondly, PANI enhanced the formation of a dense and adherent coating, which restricted the diffusion of water and oxygen from the environment onto metal surfaces; and thirdly, PANI formed an electric field at the interface, and it restricted electrons transfer from metal to the oxidant.

After studying the PANI, in EB form or ES form (doped with CSA or phenyl sulfonic acid), epoxy blend coating on mild steel in neutral, acidic and alkaline solutions with various electrochemical methods, Talo *et al.* [158] found that EB was the most effective form of PANI for corrosion protection of mild steel in saline solution while ES was more effective in acidic solution than in neutral solution, and also the choice of protonating acid was very important. EB epoxy coating provided excellent protection even a hole was made to the coating. All results showed that PANI containing coating can and must be tailored according to the environment to achieve the optimal corrosion protection.

Salt fog exposure indicated that PANI thermoplastic primer topcoated epoxy system on carbon steel was equivalent to inorganic zinc in performance [159]. Performance of PANI containing primer CORRPASSIV[TM] with various topcoats evaluated by scanning Kelvin probe (SKP), EIS and voltammetry indicated much better corrosion protection capability of the primer with an epoxy topcoat compared to a zinc-rich primer with an acrylic topcoat [160,161]. The ability of commercial PANI from Zipperling to act as a protective coating for mild steel corrosion in saline and acid was investigated using EIS by Li *et al.* [162]. The results indicated PANI passivated the metal surface by a mediated redox reaction and reoxidized itself by dissolved oxygen. The damaged films were also repassivated by such process and especially in acids. The performance of the PANI can be further enhanced by a topcoat to increase the diffusional resistance for the corrosion species. Schauer *et al.* [163] investigated the corrosion protection of iron with commercial PANI primer (Ormecon Chemie) and found that the protection takes place in three consecutive phases, i.e. conversion of original iron oxides, the active protection action of a conductive form of PANI due to the separation of partial cathodic and anodic corrosion processes, and the action of a non-conductive form of PANI. In the first stage of protection, the oxidation state of the iron surface played an important role; in the second stage, the durability of the active protection depended mainly on barrier properties of both the topcoat and the PANI primer as well as on the resistance of ES against alkaline media; and in the third stage, only the barrier properties of the PANI primer/topcoat system were important. In a later study [164], the second stage of the protection was confirmed, and if the coating system consists of an electrochemically active PANI primer and an electroconductive topcoat, a premature delamination of the topcoat can be effectively hindered.

A variety of acids, including TSA, dinonylnaphthalene disulfonic acid, and DNNSA, doped PANI were mixed with a base solution of poly (vinylbutyral) and brush-coated onto all sides of carbon steel coupons. Corrosion protection performance was evaluated by SRET for specimen with pinholes made down to bare steel of various diameters in a solution of St. Louis municipal tap water pH 7.8, conductivity 7×10^{-4} Scm^{-1}) [165]. The results showed that the sulfonic acids and phosphonic acid doped PANI coated panels (without an epoxy topcoat) exhibited anodic activity in pinholes and cathodic activity on the conductive polymer surface, and the pinholes were observed to be passivated with time. When an epoxy topcoat was applied to the conductive polymer coat, both anodic and cathodic pinholes were observed. Sulfonic acid dopants exhibited cathodic/anodic flips with the overall galvanic activity relatively constant or increasing and the flip-flop may be due to the instability of the iron/sulfonate passive layer. Phosphonic acid dopant exhibited a decrease in galvanic activity with time indicating pinhole passivation. A 1500 h salt fog exposure test also showed the phosphonic acid doped PANI formulations greatly outperformed those prepared using sulfonic acid doped PANI in terms of subfilm corrosion spread from the scribed area.

PANI/epoxy coatings were prepared as suggested by Peltola *et al.* [166], i.e. the CSA doped PANI was mixed to the epoxy with the help of an alkyl phenol as a co-solvent and the coatings were cured with a BF$_3$-based Lewis acid on mild steel plates [167]. A noble shift of corrosion potential by about 400-600mV and a decrease of redox current density by about five orders of magnitude as compared to mild steel were simultaneously shown in the potentiodynamic polarization curves in both 0.1 M HCl and 0.6 M NaCl solutions. XPS analysis revealed that the steel surface was covered with iron oxide, fluoride and possibly also with iron hydroxide and some organo-iron compound after the electrochemical experiment.

Electrochemical behavior of a homogeneous PANI based acrylic blend film, which was formed by mixing the solution of CSA doped PANI powder and that of poly methyl methacrylate (PMMA) both in *m*-cresol and then brushed the resultant solution and evaporated in air at room temperature, used for corrosion protection of iron in aerated 1.0 M H_2SO_4 solution with or without 0.1 M NaCl solution added was presented by de Souza *et al.* [168]. The OCP of the PANI coated iron was stabilized at ca. –0.38 V for 40 days even after addition of the NaCl solution, far higher than that of the PMMA coated and uncoated iron, which was –0.56 V. Potentiodynamic curves showed that the passive region was enlarged for PANI coated iron in 1.0 M H_2SO_4 solution, and mainly the passivation potential was diminished from 0.78 V to 0.3 V indicating the formation of the Fe-CSA complex. Physical barrier character was also demonstrated by potentiodynamic curves with the addition of NaCl solution, i.e. the movement of the chloride ions inside the blend was hindered, which was in good agreement with quartz crystal microbalance experiments. All results indicated the dual protection mechanism of the blend, a passivating complex formed with the dopant anion and a simultaneous physical barrier to avoid chloride anion penetration. The similarly prepared coating was tested later for corrosion protection of Fe, Cu, Ni, Zn and Al in H_2SO_4 solution by Torresi *et al.* [169,170]. For all metals, the PANI-CSA formed smartly a second physical barrier when damage is produced on the coating surface, avoiding penetration of aggressive ions to the substrates.

Double-strand PANI, which is a molecular complex of PANI and a polyanion, was dissolved in ethyl acetate to prepare coating on aluminum alloy [171]. The effectiveness of the coating was tested by salt-spray and immersion in salt and acidic solutions. EIS measurements revealed that the polymer coated Al alloy was highly resistant to corrosion. It was found that use of double-strand PANI facilitated a paintable formulation of coating and provided good adhesion to mental surface. Furthermore, the polymer prevented corrosion not by barrier effect, but by the passive layer converted on the surface layer of the alloy.

In a paper by Wang *et al.* [49], the performances of EB, ES and the trimers of PANI for corrosion protection were studied by EIS. The results indicated PANI coatings behave as anionic membranes in anticorrosion and the PANI bipolar coatings, which were PANI anionic membranes covered by a cationic membrane, acted as "electronic barriers" to inhibit anodic reaction and maintain high resistance to ionic flow, thereby providing excellent corrosion protection.

Nanocomposite materials that consist of EB and layered montmorillonite (MMT) clay were prepared by *in-situ* polymerization of aniline with the presence of MMT. The corrosion protection effect of PANI-clay nanocomposite (PCN) coatings, which were cast from NMP solutions, at low clay loading was testified by electrochemical measurement of corrosion potential, polarization resistance and corrosion current on CRS in 5-wt% aqueous NaCl solution. The PCN coatings were found to offer better anticorrosion performance than conventional PANI coating [172].

Wang *et al.* [173] obtained both EB/DBSA/PU coating and EB/epoxy solvent free coating on mild steel. The former was prepared by means of aging the dispersion at elevated temperature, and the latter was by controlling the ratio between EB, epoxy resin and tetraethylene pentamine, a curing agent of epoxy resin. A dense rust layer under EB/DBSA/PU coating indicated this kind of coating could not prevent the mild steel from corrosion completely. However, both the facts that rust produced for EB/DBSA/PU protected mild steel in 3.5-wt% NaCl solution was 3-4 times lower than that for the steel protected only with PU coating, and the fact that rust nearly stopped growing after 2 weeks for the former and kept growing for the later, showed EB/DBSA/PU a relatively good corrosion prevention coating. For EB/epoxy coating in 3.5-wt% NaCl solution, it was surprising that only about 1-wt% EB in the coating could remarkably reduce the corrosion current and move the corrosion potential to the positive direction; and in 0.1 M HCl solution, though the corrosion potential moved slightly to negative position, the corrosion current decreased remarkably, indicating EB/epoxy did have corrosion prevention effect even in 0.1 M HCl solution. A very effective network connection of the EB in the EB/epoxy coating by SEM image explained why this coating was effective even when EB was about 1-wt% in the total solid content.

A polyamic acid/PANI blend was prepared by addition of some given amount of undoped PANI powder to a NMP solution with a 15-wt% solid content of polyamic acid. The blends as made were coated on A3 steel sheets and dried at different temperatures for imidization to get the polyimide/PANI blend films [174]. The reactions between the end group $-NH_2$ in PANI and the carboxyl group or end anhydride group in polyamic acid or between the imine group in PANI and the anhydride made these two polymers a miscible system. EIS data from 0.01 Hz to 10KHz in 3-wt% NaCl solution showed that the corrosion protection effect of the blend coatings improved slowly when PANI content increased from 0 to 2-wt% and went up sharply when PANI content increased from 3 to 5-wt%, and an excellent protective effect was achieved with 10-wt% PANI content. The reason why PANI can improve the anti-corrosion property was that, according to the authors, on one hand, the dense and non-porous blend film prevented the corrupting components from access to the underlying steel surface; on the other hand, PANI served as a corrosion inhibiting agent to scavenge any protons and fostered a local basic surface environment.

A solvent free corrosion protection coating was developed by Wang *et al.* [175], in which aliphatic polyamine-tetraethylene pentaamine was used as the "solvent" for EB as well as the hardener for epoxy resin. The corrosion protection efficiency of the epoxy resin was greatly improved with a low EB loading (0 to 2-wt%) in the blends. Equivalent circuit analysis of EIS spectra indicated that new elements such as oxide layer resistance and oxide layer capacity existed in the blend coating compared with pure epoxy coating. The great increase, ca. two orders of magnitude, of the semicircle radius from the Nyquist plots, both in high and low frequency zone, indicated the improved corrosion protection effect of the blend coating even at very low EB loading (1-wt%) as compared with the epoxy coating. The blend coating, with a thickness of 200μm, successively passed a 2000h salt fog test. A "quasi-reversible" change of oxidation state of PANI was observed for the first time when it was stirred with reduced iron powder in 3.5-wt% NaCl solution for certain period at room temperature and then determined by Uv-vis spectra in NMP solution. And it was this "quasi-catalyst" of EB that caused the formation of dense iron oxide film in the interface, providing then efficient corrosion protection on mild steel even with very low EB loading [176].

Blend coatings of poly(2,5-dimethoxy aniline) (PDMA), a kind of substituted PANI, with fluoropolymers were also investigated for protection of SS in 3.0-wt% NaCl solution [177]. Electropolymerized PDMA, poly(vinylidene fluoride) (PVDF) and poly(tetrafluoroethylene-*co*-vinylidene fluoride-*co*-propylene) (*co*-PTFE) were dissolved separately in DMF. PDMA and PVDF solutions were firstly mixed to form a binary blend, and then the binary blend was mixed with *co*-PTFE to give a ternary blend. The solutions and the final blends were deposited on emery paper (220 mesh) treated SS substrates, with the solvent evaporated at room temperature, to form the corresponding coatings. OCP measurements revealed that corrosion potentials of PDMA and the blend coated coupons were (ca. 100 mV) positive than that of the uncoated and *co*-PTFE coated coupons, indicating that PDMA was active in protecting the substrate in the ternary blends. With PDMA and the blend coatings immersed in 3.0-wt% NaCl solution for 60 min and then placed in a furnace at 200°C for 180 min, OCP measurements were taken before and after this thermal treatment. It was found that PDMA lost its stability, while the blend coatings showed little difference after the treatment, indicating that the PDMA was protected by the fluoropolymers components in the coatings. Therefore, it was stated that corrosion protection with ICP at temperature higher than 150°C was possible.

Two kinds of PMMA based composite coatings, prepared by mixing of the matrix polymer with PANI solutions that were prepared firstly by dissolving CSA or phenylphosphonic acid (PPA) doped PANI powders in either m-cresol or 1,1,1,3,3,3-hexafluoro-2-propanol with vigorous sonication and then filtered to eliminate solid residue, were tested for corrosion protection of iron in aerated 1.0 M H_2SO_4 solution with or without the addition of 0.1 M NaCl [178]. The results revealed that the PANI was reduced and the anions were released concomitantly owing to the redox reaction between PANI and iron, and then the anions formed passivating complex with iron cations and protected the metal from corrosion, indicating the doped PANI acts as a reservoir for smart releasing of anions on demand as stated in [152,153,170]. Higher passivating character was observed for the PPA doped PANI coating than that of the CSA based coating, indicating the chemical nature of the dopant is a crucial choice in determining the protective performance of the conducting coatings.

3.1.3. Dispersion Prepared PANI Composite Coatings

According to the experimental and thermodynamical considerations, Wessling *et al.* [179-181] insisted that ICPs cannot be truly soluble, or solutions of ICPs cannot exist and the "solutions" most scientists referred to, for example solution of EB in NMP, solution of CSA doped PANI in *m*-cresol, etc. are "dispersions" in their opinion. And with the "dispersion" method, CORRPASSIV™ coating systems for various industrial applications have been developed [182]. An excellent paper [181] is recommended for anyone who want to better understand the "solutions" and "dispersions" proposed and debated by Wessling, in which basic thermodynamic aspects of solutions and dispersions in general were presented, and many quantitative or semi-quantitative arguments were developed concerning solubility parameter, surface tension, free energy of dissolution, solution enthalpy, lattice energy, etc.; also, numerous literatures reporting about various different experimental studies were summarized and interpreted. But here in this chapter, the dispersion we referred to is not the one as Wessling *et al.* referred to, it is simply the mixing, by various procedures such as beads milling, shear blending, of the ICP powders with conventional polymers with or without solvents. Coatings based on this approach bear exceptional interest from a practical standpoint.

With PANI chemically polymerized in HCl media at −5°C and pulverized for making into powder, PANI coating was formulated by dispersing the as-prepared PANI powder in a medium oil alkyl resin with the help of beads mill. The required properties of the coating, mainly drying, leveling etc., were adjusted by adding different additives through make up stage and four coatings with varying solid percentage of PANI were formulated by Rout *et al.* [183]. All electrochemical studies were carried out in a 3.5-wt% NaCl solution for both coated and bare steel samples. Tafel plots showed a smaller current density (2.8×10^{-7} Acm^{-2}) and a more nobler potential (−527 mV) for the PANI coated steel compared to the bare steel (current density: -7.8×10^{-6} Acm^{-2} and potential −557 mV). And the corrosion rate for the PANI coated steel was significantly lower (0.132 milli-inches per year (MPY)) than that of the bare steel (0.606 MPY). Potentiodynamic polarization measurements indicated the tendency for the formation of passive oxide layer in case of PANI coated samples and the time required to form the passive layer for PANI coated samples was far less than the bare and Zn-coated ones. EIS measurements revealed a continuous charge transfer reaction across the metal-coating interface for the PANI coated steel thereby forming a passive oxide layer. SEM studies showed a compact oxide layer was formed on the PANI coated steel, which acted as passive protection layer, while porous and flaky oxide layer was formed on bare steel which is more prone to water diffusion and accelerated corrosion.

Polyaniline (PANI) doped with CSA and DBSA were blended in PMMA and epoxy, respectively, and poly-o-ethoxyaniline (PEtOAN) doped with HCl, HNO_3, and HBF_4 were in water soluble epoxy to prepare PANI blend coatings on iron [184]. It was found, from both potentiodynamic polarization measurement and OCP measurement in a 3-wt% NaCl solution, that all the blend coatings containing PANI and PEtOAN appeared to show better corrosion protection efficiency than conventional polymer only coating. The HNO_3 and HBF_4 doped PEtOAN blended in water soluble epoxy turned out to be excellent coating materials for corrosion protection of iron as per the authors. Polyvinylbutyral coatings containing dispersed PANI-TSA inhibited effectively the cathodic disbondment when Cl^- contacts the iron surface and the PANI-TSA was reduced in the same time [185]. The reduced PANI-TSA would be reoxidized by atmospheric oxygen and converted into EB form with coating alkalization. Therefore, if no additional PANI-TSA is provided, there is only the slowing up of the disbondment, but no prevention.

In a paper by Samui *et al.* [186], ingredients of HCl doped PANI (PANI-HCl) powder, which was chemically oxidative synthesized by following the method of Chiang and MacDiarmid [187], titanium dioxide and dioctyl phthalate were added to a pre-dissolved styrene-butyl acrylate copolymer solution of xylene and the mixture was ball-milled for 4 h and filtered through fine cotton cloth to prepare PANI-HCl containing paints. The paints with various PANI-HCl loadings were applied on degreased clean sand-blasted mild steel panels of various sizes with thickness maintained at 80 ± 5 µm. Before studying the properties, all painted panels were kept at ambient temperature under 50-60% relative humidity (RH) for 1 month. Corrosion protection studies were conducted under both atmospheric and underwater conditions. Accelerated weatherometer study, in which the samples were exposed alternatively to condensing and U-V exposure for 2 h and 4 h, respectively, with temperature maintained at 40-60°C and RH maintained at 100%, showed that no trace of any corrosion was observed after 1200 h of exposure but loose gloss of the surface, indicating the paint was suitable for long term protection in spite of its poor gloss retention. Humidity cabinet exposure, in which the RH was maintained at ca. 100% and the temperature cycled over a range 42-48°C, showed that at 20 parts loading of PANI-HCl, corrosion occurred earlier to systems with lower PANI-HCl loading; at loading of 0.1 and 0.5 parts, corrosion initiated only after 1670 h; while for the control panel (0 parts PANI-HCl), corrosion occurred after 1270 h. Salt spray exposure and sea water immersion both showed that at higher PANI-HCl loading the protection performance of the paints was inferior to that at lower loading and the corrosion protection was maximum at 0.1 parts PANI-HCl loading. The poor performance of the higher PANI-HCl loading paints (10 and 20 parts PANI-HCl loading) was due to their higher water vapor permeability that increased with the PANI-HCl loading. The results of potentiodynamic measurements in 3.5-wt% aqueous NaCl solution are listed in **Tab. 5.3**. Higher corrosion potential values are observed for lower PANI-HCl loadings. Corrosion current is significantly higher for 20 parts PANI-HCl containing paint compared to other systems and decreases with PANI-HCl loading. Polarization resistance shows a maximum at lowest PANI-HCl loading and declines as the loading is increased. All above results indicated that lower

PANI-HCl containing paints protect mild steel better than paints with higher PANI-HCl loading. OCP measurement in 3.5-wt% aqueous NaCl solution also substantiated the above results, which showed that paint with 0.1 parts PANI-HCl exhibited less negative potential compared to other systems. The paint system showed appreciable corrosion resistance without any top barrier coat.

Tab. 5.3. Data on potentiodynamic measurements on mild steel coated with PANI-HCl paints. Reproduced with permission from A.B. Samui *et al.*; Prog. Org. Coat., 47, 1 (2003). Copyright © 2003, Elsevier Science Ltd.

Sample	PANI-HCl (parts)	Corrosion potential (mV)	Corrosion current × 10^8 (A/cm^2)	Polarization resistance × 10^{-6} (Ω cm^2)	Corrosion rate (MPY)
PANI 0	0.0	−413.8	1.9020	0.6010	0.009
PANI 1	0.1	−283.6	0.0224	139.6000	0.000
PANI 2	0.5	−347.6	1.9500	0.0874	0.018
PANI 3	1.0	−351.8	0.6377	2.6230	0.008
PANI 4	3.0	−382.0	0.4967	7.3980	0.006
PANI 5	5.0	−364.6	1.3820	0.9774	0.017
PANI 6	10.0	−347.9	1.2780	0.7202	0.016
PANI 7	20.0	−474.3	25.2300	2.0300	0.314
Coal tar epoxy	0.0	−524.7	19.2700	0.4164	0.088

Trieu *et al.* [188] prepared MMT-PANI nanocomposite by *in-situ* polymerization and dispersed the composite in an epoxy resin to form a film of clay-PANI on steel surfaces. Potentiodynamic polarization measurements in 3-wt% NaCl solution showed that the current density of the as-coated steel was significantly lower than that of steel coated with epoxy primer containing 40-wt% zinc chromate. Together with the EIS testing, it was indicated clearly this nanocomposite has good corrosion protection efficiency.

In Cook *et al.*'s group [189], DPA doped PANI [147] was mixed with a solid acrylic resin to form a 1-wt% concentration of PANI additive. The mixture was added to a quantity of xylene to produce a 25-vol% dispersion. The dispersion was applied to low carbon steel coupons after a 4 h's ultrasonic exposure and dried. After artificial defects were introduced carefully into the polymer coated steel coupons, corrosion tests were carried out in 0.1 M NaCl solution at room temperature. Anodic polarization profiles showed that significant inhibition was observed for the DPA added coating, in which the diffusion-limited current is approximately a factor of 10 lower than that of the control, i.e. the bare steel. Moreover, the PANI-DPA added coating and a commercially obtained polymer, Ormecon^{TM} (ORM) from Zipperling added coating provided lower level of protection than that of the DPA only added one. And more importantly, no active-passive behavior was observed and the corrosion potential and Tafel slopes were independent of coating additive, which strongly suggested that the protection was of inhibitory action rather than passivating effect. The polarization resistance suggested that all additives provided effective inhibition under free corrosion conditions. These data are partly inconsistent with that of [143], but the comparison between the two studies is not possible due to the different formulations and application method. Cathodic polarization profiles showed that both PANI-DPA and DPA inhibit the cathodic kinetics. Polarization resistance values also suggest that only PANI-DPA and DPA containing coatings offer effective inhibition under free corrosion conditions. The study exhibited that the ES formulation protected the corrosion by using an inhibitory effect of the released dopant anions, in this case, the DPA anions on the reduction of PANI-DPA to an LB form, and the LB in contact with the steel was assumed remained LB, not like the assumption by others [150,162] that LB underwent a subsequent re-oxidation by dissolved oxygen.

TSA doped PANI (PANI-TSA) was admixed with polyvinylbutyral (PVB) solution in ethanol using shear blender to form dispersion. The ethanolic dispersions were bar cast on to cleaned zinc and iron substrates and air-dried to give a coating of 30 ± 5 µm. The coatings were partially peeled to give a defect and the delamination experiments were carried out at 20°C in 0.86 M NaCl solution at pH 6.5 using SKP [190]. It was found that the delamination rates of the coatings decreased monotonically with increasing PANI-TSA volume fraction (ϕ) and no delamination was observed over a period of 48 h when (ϕ) < 0.2. Zinc oxide layers developed at the coating/substrate interface and completed their growth within 6 h of coating application when the relative humidity was larger than 95%, and the thickness attained was proportional to ϕ. Together with other experiment results, it was proposed that the inhibition arises principally as a result of the zinc oxide layer blocking cathodic oxygen reduction at the zinc-coating interface, and this effect was magnified and prolonged by the pH buffering action of PANI-TSA in the coatings, or to say PANI-TSA absorbed the OH^- by produced cathodic oxygen reduction and prevented or delayed therefore the alkaline dissolution of the zinc oxide layer.

Stable nano-sized PANI waterborne latexes were prepared by emulsion polymerization of aniline in micellar solution of sodium dodecylbenzene sulfonate, DBSA, or sodium dodecylsulfate, and the latexes exhibited good direct film formability [191]. Topcoated with acrylic resin, the latexes coated iron plates were tested for corrosion protection in 5% NaCl solution. The anti-corrosion performance of the coatings was affected by the particle size and the film formability of the latexes. The smaller the particles size and the better the film formability, the better the anti-corrosivity of the coatings. The better anti-corrosivity achieved implied that the latexes can be directly applied as anti-corrosion coatings of metals.

We also investigated the corrosion protection performance of epoxy coating with dispersed EB particles on CRS coupons. In one method [192], the chemically prepared EB particles were mixed with epoxy resin and a proper solvent and ground in a planetary ball mill with ceramic grinding media for 1.5 h at a rotating speed of 500 rounds per minute. A dispersion of EB, with EB diameters between 30-200 nm with the most of ca. 100 nm, was obtained and mixed with an aromatic curing reagent where-after. Brushed on CRS coupons and cured, EB coatings were prepared. Upon electrochemical measurements in 3.0-wt% aqueous NaCl solution, corrosion potentials of epoxy coated coupons decreased to a constant value (–590 mV *vs.* Ag/AgCl) slightly higher than that of bare CRS (–649 mV *vs.* Ag/AgCl) in 48h due to the only physical barrier effect. Coatings containing EB less than 10-wt% acted similar to epoxy coating. For coatings with EB contents equal to or exceed 10-wt% (including 10-wt%, 15-wt% and 20-wt%), their corrosion potentials decreased slowly with immersion time and stabilized at a higher value (–433 mV *vs.* Ag/AgCl) (**Fig. 5.8**), which was equivalent to that of the EB powder passivated CRS. Together with the cyclic voltammetry study, it was demonstrated that (i) the oxidation-reduction capability of EB still performed in cured epoxy coating and (ii) a minimum loading of EB, 10-wt%, was needed to achieve good anti-corrosive effect. In another method [193], EB coated solid epoxy resin particles were prepared by chemical polymerization of aniline with the presence of the epoxy particles, and the EB coated particles were mixed with a proper solvent and ground in a planetary ball mill with ceramic grinding media for 1.0 h at a rotating speed of 500 rounds per minute. Coating preparation and electrochemical tests

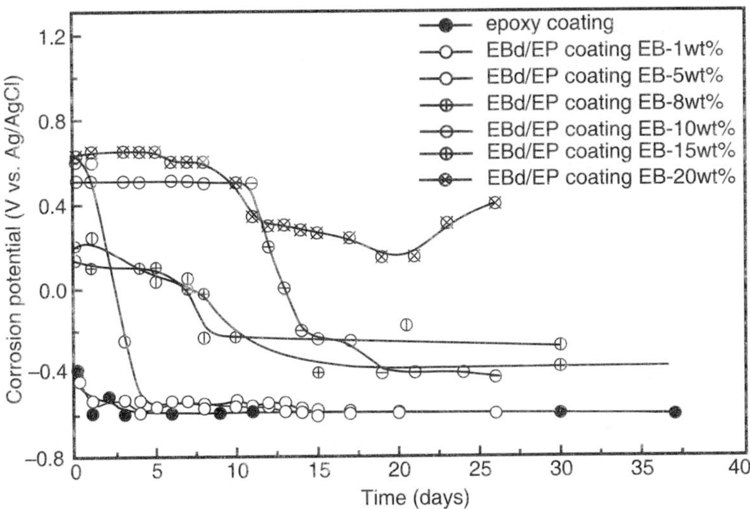

Fig. 5.8. Plots of corrosion potential against immersion time in 3-wt% aqueous NaCl solution. Reproduced with permission from X.L. Jing *et al.*; J. Chin.Soc.Corros.Prot., 24(5), 301 (2004). Copyright © 2004, The Chinese Society for Corrosion and Protection.

were similar to [192]. The overall behavior of the corrosion potential with immersion time of this EB coating was similar to that of directly ball milled EB containing coating [192], except the slower decrease of potential to a constant value and a minimum EB loading of 1.27-wt% for good corrosion protection (**Fig. 5.9**), ninefold reduced as compared to [192], which was due to the well dispersed EB particles (diameter of 20-50 nm as shown by transmission electron microscopy (TEM) images) in this method. In another one of our paper [194], EB coated silicon dioxide particles were prepared through chemical oxidative polymerization with the presence of silicon dioxide particles. The particles passivated the CRS surfaces as the EB did, and a noble shift of the corrosion potential of ca. 200 mV was observed when the feed weight ratio of silicon dioxide to aniline equals to 400. Dispersion of these particles into epoxy coatings significant lyenhanced the corrosion protection performance of the coatings, and this may pose an alternative approach to depress further the cost of the PANI coatings. However, these fascinating effects vanished for particles with further decreased aniline feed ratio in their preparation processes.

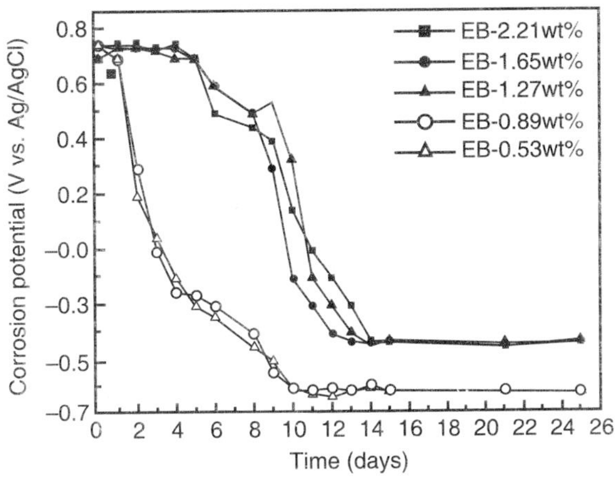

Fig. 5.9. Plots of corrosion potential against immersion time in 3.0 wt% aqueous NaCl solution. Reproduced with permission from Y.Y. Wang and X.L. Jing; Polym. J., 36(5), 374 (2004). Copyright © 2004, The Society of Polymer Science, Japan.

Moreover, low percentages of PANI (0.2, 0.4 and 0.6-wt%) as additives improved not only the corrosion resistance of the alkyd resin coatings, but also the degradation resistance of the coatings according to both the field tests in a sodium hydrogen sulfite solution acidulated with sulfuric acid to pH of 3.2 and laboratory tests in a 3-wt% NaCl solution [195]. The corrosion degree of the alkyd coating containing 0.6-wt% of PANI was about 0.5-1.0%, far less than that of the pure alkyd coating and the vinyl copolymer, acrylic blend and phenoxy coatings. Phenomena like fissures, pulverization and discoloration observed for the alkyd coating are clearly avoided in presence of PANI, indicating improved degradation resistance of the coating.

3.1.4. Electrodeposited PANI Coatings

As assumed in literature [196], an efficient protection cannot be obtained by casting of a blend because of the need to have a good quality interface between the passive layer and PANI. Even if some passivation of the iron surface can be obtained by this simple contact, it cannot be controlled with enough rigors. Note this argument, there comes the investigation of electrodeposited PANI on various metals and its corrosion protection performance.

Successive phenomena were observed for PANI electrodeposition on mild steel in phosphoric acid electrolyte [197], viz. from the beginning, a small and slow loss of oxidation level (shown by optical spectroscopy); just before the breakdown, a delocalization of carriers with withdrawal of bipolaronic sites (shown by Raman resonance) and exactly during the potential fall, a reduction of PANI to randomly protonated lecoemeraldine (shown by optical multichannel analysis). The protection brought by PANI depended on the underlying layer that stabilizes through a "quasi-potentiostatic" effect. Also PANI potentiostatically electrodeposited on iron in sulfuric acid medium was used to protect the electrode from corrosion [198] and the PANI film can exchange electrons through the passive layer on iron and thereby protect the passive layer from destruction at least until the polymer reduction was completed [199]. The acids used for PANI electrodeposition were influent on the protection offered by the polymers. Some more investigations about PANI electrodeposition in sulfuric acid, phosphoric acid etc. were carried out later and similar results were achieved. Iron was passivated during the electropolymerization of PANI film; the subsequent protection was dependent on the nature of this passive layer and PANI acted as a stabilizer of this layer [200], e.g. the presence of phosphoric acid strengthened the passive layer, which was expected to be a phosphate/oxide bilayer strongly protective film, and a copolymer formed with the insertion of metanilic acid improved further the protective property [201]. The passive layer broke down when the PANI film lost its reoxidability and the step was preceded by slow modification in the polaron distributions [202]. On the basis of these earlier works on corrosion protection studies of PANI for iron, the same group [203] later prepared potentiostatically both PANI and sulfonated polyaniline (SPANI) films on iron in 1.7 M H_3NO_3 containing 0.25 M aniline, or 0.25 M (aniline + amino benzenesulfonic acid (ABSA)), with the maximum ABSA/aniline ratio of 1, and found from OCP measurement (in 0.1 M K_2SO_4 + H_2SO_4, pH 1) that the protection was better when SPANI film was used instead of PANI. The SPANI film was a copolymer of aniline and ABSA, also called metanilic acid. Roman spectra indicated that the protection breakdown was owing to the impossibility to reoxidize the reduced polymers as found in [202], and the better protection of SPANI was due to, on one hand, the presence of partially oxidized sulfonated units even in cathodic polarization, i.e., the hindrance to achieve a full reduction of SPANI; on the other hand, the cross-linkages in SPANI, which can limit the possibility of ingress of aggressive anions.

Note that the decaying time of OCP of the PANI film deposited in phosphoric acid was four times longer than that of the PANI formed in oxalic acid [197,198], Moraes *et al.* [204] electrodeposited PANI on austenitic SS in electrolyte of phosphate-based buffer solutions of pH between 1.7 and 2.2, whose main advantage, as evidenced, was to control the pH at the polymer/electrolyte interface, promoting a very uniform film growth. The surface structure of the PANI films depended on the preparation conditions in particular on the pH of the phosphate buffer solution and the most uniform thin film was formed from 1.0 M (pH = 1.7) phosphate buffer solution containing 0.1 M aniline. The corrosion potentials of all the PANIs, both doped and undoped forms, deposited steels in 3-wt% NaCl aqueous solution shifted to noble values as compared to the corrosion potential of the bare steel. A linear decrease of corrosion potential with increasing pH of the deposition buffer solution was observed for the doped PANI, and a reversed relation was observed for the undoped PANI. During the electrodeposition, the phosphate diffused into the PANI film in order to compensate the charge from the polymer oxidation. Therefore, at high phosphate concentration, the species entered more in the polymer chain and increased the doping level; however, when they contact with the metal, these species produced an insoluble phosphate layer as well as an increase in the number of binding sites on the metal surface. This layer had a strong passivation characteristic and contributed effectively to the shifts of corrosion potential to more positive values. On the other hand, low phosphate concentration formed mainly soluble species and led to the formation of weak passivation layer, and consequently to a small shift. For the PANI electropolymerized in 0.6 M phosphate buffer solution (pH = 1.8), a remarkable potential shift of 0.176 V was observed for the doped PANI as compared with the bare SS, and a shift of 0.172 V for the undoped PANI from the potentiodynamic polarization in 3-wt% NaCl aqueous solution. The shifts were smaller in magnitude than that obtained for SS covered with chemically prepared PANI, which was cast of EB from NMP solution on SS [134]. The reason for this behavior was not clear, but the authors expected that the electropolymerization would promote a change on the metallic surface and therefore a less noble shift.

As exhibited by Camalet *et al.* [205], the galvanostatic deposition of PANI on mild steel from oxalic acid proceeded in two stages, implying first the dissolution and the progressive passivation of the substrate and then deposition of a strongly adherent PANI film. The film showed the same features as classical PANI but appeared to be less oxidized, indicating that the underlying metal could have a reducing effect on the deposited PANI film. The corrosion protection effect in an acidic solution (0.4 M NaCl + 0.1 M HCl) of the PANI coated iron samples was measured by the amount of Fe^{2+} passed into the solution, and the 1.5 μm PANI was more effective than a 3 μm PPY film. Later, the same group electrodeposited PANI films on mild steel in aqueous TSA, not in oxalic acid, due to the advantage of diminished metal oxidation with high current density and higher yield of polymer from the industrial point of view. Corrosion resistance of the PANI-coated samples were evaluated by measuring the amount of Fe^{2+} after the samples were dipped in a solution of 0.1 M HCl + 0.4 M NaCl for 5 hours and the results showed that the corrosion current was divided nearly by ten, i.e. from 1.3 $mAcm^{-2}$ for the uncoated mild steel to the 0.15 $mAcm^{-2}$ for the coated one, suggesting the polymer coating could be used for corrosion protection [206]. A similar result for iron dipped in an acidic medium containing dissolved poly (*o*-methoxyaniline) was also obtained by Sathiyanarayanan *et al.* [207], in which an inhibition efficiency of nearly 80-88% was observed even at 25 ppm concentration of the polymer. The strong adsorption of polymer following the Temkin adsorption isotherm was largely responsible for its inhibitive action according to the authors.

Smooth, well adhering PANI coatings were also obtained electrochemically in oxalic acid solutions on iron disc electrodes by Sazou *et al.* [208]. The PANI coatings exhibited good stability on exposure in air and various corrosive media, including 0.5 M H_2SO_4, 2 M H_2SO_4, 2 M H_2SO_4 + 0.03 M Br^-, and 2 M H_2SO_4 + 0.05 M Br^-. Preliminary tests, such as OCP and anodic polarization, showed the inhibitive properties of the PANI coatings seem to be good, which might be attributed to a stabilization of the passive oxide film of iron in acidic corrosive media. In the studies by Özyýlmaz *et al.* [209,210], homogenous and strongly adherent PANI films with different thickness were prepared on SS (316L) and copper electrodes using cyclic voltammetry technique in oxalic acid solution. EIS measurement in 0.1 M HCl solution and 3.5-wt% NaCl solution revealed that the films exhibited significant barrier effect against the attack of corrosive agents and the effect was affected by the synthesis condition, i.e. the different potential ranges adopted in electrochemical polymerization. The PANI coating exhibited important catalytic effect on the formation of a protective oxide layer on the surface, which protected the metal from corrosion [210]. Although the corrosion resistance property exhibited by electrodeposited PANI coating, its lifetime was limited. However, owing to its catalyzing passivation effect, the electrodeposited PANI decreased the porosity of the plated nickel coating, which then reduced drastically the corrosion rate of the mild steel and provided much longer time of protection [211].

Polyaniline (PANI) coatings deposited on steel from oxalic acid were also tested for corrosion protection in high pH media, 0.1 M NaOH and 0.1 M NaOH containing 5 gL^{-1} NaCl [212]. The iron oxalate coating, which was formed on the steel prior to aniline polymerization, exhibited less corrosion rate than the bare steel in the chloride free solution, owing to the barrier effect that slowing or inhibiting the dissolution and transport of iron to the oxalate/solution interface or the diffusion of OH^- to the iron/iron oxalate interface. However, the iron oxalate coating showed greater corrosion rate than the bare steel in chloride containing solution, indicating that the chloride ion could possibly penetrate the oxalate layer and initiate corrosion of the underlying steel. In the case of the PANI deposited sample, the corrosion rate was the lowest even in the chloride containing solution. The reason lies in the redox capability of PANI, which, in EB form due to the high pH of the solution, was reduced to lower oxidation state from the electrons produced in the oxidation of Fe to $Fe(OH)_3$ and was oxidized back to EB form by the oxygen in the solution, yielding quickly the protective oxide that prevents the further corrosion of the steel [135]. In another study [213], PANI was electrodeposited on iron from oxalic acid and corrosion tests in acidic solutions showed little effect of the polymer (ES) for iron, while in an alkaline borate solution, the polymer (dedoped to EB) exhibited a clear beneficial effect on the local breakdown of iron by chloride anions, and the positive effect of the EB was resulted from both the oxide formed on the surface of the metal [135] and the barrier property of the EB film. The inferior protection effect of PANI coating to that of the PPY coating, which was electrodeposited from phosphoric acid, was associated with the instability of the initial iron-oxalate layer than that of the iron-phosphate layer formed prior to electrodeposition [214].

Based on their experiments, Popoviæ *et al.* [215] found that the electrolyte based on sodium benzoate was the most promising one for the electrochemical synthesis of PANI for corrosion protection of mild steel. Corrosion of the as-prepared mild steels with PANI films in thickness of 1.5 µm were investigated by the means of *o*-phenantroline method, in two different corrosion media 0.1 M H_2SO_4 and in 0.5 M NaCl + 1m M HCl. The edges of the steels covered with PANI were protected by epoxy resin. Hydrogen evolution reactions were observed for all samples after their immersion. However, during the course, the rate of hydrogen evolution reaction decreased especially for the PANI coated samples, similar to the report by Camalet *et al.* [216]. After 120 h of immersion, the average corrosion current density in sulfuric acid media was 26 and 107 μAcm^{-2} for coated and uncoated samples, and in chloride containing media, 2.5 and 60 μAcm^{-2}, respectively, indicating the remarkable corrosion protection efficiency of the PANI coatings. Atmospheric corrosion, which was investigated by visual observation, also showed that PANI, though partially coated, protected the steels from corroding. It was therefore concluded by the authors that the PANI films electrodeposited from benzoate containing solutions on mild steel are potentially good anticorrosion coatings. In another study [217], the same group demonstrated that the electrodeposited PANI coating could be used for modification of mild steel prior to epoxy coating deposition and an increased corrosion protection was obtained for the PANI/epoxy coating system in the same solution.

After study of the corrosion protection performance of electrosynthesized PANI coatings on steel samples in sulphuric and phosphoric acids, Kraljiæ *et al.* [218] pointed out that PANI layers, which were possible to be electrosynthesized on SS and low alloy steel electrodes, protect SS from corrosion in sulphuric and in phosphoric acid solutions by stabilizing the oxide layer formed on the steel surface during PANI polymerization, and thus prevent the metal dissolution process. The protection depended both on the quality of the oxide film and the amount of PANI, which was usually given as the thickness. The thicker the PANI film, the longer the time of protection for SS. However, the time of protection for low alloy steel was much shorter. The layer deposited in a phosphate solution appeared to have better protective property than the layer deposited in a sulphate solution. But the protection time for the PANI deposited from phosphate solution in 0.1 M HCl was significantly shorter. EIS measurements showed that PANI layer in emeraldine state prevents the dissolution of steel, and the capacitive behavior of PANI layer prevails in the passive potential region. At the potential values of leucoemeraldine state, PANI did not prevent but slightly hindered the metal dissolution process.

In another paper, Malik *et al.* [219] showed that electrodeposited PANI alone on SS was not effective enough for long-term stabilization of the passive layer on SS in strong acid solution (2 M H_2SO_4) at steady state. While with the incorporation of platinum microparticles (with average size of 50-100 nm) at a fairly low level (ca. 25 μgcm^{-2}), the reduction of oxygen, which can be coupled to the self-passivation of steel, was fast enough to replenish the PANI charge consumed by the oxidation of SS, and to poise the steel's potential well within the passive region for an indefinite time.

Shah *et al.* [220] reported the electrochemical synthesis of poly (*N*-ethyl aniline) (PNEA) coatings on Al alloy from aqueous solution of oxalic acid by using cyclic voltammetry. DC polarization studies were done to evaluate the corrosion resistance offered by the film in an aqueous 3.5-wt% NaCl solution. The corrosion potential of the coated sample increased from –234 mV for the uncoated control to –125 mV and the corrosion current decreased from 1.97 μA to 0.325 μA. The corrosion rate of the metal was reduced by about one order of magnitude and was not significantly influenced by the thickness of the PNEA coatings. Stable PDMA film was also electrodeposited on 304 SS in an oxalic acid electrolyte and tested in 0.5 M HCl for corrosion protection [221]. The enhanced OCP of the PDMA coated SS samples, though fluctuated between –150 mV and 250 mV remained at the level for 3 weeks without decreasing to that of the bare SS sample (–320 mV), indicating promising corrosion protection capability of the polymer. However, only when the polymer is mixed with paints or other polymeric materials that provide good adhesion and durability, as stated by the authors, could the practical application of the polymer for metallic corrosion protection be achieved due to the difficulty in preparing an even and durable coverage of the film by electrochemical method.

PANI was also electrodeposited on aluminum alloy Al-2024-T3 from aqueous oxalic acid [222]. Polarization studies indicated that a significant reduction in corrosion current and corrosion rate were obtained with the PANI coating, and the corrosion current, corrosion rate and corrosion potential were all dependent on the current density and reaction time used for the electrodeposition of PANI. In the process of PANI deposition, an Al-oxalate layer was firstly deposited on the Al alloy surface followed by the PANI coating.

PANI coatings electrodeposited on pre-treated aluminum were found more resistant to corrosion in 0.1 M NaCl solution than the coatings prepared on untreated aluminum [223]. The pre-treatment was carried out by submitting aluminum to anodic galvanostatic activation in 0.1 M HNO_3 with 0.1 M aniline at 20 $mAcm^{-2}$ for 5 min, similarly to the procedure used by Hulser *et al.* [96] in pre-treatment of aluminum for PPY electrodeposition. PANI was electrodeposited successively on the pre-treated aluminum by cyclic voltammetry in 0.5 M H_2SO_4 containing aniline. Pits of 10-30 μm in diameter were generated in the pre-treatment and were three orders of magnitude larger than the pores (1-3 nm in diameter) of the porous layer formed during a simple anodization to passive Al_2O_3. The over-oxidized PANI filled in the pits and pores promoted the formation of a dense PANI in the aluminum surfaces, which significantly decreased the corrosion current density as compared with bare aluminum.

Herrasti *et al.* [224] electrodeposited films of PANI, poly-*o*-toluidine (P-*o*-tol) and a composite film of the two (P-*o*-tol-ani) on type SS-304 and investigated their corrosion protection performance in 3-wt% NaCl solution and 1 M H_2SO_4. It was found from the OCP measurements that the composite film yields better performance than single PANI or P-*o*-tol films, owing to the cement and cross-linker effect of the composite which prevent the entry of protons and other ions and consequently enhanced the corrosion protection. Potentiodynamic polarization results listed in **Tab. 5.4** show a noble shift for the polymer coatings, which indicate an inhibition of the corrosion process. The increase of the corrosion current corresponded to the oxidation of the polymer and was therefore not strictly the corrosion current. Microstructure studies of the samples after 15 days of immersion in 1 M H_2SO_4 exhibited that cracks showed on the PANI films which allowing the electrolyte solution to access the underlying steel; and holes showed across the P-*o*-tol films with the polymer continues to cover the steel inside the holes; no cracks or holes showed on the composite films which indicate again the composite renders good corrosion protection. Together

Tab. 5.4. Corrosion potentials and currents for bare and coated steel with different polymers. Reproduced with permission from P. Herrasti *et al.*; J. Appl. Electrochem., 33(6), 533 (2003). Copyright 2003 Kluwer Academic Publishers.

Substrate	Corrosion potential (in mV vs. Ag/AgCl)	Corrosion current (Acm^{-2})
Stainless	−255	1.9×10^{-6}
PANI	200	1.12×10^{-3}
P-*o*-tol	300	4.51×10^{-5}
P-*o*-tol-ani	370	1.25×10^{-3}

with the microhardness measurement which revealing nonelastic films in all cases and low hardness values that increased from PANI to P-*o*-tol to the composite, the best results, both the corrosion protection and the mechanical properties, were found in the case of the P-*o*-tol-ani composite film.

A polyaniline-thiokol rubber (PANI/TR) composite coating was electrodeposited from a nonaqueous solution containing aniline, acetonitrile, trifluoroacetic acid, trichloroacetic acid and thiokol rubber onto mild steel by Ding *et al.* [225] to overcome the bad adhesion of PANI coating chemically or electrochemically synthesized and to meet, to a certain extent according to the authors, industrial demands. The PANI/TR composite coating exhibited a dense and continuous network structure, unlike the loose reticular structure of the PANI coating, which impedes the diffusion of water and ions. The TR cross-linked the PANI particles into a kind of monolith and in turn enhanced the strength and adhesion of the coating. XPS showed that the TR content in the PANI/TR coating decreased on going from the surface to the substrate while PANI changed conversely during electrodeposition and therefore formed a highly effective barrier against water and oxygen. OCP measurements in 0.1 M HCl showed that the PANI/TR coated specimens exhibited not only a more noble corrosion potential (100 mV), but also a duration of 50 min as compared with the 15 min of PANI coated ones until the exfoliation begins; while in 5-wt% NaCl solution, although a more noble corrosion potential up to 120 mV was observed for the PANI/TR coated specimen, both PANI/TR and PANI coated specimens retained relatively stable corrosion potential for a long period. Compared with the results in 0.1 M HCl, it was concluded that H^+ is the main corrosive agent for the PANI coating and therefore acidic environment may be incompatible with the PANI application according to the authors. Corrosion currents for the PANI/TR and PANI coated samples were found to be 0.003 µA and 12.6 µA respectively according to the Tafel plots. Upon exposing the samples directly to laboratory air after the electrochemical tests, very small or no corrosion was observed beneath the PANI/TR coating even after 2 weeks while the PANI coating cracked and the end face of the steel began to rust after only 2 days. EIS measurements in 5-wt% NaCl solution indicated the impedance of the PANI/TR coated specimen was 6.3 times higher than that of PANI coated ones in the whole frequency range (from 0.05 Hz to 100 kHz), reflecting the results of the previous experiments that the PANI/TR composite coating exhibited greatly improved corrosion protection performance as compared with PANI, in addition to the improved adhesion performance.

Galkowski *et al.* [226] prepared a composite PANI film containing Prussian blue (PB) and hexacyanoferrate anions on SS by potential cycling the electrode in a solution containing 0.02 M aniline + 0.01 M $K_3[Fe(CN)_6]$ + 2 M H_2SO_4 + 0.5 M KNO_3 and tested its protection efficiency of SS against pitting corrosion in strongly acid media (0.4 M H_2SO_4 + 0.4 M HCl). The inner portion of the composite film, that is adjacent to the steel surface, seemed to be dominated by polynuclear PB, which was due to the interaction of $Fe(CN)_6^{3-/4-}$ with surface Fe^{2+}, whereas the outermost portions of the film were likely to contain hexacyanoferrate anions. The formed PB was stable in acid and chloride media, which led to the stabilization of passive state of the steel; hexacyanoferrate ions presented in the composite film strongly inhibited transfer of iron and chromium cations into the solution even in chloride containing aggressive media. By neutralizing the positive charge of PANI matrix with inorganic anions, its anion exchange properties were reversed or minimized, and the access of pitting causing chlorides was blocked. Even if some Cl^- reached the steel surface and the breakdown of passive film occurred, any iron and chromium ions appearing in the pits immediately formed PB. Accordingly, the PANI based composite film was found to show promising protective properties of SS against pitting corrosion in chloride containing acid media.

Inorganic fillers, including carbon black, carbon fibers, SiO_2, yttria stabilized zirconia (YSZ), CeO_2, clay, etc. were incorporated electrochemically into PANI coatings on Al 2024 by Shah *et al.* [227]. Tafel tests, which were carried out in 3.5-wt% NaCl solution, indicated that the presence of clay and ion exchange clay, YSZ, SiO_2, and CeO_2 in the films improved the corrosion resistance of PANI on Al 2024. However, fillers such as TiO_2 and WO_3 did not show any such effect.

3.1.5. Soluble PANI for Corrosion Inhibition

In addition to the above mentioned PANI and PANI containing protective coatings, the inhibitive performance of PANI and its derivatives were also studied and it was showed that the soluble PANI and its derivatives are very effective corrosion inhibitors for iron and steel in acidic chloride solution. For example, Sathiyanarayanan *et al.* [228] studied the corrosion inhibition property of iron in 1 M HCl solution by a soluble PANI, poly (*o*-ethoxyaniline) (PEA), which was prepared by chemical oxidative polymerization of *o*-ethoxy aniline in either sulphamic acid or hydrochloric acid. Electrochemical techniques like Tafel linear extrapolation, linear polarization resistance method, EIS and direct weight loss method were used and the results showed that the soluble polymer can be effectively used as a corrosion inhibitor for iron in acidic chloride solutions. The better corrosion inhibition was from the strong adsorption of the polymer, which was facilitated by both the co-existence of delocalized π-electrons, and the quaternary ammonium nitrogen in the polymer. A double layer capacitance study indicated that the adsorption of the polymer obeys the Temkin isotherm. Later, similar results were obtained by the same group with poly (*o*-methoxyaniline) on iron [207] and with a completely water soluble and a commercially available PANI on mild steel in HCl [229].

A completely water soluble poly (styrene sulphonic acid) doped PANI was also evaluated for the corrosion inhibition of mild steel in 1 M HCl [230]. Weight loss measurements showed that a concentration as low as 70 ppm of the polymer in 1 M HCl give an inhibition efficiency of 90%. Galvanostatic polarization studies indicated that the polymer inhibits the corrosion of mild steel in 1M HCl by a mixed type of control, i.e. both anodic and cathodic, but anodic control to a greater extent. The polymer also brought down the extent of hydrogen permeated through steel according to the hydrogen permeation studies. The effective corrosion inhibition of mild steel in 1 M HCl by the polymer, also showed by AC impedance measurements, was attributed to the formation of protective film on the metal surface due to the well adsorption of the polymer on the metal surface, which was facilitated by the greater availability of π-electrons in the aromatic ring of the polymer, and the adsorption of the polymer on the metal surface obeys Temkin's adsorption isotherm.

3.2. Polypyrrole (PPY)

Polypyrrole (PPY) film or coating used for metallic corrosion protection was mostly electrodeposited on various metals, such as iron, steel, etc. from acidic, alkaline and neutral electrolytes. Film preparing approaches including solution casting or spinning were also adopted with soluble PPY, e.g. various substituted PPYs. We reviewed here the metallic corrosion protection properties and performances of various PPY coatings also according to their preparing approaches in the order from electrodeposition approach to solution and composite approaches.

3.2.1. Electrodeposited PPY Coatings

Since Beck *et al.* [231] are the first to show that it is possible to obtain adherent, smooth PPY film on iron by electrodeposition from aqueous solutions of pyrrole and oxalic acid, which proceeds in three distinct stages, i.e. dissolution of steel, passivation of steel and finally deposition of PPY [232], many papers have been focused on the corrosion protection of iron, steel etc. by electrodeposited PPY and its derivatives, as well as on the formation and electrochemical behavior of PPY coatings on steel substrates [233].

Polypyrrole (PPY) films with different thicknesses were electrodeposited on mild steel from aqueous solution of 0.1 M pyrrole and 0.1 M oxalic acid at a constant current density by Krstajiæ *et al.* [234] and all their corrosion investigation experiments were performed in 0.1 M H_2SO_4 solution. The corrosion potential *vs.* time dependence indicated that the corrosion potentials of the deposited samples shifted to more negative value and the thinner the PPY coating, the shorter the time to achieve the steady-state value. Note that the steady-state corrosion potentials were very near the corrosion potential of the bare mild steel, PPY coating cannot provide anodic protection of mild steel in sulfuric acid. The PPY coating was rapidly undoped upon contacting with the electrolyte and the process was driven by the anodic dissolution of metal through the pores of the coating. After the short time, the total impedance of the above corrosion system increased with time, suggesting that part of the coating was reduced. The degree of reduced coating increased during prolonged exposure time, lowering the coating conductivity and the corrosion rate of the mild steel. The corrosion rate of the mild steel was reduced by at least a factor of 20 with the PPY coating. The same results were reported later by Grgur *et al.* [235]. In a more recent study [236], duplex PPY films, i.e. the inner film was doped with tetraoxalate anions and the outer doped with dodecyl sulfate anions, were electrodeposited on iron and tested for corrosion protection in 3-wt% NaCl solution. The films exhibited an increase in protection time by one order of magnitude in comparison to the oxalate doped PPY film with the same total thickness.

Ocón *et al.* [237] electrodeposited PPY films on mild steels from aqueous Na_2SO_4 solution and evaluated the corrosion protection performance of the PPY, doped and undoped, coated samples in 3-wt% solution. OCP-time measurement showed that no protection can be achieved when the film thickness is less than 17.3 μm; while if the thickness was no less than 17.3 μm, the OCP reached a pseudo-plateau of –550 mV (*vs.* Ag/AgCl) for the doped PPY and of –500 mV (*vs.* Ag/AgCl) for the doped PPY after 72 h immersion. However, the OCP values did not indicate the formation or maintenance of the passive layer. Potentiodynamic polarization revealed both a positive shift of ca. 200 mV in corrosion potential and higher exchange current densities, which include not only the corrosion current density, for both the doped and undoped PPY films as compared with the bare steel. Together with the EIS results, it was concluded that undoped PPY is the best coating in terms of corrosion protection properties for reactive metals. The reason lies in that the undoped PPY can be reoxidized by the oxygen in the solution of air and therefore the time of protection depended on the relative reaction rate, as well as the available amount of doped PPY, and in that the undoped PPY is more compact, which blocks the access of the corrosion media to the substrates. Latter, after studying the influence of the preparation methods on the morphology, mechanical properties (Universal microhardness, Young's modulus and elastic recovery) and corrosion protection properties of PPY films, the same group [238] pointed out that current constant method is the best method to obtain hard, elastic and compact PPY films, as compared with the constant potential and cyclic voltammetry methods, providing good protection for steel in highly aggressive media (0.5 M H_2SO_4). Thermal treatment of the films at 80°C for 1 day leads to a compaction of the film without degradation, improving further the quality of the corrosion protection coating, especially in non-quiescent corrosion media.

Polypyrrole and poly (*N*-methylpyrrole) were electrodeposited on low carbon steel samples in oxalic acid solution with varied pH values. Electrochemical tests including DC polarization and EIS were performed to evaluate the corrosion protection properties of the films in 1 M NaCl solution [239]. Note that PPY coated steel possesses significantly improved corrosion potential and sharply reduced corrosion current than bare steel and iron (II) oxalate films coated steel, PPY can act as a protective layer on steel and improve the overall corrosion protection performance. On the other hand, poly (*N*-methylpyrrole) coating showed higher corrosion current and higher corrosion rate than PPY. EIS results suggested the presence of a composite coating comprised a passive film and a PPY coating, which prevented corrosion by isolation and by charge transfer mechanism. Both the DC polarization and EIS results showed that PPY coatings formed at pH 2.4 exhibited the smallest corrosion rate, and the corrosion resistance of poly (*N*-methylpyrrole) coating was lower than that of PPY coatings. Later, Su *et al.* [240] electrodeposited PPY and poly (*N*-methylpyrrole) coatings on low carbon steel substrates from oxalate solutions with pH from 1 to 10 and evaluated their corrosion protection performance in 1 M NaCl solution by means of DC polarization test. With study of their electrodeposition mechanism, the authors found that a $FeC_2O_4.H_2O$ layer was first deposited on the steel surface and the steel surface was passivated, then the formed $FeC_2O_4.H_2O$ layer decomposed with the electrodeposition of PPY, that is, the PPY was deposited directly on steel surface instead of on $FeC_2O_4.H_2O$ surface, differing from the previous studies [232,239] that PPY was deposited on $FeC_2O_4.H_2O$ surface. Tafel plots showed that the electrodeposited iron (II) oxalate coatings had lower corrosion potential (–698.8 mV), higher corrosion current (1.44×10^{-4} $mAcm^{-2}$), and higher corrosion rate than the bare steel (corrosion potential: 600.6 mV, corrosion current density: 9.05×10^{-5} $mAcm^{-2}$), indicating that they cannot protect steel from corrosion. However, all PPY coatings greatly increased the corrosion potential and reduced the corrosion current. The significantly reduced corrosion rate indicated the PPY coating acts as a protection layer on steel substrate and improves the overall corrosion performance. The protection performance of the PPY coatings was also influenced by pH values of the forming electrolytes. For example, the PPY coating formed at pH 2.4 showed the best protection performance with the corrosion potential of –101.9 mV and corrosion current density 5.45×10^{-6} $mAcm^{-2}$. Compared with PPY coatings, the poly(*N*-methylpyrrole) coatings only slightly improved the corrosion potential. Although the reduced corrosion current and corrosion rate of the coated steel, the protection performance of poly (*N*-methylpyrrole) coating was inferior to PPY coating. The conclusion by the paper showed that electrodeposited PPY coating was a good corrosion inhibiting candidate and its corrosion protection performance, adhesion were influenced by the electrodepositing parameters. Identical conclusions were inferred by Martins *et al.* [241] with PPY coatings electrodeposited from tartrate containing aqueous solution on zinc-coated steels, which were tested in 3-wt% NaCl, 0.1 M HCl and 0.1 M H_2SO_4 solutions at room temperature. Furthermore, it was stated that ultrasound has important effect on the formation and properties of the PPY coatings and the coatings as prepared are more compact and adherent allowing a better protection of zinc substrates against corrosion.

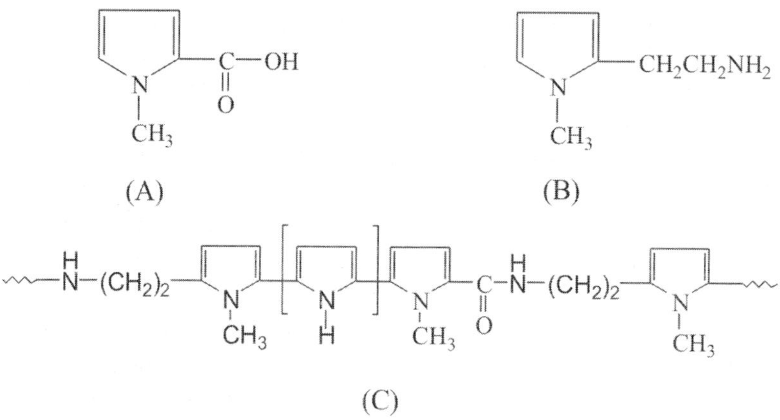

Fig. 5.10. Structures of *N*-methylpyrrole-2carboxylic acid (A), *N*-methyl-2(2-ethylamino) pyrrole (B) and schematic structure of modified PPY (C). Reproduced with permission from C.A. Ferreira *et al.*; J. Appl. Electrochem., 259 (1999). Copyright © 1999, Springer Verlag.

Polypyrrole films and modified PPY films, obtained by oxidation of a mixture of pyrrole, *N*-methylpyrrole-2-carboxylic acid and *N*-methyl-2(2-ethylamino)pyrrole (**Fig. 5.10**) in a ratio of 1/0.1/0.1 at a total concentration of 0.1 M, were electrosynthesized on mild steel and zincated steel coupons. The polymer obtained was heated at 180°C in order to cause chain coupling condensation of $-NH_2$ and $-COOH$ groups (**Fig. 5.10**) and all coatings were reduced and dedoped before any corrosion test [242]. Corrosion tests of the films in 0.5 M NaCl solution showed that corrosion potential of PPY coated mild steel decreased from –300 to –580 mV in about 40 h and then remained fairly constant, higher than that of bare mild steel, which drops rapidly from –575 to –625 mV in a few minutes and reached a pseudo steady state at about –675 mV. For the modified PPY coated samples, no corrosion was observed after eight days' immersion and the potential decreased slowly from –175 to –520 mV. Although a reddish corrosion product appeared at some points after longer time, no visible general corrosion was detected when the film was peeled off. Salt spray tests applied for 500 h showed that PPY coatings appeared to be severely damaged, while the modified PPY coatings were less attacked. When covered with a cataphoretic paint layer, the modified PPY coatings were as good as phosphating coatings and better than the PPY coatings, indicating dramatic improvements can be achieved if the PPY is suitably modified. All results proved that after they have been reduced and dedoped, conductive polymers might constitute efficient coatings against corrosion of iron or mild steel.

Polypyrrole films on mild steel samples, un-pretreated or pretreated by mechanical polishing or by chemical treatment in HNO_3, were also electrodeposited from aqueous solution of 0.1 M pyrrole and 0.1 M *p*-toluenesulphonate [243]. Electrochemical corrosion measurements were performed in 0.1 M NaCl solution at room temperature. A noble shift of 300-400 mV in corrosion potential was observed for the PPY coated treated or un-treated steel samples from the potentiodynamic polarization curves. Besides, a decrease of the redox current density as much as four orders of magnitude was measured for all the PPY coated samples. The greatest noble shift of the corrosion potential was observed for the PPY coatings on the chemically treated steel, where no pores was observed as compared with the untreated or polished steel samples.

Strongly adherent and smooth PPY layers were electrodeposited on iron by Nguyen Thi Le and his co-workers [244] in potassium tetraoxalate (0.05 M) and pyrrole (0.1 M) solution. After the coating process, the samples were rinsed in demineralized water and then dipped at open circuit in a 3-wt% NaCl solution. The OCP measurements showed that the corrosion potentials increased between –0.35 and –0.30 V (*vs.* saturated sulfate electrode (SSE)) in the early stages of immersion, which indicated the samples were in passive state and the metals were protected; then the corrosion potentials decreased to a pseudo-plateau around –0.5 V (*vs.* SSE), the value close to the reactivation potential of the reduction of the passivating oxide into a valence 2 iron. Finally, the plateau decreased sharply towards the corrosion potential of uncoated iron, indicating the metal was no longer protected as some blistering was detected and the PPY layer was dedoped. The best protection time by PPY film was 7 days in 3-wt% NaCl solution, which were two magnitudes lower than that in a H_2SO_4 solution (pH = 2). Together with the EIS study, it was suggested that the loss in protection efficiency was mainly ascribed to interchange between the counter-ion of the film and chloride ions of the solution, and therefore the nature of the counter-ion inside the film determine the efficiency of the coating. It was also suggested that a significant improvement to the protective characteristics could be gained by preventing the chloride ingress in the film. In a later study [245], Raman spectroscopy was adopted to analyze both the PPY films during the corrosion tests and the deliberately made scratches to understand the protection mechanism and the protection loss afforded by PPY film against iron corrosion in chloride neutral media. The results were perfectly consistent with the early study [244]. It was also confirmed that during the whole higher potential plateau, the PPY film kept its full doping level and maintained the iron film in passive state, confirming the anodic galvanic mechanism proposed in [244]. In the scratches, the doping level of the PPY films were non-uniform, which indicated that the oxidizing capacities of the film were not fully used, probably because when the potential decreased, the oxalate mobility became very low. The presence of sodium oxalate in the pits indicated indirectly the increased mobility of the cations at more cathodic potentials.

On the basis of their early work [244], the same group [246] electrosynthesized composite polyanions/PPY films on iron after electrodeposition of a first oxalate-doped PPY film, and after the coating synthesis, the samples were rinsed in demineralized water and then dipped at open circuit in a 3-wt% NaCl aqueous solution. The OCP curves corresponding to different counter-ions, i.e. oxalate, paratoluenesulphonate (tosylate) and polystyrenesulfonate (PSS), which were used as synthesis media, had similar forms featuring two potential plateaus before the potential dropped to the iron corrosion potential and were similar to [244]. In a first plateau, the potential increased slightly between –0.25 V and –0.2 V (*vs.* SSE), which confirmed that iron was in the passive state. A second plateau was seen around –0.4 to –0.5 V (*vs.* SSE) with the potential decreased. Finally, the potential decreased sharply towards the corrosion potential of uncoated iron, the metal was then no longer protected. The OCP curves also showed that the total protection time increases with the counter-ions size, that is, oxalate < tosylate < PSS. This improvement was mostly due to the longer duration of the first plateau where exchange between chloride ions and counter-ions occurred. The protection time of iron by using of PPY coatings as primer coating varied between 50 and 170 h, depending on the films, and was much longer than a chromatation layer primer. The duration of the first plateau was twenty times longer with the mixed oxalate/PSS-doped PPY film, but the duration of the second plateau was smaller with mixed oxalate/tosylate or PSS-doped PPY film than that with "pure" oxalate-doped PPY film due to the lower quantity of the oxalate anions, which allowing autorepair of the passive layer, in the mixed films.

Fenelon *et al.* [247] electropolymerized a PPY layer for the first time on a CuZn electrode, not like the literature documented iron or iron-alloys electrodes [214], from a sodium oxalate solution with a pH of 7.8 and assessed the corrosion protection properties of this layer in a 0.1 M NaCl solution, adjusted to pH value of 3.5 or 7.8 using either NaOH or HCl. The growth of the PPY film was facilitated by the initial oxidation of the oxalate solution to generate a zinc-containing copper (II) oxalate pseudopassive layer. The layer was sufficiently protective to inhibit further dissolution of the CuZn electrode, and sufficiently conductive to enable the electropolymerization of pyrrole at the interface and the generation of an adherent PPY film. For the uncoated CuZn electrode, the OCP reached near-constant value after a short immersion time, ca. 6×10^3 s; while for the coated electrode, the OCP decayed slowly and reached near steady-state value after ca. 2×10^6 s. At this point the differences between the OCP of the coated and uncoated electrodes were ca. 60 mV and 34 mV in the solution with pH 7.8 and pH 3.5, respectively. The difference in breakdown potential of the coated and uncoated electrodes was 300 mV from the anodic polarization of the electrodes following the 14 h immersion period in the pH 7.8 solution. The apparent stability of the PPY layer on CuZn electrode was attributed to electron transfer between copper and the N^+ of PPY to form a stable Cu-PPY complex [248], thus preventing nucleophilic attack on the positively charged nitrogen, for example, by H_2O. On prolonged immersion in the chloride containing solutions, the PPY film lost some of its corrosion protection property, but it still remained stable, offering good corrosion protection to the CuZn substrate. A homogenous and adherent PPY film was generated electrochemically at the CuNi electrode by the same group [249] in a near-neutral sodium oxalate solution containing Cu^{2+} cations that promoted the oxidation of the polymer units during polymerization process and increased the adherence of the PPY film at the electrode surface. Corrosion protection properties of the PPY coated CuNi electrodes were assessed in 0.1 M NaCl solutions of pH 7.8 and 3.5. In the near-neutral solution, the corrosion potential of the PPY coated electrode was approximately 450 mV more noble than that of the uncoated one, and the potential can be increased to 800 mV or more before the anodic current reached a 1 $mAcm^{-2}$ level. The OCP of the PPY coated electrode initially reached 500 mV and then decayed slowly to a near-steady-state value that was 180 mV higher than that of the uncoated one. The near-steady-state OCP value remained constant and differed from the iron-coated electrodes [246], which dropped to the OCP value of the uncoated electrode. In the more aggressive acidic solution, the corrosion potential was increased by almost 400 mV and the potential at which the current adopted a 1 $mAcm^{-2}$ level, by over 650 mV compared to the uncoated electrode. The OCP of the PPY coated electrode became near constant shortly after immersion in the acidic solution and it was 50 mV lower than that of the uncoated electrode. Anodic polarization of both the systems after 14h immersion period showed that the PPY coated electrodes can be polarized to relatively high anodic potentials before high anodic currents were measured; moreover, some of the anodic currents measured for PPY coated electrodes was due to the electroactivity and oxidation of the PPY and not due to dissolution of the substrate. These PPY films exhibited significant corrosion protection properties in both acidified and neutral 0.1 M NaCl solutions for CuNi electrodes.

Though tremendous papers have been published on the electrochemical and electrocatalytic properties of polyoxometalates doped conducting polymers [250], Cheng *et al.* [251] investigated for the first time the spectroscopic character, morphology and solubility of the polyoxometalate doped PPY films, which were prepared by electrochemical polymerization from fresh acetonitrile +2% H_2O (v/v) solutions containing pyrrole and polyoxometalate anions. Note that the hybrid films showed a flat surface morphology, good adhesion, good wear resistance, good stability in different solvents and a hardness of 2H as evaluated through the standard ASTM T33 63-74, they were suitable for metal coating applications and corrosion-wear preventions according to the authors.

Based on and different from the previous works [10,219], in which noble metal microparticles were incorporated into conducting polymer films by electrolysis, Hammache *et al.* [252] prepared Cu incorporated PPY films (PPY/Cu) by a cementation process, i.e., immersing the PPY coated iron substrates, which were previously electropolymerized on iron substrates from aqueous oxalic acid solution, in a solution containing 1×10^{-5} M of $CuSO_4$ in 0.1 M Na_2SO_4 + H_2SO_4, and evaluated the corrosion protection performance of the PPY/Cu films in a corrosive solution (3-wt% NaCl + HCl, pH = 1). According to the authors, the PPY electropolymerization was characterized by two distinct stages, i.e. first the dissolution and the progressive passivation of the substrate and then deposition of PPY film, differing from the three stages reported by Beck *et al.* [231,232]. For PPY coatings of different thickness (obtained at different current densities), the OCP curves showed a starting corrosion potential value of about 200 mV, which corresponds to the potential of the redox process in PPY film; then a slightly decrease of potential till a pseudo-plateau around 100 mV, indicating the iron was in its passive region; finally the decrease of potential plateau to the corrosion potential of the uncoated iron. The protection time was determined by the elapsed time between the immersion of the samples into the corrosive medium and the end of the plateau. The OCP curves also showed that the PPY film with a thickness of 16.6 µm, which was electrodeposited with current density of 2 $mAcm^{-2}$, was more effective than other films with thickness more or less. By incorporation of Cu microparticles into the PPY films, the corrosion protection of iron was significantly improved. For example, at the maximum benefit Cu amount into PPY film, i.e. 0.4 $mgcm^{-2}$, the stability of the PPY film was enhanced by a factor of 40 comparatively to the non-incorporated PPY film. The Tafel explorations showed that the PPY and PPY/Cu caused extremely large positive displacement of around 700 mV in the corrosion potential relative to the bare iron; and also showed substantial reduction in corrosion current for the coated samples (PPY and PPY/Cu) with respect to uncoated samples. But on long time (until the time the electrode potential decreased to the corrosion potential of uncoated iron) immersion of all the PPY and PPY/Cu coated samples, the polymer morphology changed, as reported by Herrasti *et al.* [5] that holes appeared on the polymer surface. All results indicated that the PPY/Cu coating appeared to be more effective than the PPY coating, and allowed to maintain the corrosion potential in the passive range for a long time in strong aggressive solution, and also the level of protection depended on the characteristic and form of the PPY/Cu and the nature of corrosion environment.

Jiang *et al.* [93] prepared PPY films on AZ91 magnesium alloys (with composition of Al 9-wt%, Zn 1-wt% and Mg the rest) by electrochemical polymerization of pyrrole from aqueous solutions of 0.2 M pyrrole and 0.2 M KOH. OCP and potentiodynamic measurements were performed in 3.5-wt% NaCl solution. It was showed in OCP curves that the potential of bare AZ91 magnesium alloy in 3.5-wt% NaCl solutions was almost no change after certain time (ca. −1670 V). But potential of PPY deposited AZ91 magnesium alloy first increased, and then decreased, also thirdly as bow increased and remained constant (ca. −1550 V). The positive open potentials of PPY deposited AZ91 magnesium alloys indicated that anti-corrosion of AZ91 magnesium alloys improved with PPY deposition. The potentiodynamic polarization curves for PPY deposited AZ91 magnesium alloys indicated that corrosion protection of PPY coatings prepared in alkaline solutions was low to protection of PPY/paint [253]. The value of corrosion protection observed in this measurement was different from that above because of the different thickness of PPY coatings. It was possible that the porosity of PPY coatings gave rise to corrosion ions permeation and diffusion, and therefore the more compact PPY is required to prepare for better protection efficiency.

Polypyrrole coating was also electrodeposited on copper from a solution of 0.1 M pyrrole and 0.125 M sodium oxalate with pH of 7.6 and the corrosion performance of the coated samples was assessed in 0.1 M NaCl solutions, with pH value between 3.5 and 7.8 adjusted by NaOH or HCl [254]. In acidic NaCl solution (pH 3.5), the corrosion potentials adopted by the "thin" and "thick" PPY films were 0 and 290 mV, respectively. Moreover, the PPY films remained protective at much higher potentials. The thin PPY film broke down at 470 mV, but the thick PPY made it difficult to mark the onset of copper dissolution as the current-potential response was governed mainly by the activity of the PPY and not the copper substrate. In neutral NaCl solution (pH 7.8), the corrosion potentials of the PPY coated samples were significantly higher at 50 mV for the "thin" polymer and 225 mV for the "thick" polymer. Similarly, the polymer exhibited good corrosion protection properties on polarization in the anodic direction with breakdown of the "thin" PPY coating occurring at 500 mV and no evidence of breakdown for the "thick" polymer. The high currents measured for the "thick" polymer appeared to be associated with the conducting properties of the polymer and not the dissolution of the underlying copper substrate, and this was supported by the addition of a copper-sensitive indicator, Murexide, which remained violet for the absence of Cu^{2+} in solution. The OCP of the uncoated copper was independent of time and remained constant at ca. −140 mV in acidic solution and −180mV in neutral solution, while the OCPs of PPY coated coppers were significantly higher in both acidic and neutral solutions. For example, in acidic solution, the OCP of the PPY coated copper was ca. 200 mV in early stages of immersion and then decayed to 0 V after 14 h for the "thick" polymer and extended between 0 and −40 mV for the "thin" polymers. The resistance of the copper-chloride corrosion film and the polarization resistance, from the EIS data, remained stable for period exceeding 1000 min, indicating that the PPY-copper composite was stable in these highly aggressive solutions.

By applying cyclic voltammetry, Tüken *et al.* [255,256] electrosynthesized a very adherent and highly conductive PPY film on mild steel from a 0.3 M oxalic acid solution. A topcoat of polyphenol was also electrodeposited on the PPY film. Corrosion tests, including EIS, anodic polarization curves and OCP *vs.* time curves, in 0.1 M Na_2SO_4 solution and 0.05 M H_2SO_4, all indicated that the topcoated film provide better protection for longer periods than the single PPY film. For example, OCP of the single PPY coated sample decreased to that of bare steel (ca. −0.560 V) after 9 h of immersion in Na_2SO_4 solution, while for the topcoated sample, it was +0.05 V after 10 h and −0.45 V even after 45 h [255]. In the case of the acidic medium [256], a protection efficiency of 98.3% was determined after 340 h of exposure. It was then concluded that the thin polyphenol topcoat decreased the permeability of the coating and water mobility within the pores, viz. improved the barrier effect of the coating. When the barrier effect was diminished, the pronounced anodic protection of the underlying PPY coating was brought into effect. The same barrier effect was observed by the group with the similarly prepared PPY coatings without any topcoat for copper and brass in 0.1 M H_2SO_4 [257] and with the similar PPY as well as the PANI coatings on nickel coated mild steels in 3.5-wt% NaCl solution [258]. A better result was obtained on copper than on brass owing to the better adhesive interaction between PPY film and the oxidized copper surface. Besides, PPY film was more effective than PANI on nickel coated mild steel [258]. In a more recent paper, the same group [259] prepared a double layer system, i.e. a PPY primer and a polyindole topcoat, on mild steel electrode and tested the corrosion protection performance of the system in 3.5-wt% NaCl solution. The thin PPY film was obtained from oxalic acid solution by using cyclic voltammetry technique. The polyindole, which is one of the ICPs and can be easily electropolymerized by anodic oxidation of indole in various electrolytes, though no adherent and corrosion protective polyindole film can be obtained on Fe electrode in $LiClO_4$ containing acetonitrile solution [260], was electrosynthesized on the thin PPY coated mild steel electrode by using the same technique from $LiClO_4$ containing acetonitrile medium. OCP of the sample was +0.226 V (*vs.* Ag/AgCl) after initial immersion, and –0.248 V, –0.185 V, –0.46 V and –0.413 V (*vs.* Ag/AgCl) after 8 h, 27 h, 96 h and 150 h of exposure respectively; finally, –0.580 V (*vs.* Ag/AgCl), almost the OCP of the bare steel sample (–0.613 V *vs.* Ag/AgCl). The increases of OCP at 27 h and 96 h were explained by a catalytic effect of the polymer coating for the formation of protective ferric compounds and the repair of passivity that had been broken by chloride ions. Much lower current and noble corrosion potential were also found for the coated sample than for the bare steel in anodic polarization curves. Together with the EIS results, it was concluded that the PPY primer and polyindole topcoat system exhibited excellent barrier effect and was able to provide anodic protection for mild steel.

Polypyrrole coatings containing hexacyanoferrate anions were electrodeposited galvanostatically on SS from aqueous solutions containing pyrrole, potassium hexacyanoferrate (II) and traces of free CN^- ions and tested for corrosion protection in highly aggressive medium (0.4 M H_2SO_4+0.4 M HCl) [261]. As that happened in the hexacyanoferrate anions containing PANI coatings [226] on SS, sparingly soluble and physicochemically stable PB was formed at the polymer/SS interface, resulting in the stabilization of passive state of the steel. Pitting causing anions such as Cl^- were blocked by the introduced hexacyanoferrate anions in the coatings and corrosion was largely suppressed. When some Cl^- anions reach the steel surface and the breakdown of passive layer occurred unfortunately, any iron or chromium ions generated at the pits were likely to form PB with the $Fe(CN)_6^{3-/4-}$, indicating promising candidates of the coatings in the protection of SS against pitting corrosion in chloride containing acid medium.

3.2.2. Solution Prepared PPY Coatings

Poly(3-octyl pyrrole) (POP) was synthesized electrochemically on a platinum electrode from a mixed solvent of CCl_4 and CH_2Cl_2 containing 0.1 M monomer (3-octyl pyrrole), 0.1 M tetrabutylammonium perchlorate and 0.025 M *p*-tetrabutylammonium *p*-toluenesulfonate. The soluble fraction of the electrosynthesized polymer was recovered from the liquor and dissolved in a mixed solution of CCl_4 and CH_2CN for cast coating of CRS, aluminum alloy and pure aluminum substrates. The coatings were dried overnight and no topcoat was employed. Scanning vibrating electrode technique (SVET) was used to probe the interactions between POP and the substrates. A defect extending to the metal surface with area ranging from 0.1 to 0.3 mm^2 was introduced to each sample. The immersion solution for Al 2024-T3 samples was dilute Harrison solution and for CRS was 3-wt% NaCl solution. All scans were initiated within 5 min of immersion and data were collected every 20 min for the duration of the experiment (typically 20 h). On CRS, the POP coating delayed the onset of corrosion within the defect, and the delay was significantly longer, typically ca. 3 h, as compared to ca. 40 min for chromated epoxy coating. Oxidation was observed within the defect and reduction occurred rather uniformly across the conducting polymer surface. The defect was protected by a mechanism involving formation and/or stabilization of a passive layer in the defect. On Al 2024-T3, no oxidation current was observed within the defect, viz. only reduction current. The oxidation process involved removal of metal from copper-rich regions of the alloy surface beneath the POP coating, and the oxidation was not driven by the conducting polymer, but was coupled to the reduction process occurring in the defect and also in the POP surface. On pure aluminum, a significant time delay was observed as for the alloy, at which point only reduction in the defect area. And in contrast to the very localized oxidation current of the alloy, the oxidation current in aluminum was much more distributed across the substrate surface. When the coating was removed, no pitting was observed and the surface was shiny, indicating a rather uniform oxidation of the aluminum surface. Although the details were presented further in the paper, the observations on pure aluminum support the notion that copper-rich regions in the alloy are responsible for the localized oxidation current for the POP coated Al 2024-T3 alloy [262]. In another work by the same group [263], POP and poly (3-octadecyl pyrrole) (PODP) were coated onto aluminum 2024-T3 by the similar procedure to [262] except with a polyurethane topcoat. EIS, during both long-term immersion and Prohesion® exposure, ENM, polarization and SRET were utilized to examine the corrosion protection performance of the coatings in a corrosive medium of dilute Harrison's solution. After 600 days' immersion of the samples, EIS data were collected at the OCP for increasing times of immersion. Compared with the Cr-epoxy coating, i.e. the control sample, the POP coating exhibited the least change and greatest inhibition of blister development during immersion by visual observation. The control sample had high impedance at the start of immersion, but showed a drastic failure at 385 days, with rather larger blisters on the samples. The PODP coated samples started to show signs of failure with the decrease of impedance at day of 376, while the POP coated samples continued to maintain the initial impedance value. ENM tests revealed that the value of R_n remained between 10^4 and 10^5 Ω, throughout the immersion for POP coated samples; PODP coated samples also had value of R_n between 10^4 and 10^5 Ω, but had large fluctuations as compared with the stable one of POP. The differences, according to the authors, were dependent on the stability of the adherent oxide layer. A less adherent oxide layer would fail and subsequently reform, giving rise to fluctuations in R_n. The POP system would form an extremely stable, tightly adhering oxide layer, thus providing better corrosion protection to the underlying substrate. An increase trend of pH was noted during the immersion, which as proposed was due to aluminum hydroxide that is slightly soluble in water and thus increased the pH value, not like the decrease of pH for PANI [149], which was attributed to the dedoping of the PANI. During the Prohesion® cyclic exposure of scribed samples, the control sample showed an initial decrease in impedance after 117 h, while the PODP system was able to maintain the initial impedance value. Even more promising was the POP system, which showed an increase in impedance after 117 h of Prohesion® exposure, which was due to the growth in the adherent oxide layer. The adhesion of the coatings also played an important role in their protection performance, which made the PODP coating not as outstanding as POP coating.

3.2.3. PPY Composite Coatings

PPY composite coatings were formed after double immersion of magnesium samples in a polyacrylic-PPY matrix solution, which was given by using *in-situ* polymerization of pyrrole in an aqueous polyacrylic emulsion [264], for 10 s with a time interval of 5 min and finally dried at 80°C for 15 min. PPY powders were also blended with polyacrylic emulsion to form PPY containing coatings. Immersion tests were performed in a 1-wt% NaCl solution at neutral pH of 6.9 for 9 days and the corrosion rate was monitored every second day by determination of the Mg^{2+} concentration in solution. For the *in-situ* prepared coatings, corrosion rate reduction of 70% can be obtained for the optimized synthesis conditions and the corrosion rate showed weak dependence on coating thickness, indicating that the bulk properties of the coatings were not of great importance. The inhibition properties of the coatings were attributed to the formation of a thin, passive layer in between the coatings and the surface metal oxides. The layer was formed from PPY polymer or oligomers strongly chemisorbed onto the metal oxides, and the structure was further stabilized by polyacrylic polymer chains interacting as dopants of PPY. On the contrary, the addition of PPY in the form of pigment had very limited effect on the corrosion rate, though 20 μm in thickness compared with the 2-5 μm of the *in-situ* prepared coatings, indicating that the protective action of PPY was minimized when the polymers are not in the physicochemical form to interact with the metal interface.

Polypyrrole containing acrylic paint (Trade name: Acrolak) was made by thoroughly mixing of the PPY powder at a given concentration with an acrylic paint, and coated then on magnesium alloy DTD 118 (with composition of Ca 0.15-wt%, Zr 0.06-wt% Mn 1.64-wt% Zn 0.01-wt% with Mg the remainder) with average coating thickness of 120 μm for corrosion protection of the alloy [253]. PPY powder was produced by chemical oxidative polymerization using ferric chloride ($FeCl_3$) as an oxidant and freshly distilled pyrrole and doped with 5-sulfosalicylic acid. The corrosion electrolyte was 3.5-wt% NaCl solution for the OCP measurements and 0.1 M NaCl solution for potentiodynamic scanning and EIS measurements. For OCP measurements and potentiodynamic scan tests, each coated specimen was drilled with a 0.6 mm diameter pinhole through to the magnesium surface, prior to testing, and for EIS measurements, was of free from any pinhole. Salt spray tests showed the PPY containing (10 and 20 parts) Acrolak coated samples exhibited minimal corrosion in the defect and a few moderate blister after exposure to salt spray for 1000 h, while the control samples (coated with Acrolak only) exhibited large blisters with white products underneath and failed completely at 200 h. Corrosion was also insignificant in the Acrolak containing PPY (5 parts) coated samples but more blisters were observed after1000 h exposure. Anodic potentiodynamic data showed that the E_{pit}-E_{OCP} value, which is an indication of the effectiveness of the resistance to corrosion with a high value indicating a coating with more effective resistance, was four times higher for the Acrolak containing PPY (20 parts) coated samples (223 mV) than that obtained for the control samples (57 mV) under the same testing conditions. It appeared that the anodic activity of the magnesium sample was significantly reduced when the PPY coatings were used. The reduction of anodic activity measured in the anodic potentiodynamic scans for the PPY coated samples may be due to a reduction in the corrosion rate of the magnesium substrate, although a reduction of the anodic reaction rate at the PPY particles was also possible. The E_{pit}-E_{OCP} values increased very rapidly with PPY content in Acrolak from 0 to 10 parts and tapered off, indicating the optimum quantity of PPY powder, which provides an effective corrosion protection and maintains good mechanical properties of the coating, is around 10 parts in Acrolak. The data obtained from the cathodic potentiodynamic scans and the EIS for the PPY containing paint coated samples showed large cathodic current densities (approximately 2 orders of magnitude higher than the control sample) and overall low impedance values, respectively, and contradicted the results of the salt spray test, immersion test, OCP and anodic potentiodynamic measurements. All these tests indicated that the corrosion protection by the coating containing PPY powder was clearly superior to the control sample. The high cathodic current densities were primarily due to the cathodic reaction(s) at the PPY particles and not at the substrate surface. Such cathodic reaction(s) and the relatively high conductivity of the PPY led to an overall decrease in the measured impedance values in the EIS scans.

Polypyrrole and PPY/carboximethylcellulose (PPY/CMC) composite coatings were electrochemically deposited on SS in 0.25 M pyrrole, 0.1 M $LiCO_4$ and 0.25 M pyrrole, 0.1 M CMC aqueous solutions, respectively by Herrasti *et al.* [5] and the electrochemical corrosion measurements were performed in 3-wt% NaCl solution at room temperature. Polarization curves showed that noble shifts in corrosion potential for PPY and PPY/CMC covered surfaces were 90 mV and 250 mV respectively. The OCP results also showed that the potential was stable with a lighter decrease in the case of SS covered with PPY and PPY/CMC compared with bare SS where the OCP presented a decreasing trend. Potentiostatic method was better than galvanostatic method to deposit more stable films against corrosion. PPY films deposited at low current density were more uniform and more stable against corrosion and better than those deposited at high current density. The PPY/CMC composite coating was more stable than PPY due to a more compact structure. But during long time immersion, the morphology of the coatings changed, appearing holes on the surface of the polymer surface, and the change was drastic for PPY coating.

Polypyrrole/inorganic particles composited films were also investigated for corrosion protection of metals. Lenz *et al.* [265] galvanostatically synthesized PPY/TiO_2 composite films on AISI 1010 steel sheets in 0.1 M oxalic acid (pH = 2.0) and all films contain 14-wt% titanium dioxide. The dedoping process of the films was performed by simple immersion after polymerization in 0.1 M NaOH applying –0.7 V for 15 min. Corrosion protection performance of the films, both in doped and dedoped forms, was evaluated by salt spray test in 3.5-wt% NaCl solution. For the steel coated by PPY/TiO_2 composite film, weight loss test showed that there was at first an acceleration of the oxidation process, probably due to initial iron oxidation through the pores/or defects of the dedoped film, and no further corrosion was detected after ca. 96 h and the corrosion process stabilized at 12.5 mg of iron per liter of NaCl solution; for the steel coated with dedoped PPY film, the initial oxidation rate was slightly higher than that for PPY/TiO_2 composite film, thereafter, the corrosion rate was slowed down but no stabilization was observed and after 7 days the samples released 25 mg of iron per liter of NaCl solution. The different behavior was explained by the low porosity of the PPY/TiO_2 composite films due to the filling of polymer pores by TiO_2 particles as observed by SEM images. The same experiments performed with the doped PPY and PPY/TiO_2 composite films showed slightly greater weight losses than the dedoped samples; however as mentioned above, the presence of TiO_2 decreased the corrosions. Salt spray test performed on doped samples showed that without an X-cut, the PPY/TiO_2 samples exposed for 5 h and 24 h, respectively, displayed a reduction of ca. 95% and 30% in the corroded area as compared to the PPY samples; with an X-cut, all the samples showed intense corrosion and the corroded area was always larger for the PPY samples as compared with the PPY/TiO_2 samples.

With oxidant of APS and dopant of DBSA, Yeh *et al.* [266] *in-situ* prepared PPY/MMT nanocomposites by chemical oxidative polymerization. After dissolution of the PPY/MMT powders in chloroform and filtered through a Teflon membrane with pore size of 1μm, the solution was then dip-coated onto CRS coupons and followed by drying in air for 12 h to give a coating of ca. 60 μm thickness. The control coating, i.e. PPY coating, was prepared by the similar procedure. The results of the electrochemical corrosion measurements in 5-wt% NaCl aqueous solution at room temperature are listed in **Tab. 5.5**. It can be seen that the corrosion potentials of the PPY/MMT composite coatings are higher than that of PPY coating, also the corrosion currents of the PPY/MMT coatings decrease with the MMT loading. The advanced corrosion protection effect of the PPY/MMT coatings compared to the pristine PPY coating resulted from the increased tortuosity of the diffusion pathway of oxygen and water when the silicate nanolayers of MMT were dispersed in PPY matrix.

Tab. 5.5. Electrochemical corrosion measurement results of the PPY/MMT coatings. Reproduced with permission from J.M. Yeh *et al.*; J. Appl.Polym. Sci., 88(14), 3264 (2003). Copyright © 2003, John Wiley & Sons, Inc.

Compound code[a]	Inorganic content in product (wt%)	Corrosion potential (mV)	Polarization resistance ($K\Omega \times cm^2$)	Corrosion current ($\mu A/cm^2$)	Corrosion rate (MPY)
Bare	—	−641	0.80	44.4	86.140
PPY	0	−546	17.96	2.75	0.064
PPYCL1	2.3	−474	24.51	1.87	0.044
PPYCL3	5.6	−470	50.43	0.87	0.020
PPYCL5	8.0	−421	68.71	0.54	0.013
PPYCL10	22.1	−351	120.38	0.02	0.001

[a] PPYCL denotes the composite coating and the number denotes the MMT content in feed composition (pyrrole and MMT).

An oxide/PPY composite film was electrosynthesized on iron from a solution containing 0.1 M pyrrole, 0.05 M potassium tetraoxalate, and the oxide in suspension, i.e. magnetite (Fe_3O_4) and hematite (Fe_2O_3) [267]. OCP curves of all the films, including the PPY film, in 3-wt% NaCl solution featured by two potential plateaus before the potential dropped down the iron corrosion potential [244]. In the early stage, the plateau was –0.3 V (*vs.* SSE) for the PPY without oxide and –0.16 V and –0.24 V (*vs.* SSE) for PPY/ Fe_3O_4 and PPY/ Fe_2O_3 composite films, respectively, indicating the iron was in the passive state. Then the potential decreased to a second plateau, with the solvent exchange and the exchange between the PPY oxalate counterions and chloride ions. The chloride ions which have reached the metal involved the oxidation of the metal and then partial polymer reduction; the reduction released oxalate anions which ensure the repairing of the passive layer by forming ferrous oxalate precipitate at the metal/polymer interface. Finally, when the chloride anion concentration was too large at polymer/iron interface or when oxalate anions were more available, the second plateau was followed by a sharp decrease toward the corrosion potential of uncoated iron. The improvement of the protection efficiency of the composite films, i.e. the longer duration of the first plateau, was due to the presence of the oxides, which allow the polymer to remain in its oxidized state using faradaic charge, and consequently to maintain the passive state of the metal. The better efficiency of the PPY/Fe_3O_4 composite film as compared with the PPY/Fe_2O_3 composite film resulted from, on one hand, the higher amount of Fe_3O_4 oxide incorporated in the film during the synthesis; on the other hand, the increased number of defects and therefore the weakness of the coating due to the incorporation of the big agglomerates like those formed by Fe_2O_3, not from the higher oxidation state of iron in the oxide Fe_2O_3 (+3) than in the oxide Fe_3O_4 (+2.6). A dispersion of the results concerning the corrosion protection tests, ca. 20% in the standard deviation, showed that all the synthesis parameters were not well controlled and more investigations are still in progress according to the authors.

In a recent work by Paloumpa *et al.* [268], PPY composite coatings containing titanium and zinc oxides were chemically prepared on Al panels by using an aqueous solution of pyrrole, fluoro-zirconic acid, fluoro-titanic acid, zinc oxide and phosphoric acid. The coatings were formed after simple immersion of the panels in the solution for 6 min. Several methods, mainly photo-emission electron microscopy (PEEM) and SEM, were employed in studying the corrosion process of the samples and it was found that the composite coatings exhibited advanced corrosion resistance in 5.0-wt% NaCl solution acidified with acetic acid (pH4) under usual corrosion conditions, and lost their durability only under extreme conditions of low pH and electric current flow. It was also found that PPY was not adhesive on iron as on Al and therefore traces of the metal caused a homogeneity in the PPY coating, which was fatal for the protection performance of the film, not only due to the exposed iron phases where the corrosion begins, but also to the fact that the iron avoids zinc to protect aluminum by its cathodic mechanism.

A composite film containing a first layer of poly(1,5-diaminonaphthalene) (PDAN) at the metal/film interface and a second layer of PPY was electro-prepared by Nguyen and his co-workers [269] on iron recently. The PDAN layer enhanced the adhesion of the composite coating as compared with PPY film alone. OCP measurements in 0.1 M K_2SO_4 solution with pH adjusted to 4 by H_2SO_4 showed that in the case of the bare iron, the sample became active in a few seconds with OCP dropped to ca. –0.7 V (*vs.* SSE); in the case of the PDAN coating, the sample became active within a couple of minutes; while in the case of composite coating, the sample reached the corrosion potential at –0.9 V (*vs.* SSE) only after 50 days, far better than the 60h of the PPY coating, which was attributed, on one hand, to the better adherence of the film on the iron surface; on the other hand, to the higher oxidation state of the composite coating, which remained higher during protection, than the PPY coating, which was progressively reduced during protection, as evidenced by the *in situ* Raman spectroscopy.

3.3. Simultaneous Use of Polyaniline (PANI) and Polypyrrole (PPY)

Tan *et al.* [270] galvanically deposited three different types of multilayered conducting coatings, consisting of combinations of the conducting PANI and PPY onto both carbon steel and SS. The first, PANI/PPY coating, was prepared by firstly a deposition of PANI from 0.1 M aniline in either 0.05 M H_2SO_4 for SS or 0.3 M oxalic acid for carbon steel and secondly deposition of PPY over the PANI film from 0.1 M pyrrole in acetonitrile and 0.5 M $LiClO_4$. The second, PPY/PANI coating, was prepared by an identical procedure and condition to those for the PANI/PPY coating except that the order in which the polymers were coated was reversed. The last, PANI-PPY coating, was prepared by deposition of copolymer from a mixture of aniline and pyrrole, both in 0.1 M, in either 0.05 M H_2SO_4 for SS or 0.3 M oxalic acid for carbon steel. Electrochemical measurements were carried out in an aqueous electrolyte of 0.028 M NaCl solution (1000 ppm Cl^-).

For carbon steel, although all three coatings caused small (<100 mV) positive shifts in the corrosion potential as compared to bare carbon steel, the Tafel extrapolations revealed that only the PANI-PPY coating gave a reduction, which is about 30%, in the general corrosion rate. PANI-PPY coating also encouraged passivation giving rise to lower currents in the so-called passive region than that of the bare steel, which suggest that the coating was able to provide corrosion protection to carbon steel in a manner akin to anodic protection or anodic inhibitors. In the meantime, PANI-PPY coating caused negative shifts in both pitting potential and the repassivation potential, indicating a reduced resistance to pitting corrosion. Both the PPY/PANI and PANI/PPY coatings caused increased corrosion current density probably due to the improved kinetics for cathodic reaction. These two coatings also failed to encourage any passivation of the carbon steel which was particularly surprising since a passive layer of Fe (II) oxalate has been reported to form on carbon steels when PANI was electrodeposited from oxalic acid [205]. And it was concluded then that for carbon steels the performance of the multilayered coatings was not sufficiently better than for single PANI coatings to justify their more complicated deposition procedures.

For SS, potentiodynamic polarization curves revealed that the all coatings were much more successful than they did on carbon steel. The PPY/PANI coating caused an extremely large positive shift of 750 mV in corrosion potential, relative to the value for the bare SS, and reduced the general corrosion rate by a factor of ca. 2000. While the PANI/PPY coating and the PANI-PPY coating did not appear to be successful at protecting SS from corrosion. Despite the noble shift of corrosion potential by more than 200 mV, these two coatings led to higher corrosion current densities. Substantial positive shifts in the pitting potential, as compared to the bare SS, were observed for all the coatings. The performance of the PANI/PPY coating on SS-304 was comparable with more resistant (and more expensive) molybdenum containing SS-316, and the PPY/PANI coated SS specimens performed even better as per the authors. It was thus suggested, as compared with the early work by Santos *et al.* [134] in which no improvement in the pitting potential was observed for SS-304 when coated with a single PANI coating, that the multilayered coatings may be better chemical barriers, trapping chloride and preventing it from attacking the SS, or that a second electronic barrier (akin to a Schottky barrier) was developed at the PPY/PANI interface. Such a barrier would not be expected for the PANI-PPY coating, but it still provided better pitting corrosion than single polymer coatings. It was summarized that PPY/PANI coating provided excellent protection to both localized and general corrosion of SS-304, whereas the PANI/PPY coating and the PANI-PPY coating provided only moderate protection against localized corrosion. Nevertheless, all the PANI/PYY coatings performed better than previously reported single PANI coating [134]. PPY/PANI coating, by far the best performed one, had a closely packed morphology and the strongest adhesion (0.2 N) to the SS, although less than half the value of a typical paint film. And all results suggested, according to the authors, that the ability of a conducting polymer film to act as electronic and chemical diffusion barriers was more important in providing corrosion protection than its ability to act as a physical barrier.

In Iroh's group, PANI-PPY composite coatings of various compositions were electropolymerized on both Al 2024 [227] and low carbon steel [271] substrates in oxalic acid. The corrosion performance of the samples was evaluated by using DC polarization tests in neutral 3.5-wt% NaCl solution and the corrosion rate was calculated by averaging of the results using a Model 352 Softcorr III software. In the case of Al substrate, the more the initial aniline concentration in the feed composition, which varied from 1:9 to 9:1 (aniline and pyrrole), the better the corrosion resistance of the obtained composite coating. And the best coating, the one with feed composition of 10% pyrrole and 90% aniline, reduced the corrosion rate by 1.5 orders of magnitude as compared the one with feed composition of 90% pyrrole and 10% aniline. The results agreed well with the morphology analysis of the coatings, as the coatings formed with higher aniline concentration were dense and compact than the ones with higher pyrrole concentration. In the case of steel substrate, the composite coating, which was formed using equal molar feed ratio of aniline to pyrrole, showed dense and compact structure, as compared with the granular morphology of PANI coating and globular one of PPY coating, and exhibited better corrosion protection performance than both PANI and PPY coatings. Furthermore, corrosion resistance of the coatings can be enhanced by the increase of the adhesion strength. In a later study [272], the same group electrodeposited PANI, PPY, PNEA, poly(*o*-anisidine) and PANI-PPY composite coatings on Al-2024 alloy in a similar procedure. Topcoated with a polyimide in thickness of 40 µm by solution cast, the samples were subjected to DC polarization test in a 3.5-wt% NaCl solution and EIS test in a Harrison solution. Both tests indicated that the PANI-PPY coating, together with the PANI, PNEA and poly(*o*-anisidine) coatings, performed better than PPY coating. The superior corrosion protection resulted from both the good barrier property of the coatings and the good adhesion between the topcoat and the primer layer, which prevented the seeping of the electrolyte inside. While for the PPY coating, the poor corrosion protection was attributed to its poor adhesion onto the Al-2024 substrate as shown in the peel tests. PANI-PPY composite coating was also electrodeposited on aluminum from a tosylic acid solution [214]. Anodic polarization in acidic chloride containing solution exhibited higher corrosion potentials and lower anodic currents of the coated aluminum, indicating that the composite coating protects the underlying substrate from corrosion.

Note that stable PANI film could not be prepared on mild steel by direct electrochemical deposition from acetonitrile-$LiClO_4$ medium, Tüken *et al.* [273] electrodeposited a very thin PPY layer, as well as a PANI layer, on mild steel from aqueous oxalic acid solution first, which were considered primer coatings and prevented the dissolution of the mild steel during the following electrochemical deposition of the PANI topcoat from the acetonitrile-$LiClO_4$ medium. Corrosion behavior of the coatings obtained, i.e. the PPY/PANI coating as well as the PANI/PANI coating, were studied in 3.5-wt% NaCl solution. Both the anodic polarization and the OCP-time curves indicated that PPY/PANI coating provided much longer time of protection, more noble shift of corrosion potential and more reduced corrosion current, as compared with the bare steel or even the PANI/PANI coating, and the reason was attribute to the better barrier property of the PPY/PANI coating, which played a more important role in the first hours of immersion but decreased dramatically after 24h of immersion. A protection efficiency of 97.0% was determined for the PPY/PANI coating from EIS study after 240 h of immersion.

4. CORROSION PROTECTION MECHANISMS OF ICPs

The corrosion protection mechanism documented for PPY is mainly the same passivation as for PANI, and is not so many in quantity and diversity as that for PANI, therefore we present here the corrosion protection mechanisms of ICPs exclusively with PANI.

4.1. Passivation Mechanism

4.1.1. Metal Oxide

Deng *et al.* [10] first proposed that ICP coatings, poly (3-methylthiophene) in their case, stabilize the potential of the metal, e.g. titanium, in a passive regime and maintain a protective oxide layer on the metal. Oxygen reduction on the polymer coating replenishes the polymer charge consumed by the metal dissolution, thereby stabilized the potential of the exposed metal in the passive regime and minimized the rate of metal dissolution (**Scheme 5.1**). The reduction rate was improved dramatically by incorporating Pt particles on the conductive film and fast enough to balance the passivation current on the metal and to poise the potential for an indefinite time.

$$(1/n)M + (1/m)ICP^{m+} + (y/n)H_2O \rightarrow (1/n)M\,(OH)_y^{(n-y)} + (1/m)ICP^0 + (y/n)H^+$$

(ICP reduction/metal oxidation)

$$(m/4)O_2 + (m/2)H_2O + OCP^0 \rightarrow ICP^{m+} + mOH^-$$

(ICP oxidation by molecular oxygen)

Scheme 5.1. Reaction scheme of ICP with metal.

Based on the "grain-boundaries pores" model [274] which has been proposed for the humidity-induced corrosion of steel in presence of Fe_2O_3/Fe_3O_4, Fahlman *et al.* [135] proposed a model for corrosion protection of Fe and CRS by EB. By donating electrons into the LUMO (quinoid) level of EB, a positive charge will initially be formed on the outer surface of the Fe/CRS metal at the EB interface. Since iron and steel are metals, the charge spread quickly to all the surfaces of the sample, creating a positively changed surface region of $\sim 0.5^{-1}$ Å thick. Due to the semimetal conductivity ($\sim 10^2$ Scm^{-1}) of the Fe_3O_4 layer and the semiconductor of Fe_2O_3 layer, a charged layer was built up at the Fe_2O_3/Fe_3O_4 interface instead of the Fe_3O_4/Fe contact and the electronic structure of the Fe_3O_4 in this region was stabilized due to the electron deficiency, changing the chemical potential. This in turn made it harder to (further) oxidize the iron ions and hence the steel became more resistant to corrosion. The charged layer at the Fe_2O_3/Fe_3O_4 interface hindered the diffusion of Fe^{2+} ions up through the bulk to the interface and O^{2-} ions from the surface down to the interface, though water molecules were not prevented from diffusing down to the metal interface causing formation of Fe_3O_4, the rate of corrosion was much slower than that at the surface region due to the space constraints.

After thorough investigation of the passivation of iron with PANI, Wessling [275] pointed out that the passivation of iron is a complex multistep process. The first step of the interaction between PANI and iron was an etching step with few microns of iron (and "dirt") removed; after that, the fresh iron surface was coated with an iron oxide layer, which prevents the metal from being corroded. The PANI layer intervened as a redox catalyst **(Fig. 5.11)** in the reaction between oxidizable metals and oxygen/water in forming of the passivating oxide layer [180,196].

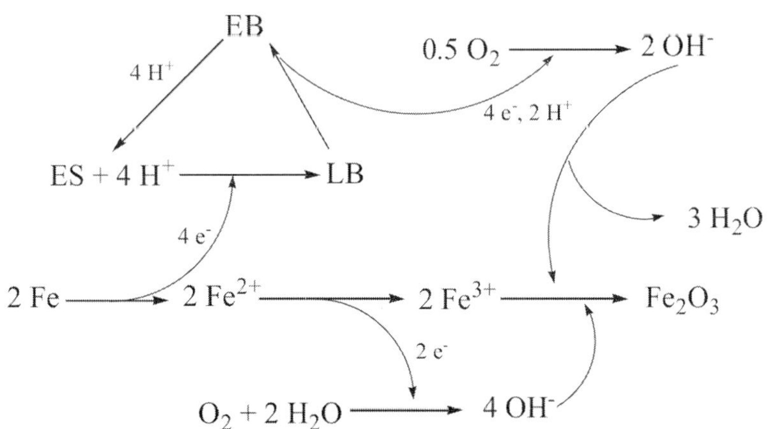

Fig. 5.11. Reaction mechanism for the passivation of iron using PANI coating. Reproduced with permission from B. Wessling; Synth.Met., 93(2), 143 (1998). Copyright © 1998, Elsevier Science Ltd.

The passivation of metal through a redox process was also found by Ahmad and MacDiarmid [1] with EB, in which the SS surface was superficially oxidized by EB and EB was itself reduced to LB. The LB was oxidized back to EB by air and EB healed the pitting corrosion in the same way. In another paper, Beard *et al.* [276] reported the redox chemistry between EB and CRS by using XPS. The EB coating cast from NMP solution onto CRS was relatively reduced and can be reduced further with heating in the absence of air (165°C, in ultrahigh vacuum). All the findings are consistent with the hypothesis that EB can catalyze or otherwise enable passivation of steel surface.

More recently, Zhu *et al.* [277] galvanically coupled a separate protonated PANI film with SS or carbon steel to investigate the corrosion protection mechanism of PANI on ferrous metals and found that PANI is capable to passivate ferrous metals in non-chlorine acid environment due to its more positive OCP and greater initial current in cathodic polarization. More importantly, the passivation of the metals resulted from the excess oxidative charge that stored in the PANI film and released during cathodic polarization, and had no relation with the oxygen reduction as stated by Wessling [275]. The excess oxidative charge and the equilibrium activity of the protonated PANI with the acidic electrolyte were the persistent driving force for ferrous metal passivation. However, if the PANI or PPY coating remained in the oxidized state, no further passivation can be achieved even if the charge stored is still remaining [278].

4.1.2. Metal/dopant Complex

After investigation of the corrosion protection behavior of various acids doped PANIs by using SRET, Kinlen *et al.* [165] found that PANI coatings shift the potential of the steel positive and

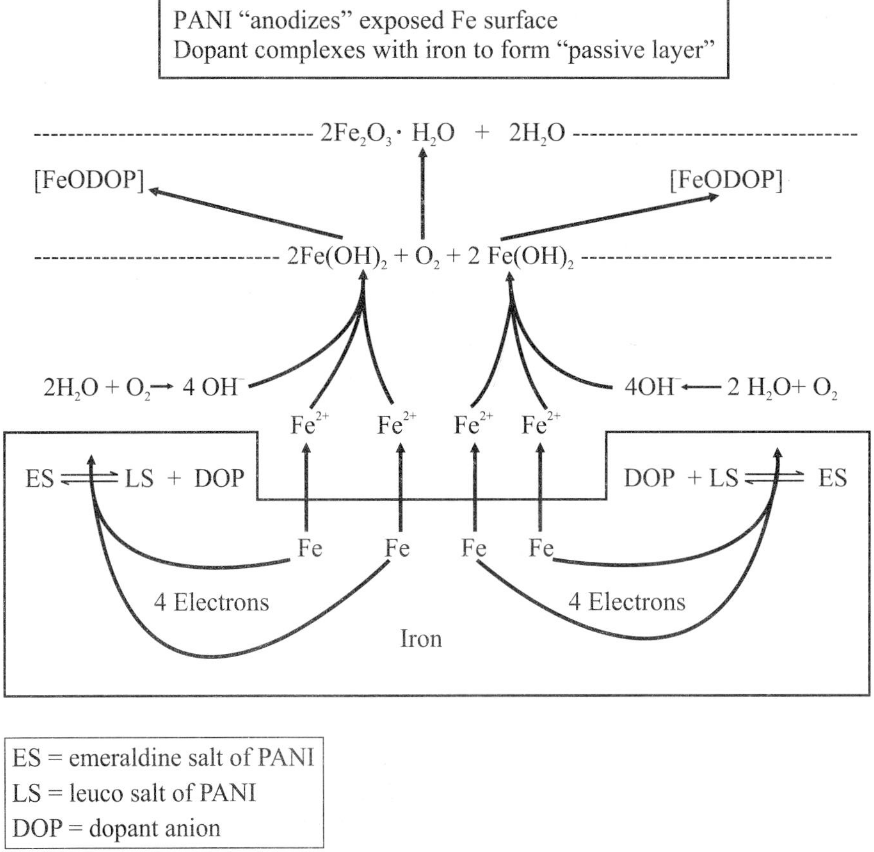

Fig. 5.12. Mechanistic view of pinhole passivation. Reproduced with permission from P.J.Kinlen *et al.*; J. Electrochem.Soc., 146(10), 3690 (1999). Copyright © 1999, The Electrochemical Society, Inc.

result in the formation of an iron/dopant complex at the coating/steel interface. For scribed coating, the author proposed that Fe^{2+} initially reacted with dopant anion in an anodic regime and the insoluble iron-dopant salt formed effectively passivated the pinholes. The phosphonic acid doped PANI formulations significantly outperformed those prepared using sulfonic acid doped PANI, indicating that the dopant of PANI affects greatly the passivity of steel, for example in this case, the iron/phosphonate complex is more effective in passivating the steel than the iron/sulfonate complex. The same mechanism was also mentioned in ref. [168,178] and the iron/PPA complex exhibited higher passivating character as compared with that of the iron/CSA complex in the case of [178] as illustrated in **Fig. 5.12**.

4.1.3. Metal-PANI Complex

Kinlen *et al.* [159] found that a Fe-PANI complex was formed when PANI was coated onto a steel surface. The Fe-PANI complex had an oxidation potential 250 mV more positive than PANI and catalyzed oxygen reduction at a higher rate than PANI alone and thereby increased the protection efficiency.

4.2. Electronic Barrier Mechanism

Both conductive and nonconductive PANI coatings as well as aniline trimer coatings exhibited anionic nature and, in this sense, behaved as electronic barriers to cations in corrosion protection. Covered by a cationic coating, the obtained PANI bipolar coatings acted as "electronic barriers" (to both cations and anions) to inhibit anodic reaction and maintained high resistance to ionic flow, thereby providing excellent corrosion protection [49]. The same mechanism was also proposed for PPY films [236].

4.3. Multistage Conditional Protection (MCP) Mechanism

Schauer and his co-workers [163] advanced a multistage conditional protection (MCP) mechanism, which consists of three stages of protection. In the first stage, PANI reacted with the Fe substrate by formation of Fe_3O_4 and γ-Fe_2O_3 as precursors of the passive layer. At this preliminary stage, the metal passivation was initiated and the basis for active protection was laid. In the second stage, i.e. the active protection stage, as soon as the water and corrosive ions diffused through the coating and contacted the metal, electrons from the anodic partial reaction were intercepted and transported by PANI to interface II (**Fig. 5.13**) and reacted there with oxygen to form hydroxyl ions. It was the presence of oxygen concentration gradient across the coating imparted by a proper topcoat and the limited supply of electrons at the metal surface that promoted the cationic reaction at the interface II rather than at the interface I, i.e. spatial separation of cationic and anodic partial reaction that favor the conditions for the maintenance of a passive state and formation of stable oxides such as Fe_3O_4, γ-Fe_2O_3 and γ-Fe_2O_3. With degradation of the coating due to factors such as a topcoat with insufficient barrier properties, deprotonation of ES and its transition to EB occurred, which in turn changed the active protection of ES to a barrier protection of EB due to its high resistance and sealing effect [279]. In this stage, the substrate can still be efficiently protected for a long time. Both the first and the second stage of the MCP mechanism of PANI were confirmed according to a later paper [164] of the same group.

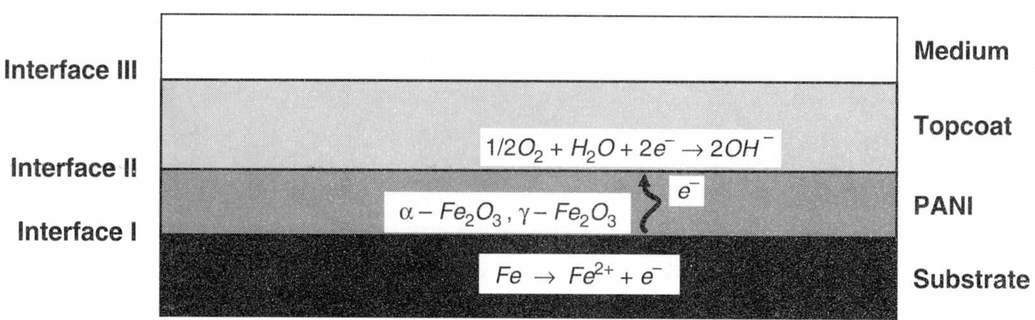

Fig. 5.13. Schematic representation of the MCP mechanism of PANI in phase of active corrosion protection. Reproduced with permission from T. Schauer *et al.*; Prog. Org. Coat., 33(1), 20 (1998). Copyright © 1998, Elsevier Science Ltd.

5. CONCLUSIONS AND FUTURE PROSPECTS

ICPs are excellent candidate materials, mostly in the form of coatings, for corrosion protection or inhibition of metals, including Fe, Cu, Ag, Al, to mention just a few, and some alloys, in various corrosive media, for example, in acidic, neutral or alkalic solutions, and in air. The advantages of adopting ICPs as corrosion protection coatings cover the higher protection effectiveness, good environmental compatibility and stability, light weight, versatility for processing, ease of synthesis, and so on. However, significant research efforts are still needed to optimize the synthetical properties of the ICPs based corrosion protection coatings, such as the nature (oxidation and protonating state, molecular weight, crystallinity, morphology, etc.) and doptant of the ICPs, preparing approaches, adhesion strength, lifetime of the coatings, and to realize wider application of the coatings. Moreover, and perhaps the most important one from the scientific point of view, the actual mechanism by which the ICPs protect metals from corroding need to be elucidated, although several mechanisms have been proposed.

ACKNOWLEDGEMENTS

The authors gratefully acknowledge financial support from the National Natural Science Foundation of China (No. 59903005), and that from the Key Technologies R&D Programme of Shaanxi Province (2003K07-G10), China.

Appendix

List of Terms and Abbreviations

ABSA	amino benzenesulfonic acid		PEEM	photo-emission electron microscopy
CMC	carboximethylcellulose		PEtOAN	poly-*o*-ethoxyaniline
CRS	cold rolled steel		PI	Polyimide
CSA	camphor sulfonic acid		PMAS	polymethoxyaniline sulphonic acid
DBSA	dodecylbenzenesulfonic acid		PMMA	poly methyl methacrylate
DDPH	4-dodecylphenol		PNEA	poly (*N*-ethyl aniline)
DFT	density functional theory		PODP	poly(3-octadecyl pyrrole)
DMF	N, N-dimethylformamide		POP	poly(3-octyl pyrrole)
DMSO	Dimethyl sulfoxide		PPA	Phenylphosphonic acid
DNNSA	dinonylnapthalene sulfonic acid		PPP	Polyparaphenylene
DOP	dopant anion		PPS	polyparaphenylene sulphide
DPA	*n*-dodecylphosphonic acid		PPV	polyparaphenylene vinylene
DSA	dodecylsulfonic acid		PPY	Polypyrrole
EB	emeraldine base		PSS	polystyrenesulfonate
EIS	electrochemical impedance spectroscopy		*co*-PTFE	poly(tetrafluoroethylene-*co*-vinylidene fluoride-*co*-propylene)
ENM	electrochemical noise method			
ES	emeraldine salt		PTH	polythiophene
HCl	hydrochloric acid		PU	polyurethane
ICP	intrinsically conducting polymer		PVB	polyvinylbutyral
LB	leuco base		PVDF	poly(vinylidene fluoride)
LS	leuco salt		RH	relative humidity
MCP	multistage conditional protection		SCE	saturated calomel electrode
MMT	Montmorillonite		SEM	scanning electron microscopy
MPY	milli-inches per year		SKP	scanning Kelvin probe
NMP	*N*-methylpyrrolidone		SPANI	sulfonated polyaniline
NSA	naphthalenesulfonic acid		SRET	scanning reference electrode technique
NTPA	nitrilotri (methyl phosphoric acid)		SS	stainless steel
OA	oxalic acid		SSE	saturated sulfate electrode
OCP	open circuit potential		SVET	scanning vibrating electrode technique
PA	polyacetylene		TA	tartaric acid
PANI	Polyaniline		TEM	transmission electron microscopy
PB	Prussian blue		TGA	thermogravimetric analysis
PCN	PANI-clay nanocomposite		TSA	*p*-toluenesulfonic acid
PCS	plain carbon steel		XPS	X-ray photoelectron spectroscopy
PDAN	poly(1,5-diaminonaphthalene)		YSZ	yttria stabilized zirconia
PDMA	poly(2,5-dimethoxy aniline)			

REFERENCES

1. N. Ahmad and A.G. MacDiarmid; Synth.Met., 78(2), 103 (1996).
2. A. Amirudin and D. Thierry; Prog.Org.Coat., 26(1), 1 (1995).
3. G.P. Bierwagen; Prog.Org.Coat., 28(1), 43 (1996).
4. W. Funke; in: Polymeric Materials for Corrosion Control (Ed. R.A. Dickie and F.L. Floyd), American Chemical Society Symposium, USA, 1986, p. 222.
5. P. Herrasti and P. Ocón; Appl.Surf.Sci., 172, 276 (2001).
6. D.W. DeBerry; J. Electrochem.Soc., 132(5), 1022 (1985).
7. Anonymous; Adv.Mater.Processes, 158, 28 (2000).
8. S.M. Cohen; Corros., 51(1), 71 (1995).
9. R.L. Twite and G.P. Bierwagen; Prog.Org.Coat., 33(2), 91 (1998).
10. Z. Deng, W.H. Smyrl and H.S. White; J. Electrochem.Soc., 136(8), 2152 (1989).
11. Z. Deng and W.H. Smyrl; J. Electrochem.Soc., 138(7), 1911 (1991).
12. S. Ren and D. Barkey; J. Electrochem.Soc., 139(4), 1021 (1992).
13. G. Kousik, S. Pitchumani and N.G. Renganathan; Prog.Org.Coat., 43(4), 286 (2001).
14. T. Tüken, B. Yazýcý and M. Erbil; Prog.Org.Coat., 51(3), 205 (2004).
15. T. Tüken, B. Yazýcý and M. Erbil; Prog.Org.Coat., 53(1), 38 (2005).
16. A.G. MacDiarmid; Synth.Met., 125(1), 11 (2001).
17. H. Shirakawa, E.J. Louis, A.G. MacDiarmid, C.K. Chiang and A.J. Heeger; J. Chem.Soc., Chem.Commun., 578 (1977).
18. L.H. Dao, M.T. Nguyen, J.W. Chevalier, T.N. Do, J.Y. Bergeron and R. Paynter; Annual Technical Conference (ANTEC), Conference Proceedings 37, 861 (1991).
19. P. Somani, A.B. Mandale and S. Radhakrishnan; Acta Mater., 48(11), 2859 (2000).
20. A. Harlin and P. Nousiainen; Chem.Fibers Int., 53(1), 42 (2003).
21. Y.Y. Wang and X.L. Jing; Polym.Adv.Technol., 16(4), 344 (2005).
22. J. Lu, N.J. Pinto and A.G. MacDiarmid; J. Appl.Phys., 92(10), 6033 (2002).
23. P. Zarras, N. Anderson, C. Webber, D.J. Irvin, J.A. Irvin, A. Guenthner and J.D. Stenger-Smith; Radiat.Phys.Chem., 68(3-4), 387 (2003).
24. A.J. Heeger; Synth.Met., 125(1), 23 (2001).
25. A.G. MacDiarmid; Angew.Chem.Int.Ed., 40(14), 2581 (2001).
26. A.J. Heeger; Angew.Chem.Int.Ed., 40(14), 2591 (2001).
27. H. Shirakawa; Angew.Chem.Int.Ed., 40(14), 2575 (2001).
28. H. Letheby; J. Chem.Soc., London 15, 161 (1862).
29. A.G. MacDiarmid, J.C. Chiang, M. Halpern, W.S. Huang, S.L. Mu, N.L.D. Somasiri, W. Wu and S.I. Yaniger; Mol.Cryst.Liq.Cryst.Sci., 121, 173 (1985).
30. E.M. Geniès, A. Boyle, M. Lapkowski and C. Tsintavis; Synth.Met., 36(2), 139 (1990) and the references therein.
31. N. Gospodinova and L. Terlemezyan; Prog.Polym.Sci., 23(8), 1443 (1998).
32. K. Tzou and R.V. Gregory; Synth.Met., 47(3), 267 (1992).
33. Z. Sun, Y. Geng, J. Li, X. Jing and F. Wang; Synth.Met., 84(1-3), 99 (1997).
34. H. Kuzmany and N.S. Sariciftci; Synth.Met., 18, 353 (1986).
35. Y. Ding, A.B. Padias and H.K. Jr. Hall; J. Polym.Sci., Part A: Polym.Chem., 37(14), 2569 (1999).
36. K.G. Conroy and C.B. Breslin; Electrochim.Acta, 48(6), 721 (2003).
37. M. Millard; in Techniquea and Applications of Plasma Chemistry (Ed. J.R.Hollahan and A.T.Bell), Chapter 5, Wiley, New York, (1974).
38. L.G. Paterno, S. Manolache and F. Denes; Synth.Met., 130(1), 85 (2002).
39. G.J. Cruz, J. Morales, M.M. Castillo-Ortega and R.Olayo; Synth.Met., 88(3), 213 (1997).
40. C. Liao and M. Gu; Thin Solid Films, 408(1-2), 37 (2002).
41. Y. Chen, E.T. Kang and K.G. Neoh; Appl.Surf.Sci., 185(3-4), 267 (2002).
42. J. Gong, X.J. Cui, Z.W. Xie, S.G. Wang and L.Y. Qu; Synth.Met., 129(2), 187 (2002).

43. J.X. Huang, J.A. Moore, J.H. Acquaye and R.B. Kaner; Macromolecules, 38(2), 317 (2005).
44. A.G. MacDiarmid and W.S. Huang; J. Chem.Soc., Faraday Trans., 82, 2385 (1986).
45. A.G. MacDiarmid, S.K. Manohar, J.G. Masters, Y. Sun, H. Weiss and A.J. Epstein; Synth.Met., 41(1-2), 621 (1991).
46. J.G. Masters, Y. Sun and A.G. MacDiarmid; Synth.Met., 41(1-2), 715 (1991).
47. J.P. Travers, F. Genoud, C. Menardo and M. Nechtschein; Synth.Met., 35(1-2), 159 (1990).
48. J.G. Wang; Synth.Met., 132(1), 49 (2002).
49. J.G. Wang; Synth.Met., 132(1), 53 (2002).
50. H. Namazi, R. Kabiri and A. Entezami; Eur.Polym.J., 38(4), 771 (2002).
51. M.E. Leyva, G.M.O. Barra, M.M. Gorelova, B.G. Soares and M. Sens; J. Appl.Polym.Sci., 80(4), 626 (2001).
52. S. Angappane, N. Rajeev Kini, T.S. Natarajan, G. Rangarajan and B.Wessling; Thin Solid Films 417(1-2), 202 (2002).
53. R. Gangopadhyay and A. De; Synth.Met., 132(1), 21 (2001).
54. S.K. Dhawan, N. Singh and S. Venkatachalam; Synth.Met., 125(3), 389 (2002).
55. P. Ghosh, S.K. Siddhanta, S.R. Haque and A. Chakrabarti; Synth.Met., 123(1), 83 (2001).
56. M. Okubo, S. Fujii and H. Minami; Colloid.Polym.Sci., 279, 139 (2001).
57. H. Xia and Q. Wang; Chem.Mater., 14(5), 2158 (2002).
58. B.H. Kim, J.H. Jung, S.H. Hong, J. Joo, A.J. Epstein, K. Mizoguchi, J.W. Kim and H.J. Choi; Macromolecules, 35(4), 1419 (2002).
59. B. Feng, Y. Su, J. Song and K. Kong; J. Mater.Sci.Lett., 20(4), 293 (2001).
60. W. Jia, E. Segal, D. Kornemandel, Y. Lamhot, M. Narkis and A. Siegmann; Synth.Met., 128(1), 115 (2002).
61. Z.F. Li and E. Ruckenstein; Macromolecules, 35(25), 9506 (2002).
62. R. Nagarajan, S. Roy, J. Kumar, S.K. Tripathy, T. Dolukhanyan, C. Sung, F. Bruno and L.A. Samuelson; J. Macromol.Sci., Pure Appl.Chem., 38(12), 1519 (2001).
63. S.Roy, J.M.Fortier, R.Nagarajan, S.Tripathy, J.Kumar, L.A.Samuelson and F.F.Bruno; Biomacromolecules 3(5), 937 (2002).
64. J. Anand, S. Palaniappan and D.N. Sathyanarayana; Prog.Polym.Sci., 23(6), 993 (1998).
65. K. Kaneto, Y. Sonoda and W. Takashima; Jpn.J.Appl.Phys., Part 1, 39(10), 5918 (2000).
66. C.W. Lin, Y.L. Liu and R. Thangamuthu; Sens.Actuators, B, 94(1), 36 (2003) and the references therein.
67. L. Ruangchuay, A. Sirivat and J. Schwank; Talanta, 60(1), 25 (2003).
68. B.J.V. Tongol, C.A. Binag and F.B. Sevilla III; Sens.Actuators, B, 93(1-3), 187 (2003).
69. Y.C. Liu and B.J. Hwang; Thin Solid Films 339(1-2), 233 (1999).
70. M. Gao, L. Dai and G.G. Wallace; Synth.Met., 137(1-3), 1393 (2003).
71. D.S. Lee, D.D. Lee, H.R. Hwang, J.H. Paik, J.S. Huh, J.O. Lim and J.J. Lee; J.Mater.Sci.-Mater.Electron., 12(1), 41 (2001).
72. C. Zhao, H. Wang and Z. Jiang; Appl.Surf.Sci., 207(1-4), 6 (2003).
73. M.I. Sanchez de Pinto, H.T. Mishima and B.A. Lopez de Mishima; J. Appl.Electrochem., 27(7), 831 (1997).
74. S.S. Pandey, W. Takashima, M. Fuchiwaki and K. Kaneto; Synth.Met., 135-136, 59 (2003).
75. K. Tada, Y. Kato and M. Onoda; Synth.Met., 135-136, 101 (2003).
76. D. Zhou, G.M. Spinks, G.G. Wallace, C. Tiyapiboonchaiya, D.R. MacFarlane, M. Forsyth and J. Sun; Electrochim.Acta, 48(14-16), 2355 (2003).
77. W. Takashima, S.S. Pandey and K. Kaneto; Synth.Met., 135-136, 61 (2003).
78. N.S. Allen, K.S. Murray, R.J. Fleming and B.R. Saunders; Synth.Met., 87(3), 237 (1997).
79. G.R. Goward, F. Leroux and L.F. Nazar; Electrochim.Acta, 43(10-11), 1307 (1998).
80. J.O. Iroh and K. Levine; J. Power Sources, 117(1-2), 267 (2003).
81. H. Kim and W. Chang; Synth.Met., 101(1), 150 (1999).
82. V. Syritski, A. Öpik and O. Forsén; Electrochim.Acta, 48(10), 1409 (2003).

83. K. Hosono, I. Matsubara, N. Murayama, W. Shin, N. Izu and S. Kanzaki; Thin Solid Films, 441(1-2), 72 (2003).
84. M. Omastová, M. Trchová, J. Kovárová and J. Stejskal; Synth.Met., 138(3), 447 (2003).
85. C. He, C. Yang and Y. Li; Synth.Met., 139(2), 539 (2003).
86. Y.Y ang, J. Liu and M. Wan; Nanotechnology, 13(6), 771 (2002).
87. S. Cuenot, S. Demoustier-Champagne and B. Nysten; Phys.Rev.Lett., 85(8), 1690 (2000).
88. K.K. Kanazawa, A.F. Diaz, W.D. Gill, P.M. Grant, G.B. Street, G.P. Gardini and J.F. Kwak; Synth.Met., 1(3), 329 (1979).
89. Y. Li and J. Quyang; Synth.Met., 113(1-2), 23 (2000).
90. J.K. Xu, G.Q. Shi, L.T. Qu and J.X. Zhang; Synth.Met., 135-136, 221 (2003).
91. S.B. Saidman; J. Electroanal.Chem., 534(1), 39 (2002).
92. S.B. Saidman; Electrochim.Acta, 48(12), 1719 (2003).
93. Y.F. Jiang, X.W. Guo, Y.H. Wei, C.Q. Zhai and W.J. Ding; Synth.Met., 139(2), 335 (2003).
94. M. Bazzaoui, L. Martins, E.A. Bazzaoui and J.I. Martins; Electrochim.Acta, 47(18), 2953 (2002).
95. S.B. Saidman and J.B. Bessone; J. Electroanal.Chem., 521(1-2), 87 (2002).
96. P. Hulser and F. Beck; J. Appl.Electrochem., 20(4), 596 (1990).
97. K. Idla, O. Inganäs and M. Strandberg; Electrochim.Acta, 45(13), 2121 (2000).
98. T. Zalewska, A. Lisowska-Oleksiak, S. Biallozor and V. Jasulaitiene; Electrochim.Acta, 45(24), 4031 (2000).
99. A.C. Cascalheira, S. Aeiyach, P.C. Lacaze and L.M. Abrantes; Electrochim.Acta, 48(17), 2523 (2003).
100. M. Bazzaoui, J.I. Martins, E.A. Bazzaoui, T.C. Reis and L. Martins; Pyrrole electropolymerization on copper and brass in a single-step process from aqueous solution, J.Appl.Electrochem., 34(8), 815 (2004).
101. M. Bazzaoui, L. Martins, E.A. Bazzaoui and J.I. Martins; J. Electroanal.Chem., 537(1-2), 47 (2002).
102. D. Cossement, F. Plumier, J. Delhalle, L. Hevesi and Z. Mekhalif; Synth.Met., 13893), 529 (2003).
103. A. Deronzier and J.C. Moutet; Coord.Chem.Rev., 147, 339 (1996).
104. W.M. Sigmund, G. Weerasekera, C. Marestin, S. Styron, H. Zhou, M.Z. Elsabee, J. Ruehe, G. Wegner and R.S. Duran; Langmuir 15(19), 6423 (1999).
105. N. Costantini, S. Capaccioli, M. Geppi and G. Ruggeri; Polym.Adv.Technol., 11(1), 27 (2000).
106. R.L. McCarley and R.J. Willicut; J. Am.Chem.Soc., 120(36), 9296 (1998).
107. L.X. Wang, X. G.Li and Y.L. Yang; React.Funct.Polym., 47(2), 125 (2001) and the references therein.
108. L. Cen, K.G. Neoh and E.T. Kang; Biosens.Bioelectron., 18, 363 (2003).
109. Y. Shen and M. Wan J. Appl.Polym.Sci., 68(8), 1277 (1998).
110. J. Joo, J.K. Lee, J.K. Hong, J.S. Baeck, W.P. Lee, A.J. Epstein, K.S. Jang, S.J. Suh and E.J. Oh; Macromolecules 31(2), 479 (1998).
111. Q. Wu and X. Xie; Mater.Chem.Phys., 77(3), 621 (2003).
112. D.P. Park, J.H. Sung, S.T. Lim, H.J. Choi and M.S. Jhon; J. Mater.Sci.Lett., 22(18), 1299 (2003).
113. Y.C. Liu; Mater.Chem.Phys., 77(3), 791 (2003).
114. Y. Lu, A. Pich and H. Adler; Synth.Met., 135-136, 37 (2003).
115. P. Dutta and S.K. De; Synth.Met., 139(2), 201 (2003).
116. D.P. Park, J.W. Kim, F. Liu, H.J. Choi and J. Joo; Synth.Met., 135-136, 713 (2003).
117. D. Kincal, A. Kumar, A.D. Child and J.R. Reynolds; Synth.Met., 92(1), 53 (1998).
118. W. Chen, X. Li, G. Xue, Z. Wang and W. Zou; Appl.Surf.Sci., 218(1-4), 215 (2003).
119. C.A. Ferreira, S.C. Domenech and P.C. Lacaze; J. Appl.Electrochem., 31, 49 (2001).
120. L. Hao, C. Zhu, C. Chen, P. Kang, Y. Hu, W. Fan and Z. Chen; Synth.Met., 139(2), 391 (2003).
121. H.B. Sonmez, G. Sonmez, B.F. Senkal, A.S. Sarac and N. Bicak; Synth.Met., 135-136, 807 (2003).
122. Y. Wei, J. Tian and K.F. Hsueh; Polym.Mater.Sci.Eng., 71, 586 (1994).
123. A. Levent, J. Hacaloglu, L. Toppare and Y. Yagci; Synth.Met., 135-136, 457 (2003).
124. M. Sacak, U. Akbulut and D.N. Batchelder; Polym., 39(20), 4735 (1998).
125. J.K. Avlyanov; Synth.Met., 102(1-3), 1272 (1999).
126. A.J.G. Zarbin, M.A. De Paoli and O.L. Alves; Synth.Met., 99(3), 227 (1999).

127. S. Sakkopoulos, E. Vitoratos, J. Grammatikakis, A.N. Papathanassiou and E. Dalas; J. Mater.Sci., 37(14), 2865 (2002).
128. T. Kojima, H. Hayakawa, M. Tsuchiya and Y. Kai; J. Appl.Polym.Sci., 55(13), 1855 (1995).
129. Y. Wei, J. Wang, X. Jia, J.M. Yeh and P. Spellane; Polym.Mater.Sci.Eng., 72, 563 (1995).
130. Y. Wei, J. Wang, X. Jia, J.M. Yeh and P. Spellane; Polymer, 36(23), 4535 (1995).
131. M. Fahlman, H. Guan, J.A.O. Smallfield and A.J. Epstein; Iron/polyaniline interface and its effect on corrosion protection of iron and cold rolled steel in aqueous and salt environments.Annual Technical Conference-ANTEC, Conference Proceedings 2, 1238 (1998).
132. D.A. Wrobleski, B.C. Benecewicz, K.G. Thompson and C.J. Bryan; Polym.Prep., 35, 265 (1994).
133. A. Mirmohseni and A. Oladegaragoze; Synth.Met., 114(2), 105 (2000).
134. J.R. Jr. Santos, L.H.C. Mattoso and A.J. Motheo; Electrochimica Acta 43(3-4), 309 (1998).
135. M. Fahlman, S. Jasty and A.J. Epstein; Synth.Met., 85(1-3), 1323 (1997).
136. R. Gašparac and C.R. Martin; J. Electrochem.Soc., 148, B138 (2001).
137. A.A. Pud, G.S. Shapoval, P. Kamarchik, N.A. Ogurtsov, V.F. Gromovaya, I.E. Myronyuk and Y.V. Kontsur; Synth.Met., 107(2), 111 (1999).
138. A.J. Epstein, J.A.O. Smallfield, H. Guan and M. Fahlman; Synth.Met., 102(1-3), 1374 (1999).
139. Y. Wei, J.M. Yeh, J. Wang, X. Jia, C. Yang and D. Jin; Polym.Mater.Sci.Eng., 74, 202 (1996).
140. L.T. Sein Jr., Y. Wei and S.A. Jansen; Comput.Theor.Polym.Sci., 11(2), 83 (2001).
141. L.T. Sein Jr., Y. Wei and S.A. Jansen; Synth.Met., 143(1), 1 (2004).
142. M.C. Bernard, C. Deslouis, T. El Moustafid, A. Hugot-Le Goff, S. Joiret and B. Tribollet; Synth.Met., 102(1-3), 1381 (1999).
143. W.K. Lu, R.L. Elsenbaumer and B. Wessling; Synth.Met., 71(1-3), 2163 (1995).
144. I. Kulszewicz-Bajer, M. Zagórska, A. Bany and L. Kwiatkowski; Synth.Met., 102(1-3) 1385 (1999).
145. S.F. Cogan, M.D. Gilbert, G.L. Holleck, J. Ehrlich and M.H. Jillson; J.Electrochem.Soc., 147(6), 2143 (2000).
146. V. Brusic, M. Angelopoulos and T. Graham; J. Electrochem.Soc., 144(2), 436 (1997).
147. A. Cook, A. Gabriel and N. Laycock; J. Electrochem.Soc., 151(9), B529 (2004).
148. D.F. Tallman, Y. Pae and G.P. Bierwagen; Corros., 56, 401 (2000).
149. D.E. Tallman, Y. Pae and G.P. Bierwagen; Corros., 55, 779 (1999).
150. G.M. Spinks, A.J. Dominis and G.G. Wallace; Corros., 59(1), 22 (2003).
151. A.J. Dominis, G.M. Spinks and G.G. Wallace; Prog.Org.Coat., 48(1), 43 (2003).
152. M. Kendig, M. Hon and L. Warren; Prog.Org.Coat., 47(3-$, 183 (2003).
153. M. Kendig and M. Hon; Corros., 60(11), 1024 (2004).
154. N.A. Ogurtsov, A.A. Pud, P. Kamarchik and G.S. Shapoval; Synth.Met., 143(1), 43 (2004).
155. W.S. Araujo, I.C.P. Margarit, M. Ferreira, O.R. Mattos and P. Lima Neto; Electrochim.Acta, 46(9), 1307 (2001).
156. L.M. Liu and K. Levon; J. Appl.Polym.Sci., 73(14), 2849 (1999).
157. T.P. McAndrew; Trends Polym.Sci., 5, 7 (1997).
158. A. Talo, O. Forsén and S. Yläsaari; Synth.Met., 102(1-3), 1394 (1999).
159. P.J. Kinlen, D.C. Silverman and C.R.J effreys; Synth.Met., 85(1-3), 1327 (1997).
160. B. Wessling and J. Posdorfer; Electrochim.Acta, 44(12), 2139 (1999).
161. J. Posdorfer and B. Wessling; Fresenius; J. Anal.Chem., 367, 343 (2000).
162. P. Li, T.C. Tan and J.Y. Lee; Synth.Met., 88(3), 237 (1997).
163. T. Schauer, A. Joos, L. Dulog and C.D. Eisenbach; Prog.Org.Coat., 33(1), 20 (1998).
164. T. Schauer, H.W. Greisiger and C.D. Eisenbach; Polym.Prep., 41(2), 1783 (2000) and the references therein.
165. P.J. Kinlen, V. Menon and Y. Ding; J.Electrochem.Soc., 146(10), 3690 (1999).
166. J. Peltola, Y. Cao and P. Smith; Adhes. Age, 38(5), 18 (1995).
167. A. Talo, P. Passiniemi, O. Forsén and S. Yläsaari; Synth.Met., 85(1-3), 1333 (1997).
168. S. De Souza, J.E. Pereira Da Silva, S.I.Cordoba De Torresi, M.L.A. Temperini and R.M. Torresi; Electrochem.Solid-State Lett., 4(8), B27 (2001).

169. J.C. Seegmiller, J.E. Pereira Da Silva, D.A. Buttry, S.I. Cordoba De Torresi and R.M. Torresi; J. Electrochem.Soc., 152(2), B45 (2005).

170. R.M. Torresi, S. De Souza, J.E. Pereira Da Silva and S.I. Cordoba De Torresi; Electrochim.Acta, 50(11), 2213 (2005).

171. R. Racicot, R. Brown and S.C. Yang; Synth.Met., 85(1-3), 1263 (1997).

172. J.M. Yeh, S.J. Liou, C.Y. Lai, P.C. Wu and T.Y Tsai; Chem.Mater., 13(3), 1131 (2001).

173. X.H. Wang, J. Li, J.Y. Zhang, Z.C. Sun, L. Yu, X.B. Jing, F.S. Wang, Z.X. Sun and Z.J. Ye; Synth.Met., 102(1-3), 1377 (1999).

174. L. Lei, M. Wu, H. Shu, H. Li, Q. Qiao and Y. Du; Polym.Adv.Technol., 12(11-12), 720 (2001).

175. X. Wang, Y. Wang, J. Lu, J. Li, X. Jing and F. Wang; Polym.Prep., 41(2), 1752 (2000).

176. J.L. Lu, N.J. Liu, X.H. Wang, J. Li, X.B. Jing and F.S. Wang; Synth.Met., 135-136, 237 (2003).

177. G. Alvial, T. Matencio, B.R.A. Neves and G.G. Silva; Electrochim.Acta, 49(21), 3507 (2004).

178. J.E.P. Da Silva, S.I.C. De Torresi and R.M. Torresi; Corros.Sci., 47(3), 811 (2005).

179. B. Wessling; Synth.Met., 102(1-3), 1396 (1999).

180. B. Wessling; Synth.Met., 93(2), 143 (1998).

181. B. Wessling; Conductive Polymer/solvent systems: solutions or dispersions? Proceedings of 3rd BPS (Bayreuth Polymer & Materials Research Symposion), Bayreuth, Germany (April 1997).

182. B. Wessling and J. Posdorfer; Synth.Met., 102(1-3), 1400 (1999).And the references therein.

183. T.K. Rout, G. Jha, A.K. Singh, N. Bandyopadhyay and O.N. Mohanty; Surf.Coat.Technol., 167(1), 16 (2003).

184. J.H. Huh, E.J. Oh and J.H. Cho; Synth.Met., 137(1-3), 965 (2003).

185. R.J. Holness, G. Williams, D.A. Worsley and H.N. McMurray; J. Electrochem.Soc., 152(2), B73 (2005).

186. A.B. Samui, A.S. Patankar, J. Rangarajan and P.C. Deb; Prog.Org.Coat., 47(1), 1 (2003).

187. J.C. Chiang and A.G. MacDiarmid; Synth.Met., 13(1-3), 193 (1985).

188. Q.D. Trieu, N.D. Nghia, N.H. Minh, N.T.T. Thuy, T.T.X. Hang, V.K. Oanh and T.A. Truc; Studies on syntheses and properties of nanocomposite from clay-polyaniline. IUPAC World Polymer Congress 2002, 39th International Symposium on Macromolecules, Beijing, China (July 7-12, 2002) p.1077.

189. A. Cook, A. Gabriel, D. Siew and N. Laycock; Curr.Appl.Phys., 4(2-4), 133 (2004).

190. G. Williams, R.J. Holness, D.A. Worsley and H.N. McMurray; Electrochem.Commun., 6(6), 549 (2004).

191. X.G. Li, M.R. Huang, J.F. Zeng and M.F. Zhu; Colloids Surf., A, 248(1-3), 111 (2004).

192. X.L. Jing, Y.Y. Wang and J.F. Qiang; J. Chin.Soc.Corros.Prot., 24(5), 301 (2004).

193. Y.Y. Wang and X.L. Jing; Polym.J, 36(5), 374 (2004).

194. X.L. Jing, Y.Y.Wang and D.H. Zhang; J. Mater.Eng., 248, 20 (2004).

195. J.I.I. Laco, F.C. Villota and F.L. Mestres; Prog.Org.Coat., 52(2), 151 (2005).

196. B. Wessling; Mater.Corros.47, 439 (1996).

197. M.C. Bernard, A. Hugot-Le Goff, S. Joiret and P.V. Phong; Synth.Met., 119(1-3), 283 (2001).

198. M.C. Bernard, A. Hugot-Le Goff and S. Joiret; Synth.Met., 102(1-3), 1383 (1999).

199. M.C. Bernard, A. Hugot-Le Goff, S. Joiret, N.N. Dinh and N.N.Toan; J.Electrochem.Soc., 146(3), 995 (1999).

200. M.C. Bernard, S. Joiret, A. Hugot-Le Goff and P.V. Phong; J. Electrochem.Soc., 148(1), B12 (2001).

201. M.C. Bernard, S. Joiret, A. Hugot-Le Goff and P.D. Long; J. Electrochem.Soc., 148, B299 (2001).

202. M.C. Bernard, S. Joiret, A. Hugot-Le Goff and P.V. Phong; J. Electrochem.Soc., 148, B304 (2001).

203. M.C. Bernard, S. Joiret and A. Hugot-Le Goff; Russian J.Electrochem., 40, 235 (2004).

204. S.R. Moraes, D. Huerta-Vilca and A.J. Motheo; Prog.Org.Coat., 48(1), 28 (2003).

205. J.L. Camalet, J.C. Lacroix, S. Aeiyach, K. Chane-Ching and P.C. Lacaze; J. Electroanal.Chem., 416(1-2), 179 (1996).

206. J.L. Camalet, J.C. Lacroix, S. Aeiyach and P.C. Lacaze; J. Electroanal.Chem., 445(1-2), 117 (1998).

207. S. Sathiyanarayanan, K. Balakrishnan, S.K. Dhawan and D.C. Trivedi; Electrochim.Acta, 39(6), 831 (1994).

208. D. Sazou and C. Georgolios; J. Electroanal.Chem., 429(1-2), 81 (1997).
209. A.T. Özyılmaz, M. Erbil and B. Yazıcı; Prog.Org.Coat., 51(1), 47 (2004).
210. A.T. Özyılmaz, T. Tüken, B. Yazıcı and M. Erbil; Prog.Org.Coat., 52(2), 92 (2005).
211. A.T. Özyılmaz, G. Kardas, M. Erbil and B. Yazıcı; Appl.Surf.Sci., 242(1-2), 97 (2005).
212. N.M. Martyak; Mater.Sci.Eng., A, 371(1-2), 57 (2004).
213. A.M. Fenelon and C.B. Breslin; Surf.Coat.Technol., 190(2-3), 264 (2005).
214. C.B. Breslin, A.M. Fenelon and K.G. Conroy; Mater.Des., 26(3), 233 (2005).
215. M.M. Popoviæ and B.N. Grgur; Synth.Met., 143(2), 191 (2004).
216. J.L. Camalet, J.C. Lacroix, T.D. Nguyen, S. Aeiyach, M.C. Pham, J. Petitjean and P.C. Lacaze; J. Electroanal.Chem., 485(1), 13 (2000).
217. M.M. Popoviæ, B.N. Grgur and V.B. Miškoviæ-Stankoviæ; Prog.Org.Coat., 52(4), 359 (2005).
218. M. Kraljiæ, Z. Mandiæ and Lj. Duiæ; Corros.Sci., 45, 181 (2003).
219. M.A. Malik, M.T. Galkowski, H. Bala, B. Grzybowska and P.J. Kulesza; Electrochim.Acta, 44(12), 2157 (1999).
220. K. Shah and J. Iroh; Synth.Met., 132(1), 35 (2002).
221. P. Yin and P.A. Kilmartin; Curr.Appl.Phys., 4(2-4), 141 (2004).
222. K.G. Shah, G.S. Akundy and J.O. Iroh; J. Appl.Polym.Sci., 85(8), 1669 (2002).
223. D. Huerta-Vilca, S.R. De Moraes and A. De Jesus Motheo; Synth.Met., 140(1), 23 (2004).
224. P. Herrasti, P. Ocón, A. Ibáñez and E. Fatás; J. Appl.Electrochem., 33(6), 533 (2003).
225. K. Ding, Z. Jia, W. Ma, D. Jiang, Q. Zhao, J. Cao and R. Tong; Prot.Met., 39, 71 (2003).
226. M. Galkowski, M.A. Malik, P.J. Kulesza, H. Bala, K. Miecznikowski, R. Wlodarczyk, L. Adamczyk and M. Chojak; J. Electrochem.Soc., 150(6), B249 (2003).
227. K. Shah, Y. Zhu, J.O. Iroh and O. Popoola; Surf.Eng., 17(5), 405 (2001).
228. S. Sathiyanarayanan, S.K. Dhawan, D.C. Trivedi and K. Balakrishnan; Inf.Manage., 23, 1831 (1992).
229. S. Sathiyanarayanan, S.K. Dhawan, K. Balakrishnan and D.C. Trivedi; Bull.Electrochem., 39, 109 (1999).
230. R. Manickavasagam, K. Jeya Karthik, M. Paramasivam and S. Venkatakrishna Iyer; Anti Corros. Methods Mater., 49(1), 19 (2002).
231. F. Beck and R. Michaelis; J. Coat.Technol., 64(808), 59 (1992).
232. F. eck, R. Michaelis, F. Schloten and B. Zinger; Electrochim.Acta, 39(2), 229 (1994).
233. W.J. Hamer, L. Koene and J.H.W. de Wit; Mater.Corros., 55(9), 653 (2004).
234. N.V. Krstajiæ, B.N. rgur, S.M. Jovanoviæ and M.V. Vojnoviæ; Electrochim.Acta, 42(11), 1685 (1997).
235. B.N. Grgur, N.V. Krstajiæ, M.V. Vojnovic, Æ Laænjevac and Lj. Gajiæ-Krstajiæ; Prog.Org.Coat., 33(1), 1 (1998).
236. N.T.L. Hien, B. Garcia, A. Pailleret and C. Deslouis; Electrochim.Acta, 50(7-8), 1747 (2005).
237. P. Ocón, A.B. Cristobal, P. Herrasti and E. Fatas; Corros.Sci., 47(3), 649 (2005).
238. P. Herrasti, L. Diaz, P. Ocón, A. Ibáñez and E. Fatas; Electrochimi.Acta, 49(22-23), 3693 (2004).
239. J.O. Iroh and W. Su; Electrochim.Acta, 46(1), 15 (2000).
240. W. Su and J.O. Iroh; Synth.Met., 114(3), 225 (2000).
241. J.I. Martins, T.C. Reis, M. Bazzaoui, E.A. Bazzaoui and L. Martins; Corros.Sci.46(10), 2361 (2004).
242. C.A. Ferreira, S. Aeiyach, A. Coulaud and P.C. Lacaze; J.Appl.Electrochem., 29(2), 259 (1999).
243. J. Reut, A. Öpik and K. Idla; Synth.Met., 102(1-3), 1392 (1999).
244. H. Nguyen Thi Le, B. Garcia, C. Deslouis and Q. Le Xuan; Electrochimi.Acta, 46(26-27), 4259 (2001).
245. H. Nguyen Thi Le, M.C. Bernard, B. Garcia-Renaud and C. Deslouis; Synth.Met., 140(2-3), 287 (2004).
246. H. Nguyen Thi Le, B. Garcia, C. Deslouis and Q. Le Xuan; J. Appl.Electrochem., 32(1), 105 (2002).
247. A.M. Fenelon and C.B. Breslin; J. Electrochem.Soc., 150(11), B540 (2003).
248. Y.C. Liu and B.J. Hwang; J. Electroanal.Chem., 501(1-2), 100 (2001).
249. A.M. Fenelon and C.B. Breslin; Corrosi.Sci., 45(12), 2867 (2003).
250. E. Coronado and C.J. Gomez-Garcia; Chem.Rev., 98(1), 273 (1998).

251. S.A. Cheng and T.F. Otero; J. Mater.Sci.Lett., 20(11), 1055 (2001).
252. H. Hammache, L. Makhloufi and B. Saidani; Corros.Sci., 45(9), 2031 (2003).
253. V.T. Truong, P.K. Lai, B.T. Moore, R.F. Muscat and M.S. Russo; Synth.Met., 110(1), 7 (2000).
254. A.M. Fenelon and C.B. Breslin; Electrochim.Acta, 47(28), 4467 (2002).
255. T. Tüken, B. Yazýcý and M. Erbil; Prog.Org.Coat., 50(2), 115 (2004).
256. T. Tüken, G. Arslan, B. Yazýcý and M. Erbil; Corros.Sci., 46(11), 2743 (2004).
257. T. Tüken, B. Yazýcý and M. Erbil; Prog.Org.Coat., 51(2), 152 (2004).
258. T. Tüken, A.T. Özyýlmaz, B. Yazýcý, G. Kardas and M. Erbil; Prog.Org.Coat., 51(1), 27 (2004).
259. T. Tüken, M. Düdükcü, B. Yazýcý and M. Erbil; Prog.Org.Coat., 50(4), 273 (2004).
260. D. Sazou; Synth.Met., 130(1), 45 (2002).
261. M.A. Malik, R. Wlodarczyk, P.J. Kulesza, H. Bala and K. Miecznikowski; Corros.Sci., 47(3), 771 (2005).
262. J. He, V.J. Gelling, D.E. Tallman, G.P. Bierwagen and G.G. Wallace; J. Electrochem.Soc., 147(10), 3667 (2000).
263. V.J. Gelling, M.M. Wiest, D.E. Tallman, G.P. Bierwagen and G.G. Wallace; Prog.Org.Coat., 43(1-3), 149 (2001).
264. A. Yfantis, I. Paloumpa, D. Schmeiâer and D. Yfantis; Surf.Coat.Technol., 151-152, 400 (2002).
265. D.M. Lenz, M. Delamar and C.A. Ferreira; J.Electroanal.Chem., 540, 35 (2003).
266. J.M. Yeh, C.P. Chin and S. Chang; J. Appl.Polym.Sci., 88(14), 3264 (2003).
267. B. Garcia, A. Lamzoudi, F. Pillier, H. Nguyen Thi Le and C. Deslouis; J. Electrochem.Soc., 149(12), B560 (2002).
268. I. Paloumpa, A. Yfantis, P. Hoffmann, Y. Burkov, D. Yfantis and D. Schmeißer; Surf.Coat.Technol., 180-181, 308 (2004).
269. T.D. Nguyen, M.C. Pham, B. Piro, J. Aubard, H. Takenouti and M. Keddam; J. Electrochem.Soc., 151(6), B325 (2004).
270. C.K. Tan and D.J. Blackwood; Corros.Sci., 45(3), 545 (2003).
271. R. Rajagopalan and J.O. Iroh; Surf.Eng., 18(1), 59 (2002).
272. K. Shah and J.O. Iroh; Surf.Eng., 20(1), 53 (2004).
273. T. Tüken, A.T. Özyýlmaz, B. Yazýcý and M. Erbil; Appl.Surf.Sci., 236(1), 292 (2004).
274. J. Robertson; Corros.Sci., 29(11-12), 1275 (1989).
275. B. Wessling; Adv.Mater., 6(3), 226 (1994).
276. B.C. Beard and P. Spellane; Chem.Mater., 9(9), 1949 (1997).
277. H. Zhu, L. Zhong, S. Xiao and F. Gan; Electrochim.Acta 49(28), 5161 (2004).
278. T.D. Nguyen, T.A. Nguyen, M.C. Pham, B. Piro, B. Normand and H. Takenouti; J. Electroanal.Chem., 572(2), 225 (2004).
279. T.P. McAndrew, S.A. Miller, A.G. Gilicinski and L.M. Robeson; Polym.Mater. Sci. Eng., 74, 204 (1996).

6

Inorganic Polymers for Advanced Applications

Bo-Hye Kim[1] and Hee-Gweon Woo[*2]

[1]Department of Korean Medicinal Supply, Dongshin University, Naju—520 714, Korea.
[2]Nanotechnology Research Center and Institute of Basic Sciences,
Department of Chemistry, Chonnam National University, Gwangju—500 757, Korea.
*Corresponding author's email: hgwoo@chonnam.ac.kr

Contents

Summary: Main group metalloid-containing inorganic polymers, which are made up of boron (group 13: polyborazines), silicon/germanium/tin (group 14: Polysilanes, polygermanes, and polystannanes) and phosphorous (group 15: polyphosphazenes) atoms, are reviewed as advanced materials with heavy emphasis on silicon-containing polymers with quite unusual properties. Catalytic coupling of hydrosilanes with molecules having active E′-H bonds (E′ = Si, O, N) or vinylic bonds through Si-Si/Si-O/Si-N dehydrocoupling, redistributive coupling, and multiple addition processes are described in this article as selective examples of our recent research developments on specialty materials of inorganic polymers.

1. INTRODUCTION

Polymers can be classified as organic polymers and inorganic polymers. The backbones of organic polymers consist mainly of carbon atoms linked together or separated by heteroatoms such as oxygen or nitrogen [1]. Organic polymers used heavily in modern daily life tend to lose their advantageous properties (e.g., light weight, flexibility, fabricability, elasticity, mechanical strength, etc.) in the presence of oxygen, ozone, ultraviolet radiation or at high/low extreme temperatures. Most organic polymers hardly decompose in nature and burn with the release of toxic chemicals as well, leading to serious environmental pollution. Furthermore, the availability of raw materials for organic polymers is limited by the anticipated shortage of petroleum/coal natural resources. Therefore, research on the design, synthesis, characterization, and applications of inorganic polymers or inorganic-organic hybrid polymers (or organometallic polymers) is needed to avoid these problems, leading to the development of novel advanced materials for specialty applications [2]. Inorganic polymers have main group metalloids (e.g., Si, B, P, etc.), main group metals (e.g., Mg, Al, etc.) or transition metals (i.e., d-block or f-block elements) within the polymers. Main group metal containing and transition metal-containing inorganic polymers have been heavily reviewed by many authors [3] and are beyond the scope of this article. Main group metalloid-containing inorganic polymers are made up of mainly boron (group 13 element; e.g., polyboranes, polyborazines, polycarboranes, etc.), silicon/germanium/tin (group 14 elements; e.g., polysilanes, polycarbosilanes, polysiloxanes, polysilathianes, polysilazanes, polygermanes, polystannanes, etc.), phosphorous (group 15 element; e.g., polyphosphazenes), and sulfur (group 16 element; e.g., polysulfur, polysulfur nitrides) atoms, which do not originate from petroleum/coal resources [1b,2]. Those main group metalloid-containing inorganic polymers possess unusual properties. Especially, silicon-containing polymers exhibit quite unusual properties as advanced materials [4].

For the heavier elements in group 14 the formation of A-A bonds to produce a long chain polymer has proven to be difficult because stable unsaturated A = A species, analogues of vinyl compounds, can be prepared only in the combination with sterically bulky substituents, which will deter their polymerization [1,4-6]. Wurtz-type coupling of group 14 dihalides using an alkali metal dispersion has widely been used in industry, but is quite problematic. As an alternative to the Wurtz-type coupling of group 14 dihalides, dehydrocoupling and redistributive coupling of group 14 hydrides to their polymers are useful routes to the synthesis of A-A bonds. Group 14 hydrides (hydrosilanes, hydrogermanes, hydrostannanes) possess an A-H bond (A = Si, bond energy of 320 kJ/mole; A = Ge, 289 kJ/mole; A = Sn, 263 kJ/mole) that is more reactive than the C-H bond of hydrocarbons (bond energy of 416 kJ/mol) [7]. In the following sections, this article mainly describes some of the recent advances that have been made by us with respect to group 14 element-containing advanced polymers prepared by dehydrocoupling, redistributive coupling, and multiple addition of group 14 hydrides. At the same time some selective examples of recent research developments made by many researchers in specialty inorganic polymers (i.e., polyborazines for group 13 element and polyphosphazenes for group 15 element) containing groups 13 and 15 elements are also described.

2. GROUP 14 ELEMENT-CONTAINING INORGANIC POLYMERS: POLYSILANES, POLYGERMANES AND POLYSTANNANES

2.1. Dehydrocoupling of Group 14 Hydride to Polymers

2.1.1. Linear-selective Dehydrocoupling of Hydrosilanes to Polysilanes

Polysilanes are an interesting type of inorganic polymers with a great deal of applications in ceramics/ composites, photoelectronics, photoresistors, and nonlinear optics [8,9]. The peculiar optoelectronic properties of polysilanes are due to sigma-conjugation of the silicon atoms in the polymer chain, varying with the molecular weight, conformation, and substituents of the polymer [10a]. Wurtz-type coupling of dihalosilanes with an alkali metal dispersion in toluene- or xylene-refluxing temperature has some serious problems: (1) intolerance of certain functional groups, (2) lack of reproducibility, and (3) limitations for controlling stereochemistry and molecular weight distribution. Some improvements on the Wurtz-type coupling of dihalosilanes have been made by tuning several factors under ultrasonic activation [10b]. As an alternative synthetic method, the dehydrocouping of hydrosilanes, catalyzed by transition metal group 4 metallocene such as dimethyltitanocene or dimethylzirconocene, was first reported by Harrod and coworkers (**eq. 6.1**).

$$ n\ PhSiH_2 \quad \xrightarrow[\text{Cp}_2\text{TiMc}_2]{-\ H_2} \quad \left(\begin{array}{c} H \\ | \\ Si \\ | \\ Ph \end{array} \right)_n \tag{6.1} $$

However, the dehydrocoupling of hydrosilanes generally produces a mixture of linear polymers and cyclic oligomers, resulting in decrease of polymer molecular weights and improper molecular weight distribution. Following studies were intensively made with great efforts by many researchers worldwide [11-13]. For the selective production of linear polymer the careful design of new group 4 metallocene catalytic systems is important with proper tuning of other factors such as addition rate/ order of reagents, reaction temperature, etc. [12a]. Linear high molecular weight polysilanes can be used as precursors for making functional polysilanes by introducing functional groups on the linear polysilane. The synthesis of high molecular weight polyphenylsilanes with number-average molecular weight (M_n) of ca. 5300 and 4700 g/mol was reported by careful control of dehydrocoupling reaction conditions of phenylsilane using zirconocene-based catalysts [12a,14]. Tanaka [15] and Harrod [16] groups also prepared polyphenylsilanes with M_n of ca. 4600 and 7300 g/mol, respectively, from the dehydrocoupling of phenylsilane by using the zirconocene-based combination catalysts of $[Me_2N(CH_2)_3\text{-}H_4C_5](Me_5C_5)ZrCl_2/2MeLi$ and $Cp(Me_5C_5)ZrCl_2/2n\text{-}BuLi/(C_6F_5)_3B$, respectively. A recent survey of the catalytic dehydrocoupling of hydrosilanes under the influence of a range of early and late transition metal complexes appeared [13b].

Woo and co-workers recently developed a facile, highly linear-selective dehydrocoupling catalyst system of phenylsilane: Cp'_2MCl_2/Hydride (Cp' = C_5H_5 or C_5Me_5; M = Ti, Zr, Hf; Hydride = Red-Al, Selectride, Super Hydride) combination catalysts (**eq. 6.2**) [11,17].

$$n\ \text{PhSiH}_3 \xrightarrow[\substack{\text{Cp}_2\text{MCl2/Red-Al} \\ (\text{M = Ti, Zr, Hf})}]{-\ H_2} \left(\begin{array}{c} \text{H} \\ | \\ \text{Si} \\ | \\ \text{Ph} \end{array}\right)_n \tag{6.2}$$

The term "composite materials" may, perhaps, be simply defined on the basis of the classical definition of a "composite material" as given in "Longman's Dictionary" "something combining the typical or essential characteristics of individuals making up a group". This short statement grasps the very essence of this new material, which has recently emerged as a leading contender for numerous applications in the automotive, aerospace, electronic and wears industry [1].

Woo's combinative catalyst system of Cp'$_2$MCl$_2$/Hydride is different from the catalyst systems using Cp'$_2$MCl$_2$/2 alkyllithiums of Corey, Tanaka, and Harrod. Real catalytic species in the dehydrocoupling of hydrosilanes could be a metallocene hydride based on a sigma-bond metathesis mechanism [12,13b]. Inorganic hydrides effectively produce a metallocene hydride whereas alkyllithium can produce a metallocene hydride via a complex process (e.g., reductive elimination of metallocene alkyls or reaction with hydrosilane). Hence, no appreciable induction period is observed for the Cp'$_2$MCl$_2$/Hydride combination catalyst. The molecular weight distributions measured from the GPC traces were bimodal, indicating the presence of linear polysilanes and cyclic oligosilanes. The formation of cyclic oligosilanes can be estimated by integration of the GPC peaks. The peaks corresponding to SiH in the ^{1}H NMR spectrum can visually be separated as linear polysilane (δ 4.2~4.8 ppm range) and cyclic oligosilane (δ 4.9~5.3 ppm range). Thus, the formation of the cyclic oligosilanes is also estimated by integration of the ^{1}H NMR peaks, and is used as a means of crosschecking the cyclic/linear ratio. Woo *et al.* also screened the other group 4 metallocene-based combination catalysts for the dehydrocoupling of phenylsilane under various reaction conditions. The dehydrocoupling of phenylsilane with Cp'$_2$MCl$_2$/Red-Al combination catalysts rapidly produces linear polyphenylsilanes. The linear selectivity increases in the order of Cp$_2$Ti (64%) < Cp$_2$Zr (92%) < Cp(C$_5$Me$_5$)Zr (95%) < Cp$_2$Hf (99%) < Cp(C$_5$Me$_5$)Hf (<99%) [17b]. The higher linear-selectivity of the hafnocene relative to the zirconocene is likely due to the lower intrinsic dehydrocoupling activity (originating from stronger Hf-H and Hf-Si bond strengths). The lower linear-selectivity of titanocene relative to zirconocene and hafnocene is probably due to the combined effect of greater intrinsic dehydrocoupling activity (stemming from much weaker Ti-H and Ti-Si bond strengths) and much smaller atomic size (overriding steric crowding around the metal center) of Ti [18]. The change in linear selectivity is more pronounced than in other catalytic combination systems: Cp$_2$TiCl$_2$/2MeLi (55%) < Cp$_2$ZrCl$_2$/2MeLi (75%), Cp$_2$TiCl$_2$/2n-BuLi (75%) < Cp(C$_5$Me$_5$)ZrCl$_2$/2n-BuLi (80%) < Cp$_2$HfCl$_2$/2MeLi (85%) [16,18]. The coordinating environment around the metal center of the Cp$_2$MCl$_2$/Red-Al combination catalysts could be different from other catalytic systems such as Cp$_2$MCl$_2$/2R'Li [14,15,18,20] Red (or Vitride; sodium bis(2-methoxyethoxy)aluminum hydride; Na[H$_2$Al(OCH$_2$CH$_2$OMe)$_2$], soluble in toluene) will be quantitatively converted into Na[Cl$_2$Al(OCH$_2$CH$_2$OMe)$_2$] after reacting with dichlorometallocene. The coordinating structure of the present catalytic system could be similar to the zwitterionic structure of the Cp$_2$ZrCl$_2$/2n-BuLi/ (C$_6$F$_5$)$_3$B catalytic system [16,19]. The Na[Cl$_2$Al(OCH$_2$CH$_2$OMe)$_2$] moiety may influence by simply coordinating to the metal through an H, Cl-bridge or H, OMe-bridge between the group 4 metal and Al metal. The favorable steric demands imposed by the Cp ring and cocatalyst moiety could prevent the formation of the inactive dimer of metallocene hydride and could suppress the cyclic oligomer formation by chain cleavage reaction as well, leading to greater chain elongation [15,16]. However, an overriding steric demand should result in low dehydrocoupling activity. The order of dehydrocoupling activity for the various zirconocenes was found to be the same as the sequence of Tilley [12] and Harrod [16]: $(C_5Me_5)_2Zr \ll Cp_2Zr < Cp(C_5Me_5)Zr$. The $(C_5Me_5)_2ZrCl_2$/Red-Al combination catalyst thus slowly produces a mixture of dimer, trimer, and tetramer. In addition, the order of linear-selectivity for the dehydrocoupling of phenylsilane catalyzed by hydrides with Cp_2ZrCl_2 was found to be Super Hydride (82%) < N-Selectride (88%) < Red-Al (92%) [17c]. The dehydrocoupling of $PhCH_2SiH_3$ yields only low-molecular weight oligomers because an alkylsilane is generally less reactive than an arylsilane [20]. The increase in molecular weight of polymer is possible by higher catalyst concentration (1 mol% *versus* 5 mol%), but it is affected little by longer reaction times (1 day *versus* 5 days) [17b]. Linear selectivity and molecular weights decrease with addition of solvent and with heating, which was similarly observed in other catalytic systems [12a,16]. This is because dilution and heating could hamper the tight coordination of the $Na[Cl_2Al(OCH_2CH_2OMe)_2]$ moiety to the metal center. Interestingly, linear selectivity and molecular weights decrease drastically by adding 4 Å molecular sieve (MS 4Å). The interaction of $Na[Cl_2Al(OCH_2CH_2OMe)_2]$ moiety with MS 4Å might prevent the close coordination of $Na[Cl_2Al(OCH_2CH_2OMe)_2]$ moiety to the metal center [17b]. An exact molar ratio of Red-Al to dichlorometallocene is necessary to replace both chlorines to attain high reactivity. The inactivity observed for higher molar ratios of Red-Al to dichlorometallocene could be attributed to over-complexation of Red-Al moieties to the metal, blocking the empty coordination site necessary for the dehydrocoupling of silane [18]. The poor activity is observed for lower molar ratios of Red-Al to dichlorometallocene. All the experimental results described above firmly suggest that better catalysts affording higher linear-selectivity and higher-molecular-weight polymer can be properly designed by tuning the steric and electronic character of the catalyst environment, including metal, ligand and co-catalyst moieties.

It is interesting to note that the dehydrocoupling of *p*-fluorophenylsilane using Cp_2ZrCl_2/Red-Al produced soluble amorphous polysilane (ca. 75%) and sparingly soluble crystalline polysilane (ca. 25%) in toluene and chloroform. The two polysilanes are soluble in THF and pyridine. In comparison, the dehydrocoupling of *p*-fluorophenylsilane using Cp_2TiCl_2/Red-Al gives soluble polysilane only [17c]. The crystalline polysilane may have interactions [17b] in the polymer molecules either between Si and F with high possibility or between F and phenyl ring with low possibility according to the medium level of MO calculation.

Linear high molecular weight polysilanes can be used as precursors for making functional polysilanes by introducing functional groups on the linear polysilane. The Si-H bonds in the backbone chain of poly(hydrophenylsilane) are transformed to Si-Cl bonds using a mild chlorinating reagent, CCl_4. The Si-Cl bonds in the poly(chlorophenylsilane) can be replaced by various nucleophiles such as cyclopropyl, epoxy, aziridinyl, pyridyl, bipyridyl, phosphinyl, poly(ethylene oxide), thiol, etc. to give new functional polymers which can be used for applications in sensors, ion-exchange resins, batteries, drug delivery, metal nanomaterials preparation, etc. [17e].

2.1.2. Homodehydrocoupling and Codehydrocoupling of 1,1-dihydrotetraphenylsilole and 1,1-dihydrotetraphenylgermole to Electroluminescent Polymers

Polysilanes [21-23] with low oxidation potentials and a high-lying HOMO exhibit unusual optoelectronic properties due to sigma-conjugation along the silicon backbone chain in the polymer [10,24] Siloles (silacyclopentadienes), with low reduction potential and low-lying LUMO's, have attracted considerable attention because of their peculiar electronic properties [25,26]. They may be used for electron-transporting materials in devices [27]. A silole does not luminesce in diluted solution but does luminesce in concentrated solution, implying that polysiloles could exhibit different luminescent behavior from that of monomeric siloles [25]. Polysiloles, which are expected to have hybrid properties of polysilane and silole by nature in the structure, can be prepared by 1,1- or 2,5-coupling reactions of siloles using various synthetic coupling methods [28]. Electroluminescent poly(silole-*co*-silane)s have also been synthesized in several laboratories [29]. West *et al.* reported recently the synthesis of polysiloles and polygermole (M_w ca. 5200-5700 g/mol) that have methoxy end groups in 30-37% yield by heterogeneous Wurtz 1,1-dehydrocoupling of 1,1-dichlorotetraphenylsilole with 2.0 equivalents of Li, Na, K metal in refluxing THF for 3 days (**eq. 6.3**) [28a]. Tamao and collaborators had earlier reported the Wurtz coupling synthesis of polysiloles [28d].

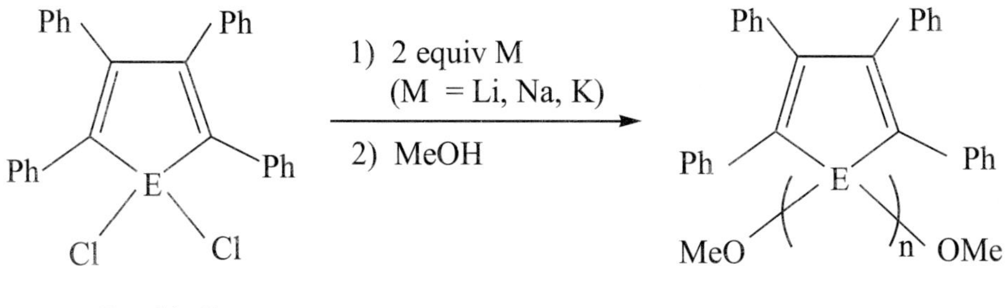

$$\text{(6.3)}$$

$$E = Si, Ge$$

As an alternative to the Wurtz coupling of 1,1-dichlorosilole, the homogeneous dehydrocoupling methodology was demonstrated in Tanaka's earlier report of the dehydrocoupling synthesis of poly(dibenzosilole) [30a]. Trogler and co-workers recently reported the 1,1-dehydrocoupling of 1,1-dihydrotetraphenylsilole (**1**) to an electroluminescent polysilole (**2**) (M_w ca. 4000-6000 g/mol), having hydrogen end groups, in high yield 80-90% yield using 1 mol% of the late transition metal complexes [(Ph$_3$P)$_3$RhCl and ((Ph$_3$P)$_4$Pd)] as catalysts [29b,30b]. Similarly, Woo and co-workers prepared **2** (M_w ca. 5500-6200 g/mol) in > 95% yield by dehydrocoupling of **1** using [RhCl(COD)$_2$]$_2$ and Pt(PMe$_3$)$_4$ as catalyst (**eq. 6.4**) [31].

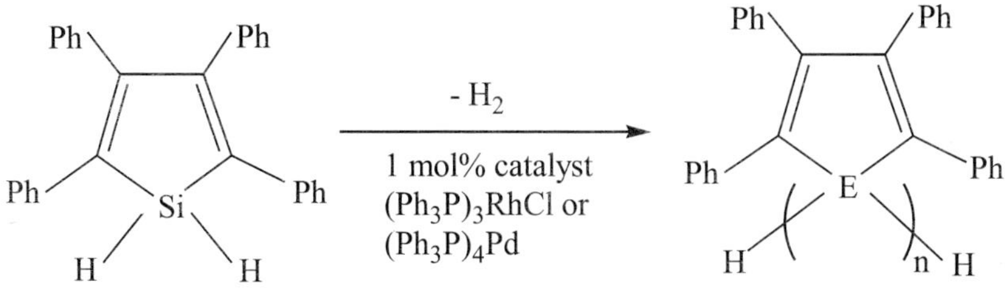

$$\text{(6.4)}$$

1

2

The same authors also synthesized poly(tetraphenylgermole) (M_w ca. 5800-6500 g/mol) in > 92% yield by dehydrocoupling of 1,1-dihydrotetraphenylgermole using the same catalysts. The UV-VIS spectrum of poly(tetraphenylgermole) shows an absorption at 377 nm, which is assignable to both the s-s* transition of the Ge-Ge backbone chain and p-p* transition of the germole ring. The polygermole is intensively photoluminescent, emitting green light at 487 nm [31]. The hydrogen end groups of the polysiloles and polygermoles are then transformed to other useful functional groups by various chemical reactions [31].

As an alternative to the heterogeneous Wurtz reductive coupling of dichlorosilole and the homogeneous late-transition-metal-complex-catalyzed dehydrocouplig of **1**, Woo *et al.* also synthesized **2** in high yield by homogeneous dehydrocoupling of **1** under mild conditions, catalyzed with inorganic hydrides such as Selectrides {MB[CH(CH$_3$)C$_2$H$_5$]$_3$H; M = Li, Na, K}, Red-Al {Na[H$_2$Al (OCH$_2$CH$_2$OCH$_3$)$_2$]}, and Super-Hydride [LiB(C$_2$H$_5$)$_3$H] as shown in **eq. 6.5** [32a,b].

$$1 \xrightarrow[\substack{< 50 \text{ mol\% of} \\ \text{Hydride, THF} \\ 25°C, 24 \text{ h}}]{-\text{H}_2} 2 \tag{6.5}$$

Dehydrocoupling of **1** catalyzed by < 50-mol% (*i.e.*, M-H/Si-H = 0.5) of Red-Al yielded **2** as light yellow powders. Polymers with molecular weights (M_w) of 4600 and 4100 g/mol were isolated in 86% and 78% yields when 15-mol% and 50-mol% of Red-Al were used, respectively. Polymer yields and molecular weights when 15-mol% of Red-Al was used were higher compared to polymers obtained when 50-mol% of Red-Al was used. Products from the reaction of **1** with 15-mol%, 25-mol%, and 50-mol% Red-Al were separated by preparative GPC and were characterized by NMR spectroscopy. Shorter oligomers such as silole dimer or trimer were not found in products. However, when 100 mol% of Red-Al (*i.e.*, M-H/Si-H = 1) was used, the formation of silole dianion **3** [33] was observed without forming **2** (**eq. 6.6**).

$$1 \xrightarrow[\substack{100 \text{ mol\% of} \\ \text{Red-Al, THF} \\ 25°C, 1 \text{ h}}]{-\text{H}_2} \mathbf{3} \text{ (Na)}_2 \tag{6.6}$$

In the similar manner, the dehydrocoupling of **1** using 15 mol% of Selectrides and Super-Hydride at 25°C for 24 hours produced **2** in 77-78%-isolated yield. The molecular weight (M_w) and polydispersity index (PDI) of all the polysiloles were in the range of 4300-5800 g/mol and 1.1-1.2, respectively. Polymerization yield and polymer molecular weight increased in the order of L-Selectride < N-Selectride < K-Selectride. This trend seems to be closely related to the ionic character of the Selectrides. The polymerization yields were almost equal for Red-Al, K-Selectride, and Super-Hydride, but the molecular weight increased in the order of Red-Al < K-Selectride < Super-Hydride. Like the polysiloles prepared by West and co-workers [28a], these polysiloles have a characteristic UV absorption around 300 nm, assigned to the σ-σ^* transition of the Si-Si backbone chain. They are photoluminescent, emitting green light at 520 nm when the excitation is at 330 nm. These polysiloles are strongly electroluminescent around 520 nm. The similar dehydrocoupling of 1,1-dihydrotetraphenylgermole with the hydrides produced polygermole in high yield. Furthermore the co-dehydrocoupling of **1** and 1,1-dihydrotetraphenylgermole (with varying the mixing mole ratio) with the hydrides produced poly(silole-*co*-germole)s in high yield [31].

For the dehydrocoupling reaction of **1** to **2**, K-Selectride and Super-Hydride were the most active catalysts examined. **2** was prepared in high yield directly from the reaction of 1,1-dichlorotetraphenylsilole (instead of **1**) in the presence of < 1.5 equiv of Red-Al (instead of < 0.5 equiv) [32c]. Unlike in the case of late-transition-metal-complex-catalyzed dehydrocoupling (generally proceeded by oxidative addition/reductive elimination processes) [30,31], catalysis for the conversion of **1** to **2** by early transition metal complexes, Cp$_2$MCl$_2$/Red-Al (M = Ti, Zr, Hf) [17] is ineffective in the dehydrocoupling of **1** because **1** is sterically hindered, considering the four-centered transition state in the sigma-bond metathesis mechanism [12]. Woo group proposed a mechanism involving the preferential attack of a hydride ion on either the silicon atom or silole ring of **1** to form an activated anionic intermediate such as a pentacoordinated sigma-complex or pi-complex [32].

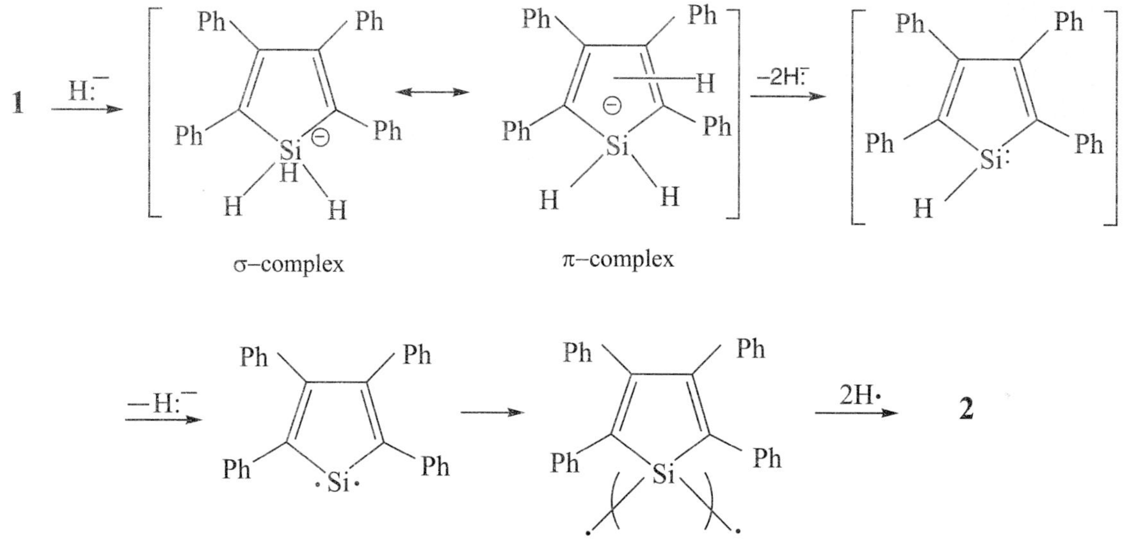

Fig. 6.1. Possible mechanism for the formation of polysilole 2 from the dehydrocoupling of 1.

The activated anionic intermediate could lose both a dihydrogen molecule and a hydride ion (this hydride may participate again in the catalytic cycle) sequentially to form a silylene type of silole. If the activated anionic intermediate accepts another hydride ion, a silole dianion **3** will be formed after losing two dihydrogen molecules. The silylene type of silole will then either self-couple or keep inserting into the Si-H bond of **1**, forming **2** (**Fig. 6.1**). For the homodehydrocoupling of 1,1-dihydrogermole to polygermole or co-dehydrocoupling of 1,1-dihydrogermole with 1,1-dihydrosilole to poly(silole-*co*-germole)s the same mechanism in **Fig. 6.1** should be applied.

2.1.3. Si-O/Si-Si Dehydrocoupling of Hydrosilane with Alcohol to Poly(alkoxysilane)s

A wide range of catalysts (e.g., acids, bases, and homogeneous/ heterogeneous transition metal catalysts) have been used for the Si-O dehydrocoupling of alcohols with silanes [13a,34,35a]. The Si-O dehydrocoupling of bis-hydrosilanes with diols, catalyzed by rhodium complex, yielding polysiloxanes was reported [35b,c]. The Si-S dehydrocoupling of hydrosilanes with dithiols to produce polysilathianes was also reported [35d]. The polysiloxanes and polysilathianes are beyond the scope of this article and thus are not covered here. Si-Si dehydrocoupling of hydrosilanes with late transition metal complex catalysts produces a mixture of oligomers along with significant amounts of redistributed by-products [13b]. Harrod *et al.* reviewed the recent dehydrocoupling of hydrosilanes with alcohols [11]. The transition metal complexes of group VIII (Ni, Co, Rh, Pd, Ir, Pt, etc.) have been extensively applied in the catalytic dehydrocoupling of hydrosilanes with various nucleophilic reagents [36]. A recent survey was conducted on the catalytic dehydrocoupling of hydrosilanes in the presence of a range of early and late transition metal complexes 13b].

Numerous studies were reported either on the alcoholysis of hydrosilanes (i.e., Si-O coupling) or on the dehydropolymerization of silanes (i.e., Si-Si coupling) under the influence of various transition metal complex catalysts [13,34,35a-c,36]. Poly(alkoxysilane)s can be used as important precursors for preparing interesting polysilane-siloxane hybrids by sol-gel methods in the presence of acid/base catalyst in the reverse micellar environment [37a]. Woo and co-workers first described the combinative Si-Si/Si-O dehydrocoupling reaction of hydrosilanes with alcohols (1:1.5 mole ratio) at 50°C, catalyzed by Cp_2MCl_2/Red-Al (M = Ti, Zr) and Cp_2M' (M = Co, Ni), producing poly(alkoxysilane)s in one pot in high yield (**eq. 6.7**).

$$\text{PhSiH}_3 + \text{EtOH} \quad \xrightarrow[\substack{\text{or} \\ Cp_2M'\ (M' = \text{Co Ni}) \\ 50°C}]{\substack{Cp_2MCl_2/\text{Red-Al} \\ (M = \text{Ti, Zr}),\ 50°C}} \quad \text{EtO}\left(\underset{\underset{\text{Ph}}{|}}{\overset{\overset{\text{OEt}}{|}}{\text{Si}}}\right)_n\!\!\text{OEt} \quad + \quad \text{PhSi(OEt)}_3 \tag{6.7}$$

The hydrosilanes are $p\text{-C}_6\text{H}_4\text{SiH}_3$ (X = H, CH_3, OCH_3, F), $\text{PhCH}_2\text{SiH}_3$, and $(\text{PhSiH}_2)_2$. The alcohols are MeOH, EtOH, *i*-PrOH, PhOH, and $CF_3(CF_2)_2CH_2OH$. The weight average molecular weights (M_w) of the poly(alkoxysilane)s ranged from 600 to 8000 g/mol. However, Cp_2M' (M´ = Co, Ni) have no catalytic activity toward Si-Si dehydrocoupling of primary hydrosilanes in inert atmosphere [37b]. Interestingly, the dehydrocoupling reactions of phenylsilane with ethanol (1:1.5 mole ratio) using Cp_2HfCl_2/Red-Al, and phenylsilane with ethanol (1:3 mole ratio) using Cp_2TiCl_2/Red-Al gave only triethoxyphenylsilane as product (**eqs. 6.8a & 6.8b**) [38].

$$\text{PhSiH}_3 \;+\; 1.5\ \text{EtOH} \quad \xrightarrow[50°C]{Cp_2HfCl_2/\text{Red-Al}} \quad \text{PhSiH}_3 \tag{6.8a}$$

$$\text{PhSiH}_3 \;+\; 3.0\ \text{EtOH} \quad \xrightarrow[50°C]{Cp_2HfCl_2/\text{Red-Al}} \quad \text{PhSiH}_3 \tag{6.8b}$$

Similarly, the combinative Si-Si/Si-O dehydrocoupling reactions of mixed hydrosilanes with alcohols (two different hydrosilanes were used in the same mole ratio; 0.5:0.5:1.5 mole ratio) at 50°C, catalyzed by Cp_2MCl_2/Red-Al (M = Ti, Zr) and Cp_2M' (M' = Co, Ni), producing copoly(alkoxysilane)s in one-pot in high yield (**eq. 6.9**) were reported [37a].

$$0.5\ RSiH_3\ +\ 0.5\ R'SiH_3\ +\ EtOH \xrightarrow[\substack{\text{or}\\ Cp_2M'\ (M' = Co\ Ni)\\ 50°C}]{\substack{Cp_2MCl_2/\text{Red-Al}\\ (M = Ti,\ Zr),\ 50°C}} EtO\left(\underset{\underset{R}{|}}{\overset{\overset{OEt}{|}}{Si}}\right)_x\left(\underset{\underset{R'}{|}}{\overset{\overset{OEt}{|}}{Si}}\right)_{n-x}OEt$$

$$+\ R'Si(OEt)_3\ +\ RSi(OEt)_3 \tag{6.9}$$

In the similar manner, Woo and co-workers performed the combinative Si-Si/Si-O dehydrocoupling reactions of hydrosilanes with mixed alcohols (two different alcohols were used in the same mole ratio; 1:0.75:0.75 mole ratio) at 50°C, catalyzed by Cp_2MCl_2/Red-Al (M = Ti, Zr) and Cp_2M' (M' = Co, Ni), producing copoly(alkoxysilane)s in one-pot in high yield [37a]. The bonding characters [mixing interaction between σ (silicon) orbital and n (oxygen) orbital] of Si-O bonds in poly(dialkoxysilylene)s were studied in detail using various levels of molecular orbital calculations [37c].

2.1.4. Si-N Dehydrocoupling of Poly(hydrosilane) with Polyborazine Additive

An exhaustive review on the Si-N dehydrocoupling of hydrosilanes with various amines (including ammonia, alkyl amines, and aromatic amines, hydrazines), catalyzed by early and late transition metal complexes, appeared [11]. Eisen and co-workers reported the Si-N dehydrocoupling of phenylsilane with various amines in the presence of $[(Et_2N)_3U[BPh_4]$ [39a]. The B-N dehydrocoupling of borazine and poly(vinylborazine) was reported by Sneddon and co-workers [39b]. No Si-B dehydrocoupling of hydrosilane with borazine has been reported up to date although it may possibly occur.

Since silicon-containing polymers are generally used for preparing non-oxide ceramics, various polymeric precursors with different structures and compositions have been designed [39c]. Preceramic polycarbosilane (PCS), used for preparing commercial Nicalon fiber, is synthesized by the poorly efficient thermolysis of polydimethylsilane (PDMS) due to the formation of hazardous gaseous compounds (leading to low yield) and irregularity of the composition of the resulting PCS [39d] although some modifications were made by adding additive such as polyborodimethylsiloxane and zeolite ZSM-5 [39e,f]. As an alternative precursor for SiC ceramics, Woo and co-workers studied polyphenylsilane (PPS,$[-Si(Ph)H-]_n$) and polyvinylsiane (PVS, $[CH_2CH(SiH_3)]_n$). An ideal preceramic precursor for Si-C ceramics should show: (1) high ceramic residue yield, to minimize the cost and volume change associated with pyrolytic conversion to ceramics and consequently to maximize the control of porosity and densification, and (2) a suitable processability (*i.e.*, viscosity) allowing it to be shaped into the needed forms prior to pyrolysis. In this context, it is disadvantageous that the low viscosity PPS and PVS undergo drastic mass loss during pyrolysis, which leads to low ceramic residue yields (20-40 wt%) [39g,40a]. To improve the ceramic residue yields and processibilities of PPS and PVS, the Seyferth group tried to use many additives including transition metal complex catalysts, urea, and decaborane [39g,h]. Woo and collaborators also reported the increase of thermal stability (*i.e.*, increase of TGA ceramic residue yield) of PCS and oligocarbosilane (OCS) by dehydrocoupling of Si-H bonds in the PCS and OCS using group 4 and 6 transition metal complex catalysts [40b,c]. Woo *et al.* described the additive effect of polyborazine (PBN) on the improvement of ceramic residue yield of the polysilane, as depicted in (**eq. 6.10**) [41].

$$(R = H, \text{ borazine}) \tag{6.10}$$

The ceramic residue yield of PPS increases from an original 39 wt% to 65 wt%, and PVS from an original 26 wt% to 64 wt% by simply heating with 1 wt% PBN at 70°C for < 6 hours. Furthermore, the low viscosity PPS and PVS are transformed into highly viscous polymers, which are suitable for hand drawing into green fibers. Analysis using NMR spectroscopy suggests that Si-N dehydrocoupling of SiH_3 in PVS and Si-H in PPS with PBN is responsible for the improved ceramic residue yields [41a]. When the adding amount of PBN increases, the B, N-incorporated SiC ceramic composites, which can be used for diverse materials applications, are obtained. The reverse reaction is interesting. Polyhydrosilanes can be also used for a cross linking agent for PBN. When the using amount of polyhydrosilane increases, the SiC-incorporated BN ceramic composites are formed [41b].

2.1.5. Dehydrocoupling of Hydrogermanes and Hydrostannanes to Polymers

Recent reviews appeared on the catalytic dehydrocoupling of hydrogermanes and hydrostannanes to polygermanes and polystannanes promoted by transition metal complex catalysts [11,42]. As in the case of polysilanes, polygermanes and polystannanes have been conventionally prepared by the Wurtz-type dehalocoupling of halogermanes and halostannanes using alkali metals (e.g., Li, Na, K) or alkaline earth metal (Mg) [43]. The catalytic dehydrocoupling of $PhGeH_3$ to a cross-linked gel type of polygermane using Cp_2TiMe_2 catalyst was first reported [44a]. The formation of the highly cross-linked insoluble gel product might be caused by further dehydrocoupling of backbone Ge-H bonds or by partial redistribution of $[PhGeH]_n$ moiety to $[Ph_2Ge]_x[GeH_2]_{n-x}$, followed by further dehydrocoupling of $[GeH_2]_{n-x}$ moiety. The first one seems to occur with higher possibility because the catalytic dehydrocoupling of Ph_2GeH_2 gives a mixture of oligomer and $(Ph_2GeH)_2$ using Cp_2TiMe_2 catalyst [44a]. Tanaka *et al.* synthesized partially cross-linked soluble poly(phenylgermane) in 67% isolated yield by the catalytic dehydrocoupling of $PhGeH_3$ using $CpCp'ZrCl_2/2n$-BuLi (Cp' = Cp or Cp^*) combination catalyst, followed by precipitation of the polymer solution in THF with pentane [44b]. Similarly, Woo and co-worker prepared soluble poly(phenylgermane) and insoluble poly(phenylgermane) by the catalytic dehydrocoupling of $PhGeH_3$ using Cp_2MCl_2/Red-Al (M = Ti, Zr, Hf) combination catalyst [44c]. From the catalytic reaction the percentage of the soluble poly(phenylgermane) decreased in the order of Hf (89%) > Zr (82%) > Ti (25%) probably because the larger size of zirconium and hafnium could provide better coordination space for polygermane chains than the smaller size of titanium.

Tilley and co-workers prepared linear polystannanes (M_w, ca. 70,000 g/mol) by dehydrocoupling of R_2SnH_2 (R = n-Bu, n-Hex, n-Oct) using $CpCp^*Zr[Si(SiMe_3)_3]$ Me catalyst [45]. The authors studied the properties of sigma-conjugated polystannanes such as low band gap [46a]. s-s* transition energy [46b] and order-disorder phase transition [46c]. The dimerization of R_3SnH (R = Ph, n-Bu) by a Fe-Pd heterobimetallic alkoxysilyl and siloxysilyl complex was reported [47]. The mechanism for the dehydrocoupling of hydrogermanes and hydrostannanes by the group 4 metallocene catalysts could be similar to the mechanism for dehydrocoupling of hydrosilanes by those catalysts [11,12]. Tilley and Neale recently proposed a modified mechanism comprising steps of σ-elimination and s-bond metathesis based on their kinetic study of hafnium hydrostannyl complexes by ^1H NMR spectroscopy [48].

2.2. Redistributive Coupling of Group 14 Hydride to Polymers

2.2.1. Desilanative Coupling of Multisilylmethanes to Oligomers

There are many types of hydride compounds, which are Lewis bases. Inorganic hydrides are generally soluble in polar solvent (e.g., THF) because of their ionic or polar nature. Nonetheless, the toluene solution of Red-Al (or Vitride; sodium bis(2-methoxyethoxy)aluminum hydride) commercially available efficiently catalyzes the polymerization of lactams and olefins and the trimerization of isocyanates [49]. Corriu and co-workers reported the desilanative proportionation reactions of di- and trihydrosilanes to the both more substituted silanes and less substituted silanes [50a] and the desilanative oligomerization of disilanes [50b] with exchanging their substituents (**eqs. 6.11 & 6.12**), promoted by inorganic hydrides (e.g., NaH, KH, *etc*) in THF.

$$3\ RSiH_3 \xrightarrow[-SiH_4]{H:^-} RSiH_2 \xrightarrow[-SiH_4]{H:^-} R_3SiH \tag{6.11}$$

$$R_3Si\text{-}Si(Me)_2H_3 \xrightarrow{H:^-} R_3SiH\ +\ (Me_2Si)_6\ +\ (Me_3Si(Si_6\ R_3Si\text{-}Si(Me_2)_nH \tag{6.12}$$

The authors proposed a mechanism involving a reactive pentacoordinated hydrosilyl anion [50c], that is formed by addition of hydride (H$^-$) on the silanes, for the redistribution reactions. They suggested that the reactivity increases hexacoordinated silicon > pentacoordinated silicon > tetracoordinated silicon.

Similarly Riviere and collaborators reported the preparation of oligophenylgermane $(PhHGe)_n$ by redistributive coupling of $(PhH_2Ge)_2$ under the influence of a Lewis base PhH_2GeLi [51]. The strong Lewis acid $AlCl_3$ also catalyzes a redistribution reaction of hydroarylsilane to quaternary arysilane with giving off pyrophoric SiH_4 [52]. Woo *et al.* reported the intriguing desilanative coupling of bis- and tris(silyl)methanes to oligomers, catalyzed by Red-Al [53]. The bis- and tris(silyl)methanes are multisilylmethanes which have reactive Si-C-Si linkages. Thus, 2-phenyl-1,3-disilapropane (**4**) undergoes desilanative coupling at ambient temperature in the presence of 1 mole% Red-Al (3.4 M solution in toluene) within 1 hour to produce benzylsilane as the major product and as yet uncharacterized high-boiling oligomers as minor products with SiH_4 gas, as shown in **eq. 6.13**. Similarly, the bis(silyl)methane, 1-phenyl-3,5-disilapentane (**5**), is quantitatively converted to methylphenethylsilane and phenethylsilane (7:3 ratio, identified by GC/MS analysis) as the major product and uncharacterized high-boiling oligomers as minor products with SiH_4 and $MeSiH_3$ gases (**eq. 6.14**). Likewise, the tris(silyl)methane, 1-phenyl-4-silyl-3,5-disilapentane (**6**), is quantitatively transformed to (**5**), methylphenethylsilane, and phenethylsilane (3:12:4 ratio identified by GC/MS analysis) as major products and uncharacterized high-boiling oligomers as minor products along with the formation of SiH_4, $MeSiH_3$, and $CH_2(SiH_3)_2$ molecules (**eq. 6.15**) [53].

$$PhCH(SiH_3)_2 \xrightarrow{\text{Red-Al}} PhCH_2SiH_3 \; + \; SiH_4 \; + \; \text{Oligomers} \qquad (6.13)$$

4

$$Ph(CH_2)_2SiH_2CH_2SiH_3 \xrightarrow{\text{Red-Al}} Ph(CH_2)_2SiH_2CH_3 + \; Ph(CH_2)_2SiH_3$$

5

$$+ \; SiH_4 \; + \; CH_3SiH_3 \; + \; \text{Oligomers} \qquad (6.14)$$

$$Ph(CH_2)_2SiH_2CH(SiH_3)_2 \xrightarrow{\text{Red-Al}} \mathbf{5} \; + \; Ph(CH_2)_2SiH_2CH_3 \; +$$

6

$$Ph(CH_2)_2SiH_3 \; + \; SiH_4 \; + \; CH_3SiH_3$$

$$+ \; CH_2(SiH_3)_2 \; + \; \text{Oligomers} \qquad (6.15)$$

Benzylsilane does not undergo further appreciable redistribution (*i.e.*, to dibenzylsilane and tribenzylsilane; such redistribution should be possible at high temperature because the cleavage of less reactive Si–C bond needs higher energy than that of more reactive Si-C-Si bonds) under the mild redistribution conditions. Methylene bridges between two silicon atoms are more readily deprotonated by strong organometallic bases than are methyl groups bonded to a single silicon atom because two silicon centers jointly can stabilize the resulting anion better than one silicon center via $p\pi$-$d\pi$ interaction between p-orbital of carbon and d-orbital of silicon [54] (but other explanation based on some molecular orbital calculation may be possible [37c]), but cleavage of the Si-C bond takes place normally under extremely harsh conditions[55]. Methylphenethylsilane can be obtained *via* Si-C bond cleavage of **4**, evolving SiH_4 gas in the redistribution reaction of **5**. However, phenethylsilane cannot be obtained *via* the direct Si-C bond cleavage of methylphenethylsilane, but *via* the direct Si-C bond cleavage of **5** with CH_3SiH_3 gas evolution. Methylphenethylsilane was obtained in higher yield than phenethylsilane, implying that SiH_4 formation is easier than CH_3SiH_3 formation, likely due to less steric hindrance exerted upon adding hydride to the silanes. On the other hand, methylphenethylsilane can be formed *via* consecutive Si-C bond scissions of **6**, giving off SiH_4 gas in the redistribution reaction of **6**. However, phenethylsilane cannot be obtained by the direct Si-C bond cleavage of methylphenethylsilane, but can be obtained either by the direct Si-C bond cleavage of **5** with release of CH_3SiH_3 or by the direct Si-C bond cleavage of **6** with the formation of $CH_2(SiH_3)_2$. The yield for methylphenethylsilane was higher than for phenethylsilane, suggesting that SiH_4 formation is easier than $CH_2(SiH_3)_2$, because of different steric hindrance upon the addition of hydride to the silane. The as yet uncharacterized high-boiling oligomers could be obtained only during the redistribution process because the reactions of benzylsilane, methylphenethylsilane, and phenethylsilane with Red-Al do not yield oligomeric mixture of silanes under the reaction conditions. The oligomers might be oligocarbosilanes or a mixture of oligocarbosilanes and oligosilanes.

Based on the experimental results described above Woo *et al.* proposed a plausible mechanism involving the preferential addition of the hydride to the silicon at the less hindered site, forming an active intermediate and SiH_4 gas [53]. The α-silyl carbanion of pentacoordinated anionic species then collapses to give an α-silyl carbanion. The a-silyl carbanion may then pick up a hydrogen from a hydrogen source around (*e.g.*, silane or solvent) to form a silane or may lose a hydride to produce a silene, associating to produce some oligo(carbosilane)s, or isomerizing to a silyl anion. The silyl anion may lose a hydride to give an unstable silylene, which can add to silane or an associate, producing some linear or cyclic oligosilanes. The regenerated hydride may add to **4** to participate again in the catalytic cycle.

$$\text{copolymer} \quad + \quad PhCH_2SiH_3 \quad +$$
$$\mathbf{7}$$

$$(PhCH_2)_2SiH_2 \quad + \quad H_3SiCH(Ph)SiH_2CH_2Ph$$
$$\mathbf{8} \qquad\qquad\qquad \mathbf{9}$$

$$\mathbf{4} \quad \xrightarrow[\substack{Cp_2MCl_2/\text{Red-Al} \\ (M = Ti,\ Zr,\ Hf)}]{-H_2/-SiH_4}$$

$$+ \quad H_3SiCH(Ph)SiH_2CH(Ph)SiH_3 \quad +$$
$$\mathbf{10}$$

$$H_3SiCH(Ph)SiH_2SiH_2CH(Ph)SiH_3 \qquad\qquad (6.16)$$
$$\mathbf{11}$$

Very interestingly the catalytic reactions of **4** with Red-Al, $Cp_2MCl_2/$Red-Al, and $Cp_2MCl_2/2n$-BuLi (M = Ti, Zr, Hf) yield different results. The redistribution-dehydrocoupling of **4** catalyzed by $Cp_2MCl_2/$Red-Al gives polymer (M_w = 550 for Ti; 750 for Zr; 2040 for Hf) with low TGA ceramic residue yield (*ca.* 3% at 800°C for Ti, Zr, Hf) in moderate yield (20% for Ti; 25% for Zr; 28% for

Hf) [56,57] as shown in **eq. 6.16** whereas the redistributive coupling (**eq. 6.13**) of **4** catalyzed by Red-Al gives oligomer (M_w < 500) in very low yield (less than 3%) [53].

A plausible mechanism for the redistribution-dehydrocoupling of **4** with Cp_2ZrCl_2/Red-Al is depicted in **Fig. 6.2** [57]. The mechanism involves the preferential attack of the hydride on the less hindered silicon with formation of a pentacoordinated anionic species, which collapses to give an α-silyl carbanion intermediate and SiH_4 gas. The α-silyl carbanion may then take up a hydrogen from the hydrogen source (*e.g.*, silane or solvent) to yield **7** or may lose a hydride, participating again in the catalytic cycle to produce phenylsilene. All attempts for trapping silene are unsuccessful. The reason could be probably due to the active Si-H bonds abundant in the reaction mixture. Compound **4** may add to phenylsilene to give **10**, which will become **8** and **9** by sequentially losing SiH_4 gas. The silanes might then undergo catalytic dehydrocoupling to produce a copolymer [56,57].

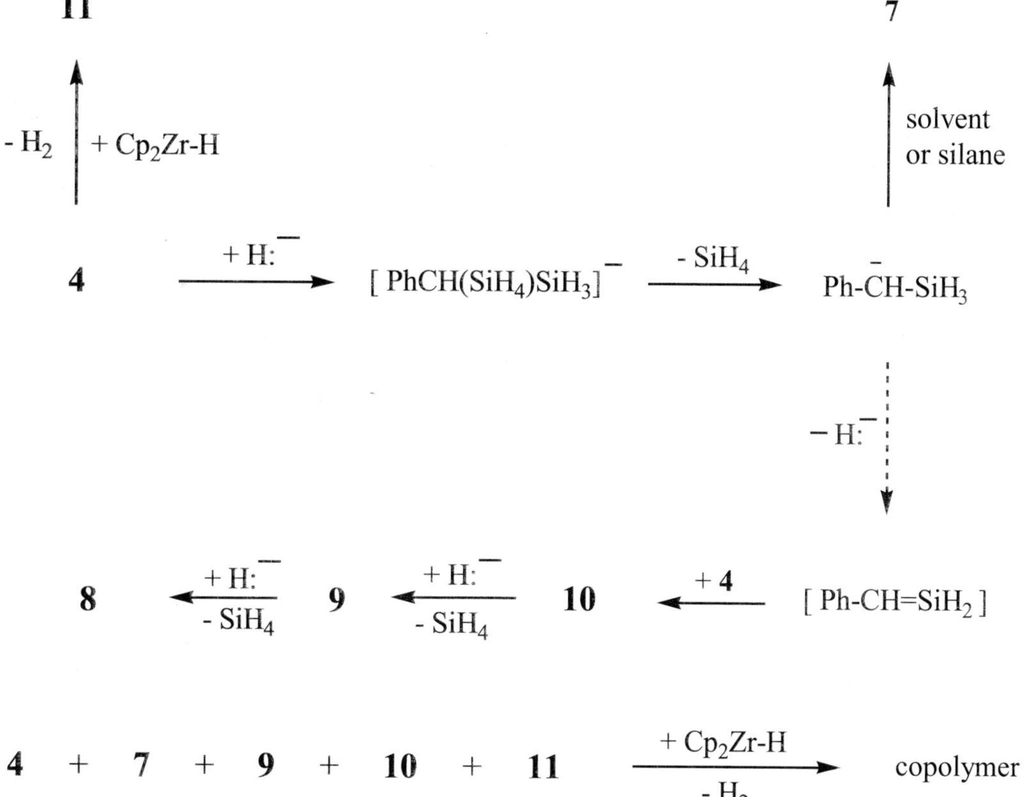

Fig. 6.2. **Plausible mechanism for the redistribution-dehydrocoupling of 4 catalyzed by Cp₂ZrCl₂/Red-Al.**

By comparison with the reactions as shown in **eqs. 6.15 and 6.16**, dehydrocoupling (without redistribution) of **4** catalyzed by Cp_2MCl_2/2*n*-BuLi gives mostly highly cross linked polysilanes along with soluble oily polymers as minor products (**eq. 6.17**) [56-58].

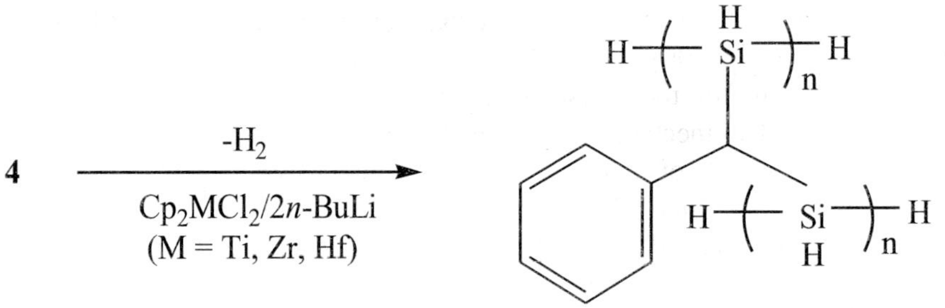

$$4 \xrightarrow[\substack{Cp_2MCl_2/2n\text{-}BuLi \\ (M = Ti, Zr, Hf)}]{-H_2}$$

(6.17)

Insoluble polymers are isolated in 82% yield for Ti, 95% yield for Zr, and 80% yield for Hf. TGA ceramic residue yields are measured to be 72% for Ti, 73% for Zr, and 74% for Hf. The weight average molecular weights of the oily polymers are 4120 for Ti, 9020 for Zr, and 5010 for Hf. The TGA ceramic residue yields of the soluble oily polymers are ca. 14%. The dehydrocoupling mechanism of **4** could be similar to the sigma-bond metathesis for the dehydrocoupling of phenylsilane [11,12].

2.2.2. Redistributive coupling of bis(silyl)phenylenes to hyperbranched polymers

As described in detail in the above section, the desilanative coupling and desilanative-dehydrocoupling of **4** possessing a Si-C-Si connection (*i.e.*, 1,1-bissilyl connection) in the molecule, catalyzed by Red-Al and Cp_2MCl_2/Red-Al (M = Ti, Zr, Hf), produce oligomers and polymers (**eqs. 6.13 & 6.16**), respectively [53,56-58]. In comparison, redistributive coupling of $1,4\text{-}C_6H_4(SiH_3)_2$ **12** using 2 mol% Red-Al gave 33% of soluble polysilane (M_w: 2800; M_w/M_n = 1.2) and hyperbranched polysilane (TGA ceramic residue yield: 63%) which is insoluble in organic solvents because of the high extent of crosslinking [59]. Similarly, redistributive coupling of **12** in the presence of 2 mol% inorganic hydride (N-Selectride, Super-Hydride) afforded soluble polysilane in moderate yield (30%, M_w = 3100 for N-Selectride; 27%, M_w = 3500 for Super-Hydride) and hyperbranched polysilane (TGA ceramic residue yield = 83% for N-Selectride; 67% for Super-Hydride) as shown in **eq. 6.18**.

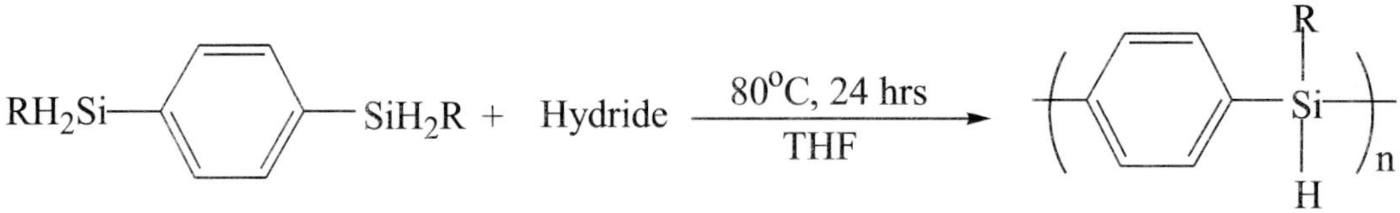

$$RH_2Si\text{—}\langle C_6H_4 \rangle\text{—}SiH_2R + \text{Hydride} \xrightarrow[THF]{80°C, 24\ hrs} \left(\langle C_6H_4 \rangle\text{—}\underset{H}{\overset{R}{Si}} \right)_n$$

Hydride = Red-Al, N-Selectride, Super-Hydride

R = H, **12**; Me, **13**; CH=CH$_2$, **14**

(6.18)

The redistributive coupling of $1,4\text{-}C_6H_4(SiH_2Me)_2$ **13** using 2 mol% Red-Al gave 70% of soluble polysilane (M_w: 1500; M_w/M_n = 1.5) and hyperbranched polysilane (TGA ceramic residue yield: 70%) [59]. Similarly, redistributive coupling of **13** under the influence of 2 mol% inorganic hydride (N-Selectride, Super-Hydride) afforded soluble polysilane in moderate yield (58%, M_w = 2100 for N-Selectride; 40%, M_w = 2500 for Super-Hydride) and hyperbranched polysilane (TGA ceramic residue yield: 75% for N-Selectride; 80% for Super-Hydride). The redistributive coupling of

1,4- $C_6H_4(SiH_2CH=CH_2)_2$ **14** with 2 mol% Red-Al yielded 70% of soluble polysilane (M_w: 1500, M_w/M_n = 1.5) and hyperbranched polysilane (TGA ceramic residue yield: 63%) [59]. Similarly, redistributive coupling of **14** in the presence of 2 mol% inorganic hydride (N-Selectride, Super-Hydride) afforded soluble polysilane in moderate yield (58%, M_w = 2100 for N-Selectride; 40%, M_w = 2500 for Super-Hydride) and hyperbranched polysilane (TGA ceramic residue yield: 75% for N-Selectride; 80% for Super-Hydride). Thus, steric hindrance on the silicon center increases the portion of soluble polymer as expected because it will retard the extensive redistributive coupling to form hyperbranched polysilane.

The dehydrocoupling of **12** using 1 mol% by Cp_2MCl_2/Red-Al (M = Ti, Zr, Hf) catalyst produces soluble polysilane (in 25%, M_w = 2600 g/mol, M_w/M_n = 1.1 for Ti; in 15%, M_w = 3300 g/mol, M_w/M_n = 1.2 for Zr; in 15%, M_w = 3600 g/mol, M_w/M_n = 1.1 for Hf) and hyperbranched polysilane (TGA ceramic residue yield: *ca.* 60% for all Ti, Zr, Hf) which is insoluble in organic solvents because of the high extent of crosslinking [60]. The dehydrocoupling of **12** with 1 mol% $M'(CO)_6$ (M' = Cr, Mo, W) catalyst also yields soluble polysilane (in 65%, M_w = 4100 g/mol, M_w/M_n = 2.1 for Cr; in 53%, M_w = 5100 g/mol, M_w/M_n = 2.4 for Mo; in 40%, M_w = 5900 g/mol, M_w/M_n = 2.0 for W) and hyperbranched polysilane (TGA ceramic residue yield: ca. 70% for Cr, Mo, W) [60]. The dehydrocoupling of **13** in the presence of 1 mol% Cp_2MCl_2/Red-Al (M = Ti, Zr, Hf) catalyst gives soluble polysilane (in 95%, M_w = 1300 g/mol, M_w/M_n = 1.4 for Ti; in 90%, M_w = 1400 g/mol, M_w/M_n = 1.1 for Zr; in 93%, M_w = 2400 g/mol, M_w/M_n = 1.1 for Hf) [60]. The dehydrocoupling of **13** with of 1 mol% $M'(CO)_6$ (M' = Cr, Mo, W) catalyst affords soluble polysilane (in 75%, M_w = 2500 g/mol, M_w/M_n = 1.3 for Cr; in 80%, M_w = 2700 g/mol, M_w/M_n = 1.1 for Mo; in 50%, M_w = 2720 g/mol, M_w/M_n = 1.1 for W) **(eq. 6.19)** [60].

$$RH_2Si\text{—}C_6H_4\text{—}SiH_2R \xrightarrow[\substack{Cp_2MCl_2/Red\text{-}Al \\ (M = Ti, Zr, Hf) \\ or\ M'(CO)_6 \\ (M' = Cr, Mo, W)}]{60^{\circ}C,\ 24\ hrs} \left(\text{—}C_6H_4\text{—}\underset{\overset{|}{R}}{\overset{\overset{R}{|}}{Si}}\text{—}\underset{\overset{|}{R}}{\overset{\overset{R}{|}}{Si}}\text{—}\right)_n$$

R = H, **12**; Me, **13**

$$(6.19)$$

In the similar manner, other disilanes such as bis(1-sila-3-butyl)benzene [57,58], 2,5-disilaoct-7-ene,2,5-disilahexane [61] undergo dehydrocoupling using group 4 metallocene combination catalysts such as Cp_2MCl_2/Red-Al (M = Ti, Zr, Hf) and $Cp_2ZrCl_2/2n$-BuLi. Thus, types of substituent on silane and catalyst can make a pronounced difference in the coupling processes.

2.2.3. Demethanative Coupling of Tertiary Germanes to Polygermanes

As described in sections 2-1-5 and 2-2-1, polygermanes can be prepared: (1) by the Wurtz coupling of chlorogermanes with alkali metals in toluene-refluxing temperature, (2) by the dehydrocoupling of hydrogermanes, and (3) by redistributive coupling of $(PhH_2Ge)_2$ under the influence of a Lewis base PhH_2GeLi [51]. Berry and co-workers reported an extraordinary demethanative coupling of tertiary germanes to polygermanes (M_w = 5,000 ~ 10,000 g/mol by light scattering measurements), catalyzed by $Ru(PMe_3)_4X_2$ (M = $GeMe_3$, Me) as shown in **eq. 6.20** [62].

$$\text{HGeMe}_2\text{Ar} \quad \xrightarrow[\substack{25^{\circ}\text{C, neat}}]{\substack{\text{Ru(PMe}_3)_4\text{X}_2 \\ (\text{X = Me, GeMe}_3)}} \quad \text{CH}_4 \;+\; \text{H}\!\left(\!\!\begin{array}{c}\text{Me}\\|\\\text{Ge}\\|\\\text{Ar}\end{array}\!\!\right)_{\!\!n}\!\!\text{CH}_3 \qquad (6.20)$$

The hydrogermanes should have at least one methyl group in the molecule for demethanative coupling: HGeMe_2Ar (Ar = methyl, phenyl, *p*-tolyl, *p*-fluoro, *p*-trifluorotolyl, *p*-anisyl, *m*-xylyl). Comparison of the properties of polygermanes prepared by catalytic demethanative coupling and Wurtz coupling of MeArGeCl_2 with sodium shows no significant differences. Furthermore, in principle, the demethanative coupling (i.e., removal of CH_4 gas) is similar to dehydrocoupling (i.e., removal of H_2 gas). According to the sigma-bond metathesis mechanism through four-center transition states the dehydrocoupling of a tertiary germane should produce a digermane if the tertiary germane is not sterically demanding. If the tertiary germane is sterically demanding, no dehydrocoupling should take place [11,12]. The authors suggested a mechanism for the demethanative coupling where Ge-C bond cleavage and Ge-Ge bond formation are attained by sequential α-CH_3 and germyl to germylene migration steps *via* an intermediacy of metal-germylene species [62b].

2.2.4. Redistributive Coupling of Hydrostannanes to Highly Branched Polystannanes

As described in section 2-1-5, polystannanes, prepared by the Wurtz coupling of chlorostannanes with alkali metals and by the dehydrocoupling of hydrostannanes, exhibit intriguing physical properties. Sita *et al.* reported a peculiar polymerization of $n\text{-Bu}_2\text{SnH}_2$ with a $\text{RhH(CO)(PPh}_3)_3$ catalyst in neat or in toluene (**eqs. 6.21a & 6.21b**) [63].

$$n\text{-Bu}_2\text{SnH}_2 \begin{cases} \xrightarrow[\text{neat}]{\text{RhH(CO)(PPh}_3)_3} (n\text{-Bu}_2\text{Sn})_5 \;+\; (n\text{-Bu}_2\text{Sn})_6 & (6.21a) \\[2em] \xrightarrow[\text{in toluene}]{\text{RhH(CO)(PPh}_3)_3} (n\text{-Bu}_2\text{Sn})_{q\text{-x-y}}(n\text{-Bu}_3\text{Sn})_x(n\text{-BuSn})_y & (6.21b) \end{cases}$$

$$q > x, y$$

$$31\% \text{ isolated yield}$$

Normal dehydrocoupling reaction occurs to give only the cyclic oligomers (Sn_5 and Sn_6) when the reaction was carried out in neat hydrostannane with fast addition of the catalyst (**eq. 6.21a**). However, redistributive dehydrocoupling takes place to afford high molecular weight, moderately branched polystannane [n-Bu$_2$Sn]$_{q-x-y}$(n-Bu$_3$Sn)$_x$(n-BuSn)$_y$, q > x, y; M_w = 50240 g/mol, M_w/M_n = 1.43] in 31% yield without appreciable formation of cyclic oligomers when the reaction is carried out in toluene with slow addition of the catalyst (**eq. 6.21b**). The high molecular weight branched polystannane of Sita is different from the high molecular weight linear polystannane [H(n-Bu$_2$Sn)$_m$H; M_w = 33430 g/mol, M_w/M_n = 2.26] of Tilley that was produced by dehydropolymerization of n-Bu$_2$SnH$_2$ using Cp$_2$ZrMe$_2$ catalyst [43c,45], judged by ^{119}Sn NMR, UV-VIS, and GPC analytical techniques. The redistribution could occur *via* a well-known process with intermediacy of metal-stannylene species.

Woo and collaborators also described such unusual redistributive dehydrocoupling of n-Bu$_2$SnH$_2$ with Cp$_2$MCl$_2$/Red-Al (M = Ti, Zr, Hf) combination catalyst [64a]. The redistributive dehydrocoupling is performed at ambient temperature for 30 min, producing soluble cyclic oligomers (Sn$_5$ and Sn$_6$) and insoluble hyperbranched polystannane [n-Bu$_2$Sn]$_{p-x}$(n-BuSn)$_x$] as depicted in **eq. 6.22**.

$$n\text{-Bu}_2\text{SnH}_2 \quad \xrightarrow[\text{neat, 25}^\circ\text{C, 30 min}]{\begin{array}{c}\text{Cp}_2\text{MCl}_2/\text{Red-Al}\\(\text{M = Ti, Zr, Hf})\end{array}} \quad \begin{array}{l}(n\text{-Bu}_2\text{Sn})_5 + (n\text{-Bu}_2\text{Sn})_6 + \\[6pt] (n\text{-Bu}_2\text{Sn})_{q-x}(n\text{-BuSn})_x\end{array}$$

$$(6.22)$$

The insoluble products (TGA ceramic residue yield: ca. 33%) are obtained in moderate yield (for Ti, 39%; for Zr, 27%; for Hf, 26%) as a yellow solid. In the same manner the redistributive dehydrocoupling of Bu$_2$SnH$_2$ with M'(CO)$_6$/Red-Al (M' = Cr, Mo, W) at ambient temperature for 16 hours afford soluble oligostannanes and insoluble hyperbranched polystannanes. The insoluble product (TGA ceramic residue yield: ca. 33%) is obtained in moderate yield (for Cr, 40%; for Mo, 48%; for W, 39%) as a yellow solid. The same authors also describe the redistributive dehydrocoupling of Bu$_3$SnH with Cp$_2$MCl$_2$/Red-Al (M = Ti, Zr, Hf) and M'(CO)$_6$/Red-Al (M' = Cr, Mo, W) at 70°C (for group 4 catalyst) or 90°C (for group 6 catalyst) for 3 days to give soluble oligomer (mainly, Sn$_2$) and insoluble hyperbranched polystannane (TGA ceramic residue yield = ca. 33%; for Ti, 17%; for Zr, 14%; for Hf, 13%; for Cr, 23%; for Mo, 23%; for W, 7%) [64b].

2.3. Hydrosilapolymerization, Hydrogermapolymerization and Hydrostannapolymerization of Group 14 Hydride on Vinyl Monomers

The addition of Si-H bond to unsaturated bonds such as alkenes (including C = C, C = O, C = N, C = metal, etc.) and alkynes (including C $\equiv$ C, C $\equiv$ N, C $\equiv$ metal, etc.) is termed *hydrosilation* (or *hydrosilylation*) and is promoted by many homogeneous transition-metal complex catalysts and heterogeneous supported metal catalysts [65-78]. Catalytic hydrosilation is a complex process and the overall mechanism is not well understood. Chalk and Harrod proposed a simplified mechanism for transition metal complex-catalyzed hydrosilation involving coordination of alkenes to a coordinatively unsaturated transition metal hydride residue, followed by metal-hydride insertion, oxidative addition of the silane to the metal alkyl, then reductive elimination of the alkylsilane to regenerate the metal-hydride [68]. Nonetheless, some experimental observation (include an induction period for many precatalysts and the formation of vinylsilanes) cannot be explained well by the Chalk-Harrod mechanism [69]. An alternative mechanism to the Chalk-Harrod mechanism involves insertion of the alkene into the M-Si bond instead of insertion of the alkene into the M-H bond [70]. Alternative mechanisms were also proposed by several researchers on the basis of their studies with their own catalytic systems [71].

The common hydrosilation of olefins is the single addition of Si-H bond to vinyl derivatives in the presence of organic or inorganic catalysts by adding more excess hydrosilanes than olefins. Versatile silicon-containing polymers can be prepared by continuous multiple hydrosilation:

1. Hydrosilation of $CH_2 = CH(CH_2)_x SiR_2 H$ (x = 0~2; R = H, Me, OMe, Cl).
2. Hydrosilation of $[CH_2 = CH(CH_2)_x]_2 Y$ (x = 0~2; Y = CH_2, CMe_2, $SiMe_2$, $GcMe_2$, phenylene, etc.) with $R_2 SiH_2$ or $HSiMe_2\text{-}X\text{-}SiMe_2 H$ [X = $(SiMe_2\text{-}O)_n$, CH_2, CMe_2, $SiMe_2$, $GeMe_2$, phenylene, etc.].
3. Hydrosilation of olefin with $(RSiH)_n$.
4. Hydrosilapolymerization of vinyl monomer with hydrosilane or $(RSiH)_n$.

For example, a series of new 3,3,3-trifluoropropyl substituted copoly(carbosiloxane)s were prepared by step-growth hydrosilation copolymerization of 1,9-dihydrido-1,1,3,5,7,9,9-heptamethyl-3,5,7-tris(3′,3′,3′-trifluoropropyl)pentasiloxane with various α,ω-divinylsilanes and siloxanes using Karsted's catalyst, Pt-divinyltetramethyldisiloxane [79a]. Poly(arylene-1,2-dioxy-*co*-oligodimethylsiloxanylene)s were synthesized by dehydrosilation condensation copolymerization of o-quinones with α,ω-dihydridooligodimethylsiloxanes catalyzed by $H_2(CO)Ru(PPh_3)_3$ [79b]. Woo and co-workers prepared new preceramic polymers (for making SiC ceramics) such as a copolymer of polycarbosilane-dichloromethylvinylsilane and a copolymer of polycarbosilane-γ-methacryloxy-propyltrimethoxysilane by hydrosilation copolymerization under the influence of Pt-based catalyst [79c,d].

Green photopolymerization technology is widely used commercially in surface coatings, photoresists, adhesives, and holography [81]. Woo and co-workers reported a novel consecutive multiple addition (i.e., hydrosilapolymerization) of vinyl monomers to hydrosilanes thermally and photochemically to produce the first organic polymers with reactive hydrosilyl end groups by the addition of excess olefin with respect to hydrosilane [80a-m].

Only a few vinyl monomers, including methyl methacrylate (MMA) and styrene, absorb the most convenient wavelength of light (250-500 nm) for common experimental work of photopolymerization. The bulk photopolymerization of MMA with *para*-substituted phenylsilanes such as $p\text{-}F\text{-}C_6H_4SiH_3$ (**15**), $p\text{-}H_3C\text{-}C_6H_4SiH_3$ (**16**), and $p\text{-}H_3CO\text{-}C_6H_4SiH_3$ (**17**) produces poly(MMA)s containing the respective silyl moiety as an end group [81]. Poly(MMA)s possessing the $p\text{-}F\text{-}C_6H_4SiH_2$ moiety as an end group with weight average molecular weight (M_w) of 6200-22020 and TGA residue yields of 12-73% were prepared in 24-85% yields by 300 nm UV light-initiated bulk polymerization of MMA with different molar ratios of **15** (MMA : **15** = 9:1 through 3:7) as shown **eq. 6.23**.

$$\textbf{15} \ + \ n \ \diagup\!\!\diagdown \begin{smallmatrix} Me \\ \\ COOMe \end{smallmatrix} \xrightarrow{h\nu} \ F\text{-}C_6H_4SiH_2 \left(\diagup\!\!\diagup \!\!\underset{COOMe}{\overset{Me}{\diagdown}}\!\! \right)\!\! H \tag{6.23}$$

Similarly, poly(MMA)s possessing p-H$_3$C-C$_6$H$_4$SiH$_2$ end groups with weight average molecular weights (M_w) of 8130-28090 and TGA residue yields of 12-67% were prepared in 30-93% yields by the bulk polymerization of MMA with different molar ratios of **16**. For the hydrosilanes, the polymerization yields and the polymer molecular weights decreased, while the TGA residue yields and the relative intensities of Si-H IR stretching bands increased as the relative silane concentration compared to MMA increased. The highest polymerization yield and polymer molecular weight were obtained for MMA:hydrosilane = 9:1, but the highest TGA residue yield was obtained for MMA:hydrosilane = 3:7. The polymerization yields and polymer molecular weights of MMA with **15-17** increased in the order of **17** < **15** < **16**.

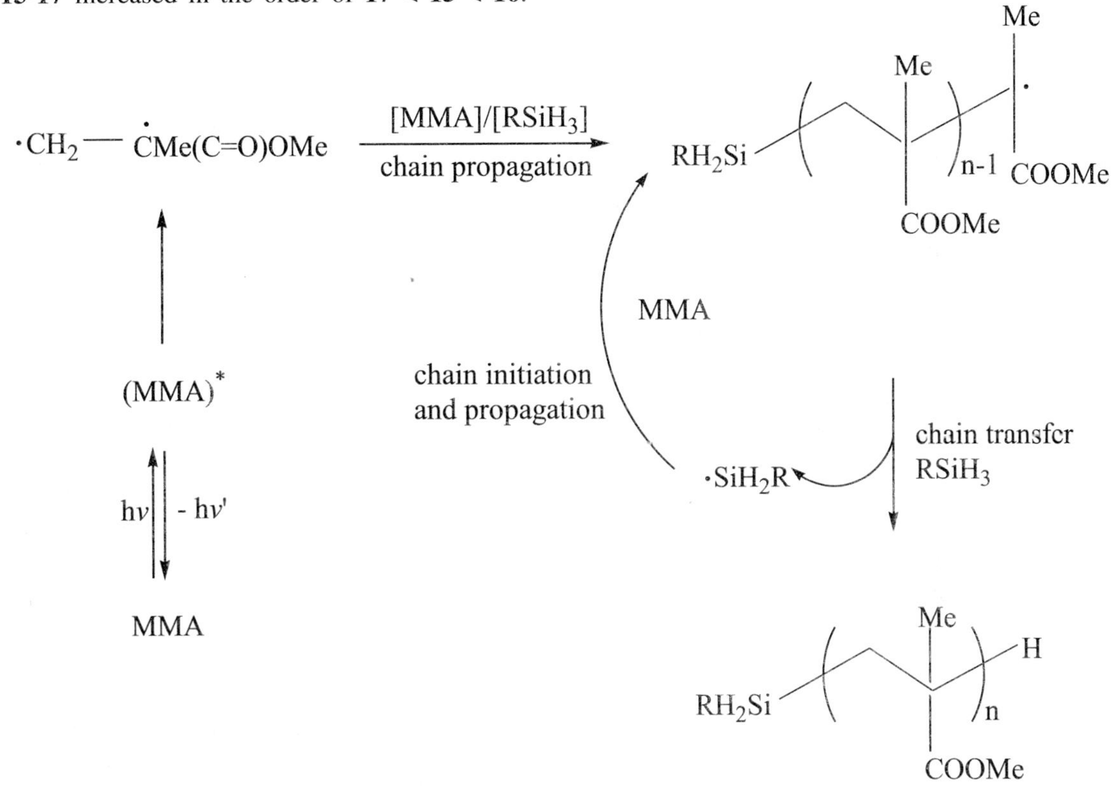

Fig. 6.3. Postulated mechanism for the hydrosilapolymerization of MMA on RSiH$_3$ under the influence of UV light.

The hydrosilapolymerization mechanism (**Fig. 6.3**) for the photopolymerization of MMA with **15-17** could be similar to those with other hydrosilanes. Light absorption of a MMA molecule is well known to produce an excited singlet state of MMA which may then either fluoresce with a return to the ground state of MMA or may be converted to a long-lived triplet excited state, the diradical of the MMA monomer [82]. Addition on another MMA by this diradical gives a new diradical of MMA dimer which either reverts to the ground state or continues attacking other MMAs to produce poly(MMA)s. Under neat conditions the latter will be a predominant process to produce poly(MMA) radicals. At high [MMA]/[silane] ratios, chain propagation will be able to compete with chain transfer over the poly(MMA) radicals. However, chain transfer will eventually rule over chain propagation with increasing silane concentration. Chain transfer could produce a silyl radical, which, in turn, leads to chain initiation, resulting in the production of poly(MMA) containing silyl end groups. Higher concentrations of hydrosilane over MMA will produce shorter chain lengths of poly(MMA), which contributes to the increase in thermal stability of poly(MMA).

The increase of Si-H contents with respect to poly(MMA) moiety will result in the higher possibility of high temperature cross-linking hydrosilation with C=O groups in poly(MMA), which will give higher thermal stability (*i.e.*, higher TGA ceramic residue yield). The reported hydrosilanes include $PhSiH_3$, $PhCH_2SiH_3$, $PhMeSiH_2$, Ph_2SiH_2, $PhSiH_2SiH_2Ph$, $PhCH(SiH_3)_2$, and $1,4\text{-}C_6H_4(SiH_{3-x}Me_x)_2$. The reported monomers include MMA, AA (acrylic acid), MA (methacrylic acid), HEMA (2-hydroxyethyl methacrylate), and styrene. The hydrosilapolymerization yield of styrene was lower than those of MMA, MA and HEMA. No appreciable thermal and photopolymerizations of the vinyl monomers with $CH_2(CH_2)_5SiH_3$, $Ph(CH_2)_2SiH_3$, Ph_3SiH, and $(mesityl)_2SiH_2$ were observed, probably because of their steric and/or electronic effects. Therefore, the types of vinyl monomer and hydrosilane appear to be important for successful hydrosilapolymerization.

Chatgilialoglu and co-workers reported the reactivity study of alkyl peroxy radicals toward poly(hydrosilane)s [83]. The Woo group extended the above hydrosilapolymerization methodology to poly(hydrophenylsilane)s instead of monomeric hydrosilanes for first making novel inorganic-organic hybrid copolymers. Such inorganic-organic hybrid graft copolymers can be used for many applications in contact lens, paints, etc. Poly(*p*-fluoro-substituted phenylsilane), $H[(p\text{-}F\text{-}C_6H_4)SiH]_nH$, ($M_w = 3300$; $M_w/M_n = 1.83$) was prepared by dehydropolymerization of $p\text{-}F\text{-}C_6H_4SiH_3$ using the Cp_2ZrCl_2/Red-Al combination catalyst. In a typical photopolymerization experiment, a quartz tube (15 mm × 120 mm) charged with MMA, poly(*p*-fluoro-substituted phenylsilane), and toluene (1 mL) was degassed, sealed, and irradiated with UV-light (280 nm) for 6 hours. Precipitation with *n*-hexane and drying in vacuo give poly(*p*-fluoro-substituted phenylsilane)-*g*-poly(MMA), which is an inorganic-organic hybrid graft copolymer **(eq. 6.24)** [84a].

$$\text{(6.24)}$$

In this photopolymerization, various molar ratios of polysilane:MMA are used: 1:1, 1:5, 1:10, 1:20, 1:50. Similarly, in the graft photopolymerization, while the polymer molecular weights and the polymerization yields decrease, the relative intensities of Si-H IR stretching bands and the TGA residue yields increase with the increase of hydrosilane molar ratio over MMA. The higher contents of Si-H moieties remaining in the graft copolymer backbone will result in the higher possibility of high temperature cross-linking hydrosilation with C=O groups in the graft chain of poly(MMA), which will give higher thermal stability (*i.e.*, higher TGA ceramic residue yield). For the other vinyl monomers such as MA and styrene, similar trends are observed [84b]. Similar trends are also observed for the copolymerization of MMA-MA, MMA-styrene, and MA-styrene [84c]. Furthermore, for various other poly(hydroarylsilane)s prepared by Cp_2MCl_2/Red-Al (M = Ti, Zr, Hf) combination catalyst, similar trends are found [84c]. The hydrosilapolymerization mechanism of vinyl monomers on poly(hydrosilanes) should be similar to the that of vinyl monomers on hydrosilanes (**Fig. 6.3**).

The addition of Ge-H bond and Sn-H bond to unsaturated bonds is termed hydrogermation (or hydrogermylation) and hydrostannation (or hydrostannylation), respectively, and is promoted by many homogeneous transition-metal complex catalysts and heterogeneous supported metal catalysts [85-88]. (Germyl)stannanes and digermanes may be used for germylstannation and double germation, respectively [87]. Catalytic hydrogermation and hydrostannation could be complex processes and the overall mechanisms are not well understood although plausible mechanisms could be similar to the Chalk and Harrod mechanism or modified Chalk-Harrod mechanism [68-70]. Hydrogermapolymerization (and hydrostannapolymerization) of hydrogermanes (and hydrostannanes) on MMA and MA gave similar trends as the hydrosilapolymerization of MMA and MA [89]. The reactivity increases in the order of Si-H < Ge-H < Sn-H which is the reverse order of bond strength if their steric bulkinesses are same. Nonetheless, the hydrogermapolymerization/ hydrostannapolymerization mechanisms of vinyl monomers on poly(hydrosilane)s should be similar to the that of vinyl monomers on hydrosilanes depicted in **Fig. 6.3**.

3. GROUP 13 ELEMENT-CONTAINING INORGANIC POLYMER: POLYBORAZINES

Boron nitride (BN) is one of technologically important non-oxide ceramics (e.g., aluminum nitride AlN, silicon nitride Si_3N_4, silicon carbide SiC) whose synthesis and properties have been intensively studied [90]. Boron nitride has different polymorphic forms: hexagonal (h-BN), cubic (c-BN), rhombohedral (r-BN), or würtzite (w-BN). In particular, the graphite-like anisotropic structure of h-BN provides unique electrooptical and thermomechanical properties such as low thermal expansion coefficient, great resistance toward thermal shocks/oxidation, and high tensile strength/Young's modulus [91]. Thus h-BN has been used for many specialty applications [92]. These advanced properties are heavily dependent on their preparation method, purity, crystallinity, and microstructure. BN ceramic powders were traditionally prepared by co-thermolysis of cheap boric acid and urea. Most of commercially available BN ceramic goods are prepared by hot-pressing technique of BN powder or CVD process. For the purpose of reinforced composite preparation BN ceramic fiber is needed. The BN ceramic fiber can be prepared by the polymer-derived ceramic route [91]. Polyborazines are frequently used as preceramic polymer precursors for the preparation of boron nitride fiber or composites [93].

Polyborazines can be prepared from the homodehydrocoupling of borazine derivatives or hetrodehydrocoupling of borazine derivatives with ammonia or coupling of aminoboranes with ammonia [91,92,94]. Simple borazine ring, a cyclic trimer $[BH = NH]_3$, has a formula of $B_3N_3H_6$ whose structure is similar to the delocalized benzene. Borazine has polarity compared to benzene (which is composed of carbon atoms with same electronegativity) because nitrogen (boron) is more (less) electronegative than carbon [95]. Thus, borazine should be more reactive than benzene toward dehydropolymerization. Polyborazine $(B_3N_3H_x)_n$ can be made by intermolecular dehydro-coupling between N-H and B-H bonds in borazines (**eq. 6.25**).

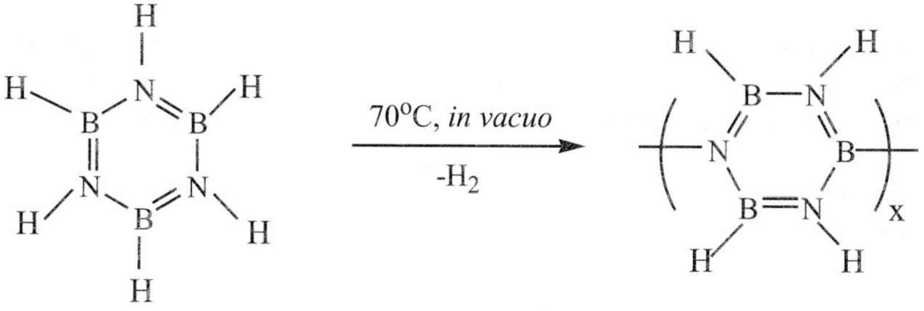

$$\text{(6.25)}$$

R = H or Borazinyl

Other polymeric precursors include polyborazines with (a) B-B linkages or (b) B-N linkages and polyborazines with (c) NR bridge or (d) BR bridge [2a,96].

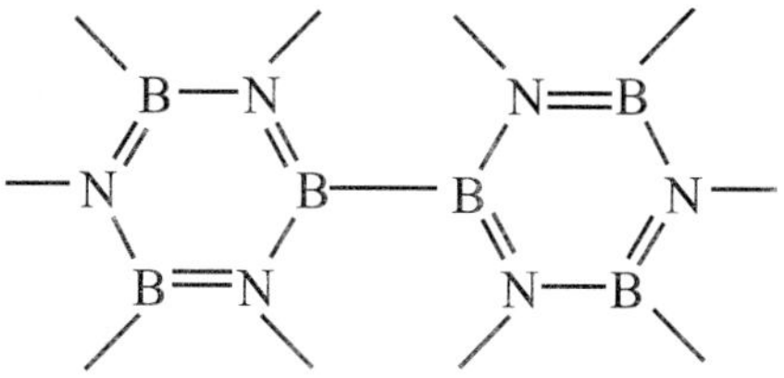

(a)

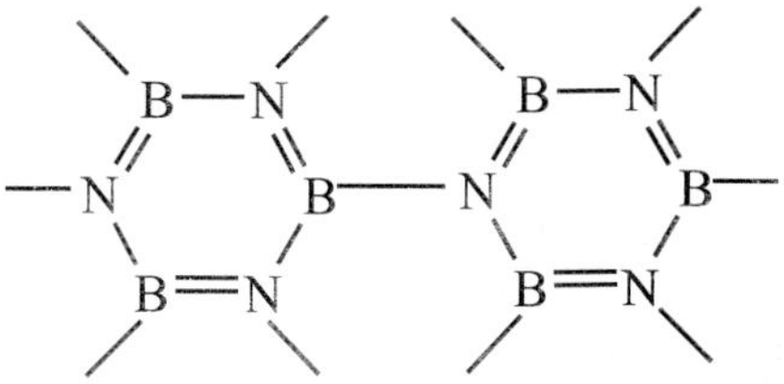

(b)

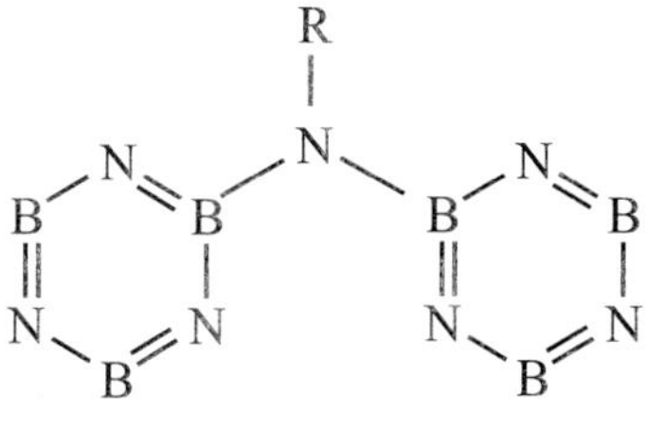

(c)

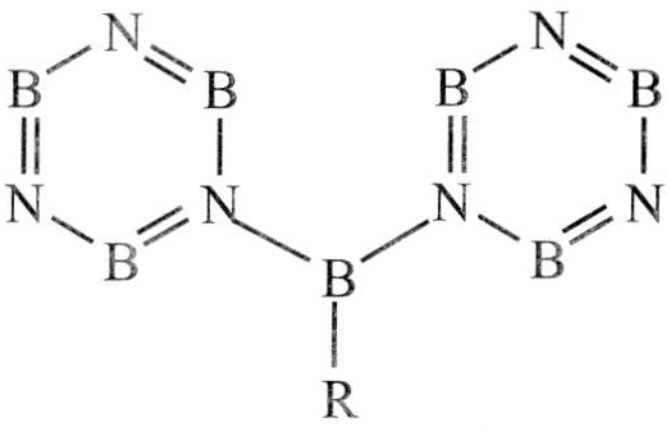

(d)

Borazine rings can also be pendant groups of polymers such as poly(vinylborazine) and borazine-substituted polysilazane [96]. Woo and co-workers recently reported that polyborazine can be used as an effective cross-linking agent for polysilane by Si-N coupling between Si-H (of polysilane) and N-H bonds (of polyborazine) [41a]. When the adding amount of PBN increases, the B, N-incorporated SiC ceramic composites are obtained. Polyhydrosilanes can also be used as a cross linking agent for PBN. When the using amount of polyhydrosilane increases, the SiC-incorporated BN ceramic composites are formed [41b]. Therefore, this methodology is a good way to make ceramic composites in high yield. Polyboranes and polycarboranes are well reviewed in the literatures [95] and thus are not included here.

4. GROUP 15 ELEMENT-CONTAINING INORGANIC POLYMER: POLYPHOSPHAZENES

Polyphosphazenes with [$-N = PR_2-$] repeating units attract many interests in material science [97]. Polyphosphazenes can be prepared by (1) the thermal ring-opening polymerization of cyclic trimer [$N = PR_2]_3$ over 200°C, (2) the substitution reactions with diverse nucleophiles on poly(dichlorophosphazene) [$N = PCl_2]_n$, and (3) the condensation reactions of phosphazene derivatives with good leaving groups (($(CH_3)_3Si$- and $-Cl$ or $-OCH_2CF_3$) such as ($(CH_3)_3Si$-$N = PR_2(OCH_2CF_3)$) and $Cl_3P = NSi(CH_3)_3$ [1b,102,103]. Oligomeric [$N = PCl_2]_x$ (x = 3, 4,....) with various ring sizes are classically prepared from the reaction of phosphorus pentachloride with ammonium chloride. The main product is white crystalline hexachlorocyclotriphosphazene [$N = PCl_2]_3$, which melts at 114°C and is a useful precursor for ring-opening polymerization to polyphosphazene unlike borazine, which is known to undergo the B-N dehydrocoupling to polyborazine. Hexafluorocyclotriphosphazene or partially substituted cyclotriphosphazenes can also be used instead of hexachlorocyclotriphosphazene as a precursor for ring-opening polymerization. The ring-opening polymerization mechanism of hexachlorocyclotriphosphazene is not fully clear. Experimental evidence highly suggests that an ionic mechanism is more possible than a radical mechanism. The catalytic species could be (pentachlorocyclotriphosphazene)$^+$Cl$^-$ formed from the ionization of a P-Cl bond of hexachlorocyclotriphosphazene. The phosphonitrilic cation then could keep attacking the nitrogen atoms of other hexachlorocyclotriphosphazene monomers to produce polyphosphazene [1b].

Polyphosphazenes exhibit low T_g values and elasticity because of large bond angles and substitution-free at nitrogen leading to high degree of chain movement, which is similar to the case of polysiloxanes. The property scope of polyphosphazene greatly counts on the type of side groups (R) from hydrophilic, hydrophobic, electroredox, coordinative to elastomeric character. Thus, polyphosphazene can be used as electrode mediator, polymeric drugs, hydrogels, liquid crystalline materials, non-burning fibers, biomaterials, low-temperature elastomers, semiconductors, composites, etc. Allcock and co-worker reported the synthesis and applications of oligoalkoxy-substituted polyphosphazenes, [$N = P(OC_6H_5)(OCH_2CH_2)_xOCH_3]_n$ and {$N = P[OC_6H_4(OCH_2CH_2)_xOCH_3]_2\}_n$, as solid polymer electrolytes to be used for rechargeable battery [98]. The phenoxy group improves mechanical strength and the oligoethyleneoxy group promotes lithium ion transport. Qin *et al.* synthesized polyphosphazenes with the indole-based chromophore (nitro-indole or sulfonyl indole) and charge-transporting carbazoyl side groups. These polyphosphazenes possess the second-order nonlinear optical property [99]. Sohn and collaborators described that polyphosphazene-platinum conjugates [$N = P(MPEG)_x(GlyGluPt(dach))_{2-x}]_n$ (MPEG = methoxy-poly(ethylene glycol); GlyGlu = glycyl-L-glutamic dipeptide; dach = trans-(±)-1,2-diaminocyclohexane) have selective targeting antitumor activity by attaining favorable pharmacokinetic profile and a longer blood circulation time [100]. Polyphosphazenes with coordinating ligands such as nitrile, pyridine or phosphine can be used as a supporting material for a transition metal complex. For example, phosphazene copolymers{[$N = P(O_2C_{12}H_8)]_{0.5}[N = P(O-C_6H_4-CO_2Pr^n)(O-C_6H_4-L)]_{0.5}\}_n$ (L = CN, PPh$_2$) and {[$N = P(O_2C_{12}H_8)]_{0.6}[N = P(O-C_6H_4-CO_2Pr^n)(O-C_5H_4N)]_{0.4}\}_n$ react with W(HOMe)(CO)$_5$ to give W(CO)$_5$-coordinated complexex [101a]. Similarly the phosphazene copolymers coordinated to CpMn(CO)$_2$ moiety have been reported [101b]. These metal complex-coordinated phosphazene copolymers show high glass transitions and are quite stable. Polyphosphazenes with sulfonamide side groups showing good proton conductivity have been developed for fuel cell application. Thus, $[N = P(OC_6H_4CH_3)(OC_6H_4-SO_2-NNa-SO_2CF_3)_{0.34n}[N = P(OC_6H_4CH_3)_2]_{1.66n}$ can be employed for application in fuel cells [102]. Amphiphilic biodegradable polyphosphazene block copolymers forming spherical micellar aggregates with 100~120 nm diameters have been reported [103]. The amphiphilic polyphosphazenes such as poly[bis(ethylglycinat-*N*-yl)phosphazene]-*b*-poly(ethylene oxide) and poly[bis(trifluoroethoxy)phosphazene]-*b*-poly(ethylene oxide) are prepared by the cationic polymerization of $Cl_3P = NSi(CH_3)_3$ at room temperature. Therefore, in principle, any interesting functional group (e.g., proteins, poly nucleic acids, antibodies, organometallic complex, nanometallic materials, carbon nanotubes, etc.) can be introduced to polyphosphazene backbone by using the synthetic methods described in this section.

5. APPLICATIONS

Some selective inorganic polymers are reviewed as advanced materials in this chapter: i.e., polysilanes, polygermanes, polystannanes, polyborazines, and polyphosphazenes. Polysilanes are used for a great deal of applications in ceramics/composites, photoelectronics, photoresistors, nonlinear optics, and more. The peculiar optoelectronic properties of polysilanes are due to sigma-conjugation of the silicon atoms in the polymer chain, varying with the molecular weight, conformation, and substituents of the polymer. Polygermanes and polystannanes possess similar properties of polysilanes. Polyborazines are mainly used as preceramic polymers for preparing boron nitride ceramics-composites. Polyphosphazenes are used as electrode mediator, polymeric drugs, hydrogels, liquid crystalline materials, nonburning fibers, biomaterials, low-temperature elastomers, semiconductors, composites, etc.

6. CONCLUSIONS AND FUTURE PROSPECTS

Main group metalloid-containing inorganic polymers, which are made up of boron (group 13), silicon/germanium/tin (group 14) and phosphorous (group 15) atoms, are reviewed as advanced materials. Especially, silicon-containing polymers with quite unusual properties are heavily described in this article as specialty materials. Group 14 hydrides possessing reactive E-H (E = Si, Ge, Sn) bonds can be used in synthesizing new functional materials with interesting physical properties under the influence of various promoters such as organometallic complexes, inorganic hydrides, heat, and UV-irradiation. Catalytic coupling of hydrosilanes with molecules having active E'-H bonds (E' = Si, O, N) or vinylic bonds thru Si-Si/Si-O/Si-N dehydrocoupling, redistributive coupling, and multiple addition processes are described in this article as selective examples of our recent research developments. Polyborazines and polyphophazenes are also described as specialty materials. However, due to the limits of space and interests, other types of inorganic polymers such as main group metal-containing polymers and transition metal-containing inorganic polymers are not described in this article. For the same reason, some main group metalloid-containing inorganic polymers (e.g., polysiloxane, polysilazane, polysilathianes, polyboranes, polycarbosilanes, polysulfur, polysulfur nitride, etc.) are not described here either.

ACKNOWLEDGEMENTS

We thank our co-workers (graduate students, technical supporting staffs, postdoctoral associates, research professors, visiting scientists and collaborating researchers), who appear as coauthors in the references cited. H.G. Woo is indebted to the Korea Science and Engineering Foundation (Grant Number R01-2001-000-00053-0; 2005) for its generous support of his research reported herein. B.H. Kim is grateful to the Korea Research Foundation (Grant No. KRF-2004-075-C00007; 2005).

REFERENCES

1. (a) G. Odian; Principles of Polymerization, 3rd ed., John Wiley & Sons, New York (1991). (b) H.R. Allcock and F.W. Lampe; Contemporary Polymer Chemistry, 2nd ed., Chapter 9, Prentice-Hall Inc., New York (1990).

2. (a) Inorganic and Organometallic Polymers II: ACS Symposium Series 572 (Ed. P. Wisian-Neilson, H.R. Allcock and K.J. Wynne), American Chemical Society, Washington, D.C. (1994). (b) Design, Activation, and Transformation of Organometallics into Common and Exotic Materials; NATO ASI Series E, no.141 (Ed. R.M. Laine), Martinus Nijhof, Amsterdam, (1988). (c) D.P. Gates; Annu.Rep.Prog.Chem., Sect.A, 99, 453-466 (2003).

3. (a) F. Ciardelli, E. Tsuchida and D. Wohrle; Macromolecule-Metal Complexes, Springer-Verlag, Berlin (1996). (b) C.U. Pittman, Jr., C.E. Carraher, Jr., M. Zeldin, J. Sheats and B.M. Culbertson; Metal-Containing Polymeric Materials, Plenum Press, New York (1996).

4. (a) Silicon in Polymer Synthesis (Ed. H.R. Kricheldorf), Springer-Verlag, Berlin (1996). (b) Silicon-containing Polymers (Ed. R.G. Jones), The Royal Society of Chemistry, Cambridge (1995).

5. (a) M. Regitz and P. Binger; Angew.Chem.Int.Ed., 27, 1484 (1988). (b) R. West; Angew.Chem.Int.Ed., 26, 1201 (1987).

6. K. Sakamoto, M. Yoshida and H. Sakurai; Macromolecules, 23, 4494 (1990).

7. (a) R. Walsh; *Acc.Chem.Res.*, 14, 246 (1981). (b) M.J.S.Dewar; Organomotallics, 5, 375 (1986). (c) R.A Jackson; J. Organomet.Chem., 166, 17 (1979).

8. R. West; J. Organomet.Chem., 300, 327 (1986).

9. Silicon-Based Polymer Science, Advances in Chemistry Series 224 (Ed. J.M. Ziegler and F.W.G. Fearon), American Chemical Society, Washington, D.C. (1990).

10. (a) R.D. Miller and J. Michl; Chem.Rev., 89, 1359 (1989) and references cited therein. (b) K.Matyjaszewski, Y.L. Chen and H.K. Kim; in Inorganic and Organometallic Polymers (Ed. M. Zeldin, K.J. Wynne and H.R. Allcock) ACS Symposium Series 360, p. 78, American Chemical Society, Washington D.C. (1988).

11. F. Gauvin, J.F. Harrod and H.-G. Woo; Adv.Organomet.Chem., 42, 363 (1998) and references cited therein.

12. (a) T.D. Tilley; Acc.Chem.Res., 26, 22 (1993) and references cited therein. (b) T.D.Tilley; Comments Inorg.Chem., 10, 37 (1990).

13. (a) J.Y. Corey; Adv.Silicon Chem., 1, 327 (1991) and references cited therein. (b) J.Y. Corey; Adv.Organomet.Chem., 51, 1 (2004) and references cited therein.

14. J.P. Banovetz, R.M. Stein and R.M. Waymouth; Organometallics, 10, 3430 (1991).

15. N. Choi, S.-Y. Onozawa, T. Sakakura and M. Tanaka; Organometallics, 16, 2765 (1997).

16. V.K. Dioumaev and J.F. Harrod; Organometallics, 13, 1548 (1994).

17. (a) H.-G. Woo, S.-Y. Kim, M.K. Han, E.J. Cho and I.N. Jung; Organometallics, 14, 2415 (1995). (b) H.G. Woo and S.-J. Song; Chem.Lett., 457 (1999). (c) H.-G. Woo and B.-H. Kim, manuscript in preparation. (d) R. Resel, G. Leising, F. Lunzer and Ch. Marschner; Polymer, 39, 5257 (1998). (e) Woo, H.-G.; Kim, B.-H., manuscript in preparation.

18. H. Li, F. Gauvin and J.F. Harrod; Organometallics, 12, 575 (1993).

19. V.K. Dioumaev and J.F. Harrod; Organometallics, 15, 3859 (1996).

20. W.H. Campbell, T.K. Hilty and L. Yurga; Organometallics, 8, 2615 (1989).

21. (a) C. Aitken, J.F. Harrod and U.S. Gill; Can.J. Chem., 65, 1804 (1987). (b) J.F. Harrod and S.S. Yun; Organometallics, 6, 1381 (1987). (c) C. Aitken, J.-P. Barry, F. Gauvin, J.F. Harrod, A. Malek and D. Rousseau; Organometallics, 8, 1732 (1989). (d) J.F. Harrod, T. Ziegler and V. Tschinke; Organometallics, 9, 897 (1990). (e) H.-G. Woo, J.F. Harrod, J. Henique and E. Samuel; Organometallics, 12, 2883 (1993). (f) J. Britten, Y. Mu, J.F. Harrod, J. Polowin, M.C. Baird and E. Samuel; Organometallics, 12, 2672 (1993).

22. (a) H.-G. Woo and T.D. Tilley; J. Am.Chem.Soc., 111, 3757 (1989). (b) H.-G. Woo and T.D. Tilley; J. Am.Chem.Soc., 111, 8043 (1989). (c) H.-G. Woo, J.F. Walzer and T.D. Tilley; Macromolecules, 24, 6863 (1991). (d) H.-G. Woo, R.H. Heyn, and T.D. Tilley; J. Am.Chem.Soc., 114, 5698 (1992). (e) H.-G. Woo, J.F. Walzer and T.D. Tilley; J. Am.Chem.Soc., 114, 7047 (1992). (f) J.P. Banovetz, H. Suzuki and R.M. Waymouth; Organometallics, 12, 4700 (1993). (g) T. Imori, H.-G. Woo, J.F. Walzer and T.D. Tilley; Chem.Mater., 5, 1487 (1993). (h) J.Y. Corey, J.L. Huhmann and X.-H. Zhu, Organometallics, 12, 1121 (1993).

23. (a) H.-G. Woo, S.-Y. Kim, W.-G. Kim, E.J. Cho, S.H. Yeon and I.N. Jung; Bull.Korean Chem.Soc., 16, 1109 (1995). (b) H.-G. Woo, S.-J. Song, M.-K. Han, E.J. Cho and I.N. Jung; Bull.Korean Chem.Soc., 16, 1242 (1995). (c) H.-G. Woo and S.-J. Song; Bull.Korean Chem.Soc., 17, 494 (1996). (d) H.-G. Woo and S.-J. Song; Bull.Korean Chem.Soc., 17, 1040 (1996). (e) H.-G. Woo, B.-H. Kim, S.-J. Song, M.-K. Han, S.-Y. Kim, J.-H. Kim and J.S.Lee; Bull.Korean Chem.Soc., 21, 935 (2000).

24. R. West; in Comprehensive Organometallic Chemistry II (Ed. A.G. Davis), pp. 77-110, Pergamon Press, Oxford (1995).

25. J. Luo, Z. Xie, J.W.Y. Lam, L. Cheng, H. Chen, C. Qiu, H.S. Kwok, X. Zhan, Y. Liu, D. Zhu and B.Z. Tang; J. Chem.Soc., Chem.Commun., 1740 (2001).

26. (a) H. Sohn, H.-G. Woo and D.R. Powell; J. Chem.Soc., Chem.Commun., 697 (2000). (b) H.-G. Woo, B.-H. Kim and H. Sohn; Chem.Lett., 544 (2000).

27. K. Tamao, M. Uchida, T. Izumizawa, K. Furukawa and S. Yamaguchi; J. Am.Chem.Soc., 118, 11974 (1996).

28. (a) H. Sohn, R.R. Huddleston, D.R. Powell and R. West; J. Am.Chem.Soc., 121, 2935 (1999). (b) K. Tamao and S. Yamaguchi; Pure Appl.Chem., 68, 139 (1996). (c) S. Yamaguchi and K. Tamao; J. Chem.Soc., Dalton Trans., 3693 (1998). (d) S. Yamaguchi, R.-Z. Jin and K. Tamao; J. Am.Chem.Soc., 121, 2937 (1999). (e) K. Kanno, M. Ichinohe, C. Kabuto and M. Kira; Chem.Lett., 99 (1998).

29. (a) T. Sanji, T. Sakai, C. Kabuto and H. Sakurai; J. Am.Chem.Soc., 120, 4552 (1998). (b) H. Sohn, M.J. Sailor, D. Magde and W.C. Trogler; J. Am.Chem.Soc., 125, 3821 (2003). (c) H.-G. Woo, B.-H. Kim and S.-J. Song; manuscript in preparation.

30. (a) B.P.S. Chauhan, T. Shimizu and M. Tanaka; Chem.Lett., 785 (1997). (b) W.C. Trogler, H. Sohn, S. Liu and S. Toal; Abstracts of Papers (225th ACS National Meeting, New Orleans, March 23-27, 2003), INOR-746 (2003).

31. B.-H. Kim, M.-S. Cho, J.-I. Kong and H.-G. Woo, manuscript in preparation.

32. B.-H. Kim and H.-G. Woo; Organometallics, 21, 2796 (2002). (b) H.-G. Woo, S.-J. Song, B.-H. Kim and S.S. Yun; Mol.Cryst.Liq.Cryst., 349, 87 (2000). (c) B.-H. Kim, M.-S. Cho, J.-I. Kong, H.-G. Woo, S.-W. Lee and C.-J. Lee; Mol.Cryst.Liq.Cryst., 425, 243 (2004).

33. (a) J.-H. Hong and P. Boudjouk; J. Am.Chem.Soc., 115, 5883 (1993). (b) J.-H. Hong, P. Boudjouk and S. Castellino; Organometallics, 13, 3387 (1994). (c) R. West, H. Sohn, U. Bankwitz, J. Calabrese, Y. Apeloig and T. Mueller; J. Am.Chem.Soc., 117, 11608 (1995).

34. (a) E. Lukevics and M. Dzintara; J. Organomet.Chem., 295, 265 (1985). (b) C.G. Pitt; Homoatomic Rings, Chains, and Macromolecules of Main-Group Elements (Ed. A.L. Rheingold), p. 203, Elsevier Scientific Publishing Company, Amsterdam (1977).

35. (a) T.C. Bedard and J.Y. Corey; J. Organomet.Chem., 428, 315 (1992). (b) Y. Li and Y. Kawakami; Macromolecules, 32, 6871 (1999). (c) R. Zhang, J.E. Mark and A.R. Pinhas; Macromolecules, 33, 3508 (2000). (d) J.B. Baruah, K. Osakada and T. Yamamoto; Organometallics, 15, 456 (1996).

36. (a) L.H. Sommer and J.E. Lyons; J. Organomet.Chem., 89, 1521 (1967). (b) R.J.P. Corriu and J.J. Moreau; J. Chem.Soc., Chem.Commun., 38 (1973). (c) I. Ojima, T. Kogure, M. Nihonyanagi, H. Kono, S. Inaba and Y. Nagai; Chem.Lett., 501 (1973).

37. (a) B.-H. Kim, S.-H. Park, M.-S. Kim and H.-G. Woo, manuscript in preparation. (b) B.-H. Kim and H.-G. Woo, manuscript in preparation. (c) J.R. Koe, M.Motonaga, M. Fujiki and R. West; Macromolecules, 34, 706 (2001).

38. B.-H. Kim, M.-S. Cho, M.-A. Kim and H.-G. Woo; J. Organomet.Chem., 685, 93 (2003).

39. (a) J.X. Wang, A.K. Dash, J.C. Berthet, M. Ephritikhine and M.S. Eisen; J. Organomet.Chem., 610, 49 (2000). (b) L.G. Sneddon, S. Kai, P. Fazen, A.T. Lynch, E.E. Remsen and J.S. Beck; in Inorganic and Organometallic Polymers and Oligomers (Ed. J.F.Harrod and R.M.Laine), p. 199, Kluwer Academic Publishers, Dordrecht, The Netherlands (1991). (c) G. Roewer, U. Herzog, K. Trommer, E. Müller and S. Frühauf; Structure and Bonding, 101, 59 (2002). (d) R.M. Laine and F. Babonneau; Chem.Mater., 5, 260 (1993). (e) S. Yajima, Y. Hasegawa, K. Okamura and T. Matsuzawa; Nature, 273, 525 (1978). (f) D.H. Riu, S.J. Kim, D.G. Shin, H.R. Kim and Y.H. Kim; J. Am.Ceram.Soc., 112, 5432 (2004). (g) H.-G. Woo; Research Report, Massachusetts Institute of Technology, M.A., U.S.A. (1991). (h) D. Seyferth, M. Tasi and H.-G. Woo; Chem.Mater., 7, 236 (1995).

40. (a) M.-S. Cho, B.-H. Kim, J.-I. Kong and H.-G. Woo; J. Organomet.Chem., 685, 99 (2003). (b) S.-Y. Yang, J.-M. Park, H.-G. Woo, W.-G. Kim, I.-S. Kim, D.-P. Kim and T.-S. Hwang; Bull.Korean Chem.Soc., 18, 1264 (1997). (c) H.-G. Woo, S.-Y. Yang, T.-S. Hwang and D.-P. Kim; Bull.Korean Chem.Soc., 19, 1310 (1998).

41. (a) L.-Y. Hong, F. Cao, D.-J. Kim, H.-G. Woo, B.-H. Kim, M.-S. Cho, X.-D. Li and D.-P. Kim; J. Organomet.Chem., 687, 27 (2003). (b) H.-G. Woo, B.-H. Kim, M.-S. Cho, L.-Y. Hong and D.-P. Kim, unpublished results.

42. P. Braunstein and X. Morise; Chem.Rev., 100, 3541 (2000).

43. (a) P. Trefonas and R. West; J. Polym.Sci., Polym.Chem.Ed., 23, 2099 (1985). (b) W.J. Szymanski, G.T. Visscher and P.A. Bianconi; Macromolecules, 26, 869 (1993). (c) T. Imori, V. Lu, H. Cai and T.D. Tilley; J. Am.Chem.Soc., 117, 9931 (1995). (d) S. Kashimura, M. Ishifune, N. Yamashita, H.-B. Bu, M. Takebayashi, S. Kitajima, D. Yoshiwara, Y. Kataoka, R. Nishida, S.-I. Kawasaki, H. Murase and T. Shono; J. Org.Chem., 64, 6615 (1999).

44. (a) C. Aitken, J.F. Harrod and A.J. Malek; J. Organomet.Chem., 249, 285 (1988). (b) N. Choi and M. Tanaka; J. Organomet.Chem., 564, 81 (1998). (c) B.-H. Kim and H.-G. Woo, manuscript in preparation.

45. T. Imori and T.D. Tilley; J. Chem.Soc., Chem.Commun., 1607 (1993).

46. (a) V. Lu and T.D. Tilley; Macromolecules, 29, 5763 (1996). (b) V. Lu and T.D. Tilley; Macromolecules, 33, 2403 (2000). (c) S.S. Bukalov, L.A. Leites, V. Lu and T.D. Tilley; Macromolecules, 35, 1757 (2002).

47. P. Braunstein and X. Morise; Organometallics, 17, 540 (1998).

48. N.R. Neale and T.D. Tilley; J. Am.Chem.Soc., 124, 3802 (2002).

49. (a) J. Kralicek, V. Kubanek and J. Kondelikova; German Patent, 2,301,784 (1973). (b) J. Kralicek, V. Kubanek, J. Kondelikova, B. Casensky and J. Machacek; German Patent, 2,445,647 (1976). (c) Z. Bukac and J. Sebenda; U.S. Patent, 3,962,239 (1976).

50. (a) B. Becker, R.J.P. Corriu, C. Guérin and B.J.L. Henner; J. Organomet.Chem., 369, 147 (1989). (b) B. Becker, R. Corriu, C. Guérin and B. Henner; Polym.Prepr. (Am.Chem.Soc., Div.Polym.Chem.), 28(1), 409 (1987). (c) R.J.P. Corriu; J. Organomet.Chem., 400, 81 (1990).

51. P. Riviere, J. Satge and D. Soula; J. Organomet.Chem., 72, 329 (1974).

52. J.L. Speier and R.E. Zimmerman; J. Am.Chem.Soc., 77, 6395 (1955).

53. H.-G. Woo, S.-J. Song, E.J. Cho and I.N. Jung; Bull.Korean Chem.Soc., 17, 123 (1996).

54. D. Seyferth and H. Lang; Organometallics, 10, 551 (1991).

55. A.D. Armitage; in Comprehensive Organometallic Chemistry (Ed. G. Wilkinson, F.G.A. Stone and E.W. Abel), Volume 2, Chapter 1, Pergamon Press, Oxford (1982).

56. H.-G. Woo, S.-J. Song, H. You, E.J. Cho and I.N. Jung; Bull.Korean Chem.Soc., 17, 475 (1996).

57. H.-G. Woo and S.-J. Song; Bull.Korean Chem.Soc., 17, 1040 (1996).

58. H.-G. Woo, S.-Y Kim, W.-G. Kim, S.II. Yeon, E.J. Cho and I.N. Jung; Bull.Korean Chem.Soc., 16, 1109 (1995).

59. H.-G. Woo, B.-H. Kim, M.-S. Cho and T.S. Hwang; Proceedings of International Symposium on Silicon-containing Polymers and Applications, p.49, Kanagawa, Japan (2001).

60. H.-G. Woo, B.-H. Kim and M.-S. Cho; Proceedings of International Symposium on Silicon-containing Polymers and Applications, p.53, Kanagawa, Japan (2001).

61. (a) H.-G. Woo, S.-J. Song, M.-K. Han, E.J. Cho and I.N. Jung; Bull.Korean Chem.Soc., 16, 1242 (1995). (b) H.-G. Woo, B.-H. Kim, M.-S. Cho, S.-J. Song, M.-K. Han, S.-Y. Kim, A.-Y. Sung, J.H. Kim and J.-S. Lee; Bull.Korean Chem.Soc., 21, 935 (2000). (c) R.M. Shaltout and J.Y. Corey; Organometallics, 15, 2866 (1996).

62. (a) J.A. Reichl, C.M. Popoff, L.A. Gallagher, E.E. Remsen and D.H. Berry; J. Am.Chem.Soc., 118, 9430 (1996). (b) S.M. Katz, J.A. Reichl and D.H. Berry; J. Am.Chem.Soc., 120, 9844 (1998).

63. J.R. Babcock and L.R. Sita; J. Am.Chem.Soc., 118, 12481 (1996).

64. (a) H.-G. Woo, J.-M. Park, S.-J. Song, S.-Y. Yang, I.-S. Kim, W.-G. Kim; Bull.Korean Chem.Soc., 18, 1291 (1997). (b) H.-G. Woo, S.-J. Song and B.-H. Kim; Bull.Korean Chem.Soc., 19, 1161 (1998).

65. (a) J.L. Speier; Adv.Organomet.Chem., 17, 407 (1978). (b) J.J. Pesek, M.T. Matyska and X. Pan; J. Chromatography A, 992, 57 (2003).

66. (a) J.F. Harrod and A.J. Chalk; in Organic Synthesis via Metal Carbonyls (Ed. I. Wender and P. Pino), p. 673, Wiley, New York (1977). (b) I. Ojima; in The Chemistry of Organic Silicon Compounds (Ed. S. Patai and Z. Rappoport), p. 1479, Wiley, New York (1989). (c) B. Marciniec; Comprehensive Handbook on Hydrosilylation, Pergamon Press, Oxford, England (1992).

67. J.P. Collman, L.S. Hegedus, J.R. Norton and R.G. Finke; in Principles and Applications of Organotransition Metal Chemistry, p. 564, University Science Books, Mill Valley (1987).

68. A.J. Chalk and J.F. Harrod; J. Am.Chem.Soc., 87, 16 (1965).

69. A.K. Roy and R.B. Taylor; J. Am.Chem.Soc., 124, 9510 (2002).

70. (a) F. Seitz and M.S. Wrighton; Angew.Chem.Int.Ed., 28, 762 (1989). (b) R.S. Tanke and R.H. Crabtree; Organometallics, 10, 415 (1991). (c) S.H. Bergens and P.N. Whelan; J. Am.Chem.Soc., 114, 2128 (1992). (d) T.R. Yasue; Organometallics, 15, 2098 (1996). (e) A.M. LaPointe, F.C. Rix, and M. Boorkhart; J.Am.Chem.Soc., 119, 906 (1997).

71. B. Marciniec; Appl.Organomet.Chem., 14, 527 (2000) and references cited theirin.

72. M. Rubin, T. Schwier and V. Gevorgyan; J. Org.Chem., 67, 1936 (2002).

73. (a) K. Oertle and H. Wetter; Tetrahedron Lett., 26, 5511 (1985). (b) N. Asao, T. Sudo and Y. Yamamoto; J. Org.Chem., 61, 7654 (1996). (c) J.B. Lambert, Y. Zhao and H. Wu; J. Org.Chem., 64, 2729 (1999).

74. (a) A.K. Dash, J.Q. Wang and M.S. Eisen; Organometallics 1999, 18, 4724. (b) Y. Kawanami, Y. Sonoda, T. Mori and K. Yamamoto; Org.Lett., 4, 2825 (2002). (c) J.W. Faller, and D.G. D'Alliessi; Organometallics, 21, 1743 (2002).

75. T. Takahashi, F. Bao, G. Gao and M. Ogasawara; Org.Lett., 5, 3479 (2003).

76. (a) J.F. Jensen, B.Y. Svendsen, T.V. la Cour, H.L. Pedersen and M. Johannsen; J. Am.Chem.Soc., 124, 4558 (2002). (b) T. Shimada, K. Mukaide, A. Shinohara, J.W. Han and T. Hayashi; J. Am.Chem.Soc., 124, 1584 (2002).

77. K. Itami, K. Mitsudo, A. Nishino and J.-I. Yoshida; J. Org.Chem., 67, 2645 (2002).

78. (a) L. Hao, J.F. Harrod, A.-M. Lebuis, Y. Mu, R. Shu, E. Samuel and H.-G. Woo; Angew.Chem.Int.Ed., 37, 3126 (1998). (b) J.F. Harrod, R. Shu, H.-G. Woo and E. Samuel; Can.J.Chem., 79, 1075 (2001).

79. (a) M.A. Grunlan, J.M. Mabry and W.P. Weber; Polymer, 44, 981 (2003). (b) J.M. Mabry, M.K. Runyon and W.P. Weber; Macromolecules, 34, 7264 (2001). (c) T.-S. Hwang, J.-T. Lim and H.-G. Woo; Polymer (Korea), 23, 197 (1999). (d) T.-S.Hwang, J.H.Lim and H.-G.Woo; Polymer (Korea), 22, 194 (1998).

80. (a) L.-Y. Hong, H.-G. Woo and H.-S. Ham; Bull.Korean Chem.Soc., 16, 360 (1995). (b) H.-G. Woo, L.-Y. Hong, S.-Y. Kim, S.-H. Park, S.-J. Song and H.-S. Ham; Bull.Korean Chem.Soc., 16, 774 (1995). (c) H.-G. Woo, L.-Y. Hong, S.-Y. Yang, S.-H. Park, S.-J. Song and H.-S. Ham; Bull.Korean Chem.Soc., 16, 1056 (1995). (d) H.-G. Woo, L.-Y. Hong, J.-Y. Park, Y.-T. Jeong, H.-R. Park and H.-S. Ham; Bull.Korean Chem.Soc., 17, 16 (1996). (e) H.-G. Woo, S.-H. Park, J.-Y. Park, S.-Y. Yang and H.-S. Ham; Bull.Korean Chem.Soc., 17, 373 (1996). (f) H.-G. Woo, S.-H. Park, L.-Y. Hong, H.-G. Kang, S.-J. Song and H.-S. Ham; Bull.Korean Chem.Soc., 17, 376 (1996). (g) H.-G. Woo, S.-H. Park, L.-Y. Hong, S.-Y. Yang, H.-G. Kang and H.-S. Ham; Bull.Korean Chem.Soc., 17, 532 (1996). (h) H.-G. Woo, J.-Y. Park, L.-Y. Hong, S.-J. Song, H.-S. Ham and W.-G. Kim; Bull.Korean Chem.Soc., 17, 560 (1996). (i) H.-G. Woo, E.-M. Oh, J.-H. Park, B.-H. Kim, Y.-N. Kim, C.H. Yoon and H.-S. Ham; Bull.Korean Chem.Soc., 21, 291 (2000). (j) H.-G. Woo, M.-S. Lee, Y.-J. Kim, B.-H. Kim and J.-I. Kong; Proceedings of International Symposium on Silicon-containing Polymers and Applications, p. 81, Kanagawa, Japan (2001). (k) H.-G. Woo, M.-S. Lee, Y.-J. Kim, B.-H. Kim and Y.-S. Kim; Proceedings of International Symposium on Silicon-containing Polymers and Applications, p. 85, Kanagawa, Japan (2001). (l) H.-G. Woo, B.-H. Kim, M.-S. Cho, D.-Y. Kim, Y.-S. Choi, Y.-C. Kwak, H.-S. Ham, A.-Y. Sung, D.-P. Kim and T.S. Hwang; Bull.Korean Chem.Soc., 22, 1337 (2001). (m) H.-G. Woo and B.-H. Kim; manuscript in preparation.

81. (a) G. Odian; in Principles of Polymerization, 3rd ed., pp. 222-223, Wiley, New York (1991). (b) S. Mimura, H. Naito, Y. Kanemitsu, K. Matsukawa and H. Inoue; J. Organomet.Chem., 611, 40 (2000). (c) C. Peinado, A. Alonso, F. Catalina and W. Schnabel; J. Photochem.& Photobiol.A.; Chemistry, 141, 85 (2001).

82. R.G. Norrish and J.P. Simons; Proc.Roy.Soc.(London), A251, 4 (1959).

83. C. Chatgilialoglu, V.I. Timokhin, A.B. Zaborovskiy, D.S. Lutsyk and R.E. Prystansky; J. Chem.Soc., Perkin Trans.2, 577 (2000).

84. (a) H.-G. Woo, M.-S. Cho, M.-S. Lee and M.-J. Jun; Proceedings of International Symposium on Silicon-containing Polymers and Applications, p. 97, Kanagawa, Japan (2001). (b) H.-G. Woo, S.-E. Lee, Y.-C. Kwak, B.-H. Kim and H.-S. Ham; Proceedings of International Symposium on Silicon-containing Polymers and Applications, p. 69, Kanagawa, Japan (2001). (c) H.-G. Woo and B.-H. Kim; manuscript in preparation.

85. (a) S. Tanaka, T. Nakamura, H. Yorimitsu, H. Shinokubo and K. Oshima; Org.Lett., 2, 1911 (2000). (b) J.W. Faller and R.G. Kultyshev; Organometallics, 22, 199 (2003). (c) A.E. Kadib, A. Castel, F. Delpech, P. Riviere, M. Riviere-Baudet, H. Gornitzka, P. Aguirre, J.M. Manriquez, I. Chavez and D. Abril; Inorg.Chem.Acta, 357, 1256 (2004). (c) K. Choi and J.M. Buriak; Langmuir, 16, 7737 (2000).

86. H. Kinoshita, T. Nakamura, H. Kakiya, H. Shinokubo, S. Matsubara and K. Oshima; Org.Lett., 3, 2521 (2001).

87. (a) T. Nakano, K. Ono, Y. Senda and T. Migita; J.Organomet.Chem., 619, 313 (2001). (b) H. Komoriya, M. Kako, Y. Nakadaira and K. Mochida; J. Organomet.Chem., 611, 420 (2000).

88. K. Miura, D. Wang, Y. Matsumoto, N. Fujisawa and A. Hosomi; J. Org.Chem., 68, 8730 (2003).

89. B.-H. Kim and H.-G. Woo, unpublished results.

90. (a) Ultrastructure Processing of Advanced Ceramics (Ed. J.D. Mackenzie and D.R. Ulrich), John Wiley & Sons, New York (1988). (b) D. Dou, D.R. Ketchum, E.J.M. Hamilton, P.A. Florian, K.E. Vermillion, P.J. Gradinetti and S.G. Shore; Chem.Mater., 8, 2839 (1996).

91. (a) D.P. Kim and J. Economy; Chem.Mater., 5, 1216 (1993). (b) D. Cornu, S. Bernard, S. Duperrier, B. Toury and P. Miele; J. Eur.Ceram.Soc., 25, 111 (2005).

92. F. Guilhon, B. Bonnetot, D. Cornu and H. Mongeot; Polyhedron, 15, 851 (1996).

93. (a) K.T.Moon, D.S.Min and D.P.Kim; Bull.Korean Chem.Soc., 19, 222 (1998). (b) K.T. Moon, J.K. Jeon, Y.J. Kim and D.P. Kim; Scripta Mater., 44, 2111 (2001). (c) J.G. Kho, K.T. Moon, G. Nouet, P. Ruterana and D.P. Kim; Thin Solid Films, 389, 78 (2001). (d) S. Bernard, D. Cornu, P. Miele, H. Vincent and J. Bouix; J. Organomet.Chem., 657, 91 (2002). (e) S. Bernard, F. Chassagneux, M.-P. Berthet, H. Vincent and J. Bouix; J. Eur.Ceram.Soc., 22, 2047 (2002). (f) J.A. Perdigon-Melon, A. Auroux, C. Guimon and B. Bonnetot; J. Solid State Chem., 177, 609 (2004).

94. (a) P. Miele, B. Toury, D. Cornu and S. Bernard; J. Organomet.Chem., 690, 2809 (2005). (b) B. Toury, D. Cornu, F. Chassagneux and P. Miele; J. Eur.Ceram.Soc., 25, 137 (2005).

95. I. Butler and J.F. Harrod; Inorganic Chemistry, Chapter 16, The Benjamin/Cummings Publishing Co., Inc., Redwood City, CA (1989).

96. R.T. Paine and L. Sneddon; Chemtech., July, 29 (1994).

97. (a) Phosphazenes: A World Wide Insight (Ed. M. Gleria and R. De Jaeger), Nova Science Publishers, New York (2004). (b) R.H. Neilson and P.W. Neilson; Chem.Rev., 88, 541 (1988). (c) G. Rojo, G. Martin, F. Lopez, G.A. Carriedo, F.J.G. Alonso and J.I.F. Martinez; Chem.Mater., 12, 3603 (2000).

98. H.R. Allcock and E.C. Kellam; Solid State Ionics, 156, 401 (2003).

99. Z. Li, W. Gong, J. Qin, Z. Yang and C. Ye; Polymer, 46, 4971 (2005).

100. Y.J. Jun, J.I. Kim, M.J. Jun and Y.S. Sohn; J. Inorg.Biochem., 99, 1593 (2005).

101. (a) G.A. Carriedo, F.J.G. Alonso, CD. Valenzuela and M.L. Valenzuela; Polyhedron, 25, 105 (2006). (b) G.A. Carriedo, F.J.G. Alonso, P.A. Gonzalez, C.D. Valenzuela and N.Y. Saez; Polyhedron, 21, 2579 (2002).

102. M.A. Hofmann, C.M. Ambler, A.E. Maher, E. Chalkova, Y. Zhou, S.N. Lvov and H.R. Allcock; Macromolecules, 35, 6490 (2002).

103. (a) Y. Chang, J.D. Bender, M.V.B. Phelps and H.R. Allcock; Biomacromolecules, 3, 1364 (2002). (b) Y. Chang, R. Prange, H.R. Allcock, S.C. Lee and C. Kim; Macromolecules, 35, 8556 (2002).

7

Light Weight Polymer Composite Materials for Automotive Industry

S.M. Sapuan

*Department of Mechanical and Manufacturing Engineering,
University Putra Malaysia, 43400 Serdang, Selangor, Malaysia.
Corresponding author's email: sapuan@eng.upm.edu.my*

Contents

- INTRODUCTION
- DEFINITIONS
- HISTORY OF AUTOMOTIVE POLYMER COMPOSITES
- ADVANTAGES AND DISADVANTAGES OF POLYMER COMPOSITES IN AUTOMOTIVE INDUSTRY
- DESIGN PROCESSES
- MANUFACTURING TECHNIQUES
- SELECTED AUTOMOTIVE COMPOSITE COMPONENTS
- THE APPLICATION OF COMPUTERS IN AUTOMOTIVE POLYMER COMPOSITE DEVELOPMENT: A CASE STUDY OF POLYMER COMPOSITE PEDAL BOX SYSTEM
- INTEGRATED DESIGN AND MANUFACTURING (CONCURRENT ENGINEERING) APPROACH FOR POLYMER COMPOSITE AUTOMOTIVE COMPONENTS
- MATERIAL SELECTION BY KBS SYSTEM FOR POLYMER BASED COMPOSITE BUMPER BEAM
- NATURAL FIBER POLYMER COMPOSITES IN AUTOMOTIVE INDUSTRY
- RECYCLING OF AUTOMOTIVE POLYMER COMPOSITES
- CONCLUSIONS AND FUTURE PROSPECTS
- ACKNOWLEDGEMENTS
- REFERENCES

Summary: Polymer composites are widely used in structural components in automobiles due to some advantages they offer, which are not found in other materials. These advantages include good specific strength and stiffness, part integration, ease of processing, corrosion resistance, low density, relatively low cost, light weight, aesthetically pleasing and flexibility in design. Almost all major automakers have made moves from using conventional materials such as metals and plastics to polymer composites. This article reviews various important aspects of polymer composites in automotive industry. Initially brief history and definition of automotive polymer composites are given. The benefits and disadvantages of polymer composites in automotive industry are discussed. Important issues in automotive composites such as design, manufacture, material selection, the use of computer in automotive composite development, concurrent engineering, natural fiber composites in automobiles, recycling of automotive composites and future trends are addressed.

1. INTRODUCTION

The rapid change in today's marketplace is driving industrialists to search for new methodologies and technologies to enable them to cope with market demand. This is illustrated by the demands, especially, of the automotive industry. The automotive industry has, in the past, experienced many technological changes and is currently benefiting from new advances. Continuous efforts are directed towards improving the performance of future cars, although the basic features of a car are still very much the same as they were a century ago, i.e. an internal combustion engine, four wheels, rubber tyres and steering wheel.

In the recent years, however, the new technologies and advances have been introduced. These include the use of alternative fuels, the utilization of advanced materials, the use of modern manufacturing methods and new technologies such as air bag restraint system and anti-lock braking system (ABS) etc. As a consumer product, the automobile must be made from materials that have lower cost, easy to process and which have long life and high reliability under adverse conditions. In recent years, there has been additional pressure to make vehicles lighter in the interest of fuel saving.

Improvements in materials can have a big impact on the economics of the automotive industry. One important need is the development of complete material systems that take into account the cost of manufacture along with specific design criteria. The major requirements in materials that may be considered the driving forces for automotive technology are the need to enhance reliability, low cost, functional improvement, durability and recyclability.

A vehicle consists of a variety of components made from different materials. These include steel, cast iron, engineering plastics and aluminium. Several reasons compel vehicle manufacturers to investigate the possibilities of replacing conventional materials with alternative and often novel materials. For example in the car body, such reasons are the reduction of cost, reduction of corrosion, reduction of fuel consumption through lower vehicle weight and reduction in air drag as a result of greater flexibility of design. For the engine, the reasons are the better thermal insulation, improved temperature and erosion resistance and reduction in mass of rotating and oscillating components.

To meet these demands, one of the most suitable candidate materials is the class of polymer composite materials. Polymer composites are made up of two or more constituents, which have markedly different physical and mechanical properties. The resulting material has a combination of properties.

Polymer composite is widely used in structural components in automobile due to some advantages it offers, which are not found in other materials. These advantages include good specific strength and stiffness, part integration, ease of processing, corrosion resistance, relatively low cost, lightweight, aesthetically pleasing look and flexibility in design. Almost all major automakers have made moves from using conventional materials such as metals and plastics to polymer composites. Polymer composites were used in automobile since the birth of modern polymer composite in 1940s. Chevrolet had manufactured its model called Corvette in 1953 where its body was made from glass fiber reinforced polymer composite [1]. And interestingly, up till today, the sixth generation Chevrolet Corvette cars are still being manufactured and the body panels are still made from polymer composite [2].

Nowadays, the use of polymer composites in automobile is no longer new and automakers are working hard to use polymer composites in the new areas within automobile. Efforts have been made to use non-conventional polymer composite materials such as natural fiber composites in semi-structural and non-structural automotive components. Effective use of polymer composites in automobile is largely determined from the production volume of the automobile. In the case where the number of production is small as in some sport cars, the process called hand lay-up is preferred where it is a labor intensive process while for the high production volume like what is practiced by the most of the automakers, processing such as injection and compression mouldings give the best solution.

2. DEFINITIONS

According to Morgalis [3], an automotive may be referred to as a land vehicle that is self propelled by an internal engine while a polymer composite is defined as an organic matrix reinforced with a second material in the form of fibers, fine particles etc. The various types of composites available include particles in the polymer matrix and fibers in the polymer matrix or a combination of both of them. The primary interest for automotive structures is with the fiber composites.

The manufacturing of polymer composites provides the means of improving stiffness, strength, resistance to temperature and creep of polymers while at the same time offering cost reduction. In these reinforced materials, the strength and stiffness of load bearing fibers are married to polymer matrices to provide a good combination of strength and toughness. The fibers are normally of glass, carbon or aramid. The fibers are very brittle and alone their strength and stiffness are not fully realized. The fibers consist of chopped strand mat, short fibers or of continuous fibers.

The reinforcements, whatever their form, for many manufacturing processes, is pre-impregnated with resin to form a partially cured composite called a prepreg. This form can be converted into the desired final shape during a final curing step of the fabrication process. The polymer matrix protects the fibers and transfers the load to them. There are two types of polymers used for the matrices viz. thermoplastics and thermosettings.

Thermoplastic materials can be heated and formed repeatedly and normally used for injection moulding applications. These include acrylics, nylons, polypropylenes, polycarbonates, vinyls, polystyrenes, acrylonitrile/butadiene/styrene (ABS), polytetrafluoroethylene (PTFE), polyethyleneterephalate (PET), polybutyleneterephthalate (PBT) and polysulfones. Thermosetting materials are formed through an initial heating and cure and cannot be melted and reformed once cured. These include polyesters, melamine, phenolics, vinylesters and epoxides.

Thermoplastics offer a wide range of advantages over thermosetting matrices, including unlimited shelf life, short process cycle times, their ability to be reformed, increased moisture resistance, increased toughness and impact resistance. However, there are several disadvantages associated with thermoplastic matrix composites. The present cost of the thermoplastic materials is higher than of conventional thermosetting materials.

3. HISTORY OF AUTOMOTIVE POLYMER COMPOSITES

Polymer composites have a long and honorable history in automotive industry, starting with hand lay-up bodies for limited production of sports cars upto the most up-to-date injection moulded air intake manifolds [4]. Demonstration cars with all-composite bodies have been built since the early 1900s, beginning with Henry Ford's coupe. It is followed by the Rohm & Haas acrylic Explorer, the Borg-Warner acrylonitrile-butadiene-styrene (ABS) Cycolac Research Vehicle (CRV) and the Mobay polyurethane car were built in the late 1950s and early 1960s [3].

Commercial volume all-composite Chevrolet Corvette cars were manufactured in 1953. The use of plastics, the matrix for polymer composites is found to be increasing day by day. In 1955, British cars contained about 4 lb and in 1970 some 70 lb of plastics [1]. It was reported that the Nissan was making effort to use 30% of material usage in Micra model from plastics in the year 2000 [5].

4. ADVANTAGES AND DISADVANTAGES OF POLYMER COMPOSITES IN AUTOMOTIVE INDUSTRY

Polymer composites offer a wide range of advantages to automotive industry. The most widely recognized benefit is the potential for weight saving offered by their low density. Reduced weight could lead to lower fuel consumption. Hancox [6] has summarised other benefits of composites for automotive application as follows:

- Component designs can be such that the fibers lie in the direction of the principal stresses and the amount of fiber used is sufficient to withstand the stress, thus, optimizing the material usage which is particularly applicable for continuous fiber composites and not for injection moulded short random fiber composites.
- The design and fabrication routes can be combined so that complicated shaped components can be moulded in one process rather than being assembled from components or machined from a blank, thus, potentially reducing manufacturing costs.
- Composites can be fabricated relatively easily at low pressures and temperatures of no more than 200°C, which again has implications on the cost of production.
- The excellent resistance to corrosion and other chemical environments could help manufacturers to prolong the life-time of individual components and whole vehicles.

Other advantages as enumerated by Sapuan [7] include:

- High integral strength.
- High impact resistance.
- Good dimensional stability.
- Good wear resistance under heavy loads.
- Capability to be manufactured in a variety of sizes suitable for a range of components.
- High strength to weight ratio.
- Good thermal insulation properties.
- Very good fatigue performance.
- Controlled flow for ease of processing.
- Cost effective.
- Easy to use.
- Improved vibration damping properties.
- Complex shapes can be manufactured at low cost.
- Longer life.

The disadvantages of polymer composites in automotive industry include [6]:

- Lack of information about composite properties particularly in relation to impact.
- Users are not familiar with composite materials and the benefits they can bring to the automotive industry.
- Lack of processing equipments because most of the production lines have been designed to use metals.
- Little information on design and design practices/codes and hence excessive safety factors.
- Composites are often seen as a substitute rather than a designed replacement, leading to inefficient usage.
- Difficulties of joining and repairing composites.
- Poor transverse and shear properties of unidirectional materials, however, automotive composites usually take the advantages of short random fiber composites.
- Doubts about the trade-off between lower weight and the possible extra cost of materials and fabrication.
- The stiffness of polymer composites is low in relation to metals.
- The properties can be anisotropic.
- Composites are susceptible to splitting and delamination.
- Cracks often propagate along the fiber direction.
- Composite is sometimes degraded by water and solvents, depending on the resins and fibers used.
- Carbon fiber based materials can cause corrosion of adjacent materials.
- Composites suffer from inherent variation in component properties due to the nature of the fabrication techniques.

5. DESIGN PROCESSES

The automobiles have been manufactured since the end of 19th century. They were built in a tedious and laborious manner, with little mechanical assistance. However, later, the automotive industry has been developed with the help of modern technologies and methods.

In designing automotive components from composites, it is quite different from designing with metal counterpart. Many features, not normally needed in metal, can be very important in composites such as the inclusion of ribs, bosses, undercuts and inserts. This is possible using some manufacturing processes like injection moulding and hand lay up. By doing this, the components can be strengthened and many parts can be integrated together.

Designing automotive components by simple replacement of steel parts by polymer composite materials, no doubt, yield significant weight savings, but as with many new materials, design and manufacturing problems arise. The change from relatively isotropic-homogenous steel alloys to anisotropic-inhomogenous fiber reinforced polymer composites cause these problems. As a result, it is not an easy task to replace steel by polymer composite materials [8].

Design criteria for automotive composites might not appear as stringent as designing for aerospace applications. But it is not so, because in automobiles, components and structure geometrical complexities and the varied types and directions of the load inputs into the structures lead to a complexity of design. In automobile, components with symmetrical or regular shape and loading are rarely found [9].

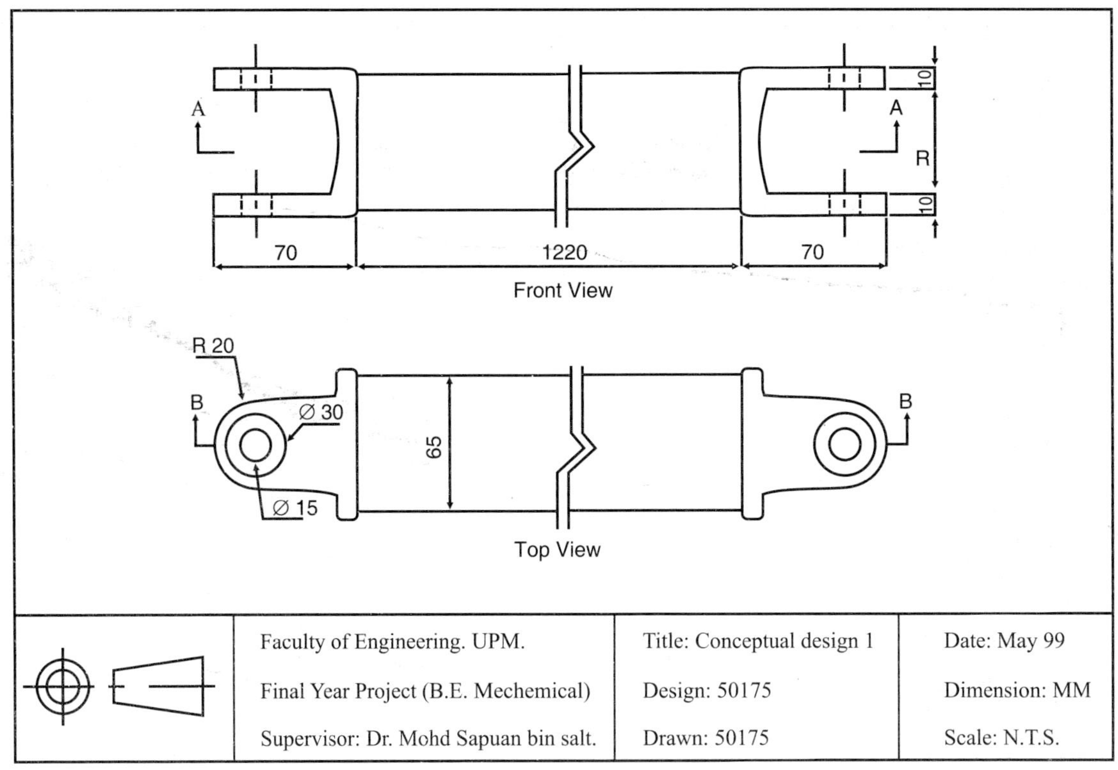

Fig. 7.1. A design concept of polymer composite automotive propeller shaft (Courtesy Mr. Saifulla Azizi).

In the design of automotive components, namely, pedal box system from polymer composites, total design route was followed in the previous publication of the author [10]. The idea is also extended to the design of composite bumper system [11]. If a product and process design are thought as distinct serial activities and not carried out in parallel, then there will be delays in bringing new products to the market. In total design activity, the activities like conceptual design and detailed design are related to each other. Conceptual design is a very important stage in total design process. Conceptual design, design concept or conceptualization is the beginning phase of the design process normally after recognition of need. Creativity is related to conceptualization because conceptualization is a means to identify solutions by considering alternatives. Creativity is a means to generate alternative solutions. In this approach, several idea generation methods like the extension of the search space, the morphological chart, the exploitation of experience, the gallery method and the voice of customers had been employed. Prior to the conceptual design, product design specification (PDS) is to be prepared. PDS is used for analysis, design, manufacturing and construction of a structure or a component in order to achieve a specified degree of safety, efficiency, performance or quality as well as a common standard of good design practice. PDS normally covers [12]:

- Product classification, scope of application, size range and condition,
- Allowable ranges of chemical composites,
- All physical and chemical properties necessary to characterize the product and
- If applicable, other requirements, such as special tolerance and surface preparation, including instruction.

Fig. 7.1 shows a concept design of polymer composite automotive propeller shaft. It can seen from this figure that, in concept design, the drawing is prepared not according to the actual scale and many details about the design are not provided. The aim is just to communicate the idea of the designer during the design process to other parties within or outside the company and normally a number of concepts are developed.

The author [13] has listed various design guides in the development of automotive components from polymer composite materials and they are, among others, given as follows:

- Integration of various information technology (IT) tools is essential because the use of IT tools like CAD, SLA, mould flow analysis, expert system and FEA and the integration of these tools is possible.
- Iterative work cannot be avoided.
- Knowledge of various fields such as material engineering, composite manufacture, design process, tooling and IT tools is essential.
- Performing total design technique.
- Carrying out comprehensive market investigation.
- Standardization of components.
- Checking for established standards like British Standard or ASTM Standards relevant to the design.
- Use of total design approach.
- Development of sound product design specification (PDS).
- Use of traditional method of design calculation to verify any findings based on modern IT tools.
- Carrying out patent search in order that no infringement of existing patented design.
- Selection of candidate materials that have comparable properties.
- Shape, color and texture can be achieved readily with composites, therefore, design one can design a component according to one's preference.
- Incorporation of features like bosses, ribs, and undercuts.

6. MANUFACTURING TECHNIQUES

Several manufacturing techniques are used to fabricate composite materials. Although many manufacturing methods can be employed to develop components from composites, only selected methods can practically be used. Filament winding, hand lay up, compression moulding, injection moulding, resin transfer moulding, pre-preg moulding and pultrusion can be considered for making automotive components.

However, the most common manufacturing techniques to produce automotive components from composite materials are injection moulding, compression moulding, resin transfer moulding and filament winding and to a lesser extent, hand lay-up for low production volume of sport car body and vacuum bag or prepreg moulding for making some prototypes. In general, the production volume is normally high in the manufacturing of automotive components while many processing methods are meant for low and medium volume production. For instance, hand lay-up process can only made custom components where the fabrication process is time consuming and involves high labor investment and specialization. This method was popular in the early stage of automotive composites in 1950s where majority of cars were sports cars and demonstration cars.

When mass-production is desired, the processing methods are not free from problems. Therefore, in the present days, the most practical solutions are to resort to employing methods like automated injection and compression mouldings where tens of thousands of components can be produced in short period of time. However, with injection moulding, the components can only be produced using short length fiber reinforced composites as opposed to long or continuous fiber composites. Obviously, many properties of short fiber composites are less superior to continuous fiber composites. However, as far as automotive application is concerned, the properties demonstrated by short fiber composites are acceptable. With compression moulding, it is possible to make components both from short and continuous fiber composites. Therefore, several body panel components such as horizontal penal like bonnet cover and boot lid and vertical panels like fender and doors are made from sheet moulding compound (SMC) or glass mat thermoplastics (GMT) that are all using compression moulding process.

There are also moves to develop components with circular in cross section such as drive shaft and propeller shaft using filament-winding process. The processing method is a medium production volume process and high investment on tooling is expected. Other methods employed include resin transfer moulding and pre-preg mouldings and the use of these methods is restricted due to many technical problems like medium tooling cost and resin and medium production time.

Whatever may be the moulded product, its quality is dependent upon the quality of the tooling and also the quality of the materials used to make the tool if the manufacturing processes involve the use of tooling. Essentially, the material should have dimensional stability, low coefficient of thermal expansion, low warpage, chemical resistance, low porosity and still be cost competitive.

6.1. Injection Moulding

Injection moulding is a process where fluid plastic/reinforced plastic material is forced or injected into a closed mould and is then allowed to solidify. Both thermoplastic and thermoset materials can be injection moulded. This technique is suitable for small to medium size products and provides greater design freedom, reduced assembly requirements and greater quality consistency since parts and mould can be designed so that no subsequent trimming or machining operation is required [14]. Injection moulding has several features that make its utilization feasible, especially for the mass production of complicated components. These features are:

- Direct path from material to finished part.
- No or minimal finishing of moulded part necessary.
- Produce parts, which are not prone to porosity.
- Process may be fully automated.
- Good repeatability of production with constant weight and identical properties.

The automotive components manufactured from injection moulding process are mainly for decorative and semi-structural components. By the introduction of reaction injection moulding (RIM) and reinforced reaction injection moulding (RRIM), more structural components can be made particularly for exterior panels such as fender. A variation of RIM is known as structural RIM or SRIM, a process, which produces the materials that possess structural-member strength [15]. SRIM is normally used to produce door trim substrates, instrument panel retainers, spare tyre covers, bumper beams, underbody structures, light truck boxes, tailgates and sun-shades etc.

Injection moulding came into significant commercial use following the development of plastic screws. The reciprocating screw type injection moulding machine has largely supplanted the early plunger type of injection moulding machine. Although, there are many variations in thermoplastic injection moulding machines, the reciprocating screw machine has become the most common type. There are variants of the thermoplastic injection moulding such as thermoplastic structural foam injection, sandwich injection moulding, hollow injection moulding, injection/compression moulding and layer injection moulding. The functions performed by the machine include heating the plastics until it is able to flow readily under pressure, pressurizing this melt to inject it into a closed mould, holding the mould closed both during the injection and while the material is solidifying in the mould and opening the mould to allow removal of the solid part.

In the injection moulding process, initially, the material from the raw state condition (usually pellet or powder form) is converted into a melt processible state. These specially formulated compounds are used in which the appropriate constituents have been optimized. Then the granulated material is plasticised by the rotation of the screw in a heated barrel. After closing the mould, which contains a cavity in the shape of the component to be moulded, the plasticised material is injected through an axial displacement of the screw. The temperature of the mould is less than that of the melt, so the mixture begins to set as soon as it comes into contact with the walls of the mould. The melt of thermoplastic materials is subsequently cooled in the mould. Finally, the mould is opened and the moulded component is ejected. During production of an injection moulded part, each step is coordinated by machine controls. In addition to the basic steps described above, other functions may take place such as retraction of the injection unit, actuation of cores and other applications.

Regardless of the material to be processed, an injection moulding machine consists of the machine frame, injection unit and clamping unit. The clamping unit serves to open and close the mould during the production cycle. The clamping unit, screw and injection unit are driven hydraulically. The pumps required to provide the necessary flow of oil are located in the machine frame and powered electrically.

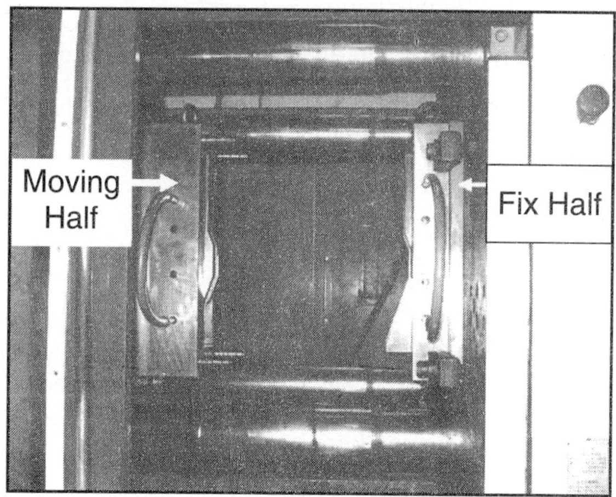

Fig. 7.2. Mould in injection moulding process (Courtesy Mrs. Sharifah Imihezri).

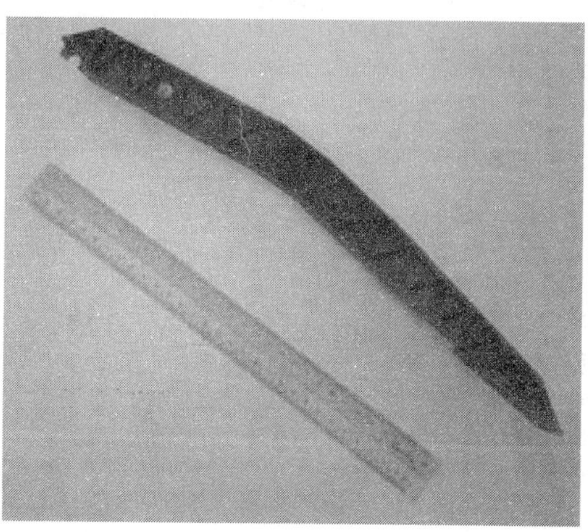

Fig. 7.3. Polymer composite automotive clutch pedal (Courtesy Mrs. Sharifah Imihezri).

To activate additional functions, such as ejectors, shutoff nozzles or core pulls, simple electromechanical or pneumatically driven devices are incorporated. To cool the molten injection plastics, the moulding compound must be cooled to below the solidification point, before the moulded part can be removed.

Imihezri *et al.* [16] carry out a simulation work to study the effect of increasing the number of gates on air traps and weld line formation in injection moulded automotive composite clutch pedal. The simulation work was carried out using Moldflow Plastic Insight (MPI) software. A mould with single gate produced more air trap compared to a double-gated mould. Dual gates mould produced longer weld line compared to single gate design. These studies are important because in injection moulding, the gate locations are decided permanently and any modification equals to increase in cost and time. **Fig. 7.2** shows a mould used in injection moulding process while **Fig. 7.3** shows a composite clutch pedal manufactured from injection moulding process.

6.2. Compression Moulding

Compression moulding or press moulding is the second major manufacturing process in automotive industry, which produces sheet-moulding compound (SMC). SMC is used in car bumpers, car body panels, grille opening panels, tailgates, heavy truck cabs and fully structural load-bearing components such as car road wheels. It involves the use of a press and a moulding compound or intermediate material in which resin, fibers, catalyst and additive are premixed in the suitable proportions. There are two types of press moulding, namely, hot press moulding and cold press moulding. Hot press moulding is the most common form of moulding process. It involves a press, heated platens and a matched tooling. Cold press moulding uses pressure alone to impregnate reinforcement laid up in a mould with catalyzed resin, which finally cured at room temperature [17].

Part configurations and cross-sections are extremely flexible, inserts and attachments may be accommodated and very high tolerances may be met. Polyesters are the predominant matrix but phenolics, melamine, silicones, diallyl phthalates and epoxies are also compression moulded. In addition to SMC, bulk/dough (BMC/DMC) or mat moulding can also be produced with compression moulding.

6.3. Resin Transfer Moulding

Resin transfer moulding (RTM) has been used to mould large non-automotive fiber reinforced parts since the mid 1970's. It is an alternative to open moulding. This method involves the transfer of resin into a closed mould into which the fiber in the form of pre-form is already placed. It is a process where a pair of matched tools is loaded with the fiber reinforcement. A catalyzed polymer is then introduced into this tool pair to impregnate the reinforcement with resin. When this polymer has cured, the tool is opened and the part is removed. The part is then post mould finished as needed.

Components like car doors and side panels are developed using this method [17]. Other components include commercial van roof, exterior body panels, leaf spring, drive shafts, pick up truck boxes and space frames.

The major advantages of this technique include closed mould, tight tolerances, provides two good surfaces, low emission of volatiles, many types of reinforcement can be used and it is possible to have inserts and cores. The developments made in the mould performance technology and mould filling techniques have reduced RTM costs considerably.

Resin transfer moulding (RTM) was employed by Harrison *et al.* [18] from Ford Motor Company, UK to produce high roof for the Ford Transit van. In this process, sandwich construction glass fiber mat reinforcements were precut and placed into closed mould as preforms and it was impregnated with a pigmented polyester resin through a carefully gated injection system.

6.4. Hand Lay-Up Technique

Hand lay up technique is one of the contact moulding processes to produce components from composite materials. It is a process in which fibers and matrix are sequentially placed into an open mould and it is the widely used method for producing composite components. It is the best known and most widely used method for producing fiber reinforced composite parts.

There is no limit to the size or complexity of structures that can be produced by this method. A wide range and variety of parts have been made, from small, non-structural, aesthetic articles up to large structures such as the shells of wing blades. The process is labor intensive, employs hand compaction only, it is a slow production rate technique and the quality control is difficult. Therefore, in automotive industry, it is only suitable for custom-made sports cars, demonstration cars and for the prototypes.

7. SELECTED AUTOMOTIVE COMPOSITE COMPONENTS

In this section, a review of selected research papers on the application and development of polymer composite automotive components is presented. The purpose is to see the main benefit i.e. reduced weight obtained from using composites in whatever components.

7.1. Drive Shaft

Two main functional requirements for power transmission rotating shafts such as drive shafts of machinery and automotive propeller shafts are the transmission of static and dynamic loads and the high fundamental bending natural frequency to avoid whirling vibration at a high rotational speed.

Long shafts made of conventional material such as aluminium and steel cannot satisfy easily these two functional requirements simultaneously because they have lower specific stiffness, which limits the magnitude of fundamental bending natural frequency. Hybrid shafts composed of aluminium and glass fiber composites can withstand more torque [19]. It is reported by Josh [20] that GNK Automotive has reduced the weight of its automotive propeller shafts by more than 5 kg per shaft through the use of glass and carbon fiber reinforced polymer composites. Weight reduction is accomplished by simplified design compared to steel counterpart. Conventional steel propeller shafts consist of two components and a center bearing, by this design, the center bearing

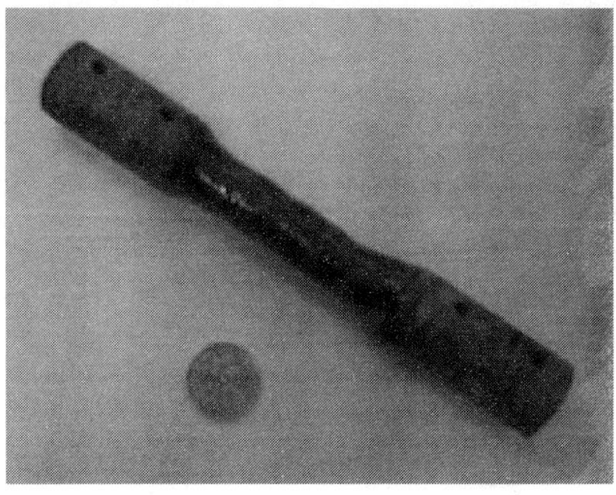

Fig. 7.4. Automotive shaft from composites (Courtesy Mr. Saad Assi Mutasher).

is removed from the unit to simplify it to one component. The benefits of this composite shaft include superior stiffness and tensile and fatigue strength in addition to weight saving. Example of automotive composite shaft is shown in **Fig. 7.4.**

7.2. Suspension

One of the latest developments in composite suspension system in USA is the work on composite springs for suspension system by Sardou *et al.* [21]. It is reported that the springs are made from thermoset epoxy and E-glass fibers, which provide longer fatigue life than that of conventional springs. The component is developed by using filament winding process.

7.3. Crossmember

Buchholz [22] reported that General Motors, USA is using a polymer composite material for the manufacture of automotive cross member. This composite component is 5 kg lighter than its steel counterpart. Compared to steel, the composite crossmember's lower and upper section wall thickness ranges from 3 to 10 mm and it does not require a stiffer plate.

7.4. Floor Pan

An automobile is a beam supported at each end by the wheels. Therefore, the automobile body must have good torsional stiffness, beam stiffness, resistance to side impact. Although steel is still widely used in automotive body panels, many automakers have developed the technology required to use polymer composites in such applications. Thermoplastic stamping is a material technology studied by Johnson *et al.* [23] to be used in the manufacture of structural components in an attempt to replace sheet moulding compound and resin transfer moulding in automotive floor pan. Various experimental work such as shear testing and compression testing were conducted on thermoplastic sheet and from these experiments, it was concluded that thermoplastic composite sheet has great potential in high volume automotive floor pan.

7.5. Bumper

Bumper is a structural component mounted across the front or rear of an automobile and its function is to absorb initial impact in a minor collision, thus protecting the body of the automobile and the main components of a bumper system is shown in **Fig. 7.5**.

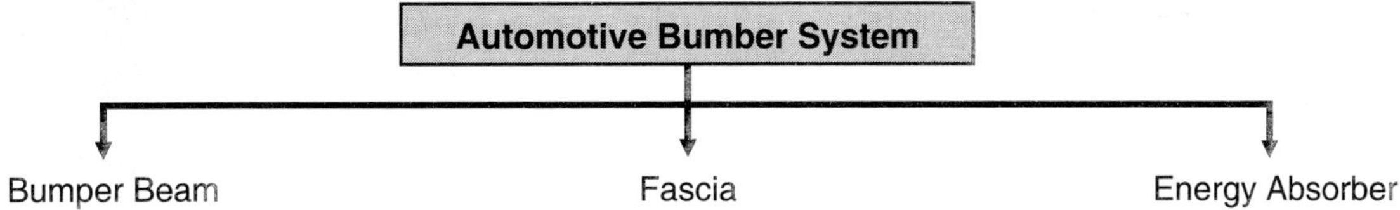

Fig. 7.5. Main components of automotive bumper system.

The bumper system is a structural component, which contribute to the crashworthiness or occupants protection during front or rear collision. There is an interest among the researcher to move from conventional material such as plastic, aluminium or steel to some cheaper materials such as polymer based composites in bumper system. Cheon *et al.* [24] developed glass fiber reinforced epoxy prepreg composite and unidirectional carbon fiber reinforced epoxy prepreg composite bumper beam for passenger cars. The box type cross section of composite bumper beam was employed to give high bending stiffness and to facilitate ease of manufacturing compared to tubular shape. The manufacturing method was vacuum bag or prepreg moulding. The total weight saving compared to metal counterpart was 30%. Minaudo *et al.* [25] developed a one piece, injection moulded, thermoplastic rear bumper system with pole impact protection. Clark *et al.* [26] described their extensive work of bumper beams using glass fiber reinforced plastics to study the stress contour in the component.

Cheon *et al.* [27] developed the composite bumper beam for passenger cars. The material used was glass fiber fabric epoxy composite material except the elbow section. Rawson [28] evaluated the performance of polyolefin in comparison with engineering thermoplastics for blow moulded bumper beams for mid size vehicles.

Ambu *et al.* [29] carried out experimental analysis of polyester SMC in FIAT-IVECO truck bumper. They analyzed the structural behavior of two experimental techniques, namely, holography interferometry and centesimal comparators. The experimental results were compared with numerical data obtained from finite element model of the bumper. **Fig. 7.6** shows the 3-D solid model of an automotive bumper beam.

Cheon *et al.* [30] reported their work on the development of glass fiber reinforced polymer composite side-door impact beams for passenger cars. Static tests were carried out to determine the optimum fiber stacking sequences and cross-sectional thickness for the composite impact beams taking into consideration of the weight saving ratio compared to the high strength steel. Dynamic tests were carried out at several temperatures using pneumatic impact tester, which was developed to investigate the dynamic characteristics of impact beams at a speed of 30 mph. They also performed finite element analysis using ABACUS software to compare the simulation results with the experimental one and good agreement was achieved.

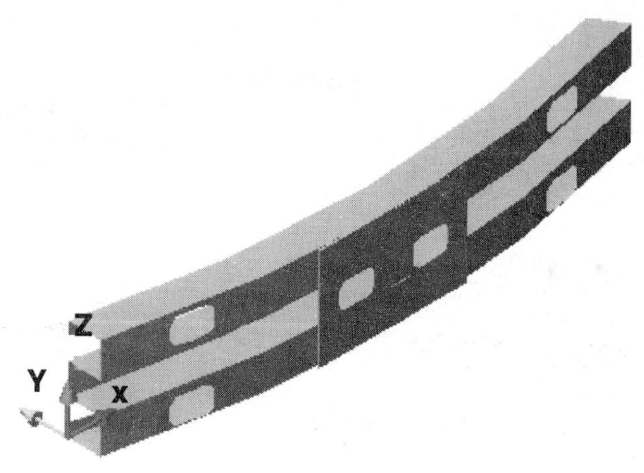

Fig. 7.6. Three-dimensional CAD automotive bumper beam (Courtesy Mr. Abdul Hadi Abdullah).

7.6. Engine Components

New materials in engines will be important to improve performance, fuel consumption, smoothness, quietness and reliability. These characteristics can be obtained from employing reciprocating parts of low friction and improved thermal control. The use of polymer composites could solve some of these problems. It is reported by Harrison [31] that injection moulded thermoplastic and thermoset polymer composites were used in intake manifolds, pump bodies, housings and cover to replace aluminium alloy castings.

7.7. Connecting Rod

Gunyaev *et al.* [32] investigated the possibility of manufacturing novel carbon fiber reinforced polymer composite (CFRP) connecting rod for car engines. Composite connecting rod offered some advantages like weight reduction, reduction in inertial forces, increase in automobile speeds and reduction in noise emissions from engines. Several studies have shown the possibility of manufacturing composite connecting rod satisfying the load requirements for engine components.

7.8. Air Intake Manifold

The function of air intake manifold is to distribute petrol and air mixtures from the carburetor as evenly as possible to each cylinder in the engine. It means that internal flow in the air intake manifold must be carefully smoothened. Increased efficiency of air intake manifold will lead to improved engine efficiency [33]. In the past various investigators have studied the polymer composite air intake manifold for automobiles. Majority of the air intake manifolds involved manufacturing by injection moulding process using low melting point fusible metal cores to produce complex internal channels. The materials are glass fiber reinforced PA 66 composites.

According to Harrison [34], composite manifolds for newer models of automobiles are being designed and manufactured from nylon-6,6. The BE91 Fiesta features a composite manifold on the new 1.25 Zetec SE engine. Some manifolds with simple shapes are now moulded in two halves and vibration welded together in order to reduce cost.

8. THE APPLICATION OF COMPUTERS IN THE DEVELOPMENT OF AUTOMOTIVE POLYMER COMPOSITES: A CASE STUDY OF POLYMER COMPOSITE PEDAL BOX SYSTEM

8.1. Computer Applications in Composite Design

The use of computers in engineering design has become an acceptable route in most engineering projects. The interest in design of composite components has caused by the advent of computer simulation packed with a multitude of analysis programs. Engineers no longer do the task of product analysis and product design separately by the task has been integrated and was performed concurrently by a sole engineer. Furthermore, these user-friendly analysis programs permit a designer to evaluate not only the form and function of a design but also the manufacturability of that design very early in the design cycle. This concurrent design approach is essential for polymer based composites since processing affects fiber orientation, which in turn greatly affects the mechanical properties of the composites.

This paperless product design approach has replaced much of the prototyping of the past. This is preferable since it reduced the rework involved in product design, time and cost of production. The use of such analysis program by product designers is one of the latest developments in the time compression trend within manufacturing.

The use of computers in the development of automotive components is considered a must activity. The trend in the recent years to developed cars is by using concurrent engineering methods as will be explained in the proceeding sections. One of the essences of concurrent engineering is the integration of information and communication technology (ICT) tools in order to facilitate integration and monitoring the various designs and manufacturing activities at various departments, localities or even countries on on-line basis. The use of advanced ICT tools such as computer aided-design (CAD), finite element analysis (FEA), expert system, mould flow analysis, materials and engineering databases, rapid prototyping (the development of rapid prototyping model is heavily dependent on IT tool) and other packages has become part and parcel of automotive composite development programs.

This section describes the use ICT tools in the design of composite pedal system. It is because, in the past 12 years, the author has been involved in the design and fabrication of polymer composite pedals. Therefore, majority of the work reported in this paper is concerned with polymer composite pedals.

8.2. Pedal Box System

A pedal box system is a component in the automobile, which receives the driver's commands centrally, and transmits them to the appropriate devices-engine, clutch or wheel brakes. In case of manual transmission, it consists of four major components, namely, a mounting bracket, a clutch pedal, a brake pedal and an accelerator pedal (**Fig. 7.7**). In recent years there has been a move to use polymer composites in car pedals. The proper selection of a material system and processes for polymer composite part depends on design and manufacturing cost, performance and size of the component. In automotive industry, glass fiber composites rather than carbon fiber composites are normally used, however, this is changing due to reduced costs of the latter. Injection and compression moulding processes are used for volume production.

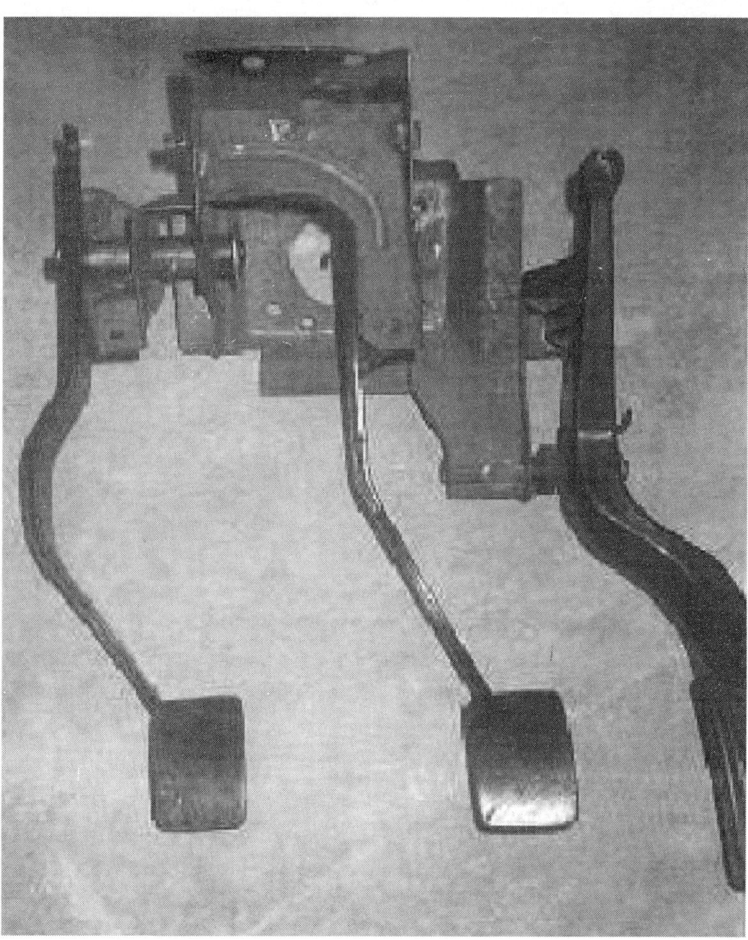

Fig. 7.7. Metallic pedal box system. Reproduced with permission from Lee *et al.*; Sudan Engg. Soc. J., 51(44), 23 (2005). Copyright © 2005, Sudan Engineering Society.

In mid 1980s, car manufacturers, plastic raw material suppliers and injection moulding companies started to use polymer composites in car pedals, Volkswagen and Ford were among the pioneers. In

1986, Ford, through its supplier, Birkbys Plastics of UK, developed polymer composite pedals for its Fiesta model.

Injection moulding is normally used in the manufacture of polymer composite pedal box system because of greater design freedom, reduced assembly requirements and greater quality consistency. Polyamides are generally used as the resins in car pedals because of higher impact strength and notched impact strength compared to some resins such as polyoxymethylene (POM) and polyesters. Polyamides possess excellent toughness and stiffness as well as the resistance towards chemicals used in automotive operation and service, e.g. greases, oils, cold cleaners etc., compared with POM and polybutyleneteraphthalate (PBT) are the factors for their selection.

The use of polymer composite pedal boxes is still in an early stage of development. It faces a lot of technical challenges and problems, although it has some advantages, especially as far as weight reduction is concerned. Polymer composites are generally more expensive than traditional materials such as steel and aluminium. There is still lack of confidence among the car manufacturers in using polymer composites as pedal materials especially in highly loaded components such as brake pedal. Until now, not a single car manufacturer, to the knowledge of the author, uses 100% polymer composite in brake pedal because of its critical nature in safety.

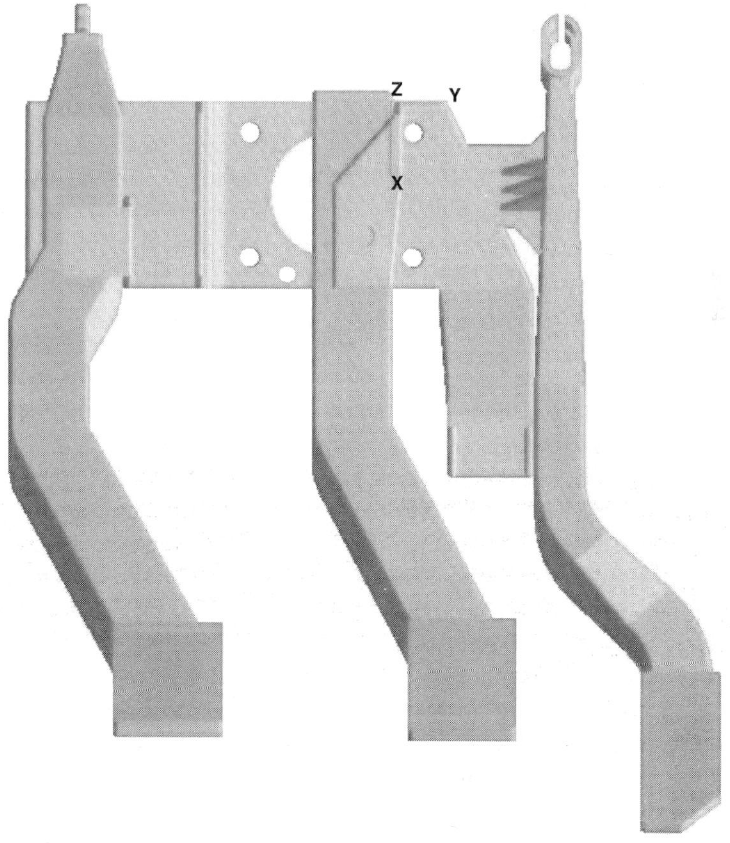

Fig. 7.8. Three-dimensional CAD automotive composite pedal box system. Reproduced with permission from Lee *et al.*; Sudan Engg. Soc. J., 51(44), 23 (2005). Copyright © 2005, Sudan Engineering Society.

8.3. Metallic Versus Composite Pedal Box System

Fig. 7.7 shows the existing metallic pedal box system. The existing accelerator pedal for Proton Wira was already made from composite material; a thicker body than the clutch and brake pedals that were made from steel can easily identify it. Hence, basically the author's design of accelerator pedal has little difference from the existing pedal. Throughout the accelerator design process, many different measurements were performed on the existing pedal and the data were used to design the new composite accelerator pedal. Both pedals have pedal pads for the driver to step his foot on it. **Fig. 7.8** shows the new design of composite accelerator pedal developed using Unigraphic CAD software system.

The pedal pad surface is grooved in the vertical direction. From the pedal pad, the pedal main body extends upwards and performed a left and right bend each. Then comes the slot hole for inserting a rod that connects the pedal to the mounting bracket. The pedal body extends upwards and incorporates a few extrusions on the left side to form a stopper surface; the pedal body is extended upward a little more and ended with a hook structure for placing the accelerator cable end. Both of the pedal bodies were using open hollow section with ribbing. The existing brake pedal is made from steel. It consists of a flat pedal pad surface. A grooved rubber pedal pad is fitted on the steel pad. Whereas the grooved surface is incorporated in the body of the composite pedal itself. From the pad location the steel brake pedal also extends upward but with a few degrees towards the left. The steel brake pedal body extends with the direction halfway of the whole length and make a small right bend to extend straight upward. The composite pedal design follows the length profile of the steel brake pedal. The only difference is that the steel pedal has a thin body and the composite pedal has a wider body with open hollow section and ribbing. At the two-thirds of the length, the steel pedal has a stopper structure welded to the pedal body. On top of the stopper structure, it has a rubber button to act as a cushion between the contacting parts. In the composite pedal design, the stopper structure is integrated in the pedal design itself and can be injection moulded with the main body of the pedal. In both the designs, the stopper structure is used to activate the brake light sensor at the mounting bracket. The steel brake pedal has a cylindrical hollow structure welded horizontally at the end of the pedal and to be used for connecting the pedal with the mounting bracket. In the composite design, the cylindrical hollow structure is not included in the design but a hole must be included for a shaft to pass through in the assembly process.

In the case of clutch pedal, the design process of the composite clutch pedal is the same as used in the brake pedal process. The composite clutch pedal design started with a pedal pad with the same size with the brake composite pedal pad. Then the body extends in the same direction as the brake pedal that is upwards with a few degrees towards the left. The body extends in this direction until about one-thirds of the whole length and turns right to extend straight upwards for a short distance. It then makes a right and left turn ending up with the body extends straight upwards again only at an offset position to the right. The body of the pedal extends upward of another 1/3 of the whole length until a hook structure is formed at the end of the extension of the pedal. The hook structure is used to connect the clutch cable of the car. In the middle of the last upward extension, a hole is made in the composite designs same as the steel pedal to allow a shaft to pass through in the assembly process. The steel pedal and composite pedal have the same body profile and the difference is that the steel pedal is a thin sheet of metal and the composite pedal is a wider body with open channel cross section with ribbing. All designs of composite pedals have the open channel cross-section facing downward with ribbing.

In the design of the composite mounting bracket, it is decided to create its own design with some important criteria. The criteria are the interface of the mounting bracket with the bulkhead of the car and the interface of the mounting bracket with the three pedals must be the same with the existing steel mounting bracket. The composite takes any shape necessary to comply with interface conditions. The composite mounting bracket is designed using a simple approach. It consists of a base plate with hole to accommodate bolt for the bulkhead and the brake pump assembly. On the base plate, a few extrusions are made at a require height and a hole is designed to have the same size with the pedal to accommodate the assembly shaft. Stopper structure was also incorporated in the design of the mounting bracket. The important design criteria to take care in the composite mounting bracket is that the position of the hole in three dimensions with reference to the base plate must coincide with the steel mounting bracket in terms of position and orientation for the assembly shaft, return spring and other minor components to be shared by both the pedals [35].

In the entire design process of the pedal box system, one important aspect that was remembered is the ultimate goal to produce a composite pedal box system that can be fitted to the existing vehicle. In order to do that, one of the ways is to design a pedal box system that has exactly the same interface as in the metal pedal box system. The two important interfaces are the interface between the driver with the pedal box which is where the pedal pad location for the foot to step on it and the interface between the pedal box with the vehicle which is actually the mounting bracket bolt hole location and brake pump location that must coincide with the bolt hole on the bulkhead of the vehicle. In order to achieve the above-mentioned criteria, the easiest approach is to design the composite pedal box system with exactly of the same size, shape and dimensions. When designing the mounting bracket, the base plate is copied from its metal counterpart with several gussets extruded out from the base plate to accommodate the three pedals. For the design of accelerator pedal, there is already an existing composite pedal produced by Proton. So the current design attempted to copy the whole sizes, shape and dimension of the sample brought from a local scrap yard.

The original metal brake and clutch pedals are in the forms of thin sheets of metal with a uniform height with some bends along the body of the pedals. Hence, it was decided to use the same height of the metal pedal with the same length and bend profile and created a bigger width for the body of the pedals. The other minor details of the design such as the hook structure for holding the clutch cable and throttle cable, the hole for the brake pump actuator, the stopper structure and the hole for placing a shaft to assemble the pedal to the mounting bracket were designed by copying the existing metal part.

The current decisions to design the pedals were applied to mimic the existing metal part with some exception where modifications have been made to accommodate the need for the pedals to be able to be fabricated using injection moulding method.

8.4. Finite Element Analysis

The main objective of finite element analysis (FEA) was to determine a meaningful structural design prediction. The results were used as guidelines for the development of pedal box system. Two major areas of work were carried out, namely, ribbing pattern determination and stress analysis in the individual parts in pedal box system.

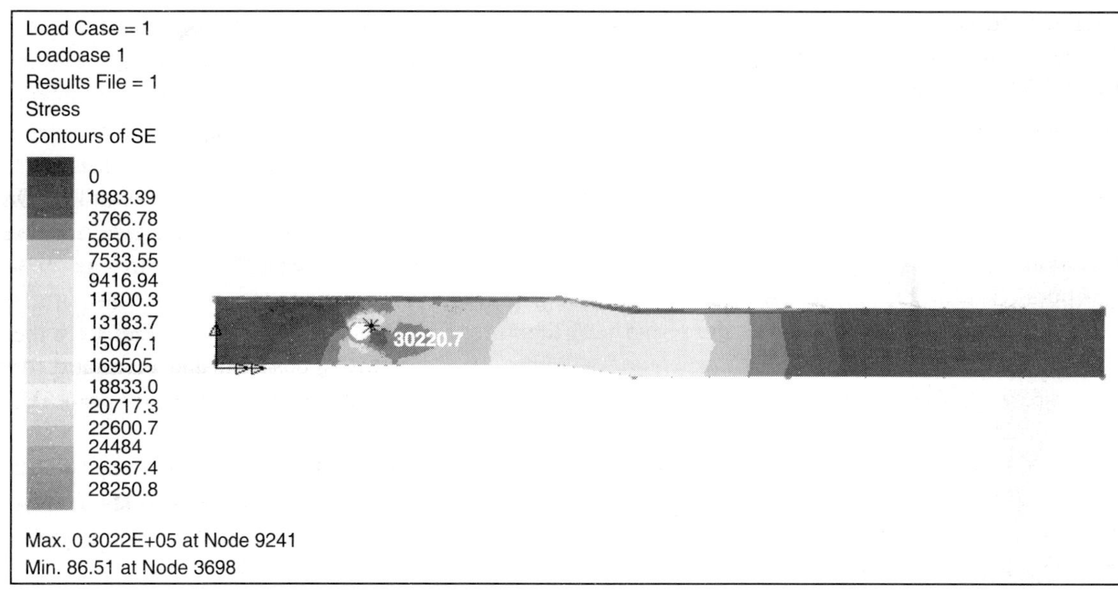

Fig. 7. 9. von Mises stress contour for 'T' profile with constant flange and web thickness of 2 mm (courtesy Mrs. Sharifah Imihezri).

A comparative study has been carried out for the two types of profiles using FEA software (LUSAS) for a constant wall thickness of 2 mm for both flange and web. The von Mises stress [36] results of the study of "I" and "T" profiles are depicted in **Figs. 7.9 & 7.10** respectively. The

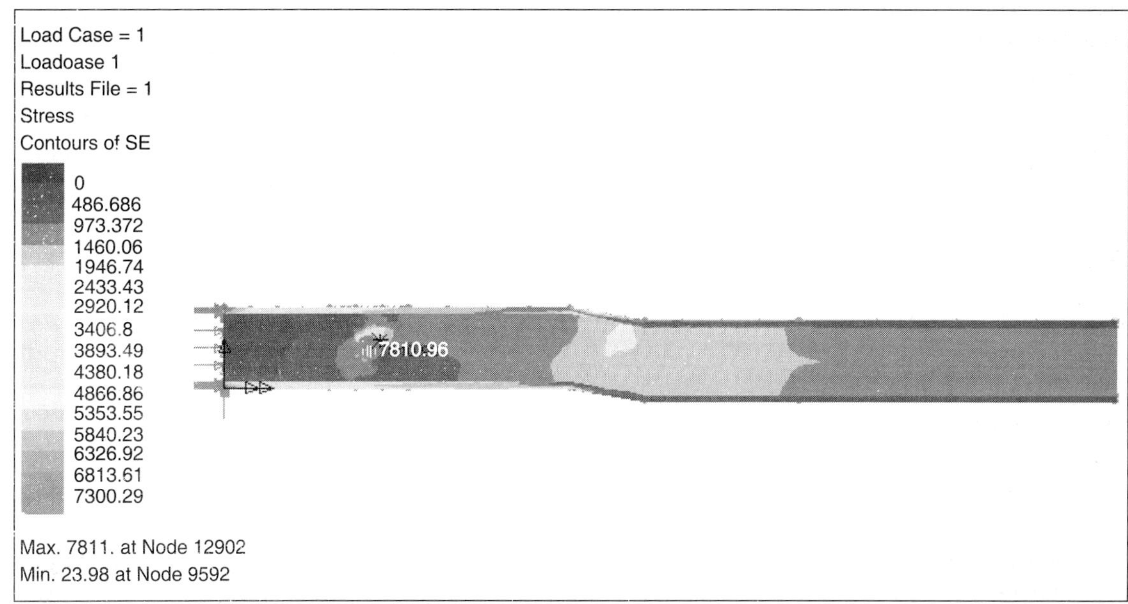

Fig. 7.10. von Mises stress contour for 'I' profile with constant flange and web thickness of 2 mm (Courtesy Mrs. Sharifah Imihezri).

material used is polyamide 6,6 reinforced with 30% short fiber reinforcement. With an applied load of 1960 N normal to the surface, the results show that higher stress occurs for a "T" profile (30.20 MPa) compared to an "I" profile cross sections (0.8 MPa). The contour plot of the von Mises stress also illustrates the regions of high stresses.

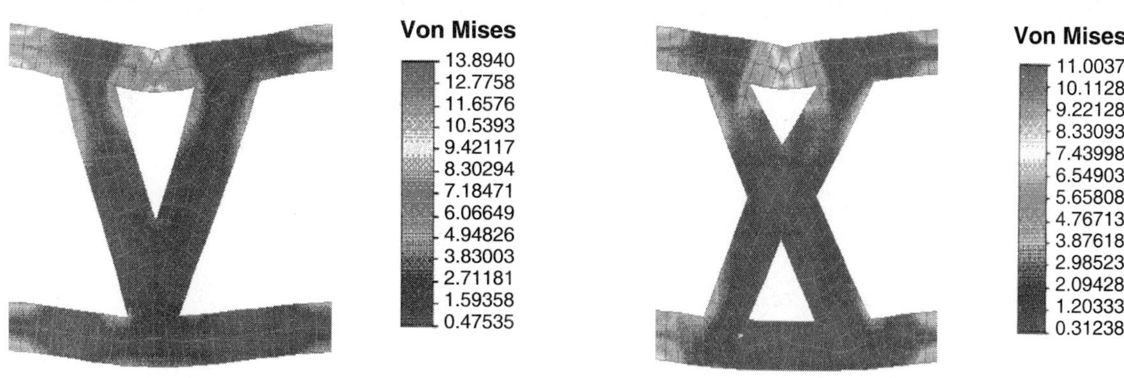

Fig. 7.11. Stress contour for "V" rib.

Fig. 7.12. Stress contour for "X" rib.

The models of "V" and "X" patterns for ribbing structures were carried out in 2-dimensional form. As can be seen in **Figs. 7.11 & 7.12**, only one element of "X" and "V" pattern was drawn to simplify the model.

The figure also shows the Von Mises stress contour plots for both types of ribbings. Maximum stress for "V" rib is 12 MPa while maximum stress for "X" rib is 15 MPa. Volume and weight calculations for similar sections showed that an "X" rib is slightly higher than a "V" rib with 360 g and 352 g respectively.

8.5. Mould Flow Analysis

The mould flow analysis of injection moulded short glass fiber reinforced nylon-6,6 composites for clutch pedal was carried out. The mould flow analysis was performed using a commercial software package Moldflow Plastic Insight (MPI). Design efficiency requires that the thickness of wall section to be decreased or increased thus reducing the product strength. In order to compensate this feature, ribbings, contours and corrugations are commonly incorporated. The addition of ribs for instance not only improves the stiffness and rigidity but also enhances mouldability since it hastens melt flow in the direction of rib. Hence, two types of ribbing patterns are considered, namely, "X" and "V" ribs. The MPI software is used to predict and simulate the fill pattern, fill time, air trap, weld line, temperature and pressure distribution for these two rib designs. The simulation study on the effects of single and double injection locations on fill time, pressure, temperature, air trap, weld line and fiber orientation was also performed for a fiber reinforced composite clutch pedal. A composite clutch pedal was fabricated using injection-moulding machine. The results are then compared and finally, the chosen rib design is incorporated as part of the composite clutch pedal design. The computer simulation presents only the elements of the "V" and "X" ribs with a standardize thickness of 2.5 mm and ran at optimum settings [37].

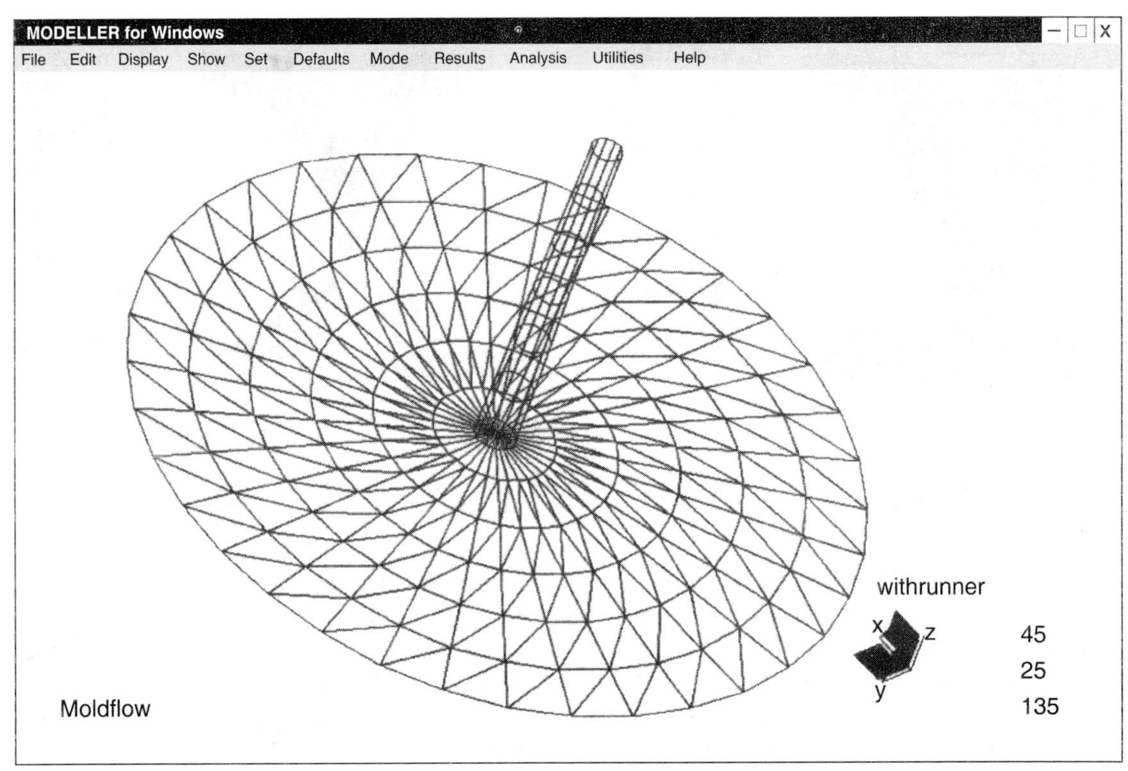

Fig. 7.13. The moulding model is created and meshed using the Moldflow Modeller software (courtesy Mr. Edisyams Zainudin).

The fill time result represents the behavior of the melt polymer at regular intervals. Thermoplastic polymer flow inside the mould is simulated using a program that calculates a flow front that grows from interconnecting nodes at each element, starting at the injection node. The cycle repeats until the flow front has fully expanded to fill the last node. One of the goals in selecting polymer injection locations is to ensure that all flow paths in the cavity fill at the same time (balanced flow paths). This prevents overpacking along the flow paths, which might otherwise fill first. The fill time analysis can be used to estimate the possible areas of short-shot, hesitation, overpacking, weld line and air traps. **Fig. 7.13** shows an example of the moulding model created and meshed using the Moldflow Modeller software. **Fig. 7.14** shows one of the capabilities of Moldflow namely the selection of material from the material database provided by Moldflow.

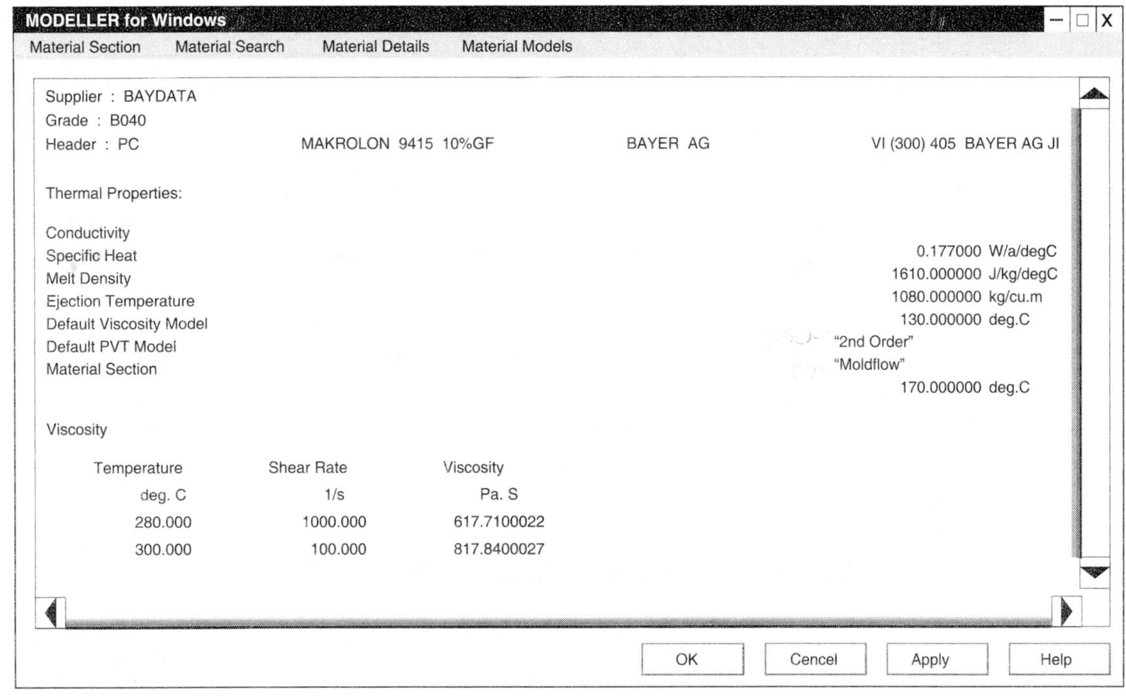

**Fig. 7.14. The material is selected from materials database, provided by Moldflow
(Courtesy Mr. Edisyams Zainudin).**

8.6. Rapid Prototyping

In the study conducted by the author on rapid prototyping, the methods called stereolithography (SLA) and 3D Printer were employed. The processes of SLA (**Fig. 7.15**) and 3D Printer (**Fig. 7.16**)

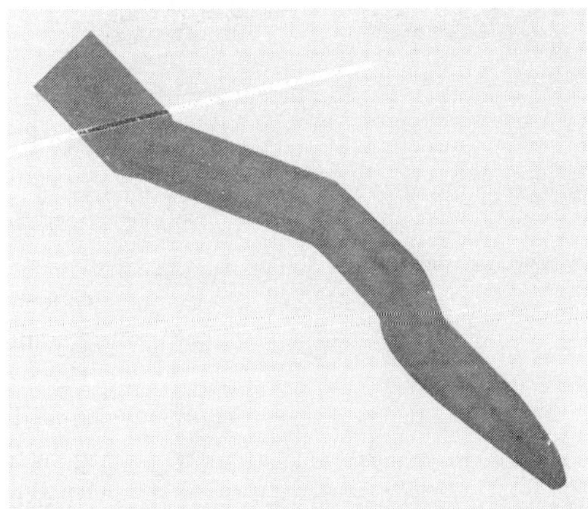

**Fig. 7.15. Rapid prototyping model (SLA) of
composite clutch pedal (Courtesy Mr. Lee Ho Boon).**

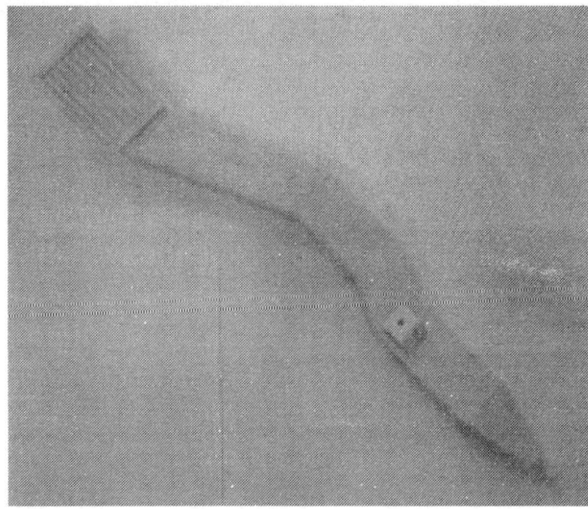

**Fig. 7.16. Rapid prototyping model (3D Printer) of
composite clutch pedal (Courtesy Mr. Lee Ho Boon).**

were used to develop the rapid prototyping models for the composite clutch pedals. The SLA 3500 rapid prototyping machine developed by 3D Systems, Inc. and 3D Printer by Z-Corp. The former was selected because it is readily available at the Universiti Putra Malaysia and the latter was sub-contracted to a local company to fabricate.

The basic operation of the SLA machine is that it has a laser, which is beamed to the container of a photosensitive resin in a specific pattern. The laser beam has a diameter of 0.20-0.30 mm. The container or the vat has a volume of 99.31 L with built envelope of 350 × 350 × 400 mm (XYZ) and maximum part weight of 56.8 kg. The resin used was RPC100ND-3 and was in transparent form.

The 3D Printer machine used has an in built envelope of 203 × 254 × 203 mm. The layer thickness was user selectable from 0.076 to 0.254 mm. The powder used was the plaster based ZP 100 powder provided by Z-Corp. The plaster-based powder delivered high-strength and detail. The starch-based powder delivered high speed at a very low cost as well as being suitable for investment casting. In this process, a cynoacrylate adhesive of medium viscosity was used to strengthen the model created.

9. INTEGRATED DESIGN AND MANUFACTURING (CONCURRENT ENGINEERING) APPROACH FOR POLYMER COMPOSITE AUTOMOTIVE COMPOSITES

Automobile manufacturers have used a more systematic way in the design of automotive components through the adoption of new approaches and powerful tools by implementing concurrent engineering. The traditional approach to the engineering design of components is highly sequential with design tasks being performed without due consideration being given to the final products. This type of engineering generally fails to consider the impact of design decisions on later design refinements and it has been termed 'over the wall' syndrome [36]. The essence of concurrent engineering is the concurrent rather than serial, execution of various phases in the product development process as depicted in **Fig. 7.17**. It entails simultaneous consideration of various product life cycle functions at the very outset of the product development in order to achieve high quality product, produced at

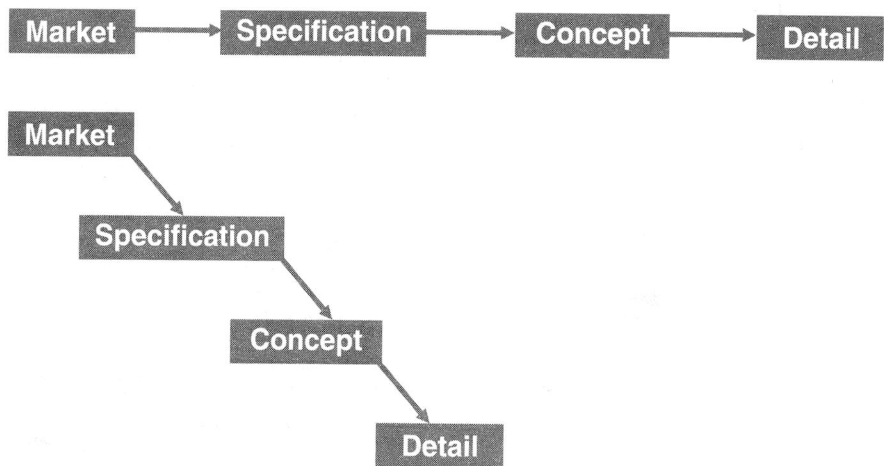

Fig. 7.17. Sequential engineering versus concurrent engineering.

shorter time and lower cost [36]. It is a way to reduce the lead time between starting a design and manufacturing a product by ensuring that design for production starts as soon after the start of the product design as possible [17].

Concurrent engineering (CE) is increasingly being used as an approach for manufacturing companies wishing to remain competitive in the modern marketplace. CE is a blending of all different areas and disciplines within a company, in order to improve the interfaces between design, engineering and manufacturing. The basic underlying philosophy is to facilitate the concurrent design of a new product as well as the manufacturing process, tooling and quality system for the production. The reduced time to market and expensive rework cost, quality improvement and teamwork are the essence of CE.

Major benefits have been gained by pioneers of the automobile industry as a result of practicing concurrent engineering over the last few decades. For instance, Henry Ford, Ransom Olds, Karl Benz and Adam Opel did not limit themselves simply to design the products but were product and process design engineers who designed the both, cars and factories that make the cars. As these industries developed and became larger, work became specialized to the point where a designer would come up with a product idea but expect a manufacturing engineer to work out how to make it [38]. Concurrent engineering (CE) has been responsible for improvements in the automotive development process. The use of CE has resulted in numerous benefits including:

- Involvement of all functions and personnel including project, designers and processes engineers.
- Engineers, cost analysts, marketing and sales personnel.
- Better processing considerations.
- Improved manufacturing launch considerations.
- Fewer revisions to the product during manufacturing.
- Improved employee involvement and satisfaction.
- Management involvement and acceptance.

In the development of automotive components within a CE environment, activities like material selection, design and analysis, components specifications, customer experience, process selection and design modification play equally important roles for the successful development of a product as presented in 'CE wheel' shown in **Fig. 7.18**.

It was pointed out by Mayer [17] that concurrent engineering (CE) is essential in polymer composite design because of the way in which materials properties are only fully realized on manufacture. Polymer composite products are made through tailor-made design, which means that while designing the products and selecting the materials, manufacturing process is studied at earlier stage of design and it is in fact an essence of concurrent engineering [39].

Concurrent engineering is a systematic procedure to optimize the manufacturing process in design of components. To get maximum benefit of concurrent engineering, the procedure is used as early as possible in the design process. The concurrent engineering results into simpler and more reliable products, which are higher in quality (Q), developed at shorter time (T) and less costly (C) to manufacture as depicted in **Fig. 7.19**.

It is reported by Quinn [40] that in composite design process, it consists of few iterative steps, in order to obtain the desired solution. The principle of design process is similar to non-concurrent engineering approach, except the geometry or shape is designed concurrently with materials selection

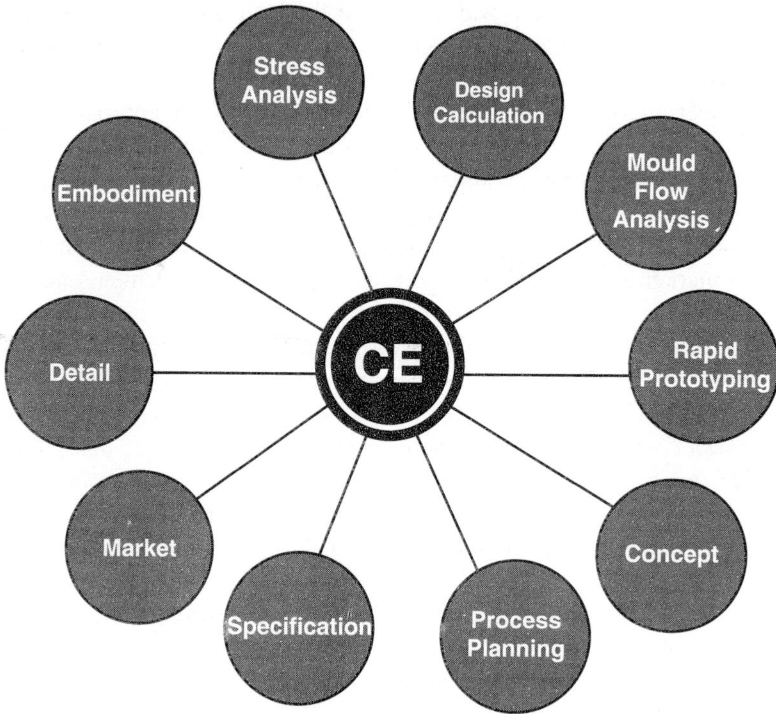

Fig. 7.18. Concurrent engineering wheel.

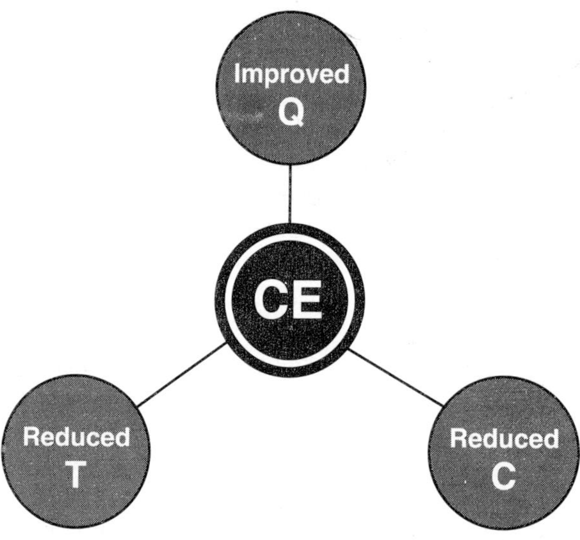

Fig. 7.19. Concurrent engineering benefits: improvement in QTC.

in this approach while the selection of materials is highly dependent on manufacturing process. In other words, component design, material selection and manufacturing method are considered early in design process, which is the essence of concurrent engineering.

It is stated by Murphy [4] that CE is possible through the use of various IT tools in the design process of composites carried out by design team, the mould maker and the processor. They are all linked together electronically so that each can make a contribution throughout the progress of the project. For instance, the use of Mold Plastic Insight enables the designer and mould maker to evaluate part geometry, design and positioning of gates and runners and prediction of mould shrinkage and stress pattern.

Schwartz [41] predicted that a small software could design both the products and the production method. Concurrent engineering of composite materials is the technique needed to meet and keep a generation ahead of competition and stay focused on better manufacturability. The work of Quinn [40] is devoted for more general purpose and what was described by Murphy [17] and Schwartz [41] can be perceived as general-purpose application.

For automotive composite development, design and manufacturing process are generally carried out in isolation [42]. In fact, many other concurrent engineering features are not fully considered in composite production such as teamwork, integration of IT tools, involvement of other personnel in

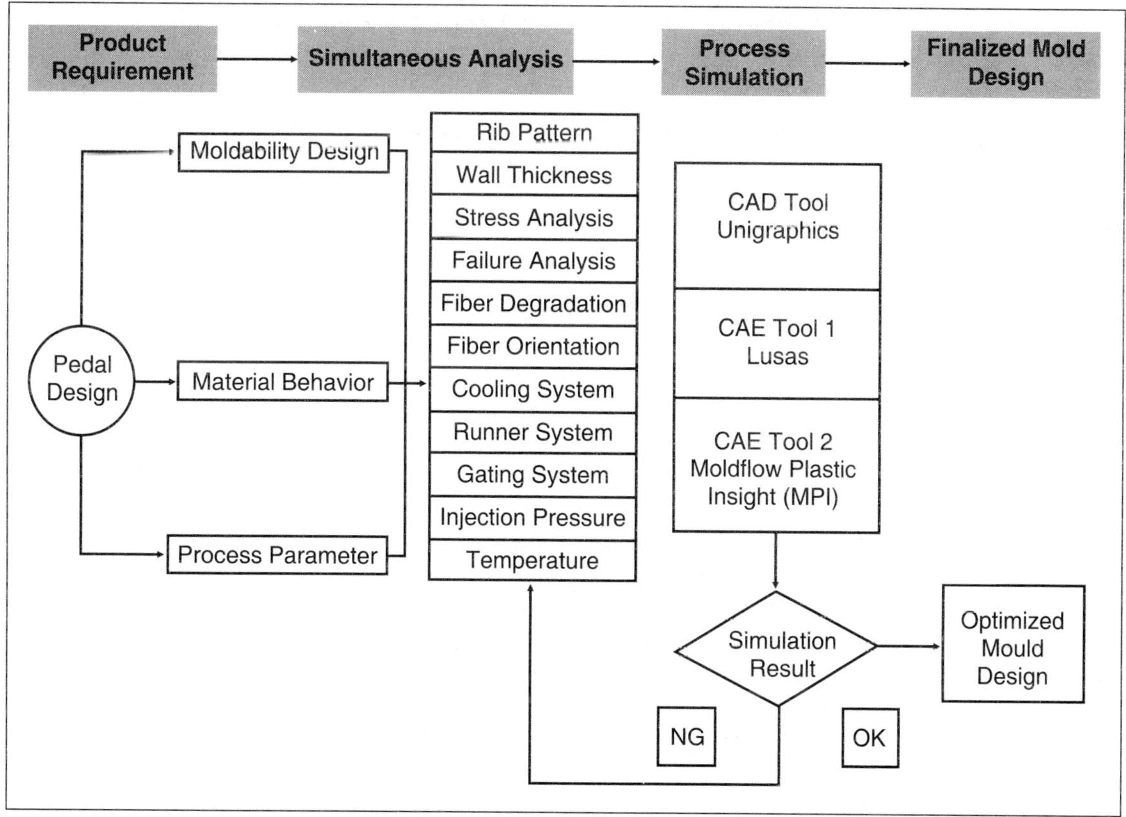

Fig. 7.20. Current concurrent engineering in composite development model (Courtesy Mrs. Sharifah Imihezri).

the organization and other issues. This situation is not only restricted to automotive composites, but also applies to automotive industry at large [43]. Recent work of the author [44] has provided some solutions to those issues. It presents the work on the use of concurrent engineering in the development of polymer composite automotive clutch pedal. It covers the use of various IT tools such as expert system, FEA, CAD, mould flow and rapid prototyping in order to carry out various activities such as material selection, total design, design analysis and mould flow analysis. The integrated IT tools enable the designer to design and manufacture automotive composite clutch pedal at higher quality and faster speed compared to its metal counterpart. By adopting composite design guides, weight saving from implementing composite materials in the clutch pedal is achieved. **Fig. 7.20** shows the flow chart of the recent work of the author on CE for automotive composites.

10. MATERIAL SELECTION BY KBS SYSTEM FOR POLYMER BASED COMPOSITE BUMPER BEAM

Polymer composites are playing an ever-increasing role in engineering applications. They are used in the manufacture of components for aircraft and automobiles besides domestic articles. Material manufacturers are trying constantly to improve and develop these types of materials. Suppliers of these materials are endeavoring to provide materials that meet customers' requirements and expectations. This is the principal reason for their rapid progress in terms of utilization and product diversification.

Material selection is an activity normally performed by design and material engineers. The purpose of material selection is to identify the materials, which after appropriate manufacturing operations, give the dimensions, shape and properties necessary for the end product or component to perform its required function at the lowest possible total manufacturing cost.

The range of materials available to a designer is vast and constantly increasing. This poses a huge challenge to designers. For instance, during the selection process, such a polymer composite must satisfy a complex set of design requirements and be at the same time cost-effective to be a successful candidate. The fundamental question is which polymer composite material is the most suitable and whether the selected material will satisfy the imposed requirements?

Currently there are thousand types of polymer composite materials that can be selected for automotive applications. Many systems are available to assist the designers to select suitable composite materials. At the most basic level, designers may work through tables of properties in various data books. However, the nature of the current data available does not meet today's advanced industry requirements, as data sheets are often incomplete and once published are difficult to update. With polymer composites, material selection has become too complicated to manage it unaided and the price of failure is too high to contemplate. These choices will not become any easier. Thus, it is hard to see how designers can optimize their choice without some form of computerized selection tool. Therefore, industrialists have implemented computer based systems and intelligent material selection tools. These design problems can be more easily and effectively tackled, if a computerized knowledge-based system (KBS) is employed to assist in the selection of materials. Artificial intelligence methodology provides computer-based tools, which can be combined with material knowledge to devise KBS for material selection. KBS is particularly suitable for the processing of unstructured scattered knowledge for the solution of complex problems. One such problem is the selection of materials like composite materials. Definitions about a knowledge-based system and its functions have been well defined by the work of [45]. A number of systems have been developed to select materials for a specific operation or a set of operations.

A KBS is a branch of artificial intelligence that uses a collection of different programming techniques and programming languages that enables computer to mimic human thinking and reasoning the process. A typical structure of KBS is shown in **Fig. 7.21**. Sapuan and Abdalla [46] developed a KBS for the selection of polymer composite materials for automotive components, in particular for pedal box system. Mechanical, physical and chemical properties together with economic and manufacturing considerations were taken into consideration during the material selection process. The KBS enabled users to select the suitable materials that satisfied all predefined criteria. In this system, the KBS shell KEE (Knowledge Engineering Environment) serves as an excellent tool to process expert knowledge. KEE was chosen because the rule-based system used for selecting materials is of pattern matchers. These pattern matchers are extremely flexible and powerful. The intelligent program of a KBS consists of an inference engine and a knowledge base. Closely associated with this intelligent program was a data or fact base.

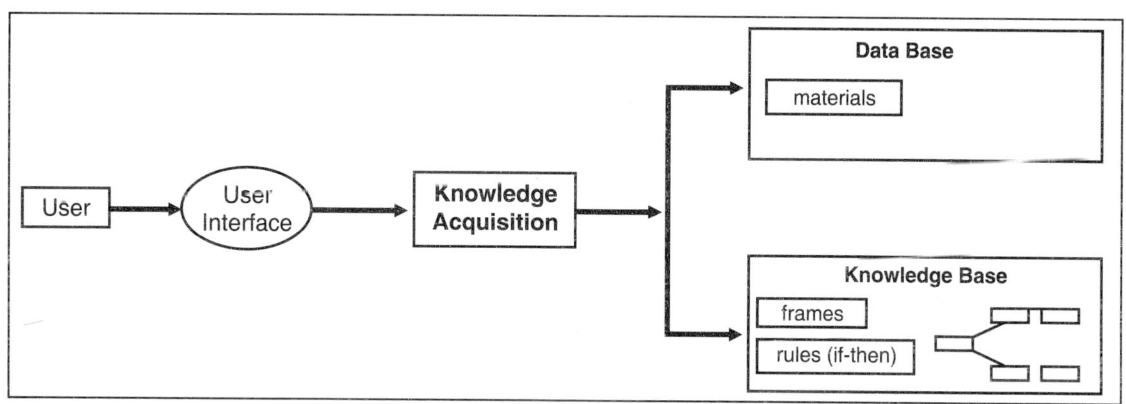

Fig. 7.21. A typical structure of a knowledge-based system (Courtesy Mr. Abdul Hadi Abdullah).

Another development of a prototype knowledge-based system (KBS) for material selection for polymer-based composite automotive bumper beam is carried out and some of the findings are presented in this section. The aim of this knowledge-based system is to select the best polymer-based composite material for bumper beam according to the required input entered by the user. The selections are based on the constraint values corresponding to the product design specifications (PDS). Product design specifications considered are mainly from mechanical and physical properties. The polymer-based composite materials properties are obtained from the material database as well as from the past experimental data compiled by other researchers. The KBS, which employs Kappa-PC 2.4 tool-kit, consists of the user interface, knowledge acquisition and inference module. All the domains are linked together by hierarchical graph. The system enables the designer to select the most appropriate material for reinforcing bumper beam according to the desired mechanical and physical properties.

10.1. Overall System Description

The system was developed using Kappa-PC tool kit developed by Intellicorp Inc. The prototype is developed using Kappa-PC 2.4 tool-kit as the interface application. The programming is a common programming language that could be used by the user by entering the value required by the system.

The proposed material selection knowledge based system for polymer based composite bumper beam consists of use interface, knowledge acquisition, inference engine, knowledge base and database. Object oriented programming techniques and rule-based techniques have been used for developing the application. Knowledge base consists of frames and rules. The rule describes the condition at which the selection procedure is carried out. It also gives the solutions for constraints. The database consists of the materials and their properties. The inference engine enables communication between the user and the knowledge base, reasons the facts and gives the solution. **Fig. 7.22** shows the simplified flow chart of the research work.

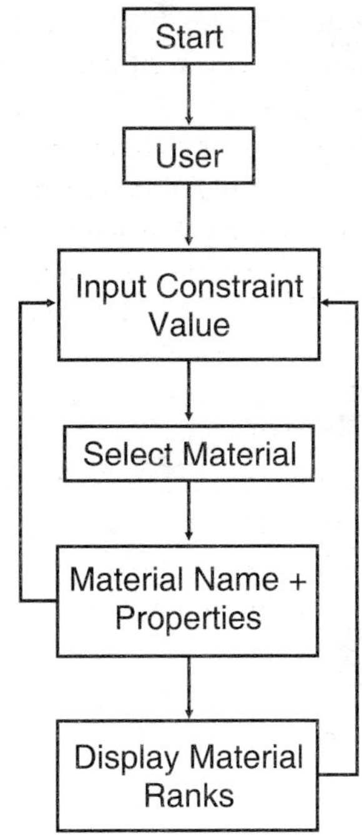

Fig. 7.22. Flow chart of material selection process (Courtesy Mr. Abdul Hadi Abdullah).

10.2. User Interface

A user-friendly interface is created using buttons, text, images and bitmaps. The user interface incorporates and organizes data that have to be evaluated for further evaluation. A session window, which is a basic component of a Kappa-PC interface is used to create an application interface. This system interface can be designed and formulated by the use of images such as buttons, text, bitmaps and transcript images. A user-friendly interface would be designed in such a way that when an image in the interface is clicked, it invokes the necessary function and starts running the inference process. **Fig. 7.23** shows a user interface, which is for material selection process for bumper beam from polymer based composite materials.

10.3. Inference Engine

The inference engine is an important element in a knowledge-based system. It finds the solution on the basis of rules by reasoning method. The reasoning method used is forward chaining. It looks through the knowledge base and finds the solution with respect to the rule constraints. The reasoning continues until a suitable result for the constraint defined in the rules is obtained.

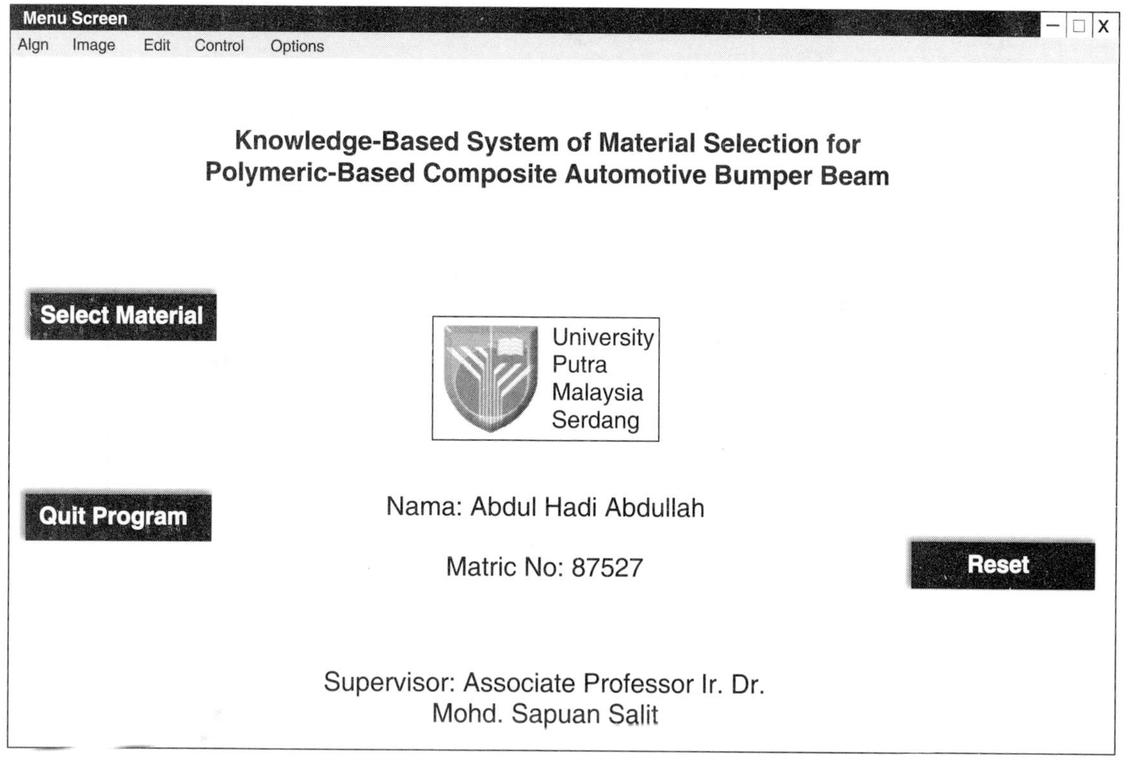

Fig. 7.23. User interface in KBS (Courtesy Mr. Abdul Hadi Abdullah).

10.4. Database Structure

The system database consists of three separate classes, namely, database for:
- Glass fiber reinforced composites.
- Natural fiber reinforced composites.
- Carbon fiber reinforced composites.

Glass and carbon fiber reinforced composites are the major types of polymer-based composites. In this research, an attempt is made to introduce natural fiber reinforced composites as an alternative. In each class, there are a variety of combinations among the reinforcements and the matrices. The properties of polymer-based composites vary depending upon the type of reinforcement, orientation of reinforcement and the volume fraction of reinforcement.

10.5. Knowledge Base Structure

The knowledge base consists of rules and techniques that are representing the knowledge in the structure. The domain of a knowledge-based system is represented as frames or objects. Object can either be classes or instances within classes [47]. The relationship between the objects was linked

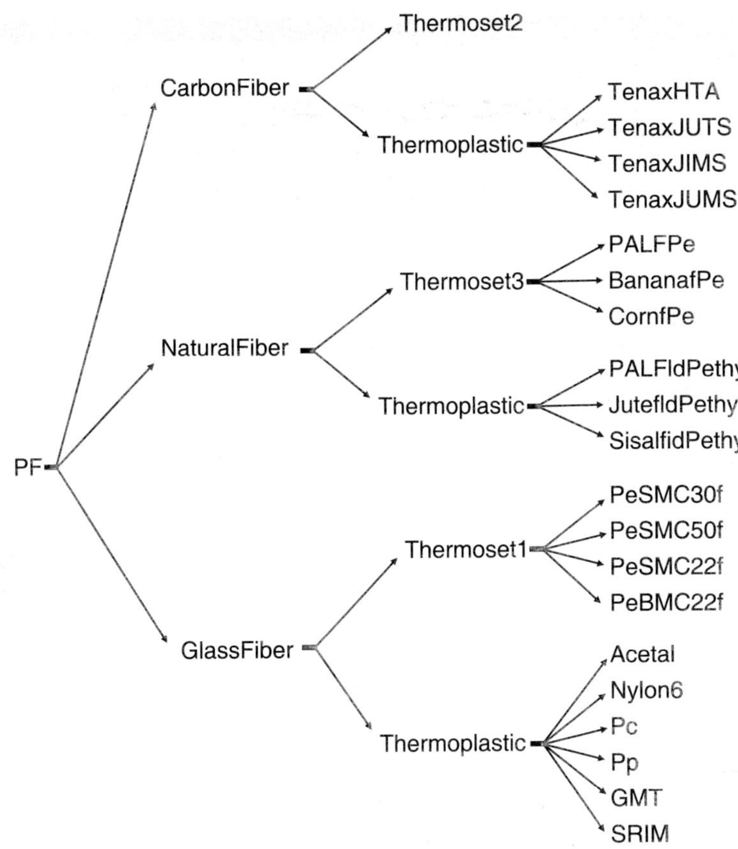

Fig. 7.24. Hierarchy graph of composite material (Courtesy Mr. Abdul Hadi Abdullah).

together by the way of hierarchy. Information related to polymer based composites and the types of reinforcement are represented as objects and their properties such as mechanical properties are used in the knowledge base. Data collected from journals and other sources and the classification are expanded with respect to reinforcement-polymer matrix combination and created as instance. The mechanical and physical properties such as material density, tensile strength, Young's modulus, flexural strength and even impact strength are created as slots. These slots are created in the parent and inherited by their children. The values for each property are stored as slot values. The slot values for each slot are stored in the slot editor window. Due to the difficulty in obtaining the suitable right mixes of various composites and to avoid the KBS from becoming too lengthy, the materials data of only 20 materials are studied and they are shown in object hierarchical graph in **Figs. 7.24 & 7.25** shows the slot and slot value window.

Bumper beam is represented in the knowledge base as object and linked in the object hierarchy. The constraint variables for the beam are represented as slots. **Fig. 7.26** shows the hierarchy relationship of the bumper system.

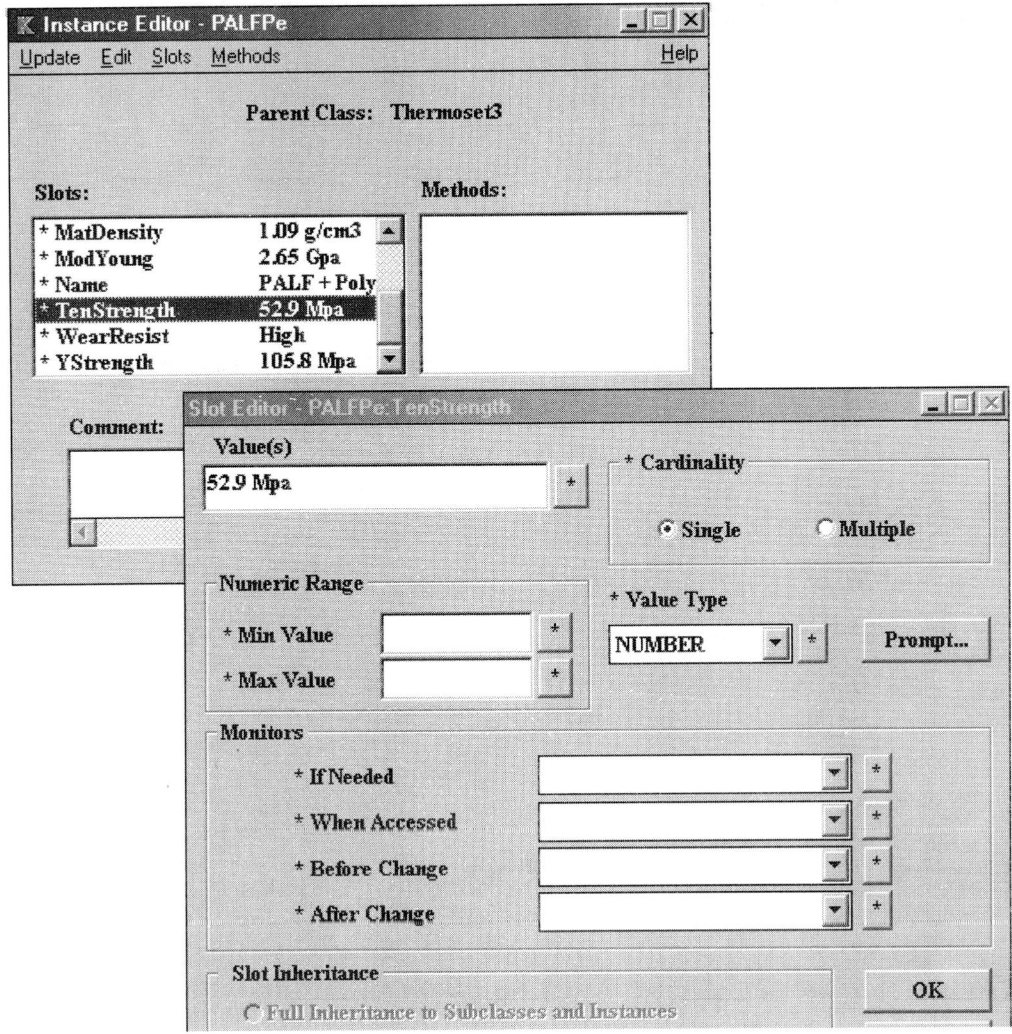

Fig. 7.25. Slot and slot value window in KBS (Courtesy Mr. Abdul Hadi Abdullah).

The sources of data that are used in this research include product design specification (PDS) values for bumper beam. The constraint values used in the rules are selected from these product design specifications. The product design specification values are obtained based on the existing bumper beam design data while some values of mechanical properties and physical properties such as flexural strength, flexural modulus, corrosion resistance, wear resistance as well as manufacturing technique are collected from material data base. **Tab. 7.1** shows the product design specification (PDS) of the automotive reinforcing bumper beam. By following some of the above steps and after carrying out some further steps (not described in this article for the sake of brevity), the user can select the most suitable composite material for the bumper beam.

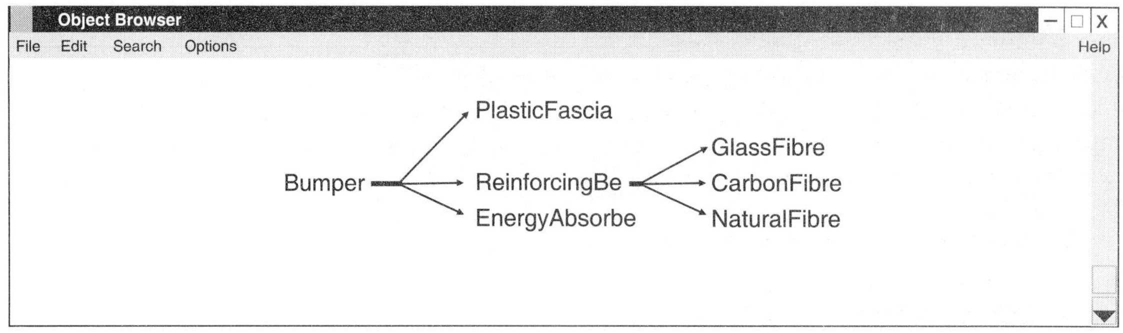

Fig. 7.26. The hierarchy relationship of the bumper system (Courtesy Mr. Abdul Hadi Abdullah).

Tab. 7.1. Product design specification (PDS) of automotive reinforcing bumper beam (Courtesy Mr. Abdul Hadi Abdullah).

Constraint variables	*Constraint values*
Material Density (g/cm^3)	< 1.90
Tensile Strength (MPa)	> 0.32
Yield Strength (MPa)	> 0.63
Young's Modulus (GPa)	> 0.02
Flexural Strength (MPa)	High
Flexural Modulus (GPa)	High
Corrosion Resistance	High
Wear Resistance	High

11. NATURAL FIBER POLYMER COMPOSITES IN AUTOMOTIVE INDUSTRY

The competition of technology among nations is to create better standard of living for the benefit of the mankind. Therefore, the mankind is seeking for the products of new design, material and production method. One of the main avenues of improvement is in the use of lightweight materials such as composite materials. However, traditionally composite materials employed well-established fibers such as glass, carbon and aramid as the composite reinforcement.

However, in many low profile applications, the use of natural fiber composites is gaining popularity due to their being cheap and easily available and having comparable mechanical properties with glass fiber composites. The use of agricultural based fiber composites is an emerging technology that is capable of replacing the conventional fiber composites such as carbon, aramid and glass fiber composites in selected non-structural applications. Among the natural fibers that are reported in literature being used as reinforcement for composites include oil palm, coconut coir and shell, rice husk, jute, henequen, cotton, baggasse, cereal straw, banana or abaca, pineapple leaf, wood, ramie, hemp, kenaf and kapok.

Malaysia is blessed with abundance of natural fiber resources and these materials have big potential to be commercialized as the substitute of conventional materials in many applications. In the past, the author has been involved in the study of properties, design and fabrication of natural fiber reinforced composite products mainly for furniture industry. Three types of natural fibers are studied, namely, coconut (including shell, coir and spathe), banana stem and pineapple leaf. The study started with carrying out determination of some important properties such as tensile, flexural, impact, electrical and thermal properties, density and morphological study using SEM of the above-mentioned fiber composites. Various different types of work were carried out based on those natural fibers such as:

1. Tensile and flexural properties of coconut spathe fiber composites.
2. Tensile and electrical properties of coconut coir composites.
3. Density, thermal insulation, tensile and flexural properties of coconut shell fiber composites.
4. Tensile, flexural, impact and morphological properties of pineapple leaf fiber composite.
5. Tensile and flexural properties as well as development of a Holy *Qur'an* stand (*rihal*), telephone stand, book shelf and coffee table from banana stem fiber composites.

Design process of making products followed total design approach to design products such as telephone stand, the Holy *Qur'an* stand (*rihal*), motorcycle pump, a small coffee table and a small bookshelf. The fabrication process is normally performed using conventional hand lay up process.

From this study, it can be concluded that natural fiber composites have great potential as an alternative material in furniture and other industries. But the present challenge is whether natural fiber composites can be used as automotive components or not (**Figs. 7.27 & 7.28**).

Fig. 7.27. Another source of natural fibre: can this new candidate be a suitable candidate for automotive component? (Courtesy Tuan Haji Nordin Osman).

Fig. 7.28. Can fibres from this wild tree be used as automotive component? (Courtesy Mr. Januar Parlaungan Siregar).

A recent trend is directed towards utilizing low cost, abundant and renewable natural fiber reinforced polymer composites in automobiles. Sufian [48] reported that plant based fiber composites are currently used in the door panels of the Mercedes-Benz E-Class. The wood fiber materials previously used for the door panels were replaced by a flax/sisal fiber mat reinforced epoxy composites. A weight saving of 20% was achieved and the mechanical properties were improved. The flax/sisal fiber composite was moulded in a complicated 3-dimensional shapes to make it more suitable for door trim panels.

A recent development reported by Buchholz [49] is on the use of natural fiber composite in seatback, rear package trays, A-, B- and C- pillar covers, center consoles, door panel inserts and headliners for heavy-duty trucks.

12. RECYCLING OF AUTOMOTIVE POLYMER COMPOSITES

Although the weight of polymer composites used is still small in comparison to total vehicle weight, the number of their applications is significant [50]. Removing these materials from cars raises the problem of recycling. The recycling of polymer composites raises some problems because of the difficulties of separating the constituent polymers. The recycling of advanced composites incorporating expensive reinforcement can be economic. The matrix can be removed by the use of solvents or by thermal oxidation to make the reinforcement available. Thermoplastic composites can be recycled using various methods previously developed. For example, most PET is reclaimed by a system that washes the PET and HDPE and separates them by the density differences (flotation or hydrocyclone process). PET recycling companies are also interested in depolymerizing PET polymers back to their monomers or oligomers in order to repolymerize them into a good grade recycled PET.

13. CONCLUSIONS AND FUTURE PROSPECTS

Polymer composites are expected to be more widely accepted in automotive industry if the total manufacturing and life-cycle cost can be reduced further. Automotive manufacturers have more confidence in using polymer composites in the components they manufactured due to intensive research in the recent years. The advantage of weight saving is the main driving force in applying these materials especially during the period of high petroleum price.

With gasoline prices so high and threatening to go higher and as per the oil price has gone up to US$65.32/barrel in September 2005, future direction in automotive industry is anticipated to be focusing on making vehicles lighter and thus more fuel-efficient [51]. Advances in nanocomposites and fillers not only allow for reduced weight but also exterior parts with high ductility and good performance with respect to coefficient of thermal expansion [52]. Nanocomposites are used in body side moulding in Chevrolet Impala and in trim and sail panel in Hummer H2 [53]. Nanocomposites are expected to play greater role in future car.

"What will motoring be like 100 years from now?", "What sort of vehicles will we be driving?", "Will there be any cars in the year 2104?". These are the questions posed by Pischetsrieder [54], Chairman of the Board of Management, Volkswagen AG when addressing the topic on future cars. Certainly no one knows about the future and only the Almighty knows best. But looking at the current trend, it is anticipated that in the future, issues like accident free motoring and balancing the individuals' need for mobility with the decline in fossil-fuel reserves are going to be the main concerns. The solutions available could be the introduction of new electronic assistants that can 'see' and enable vehicles to communicate with one another and the use of hydrogen fuels. Can polymer composites play any role in achieving those solutions? It is up to the materials engineers and scientists who are dealing with the polymer composites to provide some of the solutions.

In order to intensify the use of polymer composites in automobiles, the car manufacturers, the plastic material suppliers and the plastic moulding industry should [50]:

- Emphasize on the R&D capability that carry out thorough study on the use of this material in automobiles.
- Be working in collaboration to each other and come up with the components that could be easily produced using their existing facilities.
- Venture at more complex components once their technologies are established.
- Be exposed to these new materials and approaches by visiting and attending themselves with established overseas car manufacturers, plastic moulding companies and plastic material suppliers.
- Be ready to invest their capital in acquiring new composite processing equipments and composite moulding technologies.
- Be ready and prepared for change, especially for car manufacturers, from traditional metallic materials to polymer composites.

ACKNOWLEDGEMENTS

The author is indebted to Dr Faiz Mohammad for his invitation to write this article. Thanks are also due to Miss Qurratu Aini Mohd Sapuan and Ms Nadiah Haji Zainal Abidin for their support. The financial assistance by Universiti Putra Malaysia is highly appreciated. The contribution of fellow students and lecturers are acknowledged and some of them are mentioned here: Associate Professor Dr Megat Mohd Hamdan Megat Ahmad, Ir Mohd Rasid Osman, Associate Professor Dr Mohd Zaki Abdul Rahman, Associate Professor Dr Shamsuddin Sulaiman, Ms Sharifah Imihezri Syed Shaharuddin, Mr Mohd Nizam Suddin, Mr Shaharuddin Alimat, Mr Abdul Hadi Abdullah, Mr Shaiful Amri Mohd Khudzori, Mr Aznor Lamat, Professor Dr M. Syed Ali, Dr Kassim Ali Abbas, Associate Professor Dr Napsiah Ismail, Associate Professor Dr M.A. Maleque, Professor M. Hameedullah, Dr Liakot Ali, Mr Lee Ho Boon, Mr Saad Assi Mutasher and Mr Edi Syams Zainudin. This work is partially funded by Ministry of Science, Technology and Innovation Malaysia through IRPA scheme vote 54068.

REFERENCES

1. M.A.I. Jacobson; AA Book of the Car, p. 20, Drive Publications Limited, London (1970).
2. K. Buchholz; Automotive Engineering International, 112 (11), 75 (2004).
3. J.M. Morgalis; in International Encyclopedia of Composites (Ed. S.Lee), Vol.1, p. 99, VCH Publishers, Cambridge (1990).
4. J. Murphy; in Reinforced Plastics Handbook, p. 236, Elsevier Advanced Technology, Oxford (1994).
5. H.S. Abdalla; De Montfort University, Leicester, UK, Personal Communication, (1994).
6. N. Hancox; Materials World, 4(4), 197 (1996).

7. S.M. Sapuan; Polymer and Polymer Composites, 17(6), 421 (1999).
8. H.A. Al-Qureshi; J. Mater.Processing Technol., 118, 58 (2001).
9. P. Beardmore; in Advanced Thermoset Composites, Industrial and Commercial Applications (Ed. J.M. Margolis), p. 179, van Nostrand Reinhold Company, New York (1987).
10. S.M. Sapuan; Amer.J.Appl.Sci. 2 (2), 514 (2005).
11. S.M. Sapuan, M. Hameedullah, M.N. Suddin and N. Ismail; J. Mater.Processing Technol., 159 (2), 145 (2005).
12. M.M. Farag; Materials Selection for Engineering Design, p. 1, Prentice Hall, Hertfordshire (1997).
13. S.M. Sapuan; Proc.of the Symposium on Concurrent Engineering Manufacturing System for Polymeric-Based Composite Automotive Components (keynote address), Serdang, Malaysia (July 25, 2003), p. 1-7.
14. S.S.S. Imihezri, S.M. Sapuan, M.M.H. Ahmad, S. Sulaiman and A.K. Abbas; Proc.of the Conference on Advanced Materials (CAM 2003), Advanced Technology Congress 2003 (ATC2003), Putrajaya (May 20-21, 2003), p. 333-338.
15. T. Seagrave; Automotive Engg. International, 113 (6), 68, (2005).
16. S.S.S. Imihezri, S.M. Sapuan, S. Sulaiman, M.M.H. Ahmad, M.R. Osman and E.S. Zainudin; J. Appl.Technol., 2(2), 105 (2005).
17. R.M. Mayer; A Guide for Engineers and Designers, p.127, The Design Council, London (1993).
18. A. Harrison, M.A. Sudol, A. Priestly and S.E. Scarborough; Proc.of the Fourth International Conference on Automated Composites (ICAC95), Nottingham (September 6-7, 1995), pp. 511-525.
19. S.A. Mutasher, B.B. Sahari, A.M.S. Hamouda and S.M. Sapuan; Amer.J.Appl.Sci., Special Issue, 67 (2005).
20. K. Josh; Automotive Engineering, 105(5), 63 (1997).
21. M.A. Sardou, E.E. Dmotte, C. Zunino and P. Djomseu; Automotive Engg. International, 113(4), 62 (2005).
22. K. Buchholz; Automotive Engg. International, 113(2), 55 (2005).
23. C.F. Johnson, C.E. Wilks, C.D. Rudd and A.C. Long; Proc.of the Fifth International Conference on Automated Composites (ICAC97), Glasgow (September 4-5, 1997), p.61-73.
24. S.S. Cheon, J.H. Choi and D.G. Lee; Composite Structures, 32, 491 (1995).
25. B.P. Minaudo, J. Rawson and M. Montone; SAE Technical paper 970483 (1997).
26. C.L. Clark, C.K. Bals and M.A. Layson; SAE Technical Paper 910049, (1991).
27. S.S. Cheon, T.S. Lim and D.G. Lee; Composite Structures, 46, 267 (1999).
28. J. Rawson; SAE Technical Paper 1999-01-1015 (1999).
29. R. Ambu, F. Bertolli and F. Ginesu; Key Engg.Mater., 144, 145 (1998).
30. S.S. Cheon, D.G. Lee and K.S. Jeong; Composite Structures, 38(1-4), 229 (1997).
31. A. Harrison; Proc.of the Fifth International Conference on Automated Composites (ICAC97), Glasgow (September 4-5, 1997), p.3-12.
32. G.M.M. Gunyaev, S.M. Borovskaya and V.I. Panin; J. Automobile Engg., 208(D1), 33 (1994).
33. S.M. Sapuan, and M.R. Osman; J.Mechanical Engg., The Institution of Engineers, Bangladesh, ME27 (1&2), 39 (1999).
34. A. Harrison; in Integrated Design and Manufacture using Fibre-Reinforced Polymeric Composites, (Eds. M.J. Owen, V. Middleton and I.A. Jones), p. 149, Woodhead Publishing Limited, Cambridge (2000).
35. H.B. Lee, S.M. Sapuan, S. Sulaiman and M. Hamdan; Sudan Engineering Society Journal, 51(44), 23 (2005).
36. S.M. Sapuan; A Concurrent Engineering Design System for Polymeric-Based Composite Automotive Components, PhD Thesis, De Montfort University, UK, 1998.
37. S.S.S. Imihezri, S.M. Sapuan, S. Sulaiman, M.M. Hamdan, E.S. Zainudin, M.R. Osman and M.Z.A. Rahman, J. Mater.Processing Technol., 171(3), 358 (2006).
38. B. Evans; in Simultaneous Engineering: Integrating Manufacturing and Design (Ed. C.W. Allen), p. 3, Society of Manufacturing Engineers, Dearborn, Michigan (1990).

39. S.M. Sapuan; Proc.of the Malaysia Design Conference 2003 (Invited Lecture), Kuala Lumpur (October 8-9, 2003), p. 1-13.
40. J.A. Quinn; Composites–Design Manual, p. 1-20, James Quinn Associates Ltd, Liverpool (1995).
41. M.M. Schwartz; Composite Materials (Vol. 1): Properties, Nondestructive Testing and Repair of Polymers, Ceramics and Metal Matrices, p. 1-20, Prentice Hall PTR, Upper Saddle River, New Jersey (1997).
42. S.M. Sapuan; Invited Seminar Lecture, International Islamic University, Malaysia (12 July, 2004).
43. J.X. Gao, B.M. Manson and P. Kyratsis; J. Mater.Processing Technol., 107, 201 (2000).
44. S.M. Sapuan; Assembly Automation, 25(2), 146 (2005).
45. C.L. Dym; J. Engg.with Computers, 1(3), 9, (1985).
46. S.M. Sapuan and H.S. Abdalla; Composites Part A: Appl.Sci.Manufacturing, 29A(7), 731 (1998).
47. IntelliCorp; Kappa-PC. 2.4 User Manual (1997).
48. W. Sufian; Universiti Kuala Lumpur, Personal Communication (2005).
49. K. Buchholz; Automotive Engg.International, 113(2), 55 (2005).
50. S.M. Sapuan and Y. Nukman; J. Institution of Engineers, Malaysia, 62(2), 23 (2001).
51. Anon.; Harga minyak mentah dunia AS$65.32 satu tong sehingga 10.00 malam waktu Malaysia, Utusan Malaysia, Saturday (10 September, 2005).
52. P. Pontical; Automotive Engg.International, 113 (5), 48 (2005).
53. Anon.; Automotive Engg.International, 112(12), 42 (2004).
54. E.H.B. Pischetsrieder; Automotive Engg.International, 113(2), 84 (2005).

8

Conducting Polymer-Based Sensors

Johan Bobacka*, Tom Lindfors and Ari Ivaska

Åbo Akademi University, Process Chemistry Centre,
Laboratory of Analytical Chemistry, Biskopsgatan 8, FIN-20500 Turku-Åbo, Finland.
**Corresponding author's email: jbobacka@abo.fi*

Contents

Summary: The electrical, electrochemical and optical properties of conducting polymers can be utilized to convert chemical information (concentration, activity, partial pressure etc.) into electrical and optical signals. Immobilization of synthetic receptors and biological recognition systems in conducting polymer films allows the construction of chemical sensors and biosensors that are based on potentiometric, amperometric, voltammetric, conductimetric and optical transduction of the analytical signal. Most of the research activities on conducting polymer-based sensors have been focused on electrochemical sensors and biosensors. However, signal transduction based on changes in optical properties has also received considerable attention. The possibilities offered by conducting polymers in the area of chemical sensors are illustrated in this chapter by using various examples.

1. INTRODUCTION

Conjugated polymers have undergone a rapid development since the mid 1970s. Some of the most common conjugated polymers in their undoped (neutral) form are shown in **Fig. 8.1**. Removal of

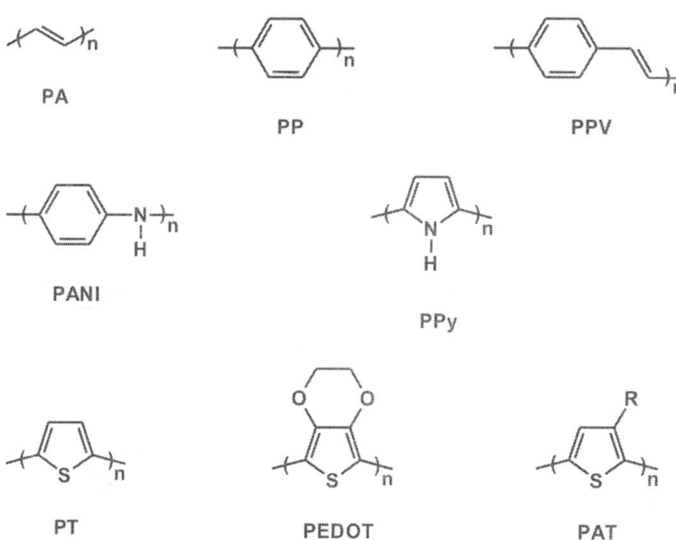

Fig. 8.1. Chemical structures of some common conjugated polymers. PA (polyacetylene), PP (polyphenylene), PPV (polyphenylenevinylene), PANI (polyaniline), PPy (polypyrrole), PT (polythiophene), PEDOT (poly-3,4-ethylenedioxythiophene), PAT (poly-3-alkylthiophene). R (alkyl) and n (number of repeating monomer units in the polymer chain).

π–electrons from a conjugated polymer backbone by chemical or electrochemical oxidation (p-doping) results in positively charged carriers (holes) that are responsible for a dramatic increase in the electrical conductivity. Removal of electrons is accompanied by insertion of charge-compensating anions, so that the material as a whole remains electrically neutral. Analogously, reduction (n-doping) corresponds to addition of electrons and the excess negative charge is compensated by cations inserted into the polymer. In their electrically conducting (doped) form, conjugated polymers are referred to as "conducting polymers" or "synthetic metals". Here the term "conducting polymer" is used exclusively for conjugated organic polymers that can be made electrically conducting by oxidation (p-doping) or reduction (n-doping) of the polymer backbone. The discovery of this particular type of conducting polymers has created a new and rapidly expanding field of research on the boundary between chemistry and physics, for which Alan J. Heeger, Alan G. MacDiarmid and Hideki Shirakawa were awarded the Nobel Prize in Chemistry in the year 2000 [1-4].

Conjugated polymers are truly multifunctional materials [5-7]. This multifunctionality is related to the chemical structure of the polymer backbone itself, including many possible substituents, and to the large variety of counterions (doping ions) or neutral molecules that can be trapped in the polymer matrix. Additionally, the material properties of conducting polymers depend on the degree of oxidation (p-doping) or reduction (n-doping) of the conjugated polymer backbone. Conducting polymers are often synthesized and characterized by using electrochemical methods [8,9]. Electrochemical doping of conjugated polymers resembles charging of a redox capacitor or battery [10] and gives rise to changes in the optical properties of the material.

Over the years the number of possible applications of conducting polymers has increased rapidly [11-13]. In their undoped state, conjugated polymers are organic semiconductors with a suitable band gap to transform electricity into light and vice versa. These features can be used to construct organic light-emitting diodes and photovoltaic devices such as solar cells. Processible conjugated polymers offer possibilities to fabricate all-polymer transistor circuits by high-resolution inkjet printing [14].

The unique electrical, electrochemical and optical properties of conjugated polymers can also be used to convert chemical information (concentration, activity, partial pressure) into electrical or optical signals. Therefore, conjugated polymers are useful as transducers in chemical sensors. One of the primary issues is to combine conjugated polymer transducers with selective molecular recognition in order to obtain durable and selective chemical sensors.

The interest in chemical sensors based on conducting polymers is reflected in the rapidly increasing number of publications in the area, as shown in **Fig. 8.2**. Different approaches to the application of conducting polymers in chemical and biochemical sensors have been explored [15-46]. Most of the research activities in this area focus on electrochemical sensors, i.e. chemical sensors and biosensors with electrochemical signal transduction. However, signal transduction based on changes in optical properties has also received considerable attention.

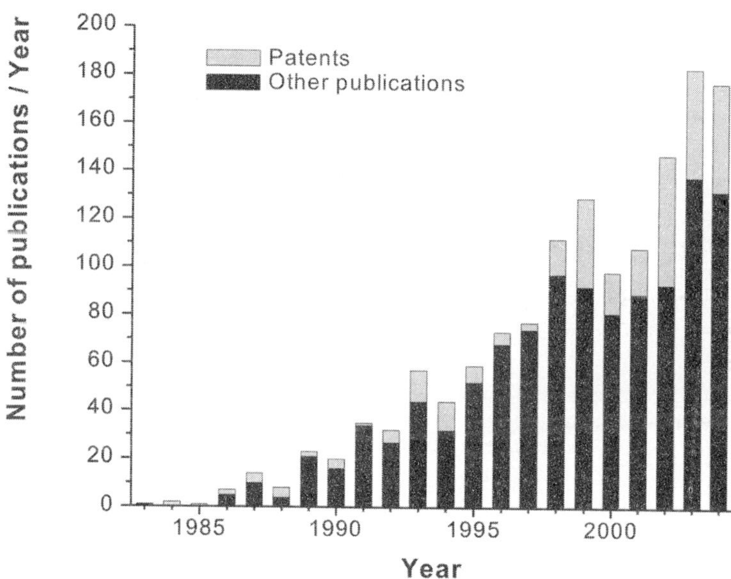

Fig. 8.2. Number of publications on chemical sensors based on conducting polymers (1983-2004). The graph is based on a literature search (4.1.2006) using the SciFinder Scholar 2004 program.

2. ELECTROCHEMICAL SENSORS

Electrochemical sensors are such chemical sensors where molecular recognition is transduced directly into an electrical signal. Conducting polymers are useful materials in electrochemical sensors for a number of reasons. Since conducting polymers are electronically conducting materials they can form an Ohmic contact to electronic conductors such as gold, platinum and carbon that have high work functions. A large variety of conducting polymers can be electrosynthesized from the corresponding monomers directly onto the electronic conductor. Since conducting polymers are electroactive materials with mixed ionic and electronic conductivity, they can transduce an ionic signal into an electronic signal, which can be utilized in ion sensors. The drawback is that conducting polymers may interact with various ions, molecules and redox species in the surrounding medium in a non-selective manner. However, the selectivity to a given analyte can be obtained by tailoring the molecular recognition properties of the conducting polymer, e.g. by using functionalized monomers and counterions or by coating the conducting polymer film by another membrane with high selectivity. Depending on the signal transduction mechanism, electrochemical sensors are usually subdivided into potentiometric, amperometric/voltammetric and conductimetric sensors.

2.1. Potentiometric Sensors

Potentiometric sensors are based on the measurement of the electrical potential of the sensor versus a reference electrode, which has a constant potential independent of the sample composition. By using an mV-meter with high input impedance, a negligible electrical current flows in the measuring circuit. The potential of the sensor is a logarithmic function of the activity of the analyte in the solution. Potentiometry is an attractive measurement technique because it is associated with small-size, portable and low-cost instrumentation. The construction of a conducting polymer-based potentiometric sensor is shown schematically in **Fig. 8.3**.

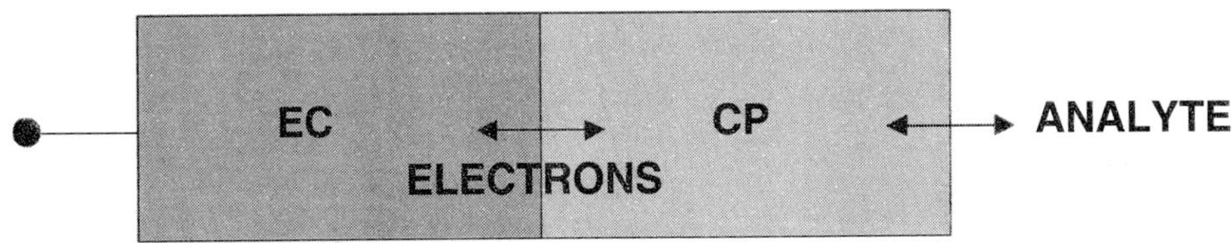

Fig. 8.3. Conducting polymer-based potentiometric sensor. Both the molecular recognition and the ion-to-electron transduction take place in the conducting polymer (CP) membrane. The electronic conductor is denoted with EC.

2.1.1. Ion Sensors

Conducting polymers offer several possibilities for the development of potentiometric ion sensors [23,41,43-45]. Oxidized (p-doped) conducting polymers have a polycationic backbone and can therefore function as anion exchangers resulting in an anionic potentiometric response [47,48]. Since different small anions, including Cl^-, Br^- NO_3^-, ClO_4^-, BF_4^- and SCN^-, may contribute to the measured potential, the anionic response of conducting polymers is generally non-selective [48-50], even though certain conducting polymers can possess selectivity towards specific ions, e.g. H^+ [51] and ClO_4^- [52]. However, the doping anion included in the polymer film during electrosynthesis may influence the structure of the conducting polymer by a process similar to molecular imprinting [53,54]. Immobilization of doping anions to obtain an excess negative charge versus the positive charge of the oxidized polymer gives a conducting polymer with cation-exchange behavior resulting in a cationic potentiometric response [55]. As a result of their electronic conductivity, conducting polymers also show a redox response [56]. The principle of the anionic, cationic and redox response for a conducting polymer is shown schematically in **Fig. 8.4**.

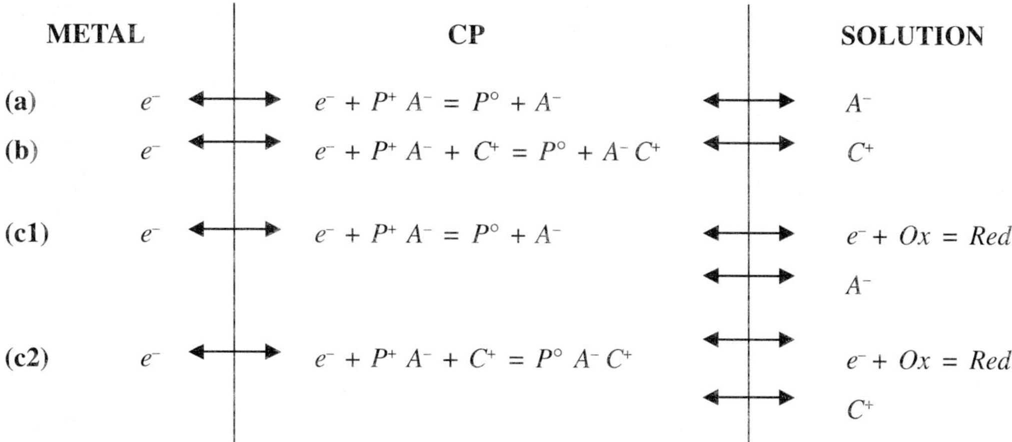

Fig. 8.4. Principles of the (a) anionic response, (b) cationic response and (c1 and c2) redox responses of conducting polymer (CP) membranes. Mobile anions are incorporated in the CP membrane in (a) and (c1), whereas (b) and (c2) describe membranes prepared with a mobile cation (immobile anion). In the solution phase, electrons originate from a redox couple: $Ox + ne^- = Red$.

The mixed anionic, cationic and redox sensitivity of conducting polymers may all contribute to the potentiometric signal, as shown both theoretically and experimentally [57,58]. Furthermore, several conducting polymers show pH sensitivity depending on the acid-base properties of the conjugated polymers [59]. The pH sensitivity of polypyrrole has been utilized in the construction of an all-solid-state pH glass electrode [60].

The potentiometric response originates partly from the chemical structure of the conducting polymer backbone and partly from the inserted doping ions. In order to enhance the selectivity to the target analyte, it is possible to modify the chemical structure of the conducting polymer by covalent binding of suitable receptors or by immobilization of functional dopants that are selective to the target analyte [61,62]. In this way conducting polymers can be functionalized by ordinary ion-complexing ligands or highly selective supramolecular receptors (ionophores) forming host-guest complexes with the target analyte [34,37,41]. Various ion-recognition sites, including complexing agents from the group of metallochromic indicators [63], sulfonated macrocycles [64,65], electrically neutral and charged carriers [66,67,68], as well as ion-exchangers [69] can be immobilized in conducting polymers to be used as potentiometric ion sensors [41]. Covalent binding of ion receptors to conjugated polymers offers excellent possibilities to construct potentiometric ion sensors where both molecular recognition and signal transduction take place even in a single macromolecule, which is of importance for the development of robust sensors that can be miniaturized.

Another possibility is to include the receptors in a separate membrane that is in contact with the conducting polymer film, as shown schematically in **Fig. 8.5**. In this configuration the conducting polymer acts as a solid-state ion-to-electron transducer and the outer membrane adds selectivity to the sensor [70]. Alternatively, the conducting polymer transducer can be integrated into the ion-selective membrane in the form a polymer blend [71]. Also in this case, the role of the conducting polymer is to convert the ionic signal into an electronic signal.

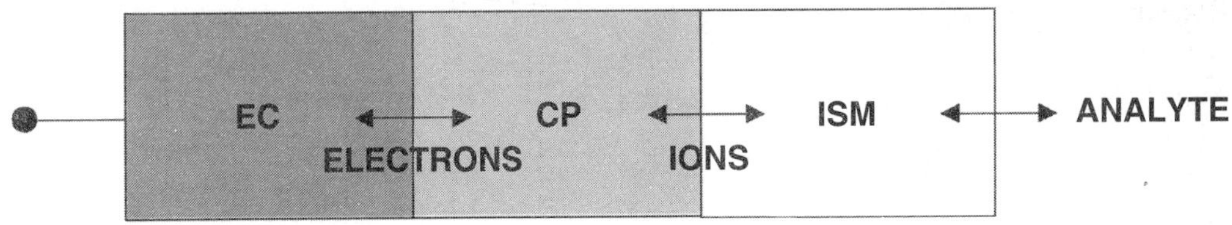

Fig. 8.5. All-solid-state ion-selective electrode using conducting polymers as ion-to-electron transducer. The molecular recognition and the ion-to-electron transduction take place in the ion-selective membrane (ISM) and conducting polymer (CP) membrane, respectively. The electronic conductor is denoted with EC.

Conducting polymers based on derivatives of polypyrrole, polythiophene and polyaniline are frequently used as ion-to-electron transducers in combination with classical ionophore-based ion-selective membranes in order to construct different types of all-solid-state ion sensors [41,43-46]. Such ion sensors are more robust and require less maintenance than conventional ion-selective electrodes with internal filling solution. All-solid-state ion sensors are therefore suitable for demanding applications, such as on-line process analysis and clinical analysis, and have been applied in flow analysis [72].

2.1.2. Gas Sensors

Conducting polymers can be applied as potentiometric sensors for gases, such as O_2, NO_x and CO_2. Potentiometric oxygen sensing can be based on oxidation of the conducting polymer by oxygen or on the formation of charge-transfer complexes between O_2 and the undoped form of conjugated polymers [73,74].

Conducting polymers are also applied in Severinghaus-type potentiometric gas sensors. In this application, both the conducting polymer-based sensor and the reference electrode are located behind a gas-permeable membrane in an internal electrolyte. Gases diffuse across the gas-permeable membrane into the internal electrolyte layer and alter the ion activity or pH, which is then recorded by the conducting polymer-based sensor. Following this principle, potentiometric sensors for NO_x and CO_2 can be based on nitrate-selective polypyrrole [75] and pH-sensitive polyaniline [76], respectively.

2.1.3. Biosensors

Potentiometric biosensors can be fabricated by immobilization of biological components such as enzymes and antibodies into a conducting polymer film. The molecular recognition of the target analyte by the immobilized biological component is transduced into an electrical potential via the conducting polymer.

Immobilization of human immunoglobulin G (IgG) in polypyrrole results in a potentiometric immunosensor for anti-IgG [77]. The complex formed between IgG and anti-IgG lowers the permeability of dissolved O_2 to the polypyrrole film, resulting in a negative shift of the electrode potential, which is proportional to the anti-IgG concentration in solution. This method utilizes the O_2 sensitivity of polypyrrole in the signal transduction process.

Potentiometric biosensors for glucose [78], urea [79,80] and creatinine [81] can be obtained by immobilization of glucose oxidase, urease and creatinine iminohydrolase, respectively, into polypyrrole. The signal transduction is based on changes in pH caused by the enzymatic reactions, which can be sensed by conducting polymers such as polypyrrole [78-81] or polyaniline [82,83]. Furthermore, polyaniline containing covalently bound boronic acid groups can work as a non-enzymatic glucose sensor [84]. The sensing mechanism is based on complex formation between phenylboronic acid and diols, which influences the acid dissociation constant (pK_a) and thereby the potential of polyaniline. Conducting polymer-based biosensors can be applied e.g. as detectors in flow injection analysis systems [85].

2.2. Amperometric and Voltammetric Sensors

Amperometric sensors are based on the measurement of the electrical current flowing through the sensor as a result of oxidation or reduction of the analyte (or a reaction product from the analyte) at a constant applied potential. Normally a potentiostat is connected to a three-electrode system in such a way that the potential is applied between the sensor and the reference electrode and the current is measured between the sensor and a counter electrode. Voltammetric sensors are based on the measurement of the electrical current versus potential during a potential scan. The current measured in amperometric and voltammetric sensors is related to the concentration of the analyte in solution. The construction of a conducting polymer-based amperometric or voltammetric sensor is principally the same as the potentiometric sensor (**Fig. 8.3**).

2.2.1. Ion Sensors

The oxidation (p-doping) or reduction (n-doping) of conducting polymers is associated with ion transfer, as shown schematically in **Fig. 8.4**. Electrochemical oxidation can take place only if charge-compensating ions are present in the solution in which the conducting polymer is immersed. This unique property can be used to detect even electroinactive ions amperometrically or voltammetrically. Consequently, conducting polymers such as polypyrrole are useful as amperometric detectors for electroinactive anions in a flow-through cell [86]. By using a carrier solution that does not contain any anions that are able to enter the polymer film, a positive potential can be applied without causing any oxidation (p-doping) of the conducting polymer and the current remains small. When the analyte is injected into the carrier stream and the analyte anions pass the detector, oxidation of the conducting polymer takes place resulting in a current peak that can be used for analytical purposes.

Electroactive ions can be oxidized or reduced directly at the conducting polymer surface, following the same principle as shown in **Fig. 8.4**. Conducting polymers can therefore be used as voltammetric sensors for a variety of electroactive ions in solution. Furthermore electroactive ions may enter the conducting polymer film, depending on the size, charge and chemical structure of the ion (guest) as well as the conducting polymer (host). This provides a means to accumulate and preconcentrate metal ions prior to electrochemical oxidation or reduction, similar to voltammetric stripping analysis [87]. Functionalization of conducting polymers by covalent binding of receptors such as crown ethers [88] or by using specific doping ions such as sulfonated calixarenes [89] can be used to influence the host-guest interactions in order to enhance the response to the target ion.

2.2.2. Gas Sensors

Amperometric gas sensors based on conducting polymers can be used to detect electroactive gases by oxidation or reduction of the gas molecules. Oxidation of SO_2 and reduction of NO_2 and O_2 can take place at a polyaniline membrane, which makes polyaniline a possible material for amperometric and voltammetric gas sensing [90]. Polypyrrole is a very promising material for amperometric detection of ammonia (NH_3) in aqueous solution [91,92]. Polypyrrole is reduced by ammonia and subsequently re-oxidized at the applied positive potential resulting in a current that is proportional to the concentration of ammonia in solution. The exact mechanism of ammonia detection by polypyrrole is still under discussion [91-93], but it has been suggested that the main oxidation product of ammonia is nitric oxide (NO) [94]. Polymer sensor arrays based on polypyrrole have also been developed for analyzing the breath of diabetic patients [95]. Another example is a voltammetric sensor for direct in vivo determination of NO that can be constructed from a carbon fiber microelectrode coated with a conducting polymer inner layer and a Nafion outer layer [96].

2.2.3. Biosensors

A large variety of biological recognition elements including enzymes, antibodies, antigens, whole cells and oligonucleotides can be immobilized in conducting polymer films to produce different types of amperometric biosensors and immunosensors [21,33,35,39,42]. In fact, electochemical immobilization of enzymes in conducting polymers is a versatile and elegant approach towards amperometric biosensors [97-99]. Electropolymerization is a one-step and fast method that allows accurate control of the spatial distribution of the enzyme in a conducting polymer film of controlled thickness.

Glucose determination is extremely important, e.g. in the case of diabetes, and therefore glucose biosensors based on glucose oxidase (GOx) have received much attention. GOx is a commercially available enzyme with relatively high chemical stability. Furthermore GOx is negatively charged at neutral pH and may act as charge compensating anion in oxidized conducting polymers. Glucose biosensors can thus be fabricated in a one-step process by electropolymerization of a monomer such as pyrrole or N-methylpyrrole in presence of GOx, which is then trapped in the polymer film [97-99]. Alternatively, GOx may be covalently bound to the conducting polymer backbone [37,100,101]. Glucose sensors based on electrospun nanofibers of polyaniline/polystyrene have also recently been reported [102].

Direct electron transfer between the redox enzyme and the conducting polymer, without any involvement of O_2 or any other mediator, would be an elegant transduction mechanism. However this mechanism is difficult to achieve because the redox site of the enzyme is located deep inside an insulating shell of proteins and is therefore difficult to access by the conducting polymer chains [35]. Therefore, glucose sensing often follows a mechanism where GOx catalyzes the oxidation of glucose by O_2, resulting in gluconolactone and H_2O_2 that can be detected electrochemically. The measured current is proportional to the concentration of H_2O_2, which in turn is proportional to the glucose concentration in the sample. The conducting polymer acting as electrical wires around the enzyme may mediate oxidation of H_2O_2. However, since H_2O_2 tends to destroy the conductivity of conducting polymers such as polypyrrole [103], it is likely that H_2O_2 is actually oxidized at the underlying electrode, which is commonly platinum. Consequently, the conducting polymer often works only as an immobilization matrix for the enzyme without taking part in the signal transduction process.

The oxidation potential of H_2O_2 is relatively high, which means that other electroactive compounds in the sample may interfere by contributing to the measured current. However, amperometric sensing can be performed at a lower positive potential by using a redox mediator (other than O_2) immobilized in the conducting polymer film. Redox mediators such as ferrocene [104], hydrazine or horseradish peroxidase (HRP) [105] can be covalently bound to the conducting polymer backbone, while negatively charge mediators such as ferrocene carboxylate can be electrostatically incorporated as doping anion in the conducting polymer film [106]. The role of the mediator is to oxidize the reduced form of GOx to its oxidized form, producing the reduced form of the mediator that is then electrochemically oxidized at a lower potential than H_2O_2.

Not only glucose sensing is of interest. Practically useful amperometric biosensors can be constructed for a large variety of analytes, including cholesterol, creatinine, ethanol, fructose, galactose, glutamate, hydrogen peroxide, hypoxanthine, lactate and urea, by immobilization of different enzymes in conducting polymer films. Immobilization of redox mediators help to reduce the contribution from electroactive interferents. However, in order to be practically useful in clinical applications the conducting polymer membrane should efficiently reject electroactive interferents such as ascorbic acid, uric acid and some drugs that may be present in biological fluids. This additional requirement can be fulfilled e.g. by immobilization of the enzyme in a permselective conducting polymer film, such as overoxidized polypyrrole, or by applying co-polymers or multi-layer membranes [35]. Additionally, covalent immobilization of glutamate oxidase to a conducting polymer can be utilized in glutamate sensing [107]. Immobilization of γ-cyclodextrin in poly(3-methylthiophene) results in a selective biosensor for dopamine, L-dopa and chloropromazine even in the presence of ascorbic acid [108]. On the other hand, an amperometric biosensor for choline makes use of multilayer films of multiwalled carbon nanotubes and polyaniline [109].

Immunosensors based on the immobilization of antibodies or antigens in conducting polymers are less common than biosensors based on immobilized enzymes [39,42]. This is particularly true in the case of amperometric and voltammetric immunosensors. Electrochemical synthesis of polypyrrole in presence of anti-human serum albumin (anti-HSA) was reported to result in an electrode that gives a voltammetric response to human serum albumin (HSA) in solution [110]. Analogously, a polypyrrole film containing anti-isoproturon allows voltammetric detection of the isoproturon pesticide in solution [111]. Furthermore, covalent binding of the *N*-2,4-dinitrophenyl-L-lysine (DNP) antigen to polypyrrole and polythiophene derivatives results in electrodes that give a voltammetric response to anti-DNP [112]. An amperometric immunosensor for vitellogenin was prepared by using poly(terthiophene) as the substrate material with horseradish peroxide and monoclonal antibody (anti-Vtg) covalently attached to this surface [113]. These results indicate that antigen (antibody) functionalized conducting polymers are able to recognize the corresponding antibody (antigen) and to transduce the antigen-antibody interaction into a voltammetric signal. However, the irreversible nature of antibody-antigen binding is a potential problem that may prevent repeated analyses with the same sensor. One way of overcoming this problem is to apply pulsed amperometric detection, where a pulsed potential waveform modulates the antibody-antigen binding so that it becomes reversible enough to allow multiple analyses with the same sensor in a flow injection analysis (FIA) system [114]. This method is suitable for conducting polymers containing different antibodies for development of amperometric immunosensors for the detection of HSA, Thaumatin, Rhesus (D) blood group antigen, halogenated organic compounds and phenols [34].

DNA sensors based on conducting polymers functionalized with single-stranded oligonucleotide (ssODN) probes are receiving increasing attention due to the utmost importance of DNA diagnostics in biotechnology and medicine, e.g. in the diagnosis of genetic diseases [115-118]. The ssODN (or ssDNA) probes can be attached to the conducting polymer film via covalent binding [115-124] or via other types of interactions [125-130]. Conducting polymer-based DNA sensors allow direct (label-free) electrochemical detection of the hybridization between the polymer-bound ssODN and a complementary ssODN in solution through the specific interaction between complementary bases, i.e. adenine-thymine and guanine-cytosine. The hybridization can be detected by methods such as voltammetry [115-121,123,125], amperometry [126,127,129] and electrochemical impedance spectroscopy [122-124,128-130]. Detection limits down to 1.6 fmol of target nucleotide in 0.1 ml (16 pM) have been reported for conducting polymer-based amperometric DNA sensors [127]. Consequently, ODN-functionalized conducting polymers are extremely promising for the fabrication of amperometric or voltammetric DNA sensors.

2.3. Conductimetric Sensors

Conductimetric sensors are based on the measurement of the electrical conductivity of the sensor material in contact with the analyte. The electrical conductivity of conducting polymers results from charge transport along individual polymer chains, charge transfer between polymer chains and possibly charge transfer between polymer aggregates or particles. This means that the conductivity is sensitive to the chemical structure (π–electron conjugation, doping) as well as the morphology of the conducting polymer, which in turn are sensitive to the chemical compounds in contact with the conducting polymer film. This makes it possible to develop conductimetric sensors based on conducting polymers.

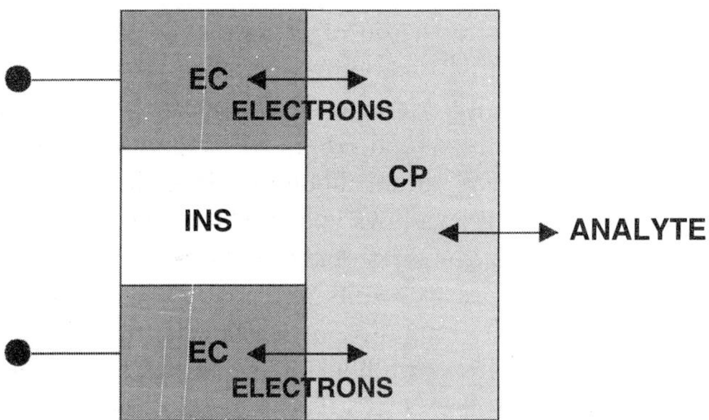

Fig. 8.6. Conducting polymer-based conductimetric sensor. The molecular recognition and the signal transduction take place in the conducting polymer (CP) membrane. The electronic conductor is denoted with EC and the electrical insulator is denoted with INS.

Conductimetric sensors do not require any reference electrode, which simplifies the experimental setup compared to potentiometric and amperometric/voltammetric sensors. Normally a thin conducting polymer film is deposited electrochemically or by solution casting (spin coating) on a pair of electrodes (two-point probe) or on four electrodes (four-point probe), e.g. in the form of interdigitated electrodes on an insulating substrate. The conducting polymer covers the electrodes and bridges the gap between the electrodes, as shown schematically in **Fig. 8.6**. The electrical resistance and the corresponding conductivity of the conducting polymer film are then obtained from dc (direct current) or ac (alternating current) measurements of current versus potential or vice versa.

By depositing the conducting polymer over two closely spaced microelectrodes it is possible to construct microelectrochemical devices for use as conductimetric sensors [131,132]. Conductimetric sensors are sometimes called "two-terminal transistors" [132] or "chemiresistors" [133], because a "transistor" is in fact a "variable resistor" [134]. In the conductimetric sensors described here the conducting polymer acts as the resistor that varies when it comes in contact with different chemical compounds (analyte).

2.3.1. Ion Sensors

Conducting polymers functionalized with ion receptors are potentially useful materials for conductimetric ion sensors [135]. Interactions between the ion and the receptor can influence the conductivity along the conjugated polymer chain via electrostatic effects or via changes in conformation (planarity) of adjacent monomer units in the polymer chain. It is interesting to note that conductimetric signal transduction of the ion recognition process can be much more sensitive than voltammetric transduction. This can be related to an inherent signal amplification of the ion recognition process [136]. In other words, a single ion-recognition event causing a change in conductivity in a single point along a conducting polymer chain acting as a molecular wire influences all electrons or holes transported through that point. Therefore a small number of ion-recognition events may cause a large change in the electronic conductivity of the material [136].

The electrical conductivity of polypyrrole is sensitive to pH. This property can be utilized to produce disposable conductimetric sensors for measurement of pH in the range of pH = 5-10 [137]. The electrical conductivity of polyaniline is also known to depend on pH [131]. In fact, polyaniline seems to be more suitable than polypyrrole for conductimetric pH sensing [138]. On the other hand, when polyaniline contains covalently bound crown ether (18-crown-6) receptors it becomes sensitive to potassium ions in solution [139]. Combinations of different conducting polymer-based sensors with pattern recognition techniques such as principal component analysis (PCA) are applied in so called "electronic tongues" that can be used e.g. to distinguish between different brands of mineral water, tea and coffee [140]. By combining an array of conductimetric sensors with mathematical signal treatment, each sensor in the array need not be highly selective to any specific analyte but each sensor should give different selectivity patterns to different ions. This is an application area where even relatively non-selective conducting polymers may prove particularly useful.

2.3.2. Gas Sensors

Interactions between conducting polymers and gases or vapors may influence the mobility and/or number of charge carriers in the conducting polymer film. Both intra-chain and inter-chain charge transport may be influenced by gases and vapors that interact with the conducting polymer chains.

This is the basis for conductimetric gas sensors. Gases that are electron acceptors (O_2, NO_2, SO_2, I_2, Br_2, Cl_2, PCl_3) or electron donors (NH_3, hydrazine, triethylamine, H_2S) may oxidize or reduce the conducting polymer, respectively, and thereby influence its electronic conductivity [141-150]. Recently, poly(3,4-ethylenedioxythiophene) nanorods have been used for conductimetric detection of both HCl and NH_3 vapors [151]. In addition to the redox properties also the acid-base properties of gases can influence the conductivity of the conducting polymer [146,147]. Furthermore, reversible charge transfer complexes may be formed between O_2 and π-conjugated polymers, which can be utilized in oxygen sensors [152]. Several conducting polymers, including polyaniline [153], polypyrrole [154], substituted polyacetylenes [155] and poly(p-diethynylbenzene) [156] show a humidity-dependent conductivity, which can be utilized in humidity sensors. Poly(3-methylthiophene) containing Pt microparticles as catalyst shows a response to O_2 and H_2 (and pH) in aqueous media [132].

Conducting polymer-based gas sensors operate at room temperature, which is an advantage compared to traditional metal oxide semiconductor gas sensors. Introducing functional groups into the conducting polymer backbone can optimize the analytical performance of conducting polymer-based gas sensors. For example, a conducting polymer containing a cobalt salen complex is promising for selective and reversible detection of NO that is an important intracellular signaling agent [157]. On the other hand, conducting polymers functionalized with copper phthalocyanine are highly sensitive to NO_2 [158].

Not only pure conducting polymers, but also composites and blends of conducting polymers and insulating polymers are used for conductimetric gas sensing applications. Composites of polypyrrole and polyamide show better sensitivity to NH_3 and CO_2 than pure polypyrrole [159]. Interestingly, a composite consisting of polyaniline in the emeraldine salt form and poly(vinyl alcohol) is selective to humidity, while a composite consisting of polyaniline in the emeraldine base form and poly(vinyl alcohol) is selective to CO_2 [160]. Furthermore, conducting polymer blends of polypyrrole with polymers such as polyvinyl acetate, polystyrene or polyvinyl chloride, are promising sensor materials for toxic gases [161]. The sensor performance of polyaniline is improved by using polyaniline nanofibers that result in a conducting polymer film with a high surface area allowing more interactions between gas molecules and the conducting polymer [162]. Thin films of vacuum-deposited nanocrystalline polyaniline are promising sensor materials for detection of carbon monoxide [163]. This shows the importance of polymer morphology in conductimetric gas sensors.

Vapors of organic solvents including methanol, ethanol, acetone, ether and toluene influence the electrical conductivity of conducting polymers [164-167]. There is a clear correlation between the conductivity change of polypyrrole and the dielectric constant of the organic solvents [168]. Polymer-vapor interactions may cause film swelling that influences the electrical conductivity of the polymer film. By using an array of poly(3,4-ethylenedioxythiophene)-based composite sensors it is possible to classify organic vapors including polar and non-polar compounds [169]. Conducting polymer foams based on polypyrrole and polyetherurethane are an order of magnitude more sensitive to amines (triethylamine, butylamine) than to alcohols (methanol) [170]. In the case of polyaniline, polar solvents result in dipolar interactions and deprotonation of the polymer while non-polar solvents influence the π-electron delocalization [171]. An optimized blend of polypyrrole and poly(methylmethacrylate) is more selective to acetone than to acetic acid and water vapor, which is of practical importance for sensing of acetone in lacquer used for room furnishing [172].

The fact that different conducting polymers are partially selective to different gases can be utilized to produce sensor arrays with specific selectivity patterns to different gases and vapors

[173-177]. Arrays of partially selective conducting polymer-based conductimetric gas sensors combined with appropriate data processing including neural networks are used to construct instruments for automatic recognition of aromas, flavors, odors and cigarette smoke [178-180]. Such instruments, which are often called "electronic noses" are commercially available and are applied in various areas including the food industry, environmental monitoring and the medical field [178,181,182]. Electronic noses based on conducting polymer sensors are potentially useful for such diverse applications as characterization of beer [173], wine [183] and olive oil [184-186], diagnosis of urinary tract infections [187] and on-line monitoring of water and wastewater quality [188].

2.3.3. Biosensors

Conductimetric biosensors based on conducting polymers are alternatives to the better-known amperometric and potentiometric ones. Conductimetric biosensors can be constructed by depositing the conducting polymer on a pair of electrodes on an insulating substrate, in the same way as for conductimetric ion and gas sensors. Thereafter, the conducting polymer layer is coated by a second layer containing the biological recognition element (enzymes, antibodies, antigens, cells), as shown schematically in **Fig. 8.7**. Biorecognition events that influence the electrical conductivity of the underlying conducting polymer film are then detected conductimetrically e.g. by measuring the current at a fixed potential between the two electrodes. Even nanowires of functionalized conducting polymers can be used for conductimetric bioaffinity sensing [189].

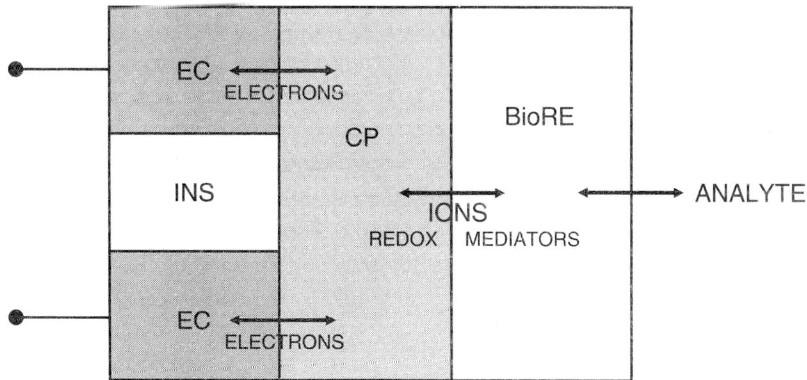

Fig. 8.7. Conducting polymer-based conductimetric biosensor. The molecular recognition and the signal transduction take place in the biological recognition element (BioRE) and conducting polymer (CP) membrane, respectively. The electronic conductor is denoted with EC and the electrical insulator is denoted with INS.

Since the electrical conductivity of polyaniline is pH dependent, enzyme reactions that result in pH changes can be monitored conductimetrically in enzyme-based biosensors. A conductimetric glucose biosensor can be constructed by depositing polyaniline on two closely spaced platinum electrodes so that polyaniline bridges the gap between the electrodes. The polyaniline film is then coated with another layer of polyaniline containing glucose oxidase [190]. In presence of glucose, the enzyme reaction produces gluconic acid that lowers the pH and thus increases the electrical conductivity of polyaniline, which is measured. Similar approaches are used to construct polyaniline-based conductimetric microsensors for urea and triglycerides [191]. Polypyrrole coated with an agarose layer containing yeast cells may be used as a simple conductimetric detector for microbial metabolism [192].

A microelectrochemical enzyme transistor (enzyme switch) responsive to glucose is based on the fact that the electrical conductivity of polyaniline depends on the degree of oxidation (p-doping). The polyaniline film is coated with a film of poly(1,2-diaminobenzene) containing glucose oxidase [193,194]. In presence of glucose the enzymatic reaction produces the reduced form of glucose oxidase that is then reoxidized by tetrathiafulvalene that works as a redox mediator transferring electrons to polyaniline. The presence of glucose is then observed as a change in electrical conductivity of polyaniline. The switch can be regenerated by electrochemical re-oxidation of polyaniline [193,194].

3. OPTICAL SENSORS

Chemical sensors in which molecular recognition is transduced into an optical signal are termed optical sensors. Optical sensors based on conducting polymers are in a relatively early stage of development and they are used more seldom than electrochemical sensors. However, it is clear that conducting polymers are potentially useful materials also in optical sensors. In their undoped state most conjugated polymers absorb electromagnetic radiation of wavelengths between 300 and 800 nm, i.e. in the UV-visible range, corresponding to band gaps between 1.5 and 4 eV. The doping process results in new energy levels in the band gap allowing electronic excitations at lower energies, resulting in absorption of electromagnetic radiation at longer wavelengths. Therefore, not only the electrical conductivity, but also the optical properties of conjugated polymers depend strongly on the degree of doping. Additionally, the optical properties depend on the chemical structure of the conjugated polymer backbone as well as on the conformation of the polymer chain. The conjugation along the polymer chain is influenced e.g. by the planarity of adjacent monomer units in the conjugated polymer chain. Consequently, chemical species that interact with the conjugated polymer can influence the optical properties of the polymer that works as an optical transducer [38]. As in the case of electrochemical sensors, the selectivity and sensitivity of optical sensors based on conducting polymers can be influenced by functionalization of the conjugated polymers by different receptor groups [34].

3.1. Ion Sensors

The optical properties of polyaniline in the visible and near-infrared region are sensitive to pH and polyaniline can be applied as an optical pH sensor within the pH range 2-12 [51,195-197]. Measurement of pH can also be conducted with polyaniline in combination with Raman spectroscopy [198]. The Raman signal of the $C = N$ stretching vibration at 1439 cm^{-1}, originating from quinoid units of the emeraldine base form of polyaniline, can be used for pH measurements between pH 3 and 6 with the 633 nm laser excitation wavelength. One advantage of this approach is that the Raman signal can be transported over long distances in optical fibers, thus allowing monitoring of pH from a remote location. Substituents on the polyaniline chain can influence the pK$_a$ value of the polymer, which influences the pH sensitivity and useful pH range of the sensor [51,199,200]. Polypyrrole can be used as an optical pH sensor in the pH range 6-12 [201]. The optical pH response of these conducting polymers originates from the acid-base properties of the conducting polymer backbone.

Some polythiophene derivatives show pH-dependent fluorescence [202]. Another approach is to covalently immobilize fluorescent indicators to the conducting polymer resulting also in a pH-dependent fluorescence signal [203]. The fluorescence is usually quenched (turned off) in most of the probes when the sensor material comes in contact with a specific analyte, e.g. Ni^{2+}, Cd^{2+}, Cr^{6+}, Mn^{2+}, Cu^{2+} [204,205], Pd^{2+} [206], Eu^{3+} [205], Fe^{3+} [207], F^- and $H_2PO_4^-$ [208]. Poly(p-phenyleneethynylene)s functionalized with a crown ether (benzo-15-crown-5) can be used for fluorescent K^+ detection [209]. The principle of fluorescence quenching in conducting polymers is based on formation of excitons (when the probe is illuminated with light, hv), which can rapidly migrate between isoenergetic sites along the polymer chain to a low energy acceptor site that quenches the fluorescence [204]. The quenching mechanism is efficient because one receptor site results in efficient quenching of several emitting units along the conducting polymer backbone. Very low analyte concentrations can therefore be measured with the fluorescence probes. It has been proposed that the turn-on fluorescent probe technique is even more sensitive than the turn-off technique [210]. The working principle of the turn-on probes is illustrated in **Fig. 8.8** [210]. Receptor molecules, which can interact with a specific analyte, are usually incorporated in the conjugated polymer matrix of the turn-on sensors. In the absence of the analyte, the fluorescence of the polymer is quenched by the relatively high-energy nonbonding electron pairs of the receptor molecules **(Fig. 8.8a)**. The HOMO level of the receptor is, however, lowered by the receptor-analyte interaction and the fluorescence is thus turned on **(Fig. 8.8b)**. The measurement of pH [210,211], Hg^{2+} [210,212] and Pb^{2+} [212] is possible with the turn-on technique.

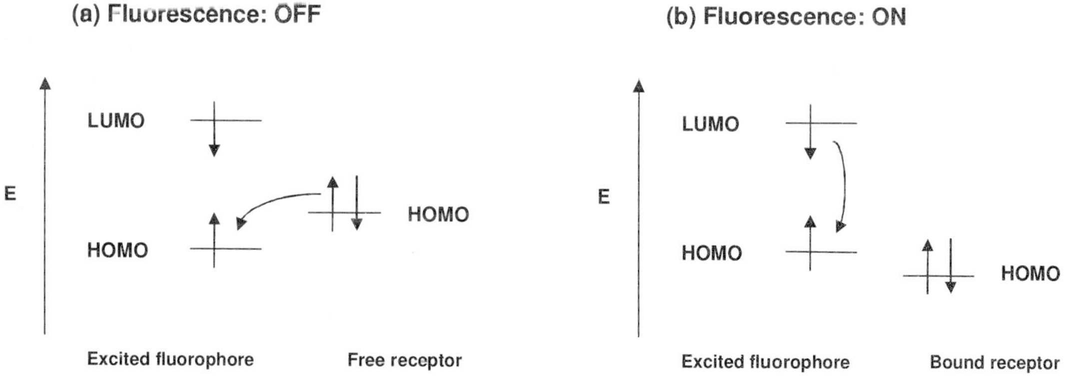

Fig. 8.8. Working principle of the fluorescence "turn-on" sensors [210]: (a) The fluorescence is quenched by the free receptor (not bound to the analyte) in the conducting polymer membrane, and (b) the fluorescence is turned on when the receptors are bound to the analyte.

Miniaturization of conducting polymer-based optical pH sensors is possible by using optical fibers. However, conducting polymers can show hysteresis and slow pH response, which may require special conditioning of this type of optical pH sensors. Hysteresis effects may also limit the practically useful pH range.

Optical sensors to different ions can be developed by utilizing conformation changes in the conjugated polymer backbone induced by complexation of different ions to receptor groups covalently bound to the conjugated polymer chain. Polythiophenes containing crown ether groups give an optical response to alkali metal ions resulting from changes in the conjugation length along the polymer backbone induced by ion complexation by the crown ether groups [213]. Similar effects can be seen in polythiophene derivatives containing oligo(oxyethylene) side groups [214]. A conducting polymer containing 3,4-ethylenedioxythiophene (EDOT) and bipyridine units in the conjugated polymer chain can coordinate protons and divalent transition metal ions, such as Fe^{2+}, Co^{2+}, Ni^{2+} and Cu^{2+}, resulting in a change in the optical absorption in the UV-visible range [215]. A similar polymer containing EDOT and 2,3-di(pyrrole-2-yl)quinoxaline (DPQ) units show an optical response to anions such as fluoride, pyrophosphate and cyanide, suggesting that these anions are bound to the DPQ group via hydrogen bonds [216]. Anion recognition by p-doped conducting polymers should be particularly favorable due to electrostatic attraction between the positively charged polymer chain and the anion, provided that sufficient selectivity can be achieved via suitable receptor groups attached to the polymer.

3.2. Gas Sensors

Optical gas sensors for NO_2 and H_2S can be fabricated from Langmuir-Blodgett films of polyaniline [217]. Sensing based on surface plasmon resonance (SPR) allow NO_2 and H_2S detection down to ca. 50 ppm (v/v). However, the SPR response of the sensor is relatively slow (30-60 min.) and not completely reversible, which are severe drawbacks of the method [217]. The optical ozone detector based on polyaniline also shows only partial reversibility, but may be useful as single-use detector of ozone leaks [218]. In contrast, polyaniline is very suitable for optical detection of ammonia (NH_3) gas down to 1 ppm (v/v) with a response time of less than 15s [219]. Interactions between NH_3 and polyaniline cause deprotonation of the emeraldine salt form of polyaniline to the emeraldine base form, which results in the recorded optical response [220]. Such optical ammonia gas sensors are clearly compatible with the conductimetric ones.

Optical sensors based on fluorescence can be extremely sensitive. Additionally, conducting polymers allow intrinsic amplification of fluorescence signals, in a similar way as for the conductimetric signal transduction, where a small number of molecular recognition events cause a large change in the recorded signal [136]. For example, the fluorescence of some conducting polymers based on iptycene-substituted poly(phenylene-ethynylene) [34], poly(p-phenylenevinylene) and poly(diphenylacetylene) [221] is effectively quenched by electron-accepting nitro-containing analytes, such as the explosive 2,4,6-trinitrotoluene (TNT) as well as 2,4- and 2,6-dinitrotoluene (DNTs). Such fluorescence-based optical sensors can be used to detect TNT down to a concentration of 10 ppb of TNT vapor. Highly sensitive detectors for TNT and other explosives is of utmost important for safety reasons, e.g. when localizing unexploded land mines in different parts of the world.

3.3. Biosensors

Cytochrome c is an electron transfer protein that causes highly efficient fluorescence quenching of the conjugated polymer poly[lithium 5-methoxy-2-(4-sulfobutoxy)-1,4-phenylenevinylene] (MBL-PPV) [222]. Quenching is observed at cytochrome c concentrations down to 10 pM. The efficient quenching is related to fast photoinduced electron transfer between cytochrome c and MBL-PPV. Additionally, the covalently bound sulfonate groups in MBL-PPV can electrostatically attract the cationic cytochrome c, which facilitates the fluorescence quenching. This means that MBL-PPV is an extremely promising material to produce a highly sensitive fluorescence-based optical biosensor for cytochrome c [222]. A fluorescent turn-on sensor for D-fructose, D-galactose and D-glucose was developed by combining the anionic sulfonated poly(phenyleneethynylene) with a cationic quencher comprising of a boronic acid functionalized benzyl viologen [223]. A strong fluorescence quenching was observed in the absence of sugar, but the addition of sugar results in even up to 70-fold increase in the fluorescence signal.

Polypyrrole containing immobilized tiron can be used as an optical biosensor for the detection of monophenolase in food products [224]. Such determinations are important in order to estimate the quality and shelf life of food products, such as apple juice.

Conducting polymers offer unique possibilities also for optical DNA sensing based on fluorescence [225-228] and absorbance in the UV-visible range [227].

4. CONCLUSIONS AND FUTURE PROSPECTS

Functionalization of conducting polymers with synthetic receptors and biological recognition systems allows the construction of novel chemical sensors and biosensors based on electrochemical and optical signal transduction. Conducting polymers are applied both as signal transducers and as sensing membranes. Representative examples include all-solid-state potentiometric ion sensors, conductimetric gas sensors and amperometric biosensors. Conducting polymers offer unique means towards signal amplification in the case of conductimetric and fluorescence-based chemical sensors. Development of conducting polymer-based DNA sensors and DNA-chips has become a rapidly expanding research area. Many of the conducting polymer-based sensors discussed in this chapter are still under development and new concepts are continuously tested all over the world. There is no doubt that conducting polymers will have a great impact on the area of chemical sensors also in the future.

ACKNOWLEDGEMENTS

Financial support from the Academy of Finland and the National Technology Agency (TEKES) is gratefully acknowledged. This work is part of the activities at the Åbo Akademi Process Chemistry Centre within the Finnish Centre of Excellence Program (2000-2011) by the Academy of Finland.

REFERENCES

1. H. Shirakawa, E.J. Louis, A.G. MacDiarmid, C.K. Chiang and A.J. Heeger; J. Chem.Soc., Chem.Commun., 578 (1977).
2. H. Shirakawa; Synth.Met., 125, 3 (2002).
3. A.J. MacDiarmid; Synth.Met., 125, 11 (2002).
4. A.J. Heeger; Synth.Met., 125, 23 (2002).
5. Handbook of Conducting Polymers (Ed. T.A. Skotheim), Marcel Dekker, New York (1986).
6. Handbook of Organic Conductive Molecules and Polymers (Ed. H.S. Nalwa), Wiley, Chichester (1997).
7. Handbook of Conducting Polymers, Second Edition (Ed. T.A. Skotheim, R.L. Elsenbaumer and J.R. Reynolds), Marcel Dekker, New York (1998).

8. Electroactive Polymer Electrochemistry (Ed. M.E.G. Lyons), Plenum Press, New York (1994).
9. G. Inzelt, M. Pineri, J.W. Schultze and M.A. Vorotyntsev; Electrochim.Acta, 45, 2403 (2000).
10. D. MacInnes, Jr., M.A. Druy, P.J. Nigrey, D.P. Nairns, A.G. MacDiarmid and A.J. Heeger; J. Chem.Soc., Chem.Commun. 317 (1981).
11. J.S. Miller; Adv.Mater., 5, 587 (1993).
12. J.S. Miller; Adv.Mater., 5, 671 (1993).
13. G.M. Spinks, P.C. Innis, T.W. Lewis, L.A.P. Kane-Maguire and G.G. Wallace; Mater.Forum, 24, 125 (2000).
14. H. Sirringhaus, T. Kawase, R.H. Friend, T. Shimoda, M. Inbasekaran, W. Wu and E.P. Woo; Science, 290, 2123 (2000).
15. A. Ivaska; Electroanalysis, 3, 247 (1991).
16. M.D. Imisides, R. John, P.J. Riley and G.G. Wallace; Electroanalysis, 3, 879 (1991).
17. G. Bidan; Sens.Actuators, B 6, 45 (1992).
18. G. Zotti; Synth.Met., 51, 373 (1992).
19. P.N. Bartlett and P.R. Birkin; Synth.Met., 61, 15 (1993).
20. P.R. Teasdale and G.G. Wallace; Analyst, 118, 329 (1993).
21. P.N. Bartlett and J.M. Cooper; J. Electroanal.Chem., 362, 1 (1993).
22. S.A. Emr and A.M. Yacynych; Electroanalysis, 7, 913 (1995).
23. M. Josowicz; Analyst, 120, 1019 (1995).
24. O.A. Sadik; Anal.Methods Instrum., 2, 293 (1995).
25. J.N. Barisci, C. Conn and G.G. Wallace; TRIP, 4, 307 (1996).
26. S.B. Adeloju and G.G. Wallace; Analyst, 121, 699 (1996).
27. M.D. Imisides, R. John and G.G. Wallace; CHEMTECH, May, 19 (1996).
28. L. Rover Jr., G. De Oliveira Neto and L.T. Kubota; Química Nova, 20, 519 (1997) (in Portuguese).
29. W. Göpel and K-D. Schierbaum; in Handbook of Organic Conductive Molecules and Polymers (Ed.H.S.Nalwa), Vol.4, p.621, Wiley, Chichester (1997).
30. B. Fabre and J. Simonet; Coord.Chem.Rev., 178-180, 1211 (1998).
31. A. Giuseppi-Elie, G.G. Wallace and T. Matsue; in Handbook of Conducting Polymers, Second Edition (Ed. T.A. Skotheim, R.L. Elsenbaumer and J.R. Reynolds), p. 963, Marcel Dekker, New York (1998).
32. T.W. Lewis, G.G. Wallace and M.R. Smyth; Analyst, 124, 213 (1999).
33. G.G. Wallace, M. Smyth and H. Zhao; Trends Anal.Chem., 18, 245 (1999).
34. D.T. McQuade, A.E. Pullen and T.M. Swager; Chem.Rev., 100, 2537 (2000).
35. F. Palmisano, P.G. Zambonin and D. Centonze; Fresenius J.Anal.Chem., 366, 586 (2000).
36. L.A.P. Kane-Maguire and G.G. Wallace; Synth.Met., 119, 39 (2001).
37. B. Fabre; in Handbook of Advanced Electronic and Photonic Materials and Devices (Ed. H.S. Nalwa), Vol. 8, p. 103, Academic Press, San Diego (2001).
38. M. Leclerc; in Sensors Update (Ed. H. Baltes, W. Göpel and J. Hesse), Vol. 8, p. 21, Wiley-VCH, Weinheim (2001).
39. A. Ramanaviciene and A. Ramanvicius, Crit.Rev.Anal.Chem., 32, 245 (2002).
40. J. Janata and M. Josowicz, Nature Materials, 2, 19 (2003).
41. J. Bobacka, A. Ivaska and A. Lewenstam; Electroanalysis, 15, 366 (2003).
42. M. Trojanowicz; Microchim.Acta, 143, 75 (2003).
43. J. Bobacka, T. Lindfors, A. Lewenstam and A. Ivaska; American Laboratory, February, 13 (2004).
44. J. Bobacka; Electroanalysis, 18, 7 (2006).
45. A. Michalska; Anal.Bioanal.Chem., 384, 391 (2006).
46. J. Bobacka; in Encyclopedia of Sensors (Ed. C.A. Grimes, E.C. Dickey and M.V. Pishko), American Scientific Publishers, in press.
47. S. Dong, Z. Sun and Z. Lu; Analyst, 113, 1525 (1988).
48. S. Dong, Z. Sun and Z. Lu; J. Chem.Soc., Chem.Commun., 993 (1988).
49. Z. Lu, Z. Sun and S. Dong; Electroanalysis, 1, 271 (1989).
50. A. Cadogan, A. Lewenstam and A. Ivaska; Talanta, 39, 617 (1992).
51. T. Lindfors and A.I vaska; J. Electroanal.Chem., 531, 43 (2002).

52. T.A. Bendikov and T.C. Harmon; Anal.Chim.Acta, 551, 30 (2005).
53. R.S. Hutchins, V. Salvadó and L.G. Bachas; Polym.Mater.Sci.Eng., 71, 658 (1994).
54. R.S. Hutchins and L.G. Bachas; Anal.Chem., 67, 1654 (1995).
55. T. Okada, H. Hayashi, K. Hiratani, H. Sugihara and N. Koshizaki; Analyst, 116, 923 (1991).
56. A. Michalska, A. Lewenstam, A. Ivaska and A. Hulanicki; Electroanalysis, 5, 261 (1993).
57. A. Lewenstam, J. Bobacka and A. Ivaska; J. Electroanal.Chem., 368, 23 (1994).
58. J. Bobacka, Z. Gao, A. Ivaska and A. Lewenstam; J. Electroanal.Chem., 368, 33 (1994).
59. Q. Pei and R. Qian; Electrochim.Acta, 37, 1075 (1992).
60. W. Vonau, J. Gabel and H. Jahn; Electrochim.Acta, 50, 4981 (2005).
61. F. Garnier; Angew.Chem., 101, 529 (1989).
62. D.J. Walton, C.E. Hall and A. Chyla; Synth.Met., 45, 363 (1991).
63. J. Migdalski, T. Blaz and A. Lewenstam; Anal.Chim.Acta, 322, 141 (1996).
64. C. Sun, J. Zhao, H. Xu, Y. Sun, X. Zhang and J. Shen; Talanta, 46, 15 (1998).
65. Z. Mousavi, J. Bobacka and A. Ivaska; Electroanalysis, 17, 1609 (2005).
66. J. Bobacka, A. Ivaska and A. Lewenstam; Anal.Chim.Acta, 385, 195 (1999).
67. T. Lindfors and A. Ivaska; Anal.Chim.Acta, 404, 101 (2000).
68. T. Lindfors and A.I vaska; Anal. Chim.Acta, 437, 171 (2001).
69. P. Sjöberg, J. Bobacka, A. Lewenstam and A. Ivaska; Electroanalysis, 11, 821 (1999).
70. A. Cadogan, Z. Gao, A. Lewenstam, A. Ivaska and D. Diamond; Anal.Chem., 64, 2496 (1992).
71. J. Bobacka, T. Lindfors, M. McCarrick, A.I vaska and A. Lewenstam; Anal.Chem., 67, 3819 (1995).
72. T. Toczylowska, R. Pokrop, A. Dybko and W. Wróblewski; Anal.Chim.Acta, 540, 167 (2005).
73. Y-B. Shim, D.E. Stilwell and S-M. Park; Electroanalysis, 3, 31 (1991).
74. J. Kankare and I.A. Vinokurov; Anal. Chem., 69, 2337 (1997).
75. E.C. Hernández, C. Mortensen and L.G. Bachas; Electroanalysis, 9, 1049 (1997).
76. G. Cui, J.S. Lee, S.J. Kim, H. Nam, G.S. Cha and H.D. Kim; Analyst, 123, 1855 (1998).
77. I. Taniguchi, T. Fujiyasu, S. Tomimura, H. Eguchi, K. Yasukouchi, I. Tsuji and M. Unoki; Anal.Sci., 2, 587 (1986).
78. M. Trojanowicz and M.L. Hitchman; Electroanalysis, 8, 263 (1996).
79. S.B. Adeloju, S.J. Shaw and G.G. Wallace; Anal.Chim.Acta, 281, 611 (1993).
80. S.B. Adeloju, S.J. Shaw and G.G. Wallace; Anal.Chim.Acta, 281, 621 (1993).
81. T. Osaka, S. Komaba and A. Amano; J. Electrochem.Soc., 145, 406 (1998).
82. A.A. Karyakin, M. Vuki, L.V. Lukachova, E.E. Karyakina, A.V. Orlov, G.P. Karpachova and J. Wang; Anal.Chem., 71, 2534 (1999).
83. A.A.Karyakin, L.V.Lukachova, E.E.Karyakina, A.V.Orlov and G.P.Karpachova; Anal.Commun., 36, 153 (1999).
84. E. Shoji and M.S. Freund; Proc.Electrochem.Soc., 2001-18, 293 (2001).
85. T. Osaka, S. Komaba, Y. Fujino, T. Matsuda and I. Satoh; J. Electrochem.Soc., 146, 615 (1999).
86. Y. Ikariyama and W.R. Heineman; Anal.Chem., 58, 1803 (1986).
87. D.M.T. O'Riordan and G.G. Wallace; Anal. Chem., 58, 128 (1986).
88. P. Bäuerle and S. Scheib; Adv.Mater., 5, 848 (1993).
89. G. Bidan and M-A. Niel; Synth.Met., 84, 255 (1997).
90. M. Fabrizio, F. Furlanetto, G. Mengoli, M.M. Musiani and F. Paolucci; J. Electroanal.Chem., 323, 197 (1992).
91. M. Trojanowicz, A. Lewenstam, T. Krawczyñski vel Kravczyk, I. Lähdesmäki and W. Szczepek; Electroanalysis, 8, 233 (1996).
92. I. Lähdesmäki, A. Lewenstam and A. Ivaska; Talanta, 43, 125 (1996).
93. I. Lähdesmäki, W.W. Kubiak, A. Lewenstam and A.I vaska; Talanta, 52, 269 (2000).
94. M. Vidotti, L.H. Dall'Antonia, S.I. Córdoba de Torresi, K. Bergamaski and F.C. Nart; Anal.Chim.Acta, 489, 207 (2003).
95. J.-B. Yu, H.-G. Byun, M.-S. So and J.-S. Huh; Sens. Actuators B, 108, 305 (2005).
96. B. Fabre, S. Burlet, R. Cespuglio and G. Bidan; J. Electroanal.Chem., 426, 75 (1997).
97. N.F. Foulds and C.R. Lowe; J. Chem.Soc., Faraday Trans., 82, 1259 (1986).

98. M. Umaña and J. Waller; Anal.Chem., 58, 2979 (1986).

99. P.N. Bartlett and R.G. Whitaker; J. Electroanal.Chem., 224, 37 (1987).

100. T. Kuwahara, K. Oshima, M. Shimomura and S. Miyauchi; Synth.Met, 152, 29 (2005).

101. T. Kuwahara, M. Shimomura and S. Miyauchi; Polymer, 46, 8091 (2005).

102. D. Aussawasathien, J.-H. Dong and L. Dai; Synth.Met, 154, 37 (2005).

103. D. Bélanger, J. Nadreau and G. Fortier; J.Electroanal.Chem., 274, 143 (1989).

104. N.C. Foulds and C.R. Lowe; Anal.Chem., 60, 2473 (1988).

105. M.A. Rahman, M.-S. Won and Y.-B. Shim; Biosens.Bioelectron., 21, 257 (2005).

106. C. Iwakura, Y. Kajiya and H. Yoneyama; J. Chem.Soc., Chem.Commun., 1019 (1988).

107. M.A. Rahman, N.-H. Kwon, M.-S. Won, E.S. Choe and Y.-B. Shim; Anal.Chem, 77, 4854 (2005).

108. D. Bouchta, N. Izaoumen, H. Zejli, M. El Kaoutit and K.R. Temsamani; Biosens.Bioelectron., 20, 2228 (2005).

109. F. Qu, M. Yang, J. Jiang, G. Shen and R. Yu; Anal.Biochem., 344, 108 (2005).

110. M.R. Smyth, E. Buckley, J. Rodriguez Flores, R. John and G.G. Wallace; in Contemporary Electroanalytical Chemistry (Ed. A. Ivaska, A. Lewenstam and R. Sara), p. 367, Plenum Press, New York (1990).

111. M.A. Petit, D. Zante, C. Colin, C. Combellas and M.-C. Hennion; Electroanalysis, 13, 856 (2001).

112. H. Korri-Youssoufi, C. Richard and A. Yassar; Mater.Sci.Eng., C 15, 307 (2001).

113. F. Darain, D.S. Park, J.-S. Park, S.-C. Chang and Y.-B. Shim; Biosens.Bioelectron., 20, 1780 (2005).

114. O.A. Sadik and G.G. Wallace; Anal.Chim.Acta, 279, 209 (1993).

115. H. Korri-Youssoufi, F. Garnier, P. Srivastava, P. Godillot and A. Yassar; J. Am.Chem.Soc., 119, 7388 (1997).

116. P. Bäuerle and A. Emge; Adv.Mater., 3, 324 (1998).

117. F. Garnier, H. Korri-Youssoufi, P. Srivastava, B. Mandrand and T. Delair; Synth.Met., 100, 89 (1999).

118. J. Wang; Chem.Eur.J., 5, 1681 (1999).

119. J. Cha, J.I. Han, Y. Choi, D.S. Yoon, K.W. Oh and G. Lim; Biosens.Bioelectron., 18, 1241 (2003).

120. H. Gu, X. Su and K.P. Loh; Chem.Phys.Lett., 388, 483 (2004).

121. S. Reisbeg, B. Piro, V. Noël and M.C. Pham; Anal.Chem., 77, 3351 (2005).

122. T.-Y. Lee and Y.-B. Shim; Anal.Chem., 73, 5629 (2001).

123. H. Peng, C. Soeller, N. Vigar, P.A. Kilmartin, M.B. Cannell, G.A. Bowmaker, R.P. Cooney and J. Travas-Sejdic; Biosens.Bioelectron., 20, 1821 (2005)

124. C. Tlili, H. Korri-Youssoufi, L. Ponsonnet, C. Martelet and N.J. Jaffrezic-Renault; Talanta, 68, 131 (2005).

125. M. Jiang and J. Wang; J. Electroanal.Chem., 500, 584 (2001).

126. J. Wang, M. Jiang and A.-N. Kawde; Electroanalysis, 13, 537 (2001).

127. E. Komarova, M. Aldissi and A. Bogomolova; Biosens.Bioelectron., 21, 182 (2005).

128. H. Cai, Y. Xu, P.-G. He and Y.-Z. Fang; Electroanalysis, 15, 1864 (2003).

129. A. Ramanaviciene and A. Ramanavicius; Anal.Bioanal.Chem., 379, 287 (2004)

130. H. Gu, X.D. Su and K.P. Lu; J. Phys.Chem.B, 109, 13611 (2005).

131. E.W. Paul, A.J. Ricco and M.S. Wrighton; J. Phys.Chem., 89, 1441 (1985).

132. J.W. Thackeray and M.S. Wrighton; J. Phys.Chem., 90, 6674 (1986).

133. J. Janata and M. Josowicz; Nature Materials, 2, 19 (2003).

134. J. Janata; Phys.Chem.Chem.Phys., 5, 5155 (2003).

135. M.J. Marsella, R.J. Newland, P.J. Carroll and T.M. Swager; J. Am.Chem.Soc., 117, 9842 (1995).

136. T.M. Swager; Acc.Chem.Res., 31, 201 (1998).

137. F. Yue, T.S. Ngin and G. Hailin; Sens.Actuators, B 32, 33 (1996).

138. A. Talaie and J.A. Romagnoli; Synth.Met., 82, 231 (1996).

139. R.B. Dabke, G.D. Singh, A. Dhanabalan, R. Lai and A.Q. Contractor; Anal.Chem., 69, 724 (1997).

140. A. Riul Jr., A.M. Gallardo Soto, S.V. Mello, S. Bone, D.M. Taylor and L.H.C. Mattoso; Synth.Met., 132, 109 (2003).

141. J.J. Miasik, A. Hooper and B.C. Tofield; J. Chem.Soc., Faraday Trans., 82, 1117 (1986).

142. G. Gustafsson and I. Lundström; Synth.Met., 21, 203 (1987).

143. T. Hanawa and H. Yoneyama; Bull.Chem.Soc.Jpn., 62, 1710 (1989).
144. N.M. Ratcliffe; Anal.Chim.Acta, 239, 257 (1990).
145. D.L. Ellis, M.R. Zakin, L.S. Bernstein and M.F. Rubner; Anal.Chem., 68, 817 (1996).
146. E. Kriván, C. Visy, R. Dobay, G. Harsányi and O. Berkesi; Electroanalysis, 12, 1195 (2000).
147. R.M.G. Rajapakse, A.D.L. Chandani Perera and H.D.S. Premasiri; J. Natn.Sci.Foundation Sri Lanka, 28, 277 (2000).
148. M.K. Ram, O. Yavuz and M. Aldissi; Synth.Met, 151, 77 (2005).
149. S. Jain, A.B. Samui, M. Patri, V.R. Hande and S.V. Bhoraskar; Sens.Actuators, B 106, 609 (2005).
150. S. Brady, K.T .Lau, W. Megill, G.G. Wallace and D. Diamond; Synth.Met, 154, 25 (2005).
151. J. Jang, M. Chang and H. Yoon; Adv.Mater., 17 1616 (2005).
152. M.S.A. Abdou, F.P. Orfino, Z.W. Xie, M.J. Deen and S. Holdcroft; Adv.Mater., 6, 838 (1994).
153. B.Z. Lubentsov, O.N. Timofeeva and M.L. Khidekel; Synth.Met., 45, 235 (1991).
154. L.S. Hwang, J.M. Ko, H.W. Rhee and C.Y. Kim; Synth.Met., 55-57, 3671 (1993).
155. Y. Li and M. Yang; Synth.Met., 129, 285 (2002).
156. M. Yang, Y. Li, X. Zhan and M. Ling; J. Appl.Polym.Sci., 74, 2010 (1999).
157. T. Shioya and T.M. Swager; Chem.Commun., 1364 (2002).
158. S. Radhakrishnan and S.D. Deshpande; Sensors, 2, 185 (2002).
159. F. Selampinar, L. Toppare, U. Akbulut, T. Yalçin and S. Süzer; Synth.Met., 68, 109 (1995).
160. K. Ogura and H. Shiigi; ACS Symposium Series, 832, 88 (2003).
161. S. Hossein Hosseini and A.A. Entezami; J. Appl.Polym.Sci., 90, 49 (2003).
162. J. Huang, S. Virji, B.H. Weiller and R.B. Kaner; Polymer Preprints, 44, 130 (2003).
163. S.C.K. Misra, P. Mathur and B.K. Srivastava; Sens.Actuators, A 114, 30 (2004).
164. P.N. Bartlett and S.K. Ling-Chung; Sens.Actuators, 19, 141 (1989).
165. P.N. Bartlett and S.K. Ling-Chung; Sens.Actuators, 20, 287 (1989).
166. L. Jiang, H.-K. Jun, Y.-S. Hoh, J.-O. Lim, D.-D. Lee and J.-S. Huh; Sens.Actuators, B 105, 132 (2005).
167. J.-S. Kim, S.-O. Sohn and J.-S. Huh; Sens.Actuators, B 108, 409 (2005).
168. J.M. Charlesworth, A.C. Partridge and N. Garrard; J. Phys.Chem., 97, 5418 (1993).
169. G.A. Sotzing, S.M. Briglin, R.H. Grubbs and N.S. Lewis; Anal.Chem., 72, 3181 (2000).
170. Y. Wang, G.A. Sotzing and R.A. Weiss; Chem.Mater., 15, 375 (2003).
171. G. Anitha and E. Subramanian; Sens.Actuators, B 92, 49 (2003).
172. L. Ruangchuay, A. Sirivat and J. Schwank; Talanta, 60, 25 (2003).
173. T.C. Pearce, J.W. Gardner, S. Friel, P.N. Bartlett and N. Blair; Analyst, 118, 371 (1993).
174. J.M. Slater, J. Paynter and E.J. Watt; Analyst, 118, 379 (1993).
175. M. Masila, A. Sargent and O.A. Sadik; Electroanalysis, 10, 312 (1998).
176. T.A. Nguyen, J.N. Barisci, A. Partridge and G.G. Wallace; Synth.Met., 137, 1445 (2003).
177. Q. Ameer and S.B. Adeloju; Sens.Actuators, B 106, 541 (2005).
178. K.J. Albert, N.S .Lewis, C.L. Schauer, G.A. Sotzing, S.E. Stitzel, T.P. Vaid and D.R. Walt; Chem.Rev., 100, 2595 (2000).
179. P. Ködderitzsch, R. Bischoff, P. Veitenhansl, W. Lorenz and G. Bischoff; Sens.Actuators, B 107, 479 (2005).
180. K.C. Persaud; Materials Today, April, 38 (2005).
181. D. Hodgins and D. Simmonds; Journal of Automatic Chemistry, 17, 179 (1995).
182. P. Greenfield; Clinica, 604, 20 (1994).
183. J.E.G. de Souza, B.B. Neto, F.L. dos Santos, C.P. de Melo, M.S. Santos and T.B. Ludermir; Synth.Met., 102, 1296 (1999).
184. R. Stella, J.N. Barisci, G. Serra, G.G. Wallace and D. De Rossi; Sens.Actuators, B 63, 1 (2000).
185. A. Guadarrama, M.L. Rodríguez-Méndez, C. Sanz, J.L. Ríos and J.A. de Saja; Anal.Chim.Acta, 432, 283 (2001).
186. M. Chimenti, D. De Rossi, F. Di Francesco, C. Domenici, G. Pieri, G. Pioggia and O. Salvetti; Meas.Sci.Technol., 14, 815 (2003).

187. A.K. Pavlou, N. Magan, C. McNulty, J. Meecham Jones, D. Sharp, J.Brown and A.P.F. Turner; Biosens.Bioelectron., 17, 893 (2002).
188. W. Bourgeois, P. Hogben, A. Pike and R.M. Stuetz; Sens.Actuators, B 88, 312 (2003).
189. K. Ramanathan, M.A. Bangar, M. Yun, W. Chen, N.V. Myung and A. Mulchandani; J. Am.Chem.Soc., 127, 496 (2005).
190. D.T. Hoa, T.N. Suresh Kumar, N.S. Punekar, R.S. Srinivasa, R. Lal and A.Q. Contractor; Anal.Chem., 64, 2645 (1992).
191. H. Sangodkar, S. Sukeerthi, R.S. Srinivasa, R. Lal and A.Q. Contractor; Anal.Chem., 68, 779 (1996).
192. E. Palmqvist, C. Berggren Kriz, M. Khayyami, B. Danielsson, P-O. Larsson, K. Mosbach and D. Kriz; Biosens.Bioelectron., 9, 551 (1994).
193. P.N. Bartlett and P.R. Birkin; Anal.Chem., 65, 1118 (1993).
194. P.N. Bartlett and P.R. Birkin; Anal.Chem., 66, 1552 (1994).
195. Z. Ge, C.W. Brown, L. Sun and S.C. Yang; Anal.Chem., 65, 2335 (1993).
196. U-W. Grummt, A. Pron, M. Zagorska and S. Lefrant; Anal.Chim.Acta, 357, 253 (1997).
197. P.T. Sotomayor, I.M. Raimundo Jr., A.J.G. Zarbin, J.J.R. Rohwedder, G. Oliveira Neto and O.L. Alves; Sens.Actuators, B 74, 157 (2001).
198. T. Lindfors and A. Ivaska; J. Electroanal.Chem., 580, 320 (2005).
199. E. Pringsheim, E. Terpetschnig and O.S. Wolfbeis; Anal.Chim.Acta, 357, 247 (1997).
200. M.D.P.T. Sotomayor, M-A. De Paoli and W.A. de Oliveira; Anal.Chim. Acta, 353, 275 (1997).
201. S. de Marcos and O.S. Wolfbeis; Anal.Chim.Acta, 334, 149 (1996).
202. Y-G. Kim, L.A. Samuelson, J. Kumar and S.K. Tripathy; J. Macromol.Sci. A, 39, 1127 (2002).
203. M. Uttamlal, W.D. Sloan and D. Millar; Polym.Int., 51, 1198 (2002).
204. Y. Zhang, C.B. Murphy and W.E. Jones Jr.; Macromolecules, 35, 630 (2002).
205. G. Ramachandran, T.A. Smith, D. Gómez and K.P. Ghiggino; Synth.Met, 152, 17 (2005).
206. H. Huang, K. Wang, W. Tan, D.An, X.Yang, S.Huang, Q.Zhai, L.Zhou and Y.Jin; Angew.Chem.Int.Ed., 43, 5635 (2004).
207. J. Na, Y.-S. Kim, W.H. Park and T.S. Lee; J. Polym.Sci.:Part A:Polymer Chemistry, 42, 2444 (2004).
208. G. Zhou, Y. Cheng, L. Wang, X. Jing and F. Wang; Macromolecules, 38, 2148 (2005).
209. J. Kim, D.T. McQuade, S.K. McHugh and T.M. Swager; Angew.Chem.Int.Ed., 39, 3868 (2000).
210. L.-J. Fan, Y. Zhang and W.E. Jones Jr.; Macromolecules, 38, 2844 (2005).
211. D.T. McQuade, A.H. Hegedus and T.M. Swager; J. Am.Chem.Soc., 122, 12389 (2000).
212. I.-B. Kim, B. Erdogan, J.N. Wilson and U.H.F. Bunz; Chem.Eur.J., 10, 6247 (2004).
213. M.J. Marsella and T.M. Swager; J. Am.Chem.Soc., 115, 12214 (1993).
214. I. Lévesque and M. Leclerc; Synth.Met., 84, 203 (1997).
215. G. Zotti, S. Zecchin, G. Shiavon, A. Berlin and M. Penso; Chem.Mater., 11, 3342 (1999).
216. D. Aldakov and P. Anzenbacher, Jr.; Polym.Mater.Sci.Eng., 88, 512 (2003).
217. N.E. Agbor, J.P. Cresswell, M.C. Petty and A.P. Monkman; Sens.Actuators, B 41, 137 (1997).
218. M. Ando, C. Swart, E. Pringsheim, V.M. Mirsky and O.S. Wolfbeis; Solid State Ionics, 152-153, 819 (2002).
219. Z. Jin, Y. Su and Y. Duan; Sens.Actuators, B 72, 75 (2001).
220. H. Hu, M. Trejo, M.E. Nicho, J.M. Saniger and A. García-Valenzuela; Sens. Actuators, B 82, 14 (2002).
221. C.-P. Chang, C.-Y. Chao, J.H. Huang, A.-K. Li, C.-S. Hsu, M.-S. Lin, B.R. Hsieh and A.-C. Su; Synth.Met., 144, 297 (2004).
222. C. Fan, K.W. Plaxco and A.J. Heeger; J. Am.Chem.Soc., 124, 5642 (2002).
223. N. DiCesare, M.R. Pinto, K.S. Schanze and J.R. Lakowicz; Langmuir, 18, 7785 (2002).
224. S. Dutta, S. Padhye, R. Narayanaswami and K.C. Persaud; Biosens.Bioelectron., 16, 287 (2001).
225. T. Livache, E. Maillart, N. Lassalle, P. Mailley, B. Corso, P. Guedon, A. Roget and Y. Levy; J. Pharm.Biomed.Anal., 32, 687 (2003).
226. G. Bidan, M. Billon, M.L. Calvo-Muñoz and A. Dupont-Fillard; Mol.Cryst.Liq.Cryst., 418, 255 (2004).
227. H.-A. Ho, M. Béra-Abérem and M. Leclerc; Chem.Eur.J., 11, 1718 (2005).
228. K.F. Karlsson, P. Åsberg, K.P.R. Nilsson and O. Inganäs; Chem.Mater, 17, 4204 (2005).

9

Photo-Converters Based on Dye-Doped Polymers

Alexander Ishchenko

Institute of Organic Chemistry, National Academy of Sciences of Ukraine,
5, Murmanskaya St., Kiev-94, 02094, Ukraine.
Corresponding author email: alexish@i.com.ua

Contents

- **INTRODUCTION**

- **THE METHODS OF INCORPORATION OF THE DYES INTO POLYMERS**

- **CHEMICAL STRUCTURE AND SPECTRAL, LUMINESCENT PROPERTIES OF DYES IN POLYMER MATRICES**

- **THE EFFECT OF ELECTRON TRANSFER ON PHOTOLUMINESCENCE SPECTRA OF DYE-DOPED PHOTO-CONDUCTING POLYMERS**

- **LASER MEDIA**

- **LUMINESCENT SOLAR CONCENTRATOR**

- **SOLAR CELLS**

- **ELECTROLUMINESCENT MEDIA**

- **CONCLUSIONS AND FUTURE PROSPECTS**

- **REFERENCES**

Summary: The main approaches to molecular design of dye-doped polymers matrices have been considered. The advantages of such matrices over polymers and dyes separately were analyzed. It was shown that these advantages can be realized in that case where the doping of polymers by low molecular-weight dyes does not essentially distort spectral and luminescent properties of the individual components of polymer composition. The particularities of these properties of dye-doped nonphotoconducting and photoconducting polymers are discussed. The influence of polymer nature and chemical constitution of organic dyes on photophysical and photochemical properties of these matrices were analyzed. The processes of dyes aggregation in polymers were characterized. Their influence on photophysical properties and photochemical stability of dye-doped polymers is discussed. The different approaches to struggle for it are offered. The main paths of energy degradation of electronic excitation in such materials, internal conversion, intersystem crossing, electron transfer and photoisomerization were analyzed. The role of cation, anion and intra-inion dyes in the processes of generation and recombination of the charge carriers in photoconductive polymers was discussed. The prospects for the applications of dye-doped polymer materials as passive Q-switches of solid-state lasers, as active laser media, as luminescent solar converters, as photovoltaic cells and as electroluminescent emitters are demonstrated.

1. INTRODUCTION

The dye-doped polymers are unique photo-converters [1,2]. Depending on structure, they may absorb and luminesce in the whole visible and near infrared region of the spectrum [3-5]. Band shape can change from being strongly structured to being completely diffuse, from the narrow to the wide. The fluorescence quantum yield (φ) and the lifetime lie within the limits of 0.01-100% and 10^{-12}-10^{-15}s respectively. Stokes shifts (SS) change from several to hundreds of nm. The spectral and luminescent properties of some dyed matrices are strongly sensitive to the change in the nature of the medium and other factors insignificantly [3-5].

The dye-doped polymers are prospective materials for photographic and photoconductor sensitizers [6], optoelectronics [2,7,8], specially for devices used in quantum electronics [9,10], electroluminescence (EL) [11,12], photovoltaic solar cells [8,13], recording and storage of the information [7,8] and so on. The advantage of the dye-doped polymers in comparison with the colored inorganic materials are their flexibility, lower density, products creation possibility with large working square, rather low cost and high radiative stability. These denominations in a combination to ease of the mechanical treatment make of them irreplaceable materials for the above-mentioned application fields.

Modern photoconducting polymers are the effective converters of electric and light energy due to the interaction between electromagnetic radiation and π-electrons of polymer conjugated bonds [14]. However, they possess less change in the range of spectral, luminescent, non-linear optical and electro-physical properties than organic dyes.

The materials based on only organic dyes are not particularly effective as photo-converters. The reason is that electronic spectra of dyes in solid state undergo essential distortion in comparison with liquid solution [15-17]. It is caused by strong intermolecular interactions between dye molecules because of the short distance between them in solid thin film. Such interactions in polymer matrices could be weakened by the incorporation of polymer molecules between dye molecules [15,17].

The dye-doped polymers enable the transformation of light radiations over a wide spectral range with higher efficiency, than polymers and dyes separately. However, these advantages can be realized only in the case of the doping of polymers by low molecular weight dyes, which do not essentially distort the spectral, luminescent and other photo-physical properties of dye-doped polymers in comparison with the individual components of this polymer composition.

2. THE METHODS OF INCORPORATION OF THE DYES INTO POLYMERS

The dyed polymers represent a solid solution of a dye in a polymer. Such solutions can be prepared by the following methods.

2.1. Slop from the Solution

The solution of polymer and dye in definite concentration is prepared in a solvent, which dissolves them well and easily evaporates on drying. If the polymer and dye require different solvents for solubility reasons, a mixture of the two can be used to prepare the solution. The solution is poured on a flush surface and solvent is then evaporated, surface can be heated if necessary, to get dyed polymer film. The film is removed by the mechanical means. In case of strong adhesion of the film to the surface, it is desirable to coat the surface with some anti-adhesive cover before processing.

The slop method allows preparing the dyed polymer film of high optical quality with homogeneous distribution of the dye. However, this method is efficient only for thin films (thickness ≤ 500 μm).

2.2. Diffusion from the Solution

The dye is dissolved in such a solvent, which does not dissolve the polymer but only swells it slightly. The polymer product is dipped in this solution and the dye diffuses into its partially swelled surface. The swelling of polymer facilitates the dye penetration into its pores. The solvent evaporates in the process of drying of the dyed polymer.

The diffusion method does not produce homogeneous distribution of the dye within the thickness of polymer film. The properties of dyed polymer films, prepared by this method, frequently, are unstable for long period of time. Besides, the polymer surface can distort and even decrepitate after the solvent evaporation.

2.3. Preparation from the Melt

The dye is incorporated in the polymer melt in an inert atmosphere. The products of the desired shape are made from polymer melt by molding under pressure, molding or extrusion. The padding of optical-mechanical processing of a product can be performed, if necessary.

The main drawback of this method is the considerable thermo-destruction of the dye during processing, as the process involves high temperatures (> 160°C). Therefore, the dye assortment by this method is limited. There are internal stresses in the polymer that lead to a distortion in the orientation of polymeric layers causing incorporation of anisotropy in the optical properties of such products [18].

2.4. Polymerization of the Dye Containing Monomer

The dye is dissolved in the monomer or oligomer, or in one of the components of the polymerization reaction mixture. In a number of cases, the addition of a solvent is possible to increase the solubility and homogenization of reaction mixture followed by the polymerization to get amorphous dyed polymer.

This method is most widely used since it has huge potential of possibilities. Depending on the type of monomer, the technique and conditions of polymerization can be selected to make the films and products in the desired form as well as polyplexes (triplexes) [19]. The latter represents a thin layer of dyed polymer interposed between two optical substrates during polymerization. Polymeric composites based on polyurethane and epoxide strongly paste together the substrates due their high adhesion abilities. Such products do not require the optical-mechanical processing any more. By this technique, the more complex pasting of polyplexes, consisting of several substrates with the polymeric interlayer between them, has become possible. Films in the triplexes (polyplexes) possess, not only a higher optical quality in comparison to the free films, but also do not distort under operation of the power light fluxes, owing to the heat tapping by the substrates from a working irradiation zone. The films in the triplexes are not contaminated with dust and are not twisted during the course of operation. Besides, it is isolated from the effect of the atmospheric humidity, as a result of the protective coatings on both the faces of the device.

It is necessary to take into account, that the dyes in polymerization reactions are not always chemically inert. Depending upon the structure, they can interact with initial radicals or macro-radicals influencing the development of a polymer chain, molecular weight, molecular weight distribution and the rate of polymerization reaction. For example, in a number of polymethine (cyanines) (dyes **1-3**), indocarbocyanine (dye **1**) is an initiator, indodicarbocyanine (dye **2**) is a regulator while indotricarbocyanine (dye **3**) is an inhibitor for the thermo-polymerization of polymethylmethacrylate (PMMA) [20]. Such influence takes place both in the presence of the initiator as well as in its absence.

Among organic dyes, polymethine dyes are of greatest interest. Depending upon the structure of heterocyclic end groups and the length of polymethine chain (n), the position, intensity and form of their spectral band, values φ and SS as well as the relaxation time of excited state may vary greatly [3,4]. No other class of organic dyes possesses such contrasting photo-physical properties. The role of cyanine dyes as photo-converters in the photo-polymerization reactions is especially very efficient [1].

Dye **1. HIC** (n = 1, R = Me, $An^- = ClO_4^-$); Dye **1a.** (n = 1, R = Bu, $An^- = BuB(Ph)_3^-$); Dye **1b.** (n = 1, R = CH_2CH_2OH, $An^- = ClO_4^-$) ; Dye **2. HIDC** (n = 2, R = Me, $An^- = ClO_4^-$); Dye **3. HITC** (n = 3, R = Me, $An^- = ClO_4^-$)

The great interest in these investigations arose after Garry Schuster and his colleagues discovered the photo-initiation of polymerization reactions by the visible light [21]. Such photo-initiation based on electron photo-transfer from the anion (An^-) to the cation (Ct^+) in ion pairs of the polymethine dyes. Alkyl (R) and aryl (Ar) borate anions (for example, dye **1a**), which have low oxidation potentials, were used as counter-ions for this purpose [21,22]. During the photo-excitation of the dye, the cation transfers to the first singlet state $S_1(Ct^+)^1$.

$$Ct^+An^- + h\nu \rightarrow (Ct^+)^1An^- \tag{9.1}$$

The presence of the unpaired electrons makes it much more reactive than in the S_0 state. This gives rise to the electron transfer from the anion to the cation with the formation of a radical pair followed by its subsequent dissociation to the neutral radicals. A distinctive feature is the formation of alkyl radicals on further decomposition.

$$(Ct^+)^1An^- \rightarrow Ct^\bullet An^\bullet \tag{9.2}$$

$$Ct^\bullet An^\bullet \rightarrow Ct^\bullet + An^\bullet \tag{9.3}$$

$$[BuB(Ph)_3]^\bullet \rightarrow Bu^\bullet + B(Ph)_3 \tag{9.4}$$

This excludes the reverse transfer of an electron in the radical pairs (**eq. 9.2**). For this reason and also on account of the low oxidation potentials of the alkyl borate anions, the yield of the free radicals during the photo-irradiation of the ion pairs of polymethine dyes is fairly high. As a result, such pairs are effective initiators of photo-polymerization, e.g., the polymerization of methyl methacrylate.

$$R^\bullet + n\,[CH_2=C(Me)(COOMe)] \rightarrow R\text{-}[CH_2\text{-}(Me)(COOMe)]_n^\bullet \tag{9.5}$$

The probability of photo-transfer of electron in the ion pairs increases with decrease in the reduction and oxidation potentials of the cation and the anion respectively, decrease in the distance between them and the decrease in the dielectric constant of the solvent as well as its solvating power [1]. The probability of the formation of an ion pair is increased as the length of the polymethine chain of the cation is increased, e.g. series of polymethine dyes (dyes **1-3**) while its reduction potential decreases [23]. It becomes more intimate as a result of the increase of the electrostatic interactions on account of the greater localization of the positive charge of the cation with increase in n [24]. The photo-transfer of electron must, therefore, be facilitated in the indotricarbocyanine dye (dye **3**) compared to the carbocyanine dye (dye **1**). In fact, it is observed in benzene during direct excitation at the absorption band of the cation even in the case of the Cl^- and I^- anions in the former [25].

The photo-transfer of electrons is also facilitated substantially when organic anions with conjugated bonds are used. Thus, in cationic-anionic polymethine dyes, it occurred even in the mono- and trimethinecyanines during direct photo-excitation [26].

The dye radicals can be formed not only in the S_1 state, but also in the triplet state (T). They can be formed both through intra-molecular electron transfer from the anion to the cation in T- state $(Ct^+)^3$ and as a result of diffusion encounters of the ion pairs [27].

$$(Ct^+)^3An^- \rightarrow Ct^\bullet An^\bullet \rightarrow Ct^\bullet + An^\bullet \tag{9.6}$$

$$(Ct^+)^3An^- + (Ct^+)^3An^- \rightarrow Ct^\bullet + (Ct^{2+})^\bullet(An^-)_2 \tag{9.7}$$

The **eq. 9.6** is realized in the case of anions with a low oxidation potential, while the **eq. 9.7** is realized for those with a high oxidation potential. The yield of the radicals is usually higher during photosensitized excitation than during direct excitation [28]. For example, it is possible to obtain radicals for the dyes **1** and dye **3** as per **eq. 9.6**, containing a difficultly oxidizable perchlorate anion [25,27].

The high probability of the transfer of an electron in the T-state compared with S_1-state is due to the smaller role of the concurrent processes in the former compared with the latter [25]. In the triplet radical pair, in contrast to the singlet one, the reverse transfer of an electron with the formation of the initial ion pair Ct^+An^- is forbidden by the spin. Inter-system crossing takes place considerably more slowly than the internal conversion. Photo-isomerization processes are much less likely in the T-state than in the S_1 state. In addition, the lifetimes of the triplet states are several orders of magnitude longer than that of the singlet states, and this increases the probability of slower electron transfer processes. Furthermore, the radicals in the T-state can also be formed by the mechanism **(eq. 9.7)**.

The possibility of photo-transfer of electrons must be taken into account during the creation of colored polymeric materials. This question is particularly important for polymethine dyes, since the main competing channel for deactivation of their electronically excited states, i.e., photo-isomerization in the polymeric matrices is excluded to a significant degree. This increases the probability of electron transfer through the S_1 state **(eqs. 9.1-9.3)**. The absence of a concurrent photo-isomerization channel creates favorable conditions for the inter-system crossing. Under the conditions of reduced oxygen diffusion in the polymer (it is often evacuated), the triplet states are slightly deactivated. As a result, the probability of the formation of free radicals **(eqs. 9.6 & 9.7)** is also increased in polymeric matrices. The photo-transfer of electrons can give rise to undesirable photochemical reactions, leading to the destruction of functional materials based on polymethine dyes.

The synthesis of dye-doped matrices by the polymerization of monomer in the presence of dye does not provide the homogeneous coloring because of incomplete and non-uniform distribution of the dye in the polymer bulk. Synthesizing structurally dyed polymers, in which dye molecules are sewed with macromolecules by chemical bonds, can eliminate this drawback. The poly-condensation and poly-addition are most prospective methods for this purpose. For example, authors have successfully synthesized modified polyurethane (PU) by attaching dye molecules through its $-NH_2$ group using epichlorohydrin as a coupling agent [28]. Cyanine dye (dye **1a**) was sewed to PU through the interaction of its hydroxyl groups with isocyanate groups of the monomer [2].

There are several ways of linking cyanine dyes covalently to the nucleophilic substrates. For example, amino and carboxyl groups are capable of linking to macromolecules that possess nucleophilic groups. Since cellulose is a polyol, it has the potential to form ester linkages upon condensing with carboxylic acids. It has been shown that N,N'-carboxyalkylcarbocyanine dyes can easily be anchored to a cellulosic substrate using Fischer's esterification conditions [29].

3. CHEMICAL STRUCTURE, SPECTRAL AND LUMINESCENT PROPERTIES OF DYES IN POLYMER MATRICES

To make a purposeful search for colored polymeric media, it is necessary to establish the dependence between the chemical structure of the dye and its spectral and luminescent properties in polymer matrices.

3.1. Spectral and Luminescent Properties of Ion Dyes

An attempt had been made to study the dependence of spectral and luminescent properties in the cation thiopyrylotricarbocyanine dyes (dyes **4-14**) [30], representatives of which are used in lasers [9]. For this purpose, PMMA films dyed with these dyes were prepared by evaporation of 1,2-dichloroethane (DCE) from the solution of DCE, PMMA and the dye. The concentration of the dye in the solvent was chosen such that the number of molecules in unit volume of the solution could easily be correlated with their concentration of the film.

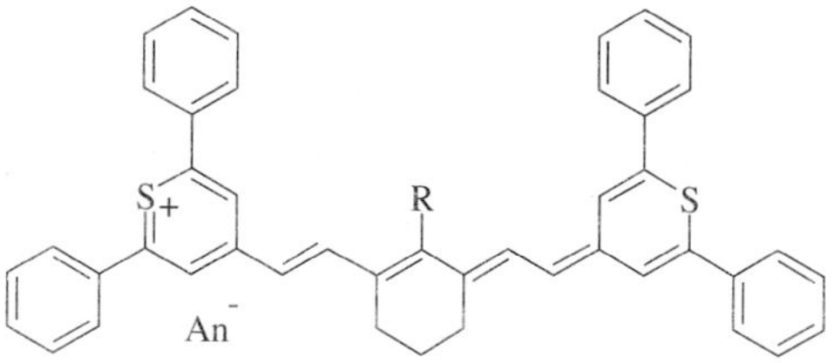

Dye **4**. (R = H, An^- = ClO_4^-); Dye **4a**. (R = H, An^- = Cl^-); Dye **4b**. (R = H, An^- = $MeC_6H_4SO_4^-$), Dye **4c**. (R = H, An^- = I^-); Dye **5**. (R = Cl, An^- = ClO_4^-); Dye **6**. (R = Me, An^- = ClO_4^-); Dye **7**. (R = iso-C_3H_7O, An^- = ClO_4^-); Dye **8**. (R = Ph, An^- = ClO_4^-)

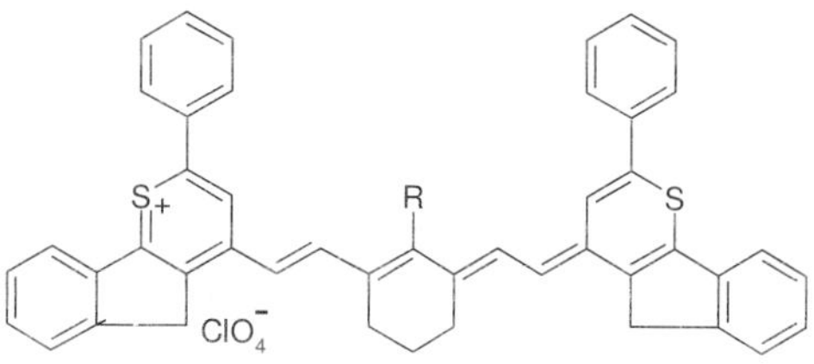

Dye **9**. (R = H); Dye **10**. (R = Cl); Dye **11**. (R = Ph)

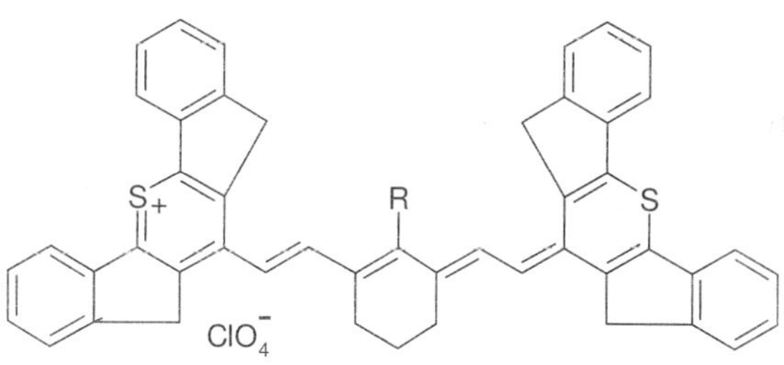

Dye **12**. (R = H); Dye **13**. (R = Cl); Dye **14**. (R = Ph)

The characteristics of the long-wave absorption band of the thiopyrylotricarbocyanines (dyes **4-11**) in DCE and in PMMA are given in **Tab. 9.1**. The values of the short- (λ_1) and long-wave (λ_2) absorption maxima as well as the ratio of the optical densities (D_1/D_2) at these maxima are also given in **Tab. 9.1**. Considering the absorption spectra of dyes **4** and dye **8**, as presented in **Fig. 9.1**, it may be seen that the positions and shapes of the absorption bands of the dyes under investigation in DCE and DCE + PMMA are practically identical. All the thiopyrylotricarbocyanines exhibit two absorption maxima in the polymer matrix similar to that in DCE. However, as compared in the latter solvent, λ_1 is appreciably shifted towards the short-wave region of the spectrum (**Tab. 9.1**). While for dyes which do not contain substituents in the meso-position of the polymethine chain (dyes **4** and **4a-c**), this effect is expressed much more strongly than for those that are substituted (dyes **5-8**). The hypsochromic shift of the short-wave absorption bands of the thiopyrylotricarbocyanines in going from DCE to PMMA cannot be due to universal interactions, since the polymer matrix possesses a larger refractive index value than that of the solvent [30]. Moreover, the mentioned interactions would affect both absorption maxima to the same degree. In our case the greatest hypsochromic shift in the λ_1 achieved was 80 nm while it was only 20 nm for λ_2. The non-uniform shift of these bands indicates that the short-wave absorption maximum of the thiopyrylotricarbocyanines in PMMA, in contrast to that in DCE, is not vibrational.

Tab. 9.1. Characteristics of long-wave absorption bands of the thiopyrylotricarbocyanines (dyes 4-11). Reproduced with permission from A.A. Ishchenko, "Structure and spectral, luminescent properties of polymethine dyes". Copyright © 1994, Naukova Dumka, Kiev.

Dye In	DCE			In PMMA		
	$\lambda_{1,\,nm}$	$\lambda_{2,\,nm}$	D_1/D_2	$\lambda_{1,\,nm}$	$\lambda_{2,\,nm}$	D_1/D_2
4	930	1035	0.30	855	1025	1.57
4a	930	1035	0.30	855	1030	1.19
4b	930	1035	0.30	855	1015	1.66
4c	930	1035	0.30	855	1030	1.75
5	960	1065	0.29	930	1065	0.76
6	950	1048	0.34	920	1040	0.73
7	925	1020	0.30	920	1015	0.57
8	940	1040	0.30	935	1040	0.56
9	940	1060	0.32	740	1040	3.91
10	970	1085	0.29	710	1085	3.07
11	950	1065	0.30	805	1060	1.52

The intensity of λ_1 of the thiopyrylotricarbocyanines (dyes **4**, **4a-c** and **8**) in the polymer matrix is increased compared with that in DCE, which is shown in the growth in the ratio of D_1/D_2. These effects are also clearly expressed for the dyes, which are not substituted in the polymethine chain (**Tab. 9.1 & Fig. 9.1**). The intensity of their λ_1 bands is increased to such a degree that D_1 exceeds D_2. In DCE, the ratio D_1/D_2 remains practically constant for all the studied thiopyrylotricarboncyanines.

In DCE, the polymethine dyes are usually completely dissociated into ions [31]. In such cases, the spectral characteristics of the dyes do not depend on the nature of the anion. In fact, λ_1, λ_2 and D_1/D_2 for the thiopyrylotricarbocyanines (dyes **4 & 4a-c**) are identical (**Tab. 9.1**). In PMMA, the spectral characteristics of the same dyes greatly differ from one another. The salt-like organic dyes, depending on the polarity of the medium, may exist in different ionic forms such as solvated ions (Ct^+S and An^-S), solvent-shared ions (Ct^+SSAn^-), contact ion pairs (Ct^+An^-) and in the form of the corresponding associates of $Ct^+S...SCt^+$, $An^-S...SAn^-$, $Ct^+SSAn^-...Ct^+SSAn^-$, and $Ct^+An^-...Ct^+An^-$

types [4] where S are solvent molecules. Ionic pairs aggregate substantially easier than ions because their aggregation is favored by electrostatic attractive forces between Ct^+ and An^- with opposite charges [4,31]. This attraction is most efficient for contact ion pairs because it is weakened by S molecules to a lesser extent than in solvent-shared pairs [4]. The lower is the dielectric permittivity ε_D and the solvating ability of a medium (nucleophilic and electrophilic solvation); the higher is the probability of formation of contact pairs in comparison with solvent-shared pairs. High-molecular compounds are characterized by a low value of ε_D. Therefore, it can be assumed that the anomalous changes in the absorption spectra of the salt-like dyes (**4-11**) may arise in going from DCE ($\varepsilon_D = 10.4$) to the less polar PMMA ($\varepsilon_D = 2.8$) due to the difference in the ionic

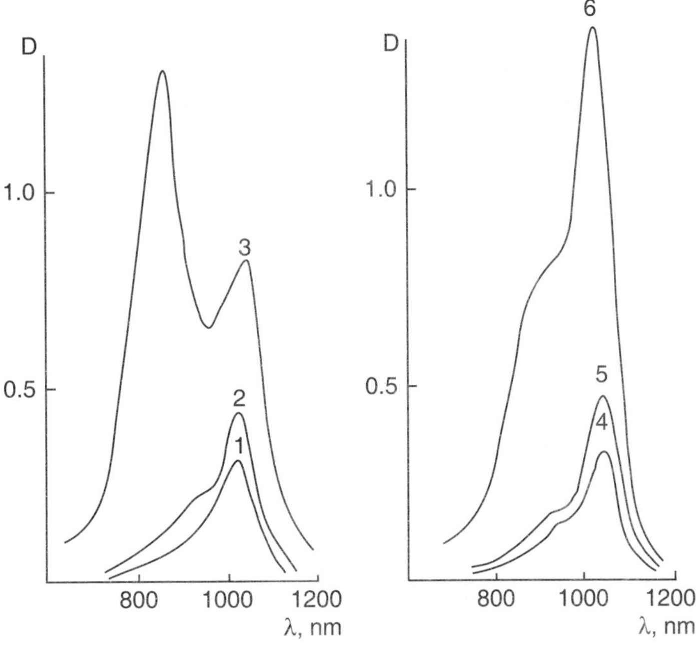

Fig. 9.1. Electronic absorption spectra of solutions of the dye 4 (curves 1-3) and dye 8 (curves 4-6) (i) in DCE (Curves 1 and 4), (ii) in DCE + PMMA (Curves 2 and 5) and (iii) in PMMA films (Curves 3 and 6). Dye concentration: 1.10^{-4} (Curves 1 and 4), 1.10^{-5} mol/liter (Curves 2 and 5), 32.10^{-7} (Curve 3) vand 17.10^{-7} mol/g of polymer (Curve 6). Layer thickness; 69 μm (Curves 1 and 4). 1012 μm (Curves 2 and 5), 75 μm (Curve 3), 60 μm (Curve 6). D: optical density. Reproduced with permission from A.A. Ishchenko, "Structure and spectral, luminescent properties of polymethine dyes". Copyright © 1994, Naukova Dumka, Kiev.

forms of thiopyrylotricarbocyanines present in these solutions.

The dependence of the spectral characteristics of the dyes on the nature of the anion is possible in two cases: (i) association of contact ion pairs and (ii) change in their structure [31]. In order to make a choice between (ii) and (i), a check of the Lambert-Beer law has been made over the concentration interval of 10^{-7}-10^{-6} mole of dye per 1 g of polymer [31]. If the first case occurs, then this law should not be obeyed, since here the constancy of the number of species in solution is broken. In the second case, the number remains constant and, therefore, it may be expected that this law will be followed. Polymeric media with dyes **4** and **4a-c**, in contrast to liquid media, do not obey the Lambert-Beer law. Hence, the hypsochromic shift of the absorption bands and the increase in intensity of the short-wave maximum in PMMA, compared with DCE are due to associations of ion pairs of these dyes.

Films of meso-substituted thiopyrylotricarbocyanines (dyes **5-8**) showed appreciably less deviation from the Lambert-Beer law and their spectra approximated to the corresponding spectra in DCE to a much greater extent than did the unsubstituted dyes **4 & 4a-c** (**Fig. 9.1**). Hence, the dyes (dyes **5-8**) exist in the polymer matrix predominantly as undissociated ion pairs. The different ability of association of the substituted and unsubstituted dyes can be due to the electronic and steric effects of the substituents. From the absorption spectra of thiopyrylotricarbocyanines (dyes **4-8**) in DCE, it may be observed that the meso-substituents can be divided into two opposite groups based on their electronic effect on the absorption maxima. One of these includes the isopropyloxy group, which causes a hypsochromic shift of the band, and the other is a chlorine, methyl and phenyl group, leading to a bathochromic shift (**Tab. 9.1**). If the aggregating ability of the dyes (dyes **5-8**) is determined mainly by the electronic effects of the substituents, then the ratio, D_1/D_2, would change in one direction when the iso-C_3H_7O group is introduced into the meso-position of dye **4**, and in the opposite direction for the introduction of Cl, Me and Ph into meso-position of the dye **4**. However, the substituents of both the types of the groups affected this value in the same direction in our studies i.e. it was decreased in comparison with dye **4**. Hence, the lower tendency towards aggregation, bringing about a weaker transformation of the absorption spectra of thiopyrylotricarbocyanines (dye **5-8**) than those of dyes **4** and **4a-c** on passing from DCE to PMMA, is due to the steric effects of substituents. The validity of this conclusion is confirmed by the fact that the ratio (D_1/D_2) regularly falls in the series in order of dye **4 > 5 > 6 > 7 > 8** i.e. with increase in the effective radius of the meso-substituents [32]. At first glance, the phenyl group is an exception to this regularity. According to the value of its effective radius (the nuclear half-thickness is intended), it should be situated between hydrogen and chlorine. However, this is valid only for the case where the phenyl group is present in the plane of the dye molecule. In the meso-position of tricarbocyanines with six-membered rings in the γ,γ'- positions of the polymethine chain, it emerges from the plane of the dye molecule [30]. In this state the phenyl ring, hinders the approach of molecules to it to a greater extent, than when lying in the plane of the dye and hence their association.

If the ability of the studied thiopyrylotricarbocyanlnes to aggregate is determined predominantly by the steric, and not by the electronic effect of the substituent in the polymethine chain, then it must also depend on the steric barriers in the heterocyclic rings. In order to confirm this assumption, we synthesized the dyes **9-14** [4]. According to structure of the polymethine chain, they were analogous respectively to the dyes **4, 5** and **8**. For the latter, all the phenyl substituents in the thiopyrylium residues are found at an angle of ~40° to the plane of the dye [4]. Their flattening should reduce the steric hindrances and thereby increase the tendency of the dye towards association. In fact, the passage from dyes **4, 5** and **8** respectively to dyes **9, 10** and **11** in PMMA is accompanied by an appreciable hypsochromic shift of the bands and by an increase in the ratio D_1/D_2 (**Tab. 9.1**). These effects are expressed for dyes **12-14** even more strongly, for which all phenyls are hardly anchored in the plane of the chromophore. Successive substitution of the hydrogen atom in the meso-position of the dye **9** and dye **12** by a chlorine atom (dye **13**) and phenyl group (dye **14**), similarly to the dyes **4, 5** and **8** results in the decrease of this ratio. The maximum value, as would be expected from all that has been discussed above, was attained for thiopyrylotricarbocyanine (dye **12**), which possesses the minimal steric hindrance among the studied compounds.

Usually the planar dyes form a sandwich of ion pairs [4,31,33]. Their formation causes the disappearance of the long-wave absorption band, appearance of the new short-wave band towards the monomer band and luminescence quenching. The presence of such effects for dyes **4** and **4a-c** testifies that their aggregates have a sandwich like structure. Sandwich of ion pairs is stabilized at both the ends of the molecule by forces of attraction between opposite charges of the counterions. Therefore, cyanines are able to associate at much lower concentrations of 10^{-5}-10^{-6} mol. litre^{-1} in weakly polar media than at the concentrations of 10^{-2}-10^{-3} mol.litre^{-1} in strongly polar media.

The tendency of cyanines to form ion pairs and consequently their associates depends on the uniformity of the charge distribution in the organic dye chromophores [33]. Therefore, varying the electronic structure of the chromophore (by replacing the hetero-residues with electron-donating or electron-accepting substituents) can also be used to control the aggregation processes in polymers. A uniformity of the charge distribution can be described by the total charges at hetero-residues (Σq_N) and in the polymethine chain (Σq_c) [33]. Since, in polymethine dye **15** based on benz[c,d]indolium, the values of Σq_N and Σq_c differ much less than in the case of thiopyrylocyanines (dye **1-14**) [33], the tendency for the association in polymer matrices is much weaker for the former than for the latter.

Dye **15**

In the case of dye **15,** the replacement of the anion alters the width of the absorption band in PMMA just by 10 cm^{-1} and the fluorescence quantum yield by 0.003% [4]. Whereas in the case of thiopyrylocyanine (dye **5**), which has the same polymethine chain structure, the corresponding changes in these quantities are several times greater.

Since the process of association of ion pairs involves a counter-ion, varying the structure of this ion can also be used to control this process. Suppression of the aggregation of ionic dyes in weakly polar media requires the use of bulky counterions with a low nucleophilicity (or electrophilicity in the case of anion polymethine dyes). The nucleophilicity can be reduced by the introduction of electron-accepting substituents into the counter-ion that reduces the charges carried by the counter-ion atoms. For example, replacing the methyl group in the $MeSO_3^-$ anion with the trifluoromethyl group (as in $CF_3SO_3^-$ anion), which weakens the nucleophilicity, strongly reduces the probability of the formation of ion pairs and their associates in cation dyes [4]. This probability is also reduced by the use of anions either with a strongly delocalized charge such as $C(CN)_3^-$ or based on any aromatic compound capable of becoming oriented parallel to the cation chromophore, for example, by the introduction of the tosylate anion [4].

Like in the case of liquid solvents, the change in polymer polarity also affects the spectral and luminescent properties of polymethine dyes. An investigation of the electronic spectra of thiopyrylotricarbocyanine (dye **8**) in copolymer films with different ratios of the components (styrene and methylmethacrylate) has been shown in **Fig. 9.2**. An increase in the content of the less polar and nucleophilic component (styrene) in the copolymer increases the intensity of the short-wavelength band and induces hypsochromic shift in it. This quenches the original luminescence band. The existence of an isobestic point, the dependence of the spectra on the nature of the anion, and the failure of the dye **5** solutions in copolymer to obey the Lambert-Beer law, all support the existence of equilibrium between ion pairs and their associates. In pure polystyrene, the equilibrium shifts almost completely towards associates **(Fig. 9.2)**. Consequently, in the matrices characterized by a very low polarity and weak nucleophilicity, even thiopyrylotricarbocyanines (dyes **5-8, 10 & 11**) with bulky meso-substituent, may preferentially assume the associated state in which a major change occurs in the spectral and luminescent properties compared with the corresponding dyes in the liquid solutions. It follows from this account that the tendency of salt-like polymethine dyes to form ion pairs and their associates can be reduced by the use of high-permittivity polymers. However, the development of such polymers is very problematic. The most widely used polymers are characterized by $\varepsilon_D <$ 10, so that dyes in the polymers are not fully dissociated. It has been proposed that the dissociation of salt-like cyanines in low ε_D matrices can be facilitated by embedding cationic polymethine dyes in polymers containing strongly nucleophilic functional groups. These groups solvate positively charged centers of the cations, separate counter-ions and decrease the formation of tight ion pairs [4]. A similar effect can be achieved in anionic dyes embedded in polymer with electrophilic groups at the expense of the electrostatic interactions of these groups with negative anion charges. Solvent-shared ion pairs have practically the same electronic spectra as solvated ions [4] and, therefore, the formation of such pairs in the polymer should not significantly change the spectral and luminescent properties of polymethine dyes compared with their liquid solutions, even if the cyanines are fully dissociated in such solutions.

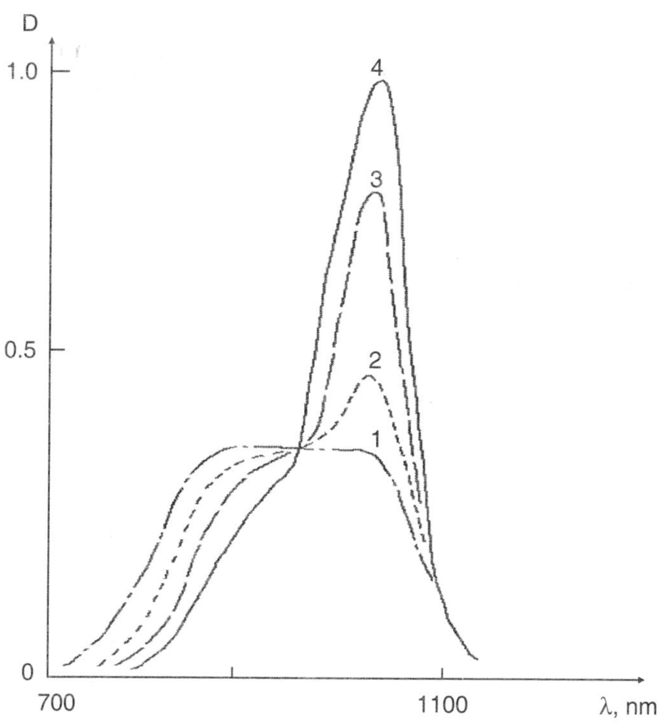

Fig. 9.2. Absorption spectra of films of dye 8 in polystyrene (1), in copolymer of methylmethacrylate and styrene in the ratio of 1:4 (2) and 4:1 (3) and in polymethylmethacrylate (4). Dye concentration −10⁻⁶ mol/g of polymer. Film thickness: 60 μm. *D*-optical density. Reproduced with permission from A.A. Ishchenko, "Structure and spectral, luminescent properties of polymethine dyes". Copyright © 1994, Naukova Dumka, Kiev.

Specific solvation of dye ions by nucleophilic (electrophilic) groups in a polymer also weakens the electrostatic and dispersion interactions between the chromophores of these ions, which additionally hinders the association of the ions. Strongly nucleophilic properties are possessed by cellulose diacetate, polyvinyl butyral, epoxides, polyimide, and polyurethane. These polymers also have higher value of e_D than PS and PMMA. Therefore, even unsubstituted at the meso-position thiopyrylotricarbocyanines (dye **4** and **4a-c**) with their very strong tendency to association do not aggregate in these polymers. For example, going from PS to more nucleophilic PMMA and polyurethane, is accompanied by the improvement of absorption spectra of these dyes. They have in PU practically the same spectra as in DCE [4].

The above-mentioned spectral regularity concerned of the ionic salt-like dyes in non-photoconducting polymers.

3.2. Spectral and Luminescent Properties of Intra-Ion Dyes

The particularities of spectral and luminescent properties of intra-ionic (nonsalt-like dye) squarilic dyes in non-photoconducting and photoconducting polymer matrices were reported in work of Skryshevski *et al.* [34].

Squarylium organic dyes (squaraines), which represent derivatives of the quadric acid, are characterized by high thermal and photochemical stability and have intensive absorption bands [35]. Because of this, they are extensively used in electrophotography, solar cells and in laser disks for recording and storage of information [35].

Squaraines, as differentiated from the above-mentioned dyes, represent intra-ionic compounds in which ionic centers are bound by chemical bonds rather than by electrostatic attraction forces. Because of this, particularities of spectral and luminescent properties of squarylium dyes are bound to be strongly pronounced in comparison with the features of ionic salt-like dyes of the corresponding structure.

Dye **16**. **HISD**

PEPK

Squaraine, **HISD** (dye **16**), was studied in comparison with a cationic cyanine, **HIDC** (dye **2**) with similar structure of end heterocyclic groups in poly-N-epoxypropylcarbazole (PEPK) and polystyrene (PS) films [36]. PEPK possesses photoconductivity and, therefore, it is extensively used as an optical recording medium in electro-photography and holography [7,36] whereas charge photogeneration in PS does not take place due to its high ionization potential [36]. PS is used as a base for a large number of organic luminophores because of its good optical and mechanical properties [37].

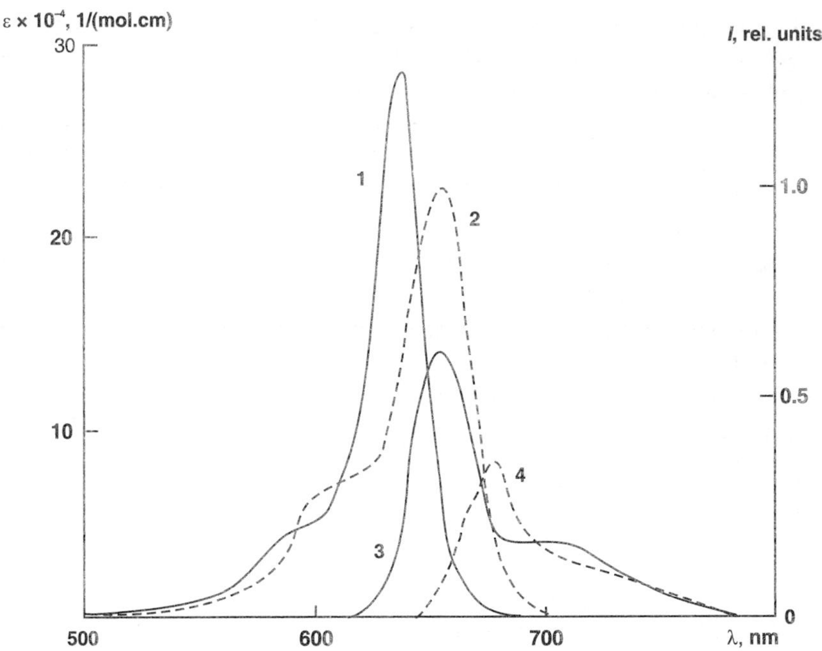

Fig. 9.3. Absorption (1,2) and luminescence (3,4) spectra of HISD (1,3) and HIDC (2,4) dyes in the methylene chloride solution with $C = 1.10^{-4}\%$ at 300 K. ε-extinction; *I*-photoluminescence intensity. Reproduced with permission from Skryshevski et al, Optics and Spectroscopy, 88(3), 352 (2000). Copyright © 2000 MAIK "Nauka/Interperiodika".

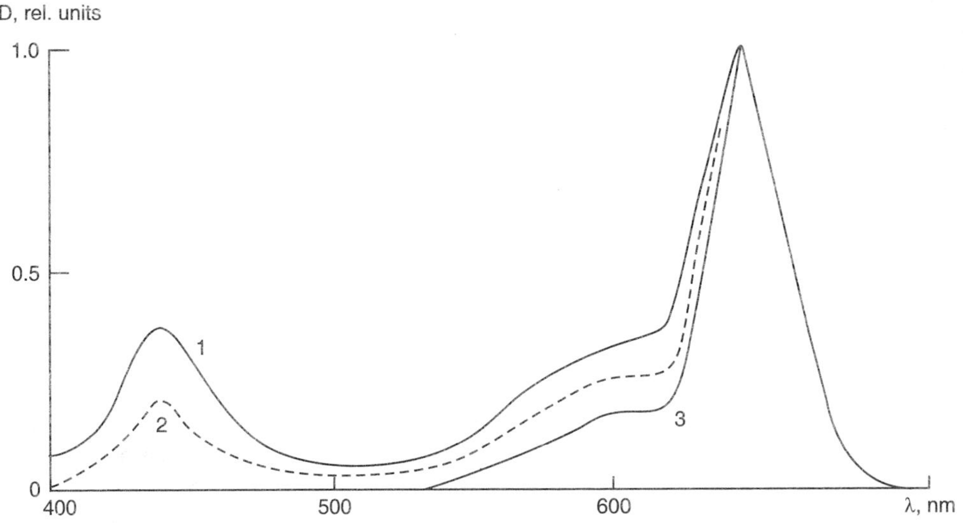

Fig. 9.4. Normalized absorption spectra of the HISD dye in PEPK films at 300 K for $C = 1.10^{-4}(1)$, 0.1(2), and 2% (*3*). *D*-optical density. Reproduced with permission from Skryshevski *et al.*, Optics and Spectroscopy, 88(3), 352 (2000). Copyright © 2000 MAIK "Nauka/Interperiodika".

The change from solutions in methylene chloride (MC) to PEPK and then to PS films, is accompanied by the bathochromic shift in the absorption band of the **HISD** (dye **16**) (**Figs. 9.3 & 9.4**). Moreover, a weak band at λ_{max} = 442 nm appears in PEPK at low concentration. An increase of the concentration (C) of **HISD** (dye **16**) in PEPK leads to a decrease of its intensity and an increase of intensity of the long-wavelength band (**Fig. 9.4**). For $C \geq 0.1\%$, the short-wavelength band vanishes. As the **HISD** (dye **16**) concentration in PEPK is increased in the range $1 \cdot 10^{-4}\% \leq C \leq 5\%$, one observes for the long-wavelength band a small bathochromic shift from 642 to 646 nm. Moreover, an increase of C has an effect on the shape of this absorption band and causes its narrowing, predominantly due to a decrease of intensity at the short-wavelength edge in the region of the vibrational peak. A similar picture is observed for PS, but the weak short-wavelength band is absent. The fact that the absorption spectra of **HISD** (dye **16**) depend on the concentration gives evidence of the association of its molecules in PEPK and PS. This statement is also supported by the fact that the positions of λ_{max} in the absorption bands of **HISD** (dye **16**) in these polymers are almost identical in spite of a strong difference of n_D of PEPK and PS, which are equal to 1.80 and 1.59, respectively. Because of a larger value of n_D of PEPK in comparison with PS, it was reasonable to expect, according to the theory of universal interactions [4], a bathochromic shift of the absorption band for PEPK in comparison with its position for PS due to an enhancement of dispersion interactions. The insensitivity of λ_{max} of **HISD** (dye **16**) to changes of n_D for the polymers under consideration suggests that the band position is predominantly determined by the interaction of monomer dye molecules with one another rather than with molecules of a medium. The absorption spectrum of the **HIDC** (dye **2**) in PEPK in the concentration range under study is virtually independent of C. The change from MC to PEPK caused a natural 300 cm^{-1} bathochromic shift of the **HIDC** (dye **2**) absorption band (for MC, n_D = 1.42), and it exceeded the shift for the **HISD** (dye **16**) whose value for $C = 1 \cdot 10^{-4}\%$ was equal to 200 cm^{-1}. Therefore, in this case, in contrast to **HISD** (dye **16**) in PEPK, the intermolecular interaction in the dye–polymer system is stronger than the interaction between monomer dye molecules, and **HIDC** (dye **2**) in PEPK exists predominantly in the monomer form.

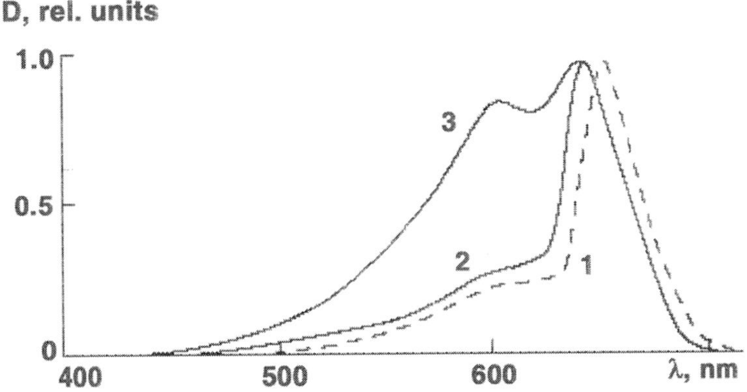

Fig. 9.5. Normalized absorption spectra of the HIDC dye in (*1*) PEPK and (*2,3*) PS films at 300 K for C = (*1*) 1%, (*2*) 0.2% and (*3*) 2%. *D*-optical density. Reproduced with permission from Skryshevski *et al.*, Optics and Spectroscopy, 88(3), 352 (2000). Copyright © 2000 MAIK "Nauka/Interperiodika".

The **HIDC** (dye **2**) in PS, as well as **HISD** (dye **16**), shows the dependence of absorption spectra on concentration. However, an increase of **HIDC** (dye **2**) concentration in PS causes a broadening of the absorption band and its small hypsochromic shift **(Fig. 9.5)**, as differentiated from **HISD** (dye **16**). This also causes an increase of intensity at the short-wavelength edge of the absorption band. It is worth noting that the change from MC to PS for **HIDC** (dye **2**), contrary to the theory of universal interactions, is accompanied by a hypsochromic instead of a bathochromic shift. For the dyes under study in polymers, the manifestation of association processes in the absorption spectra (the disappearance of monomer bands and the appearance of new distinct associate bands) is pronounced weaker in comparison with the expectations based on the classical concepts and with the results observed for dyes **4-14** [4].

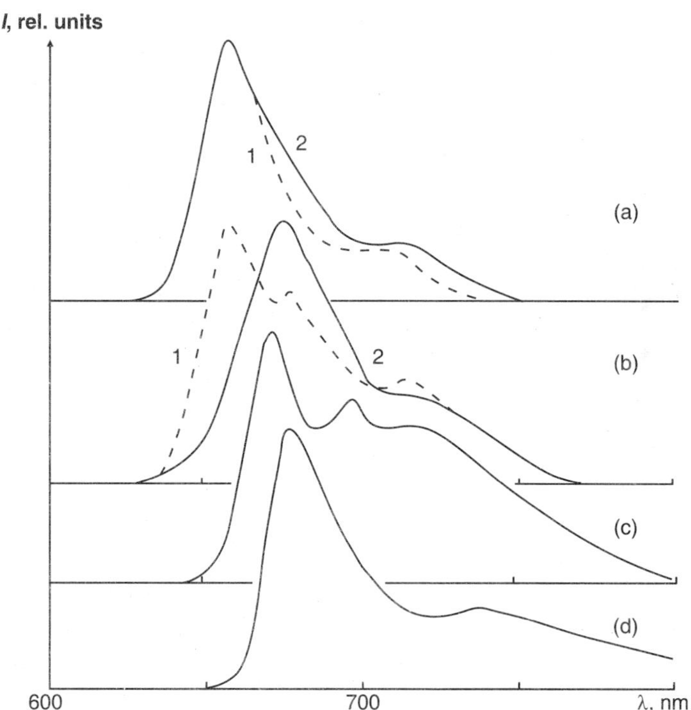

Fig. 9.6. Photoluminescence spectra of the HISD dye in PEPK films at 4.2 K for λ_{ex} = 546 nm (*1*) and 365 nm (*2*) and C = 1.10^{-4}% (a), 1.10^{-3}% (b), 0.8% (c) and 2% (d). I-photoluminescence intensity. Reproduced with permission from Skryshevski *et al.*, Optics and Spectroscopy, 88(3), 352 (2000). Copyright © 2000 MAIK "Nauka/Interperiodika".

The photoluminescence (PL) spectra measured at 300 K were unsuitable to study the picture in detail because their profiles were strongly blurred and broadened. Because of this, we studied PL in detail at low temperatures **(Figs. 9.6 & 9.8)**. In the PL spectra of **HISD** (dye **16**) in PEPK at low concentrations ($C \leq 10^{-4}$%) measured at the excitation wavelength λ_{ex} = 546 nm, which lies in the region of the long-wavelength absorption band of the dye, and the wavelength λ_{ex} = 365 nm, which lies outside the limits of this absorption band, we observed only one emission band with

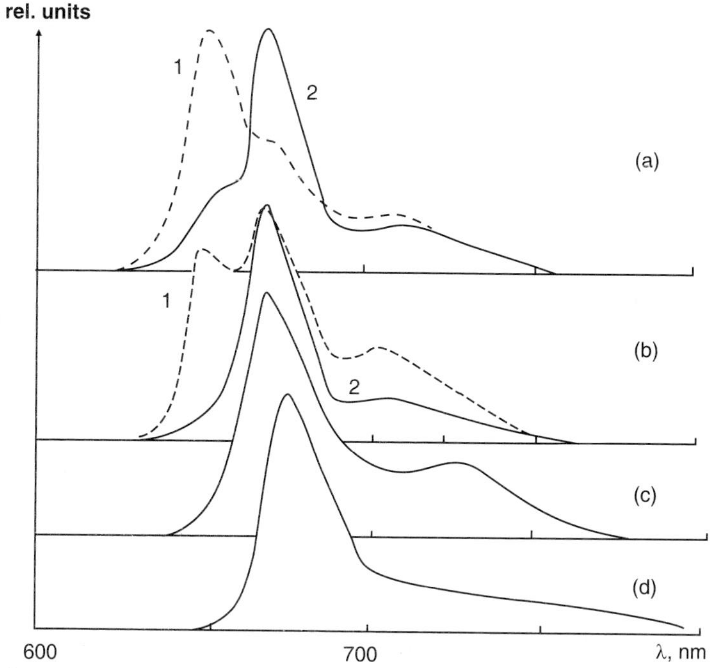

Fig. 9.7. Photoluminescence spectra of the HISD dye in PS films at 4.2 K for λ_{ex} = 546 nm (*1*) and 365 nm (*2*), *C* = 1.10⁻⁴ % (a), 1.10⁻³ % (b), 0.8% (c) and 2% (d). I-photoluminescence intensity. Reproduced with permission from Skryshevski *et al.*, Optics and Spectroscopy, 88(3), 352 (2000). Copyright © 2000 MAIK "Nauka/Interperiodika".

λ_{max} = 660 nm (**Fig. 9.6**). In the case of λ_{ex} = 365 nm, this PL band shows only broadening to a certain extent due to the long-wavelength edge. For the PS films under the same conditions, we observed two PL bands, namely, the short-wavelength band with λ_{max} = 652 nm for λ_{ex} = 546 nm and the long-wavelength band with λ_{max} = 670 nm for λ_{ex} = 365 nm (**Fig. 9.7**). Because λ_{ex} = 546 nm falls predominantly in the absorption region of **HISD** (dye **16**) monomers, it is most likely that the short-wavelength emission relates to this form. In this case, the long-wavelength emission should be assigned to the **HISD** (dye **16**) associates because their radiative level is lower than the level of nonaggregated molecules [4,38]. This identification of bands is supported by the dependence of the fluorescence spectra on the dye concentration. The proportion of monomer **HISD** (dye **16**) molecules is bound to decrease and the proportion of associated molecules to increase with increasing *C*. Indeed, the short-wavelength fluorescence band decreases and the long-wavelength band increases in intensity with increasing *C*. This picture is best pronounced for PS films (**Fig. 9.7**). Our results suggest that the decrease in the absorption observed at the short-wavelength absorption band of **HISD** (dye **16**) with increasing *C* is caused by a decrease in the content of its monomers (**Fig. 9.4**).

As the concentration *C* = 0.8% is achieved, the emission of the monomer form of **HISD** (dye **16**) in both the polymers vanishes. Light, both with λ_{ex} = 365 nm and λ_{ex} = 546 nm (and any other λ_{ex} from this range), causes only PL of associates (**Figs. 9.6 & 9.7**). A further increase in the

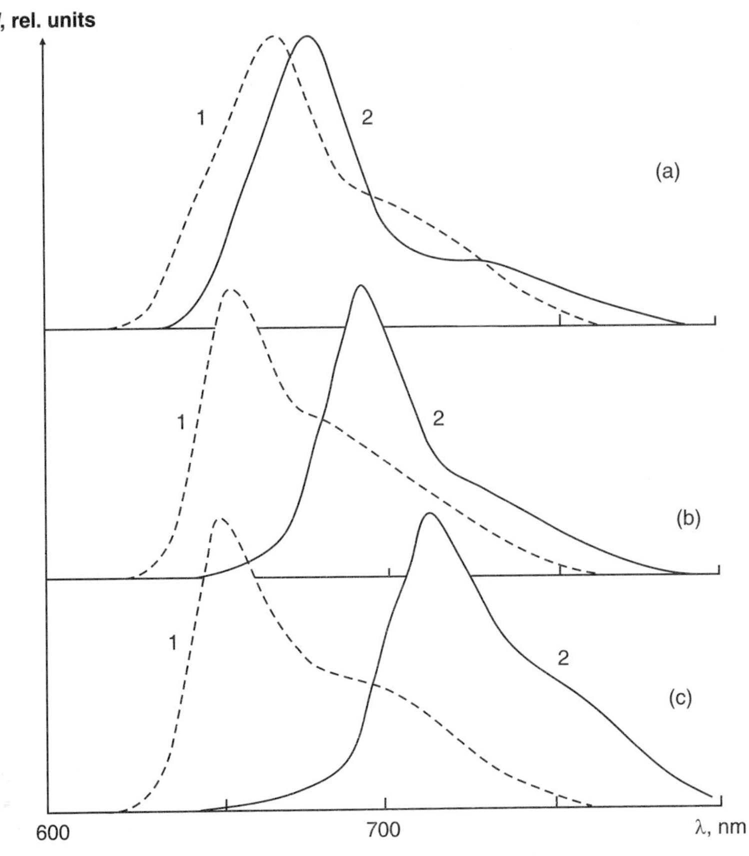

Fig. 9.8. Photoluminescence spectra of the HIDC dye in PS (*1*) and PEPK (*2*) polymer films at 4.2 K for C = 1.10^{-3}% (a), 0.2% (b) and 2% (c). I-photoluminescence intensity. Reproduced with permission from Skryshevski *et al.*, **Optics and Spectroscopy, 88(3), 352 (2000). Copyright © 2000 MAIK "Nauka/Interperiodika".**

dye concentration leads only to the ordering of the associate structure. In PEPK, associates emitting in the region of longer wavelengths (λ_{max} = 695 nm) appear **(Fig. 9.6)**. As the concentration C = 2% is achieved, its further increase causes no change in the fluorescence spectra of **HISD** (dye **16**), neither in PEPK nor in PS **(Figs. 9.6 & 9.7)**. In both the matrices, like in the case of absorption spectra, the same value of λ_{max} = 676 nm is achieved supporting the association nature of this band. This conclusion agrees with the narrowing of absorption and PL bands, which is caused by the weakening of vibronic interactions in the case of molecular aggregation of dyes [38]. The comparison of the fluorescence spectra obtained for λ_{ex} = 365 nm and the same C made it possible to compare the aggregating abilities of **HISD** (dye **16**) in PS and PEPK. In particular, for C = 1 × 10^{-4}%, we observed the appearance of an associate band for PS, as opposed to PEPK. For C = 1 × 10^{-3}%, the intensity of this band in PS exceeded, in contrast to PEPK, the intensity of the monomer form of the dye **(Figs. 9.6 & 9.7)**. Therefore, the **HISD** (dye **16**) aggregation in PS is easier than in PEPK. The PL spectra of **HIDC** (dye **2**) in PEPK and PS and their concentration dependencies sharply differ from those of **HISD** (dye **16**). In the PL spectrum of **HIDC** (dye **2**) in PEPK at $C = 1 \times 10^{-3}\%$, we observed only one band with a peak at $\lambda_{max} = 678$ nm, which had a bathochromic shift with increasing C and reached $\lambda_{max} = 718$ nm for $C = 2\%$ (**Fig. 9.8**). In the PL spectra of **HIDC** (dye **2**) in PS, we also observed only one band. However, its peak showed a hypsochromic shift from $\lambda_{max} = 672$ to $\lambda_{max} = 653$ nm with increasing C from $1 \times 10^{-3}\%$ to 2% (**Fig. 9.8**).

Squarylium dyes, owing to intra-ionic structure, are chemically bound by a nondissociating contact ionic pair in which ions always exist independent of polarity of the medium. This represents the principal difference of **HISD** (dye **16**) from **HIDC** (dye **2**). Because of this, the aggregation of **HISD** (dye **16**) molecules in PEPK and PS is stronger than that of **HIDC** (dye **2**). One can see this from the fact that **HISD** (dye **16**) is characterized by stronger changes in absorption (**Figs. 9.4 & 9.5**) and PL (**Figs. 9.6 & 9.8**) spectra in comparison with **HIDC** (dye **2**). It is reasonable to assume from the concentration dependence of the absorption spectra of **HIDC** (dye **2**) (**Fig. 9.5**) that this dye also strongly aggregates in PS. However, the band in the PL spectra hypsochromically shifts with increasing C independent of λ_{max} (**Fig. 9.8**). According to the energy migration mechanism, this feature precludes the emission of the monomer cations because the shift observed with increasing concentration of C would be bathochromic in this case, like in the PEPK matrix [39]. This picture is not typical for associates, because the "head-tail" associates are bound to give the bathochromic shift with increasing concentration, whereas the formation of "sandwiches" would cause PL quenching. The picture observed in the absorption and PL spectra can be attributed to the formation of contact (close) ionic pairs for the salt-like polymethine **HIDC** (dye **2**) in a weakly polar PS [30,31,38]. An increase in C leads to the formation of a closer ionic pair. As a result, an anion lowers the energy level of the ground anion stronger than that of the excited state, which causes a hypsochromic shift in the absorption and PL spectra. An increase in the proportion of contact ionic pairs in comparison with solvated anions or solvent-shared ionic pairs with increasing C causes a substantial change in the absorption spectra of **HIDC** (dye **2**) in PS, primarily, their broadening due to an increase in the absorption at the short-wavelength edge (**Fig. 9.5**).

An enhancement in the polarity or solvating ability of a polymer opposes the aggregation of both the cationic and squarylium dyes. We noted this when passing from PS to PEPK. This is caused by the fact that PEPK weakens the electrostatic attraction of pairs to one another independent of the prehistory of contact pair formation due to the electrostatic attractive forces of counterions or their chemical joining. This weakening is caused by the presence of polar C–N and C–O bonds in PEPK, which are capable of electrostatic interaction through their dipoles (nucleophilic or electrophilic solvation) with charged chromophore centers [4]. One of the most probable arrangements of **HISD** (dye **16**) molecules in an aggregate chain may be represented by the structure shown in **Fig. 9.9**. The conformation shown in **Fig. 9.9** for a dye molecule corresponds to the data of X-ray diffraction analysis [40]. In the associate structure proposed here, the angle between chromophore directions is close to 180°. For this angle, only one long-wavelength absorption band is bound to show itself in the absorption spectra [4,31,38]. It is precisely this picture that is observed in the cases described above. At the lower concentrations, deviations from an angle of 180° are possible when the formation of associate structure goes on. They show themselves in the appearance of the short-wavelength absorption band forbidden for the structure shown in **Fig. 9.9** and this band has a hypsochromic shift with respect to the band of monomer molecules, which was observed for **HISD** (dye **16**) in PEPK (**Fig. 9.4**).

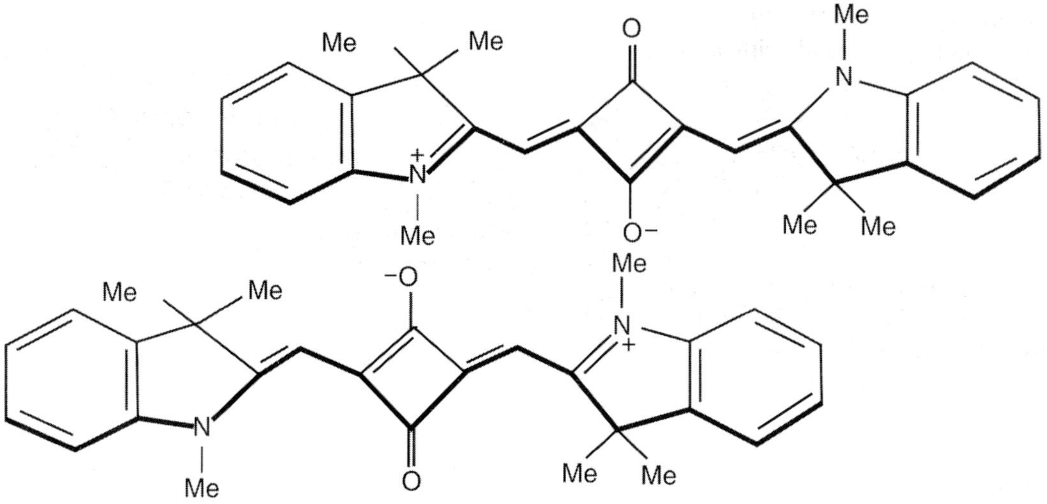

Fig. 9.9. Hypothetical arrangement of HISD molecules in the associate. Reproduced with permission from Skryshevski *et al.*, Optics and Spectroscopy, 88(3), 352 (2000). Copyright © 2000 MAIK "Nauka/Interperiodika".

The disappearance of the short-wavelength band of **HISD** (dye **16**) associates with increasing concentration gives evidence that the angle between chromophore directions in the structure is in close proximity to 180°. This structure of associates for squarylium dyes is thought to be rather advantageous because molecules are fixed there at both ends by electrostatic attractive forces between opposite charges. A negatively charged oxygen atom of one molecule is oriented above the positively charged region of the nitrogenous heterocycle of another molecule and vice versa (**Fig. 9.9**). In the PL of **HISD** (dye **16**) in PEPK in the range $1 \times 10^{-2}\% \leq C \leq 0.8\%$ when excited by light with $\lambda = 365$ nm, a wide band with $\lambda_{max} = 480$ nm appeared. A further increase in the concentration leads to its disappearance. This PL is not associated with the emission of **HISD** (dye **16**) associates which absorb at $\lambda_{max} = 442$ nm because they are characterized by the nonradiative dissipation of the electronic-excitation energy [4]. It is most likely that the band with $\lambda_{max} = 480$ nm belongs to the PEPK excimers [41]. A similar band with $\lambda_{max} = 460$ nm is also observed in the PL spectra of **HIDC** (dye **2**) in PEPK.

The spectral and luminescent properties of the **HISD** (dye **16**) intra-ionic squarylium dye and the **HIDC** (dye **2**) cation dye in polymer films are determined by their association. This is observed both in PEPK where photo-generation and transport of charges are possible and in PS where charge transport does not take place. **HISD** (dye **16**) is characterized by stronger aggregation than **HIDC** (dye **2**). This is caused by the fact that electrostatic attractive forces are of substantial importance in the association of ionic and intra-ionic dyes. As for ionic dyes, their dominant role shows itself only in the case of formation of contact ionic pairs. For this purpose, the matrix should have such properties that ionic pairs with spacing as small as possible be formed in it. Squaraines retain bipolarity (exist in the form of a chemically bound ionic pair) in any medium. Because of this, opposite charges of their molecules are always attracted to one another which leads to the dimer formation even in dilute solutions [42]. Thus, the developed criteria of suppression of dyes aggregation are fair not only for non-photoconducting polymer matrices but also for photoconducting matrices.

It should be noted that polymer matrix carry out one more very important function, it provides a rigid medium. It appears in strong change in the fluorescent ability of dyes inclined to photoisomerization. Typical representatives of this class of dyes are polymethine dyes **17** and **18**. They have a low fluorescence quantum yield in liquid solutions. It is connected with the

Dye **17**   Dye **18**. (R = Et, An⁻ = Br⁻)

formation of photo-isomers at the expense of rotation around polymethine chain bonds under the influence of electronic excitation [4,43]. In a polymer matrix, these processes are considerably delayed. Therefore, in going from liquid solution to polymer causes strong enhancement in the fluorescence. For example, value of ϕ of thiacarbocyanine (dye **17**) with open polymethine chain equals 5% in ethanol, whereas in PU matrix achieves 72%. It was revealed that a polymer enhances fluorescence much more than cyclization of chain links by the bridging groups. Dye **17**, a rigid analogue of dye **18**, has a ϕ value of only 53% in ethanol [4].

Marked increase in the fluorescence quantum yield for cyanines in polymer matrices is attained because the structure of dye molecules is rigidized virtually without changing their molecular structure. The bridging groups can strongly change it, and consequently, degree of vibronic and intermolecular interactions, essentially influencing on nonradiating processes in molecules of dye. Moreover, polymer matrices suppress not only photoisomerization but also PL quenching upon diffusion-controlled collisions with oxygen and other quenching impurities [1].

Another important advantage of polymer matrices is that the embedding of dye with open polymethine chain in polymer provides the higher value of Stokes shift than cyclization of its polymethine chain by the bridging groups. For example, in going from thiacarbocyanine (dye **17** and **18**), the SS decreases by 140 cm^{-1} in ethanol, whereas embedding dye **17** in PU even increases it by 25 cm^{-1} in comparison with ethanol solution. Consequently, polymer matrices can provide some rigidity sufficient for attaining maximum ϕ of polymethines. At the same time, the structural mobility of the excited molecules remains fairly adequate for realization of the four-level scheme of electronic transitions, which ensures high SS.

4. THE EFFECT OF ELECTRON TRANSFER ON PHOTOLUMINESCENCE SPECTRA OF DYE-DOPED PHOTO-CONDUCTING POLYMERS

As photo-conducting polymeric matrices, the oligomers of PEPK [36] of $M = 900$, anthracenyl glycidyl ether (OAGE) [44] of $M = 1100$ and PVK polymers of $M = 6\text{-}7 \times 10^3$, $1\text{-}2 \times 10^4$, 7×10^4 and 3×10^5 (PVK-I to PVK-IV respectively) were used. Polyvinylbutyral (PVB) of $M = 7 \times 10^4\text{--}10^5$, polyvinylethylal (PVE) of $M = 2 \times 10^4\text{--}10^5$, polyvinyl alcohol (PVAL) of $M > 10^5$, PS of $M = 2.3 \times 10^5$ and polyethylene (PE) of $M = 1.25 \times 10^5$ [45] were applied as non-photo-conducting polymers. The sensitizers were polymethine dyes **1-3**, **19-23** (dyes **1-3** are cationic dyes, dyes **19** and **20** are anionic dyes, dye **21** is a cation-anionic dye and dye **22** is a neutral dye) and cationic xanthene, **R6G** (dye **23**) [45].

Anionic cyanine (dye **19**) has the same chromophore as cationic dye (dye **1**), despite of the fact that the net charge of the colored ion of the former and the latter is −1 and +1 respectively. This is

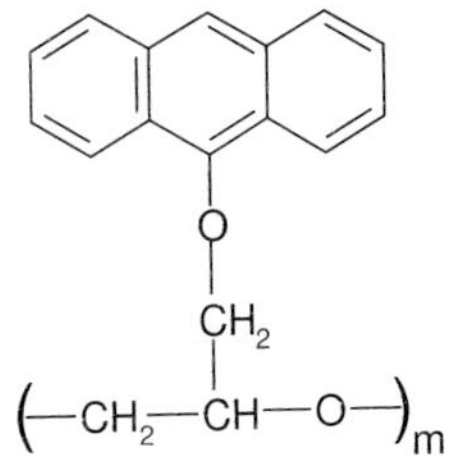

OAGE

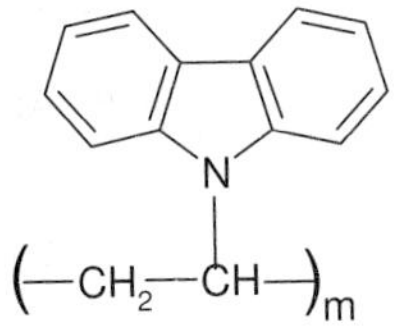

PVK

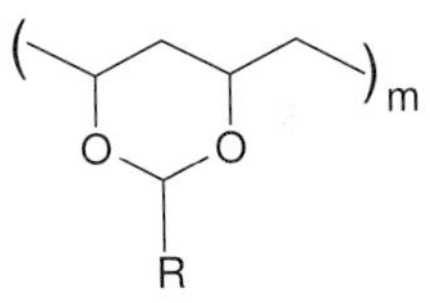

PVE (R = Me) and PVB (R = C_3H_7)

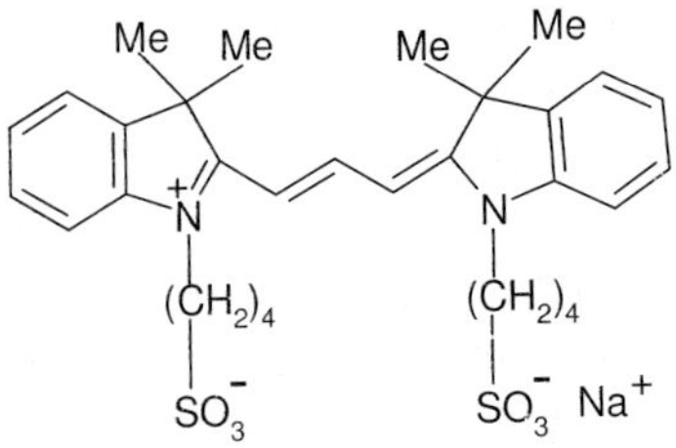

Dye **19**

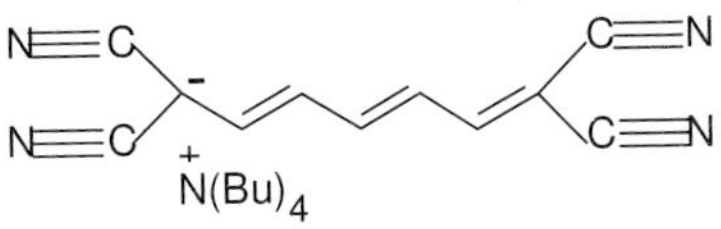

Dye **20**

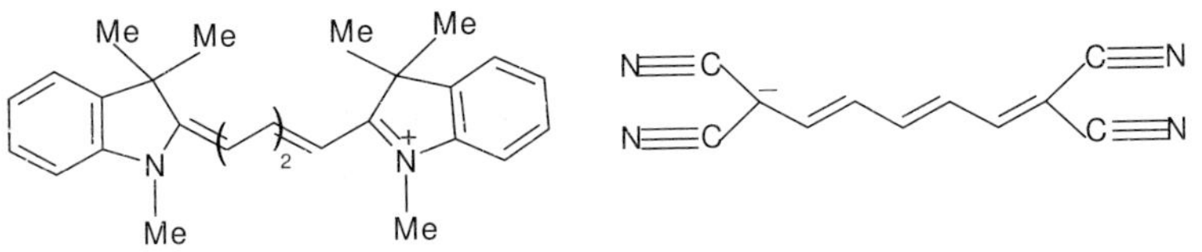

Dye **21**

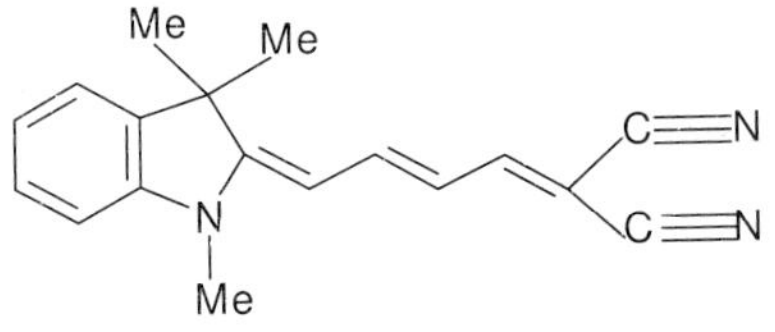

Dye **22**

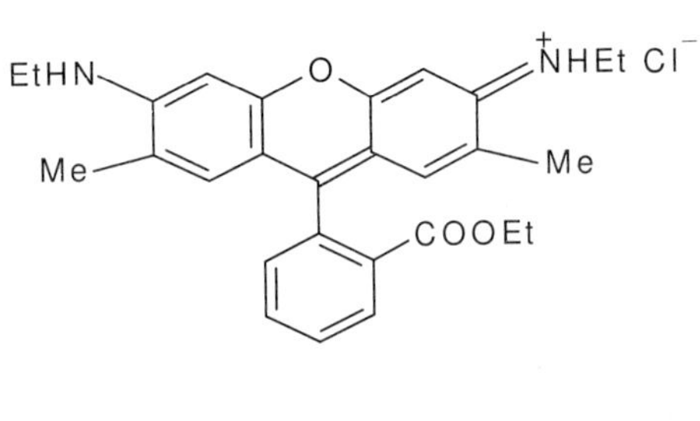

Dye **23. R6G**

due to the sulfo groups, responsible for the net negative charge, are attached to the atoms that are not involved in the conjugation chain. Cationic–anionic dye **21** is composed of the cation and anion of polymethines **2** and **20,** respectively. Neutral merocyanine (dye **22**) is a hybrid of cationic dye (dye **1**) and anionic dye (dye **20**). Thus, the change in the charge and chemical structure in the cyanine dyes **1-3** and **19-22** is provided within the same structural type of dyes, making them suitable for use in study of the photo-physical features of dyed polymers.

The specimens were prepared as free surfaces (glass substrate–amorphous polymer film or glass substrate–conducting SnO_2 layer-amorphous polymer film) and sandwiched (quartz substrate–SnO_2–amorphous polymer film–Al) structures. Dyed films were prepared by drying a polymer solution with a dye in DCE cast onto the glass substrate surface with or without a conducting layer. The films of 1μm thickness were dried in a drying oven for 3 h at 80°C for experiments.

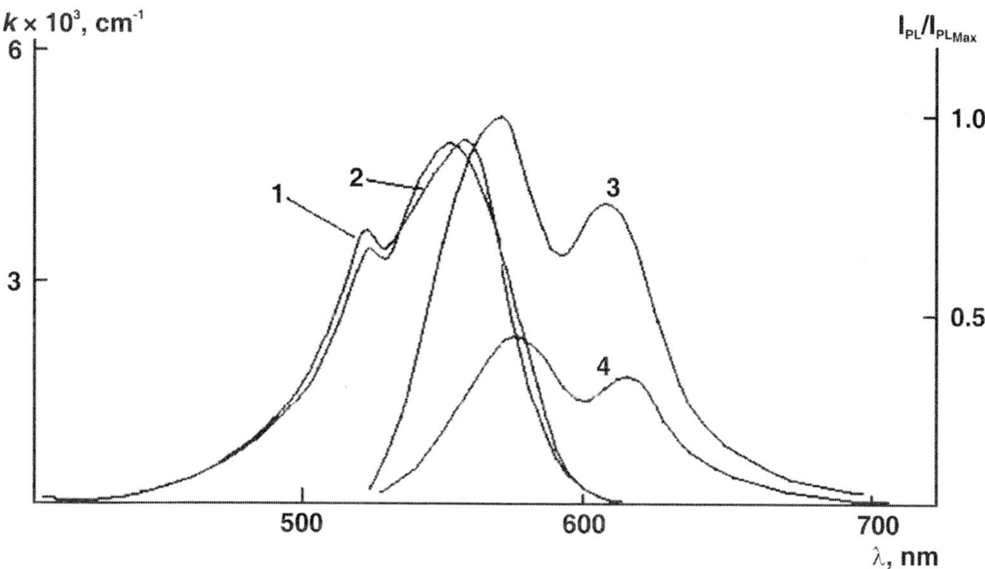

Fig. 9.10. Absorption (*1,2*) and photoluminescence (*3,4*) spectra of 1 wt% dye 1 in PEPK (*1,3*) and PS (*2,4*) films; κ-the absorption coefficient, I_{PL}-photoluminescence intensity. Reproduced with permission from Getmanchuk *et al.*, Polym. Sci., Ser.A., 44(8), 855 (2002). Copyright © 2002 MAIK "Nauka/Interperiodika".

The dye-free polymer films made from both the photo-conducting and non-conducting polymers do not absorb light in the visible region of the spectrum. When the dyes of interest are introduced into the films in an amount of 1 wt%, intense absorption bands appear in this region. For the PL of the dyed films, one PL band was observed upon excitation in their absorption bands. The envelope of the fluorescence curve is the mirror image of that of the absorption band. The replacement of the polymeric matrix PEPK by PVK or PS does not produce a substantial effect on the position of absorption and photo-luminescence bands **(Fig. 9.10)** which is due to the close values of the n_D of these polymers [45]. The absorption spectrum of dye **21** displays two bands corresponding to the absorption of colored forms of anion **20** and cation **2**, respectively. However, its PL spectrum shows absorption only in the long-wavelength region upon excitation in the short- and long-wavelength absorption bands, thus indicating excitation energy transfer from the anion to the cation. As in the case of cationic polymethine dyes (dyes **1-3**) and xanthene dye (dye **23**), the replacement of the matrix of PS, PVB, PVE, PVAL, and PE by PEPK, OAGE, or PVK leads into the enhancement in the PL upon excitation at the short-wavelength edge of the dye absorption band. None of the exciting wavelengths fall in the absorption region of polymers in this case. The PL enhancement effect is illustrated in **Fig. 9.11** which shows the ratio of PL quantum yields (φ) in dyed photo-conducting polymers to φ in PS as a function of excitation wavelength (λ_{irr}) for PL recorded at band maximum. In the photo-conducting films, φ increases, passing through a maximum, and then

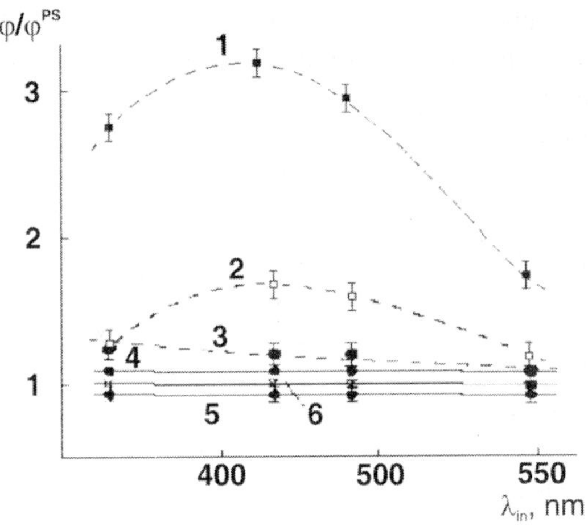

Fig. 9.11. Dependencies of relative photoluminescence quantum yield ϕ in the films of PEPK (*1*), PVK I (*2*), PVK IV (*3*), PE (*4*), PVB (*5*), PVAL (*6*), doped with 1 wt% of dye 1 on wavelength of excitation light λ_{irr} at registration of photoluminescence in band maximum (λ = 575 nm). Reproduced with permission from Getmanchuk *et al.*, Polym. Sci., Ser.A., 44(8), 855 (2002). Copyright © 2002 MAIK "Nauka Interperiodika".

decreases with decreasing φ_{irr} (**Fig. 9.11**, curves 1-3). The highest value of φ is achieved in the PEPK films. The PL intensity is proportional to the intensity of exciting light, but j does not change in this case. The luminescence enhancement effect becomes weaker with increasing the length of the polymethine chain in the series of cationic dyes (dyes **1-3**) or upon replacement of PEPK by OAGE and PVK. This effect is not observed in polymer films containing anionic dyes **19** and **20** or neutral dye **22** but is detected in the films with cation-anionic dye (dye **21**) at its cationic band.

It is remarkable that the effect for dye **21** is displayed upon excitation at the absorption region of both the cation and anion. Consequently, the anion is incapable of enhancing the intrinsic PL in anionic dye (dye **20**) but takes part in its enhancement in the cation of cyanine (dye **21**) via excitation energy transfer in the cation–anion pair. However, it should be noted that the PL intensity of cationic dyes in photo-conducting polymers depends on the nature of anion (An$^-$). For example, when ClO_4^- is used instead of I$^-$, the PL intensity of dye **1** is higher.

The external electric field has no substantial effect on the spectral line pattern of PL but quenches its intensity. The quenching effect in an electric field for the PEPK, OAGE and PVK films is enhanced with decreasing λ_{irr}. In the film samples based on PS, PVB, PVAL and PE, its influence on the PL of the dyes **1-3** and **19-23** was not observed. An electric field also does not affect the PL of polymer films doped with anionic dyes (dyes **19** and **20**). Hence, this PL quenching by an external electric field may be associated with the charge separation process of photo-generated electron–hole pairs and the decrease in the radiative recombination probability, as assumed in reference 46. No photoconductivity was detected in the sandwiched samples of PS, PVB, PVAL and PE films doped with the dyes. On the contrary, similar samples of PEPK, OAGE and PVK doped with dyes **1-3** or **21-23** exhibit photoconductivity upon their irradiation with light in the dye absorption region.

In the sandwiched samples with dye **3**, the photocurrent values are two orders of magnitude lower than those in similar samples with polymethine (dye **1**) and an order of magnitude smaller than in the samples with dye **2**. The samples doped with dyes **19** and **20** did not exhibit photoconductivity of the polymer films.

To explain the results obtained, let us analyze the role of the basic possible mechanisms of enhancement of PL of organic dyes in photoconducting polymer films.

Case 1 : The absorption of light energy by uncontrollable impurities and excitation energy transfer to dye molecules,

Case 2 : Influence of absorbed oxygen on intra-molecular intersystem crossing,

Case 3 : Two-photon absorption and emission processes in dye doped polymers,

Case 4 : The appearance of short-wavelength absorption of dye aggregates and emission from their monomer and aggregate forms,

Case 5 : Difference in the molecular mass of polymers,

Case 6 : Absorption of light energy by excimer states of PEPK, OAGE and PVK and the transfer of excitation energy to dye molecules from these states and

Case 7 : Radiative recombination of photogenerated electron-hole pairs.

The first of these mechanisms cannot be determining, as the PEPK used was subjected to special chemical purification [36]. In addition, the increase in PL intensity does not depend on its preparation procedure. Nor is the second mechanism dominant since the PL intensity enhancement effect is displayed in an almost identical manner for PS and PVAL films. Oxygen adsorption on PVAL is a few orders of magnitude smaller than that on any of the polymers studied. In contrast, the oxygen solubility in PS due to a large pore volume is so high that even photochemical reactions involving triplets do not occur in it [47].

The third mechanism cannot take place in our case since the intensity of light used to excite luminescence is too low to cause nonlinear optical processes in these polymers [9,10]. The fourth mechanism can be a subject for consideration, as the absorption spectra of dyes in PS and PE appear to depend on dye concentration. An increase in the concentration of these dyes results in broadening of the absorption band and its small hypsochromic shift [4]. In this case, the band intensity increases at the short-wavelength edge. The association of the dyes in these polymers is also indicated by the fact that passing from liquid solutions of dyes **1-3** and **19-21** in DCE to PS and PE polymer matrices is accompanied by a hypsochromic shift, rather than by the bathochromic shift as predicted by the universal interaction theory due to large values of n_D of these polymers [4,48].

The absorption spectra of **1-3** and **19-21** in PEPK and PVB are almost independent of concentration. The transition from the liquid solution to PEPK and PVB is accompanied, according to the n_D values, by a predictable hypsochromic shift in the absorption bands of the dyes. This finding suggests that dyes **1-3** and **19-21** exist primarily in the monomer state in the given matrices. Thus, if aggregation were to play a dominating part in the change in the PL intensity of the dyes, they would possess similar luminescent ability.

The role of the fifth mechanism can be associated with the fact that an increase in the molecular mass of a polymer is accompanied by a decrease in its dielectric permittivity and, hence, an increase in the formation probability of tight ion pairs of salt-like dyes [4]. In contrast to solvated ions, even an uncolored counterion can take active part in the deactivation of excited states of such pairs [49].

The influence of molecular mass is most explicitly seen in the homologous series of carbazolyl-containing polymers. The monomer units of PVK-I to PVK-IV are identical in their chemical structure. Although PEPK differs to some extent from PVK in its saturated hydrocarbon radical, it does not considerably change the molecular mass of the carbazolyl moiety. Consequently, the observed weakening of the PL enhancement effect through the series PEPK, PVK-I to PVK-IV can be due to the increase in the polymer molecular mass. Since the molecular mass of PS, PVB, PVAL or PE is much higher than that of PEPK and PVK-I, it may be assumed that this circumstance is primarily responsible for the decrease in the emissive ability on passing from PEPK to PVK-IV. The difference in the molecular mass of these two groups of polymers certainly contributes to the deactivation of the excited states of dyes. However, this contribution is not dominant, since dyes **1-3**, **21**, and **23** exhibit stronger luminescence in PVK-IV than in the photoconducting polymers examined, despite the fact that its molecular mass is higher than that of any of the latter.

The sixth mechanism is taken into consideration because of the existence of low-intensity electronic transitions in PEPK and PVK films beyond the red edge of absorption in the wavelength region of $\lambda > 400$ nm which are attributed to excimer states. In particular, we discovered such a state in PEPK [50] which is characterized by an emission band with a maximum at $\lambda \sim 400$ nm at low temperatures. Although excimer states exist in PVK films as well [51,52], the relative φ values of cationic dyes in PEPK films are much higher than those for PVK films doped with the same dyes.

To reveal the feasibility of the seventh mechanism, let us compare the plots of φ and electric field-induced variation in PL intensity versus λ_{irr} [45]. The comparison shows the existence of a correlation between the PL enhancement effect, photoconductivity and electric field quenching.

The photo-generation of charges by dye molecules at photo-generation sites occurs in two steps. In the first photo-generation step, a geminate electron-hole pair is formed after absorption of a light quantum $h\nu$ by a dye molecule (Dye). The higher occupied molecular orbitals in all the dyes, regardless of the charge sign of their ions, contain two electrons each. When a dye is excited, these electrons are decoupled because of the transition of one of them onto the lower unoccupied molecular orbital. The presence of unpaired electrons strongly increases the reactivity of the dye ion in the S_1 state. Therefore, the excited dye (Dye*) can capture a valence electron from PEPK, OAGE or PVK molecules to form a geminate pair of charges. The pair comprises a hole which represents the carbazole (Cz) radical cation $Cz^{\bullet+}$ in PVK and PEPK and the electron abstracted from a polymer molecule. In the case of cationic dyes **1-3** and **23** and cation-anionic dye **21**, a neutral radical $Dye^{\bullet}$ is formed whereas merocyanine (dye **22**) produces the radical anion $Dye^{\bullet-}$ and anionic dyes **19** and **20** give a radical dianion $Dye^{\bullet 2-}$. This allows the process of generation of geminate electron-hole pairs in doped PVK and PEPK films to be represented by the following electron-transfer reactions:

$$Cz + Dye^+ + h\nu \rightarrow Cz + (Dye^+)^* \rightarrow Cz^{\bullet+} + Dye^{\bullet} \tag{9.8}$$

$$Cz + Dye + h\nu \rightarrow Cz + (Dye)^* \rightarrow Cz^{\bullet+} + Dye^{\bullet-} \tag{9.9}$$

$$Cz + Dye^- + h\nu \rightarrow Cz + (Dye^-)^* \rightarrow Cz^{\bullet+} + Dye^{\bullet 2-} \tag{9.10}$$

In the second step of photo-generation, the hole either recombines with the electron in the same dye molecule in which it was produced (geminate recombination)

$$Cz + ... + Cz + Cz^{\bullet+} + Dye^{\bullet} \rightarrow Cz + ... + Cz + Cz + (Dye^{+})^{*}$$

$$\rightarrow Cz + ... + Cz + Cz + Dye^{+} + h\nu_{r}, \quad (9.11)$$

$$Cz + ... + Cz + Cz^{\bullet+} + Dye^{\bullet-}$$

$$\rightarrow Cz + ... + Cz + Cz + Dye^{*} \rightarrow Cz + ... + Cz + Cz + Dye + h\nu_{r} \quad (9.12)$$

$$Cz + ... + Cz + Cz^{\bullet+} + Dye^{\bullet 2-}$$

$$\rightarrow Cz + ... + Cz + Cz + (Dye^{-})^{*} \rightarrow Cz + ... + Cz + Cz + Dye^{-} + h\nu_{r} \quad (9.13)$$

or leaves the electron via transitions over neighboring Cz groups (dissociation of the pair into free charge carriers).

$$Cz + ... + Cz + Cz^{\bullet+} + Dye^{\bullet}$$

$$\rightarrow Cz + ... + Cz^{\bullet+} + Cz + Dye^{\bullet} \rightarrow Cz^{\bullet+} + ... + Cz + Cz + Dye^{\bullet} \quad (9.14)$$

$$Cz + ... + Cz + Cz^{\bullet+} + Dye^{\bullet-}$$

$$\rightarrow Cz + ... + Cz^{\bullet+} + Cz + Dye^{\bullet-} \rightarrow Cz^{\bullet+} + ... + Cz + Cz + Dye^{\bullet-} \quad (9.15)$$

$$Cz + ... + Cz + Cz^{\bullet+} + Dye^{\bullet 2-}$$

$$\rightarrow Cz + ... + Cz^{\bullet+} + Cz + Dye^{\bullet 2-} \rightarrow Cz^{\bullet+} + ... + Cz + Cz + Dye^{\bullet 2-} \quad (9.16)$$

In charge recombination according to eqn. 11-13, a valence electron from the neutral $Dye^{\bullet}$ or the charged forms $Dye^{\bullet-}$ and $Dye^{\bullet 2-}$ can transfer onto $Cz^{\bullet+}$, producing the excited forms of the dye $(Dye^{+})^{*}$, Dye^{*} and $(Dye^{-})^{*}$ respectively. After deactivation of the excited state, releasing the energy $h\nu_{r}$, this dye molecule can absorb a quantum of light again and take part in photogeneration. As a result of dissociation of an electron-hole pair according to eqn.7-9, the electrostatic interaction of a hole with an electron localized on $Dye^{\bullet}$, $Dye^{\bullet-}$ or $Dye^{\bullet 2-}$ is weakened. As electric field strength increases, the pair separation probability and, hence, the photocurrent value increase and the probability of geminate recombination as well as the intensity of recombination luminescence decrease.

Taking into account the electrostatic interaction forces between an ion and a free electron, it may be expected that the electron capture probability will dramatically decrease in the following order: cationic, neutral, and anionic dye. Another reason for its decrease is associated with an increase in energy consumption for charge separation at the formation of the neutral radical, radical anion and radical dianion **eqs. 9.8-9.11** respectively in this series dye. Therefore, anionic dyes turned out to be incapable of accepting a valence electron from PEPK, OAGE and PVK. As a result, films based on them do not display photoconductivity. Neutral merocyanine (dye **21**) has a higher ability to capture an electron. This is manifested in the onset of photoconductivity. However, this effect is still insufficient for PL enhancement.

It is known [53] that the ability of dyes to photogenerate electron–hole pairs in carbazolyl-containing polymers is determined by the relative arrangement of valence orbitals in the dye and carbazole. This relationship is also observed for the dyes examined in this work. For example, as the polymethine chain length increases in the series of cationic dyes **1-3**, the HOMO and LUMO energies become closer to one another [4] and the difference in the dye and carbazole HOMO energies decreases and can become even negative. Therefore, when dye **1** is replaced by dye **2** or **3**, the photoconductivity of PEPK, OAGE, and PVK films decreases.

Another difference of the PEPK from PVK and OAGE polymeric matrices is that in PEPK is achieved the greatest PL enhancement effect of cationic dyes as well as the presence of high-temperature recombination luminescence of charge carriers produced by photogeneration from these dyes [50]. In this case, both the effects are varied in the same manner by decreasing the excitation wavelength. As discussed previously [50], we may assume that these specific features are associated with the molecular mobility of PEPK carbazolyl groups and the presence of predimer electronic states (traps) in PEPK which are capable of capturing photogenerated holes. The recombination of these thermally detrapped holes via electron-hole pair annihilation is radiative in character. The emergence of traps in PEPK is determined by the nature of not only the polymer and colored dye cation but also by that of its anion [50]. This conclusion is consistent with the dependence of the degree of PL enhancement on the nature of An^-. Such traps appear during photogeneration of carriers, exist only in the filled state and are formed as a result of the additional turn of PEPK carbazolyl groups involved in the predimer states near An^- upon photogeneration of a hole from the photogeneration site onto these carbazolyl groups. The formation of this type of trap, the trapping of a charge carriers, degradation of the trap and charge detrapping take place at the same temperature and the activation energy of these processes is close to that of β-relaxation of PEPK. Photogenerated holes can occur in the predimer state located near An^- and can facilitate the transition of the predimer to the dimer states via the additional turn of the carbazolyl groups (β-relaxation) in the electrostatic field of An^-.

It should be noted that the usual processes connected with an electron transfer are accompanied by a quenching of the fluorescence. However in our case, the probability of formation of electron-hole pairs in these processes with much more probability is realized through triplet states of molecules than singlet states. The dye molecules, which transfer to the triplet state at the photo-excitation, do not fluoresce from S_1 state. In case of formation electron-hole pair, the part from them transfers to a singlet state during of spin conversion. The recombination electron-hole pairs happen mainly in this state with formation radiant S_1 state of a dye. Thus, the dye molecules, which should not participate in absence of charges photogeneration in the fluorescent process, deposit the contribution to a fluorescent S_1-S_0 transition. Therefore, recombination of luminescence, amplify PL (fluorescence) of dyes under study.

Thus, a filled dimeric trap for holes appears and the deficit of energy for its formation at the given temperature is compensated by polarization of the medium, the localized hole and electrostatic interaction of this hole with An^-. The hole leaves the trap during recombination or separation but the dimer will degrade via spread-turn of the carbazolyl groups and the predimer state will again be formed from the dimer state though unfilled in this case. The increase in the intensity of recombination luminescence in PEPK doped with cationic dyes with decreasing excitation wavelength seems to be determined by the increase in the probability of trapping photogenerated holes in these traps with decreasing in λirr. Since the PL enhancement and quenching effects in the external electric field of cationic dyes are stronger in OAGE and PEPK than in PVK, it may be assumed that this type of trap is also present in OAGE-based films in which anthracenyl groups possess as high an activity as do the carbazolyl groups in PEPK. This is due to the fact that OAGE and PEPK have a longer main chain as compared to PVK and anthracenyl and carbazolyl groups are not immediately linked to the chain; they are linked through bridging $-CH_2-$ and $-O-CH_2-$ groups respectively.

5. LASER MEDIA

5.1. Passive Laser Switches (Q-Switches)

Dye-doped polymer materials are of considerable interest in quantum electronics because they have a number of operational advantages compared with liquid media. First of all, they make it possible to simplify laser configurations because the system for circulating the liquid switch can be omitted. Polymers can also operate at various temperatures and in the weightless state, there is no need for contact with aggressive solvents and the reproducibility of the energy parameters of the laser output pules is higher [9,10].

Lasers which generate intensive short single pulses of approximately 10^{-8} s duration and high-energy trains of nanosecond pulses are of great interest in many fields of science and technology. Solid-state lasers with a neodymium-doped active medium are finding increasing application in technological processes used in the treatment of various materials. Some of the applications such as scribing, drilling, marking and surface doping are often performed by repetitively pulsed lasers. When materials are processed mainly by evaporation, power densities in excess 10^6 W/cm^2 and repetition rates of tens and hundreds of kilohertz are required [54]. This is achieved mainly by the Q-switching of the laser cavity. In practice, this switching is performed by a variety of methods such as acousto-optical Q-switching which is difficult to implement and requires focusing of radiation in an acousto-optical cell in order to reduce the switching time [55] which limits the permissible optical loads.

The simplest way to obtain multi-pulse generation and single pulses is to use cavities with a Q-quality adjustment. Lasers with passive cavity Q-switching are the most reliable sources of high-power coherent radiation in the near infrared. Solid-state variants of passive laser switches (PLSs) fit best the current operational requirements that laser technology components must satisfy. The application of PLSs considerably simplifies the laser design because there is no need in the use of schemes, which provide control and synchronization.

The results obtained by Q-switching with nonlinearly absorbing LiF crystals containing F_2^- color centers were reported in references 56 and 57. Such switching was performed in Nd^{3+}:YAG (yttrium-aluminium-granat) industrial laser by LiF crystals when the pumping was pulse-periodic [58]. However, they are subject to degradation with time and their absorption cross-section at the operating wavelength depends on temperature although these PLSs can have good initial parameters [59].

Phototropic Cr^{4+}:YSGG (yttrium-scandium-gallium-granat) crystals were used in order to investigate a multi-spike operation of Nd^{3+} lasers subjected to pulse-periodic pumping [60]. A passive switch based on Cr^{4+}:YAG was used to generate spiked radiation in Nd:YAG laser subjected to continuous pumping [61]. In these two cases, high photochemical stability was achieved but the Q-switching efficiency was only $\eta = 30$–50%. It should be pointed out that the peak power of multi-spike generation resulting from continuous pumping was only 10^3–10^4 W [55,56,61]. Liquid and solid organic compound solutions with phototropic properties can be used as PLSs. Solid-state PLSs are more suitable in practice. Transparent polymers have long been used as organic-dye matrices in solid-state switches for generation of single pulses [62-64]. More extensive use of polymer components in lasers is hindered by the low radiation damage threshold (RDT) of matrices made up of PS and PMMA [65], PVAL [62] and epoxy polymers [66] by the insufficient photochemical and operational stability of passive switches [67] and by the structure shortcomings of polymer film switches [65].

Specific investigations of the laser damage mechanism in polymer matrices [68,69] have established the need to use elastic materials capable of withstanding considerable elastic strains in a wide range of operating temperatures. The use of polyurethane acrylate (PUA) matrices [59,60], which retain their elastic properties even in the absence of plasticising additives, has made it possible to increase considerably the RDT of laser components. However, organic dyes embedded in matrices formed by radical hardening (polymerization) suffer photodecomposition as a result of the interaction with the chemically active free radicals generated by scattered daylight and by radiation from pump flashlamps.

However, all these advantages can be realized only if the spectral, luminescence and nonlinear optical properties of dyes do not change significantly with the adoption of polymer matrices compared with the solution of these dyes in liquids.

The first attempts to introduce cyanines into polymers have resulted into a strong distortion of these properties [70]. It is caused by the formation of dyes ion-pair associates with a sandwich structure [30]. Their formation is accompanied by a reduction in the degree of bleaching, radiation damage threshold and energy of ultrashort pulses [4,9]. It is connected to an overlap of dye associates absorption bands with analogous bands of the corresponding monomer dyes. This results in the absorption of laser radiation by the aggregates that form in the polymer passive laser switches and, consequently, in reduction in the degree of bleaching of these switches and in the energy of the ultrashort output pulses. Since the sandwich-type of associates do not emit luminescence, they convert the absorbed energy into heat causing local temperature rise that damages the dye [65]. For this reason, PMMA films based on thiopyrylotricarbocyanines (dye **4** and **4a-c**) which easily aggregate, are characterized by a low RDT, a low degree of bleaching and a low ultrashort-pulse energy [9]. These values decrease regularly with an increase of the dyes aggregation degree, i.e. with an increase in the intensity of the short-wavelength absorption [30].

Salt-like organic dyes in weakly polar media readily associate at much lower concentrations than in strongly bipolar media as mentioned above. Therefore, it is difficult to prevent the formation of associates by the variation of the concentration of dyes in polymer films, especially as passive switches are made of thin (50-100 μm) films. The necessary optical density of such films can be ensured only if the concentration of the dye is several orders of magnitude higher than the concentration used in the determination of the absorption spectra in a cell 1 cm thick. There is therefore a need for dyes that do not aggregate under these conditions. The following approaches [4] have been developed to achieve this.

Approach 1. The introduction of bulky substituents in hetero-residues and in the polymethine chain

It prevents the joining of cations to form sandwich configurations [30]. For example, the replacement of the hydrogen atom in the meso position of thiopyrylocyanine (Dye **4**) with the bulky phenyl group (dye **8**) makes it possible not only to conserve the spectral profile when a PMMA film is used instead of a liquid solvent [30] but also to increase the RDT to 400 MWcm^{-2} [71] to enhance the degree of bleaching by a factor of almost 2 and to increase the energy of ultrashort pulses by a factor of 7.5 in the case of a YAG: Nd^{3+} laser [65].

Approach 2. An increase in the uniformity of the charge distribution in the dye chromophores

As explained in reference 33, a polymer film switch based on cyanine (dye **15**) and PMMA has a service life similar to the best thiopyrylotricarbocyanines (dyes **5-8**) in spite of the fact that it does not contain any bulky substituent [72].

Approach 3. The use of the bulky counter-ions or counter-ions with delocalised charge

Replacing the perchlorate anion in the dye molecule **8** by tosylate anion doubles the relative degree of bleaching and the energy of ultrashort pulses [65]. The value of time relaxation of the first excited state hardly changes (in the case of dye **8** and **8**(tosylate) in PMMA, it is 75 ± 6 and 80 ± 9 ps, respectively. The power of ultrashort pulses generated with the aid of a polymer passive laser switch based on dye **8**(tosylate) in PMMA can reach upto 80 MW for pulses of 19 ps duration [65].

Approach 4. The use of high-permittivity polymers, containing strongly polar groups

For example, in the case of dyes of the dye **15** type the use of more polar and nucleophilic polyurethane acrylate instead of PMMA, increases the service life of passive laser switches to 10^4 pulses at any particular point when the energy of a pulse train is at least 2 mJ [73].

Thus, highly efficient dye-doped polymer passive laser switches can be made by constructing dye molecules with the highest uniformity in the charge distribution in their chromophore and in the counter-ion and by introducing bulky substituents into the dye molecules by using matrices with the highest possible polarity and with strongly nucleophilic (in the case of cation dyes) or strongly electrophilic (in the case of anion dyes) functional groups. These criteria have been used to develop prospective polymer PLSs for solid-state laser. The best results are achieved for PU compositions [19,73].

The multicomponent urethane composite as a polymer matrix is a network elastomer, representing a continuous three-dimensional entity in which all the component molecules are linked by chemical bonds of different types. The PU matrix was synthesized by means of the following polycondensation reaction of diol with diisocyanate in which R and R' are the aliphatic or aromatic hydrocarbon groups.

$$n\ HO-R-OH + n\ OCN-R'-NCO \rightarrow -[-O-R-O-CO-NH-R'-NH-CO-]_n- \qquad (9.17)$$

The hardening of this polymer composition as well as epoxy polymers is carried out by a way of a polycondensation reaction. However, contrary to epoxy polymers, it undergoes under mild conditions at the room temperature in a neutral medium. Thus, dyes of different types (polymethine, pyrromethene, oxazine, xanthene, phenalenone, coumarine and metal-organic complexes) can be incorporated in these polymers without their decomposition (total or partial) at certain stage of polymerization as well as in the wide range of concentrations (10^{-4}-10^{-7} mol/g of the polymer) without causing any aggregation.

Spectroscopic measurements demonstrate that the uncolored polymer matrix has a transparency window at 0.32-2.20 mm with an integral light transmission coefficient of T»100% in the visible and near infrared region. Variation in the relative content ratio of the polymer components alters its refractive index from 1.48 to 1.52, ensuring an almost complete immersion at the polymer-optical substrate when the switch is in the form of a triplex configuration. The well-tested method of hardening an initially liquid composite between two glass plates can be used to realize more complex variants: a pentaplex with two polymer layers between three substances; a polyplex which is a multi-layer configuration polymer layers for lasers with a high average output power. Moreover, a module containing a cavity mirror and a passive switch can be formed. The dyed polymer composite strongly bonds the optical substrates with reflecting coatings. The adhesion strength of the polymer-optical substrate (BK7 glass) joint, measured on a test device by breaking away along the normal to the contact area, exceeded 120 kg/cm^2.

The high RDT was due to the highly elastic properties of the polyurethane matrix. The relative elongation of this polymer, without incorporating any plasticizer in it, was 320% and the highly elastic state existed between –50°C and +160°C [19].

Stable and highly efficient polymer PLSs based on PU and an organic dye **BDN** (dye **24**) were developed for the neodymium lasers. The maximum energy of 0.98 J was obtained for a

Dye **24. BDN**

single pulse at passive Q-switching of these lasers. Repetition of single pulses at 50 Hz was possible. Repetitively pulsed operation with a peak output power upto 2 MW and an energy of the train of nanosecond pulses 14.1 J, were achieved in an yttrium aluminate laser for the first time [19]. Variation of the initial switch transmittance made it possible to vary the pulse (spike) duration in the range of 28-90 ns. In ruby and neodymium lasers, polymeric PLSs based on organic dyes, made it possible to realize a regime of operation with gigantic pulses packets with a Q-switching efficiency of 95% [74] and 98% [19] correspondingly. The maximum pulse repetition rate in a train was 350 kHz. The regime of gigantic pulses packets is of interest for the treatment of metals and ophthalmology in connection with the need to struggle against negative side effects brought about by the shock waves from the single powerful pulse. In the regime with spikes of pulses, the power at the individual spike does not reach the threshold for the excitation of one shock wave and the integral energy of the spikes secures the necessary effect without any side effects [19,74].

As it is necessary in the industrial processes, a variation of the PLSs transmittance and the pump energy made it possible to generate the required number of spikes from one **(Fig. 9.12a)** to the maximum possible at the fastest pumping rate **(Fig. 9.12b)** [19]. In these experiments, the maximum number of spikes in a train was 310.

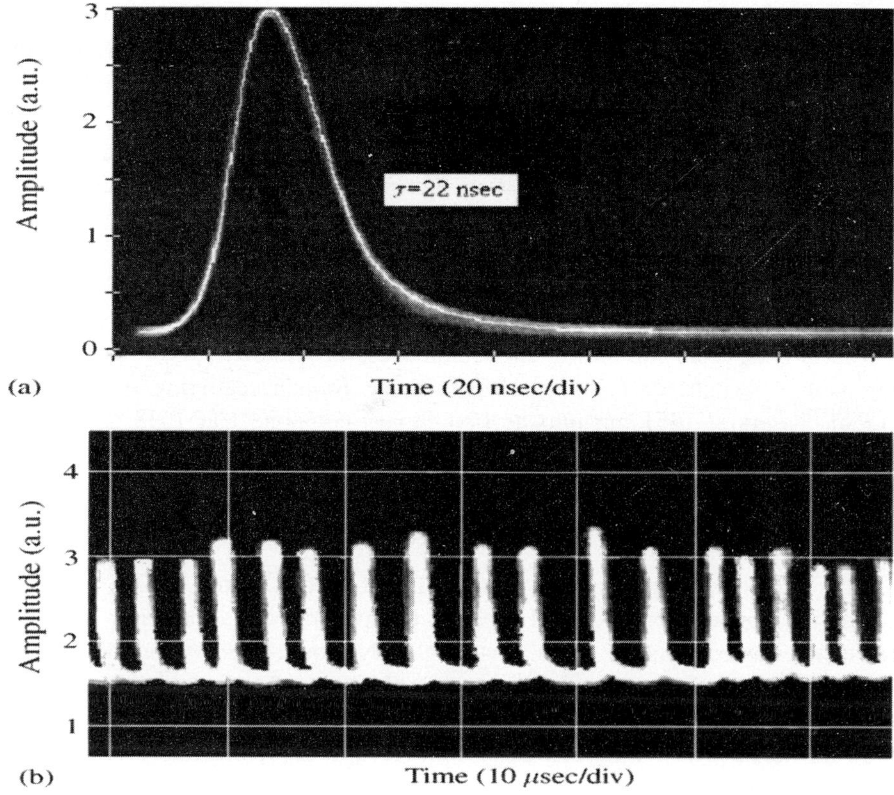

Fig. 9.12. Oscillograms of a laser single pulse (a) at polymer passive laser switches (transmittance = 20%) and of nanosecond pulse spikes (b) at such switches (transmittance = 95%). Reproduced with permission from Bezrodnyi *et al.*, Optics and Laser Technol., 34(1), 7 (2002). Copyright © 2002 Elsevier Science Ltd.

The RDT of the investigated polymer matrix was achieved 14 and 18 J/cm^2 for single pulse duration (15 and 35 ns correspondingly) and 52 J/cm^2 in multi-spike generation regime of duration of 80 ns [19].

The service life of Q-switches with λ = 1060 nm achieves 2×10^6 pulses at an operation of radiation energy of 0.9 J in a single pulse at the same point in a nanosecond operation regime [19] and 10^6 pulses at an operation of radiation energy of 6 mJ in the train in a picosecond regime [73]. For neodymium lasers on phosphate glass, the first polymer PLSs made it possible to reduce the duration of the generated pulses to a record value of 0.9 ps [75].

Thus developed Q-switches secure stable energy and time characteristics in picosecond lasers with passive negative feedback on semiconducting structures of the GaAs type, making it possible to obtain the long trains (1 μs) at the output [76]. Such trains are essential for the synchronous pumping of subpicosecond dye lasers [76]. The relatively high peak power of the pumping spikes secures stable generation in these lasers without the use of additional cascade amplification.

An important step in the development of dyed functional materials is the creation of the first polymer Q-switches based on polymethine dyes for lasers emitting in the regions of 1300 nm [77] and 1540 nm [78]. Emission in these regions is of interest for location, light guide communication and medicine. This is due to the fact that it falls into the transparency window of the atmosphere and optical fibers [79]. Radiation with $\lambda = 1300$ and 1540 nm is less hazardous than visible light for the human eye on account of the increase of the absorption coefficient of biological tissues on moving into the IR region [80]. The efficiency of Q-switches for $\lambda = 1300$ nm attained a value of 40% for a single pulse energy 0.16 J and duration of 18 ns. Passive mode locking was employed to generate ultra-short pulses of 35-60 ps duration with energy of 6-10 mJ in the train. The photostability of the polymer Q-switches at 1300 nm was two orders of magnitude higher than that of corresponding liquid switches [77].

Such polymer passive switches may turn out to be promising for use in ophthalmologic lasers. The reason is that radiation with $\lambda = 1300$ nm was recently found to give better results than with $\lambda = 633$ or 780 nm, especially in curing shortsightedness in children [81].

Polymer passive switches effectively transform the Raman scattering of the emission from $KGd(WO_4)_2$:Nd^{3+} lasers at 1351 nm into its first Stokes component at 1538 nm [78]. The output energy was 9 mJ for single pulse duration of 40 ns.

5.2. Active Laser Media of Visible Spectral Range

Dye lasers have attractive properties due to their broad spectral range tuning and high quantum efficiency. They are simple to produce and cheap in comparison with other laser sources. Two types of dye-doped active laser media are mainly used, namely, liquid-state media and solid-state media. The second type has some operational advantages relative to the first. Solid-state media have a lower refractive index dependency on temperature. In other words, their thermo-optical characteristics during lasing are better. Thus, they can be used over a wider temperature range, including low temperatures [10].

PMMA and its modifications [68,82-84], PUA [85,86] and various epoxides [87,88] are widely used as polymers for dye-doped active laser media (ALMs). The PMMA and PUA matrices are hardened by the method of radical polymerization. Free radicals that are present in the system induce side radical reactions involving dye molecules which results in the destruction of the latter. Epoxy polymers are hardened by the polycondensation reaction. However, hardeners used in these reactions possess either basic or acidic properties, which are unacceptable for organic compounds of many classes because these properties cause irreversible chemical decomposition of organic molecules. The possibility of using organic dyes doped PU in the ALMs was studied in work reported in reference 89.

The polymer matrices of ALMs experience stronger radiation and thermal loads than PLSs upon the interaction with the pump and lasing radiation. This is explained by the fact that the dye concentration used in active media is at least an order of magnitude higher than in passive media which substantially increases the probability of their aggregation. The aggregation of dyes results in the degradation of the photostability, service life and luminescent properties of dyed media [9]. As the dye concentration increases, the probability of formation of ion pairs also increases. In particular, the contact pairs are formed [4] in which electron phototransfer from an anion to a cation can occur [1] resulting in the formation of active neutral radicals.

ALMs require significantly higher power densities for their pumping than the power densities used for the PLSs bleaching. This is explained by the fact that it is necessary to produce the level population inversion as great as possible to obtain the efficient lasing. Whereas obtaining equal populations of the ground and excited levels is necessary to perform Q-switching. In addition, the energy load in ALMs is also higher than in PLSs because the radiation is absorbed at two wavelengths (pump and lasing).

It is known that the RDT of polymer matrices decreases with decreasing the pump wavelength [90]. The RDT of the PU matrix was measured at the second-harmonic wavelength of 532 nm of a single-mode Nd:YAG laser operating in a single-pulse regime. The single-pulse damage threshold was 15 J/cm^2 for pulse duration of 15 ns and the diameter of the irradiated area of 200 µm, which exceeds by a factor of 1.5 the threshold for PUA [86] measured under the same conditions.

Most laser polymer materials have intense absorption bands in the region of two-photon transitions for pump frequencies often used in experiments. For example, the energy of the electronic transition in a polymer with the 266 nm absorption band is equal to the absorbed energy of two 532 nm pump photons. Therefore, the degradation of ALMs upon high power pump can also be caused by two-photon absorption.

The purposeful studies [68,82] of the mechanism of laser damage of polymer matrices showed that it is necessary to use elastic materials, which exhibit substantial elastic deformations in a broad temperature range. Thus, the introduction of low- and high-molecular additions into PMMA made it possible to obtain the high radiation resistance that satisfied the operation parameters [68,82].

ALMs based on cross-linked PUA [85,86,91] and especially PU has highly elastic properties in a broad temperature range. Such laser elements can operate upon high-power pulsed pump without the addition of plasticisers to the polymer. The active media based on PU and PUA were fabricated in the form of a triplex [90]. Three organic dyes from three classes of lasing dyes viz. xanthene dye **R6G** (dye **23**), polymethine dye **HIC** (dye **1**) and pyrromethene 597 dye **PM597** (dye **25**) have been

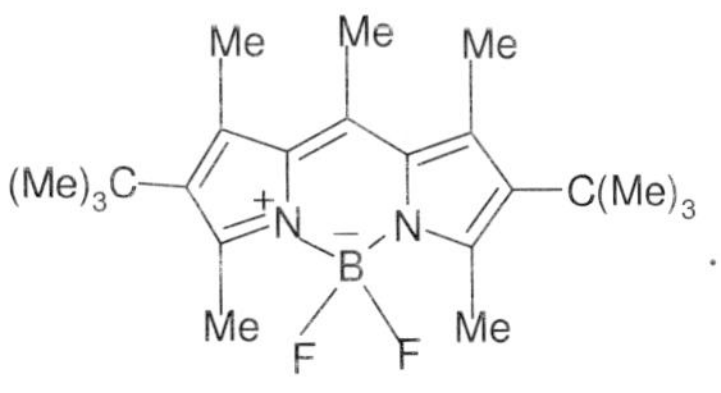

Dye **25. PM597**

chosen for these media [90]. The main operation parameters of ALMs are the operating resource (*N*), the lasing efficiency (η) and photostability. To study the photostability, we fabricated triplexes with silica substrates. The thickness of PU and PUA films doped with dyes in photostability experiments was 300 mm. The optical density (*D*) at the maximum of the main transition was about unity. Samples were irradiated by a xenon lamp which has the continuous emission spectrum in the UV and visible regions. The radiation power density on a sample was 50 mW/cm^2. The dependence of the optical density of polymer samples on the incident radiation dose (*E*) was measured at the maximum of the absorption band of the dye. **Fig. 9.13** shows the dependencies of photodissociation of dyes in polymer matrices on the irradiation energy. One can see that the photostability of dyes substantially increases on passing from PUA to PU.

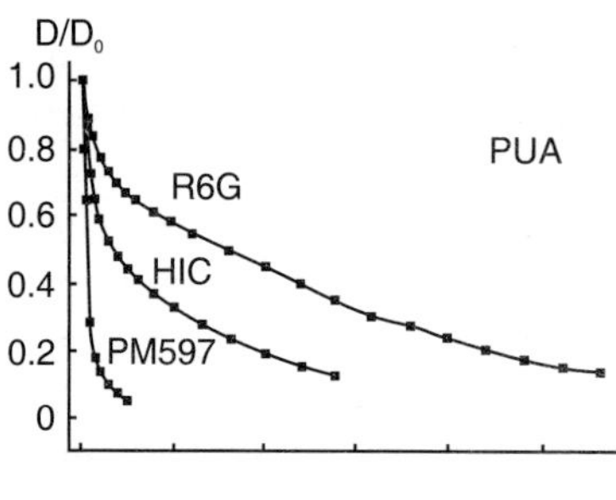
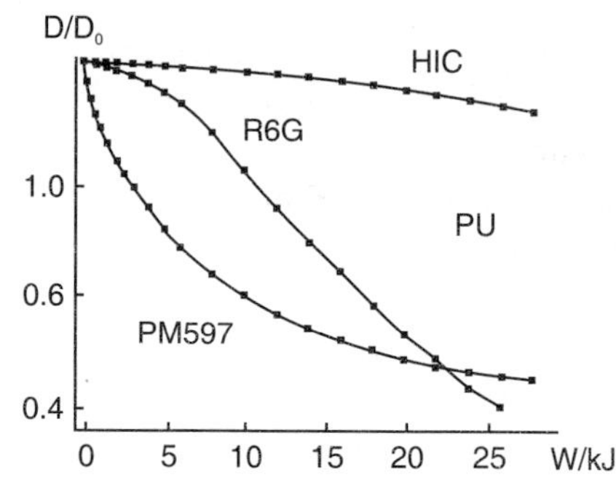

Fig. 9.13. Dependencies of the normalized optical density D/D_0 of polymer active media on the irradiation energy W for dyes R6G, HIC and PM597 in polyurethaneacrylate (PUA) and polyurethane (PU). Reproduced with permission from Bezrodnyi *et al.*, Quantum Electronics, 30(12), 1043 (2000). Copyright © 2000 Kvantovaya Elektronika and Turpion Ltd.

The **PM597** dye with the skeleton molecular structure proved to be unexpectedly most unstable in both the polymers. Whereas the **HIC** dye with an open polymethine chain which is highly sensitive to photochemical reactions, was most photostable. The **PM 597** dye exhibited rather strong absorption in the region between 320 nm and 420 nm whereas absorption of **HIC** in this region was very weak [90].

Thus, we can conclude that the photochemical dissociation of **PM597** dye exposed to the UV irradiation is mainly determined by electronic transitions to higher electronic states. This circumstance should be taken into account in the synthesis of photostable dyes. The highly excited electronic states of dyes can undergo nonradiative deactivation due to vibrational relaxation and internal conversion via high vibrational states of polymers along with the formation of free radicals of macromolecules. Because the absorption bands of dyes in the UV region are overlapped with the absorption band of PUA stronger than with that of PU, the probability of formation of free radicals in PUA should be higher which is confirmed by photodissociaiion curves shown in **Fig. 9.13**.

The resource parameters substantially improve in thick polymer layers, for example, in PMMA films [92]. The reduction of side concentration defects and weaker heating of the polymer layer may explain this. As the dye concentration decreases, the polymer layer thickness increases, resulting in the retention of the optimal optical density. The operating volume increases with increasing layer thickness. Therefore, the heat capacity and heat conduction of the working volume also increase. This prevents the overheating of this volume. Thus, the thickness of the PMMA layer in paper [92] was 20-30 mm. The high radiation resistance of both PU and PUA allowed us to perform high-power pumping of thin layers of these polymers at the same point without scanning a pump beam over their area.

The resource dependencies of the radiation conversion efficiency (η) on the number of pump pulses (N) are presented in **Fig. 9.14**. The conversion efficiency is defined as the ratio of the output energy of the dye laser to pump energy. A decrease in η for all the samples with increasing N was caused by the dye bleaching and the polymer matrix being undamaged. For the dyes studied, the operating resource in the PU matrix was greater than that in the PUA matrix.

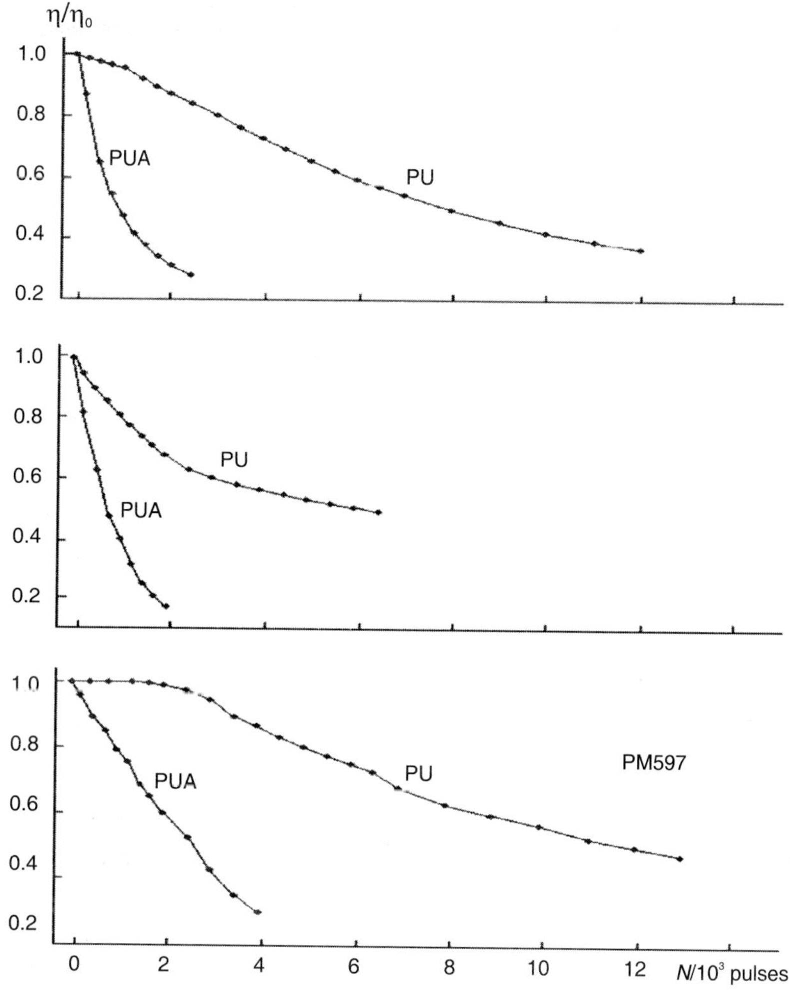

Fig. 9.14. Normalized resource dependencies of the lasing efficiency η/η_0 for dyes R6G, HIC and PM597 in polyurethaneacrylate (PUA) and polyurethane (PU). Reproduced with permission from Bezrodnyi *et al.*, Quantum Electronics, 30(12), 1043 (2000). Copyright © 2000 Kvantovaya Elektronika and Turpion Ltd.

The initial conversion efficiency (η_0) for PU elements was higher than that for PUA elements. The values of η_0 for three dyes in these polymer matrices are presented in **Tab. 9.2**. The dye destruction that we observed during fabrication of PUA elements [85] was later confirmed by the work reported in reference 93 where PMMA elements were fabricated using the radical reaction of hardening with the help of benzoyl peroxide. The products of the dye destruction absorb in the lasing region.

Tab. 9.2. Basic operation parameters of active laser media. Reproduced with permission from Bezrodnyi *et al.*, Quantum Electronics, 30(12), 1043 (2000). Copyright © 2000 Kvantovaya Elektronika and Turpion Ltd.

ALM type	$\eta_0 (\%)$	$N_{1/2}$	$E_{1/2}$ (kJ/cm²)
R6G in PUA	29	900	8.0
R6G in PU	34	8000	23.0
HIC in PUA	30	700	2.0
HIC in PU	32	6500	160.0
PM597 in PUA	58	2200	0.3
PM597 in PU	75	12200	18.5

Note: $N_{1/2}$-is the number of pulses at which h_0 decreases by half; $E_{1/2}$-is the radiation dose at which D decreases by half.

To detect the absorption related to the dye destruction products, we examined the long-wavelength parts of the absorption spectra of PUA oligomer compositions before their hardening and after hardening [90]. The bathochromic shift of the absorption band and the additional absorption in the lasing region takes place. It is caused by the dye decomposition products and affects the conversion efficiency.

The greater initial conversion efficiency in PU compared to PUA can be explained by the absence of the dye decomposition in the PU matrix during its fabrication. The **HIC** dye in the PUA matrix undergoes only weak destruction that is accompanied by a weak absorption in the lasing region [90]. This explains a small difference in η_0 for HIC in PU and PUA matrices as evident from **Tab. 9.2**.

Therefore, we can conclude that the photo-stability, the operating resource, the conversion efficiency and the uniformity of the dye concentration distribution over the thickness of the PUA matrix, are smaller than for the PU matrix. This is caused by a substantially greater decomposition of dyes in the former case. On the one hand, this occurs due to a partial decomposition of dyes in PUA during polymerization owing to the radical reaction, which favors the further activation of radical reactions upon the interaction of polymer ALMs with light. On the other hand, the advantage of PU is not only in the method of its polymerization but also in its high polarity [4]. Because the permittivity of PU is higher than that of PUA, epoxides and PMMA [4], the salt-like dyes are dissociated in PU to a greater degree than in the above-mentioned polymers. The separation of counter-ions in PU is also facilitated by the presence of many highly nucleophilic functional groups in its chains. These groups subject the dye cations to the nucleophilic solvation by producing a solvation shell around them [4]. As a result, the probability of formation of contact ion pairs in PU is much lower than that in PUA, epoxides and PMMA. Hence, the probability of the electron phototransfer and of the association of contact ion pairs due to the electrostatic attraction of their opposite charges in PU is minimal, in contrast to the above-mentioned polymers [4].

The replacement of the chloride anion in **HIC** by more nucleophilic tetrafluoroborate which has a greater electron affinity, virtually does not affect the spectral, lasing, and resource parameters of dyes in PU whereas the absorption band slightly broadens and exhibits a hypsochromic shift and the dye photostability increases in PUA. The dependence of these characteristics on the counter ion type indicates [4] to the formation of contact ion pairs in PUA in contrast to PU.

The **PM597** dye that is most unstable among the dyes studied, has the greatest operating resource (**Fig. 9.14 & Tab. 9.2**). This fact again confirms the involvement of highly lying excited estates of the **PM597** dye in processes of its photochemical decomposition. The matter is that the resource parameters were studied upon excitation only into the main absorption band whereas the photostability was examined upon excitation both into the $S_o \rightarrow S_1$ and $S_o \rightarrow S_2$ absorption bands.

The intra-ion **PM597** dye represents a chemically bonded contact ion pair, which cannot dissociate at any polarity of a polymer. However, the charges in this pair are well separated. In addition, the negative charge is localized at the borofluoride bridge because of a substantial electronegativity of fluorine atoms. Thus, the intermolecular phototransfer of an electron from boron to nitrogen in **PM597** is hindered to a greater degree than in chemically uncoupled contact pairs of cationic dyes **HIC** and **R6G** where counter ions can either approach or move away from each other depending on the polarity and solvation ability of the medium.

As a result, the **PM597** dye has a greater operating resource in PU and PUA than **HIC** and **R6G**. As in salt-like cationic dyes, the electrostatic attraction between the opposite charges of the intra-ion **PM597** dye can favor the formation of aggregates. For the reasons considered above, these processes in PU are much less probable than in PUA. This circumstance in a combination with mild polymerization provides the better resource characteristics to **PM597, HIC** and **R6G** in PU.

5.3. Active Laser Media of near IR Range

Second-harmonic generation in a neodymium laser serves as a laser pump for dye lasers in many cases [94]. Less frequently, neodymium laser third-harmonic or XeCl [95] and N_2 [96] laser generation is used for this aim. However, application of these pump sources provides effective lasing, mainly in the visible region. In the case of dye-doped active polymer media, there are some fundamental difficulties in shifting the lasing to the near IR range when using such sources. To overcome them, it is necessary to create luminophores with large Stokes shifts or dyes which intensively absorb light in the near IR and can be pumped in their second or higher electronic transitions. The development of dyes with Stokes shifts greater than about 500 nm and having a high fluorescence quantum yield is problematic [4]. Moreover, this shift is decreased in polymer matrices in comparison with liquid dye solutions.

The pumping of higher electronic transitions is of little effectiveness, mainly due to the significant decrease in dye photostability. Therewith the drop of lasing efficiency often takes place. It is caused by the increased probability of radiationless transitions as a result of participation by the highest excited state. In addition, the radiation of UV pump sources is in the absorption region of many polymers. This leads to polymer destruction and the decrease in their conversion efficiency.

There are excellent pump sources for producing tunable lasing in the near IR range (1100-1500 nm), namely, neodymium lasers ($\lambda = 1060$ nm). However, in this case some difficulties appear in obtaining generation even when liquid solutions such as active media are used [97]. Stable and effective operation of the dyed active medium into the IR region is limited by the small fluorescence lifetime and low photochemical stability of the dyes [4].

These problems were successfully decided on an example of thiopyrylo-tricarbocyanine dye (dye **5**) in PU [98]. The maximum in its main absorption band in PU lies at 1079 nm. It has a high absorption cross section in this polymer at $\lambda_p = 1064$ nm and its value is 6.3×10^{-16} cm^2. The dye **5** has a relatively large excited state relaxation time of the order of 50 ps in liquid solutions [9] and

100 ± 10 ps in PU matrix [98]. The absorption shape of this dye in the polymer did not undergo distortions **(Fig. 9.15a)** and corresponds to that in the best liquid solvents. Consequently, dye aggregation responsible for the distortion of the absorption shape is absent in the polymer used in the operating concentration range of $10^{-5} - 0.5 \times 10^{-3}$ mol/l. It is connected to the above-mentioned features of a structure of a dye and polymer.

As may be seen in **Fig. 9.15b**, the fluorescence band of the thiopyrylotricarbocyanine (dye **5**) in polyurethane as well as that of other cation tricarbocyanines in polar liquid solvents is narrower than the absorption band [99]. This is caused by the weakening of electrostatic interactions between the polar groups of the polymer and the distribution of charge in the dye chromophore in the excited state in comparison to the ground state [99]. From **Fig. 9.15**, it may be seen that the ratio between absorption and fluorescence bands allows the pumping of neodymium lasers ($\lambda = 1060$ nm) to obtain lasing beyond 1100 nm.

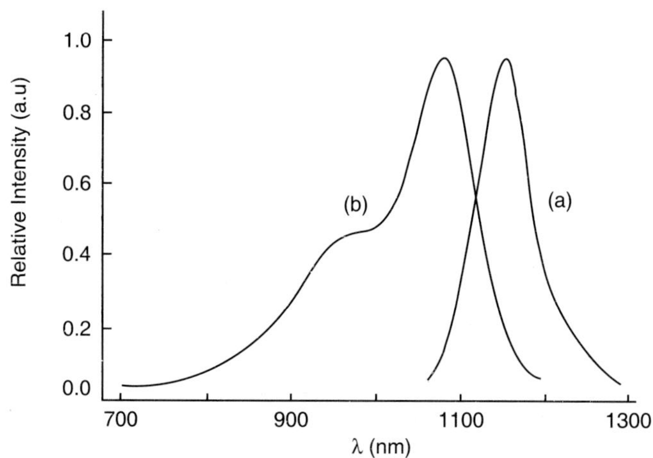

Fig. 9.15. Absorption (a) and fluorescence (b) spectra of dye 5 in polyurethane. Reproduced with permission from Bezrodnyi *et al.*, Appl.Phys., B73(3), 283 (2001). Copyright © 2001 Springer-Verlag.

The particularities of spectral, luminescence and nonlinear optics properties of thiopyrylotricarbocyanine (dye **5**) in PU are prospective to create of ALMs in near IR range. In fact, tunable lasing from a dye laser with an ALM based on dye **5** and PU has been obtained using 1060 nm pumping for the first time [98]. The conversion efficiency of 43% and the tunable range of 63 nm. Such efficiency corresponds to the best samples of a visible range.

Therefore, PU is a promising material both for PLSs and ALMs based on organic dyes. Unlike other polymers, doping of PU with dyes of any class is not accompanied with their decomposition. ALMs based on PU have a great potential for improvement of photochemical, resource and lasing parameters by optimizing dye-doped matrices.

The efficient and highly photostable laser operation from newly synthesized modified dipyrromethene. BF$_2$ complexes dyes **26-35**, analogs of commercial dye **PM567**, incorporated into solid polymeric matrices has been reported by the authors of the work published in references

Dyes **26-30** (n = 1-5) Dyes **31-35** (n = 1-5)

100 and 101 reported. All materials where the modified dipyrromethene BF_2 complexes are covalently bonded to the polymeric chains (dyes **31-35**) exhibit a photostability higher than that of the material where the same dye (dyes **26-30**) is simply dissolved in the polymer. This same effect takes place in the case of modified **R6G** molecule copolymerized with methacrylic monomers [102]. The increase in photostability was attributed to the improved dissipation of the excess of absorbed energy. The covalent linkage of the dye to the polymeric chains provides additional channels for the elimination of the absorbed pump energy that is not converted into emission which prevents the accumulation of heat in the material.

6. LUMINESCENT SOLAR CONCENTRATOR

The conceptual operation of the luminescent solar concentrator (LSC) is based on light pipe trapping of luminescence induced by the absorption of solar radiation [103,104]. A transparent material, such as PMMA or polycarbonate, is impregnated with guest luminescent absorbers such as organic dye molecules. Solar photons entering the upper face of the plate are absorbed and photons are then emitted. A large fraction of these luminescent photons will be trapped by total internal reflection, for example ~74% of an isotropic emission will be trapped in a PMMA plate having an $n_D = 1.49$. It is possible to increase the solar power density of only directional radiations by using mirrors and lenses. The collector makes it possible to use the entire radiations including the diffuse parts, which are of considerable importance in temperate zones. Successive reflections transport the luminescent photons to the edge of the plate where they can enter into an edge-mounted array of solar cells. The light amplification is given by the ratio of the plate surface to the plate edge.

The LSC offers the promise of reducing the cost of photovoltaic energy conversion by the use of high gain concentrators, which do not require tracking. A suitable materials for LSC should have (i) a high absorption intensity, (ii) a wide absorption band in order to be able to absorb a large proportion of the available light, (iii) a high φ so that the irradiated energy is lost by non-radiative processes, (iv) a large SS so that the losses by re-absorption can be kept to a minimum and (v) a high photostability.

The organic dyes have the intensive and wide absorption bands than the ions of rare-earth metals used as LSC based on inorganic glasses. The value φ in visible range of a spectrum can reach values close to 1 and considerable SS both for organic as well as inorganic compounds. The different dye-doped polymer compositions based on well-known luminophores (coumarins, rhodamines, oxazines etc.) was developed for this spectral range [37,103-106].

The concentration of the fluorescent dyes in the LSC material is very high. The absorption of sunlight increases with increasing concentration but the proportion of reabsorbed fluorescent light also increases, as the overlapping area of absorption and fluorescence bands is incremented. In this area, the luminescence can be reabsorbed by another dye molecule, which is still in the ground state [104]. To avoid this effect it is necessary to create dyes with large SS. Such compounds are necessary for one reason. The value of φ for SS for the organic materials is as a rule less than for inorganic materials in the near IR range. The most SS among known organic dyes has pyridopyrylocyanine (dye **36**), which is characterized by the greatest electronic asymmetry [107].

Dye **36**

On the basis of results of quantum-chemical calculations, it was shown that the charges and bond orders in an excited state of this dye are considerably equalized in comparison with its ground state [108]. Dye absorbs light with the geometry of a ground state but emits with the geometry of an excited state. Therefore, this explains the large SS and transformation of the light energy from the visible range to the near IR range.

Lightfastness of the dye-doped polymer is major importance for the industry. Photochemical decomposition occurs in the excited state. Thus, a dye that has returned to the ground state by luminescence will remain undestroyed until the next excitation. There are various ways to solve this problem. Optimization of the characteristics of dye-doped polymer materials is possible by sol-gel technology [109-111], the creation of mutually penetrating polymer networks [112], the covalently linking dyes to macromolecules [2] and the introduction of various types of stabilizers [113]. Among the latter, quenching agents [113] which make it possible substantially to increase the stability of colored polymers, are of great interest.

7. SOLAR CELLS

The present organic photoelectric cells have a low efficiency of light conversion in near IR spectral region. For example, one of the most effective organic solar cells based on C_{60}/polyphenylenevinylene (efficiency > 3%) do not absorb solar light in the range of maximal solar photons flux, namely, near IR region [114]. The most of organic compounds have long wavelength edge of absorption at < 600 nm [35,115]. This is one of the reasons of relatively small integral efficiency of photo-devices based on organic materials, such as photovoltaic elements, solar cells etc. The best of the now a days developed organic solar cells transform illumination at 400–600 nm spectral range only [114].

Therefore, the development of flexible organic layers, photosensitive in near IR region close to the region with maximal solar photon flux (750–850 nm), which can be used as components of organic photodevices, including solar cells, is realistic. To solve this problem, polymethine dyes can be selected [4]. These dyes have maximum of the π-π^* absorption in the 400-1600 nm spectral range. Their absorption and luminescence can be shifted over a wide spectral range by the modification of the structure of heterocyclic moieties and the length of polymethine chains. Besides, by modifying the structure of various fragments of dyes through synthetic methods, high solubility has been achieved both in the polar and in weakly polar organic solvents. Furthermore, π-π^* absorption bands generally are additionally red-shifted and broadened for films and polymer composites [4].

The results of the study on the films and polymer composites based on **HITC** (dye **3**) incorporated in polymer films of photoconducting PEPK and nonphotoconducting PVE, were presented in reference 116. This dye was selected because of its stability and good solubility in various media as well as the position of the maximum of its absorbance close to the maximum of the solar photon flux (725-775 nm) [114].

Dyed films were deposited from DCE solution by the spin-coating technique. The films obtained were 1-2 mm thick with different concentrations of **HITC** in PEPK ranging from 1 wt% to 50 wt%. The film of 50 wt% HITC in PVE was obtained to compare the photosensitivity of composites based on different classes of polymers.

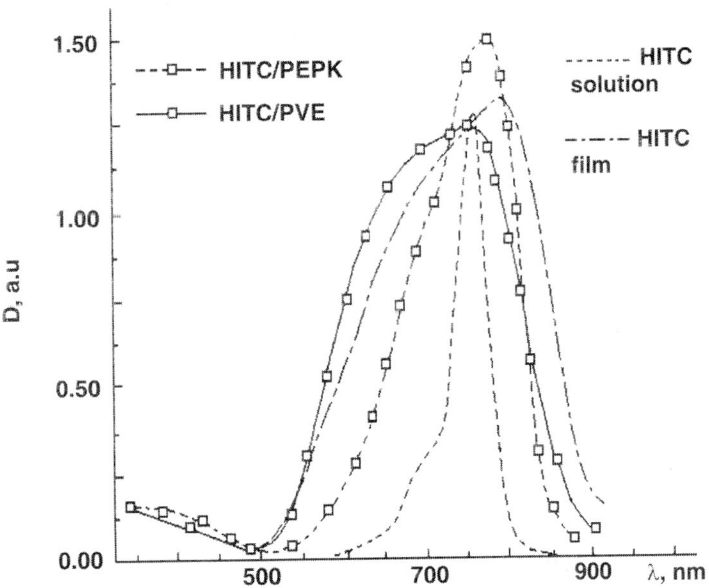

Fig. 9.16. Absorption spectra of HITC dye in different states. D-optical density. Reproduced with permission from Ishchenko *et al.*, Mater. Sci., 20(4), 5 (2002). Copyright © 2002 Oficyna Wydawnicza Politechniki Wrocławskiej.

Fig. 9.16 shows absorption spectra of the **HITC** solution in DCE, **HITC** solid film as well as dye-in-polymer composites in PEPK and PVE (30 wt.% of dye concentration). It may be seen that the absorption spectrum of HITC solution in DCE has the shape, typical of polymethine dyes, the main band with a vibronic maximum at the short-wavelength edge. When applying a polymer matrix, a red shift in the dye absorption band has been observed both in photoconducting PEPK and in non-photoconducting PVE. The feature can be explained by the increase in the dispersion interactions due to a higher refraction index of polymers as compared to DCE. The broadening of the absorption band, observed in the polymer matrices, is caused by a nucleophilic solvation of positively charged centers of **HITC** cations by polar groups of PEPK and PVE. In **Fig. 9.17**, the photovoltage spectra of **HITC** composite films in PEPK and PVE are shown. Their shapes generally correspond to the respective absorption spectra. Some minor changes in the intensity and bands position are observed.

It is obvious that the photosensitivity of the samples containing **HITC** in a photoconducting polymer (PEPK) is greater than that of the film of the dye in a nonphotoconducting polymer matrix (PVE). This fact testifies that not only the dye concentration influences the efficiency of charge generation but the type of polymer also. Charge transfer between the dye and polymer molecules is much more effective in the PEPK composite than in the PVE based film. This must result in the rise of the number of generated charge carriers.

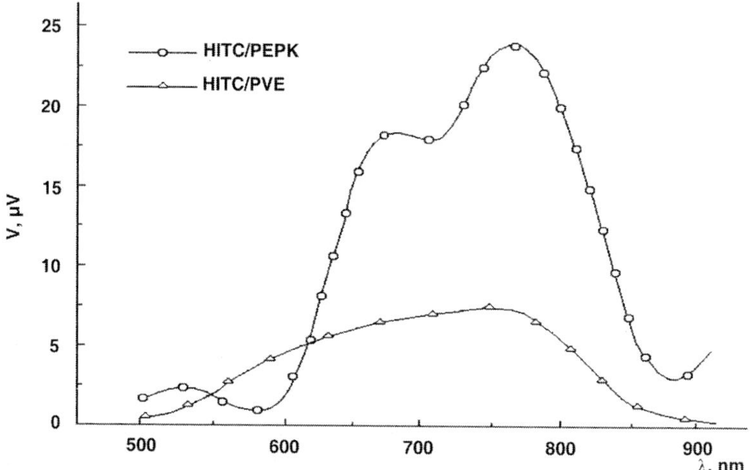

Fig. 9.17. Photovoltage spectra of HITC composite films. Reproduced with permission from Ishchenko *et al.*, Mater. Sci., 20(4), 5 (2002). Copyright © 2002 Oficyna Wydawnicza Politechniki Wroclawskiej.

In order to clarify processes in the **HITC**:PEPK composite in which more effective photogeneration takes place, the studies were conducted on the concentration dependence of the absorbance and PL under different conditions and using different sources of excitation [116].

The positions of maxima are weakly dependent on the dye concentration in the 1-20 wt% range. An additional broadening of the absorption bands occurs on increasing the dye concentration due to the formation of different types of ionic pairs by salt-like dyes in macromolecular compounds [4]. When the concentration of the dye is raised, the fraction of contact ionic pairs increases. A counter-ion on which a maximal positive charge is concentrated and being localized in the region of one of the heterocycles, destroys the electronic symmetry of the dye molecule [4]. This leads to the amplification of vibronic interaction and consequently to the band broadening.

The above-mentioned effect is maximal in solid films of dye. Not only a significant broadening of the absorption band but also a strong deformation of its shape is observed (**Fig. 9.16**). Besides, an intermolecular interaction in the dye:dye system also influences the band shape in the film due to short distances between the dye molecules.

PL spectra of **HITC,** both in the solution and in the polymer matrix, are mirror-similar to the absorption spectra [116]. The shape of PL bands remains the same for excitation in the region of the first (low-energy) electronic transition (λ_{ex} = 651 nm) as well as in the region of the second (high-energy) one (λ_{ex} = 337 nm). In the latter case, the PL intensity is weaker since the excitation energy is dissipated in the radiationless S_2–S_1 transition.

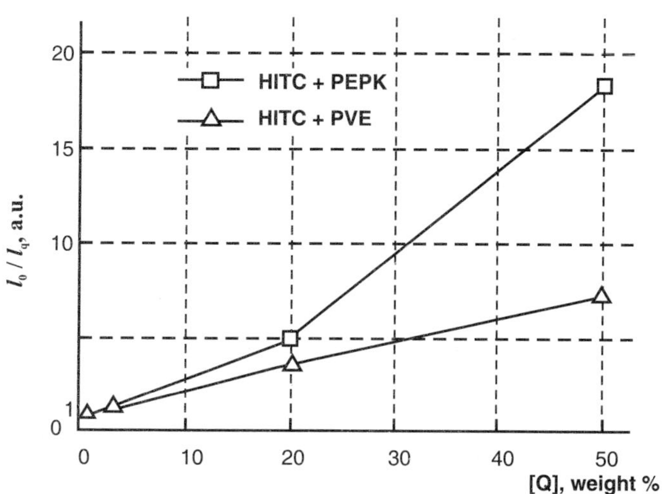

Fig. 9.18. Photoluminescence quenching plot in Stern–Volmer coordinates. I_0 and I_q is the photoluminescent intensity without a quencher (Q) and with a quencher respectively. [Q] is the quencher concentration in mol/ 1000 g of polymer. Reproduced with permission from Studzinsky *et al.*, Nonlinear Optics, Quantum Optics, 33(1-2), 151 (2005). Copyright © 2005 Old City Publishing.

A strong PL quenching occurs with the concentration rise. In **Fig. 9.18**, the Stern-Volmer plot of PL intensity dependence on **HITC** concentration is presented. It may be seen from this figure that the dependence of the PL quenching on the concentration of the dye in PEPK cannot be satisfactorily described by the linear Stern-Volmer equation **(eq. 9.18)** for the system with one luminophore and one quencher.

$$I_0/I_q = 1 + K[Q] \qquad (9.18)$$

where I_0 and I_q are the PL intensity without a quencher (Q) and with a quencher respectively, [Q] is the quencher concentration in mol/1000 g of polymer and K is the Stern-Volmer constant. Contrary to this, a linear dependence takes place in the case **HITC**-doped PVE film **(Fig. 9.18)**.

The dependence of the PL quenching on the concentration **HITC** in PEPK can be fitted in the quadratic Stern-Volmer equation, **eq. 9.19**.

$$I_0/I_q = (1 + K[Q])(1 + K'[Q]) \qquad (9.19)$$

where K and K' are the Stern-Volmer constants for different quenching processes. The fitting analysis gives the following values for these constants (K = 2.25 and K' = 9.39 in kg/mol).

Consequently, PL quenching is connected with two different mechanisms of quenching in photoconducting polymer PEPK and only with single mechanism in nonphotoconducting PVE. Concentration quenching caused by the migration of energy between dye molecules during concentration increase takes place in both the cases. The second most probable mechanism is of electron phototransfer from PEPK to the cation of the dye. PEPK should form $Cz^{\bullet+}$ cation-radical and the cation of **HITC** can form a neutral radical $Ct^{\bullet}$ **(eq. 9.8)**. Since the dye molecules at high concentrations can exist as contact anion-radical pairs $Ct^{\bullet}An^{-}$, thus formed radical-ions, $Cz^{\bullet+}$ and $Ct^{\bullet}An^{-}$, are charge carriers in the PEPK photoconducting matrix doped with **HITC** dye. Their presence explains the photovoltaic properties of **HITC:PEPK** composites. The fact that photosensitivity of the composite based on non-photoconducting PVE is much smaller than that of composites based on photoconducting PEPK **(Fig. 9.18)**, is in agreement with this model.

It is revealed that the value of photovoltage in **HITC**-doped polymer films increases on replacement of the matrices based on PVK by the iodine-containing co-polymer of PVK [117]. It is caused by the enhancement of intersystem crossing in iodine containing polymers due to the effect of internal heavy atom. As a result, the generation of charge carriers not only from singlet but also from long-lived triplet electron-hole pairs can also be observed [118]. This leads to the increase in the concentration of such charge carriers and, therefore, to the increase in the bulk photovoltage.

Thus, flexible composite films of **HITC** in photoconductive polymers have noticeable photosensitivity in the region close to the maximum of solar photon flux. Therefore, such films may be used in perspective as components of organic photosensitive devices including solar cells. In carbazol-containing polymer films and cationic polymethine dyes particularly **HITC**, behave as hole photogeneration centers [119]. The less mobile electrons that remain in the cationic dye molecules after photogeneration of the electron-hole pairs, are efficient capture and recombination centers for the charge carriers. In the films with ionic dyes, the electric charge of the colorless counterions influences the photogeneration of the mobile charge carriers. At the first step of photogeneration, the photogenerated carriers are concentrated near these counterions. The electrostatic interaction in these charge pairs determines the activation energy of the photogeneration and the electrical conductivity.

The intra-ionic dyes, for example, the synthesized merocyanine **HITB5** (dye **37**) in work reported in reference 119, exhibits positive solvatochromism. Thus, its ground state is mainly described by the neutral canonical structure while the excited state is described by a set of bipolar structures of types. Therefore, these merocyanines can be regarded as photogeneration centers of both holes and electrons. The photoconductivity increases upon replacement of a cationic dye **HITC** by an intra-ionic merocyanine **HITB5**, mainly because of photogeneration of mobile charge carriers of both the signs and a decrease in the activation energy for the photoconduction current. The decrease in the

Dye **37**. **HITB5**

activation energy for the photocurrent is due to the mobile charge carriers move away from each other during their separation in the case of an intra-ionic dye while the colorless counterion strongly holds the photogenerated charge carrier in the case of a cationic dye.

Particular attention should be placed on the extensive color and the high absorption intensity of the merocyanine **HITB5**, which is unusual for this type of dyes. The point is that merocyanines which are electron-asymmetric dyes, contain alternating single and double bonds in the polymethine chain [20] unlike symmetrical polymethines, for example **HITC** in which these bonds are largely equalized [4]. This bond alternation, similar to that in polyenes, induces a hypochromic and hypsochromic effects in the absorption spectra. Therefore, the absorption bands of merocyanines are located at shorter wavelengths than these bands for the corresponding symmetrical (parent) dyes with the same length of the polymethine chain [20,21]. As the chain length increases, the degree of alternation is also enhanced resulting in the attenuation of the vinylene shifts in merocyanines as opposed to symmetrical cyanines [4,43,115]. The merocyanine **HITB5** absorbs virtually in the same region as the cationic dye **HITC**. This substantial color deepening was attained by introducing five-membered rings into the even positions of the chromophore resulting in the bathochromic shift of the bands and the hyperchromic effect [4] and by using the thiobarbituric acid residue with a greater effective length of the chromophore as the terminal group [115].

Photoconducting polymer doped by the merocyanines of **HITB5** type are the prospective materials for photovoltaic device of near IR range due to their capacity to generate mobile carriers, both the holes and the electrons, as well as the deep color and high intensity of light absorption.

Merocyanines are the intramolecular donor-acceptor systems in which the charge transfer is realized on the chain of conjugate links. Such systems are more efficient in comparison to similar systems based on blends of the donor and acceptor. It is caused by the fact that first has considerably higher light absorption intensity than last. Recently, the interesting hybrid compound **38** with chemically linking a hole-conducting oligophenylenevinylene moiety to an electron-conducting fullerene C_{60} has been synthesized [120]. This compound represents a polymer in which the donor oligomer moiety and acceptor C_{60} are conjugated between themselves. The polymer film based on

38. Hybrid compound

hybrid compound **38** is ready by the photovoltaic element is located between two contacts of the solar battery. Such films allow avoiding any problems arising from bad contacts at the junction which take place in blends of donor and acceptor [120]. The interface of metal-organic junction is one of the important key factors in the development of the efficient solar cells.

8. ELECTROLUMINESCENT MEDIA

EL devices or light-emitting diodes (LEDs) have attracted much attention because of their high brightness, low drive voltage and a variety of emission colors [8,11,12,121]. The possibility to produce large-area display with cheap and well-established technique as well as a wide selection of emission colors, particularly in the blue region, through the molecular design of organic molecules is the main advantage of organic LEDs against those based on inorganic semiconductors. In EL devices, electric energy is transformed into light through the excitation of the organic molecules. The excitation processes involve the injection of holes and electrons from the electrodes to the organic layer and the recombination of these carriers generate excited molecules. Either low molecular weight materials or high molecular weight polymeric materials can be used as long as a material possesses appropriate carrier transport properties [8,11,12].

Developing new materials based on organic polymer films for EL devices is a currently important task because of high practical demands and considerable problems encountered in the use of existing materials [8,11]. In order to reproduce the EL effect, it is necessary to satisfy several requirements providing for the effective injection of charge carriers from electric contacts into an organic polymer film, the transport of carriers to the recombination centers in the bulk of polymer and the recombination of carriers accompanied by the emission of light quanta. The main problems in meeting these requirements are related to selecting the proper contact materials and providing conditions for the effective transport of carriers without trapping in the polymer film bulk and the effective radiative recombination.

It was shown that the recombination of carbazole cation-radical of PVK or PEPK and neutral radical dye, for example dye **1**, in electric field could be accompanied by EL [122]. The EL spectrum coincides with the PL spectrum of this dye [122]. It should be pointed out that this is **eq. 9.8** realized. The injection properties of this dye-doped polymer depend on the nature of the contacts and the dye concentration [123,124]. It was shown that the dye-doped PEPK possesses the effect of memory, namely, if primarily to irradiate a sample by visible light, the EL may appear even some hours after the electric field is switched on [122]. This effect is also caused by the ability of PEPK to form deep traps, which delay the transport of charge carriers before the operation of electric field. EL spectra for derivatives of **thiacarbocyanine10** are also observed in aromatic polyimides [125].

The results of investigations directed to the creation and characterization of new EL materials and compositions, in particular, the films of PEPK doped with a boron difluoride organic dye **39** complex was presented in work reported in reference 126. The experiments were performed on the

dye **39**

samples of sandwich structures comprising a glass substrate, a transparent conducting $SnO_2{:}In_2O_3$ (ITO) film, a dyed polymer layer (PEPK + N mass % of dye **39**) and an aluminum electrode. The dye concentration N was varied from 1 to 70 mass %. The PEPK + N mass % of dye **39** films were obtained by casting from chloroform solutions containing the corresponding amounts of PEPK and dye **39** onto the ITO-coated glass substrates followed by drying in a thermal box for 3 h at 80°C. The polymer film thickness was varied from 0.5 to 5 µm. The metal contacts were applied onto the polymer film by metal deposition under vacuum.

A complicated technology is required for the fabrication of well-known sandwich heterostructures based EL devices because they employ separate transport and emitting polymer layers and magnesium-silver cathodes [8,11,12]. The samples of the sandwich structure ITO:PEPK:Al with a dye-free PEPK layer possess low electrical conductivity, do not show any EL effect and exhibit optical absorption in the region of $\lambda < 400$ nm. In contrast, the heterostructures ITO:(PEPK + N mass % of dye **39**):Al are characterized by a significant electrical conductivity which increases with the increase in dye content (N). The optical absorption in the visible spectral range is determined by the electronic absorption of **39** showing a pronounced vibrational structure. When the dye **39** content increases above 1 mass %, the vibrational structure exhibits considerable smoothening while the absorption band broadens and shifts toward longer wavelengths. As the dye concentration N varies from 1 to 50 mass %, the wavelength corresponding to maximum absorption changes from 530 nm to 550 nm while the fluorescence peak shifts from 612 nm to 638 nm. This is the evidence of the molecular interactions arising in the polymer:dye system with aggregation of dye molecules in the limiting case [4]. An increase in the dye **39** content in the ITO:(PEPK + N mass % of dye **39**):Al structure leads to the appearance of EL. The PL and EL spectra coincide. The EL brightness can reach upto several tens of cd/m^2, which makes the emission detectable with naked eye in the usual room lighting. The experimental results presented above can be interpreted within the framework of the following model.

The electron injected from contact, occupied HOMO of a molecule **39**, forming an anion-radical **39a**. The hole injected from other contact, arises at the expense of a deficit of an electron

39a

39b

on HOMO of a carbazole fragment of PEPK, transforming it into a cation-radical Cz$^{\bullet+}$. The injected electrons and holes in volume of a polymer film under operation of an external electric field move towards each other. In case of large concentration of the moved charge carriers their occurring is rather probable. Electroneutral the singletly-excited molecule of dye **39** will be formed from two of opposite charged radicals. In it, the electron is found in LUMO and the deficit of one electron takes place on HOMO. If the relaxation of excited state happens on the radiant channel, energy of a radiated light quantum corresponds to energy of luminescent transition of dye **39**. Besides, it is possible to assume at high concentration of dye **39** that the additional contribution to derivation of its luminescent singlet state is deposited by recombination of an anion-radical **39a** and cation-radical **39b**, formed at the expense of electron recoil on the positively charged electrode. El of sandwich structures, PEPK + N mass % of dye **39,** is defined by the features of chemical constitution of the dye **39**. These features promote spin conversion and uncorrelated pairs of the injected charges with their subsequent radiant recombination. The efficiency of spin conversion is connected to the presence of the atom of boron in the structure of the dye **39**. In the region of this atom spin density of a unpaired electron and greatest negative charge in an anion-radical **39a** is localized. Such anion-radical included in composition contact electron-hole pair, differs from anions-radicals with delocalized spin and electronic density of a unpaired electron by the fact that it in the greater degree promotes to an increase of probability of charges recombination. It is connected to several reasons. At first, the point charge on small distances has larger energy of electrostatic interaction with a charge of the opposite sign than a delocalized charge. It means, that the charged atom of boron in an anion-radical **39a** promotes capture and deduction of an hole, drifting by it. Secondly, the boron atom contains the magnetoactive nucleus B^{11} with a magnetic moment of μ = 2.6885. It is coordinated with two fluorine atoms with magnetoactive nucleus F^{19} of μ = 2.6288 in structure of the dye **39**. On small distances between an anion-radical **39a** and cation-radical $Cz^{\bullet+}$, the change of spin state ion-radical pairs is possible not only at the expense of fine and hyperfine interaction of electrons with protons and nucleus of nitrogen atoms but also because of a hyperfine interaction with boron and fluorine nucleus.

It is not always necessary to create materials based on photoconducting polymers and dyes for EL devices. It is possible to use for these purposes nonphotoconducting polymers, containing bifluorophores or a mixture of luminophores, which can participate in the generation and recombination of charges. For example, the mechanical mixture of *N,N'*-diphenyl-*N,N'*-bis(3-methylphenyl)-1,1'-biphenyl-4,4'-diamine (TPD) and tris(8-quinolinolyato)aluminum (Alq) in PMMA [127]. TPD has a high hole drift mobility. Alq is a luminescent metal complex possessing electron transport properties and has been used as a luminescent layer in EL cells. PMMA is an optically and electronically inert polymer and has good film forming properties with a high glass transition temperature of 105°C [127]. Electrons and holes are transported in the polymer layer through Alq and TPD as shown in **eqs. 9.20 and 9.21**.

$$(Alq^-)^\bullet + Alq \rightarrow Alq + (Alq^-)^\bullet \tag{9.20}$$

$$(TPD^+)^\bullet + TPD \rightarrow TPD + (TPD^+)^\bullet \tag{9.21}$$

The green emission with EL of 920 cd/m^2 was achieved at a drive voltage of 17 V in the above-mentioned system [127]. Two mechanisms for the excitation of Alq are possible. One is the direct excitation of Alq by the reaction between a radical anion of Alq and a radical cation of TPD as given by **eq. 9.22**.

$$(Alq^-)^\bullet + (TPD^+)^\bullet \rightarrow Alq^* + TPD \tag{9.22}$$

The other mechanism involves the excitation of TPD by the reaction of the two charge carriers. It follows that the excited energy is transferred to an adjacent Alq molecule via Forster-type energy transfer as shown in **eqs. 9.23 and 9.24**.

$$(Alq^-)^\bullet + (TPD^+)^\bullet \rightarrow Alq + TPD^\bullet \tag{9.23}$$

$$Alq + TPD^\bullet \rightarrow Alq^* + TPD \tag{9.24}$$

In the PL of PMMA containing both the Alq and TPD, the latter mechanism of energy transfer from TPD to Alq cannot be excluded because the luminescence from Alq was observed by the excitation of TPD [127].

Other organic compounds can also carry out functions of Alq and TPD. For example, the combination of strong acceptors of fullerenes and unique converters of light energy of polymethine dyes should have major prospectives [128]. It should be noted that the molecules of the donor and acceptor could efficiently activate luminescence not only when they lie in the same polymer layer but also when the donor and acceptor are in different layers namely: acceptor in electron transport and emission layer and donor in hole transport layer [129].

9. CONCLUSIONS AND FUTURE PROSPECTS

The dye-doped polymers possess powerful potential of photophysical and photochemical properties, which are of interest for modern photo-converters. It is possible to control these properties over a wide range by changing the structure of dye and polymer, both photoconducting and non-photoconducting. Both thin-film materials and bulky products based on dye-doped polymers can easily be fabricated.

The emerging field of nanotechnology represents new steps to exploit new materials as well as new technologies in the development of efficient and low-cost electronic and photonic devices. Significant efforts have been made in recent years to investigate the photophysical and photochemical behavior of multicomponent nanostructured assemblies consisting of metals, semiconductors and photoactive dyes. The inorganic-organic hybrid assemblies are of particular interest because of their good conductivity and possibility to change their electronic properties by sized effect [130-132]. Gels of complex metal oxides or metals, in particular of SiO_2 [111], TiO_2 [130], ZrO_2 [110], V_2O_5 [131,133], Au [132] etc. are prospective materials for the use as component of multilayered or composite structures for different photonic applications. Engineering of the nanocluster surfaces with photoactive molecules can provide three-dimensional molecular arrangements around the metal containing nanoparticles. The composite materials based on dye-doped polymers and gels of metal oxides or metals are of special interest for developing efficient light energy conversion systems, optical devices and sensors.

Abbreviations

ALM (active laser medium), Alq (tris(8-quinolinolyato)aluminum), An (anion), **BDN** (bis(1-(4-dimethylaminophenyl)-2-phenyl-1,2-ethene dithiolato)nickel), Ct (cation), Cz (carbazole), D (optical density), DCE (1,2-dichloroethane), EL (electroluminescence), **HIC** (1,1',3,3,3',3'–hexamethylindocarbocyanine), HIDC (1,1',3,3,3',3'-hexamethylindodicarbocyanine), **HISD** (1,1',3,3,3',3'-hexamethylindosquaraine), **HITB5** (5-{3-[(1,3,3-Trimethyl-1,3-dihydro-2*H*_indol-2-yl-iden)ethyliden]-2-phenylcyclopent-1-en-1-ylmethylene}-2-thioxodihydropyrimidine-4,6(1*H*,5*H*)-dione), **HITC** (1,1',3,3,3',3'-hexamethylindotricarbocyanine), HOMO (highest occupied molecular orbital), ITO (SnO_2:In_2O_3), LSC (luminescent solar concentrator), LUMO (lowest occupied molecular orbital), MC (methylene chloride), OAGE (anthracenyl glycidyl ether), PE (polyethylene), PEPK (poly-N-epoxypropylcarbazole), PL (photoluminescence), PLS (passive laser switch), PMMA (polymethyl methacrylate), **PM597** (pyrromethene597), PS (polystyrene), PU (polyurethane), PUA (polyurethane acrylate), PVAL (polyvinyl alcohol), PVB (polyvinylbutyral), PVE (polyvinylethylal), PVK (polyvinylcarbazole), RDT (radiation damage threshold), **R6G** (Rodamine 6G), SS (Stokes shift), **TPD** (N,N'-diphenyl-N,N'-bis(3-methylphenyl)-1,1$'$-biphenyl-4,4$'$-diamine) and YAG (yttrium-aluminium-granat)

REFERENCES

1. A.A. Ishchenko; Theor. Exp. Chem., 34(4), 191 (1998).
2. A.A. Ishchenko; Polym. Adv. Technol., 13 (10-12), 744 (2002).
3. A.A. Ishchenko; Russ.Chemical Reviews., 60(8), 865 (1991).
4. A.A. Ishchenko; Structure and spectral, luminescent properties of polymethine dyes [in Russian], Naukova Dumka, Kiev (1994).
5. A.I. Tolmachev, Yu. L. Slominskii and A.A. Ishchenko; in NATO ASI Series.3. (Eds. S. Daehne, U. Resch-Genger and O.S. Wolfbeis), Vol.52, p. 385-415, Kluwer Academic Publishers, Dordrecht-Boston-London (1998).
6. A. Mishra, R.K. Behera, P.K. Behera, B.K. Mishra and G.B. Behera; Chem. Rev., 100(6), 1973 (2000).
7. N.A. Davidenko and A.A. Ishchenko; Theor. Exp. Chem., 38(2), 88 (2002).
8. P. Strohriegl and J.V. Grasulevicius; Adv. Mater., 14(20), 1439 (2002).
9. A.A. Shchenko; Quantum Electronics, 24(6), 471 (1994).
10. F.J. Duarte; in Tunable Lasers Handbook (ed. F.J. Duarte), p. 167, Academic Press, San Diego-New York-Boston-London-Sydney-Tokyo-Toronto (1995).
11. U. Mitschke and P. Bauerle; J. Mater. Chem., 10(8), 1471 (2000).
12. A.V. Kukhta; J. Appl. Spectroscopy, 70(2), 152 (2003).
13. C.J. Brabec and N.S. Sariciftci; Matirials Today, 3(2), 5 (2000).
14. V.D. Pokhodenko and N.F. Guba; Theor. Exp. Chem., 30(5), 241 (1994).
15. N. Derevyanko, A. Ishchenko and A. Verbitsky; Mater. Sci. 20(4), 13 (2002).
16. A.M. Bonch-Bruevich, E.N. Kaliteevskaya, V.P. Krutyakova and T.K. Razumova; Izvestiya Akademii Nauk (Russia). Ser. Fizicheskaya, 65(4), 478 (2001).
17. V.S. Manzhara, R.D. Fedorovich, A.B. Verbitsky, A.V. Kulinich and A.A. Ishchenko; Mol.Cryst.Liq. Cryst., 426, 303 (2005).
18. K.M. Dyumaev, A.A. Manenkov, A.P. Maslyukov, G.N. Matyushin, V.S. Nechitailo and A.M. Prokhorov; Interaction of laser radiation with optical polymers[in Russian]; Nauka, Moscow (1991).
19. V.I. Bezrodnyi and A.A. Ishchenko; Optics and Laser Technol., 34(1), 7 (2002).
20. T.M. Murav'eva, A.A. Ishchenko, I.E. Gorokhova, Zhurnal. Prikladnoi. Khimii., 64(12), 2520 (1991).
21. S. Chatterjee, P. Gottschalk, P.D. Davis and G.B. Schuster; J. Am. Chem. Soc, 110(7), 2326 (1988).
22. G.B. Schuster; Pure Appl. Chem., 62(8), 1565 (1990).
23. N. Tyutyulkov, J. Fabian, A. Mehlhorn, F. Dietz and A. Tadjer; Polymethine Dyes. Structure and Properties, St. Kliment Ohridski University Press, Sofia (1991).
24. I.V. Komarov, A.V. Turov, A.A. Ishchenko, N.A. Derevyanko and M. Yu. Kornilov; Dokl. Akad. Nauk SSSR, 306(5), 1134 (1989).
25. A.A. Ishchenko, N.A. Derevyanko and A.M. Vinogradov; Zh. Obshch. Khim., 67(7), 1191 (1997).
26. A.S. Tatikolov, L A. Shvedova, Kh. S. Dzhulibekov, Zh. A. Krasnaya, A.L. Sigan and V.A. Kuz'min; Izv. Akad. Nauk (Russia). Ser. Khim., 878 (1995).
27. A.K. Chibisov, G.V. Zakharova, V.L. Shapovalov, A.I. Tolmachev, Yu. L. Bricks and Yu. L. Slominskii; Khim. Vys. Energ., 29(3), 211 (1995).
28. H.-H. Wang and I.-S. Tzun; J. Appl. Polym. Sci., 73, 245 (1999).
29. 29.R.E.F.Boto, A.S. Oliveira, L.F. Vieira Ferreira and P.A. Almeida; Dyes and Pigments, 48(2), 71 (2001).

30. A.A. Ishchenko, A,F. Dokukina, Z.A. Smirnova and A.I. Tolmachev, Dokl.Akad.Nauk SSSR, 284(6), 1407 (1985).
31. A.A. Ishchenko, F.G. Kramarenko, A.G. Maydannic, S.V. Sereda and N.P. Vasilenko; J. Inf. Rec. Mater, 19(3), 207 (1991).
32. A.J. Gordon and R.A. Ford; The Chemists Companion, Wiley, New York (1972).
33. A.A. Ishchenko, Yu. L. Slominskii, A.I., M.I. Demchuk, V.P. Mikhailov and K.V. Yumashev; Optika i Spektroskopiya., 64(3), 650 (1988).
34. Yu. A. Skryshevski, N.A. Davidenko, A.A. Ishchenko, A.K. Kadashchuk and N.I. Ostapenko; Optics and Spectroscopy, 88(3), 352 (2000).
35. A. Kakuta; in Infrared Absorbing Dyes (Ed. M. Matsuoka), Plenum Press, New York and London (1990).
36. V.D. Filimonov and E.E. Sirotkina; Chemistry of Monomers based on Carbazole [in Russian], Nauka, Novosibirsk (1995).
37. B. M. Krasovitskii and B. M. Bolotin; Organic Luminescent Materials, VCH Verlagsgesellschaft GmbH, Weinheim (1988).
38. A.A. Ishchenko, I.L. Mushkalo amd N.A. Derevyanko; J. Inf. Rec. Mater., 17(1), 39 (1989).
39. N. Nizamov; Luminescence of Associated Molecules of Organic Dyes in Solutions and Films[in Russian], Zarafshan, Samarkand (1997).
40. Y. Kobayashi, M. Goto, and M. Kurahashi; Bull. Chem. Soc. Jpn., 59, 311 (1986).
41. V.I. Gadyalis, V. Yu. Krishchyunas and B.P. Krishchyunene; Litov. Fiz. Sb., 20(4), 69 (1980).
42. G.J. Ashwell, P. Leeson and G.S. Bahra; J. Opt. Soc. Am. B: Opt. Phys., 15, 484 (1998).
43. G. Bach and S. Daehne; in Rodd's Chemistry of Carbon Compounds (Ed. M.Sainsbury), Vol. IVB, p. 383, Elsevier Science, Amsterdam (1997).
44. Yu. P. Getmanchuk, I.N. Itskovskaya, D.G. Usov and N.I. Sokolov; Teor. Eksp. Khim., 37(1), 8 (2001).
45. Yu. P. Getmanchuk, N.A. Davidenko, N.A. Derevyanko, A.A. Ishchenko, L.I. Kostenko, N.G. Kuvshinskii, S.L. Studzinskii and V.G. Syromyatnikov; Polym. Sci., Ser.A., 44(8), 855 (2002).
46. J.M. Pochan, D.F. Hinman and R. Nash; J. Appl. Phys., 46(10), 4115 (1975).
47. A.V. Vannikov and A.D. Grishina; Photochemistry of Polymer Donor–Acceptor Complexes [in Russian], Nauka, Moscow (1984).
48. Ch. Reichardt; Solvents and Solvent Effects in Organic Chemistry, WILEY-VCH Verlag GmbH & Co. KGaA, Weinheim (2003).
49. M.I. Demchuk, A.A. Ishchenko, V.P .Mikhailov, and V.I. Avdeeva; Chem. Phys. Lett., 144(1), 99 (1988).
50. N.A. Davidenko, A.A. Ishchenko, A.K. Kadashchuk, N.G. Kuvshinskii, N.I. Ostapenko and Yu. A. Skryshevskii; Fiz. Tverd. Tela, 41(2), 203 (1999).
51. W. Klopffer; J. Chem. Phys., 50, 2337 (1969).
52. J.M. Pochan, D.F. Hinman and R. Nash; J. Appl. Phys., 46(10), 4115 (1975).
53. N.A. Davidenko and N.G. Kuvshinsky; J. Inf. Rec. Mater., 21, 185 (1993).
54. V.S. Kovalenko; Processing of materials by pulsed laser radiation [in Russian], Vysshaya Shkola, Kiev (1977).
55. D. Maydan and R.B. Chesler; J. Appl. Phys., 42(3), 1031 (1971).
56. T.T. Basiev, A.N. Kravets, S.B. Mirov, A.V. Fedin, VA. Konyushkin; Sov. J. Quantum Electron., 18(2), 223 (1991).
57. T.T. Basiev, A.N. Kravets, S.B. Mirov and A.V. Fedin; Sov. J. Quantum Electron., 19(7), 713 (1992).
58. T.T. Basiev, A.N. Kravets and A.V. Fedin; Sov. J. Quantum Electron., 20(6), 594 (1993).
59. N.N. Il'ichev, N.P. Kushch and A.A. Malyutin; Sov. J. Quantum Electron., 12(8), 1721 (1985).
60. I.V. Klimov, M. Yu. Nikol'skii, V.B. Tsvetkov and I.A. Shcherbakov; Sov. J. Quantum Electron., 19(7), 653 (1992).

61. Y. Shimony, Z. Burshtein, A. Ben-Amar Baranga, Y. Kalinsky and M. Strauss; IEEE J. Quantum Electron., 32(2), 305 (1996).

62. S.F. Sharlai; Zh. Prikl. Spektrosk., 13(4), 730 (1970).

63. V.I. Bezrodnyi and E.A. Tikhonov; Sov. J. Quantum Electron., 13(12), 2486 (1986).

64. V.I. Bezrodnyi and E.A. Tikhonov; Zh. Prikl. Spektrosk., 50(2), 318 (1989).

65. M.I. Demchuk, V.P. Mikhailov, V.S. Pavlovich, K.V. Yumashev, A.A. Ishchenko, E.P. Eremeeva, Z.A. Smirnova, A.I. Tolmachev; Khim. Fiz. 5(9), 1184 (1986).

66. Yu. V. Korobkin, O.I. Sidorov, V.B. Studenov; Pis'ma Zh. Tekh. Fiz., 18(16), 38 (1992).

67. V.I. Bezrodnyi, E.A. Tikhonov and N. Ya. Nedbaev; Sov. J. Quantum Electron., 13(6), 1214 (1986).

68. K.M. Dyumaev, A.A. Manenkov, A.P. Maslyukov, G.A. Matyushin, V.S. Nechitailo and A.M. Prokhorov; Izv. Akad. Nauk. SSSR. Ser. Fiz., 49(6), 1084 (1985).

69. K.M. Dyumaev, A.A. Manenkov, A.P. Maslyukov, G.A, Matyushin, V.S. Nechitailo and A.M.Prokhorov; Izv. Akad. Nauk. SSSR. Ser. Fiz., 51(8), 1387 (1987).

70. A.F. Dokukina, E.P. Eremeeva, T.F.I vanova, L.I. Kotova, M.A. Kudinova, B.D. Piterkin, Z.A. Smirnova and A.I. Tolmachev; Zh. Prikl. Spektr., 39(6), 1003 (1983).

71. M.I. Demchuk, V.P. Mikhailov, K.V. Yumashev, E.P. Eremeeva, A.A. Ishchenko, V.S. Pavlovich, Z.A. Smirnova and A.I. Tolmachev; Zh. Tekh. Fiz., 55(6), 1218 (1985).

72. A.I. Gibelev, A.F. Dokukina, E.P. Eremeeva, E.D. Isyanova, V.M. Ovchinnikov and Z.A. Smirnova; Pis'ma Zh.Tekh. Fiz., 10(3), 157 (1984).

73. V.I. Bezrodnyi, A.A. Ishchenko, L.V. Karabanova and Yu. L. Slominskii; Quantum Electronics, 25(8), 819 (1995).

74. V.I. Bezrodnyi, N.A. Derevyanko and A.A. Ishchenko; Quantum Electronics, 26(4), 345 (1996).

75. R. Grigonis, M. Eidenas, V. Sirutkaitis, V. Bezrodnyi. A. Ishchenko and Yu. Slominskii; Ultrafast Processes in Spectroscopy, 449 (1996).

76. R.A. Ganeev, K.R. Samigullin, L.A. Begishev, S.R. Kamalov, S.I. Matafonov, V.I. Redkorechev and T. Usmanov; Quantum Electronics, 27(4), 312 (1997).

77. V.I. Bezrodnyi, N.A. Derevyanko, A.A. Ishchenko and Yu. L. Slominskii; Quantum Electronics, 25(8), 823 (1995).

78. N.S. Ustimenko and A.V. Gulin; Pribory i Tekhnika Experimenta, 99 (1998).

79. E.M. Dianov, V.I. Karpov, A.S. Kurkov and V.N. Protopopov; Kvantovaya Elektronika, 22(12), 1059 (1996).

80. V.A. Berezovskii and N.N. Kolotilov; Biophysical Characteristics of Human Tissues [in Russian], Naukova Dumka, Kiev (1990).

81. E.B. Anikina, E.I. Shapiro and G.L. Gubkina; Vestnik Oftal'mol., 110(3), 17 (1994).

82. D.A. Gromov, K.M. Dyumaev, A.A. Manenkov, A.P. Maslyukov, G.A. Matyushin, V.S. Nechitailo and A.M. ProkhorovIzv; Akad. Nauk. SSSR. Ser. Fiz., 48, 1384 (1984).

83. L.K. Denisov, I.G. Kytina, V.G.Kytin, S.A. Tsogoeva, L.G. Saprykin and B.A. Konstantinov; Quantum Electron. 27(2), 115 (1997).

84. A.A. Kravchenko, A.A. Manenkov and G.A. Matyushin; Quantum Electron., 26(5), 1045 (1996).

85. V.I. Bezrodnyi, M.V. Bondar, G. Yu. Kozak, O.V. Przhonskaya and E.A. Tikhonov; Zh. Prikl. Spektrosk., 50(5), 711 (1989).

86. V.I. Bezrodnyi, O.V. Przhonskaya, E.A. Tikhonov, M.V. Bondar and M.T. Shpak; Sov. J. Quantum Electron., 9(12), 2455 (1982).

87. T.B. Bermas, Vu.S. Zaitsev, Yu. V. Kostenich, M.K. Pakter, Yu. M. Paramonov, A.N. Rubinov, A. Yu. Smirnov and T. Sh. Effendiev; Zh. Prikl. Spektrosk., 47(4), 569 (1987).

88. A.V. Bortkevich, S.A. Geidur, O.O. Karapetyan, A.R. Kuznetsov, S.M. Lan'kova, A.G. Morozov, V.E. Polyakov and V.P. Sidyakova; Zh. Prikl. Spektrosk. 50(2), 210 (1989).

89. V.I. Bezrodnyi and A.A. Ishchenko; Quantum Electronics, 30(12), 1043 (2000).

90. M.I. Aldoshin, A.A. Manenkov, V.S. Nechitailo and V.I. Pogonin; Zh. Tekh. Fiz. 49(11), 2498 (1979).

91. M.V. Bondar and O.V. Przhonskaya; Quantum Electron., 28(7), 753 (1998).

92. R.S. Anderson, R.E. Hermes, G.A. Matyushin, V.S. Nechitailo and S.C. Picarello; Proc. SPIE., 3265, 13 (1998).

93. M.D. Rahn, T.A. King, A.A. Gorman and I. Hamblett; Appl. Opt., 36, 5862 (1997).

94. A. Costela, I. Garcia-Moreno, J. Barroso and R. Sastre; Appl. Phys. B 70(3), 367 (2000).

95. S. Wu and C. Zhu; Optical Materials, 12(1), 99 (1999).

96. A. Costela, I. Garcia-Moreno, J. Barroso and R. Sastre; Appl. Phys. B 67(2), 167 (1998).

97. A. Seilmeier, W. Kaiser, B. Sens and K.H. Drexhage; Opt. Lett., 8(4), 205 (1983).

98. V.I. Bezrodnyi and A.A. Ishchenko; Appl.Phys., B73(3), 283 (2001).

99. A.A. Ishchenko, V.A. Svidro and N.A. Derevyanko; Dyes Pigments, 10(2), 85 (1989).

100. A. Costela, I. Garcia-Moreno, C. Gomez, F. Amat-Guerri, M. Liras and R. Sastre; J. Appl. Phys. B76, 365 (2003).

101. F.L. Arbeloa, J.B. Pieto, I.L. Arbeloa, A. Costela, I. Garcia-Moreno, C. Gomez, F. Amat Guerri, M. Liras and R. Sastre; Photochem. and Photobiol, 78(1), 30 (2003).

102. A. Costela, I. Garcia-Moreno, J.M. Figuera, F. Amat-Guerri, R. Mallavia, M.D. Santa-Maria and R. Sastre; J. Appl. Phys. 80, 3167 (1996).

103. J.S. Batchelder, A.H. Zewail and T. Cole; Appl. Optics, 20(21), 3733 (1981).

104. M.B. Levin, A.S. Cherkasov and V.K.Baranov; Optiko-mekhanicheskayaya promyshlennost', 47 (1988).

105. R. Raue, H. Harnisch and K.H. Drexhage; Heterocycles, 21(1), 167 (1984).

106. N.N. Barashkov, M.E. Globus, A.A. Ishchenko, I.P. Krainov, T.M. Murav'eva, V.V. Pomerantsev, V.V. Popov, O.K. Rossikhina, V.G. Senchishin, and A.V. Sidel'nikova; Zhurn. Prikl. Spektrosk., 1991, 55(6), 906 (1991).

107. N.A. Derevyanko, A.A. Ishchenko, Yu. L. Slominski and A.I. Tolmachev; Mendeleev Commun., 91 (1991).

108. A.A. Ishchenko, N.A. Derevyanko and V.A. Svidro; Dyes and Pigments, 19(2), 169 (1992).

109. M.D. Rhan and T.A. King; Apl. Optics, 34(36), 8260 (1995).

110. R. Reisfeld; J. Fluorescence, 12(3/4), 317 (2002).

111. R. Reisfeld, A. Weiss, T. Saraidarov, E. Yariv and A.A. Ishchenko; Polym. Adv. Technol., 15(10-12), 1 (2004).

112. L.M. Sergeeva and L.A. Gorbach; Usp. Khim., 65(4), 367 (1996).

113. J.C.V.P. Moura, A.M.F. Oliveira-Campos, and J. Griffiths; Dyes and Pigments, 33(3), 173 (1997).

114. J.C. Brabec, N.S. Sariciftci and J.C. Hummelen; Adv. Funct. Mater., 11, 15 (2001).

115. J. Fabian; Chem. Rev., 92(6), 1197 (1992).

116. A. Ishchenko, N. Derevyanko, Yu. P. Piryatinskii, A. Verbitsky, D. Filonenko and S. Studzinsky; Mater. Sci., 20(4), 5 (2002).

117. S. Studzinsky, V. Syromyatnikov, A. Ishchenko, N. Derevyanko, Ya. Vertsimakha and A. Verbitsky; Nonlinear Optics, Quantum Optics, 33(1-2), 151 (2005).

118. N.A. Davidenko and A.A. Ishchenko; Chem. Phys., 247, 237 (1999).

119. N.A. Davidenko, N.A. Derevyanko, A.A. Ishchenko, N.G. Kuvshinsky, A.V. Kulinich, O. Ya. Neiland and M.V. Plotniece; Russ. Chem. Bull., 53(8), 1674 (2004).

120. J.-F. Nierengarten, G. Hadziioannou and N. Armaroli; Materials Today, 4(2), 16 (2001).

121. M.A. Diaz-Garcia, S. Fernandez De Avila and M.G. Kuzyk; Appl. Phys. Lett., 80924), 4486 (2002).

122. N.A. Davidenko and A.A. Ishchenko; Adv.Mater.Opt. Electron, 8, 201 (1998).

123. N.A. Davidenko, L.I. Fenenko, A.A. Ishchenko, M. Kuzma, P.S. Smertenko and S.V. Svechnikov; Synth. Metals, 122(1), 173 (2001).

124. L. Fenenko, P. Smertenko, S. Svechnikov, O. Lytvyn, A. Ishchenko, N. Davidenko, M. Hietschold and F. Mueller; Mol.Cryst.Liq.Cryst., 427, 49 (2005).

125. E.I. Mal'tsev, D.A. Lypenko, B.I. Shapiro, M.A. Brusentseva, E.V. Lunina, V.I. Berendyaev, B.V. Kotov and A.V. Vannikov; Vysokomolekulyarnye Soedineniya. Ser.A., 41(9), 1487 (1999).

126. N.A. Davidenko and A.A. Ishchenko; Theoretical and Experimental Chemistry, 34(4), 191 (2002).

127. J. Kido, M. Kohda, K. Okuyama and K. Nagai; Appl. Phys. Lett., 61 (7), 761 (1992).

128. N.O. Mchedlov-Petrossyan, V.K.Klochkov, G.V. Andrievsky and A.A. Ishchenko; Chem. Phys. Lett., 341(3-4), 237 (2001).

129. T. Noda, H. Ogava, N. Noma and Y. Shirota; Appl. Phys. Lett., 70 (6), 699 (1997).

130. K. Hara, Y. Tachibana, Y. Ohga, A. Shinpo, S. Suga, K. Sayama, H. Sugihara, H. Arakawa; Solar Energy Materials and Solar Cells, 77. 89 (2003).

131. M.L. Rojas-Cervantes, B. Casal, P. Aranda, M. Saviron, J.C. Galvan and E. Ruiz-Hitzky; Coll. Polym. Sci., 990 (2001).

132. N. Chandzasekharan, P.V. Kamat, J. Hu and G. Jones; J. Phys. Chem. B., 104(47), 11103 (2000).

133. L.I. Fenenko, A.A. Ishchenko, A.B. Verbitsky and Ya. I. Vertsimakha; Mol.Cryst.Liq.Cryst., 426, 157 (2005).

10

Polymers in Medicine

Ayesha Ahmad[1], Faiz Mohammad[2] and Mohammad Ahmad[3*]

[1]E-9, Johri Farm, Okhla, New Delhi-110025, India.
[2]Department of Applied Chemistry, Aligarh Muslim University, Aligarh—202002, India.
[3]A-13, Medical Colony, Aligarh Muslim University, Aligarh—202002, India.*
[*]Corresponding author's email: mahmad202@yahoo.com

Contents

❑ **INTRODUCTION**

❑ **POLYMERS AND POLYMER CHEMISTRY PERTINENT TO MEDICAL APPLICATIONS**

❑ **MEDICAL APPLICATIONS**

❑ **CONCLUSIONS AND FUTURE PROSPECTS**

❑ **REFERENCES**

Summary: The applications of polymers in medicine are enormous, ranging from their use in simple articles as medical disposable items to very advanced disciplines. This chapter covers the significance and new developments in the field of biodegradable and non-biodegradable polymers for applications in medicine, such as gene delivery, drug delivery, tissue engineering, cardiovascular system, dentistry, orthopedics, nerve regeneration, ophthalmology, ear, breast implant, facial implant, gynecology, etc.

1. INTRODUCTION

Polymer science has made tremendous advances over the last 30 years, and its applications have made a significant impact in medical sciences. Polymeric materials have certain qualities, which make them superior to other metallic and non-metallic substances. They are easily available, their strength and durability can be modified to suit the varied applications. They can be moulded in different forms yet maintain the three-dimensional structure of the scaffold generated from them [1]. Their structure can be suitably modified to incorporate the desired properties for their use in living organisms such as biocompatibility [2], biodegradability [3] and biostability [4]. They can be made to be insulating or electrically conducting depending on requirement [5]. Their aesthetic is superior to other materials. They can be converted into fibers or sheath of nano-range [6-8]. They can be moulded like plastics and stretched like rubber or spun like nylon or polyester [9]. They can be used to envelope drugs and deliver it to targeted sites for treating cancer and other diseases while at the same time avoiding toxicity to other tissues of the body [10]. And above all, they can be well tolerated by living organism. Besides these important characteristics they are cheaper and cost effective. These qualities and many others, have enabled them to emerge as revolutionary material for a wide range of applications and the future holds tremendous opportunities for use of polymers.

The polymers have applications in chemistry, engineering, electronics, horticulture, environmental studies, mathematics, physics, computers, geology, health and pharmaceutical industry etc. In fact there is hardly any sphere of human activity, which has remained untouched by this material. The most promising fields in medical science and other disciplines are heavily dependent on polymers and polymer science. In tissue engineering, the polymers play a prominent role in the fabrication of 3D scaffolds [11]. The common medical uses of polymers are summarized in the **Tab. 10.1**.

Thus, the applications of polymers in the field of medicine are enormous, ranging from their use in simple articles as medical disposable items to such advanced disciplines as tissue engineering and drug delivery, which has unlimited potential in coming years, and is dependent on advances in polymer science. The most prominent applications are in the field of cardiology, dentistry, neurology, dermatology, oncology, and genetic engineering, just to name a few of them. It will be a proper statement to say that the coming years will be known as the age of specialty polymers in terms of material and applications.

Use of plastics as biocompatible material is rising rapidly as is evident from **Fig. 10.1**. The market is expected to grow at an average annual growth rate of 7.7%. The consumption of polymer will grow at an average annual growth rate of 7.9% to reach $ 11.9 billion through 2008 in United States alone. As compared to that phenomenal growth, the use of non-polymer biocompatible materials

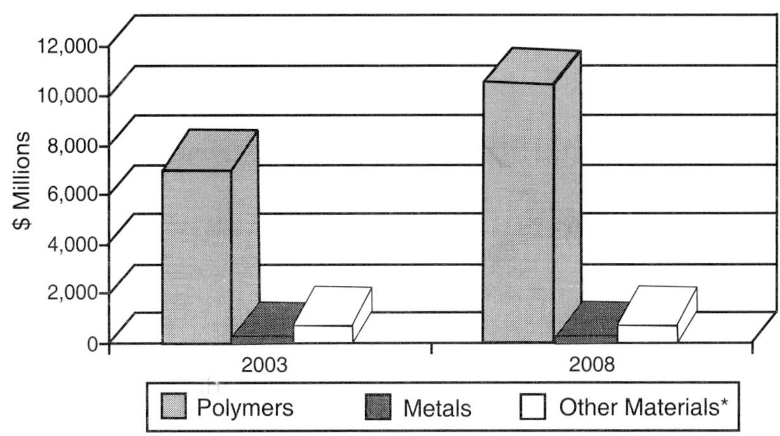

Fig. 10.1. Market size for biocompatible materials in medical devices for the years 2003 and 2008 ($ Millions). Reproduced with permission from http://www.bccresearch.com/biotech/b072N.html, Copyright © BCC Research.

Tab. 10.1. Use of polymers and polymer devices in various areas of medicine.

Area of use	Polymers and polymer devices

Cardiopulmonary and Vascular Uses	• Pacemakers
	• Heart valves
	• Intravascular stents
	• Vascular grafts
	• Oxygenators
	  — Blood membrane oxygenators
	  — Intravascular oxygenators
	• Hemoconcentrators
	• Catheters
	  — Diagnostic catheters
	  — Balloon angioplasty
	  — Other catheters
	• Heart assist devices
	  — Intra-aortic balloon pump

Cosmetic and Reconstructive Surgery	• Breast implants • Other cosmetic and reconstruction surgery

Orthopedics	• Orthopedic implants
	• Synthetic bone grafts
	• Joint lubrication
	• Fixation devices
	  — Pins
	  — Rods
	  — Plates
	  — Mesh
	  — Screws

General Surgery	• Sutures • Gloves
Dentistry	• Dentures and denture adhesives • Dental impression materials • Permanent dental implants • Crowns and bridges • Tooth regeneration
Ophthalmic	• Intraocular lens • Contact lenses • Viscoelastic substances
Renal and Urological	• Penile and testicular implants • Urinary incontinence
CNS	• Hydrocephalus shunts • Nerve regeneration
Dermatology	• Trans-dermal drug delivery patches • Skin regeneration
Systemic Drug Delivery and Oncology	• Drug encapsulation • Tissue targeting
Tissue Engineering	• Sheets of tissues and body organs
Genetic Engineering	• Treatment of metabolic disorders

(titanium, stainless steel and other metals) will be approximately $ 212 million in 2008 [12]. Nanotechnology will be a major thrust area in the field of drug delivery products, sensor and high-performance biocompatible materials that will help to produce "smart" rejection resistant implants [2].

2. POLYMERS AND POLYMER CHEMISTRY PERTINENT TO MEDICAL APPLICATIONS

Polymers have been used in the augmentation and repair of the human body with spectacular success. The specific use of the polymer depends upon mechanical and thermal properties, biodegradability and the tissue reaction to the polymer (both good and adverse) i.e. its bio-compatibility. Accordingly different polymers have been used for different purposes [3].

A polymer is a long chain macromolecule composed of many repeating units called monomers, which are bound together by suitable linkages. A polymer is generally named by prefixing 'poly' on the monomer it is synthesized from. For example, polymer derived from ethylene is poly(ethylene)— while those derived from glycolic acid and lactic acid are polyglycolide and polylactide respectively. The polymers, which exist in dextrorotary or levorotary forms, their names are prefixed by 'd' and 'l' forms or 'dl' for the recemic mix.

Polymers are classified into two broad groups, the natural polymers and synthetic polymers on the basis of their origins. Natural polymers are those obtained from natural sources while the synthetic polymers are chemical substances synthesized in the laboratory. Natural polymers can be further classified on the basis of the chemical group they belong to. Synthetic polymers offer certain advantages over the polymers obtained from natural sources. These include:

1. They possess a wider range of properties.
2. Their behavior is predictable.
3. They are free from immunogenicity.
4. Histological response to synthetic polymer is generally more predictable [10].

Some of the important examples of each group are mentioned below and given in **Tabs. 10.2 & 10.3**.

2.1. Natural Polymers

2.1.1. Proteins and Protein Based Polymers

1. Albumin.
2. Collagen.
3. Poly(amino acids).

2.1.2. Polysaccharides and Derivatives

1. **From vegetable source:** (i) agarose, (ii) alginate (marine sources e.g., algae), (iii) carbohy-drate, cellulose etc., (iv) carboxymethyl cellulose, (v) cellulose sulphate and (vi) caragenan.
2. **From human and animal sources:** (i) heparin and heparin-like glycoaminoglycane and (ii) hyaluronic acid.

Tab. 10.2. Use of individual natural and synthetic polymers in different areas of medicine.

(A) NATURAL POLYMERS

Polymer group	Uses
Proteins and Protein Based Polymers	
Albumin	• Cell and drug micro-encapsulation
Collagen	• Absorbable sutures • Sponge wound dressing • Drug delivery microspheres
Polyaminoacids	• Used as oligomeric drug careers
Polysaccharides and Derivatives (from vegetable sources)	
Agarose	• Clinical analysis • Immobilization matrix
Alginate	• Immobilization matrices for cells and enzymes • Controlled release of bioactive substances • Injectible microcapsules for treating neurodegenerative and hormone deficiency diseases
Carboxymethyl Cellulose	• Cell immobilization • Drug delivery • Dialysis membranes
Carrageenan	• Microencapsulation
Cellulose Sulphate	• Immunoisolation
Polysaccharides and Derivatives (from human and animal sources)	
Heparin and Heparin-like Glycoaminoglycans	• Antithrombotic • Anticoagulant • Surgery • Ionotropic gelatione • Capsule formation
Hyaluronic Acid	• A therapeutic agent
Polysaccharides and Derivatives (from microbial sources)	
Chitosan and Derivatives	• Biocompatible • Controlled delivery systems (e.g. gels, membranes, microspheres)
Dextran and Derivatives	• Plasma expander • Drug carrier

(B) SYNTHETIC POLYMERS

Polymer Group	Uses
Poly lactic acid, poly glycolic acid and their copolymers	• Sutures • Drug delivery systems • Tissue engineering • Biodegradable
Polyhydroxybutyrate, poly(ε-caprolactone) and copolymers	• Biodegradable • Drug delivery system • Cell microencapsulation

Polyamides (NYLONS)	• Joints • Blood vessels • Sutures • Dressings • Hemofiltration membranes (Dialysis)
Polyanhydrides	• Biodegradable • Tissue engineering • Release of bioactive molecules
Poly(ortho-esters) and polyesters	• Surface acting polymer • Drug delivery • Ophthalmology • Lungs • Kidneys • Livers • Blood vessels
Polycyanoacrylates	• Drug delivery
Polyphosphazenes	• Sutures • Catheters • Membranes
Polyvinylalcohol	• Gels and blended membranes
Polyethyleneoxide	• Highly biocompatible • Variety of biomedical applications
Polyhydroxyethylmethacrylate	• Dental implants • Bone replacement
Polytetrafluoroethylene (TEFLON)	• Vascular grafts • Clips • Sutures • Artificial vasculature
Polydimethylsiloxanes	• Implants in plastic surgery • Implants in orthopedics • Blood bags • Pacemakers
Poly(ethyleneoxide-co-propyleneoxide)	• Protein deliver • Skin treatment
Polymethylmethylcrylate (PMMA)	• Bone cement • Contact lenses
Polyvinylchloride (PVC)	• Blood vessels • Gastrointestinal grafts • Heart components
Polyurethane (used with adhesive)	• Wound dressing
Poly(2-hydroxyethylmethacrylate)	• Wound dressing
Silicon membrane on nylon velour	• Wound dressing
Polyethyleneoxide bonded to polyethylene mesh	• Wound dressing
Polyetheretherketones (PEEK)	• Dentistry (Healing caps, dentures, crowns, other structures)

Tab.10.3. Common polymer devices used in medicine.

Application type	Composition (Trade name)
Anastomosis clip	• poly(l-lactide) (Lactasorb)
Angioplastic plug	• poly(dl-lactide-co-glycolide) (Angioseal)
Anastomosis ring	• polyglycolide (Valtrac)
Dental	• poly(dl-lactide) (Drilac)
Guided tissue	• poly(dl-lactide-co-glycolide) (Resolut)
Interference screws	• poly(dl-lactide) (Sysorb), • poly(glycolide-co-trimethylene carbonate) (Endofix), • poly(l-lactide) (Arthrex), • poly(l-lactide) (Bioscrew), • poly(l-lactide-co-dl-lactide) (Phusiline), • poly(dl-lactide-co-glycolide) (Biologically Quiet)
Pins and rods	• poly(l-lactide) (Biofix), • poly(l-lactide-co-dl-lactide) (Resor-Pin)
Plates, mesh and screws	• poly(l-lactide-co-glycolide) (LactoSorb), • poly(dl-lactide) (Antrisorb) • poly(l-lactide) (Smart Screw)
Sutures	• polyglycolide (Dexon), • poly(glycolide-co-trimethylene carbonate) (Maxon), • poly(l-lactide-co-glycolide) (Vicryl), • poly(dioxanone) (PDS), • poly(l-lactide-co-glycolide) (Polysorb), • polyglycolide (PGA Suture)
Suture anchor	• poly(l-lactide) (Bio-Statak), • poly(glycolide-co-trimethylene carbonate) (Suretac)
Tack	• poly(l-lactide) (Smart Tack)

2.1.3. Microbial Polysaccharides

    (i) Dextran and its derivatives.
   (ii) Chitosan and its derivatives.

2.2. Synthetic Polymers

2.2.1. Aliphatic Polyesters

      (i) poly(lactic acid) and derivatives,
     (ii) poly(glycalic acid) and derivatives,
    (iii) polyamides(nylons),
    (iv) polyanhydrides,
     (v) poly(orthoesters),
    (vi) poly(cyanoacrylates),
   (vii) polyurethanes(thermoplastics),
  (viii) polyethylene(low density),
    (ix) polyvinylalcohol,

   (x) poly(ethyleneoxide),
  (xi) poly(hydroxyethylmethacrylate),
 (xii) poly(methylmethacrylate),
(xiii) poly(tetrafluoroethylene) (teflons) and
(xiv) polydimethylsiloxanes.

2.2.2. Environmentally Responsive Synthetic Polymers

   (i) poly(ethyleneoxide-co-propyleneoxide),
  (ii) poly(vinylmethylether) and
 (iii) poly(n-alkylacrylamides).

2.3. Biocompatible and Biodegradable Polymers in Medical Applications

The search for an ideal biocompatible biodegradable polymer has been going on for more than 100 years, but no single polymer has so far been found to fulfill all the criteria of biocompatibility and biodegradability [1].

Some polymers, such as PLA, PDA, polyanhydride, polyorthoester, polycaprolactone, polycarbonate, polyfumarate etc. are generally considered biocompatible. However, it cannot be guaranteed that they will not elicit adverse response under any circumstances. The amount of polymer used and body's ability to handle the polymer degradation product load also play a role in adverse reaction to polymers. For instance, large amounts of biodegradable polymers implanted at highly avascular sites may exceed the body's ability to clear the degradation products, leading to undesirable responses. The mechanical properties are important in places where weight bearing is required as in various orthopaedic appliances [13].

The factors, which affect the mechanical properties include [13]:
1. Hydrophilicity.
2. Crystallinity.
3. Melt point (T_m) and glass-transition temperatures (T_g).
4. Molecular weight.
5. Molecular weight distribution and end groups.

Biodegradable polymers can be either natural or synthetic. The properties of an ideal biodegradable polymer will depend upon the desired application. The general principal properties of such a substance are listed below:
1. It should have mechanical properties required for the desired use.
2. It should remain sufficiently strong until the healing process is complete.
3. It should not evoke an inflammatory or toxic response and electrical (especially for those used in the mouth) or other side effects.
4. It should be completely metabolized in the body after fulfilling its purpose.
5. It should be easily processible into the final product form.
6. It should have an acceptable shelf life.
7. It should be easily sterilized without undergoing any physical or chemical change, which will alter their mechanical properties or biocompatibility [3].

Biodegradation is accomplished by the hydrophilic unstable linkages in the backbone of the polymer. The most common chemical groups that incorporate this property are anhydrides, orthoesters and amides. Most of the commercially available biodegradable polymers are homopolymers or copolymers of glycolide and lactide. Other substances, such as copolymers of trimethylene carbonate and ε-caprolactone, have also been used in medical devices. Polydioxanone has also been used to produce bioabsorbable sutures. The factors that accelerate polymer degradation include [14]:

1. More hydrophilic backbone.
2. More hydrophilic endgroups.
3. More reactive hydrolytic groups in the backbone.
4. Less crystallinity.
5. More porosity.
6. Smaller device size.

2.4. Intelligent Polymers

Intelligent polymers are those polymers, which respond with large property changes in response to small physical or chemical stimuli, such as heat, light, sound, pH changes, etc. They are also known as "smart", "stimuli responsive", or "environmentally sensitive". The changes in the polymer in response to such stimuli may be in size, shape, consistency, physical state, etc. These changes in properties have important uses in biological systems [15].

Polymers, which change their shape in response to temperature, have been combined with polymers, which are light sensitive. Such materials have been used in intelligent suture materials and blood vessel stents in medicine [16].

Intelligent polymers may be used to deliver drugs in gastrointestinal tract. For example a pH sensitive polymer can be used to deliver bile acids (substances which digest fat). Normally bile is secreted in the upper part of intestines (duodenum) for digestion of fat. The pH in stomach does not exceed 4 while the pH intestine is 6 or more. The pH sensitive polymer with bile can be programmed to become activated at pH 5 and thus deliver bile at the appropriate site in the intestine [17].

Fusion of intelligent polymers and cells may be achieved by nanotechnology in tissue engineering. This technique has been used to engineer sheets of the cells. This is achieved in three steps. In the first step non-invasive harvest of cell cultures and cell sheets are carried out in temperature-responsive culture dishes with nano-sized thickness controlled polymeric surface modifications. In the second stage, the harvested cell sheets are transferred to other surfaces of the culture dishes (a technique called 2-D cell sheet manipulation). In the last stage the sheets can be piled up to produce structures, which have layers of cell sheets (called stratified tissue architecture). Such stratified structure is found in the tissues of liver lobule, kidneys and blood vessels. It has been claimed that these 2-D and 3-D cell sheet manipulation achieved by nano-sized thickness polymeric surface modification will become "new revolutionary tools for regenerative medicine".

Intelligent polymers may be combined with biomolecules to synthesize polymer biomolecule conjugate. The biomolecules which have been conjugated with polymers include protein, –COOH groups of aspartic or glutamic acid, –OH groups of serine or tyrosine, and –SH group of cystine residues. Usually the conjugations are carried out in a non-specific way, but it has also been possible to conjugate smart polymer to specific sites on some proteins. The ability of polymer to control protein-ligand recognition and binding process has been demonstrated in appropriately designed experiments. The conjugates of polymer-biomolecule with triggered-release of a bound ligand have applications in diagnostics, affinity separations, drug delivery and signal generation, which have important applications in medicine [18-20].

2.5. Polymers and Nanotechnology

Nanotechnology has been defined as the science and engineering involved in the design, synthesis, characterization and application of materials and devices whose smallest functional organization in at least one dimension is on the nanometer (nm) scale. A nanometer is one-billionth of a meter. It is roughly the size of a molecule itself. One can imagine a nanometer scale by comparison with a DNA molecule, which is about 2.5 nm in length, and a sodium atom, which is about 0.2 nm [21].

The idea of exploring the realm of atoms and molecules (nanotechnology) for development of products and technologies was proposed by Richard Feynman in 1959. At that time, it was difficult to imagine the impact it would have in science and technology [22].

Production of nano-materials can be done in two ways: 'top down' and 'bottom up' approaches and their combinations. In the first method, one starts with a macro level material and gradually converts it to nano-level, while in the second approach one starts at molecular level and then assembles and organizes the material into higher order structures of meso-scale and macro-scale. With the above techniques, polymer substances may be made on the form of nano-tubes or sheets of nano-range. These are used in regeneration of nerves and other structures, which are discussed elsewhere in this write-up. Nano-sized polymeric spheres are extensively used in drug delivery and gene delivery systems, which are discussed under appropriate headings [23]. As of now, nanotechnology is the most valuable advancement in science, which has its impact on 'just about everything in medicine and surgery'.

3. MEDICAL APPLICATIONS

3.1. Gene Delivery

Genes are chemical substances (DNA, RNA) located on chromosomes. It is the function of the genes, which determine the structure and function of the body. Abnormalities in gene structure adversely affect the biochemical functions in the body and result in disease processes. Such gene abnormalities may be inherited or may be acquired as a result of genetically determined mutations (sudden changes) or by physical, chemical and biological agents. With the developments in science and technology it is now possible to recognize the abnormal gene and correct the abnormalities in it.

Gene therapy is now being used to: (i) replace missing gene, (ii) deliver genes that cause destruction of cancer cells or revert them to normal tissue, (iii) deliver viral or bacterial genes as vaccines and (iv) deliver genes that promote growth of new tissues or regeneration of damaged tissues. Gene therapy has thus the most promising future in clinical medicine and therapeutics.

In gene therapy, the first step is to recognize the defective gene and then to deliver it at the appropriate site. Gene delivery is the crucial step in this process. Genes can be delivered with the help of lipids, viruses or by use of appropriate polymers. Polymers have been used with great success in this process of gene delivery.

Many risks are involved with the use of lipid and viral agents for gene delivery. Lipids can be toxic and also not very efficient. Though the virus vectors are efficient, yet they can be immunogenic. They are also not suitable for repeat doses, as the body tends to reject them. They can be potentially toxic as well. They are more difficult to make and they can get inserted into the genome of the cell and induce the development of cancer (**Fig. 10.2**). The benefits of polymers as carriers of genes are that:

1. These can be easily controlled.
2. They can be designed to be biocompatible and biodegradable.
3. They can be made non-toxic and non-immunogenic.
4. They can be made biodegradable.
5. Their designs can be improved to make them more efficient [24].

Both cationic and other polymers are used in gene delivery system, each having its own advantages. The cationic polymers are of greater significance. In fact DNA itself is a highly negatively charged polymer. Cationic polymers can form complexes with DNA and create 'packets' or 'particles'. They can be modified for attaching targeting legends to them. They can be 'eaten' (endocytosis) by cells and thus easily reach inside the cell more efficiently. Examples include Poly(lysine), Poly(ethyleneimmine), polyamidoamine, etc. The other polymers can slowly release genes (DNA) from polymer devices inside the body. These can be in the form of small particles or matrices or gels. The examples are Poly(lactide-co-glycolide) or PLGA and Polyvinylpyrrolidone (PVP) [24].

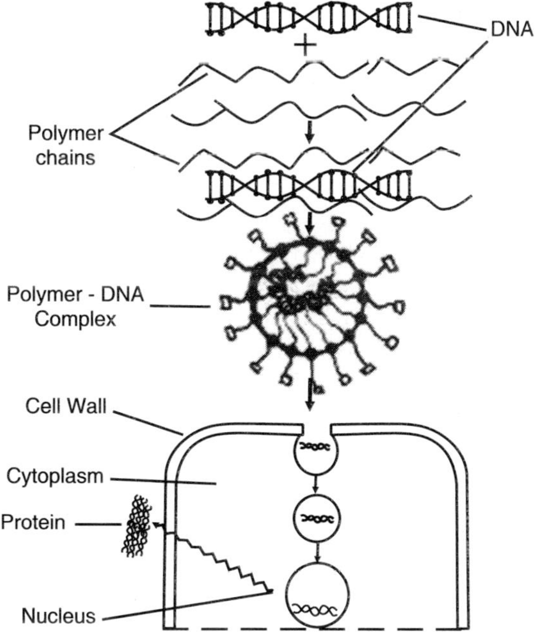

Fig. 10.2. Schematic diagram of gene delivery using polymer encapsulation. Genes are DNA molecules. These are encapsulated in polymer matrix. The combination forms a polymer-DNA complex, which can be transported through the body. On reaching the desired site, it is taken up by the cell. The gene is then released into the cell from the complex and is then taken up by the cell nucleus to produce the appropriate protein.

3.2. Drug Delivery

Polymers have come to occupy important position in delivery of drugs to targeted sites. The basic mechanism consists of encapsulating a drug in a polymer that either allows the drug to diffuse out from it or the polymer itself undergoes degradation to directly release the drug. The delivery of drug to the required site may be done locally or through blood stream. The common examples of local delivery of drug are trans-dermal patches, which release drug into the skin and drug-eluting stents, which release drug over a course of time from stents at their implantation sites. Examples of drug delivery through the blood stream include delivery to the site of infection or to a joint in inflammatory joint disease. Polymers are important aids in achieving this goal. Diffusion is the main mechanism by which drugs are released from drug eluting stents, which are used extensively in opening blocked arteries.

Trans-dermal patches are the devices, which are impregnated with some drug and can be stuck to the skin at some place on the body. These release drugs locally either gradually or through the skin in predetermined doses over a period of time, either in a sustained fashion, or in burst, or in accordance with the circadian rhythm of the body. The drug release behavior of the adhesive polymer depends upon the type of drug used and the type of polymer. Drug delivery pattern can be further improved by appropriately designing the adhesive polymer function.

Polymers have been effectively used in tissue targeting in inflammatory diseases. Inflammation is a defense process of the body in which the site of disease, as a bacterial infection, becomes red, hot and swollen because of dilatation of blood vessels in that area. This process brings a large number of white blood cells loaded with antibacterial chemicals, which help in restricting the infective process. A similar process of inflammation occurs in non-infective diseases, such as arthritis (pain and swelling of joints). While in infective process increased accumulation of white blood cells helps pooling of substances which prevent spread of infection and which kill the invading bacteria, in case of arthritis the process of inflammation does more harm than good. Biodegradable polymers have been developed which imitate white blood cells to target inflamed blood vessels. Biodegradable beads can be covered with targeting molecules. These can move through blood stream and effectively stick at the site of inflammation. Great interest has been generated in drugs made from biodegradable polymers because they can be easily prepared and have a long shelf life [25].

In patients with arthritis, the same process can be used to load the polymer particles with drugs. The drug-loaded particles mimicking the leucocytes (the white blood cells) accumulate at the site of the inflamed joint and deliver the drug there. Conventional drugs used in the treatment of arthritis while relieving the symptoms also affect other tissues in the body since they come in contact with other body tissues during transit in the body. This is responsible for the toxic effects of these drugs. Polymer assisted tissue targeting will help to minimize the toxic effects of drugs as the drug being covered in the polymer envelope, does not come in contact with other body tissues during transit to the desired site of delivery. This will help in better management of patients suffering from chronic autoimmune or degenerative diseases [25].

In certain cases the drug is required to be delivered slowly over a prolonged period to get the desired effect, for example, dipyridamole, a cardiovascular drug. This function can be achieved by the use of medicines dissolved in gel which have low diffusion rate compared to water. Studies have shown that triblock copolymer gel can be used to load dipyridamole for this purpose. It has been further observed that the drug release rate could be further tuned by altering the pH of the solution on the concentration of the copolymer. Lowering of the pH causes rapid dissolution of the gel, hence a rapid release of the drug. As against this, a slow and sustained release of the drug at the rate of 1 to 2% per hour can be achieved at physiological pH [25].

Another significance of polymers is in the delivery of hydrophobic drugs. These drugs being insoluble in water are not readily transported in the body. To overcome this problem, the drug can be encased in a larger host molecule that is soluble in water. Hyper-branched polymers can be used as host molecules for this purpose. It has been shown that these molecules can encapsulate a variety of hydrophobic drugs within their hydrophobic interior [10,26].

A further improvement in drug delivery to the targeted site can be achieved by the use of biodegradable polymers. In this case, the drug is encapsulated in a biodegradable polymer envelope, which has an appropriate ligand to target the desired tissue. On reaching the desired site, the polymer envelope gradually releases the drug over a period of time and the envelope itself gradually disintegrates in accordance with the programmed disintegration schedule. Thus the polymer envelope disappears by the time the drug is completely released as shown in **Fig. 10.3**.

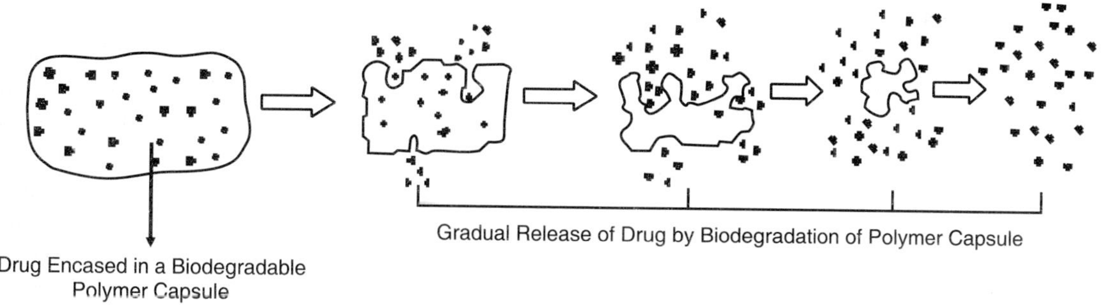

Fig. 10.3. Schematic diagram of drug delivery using biodegradable polymer envelop.

3.3. Tissue Engineering

Broadly speaking, tissue engineering comprises generating living tissues and organs for the purpose of replacement of damaged or diseased tissues. Such situations arise in replacement of diseased joint, production of dentures, replacement of skin lost by burn, injury or when removed by surgery, development of conduits for reconstruction of blood vessels, replacement of damaged heart valves, producing favorable conditions for regeneration of damaged nerves etc.

Earlier the only alternative under such situations was the use of non-biological materials such as alloys, ceramics, patches of Dacron, Dacron arteries, skin grafting, use of person's own arteries as in bypass surgery etc. The artificial materials had many drawbacks with respect to toxicity, durability, aesthetics, cost and other physical and chemical features. Use of biological material from same patients (called autografts) and from other persons (called homografts) or other species (called heterografts) was associated with morbidity, infection and rejection of tissues etc.

Tissue engineering has made available the material for replacement/reconstructive surgery and has eliminated the need of removing healthy tissues for grafting purposes. Thus the advances in tissue engineering have solved many of these problems and have opened new dimensions of research and clinical applications. A little bit of background about the principles involved in tissue engineering is essential. The body is made of microscopic units called cells. The cells start dividing and arrange themselves to form collection of cells (called tissues) and collections of tissues performing a particular function (called organs). When this job is accomplished, the cells are said to have matured. Mature cells fail to grow. Nature has programmed the cells in such a way that a cell can divide about 50 times. After that it stops growing. Thus the mature cells lose the property of multiplication. However, the immature cells, called stem cells, persist in sufficient numbers. They retain the property of dividing. The growth of cells occurs in an orderly manner. Cells destined to develop a finger will grow and arrange themselves in a manner that gives rise to formation of a finger and nothing else. Some growth factors are required to stimulate the cells to grow. The growth of cells also requires a three-dimensional framework on which it forms the particular shape of the tissue [27].

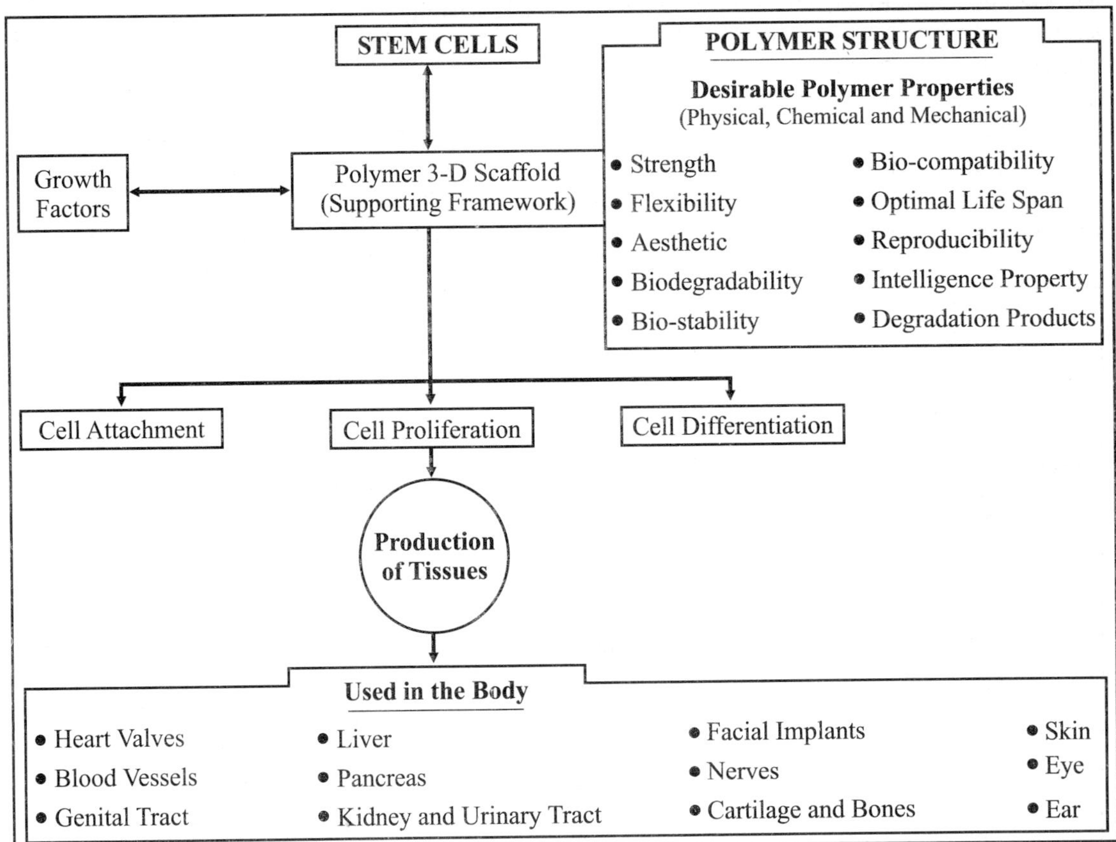

Fig. 10.4. Schematic diagram of tissue engineering. Stem cells and growth factors are mixed with an appropriate scaffold made of polymer. The cells attach themselves to the polymer scaffold and start proliferation (increase in number) and differentiation (organization to form a tissue or organ) to the desired shape. The engineered structure is then implanted on the body. The attachment, proliferation and differentiation of the cells heavily depend upon the physical and chemical properties of the polymer scaffold.

The whole process of formation of new tissues is called tissue engineering. It has been defined as "an interdisciplinary field that incorporates and applies the principles of engineering and life sciences toward the development of biological substitutes that restore, maintain, or improve tissue organ or function" [28,29].

Tissue engineering thus means regeneration of human tissues. Three things are required to grow the immature cells and form a particular tissue or organ. These include:

1. The cells which can grow (stem cells).
2. Appropriate growth factors.
3. A three-dimensional supporting framework.

It is the last requirement, where polymers play an important role. The three-dimensional structure is called a scaffold. It can be moulded according to the desired shape as per the requirement of the tissue/organ desired to be regenerated. Tissue growth factors are added to induce the cells to grow and multiply at a faster controlled rate. The physical and chemical properties of cell polymer interface strongly influence the number of cells that attach to the scaffold and their course of differentiation and growth. This technique can also be used to produce sheets of multilayered tissues for repair of defects in body organs such as in repair of certain defects in the heart (**Figs. 10.4 & 10.5**) [30].

Synthetic biodegradable polymer scaffolds, mainly comprised functional poly(lactic-co-glycolic acid) or poly (lactic acid-co-lysine) based comb-like graft copolymers with appropriate amino-acid chains. The amino acid chains impart the desired surface properties to the scaffolds without adversely affecting their mechanical characteristics. The property changes depend upon the particular peptide

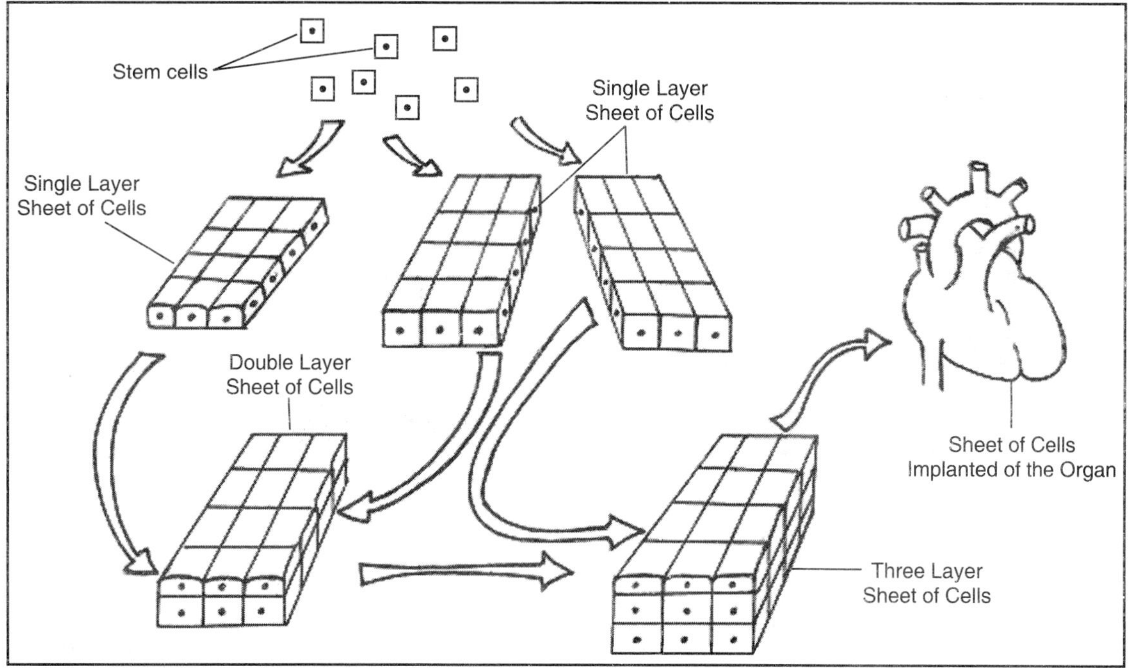

Fig. 10.5. Schematic diagram of tissue engineering to produce sheets of multilayered tissues for repair of defects in body organs.

used, the chain length' and overall composition of the polymer. The polymer chosen should have functional sites for chemical modification with peptides, so as to provide opportunity for the attachment of biologically active molecules that can actively promote favorable cell-polymer interactions [31,32].

Advances in biodegradable polymers have made it possible to design 3-dimensional structures for generating organs and tissues. Over a programmed period of time the 3-dimensional polymer structure gets degraded leaving behind the shape of tissue that was intended to be designed. Two examples, that of regeneration of blood vessels and bone, are given below.

Blood vessels: Production of autologous blood vessels (growing blood vessels from one's own tissue) is a fascinating development. There is great demand for functional, small diameter blood vessels for vascular grafts such as in bypass surgery, treatment of vascular aneurysms (abnormal dilatation of blood vessels) and replacement of damaged blood vessels in other diseases. Normal blood vessels mainly comprise muscle tissue, which make the tube, and a layer of cells which line the inner surface of the tube, called the endothelial cells. The technique of production of autologous blood vessels comprises implantation of muscle cells on a tube made of biodegradable polymer. The muscle cells loaded vascular scaffolds are placed in a bioreactor. Smooth muscles are generated in about two-month's time. Endothelial cells are then added to the tube. The polymer gets dissolved over a period of time and the regenerated blood vessel can be used in surgical procedures **(Fig. 10.6)** [33].

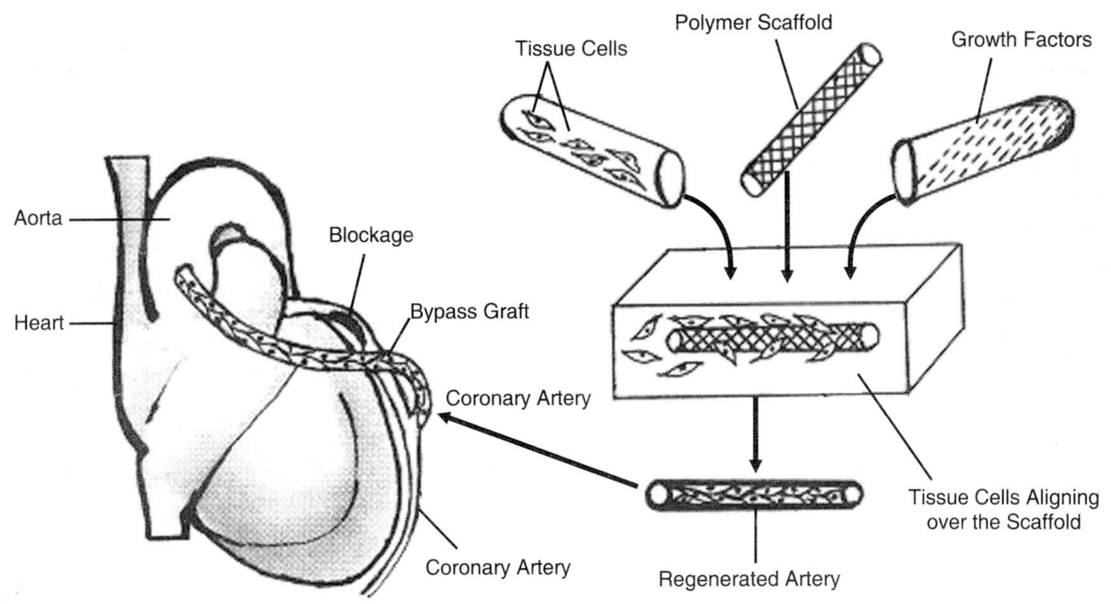

Fig. 10.6. Generation of autologous blood vessel by tissue engineering.

Bone: Bone tissue can be regenerated on a similar pattern. Appropriate three-dimensional porous biodegradable polymer scaffolds are prepared as needed. Bone marrow cells are seeded onto the scaffold and appropriate growth factors are added for promoting tissue regeneration. Changes in composition of polymer permit variation of mechanical properties over a wide range. The hydrophilicity and rate of degradation by hydrolysis can be appropriately modified for optimal tissue regeneration. Moulding and particulate leaching, freeze-drying and combinations of these methods can easily prepare porous structures of different morphologies. Composition of polymer influences bone marrow cell growth and adhesion properties, which can be optimized by surface modification with gas-plasma treatments [33].

3.4. Cardiovascular System

Diseases of cardiovascular system constitute important causes of morbidity and mortality. Approximately 20% of deaths between the ages of 36 and 74 throughout the world are due to cardiovascular lesions. There are a large number of uses of polymer substances in clinical cardiology. The important ones include intra-aortic balloons, cardiac-assist devices, vascular grafts, stents and angioplasty balloons etc. Some of these are discussed below [34].

3.4.1. Heart Valves

An important cause of cardiovascular morbidity is the involvement of heart valves in atherosclerotic, degenerative or rheumatic process. There are four valves in the heart which direct blood to flow in one direction only. These valves can become regurgitant (leaky) or stenosed (narrow), causing disturbance in normal blood flow. The significantly damaged valves have to be replaced with an artificial valve.

So far there have been two main approaches in fabricating artificial heart valves: a mechanical device or a bioprosthetic valve. The mechanical device consists of valves made of synthetic materials, while the bioprosthetic valve is constructed from the biological membranes, such as pericardium, from human or porcine sources or cadaveric aortic or pulmonary valves, some of which also have a synthetic component. Neither type of valves is fully satisfactory. The synthetic valves have a tendency to provoke thrombus formation. So these patients have to be on continuous anticoagulant therapy throughout life. The bioprosthetic valves, especially those obtained from porcine source, though less prone to thrombus formation, tend to degenerate and calcified over a period of time [35-38].

The requirements for a good artificial valve are:
1. It should not promote and propagate clot formation and embolisation.
2. It should be chemically inert so as not to react with any chemical or organelles found in blood.
3. It should not alter the physiological hemodynamics.
4. It should be durable.
5. It should not annoy the patient by the clicking sound of metallic surface impact.

The durability and thrombogenecity of the valve depends on the design of the valve and the material used. The improvement in designs comprised cage ball valve prosthesis, non-tilting discs, tilting discs, bi-leaflet discs for better hemodynamic performance.

Introduction of polymers in artificial valves has a future in solving the above two problems. It is expected that a suitable valve with the desired mechanical properties and lack of thrombotic properties can be developed. The most promising solution may be from polymeric surface treatments that will allow the device to be totally inert. The polymeric material or surface treatment with polymeric material may help to reduce mechanical noise. Polymeric scaffolds may be used for generation of tricuspid pulmonary valve. It is likely that an effective and safe polymer valve will soon become a third clinical option in cardiovascular therapeutics [39].

In early 1952, the artificial cage ball valve prosthesis used a silastic rubber ball housed in a cage. Silicone elastomer disc was developed in 1965. Delrin polymer disc came in 1975. However, Delran absorbed water and led to a change in the configuration of the disc. Because of this problem, pyrolate discs were introduced [40].

The first generation heart valves were mostly made from titanium. Susceptibility of titanium to corrosion and fatigue necessitated search for alternative materials. Polymers have helped in fabricating

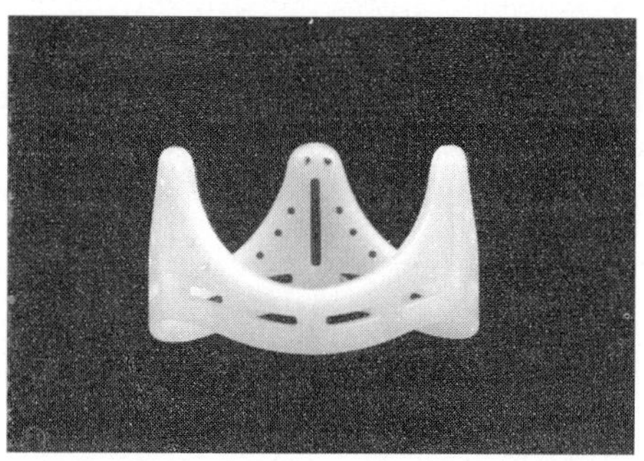

Fig. 10.7. Polymer stent for mounting heart tissue valve. Reproduced with permission from http://www.sjm.com/devices/devicetype.aspx?location=us&type=20. Copyright © 2006, St. Jude Medical Inc.

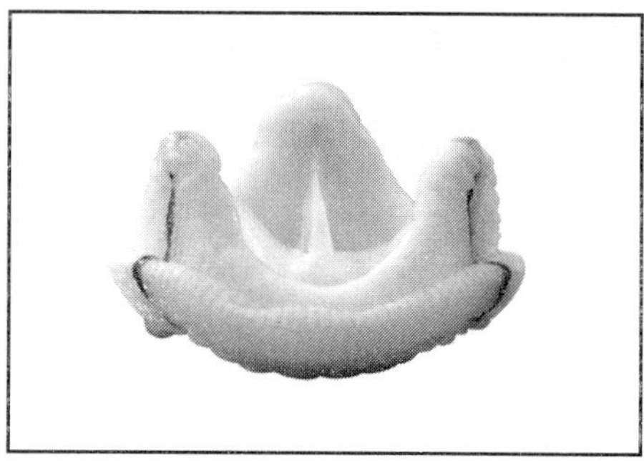

Fig. 10.8. Heart tissue valve mounted over polymer stent. Reproduced with permission from http://www.sjm.com/devices devicetype.aspx?location=us&type=20. Copyright © 2006, St. Jude Medical Inc.

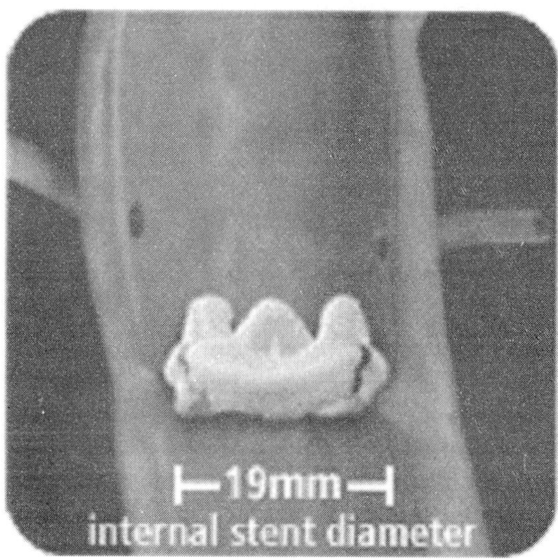

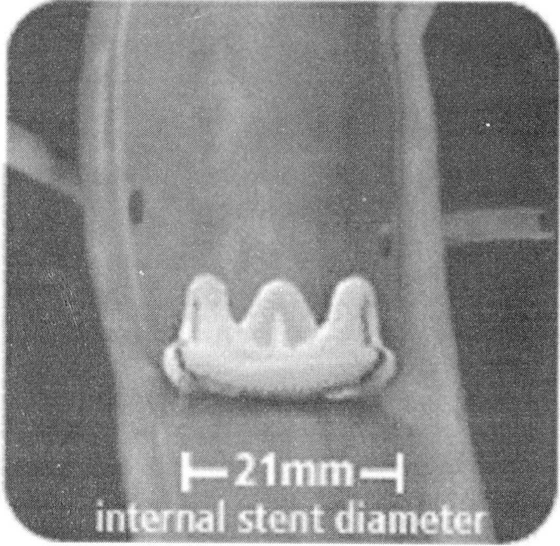

Fig. 10.9. Polymer mounted tissue valve of two different sizes in aortic position. Reproduced with permission from http://www.sjm.com/devices/devicetype.aspx?location=us&type=20. Copyright © 2006, St. Jude Medical Inc.

heart valves which are well suited to body environment as shown in **Figs. 10.7-10.9**. Polyurethane is one such material, which has been found useful for prosthetic valves. Polyurethanes consist both of hard segments that confer strength and soft segments that confer flexibility and elasticity. Thus, by altering the hard and soft segments the desired strength and flexibility can be achieved. Increasing the ratio of hard segment increases the durability of the device but its hydrodynamic function is compromised. On the other hand, with a higher ratio of soft segments the valve will have more strain accumulation, which will cause fatigue, creep or fracture leading to failure of the device [41].

Fatigue testing of valves made from polyetherurethane (PEU) and polyetherurethane urea (PEUE) has been carried out in many experiments. The valve frame and the leaflets of PEU consist of a polyether soft segment and a 4,4-diphenylmethane diisocyanate (MDI) chain extended with butanediol hard segment. The PEUE polymer also comprises polyether soft segments but the hard segment is made from MDI chain extended with ethylenediamine rather than butanediol. Thus the PEUE valves have urea groups within the polymer while PEU valve has urethane group in its structure. Since PEUE cannot be melted it is not possible to mould it into the desired shape of the specified valve model. The valve frames of PEUE have therefore to be made from PEU, which is injection mouldable. Only the leaflets are made from PEUE [40,42,43].

Besides the above features, the valve/device should be clinically inert to prevent reaction with body chemicals or organelles in blood. Though such materials are available, yet in certain areas such as articulating joints and surfaces they pose a problem. Polymeric surface treatment allows the device to become truly inert and obviate the problem of interaction with constituents and chemicals in blood. Anja Mueller of Clarkson University is working on a novel polymer that is non-sticky which can be coated on the inside of the valve that will allow the coating to adhere to the artificial valve materials and may have the potential to incorporate anticoagulants in the valve coating. This will prevent blood platelets and bacteria from sticking to heart valves installed in patients [44].

Another problem encountered in artificial valves is that of noise produced by the impact of the articulating surfaces. A clicking sound is produced by the impact of valves with occluder shuts. This noise can be greatly reduced with the use of polymeric materials or surface treatment with polymers [45].

3.4.2. Angioplasty Balloons

Another major mechanism of cardiovascular involvement is due to the condition known as atherosclerosis, which results from deposition of fatty material (athero = fat + sclerosis = hardening) in the blood vessels, especially the aorta, coronary arteries and cerebral arteries. This causes a decrease in the lumen of blood vessels and consequently an encroachment upon the blood supply to the concerned organ, resulting in angina, myocardial infarction and strokes. The blockage in the arteries can be opened by angioplasty balloons [46]. Angioplasty balloons are critical devices for treating blockages. In order to reduce surgical interventions, a procedure was developed in the 1980s to recanalize coronary vessels by using balloon dilation. The technique, percutaneous transluminal coronary angioplasty (PTCA), uses a catheter with a balloon at the distal end. The balloon is pushed forward into the stenosis and inflated to dilate the blockage [47]. The calcified deposits are thus pressed into the vessel walls under high pressure to create a wider lumen, improving blood flow **(Fig. 10.10)** [48].

Fig. 10.10. Polymer angioplasty balloon.

3.4.3. *Drug Eluting Angioplasty Stents*

Stents are devices, which prevent recoil of blood vessel after the obstructing lesion has been dilated by PTCA. The dilated blood vessels however tend to coil back. To prevent this re-coiling a metallic device, called 'stent' is deployed at that place. However, there was still high incidence of re-growth of the vessel tissue producing what is termed 'in-stent re-stenosis'. Researches in this direction established that this tissue re-growth could be prevented by local administration of tissue growth-inhibiting substances. Polymers have been found to be promising substances for this local drug delivery after PTCA (**Fig. 10.11**) [49].

The basic mechanism of drug delivery from a polymeric scaffold involves encapsulating a drug in a polymer that either allows the drug to diffuse outward from it or that undergoes degradation in order to release the drug directly. Generally, for long term applications, such as in stents, a non-erodable polymer is used. This is because the fragments that break off from the polymer coating, particularly in polymers that undergo bulk erosion, may be phagocytosed by macrophages and other lymphocytes. Phagocytosis of polymer fragments can lead to macrophage activation, which release inflammatory cytokines, causing increased lymphocyte infiltration of the site and leading to inflammation. Numerous polymer systems that seemed promising in vitro have subsequently been abandoned after in vivo studies demonstrated inflammatory responses to them. Two commonly used medicated stents are CYPHER stent and TAXUS stent. Diffusion is the principal mechanism by which drugs are released from these stents [50,51]. The CYPHER stent is composed of three layers of polymers over a frame made of laser cut 316L stainless steel. This metal stent is electropolished and coated in a primer layer of Parylene C. A mixture of polyethylene-co-vinyl acetate (PEVA) and

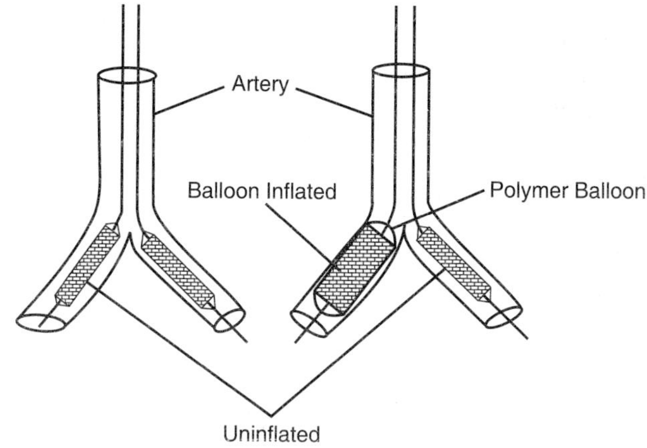

Fig. 10.11. Polymer coated metallic stents used in angioplasty. This is a schematic representation of a bifurcating lesion of two arteries. The figure on the left shows insertion of stents across the two arteries. The figure on the right shows that one stent has been dilated with a polymer angioplasty balloon.

poly n-butyl methacrylate (PBMA) is then dissolved in a solvent suitable for dissolving organic molecules. This copolymer of PBMA to PEVA is 67% PEVA, 33% PBMA. Sirolimus is then dissolved in the THF/polymer mixture and the mixture is applied to the Parylene C coated stent. Another mixture of PEVA and PBMA, without sirolimus, is dissolved in THF and applied to the stent by spraying with a fine nozzle. This outer coating prevents the so-called "burst effect" which results when drug on the surface of the polymer is rapidly released following immersion in water or another solvent. Small amount of sirolimus migrates to the final layer during this step because it dissolves in the THF and precipitates in the PEVA/PBMA outer layer. This causes a small but noticeable burst effect. The entire three-layered coating is applied to both the luminal and abluminal sides of the stainless steel stent. Finally, the stent is placed on a delivery catheter, sterilized, and packaged. The TAXUS stent also uses a similar approach [52,53].

3.4.4. Biodegradable Stents

Until now the use of polymers in Percutaneous Transluminal Coronary Angioplasty (PTCA) has been that of a drug-delivering device. Polymeric treatment of the metallic stents slowly delivers the appropriate drug over a period of time to prevent re-stenosis after PTCA. However, the metallic stent remains in place. It is not possible to remove the stent after deployment as it gets impacted in the blood vessel. It is for this reason that biodegradable stents have a scope. The advantage of biodegradable stents is that they will gradually dissolve while releasing the drug. It is expected that such drug eluting stents will perform better with respect to biocompatibility as well. They will also have sufficient strength to open and maintain the conformation of blood vessel after deployment and can be made cheaper than the conventional metal stents currently in use [54,55].

3.5. Dental Applications

Use of polymer in dentistry has found applications since the very beginning, as dental cement material and dental devices. The availability of better polymers and more insight into polymer applications have opened great hope for the use of polymer in cosmetic and regenerative dentistry in future [56]. The normal tooth structure is shown in **Fig. 10.12**.

3.5.1. Cosmetic Dentistry

The main use of dental material is in crowns, bridges, abutments, healing caps and implants [57]. Metallic alloys such as gold and titanium and ceramics such as porcelain, zirconium and alumina, have been among the most widely used materials in dental applications. Polymers form alternative materials for dental devices. The use of strong

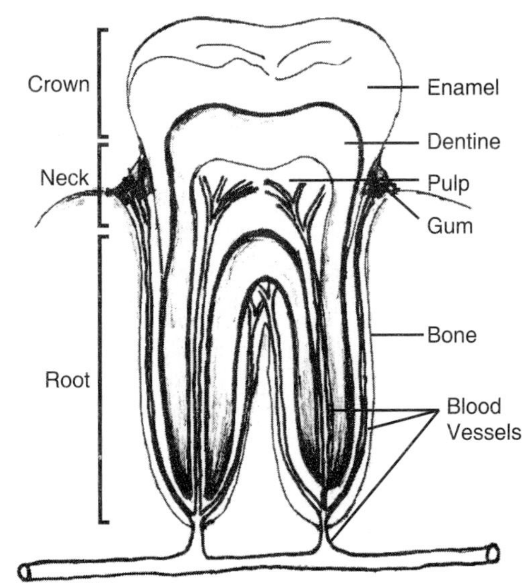

Fig. 10.12. Normal tooth structure.

polymers and composites have found a place in load bearing applications such as crowns and bridges, abutments, healing caps and implants. It is hoped that polymers will facilitate in the production of structurally superior devices and applications. Refinements in production methods, such as moulding/casting development and use of CNC/CAD have encouraged the dental device manufacturers to investigate these material alternatives. Dental healing caps, dentures, crowns and other structures, can all be fabricated from polymers [58].

Two polymeric materials best suited for dental applications belong to polyetheretherketones, namely, PEEK-OPTIMA and PEEK-CLASSIX. The former is a semi-crystalline thermoplastic material. It is used in fabrication of those devices, which are required to be implanted for a period exceeding thirty days. On the other hand, the PEEK-CLASSIX is used for applications having blood and tissue contact for less than thirty days [59]. Advantages of both these polymers are:

1. They are among the most chemically resistant polymers.
2. They can be repeatedly sterilized by conventional methods.
3. They are resistant to hydrolysis as well as to strong ionizing radiation.
4. Their strength can be greatly enhanced by addition of reinforcing fibers for applications requiring very high strength.
5. Their modulus (degree of elasticity or stiffness) may be adapted to closely match that of cortical bone.

3.5.2. Crowns, Bridges and Dentures

In fabrication of permanent crowns and bridges, the following considerations are important in selection of the material for their fabrication:

1. Strength.
2. Aesthetics.
3. Cost.
4. Biocompatibility.
5. Battery effect.

The features of substances commonly used in dental applications are summarized in **Tab. 10.4**. Gold alloys offer maximum strength but are poor in aesthetics and are expensive as well. Titanium has an advantage over gold in terms of cost. Non-precious alloys are used where cost and strength are important factors but aesthetics is not an important consideration [59].

Tab. 10.4. Property Comparison of some materials used in dental devices.

PROPERTIES	MATERIAL			
	Gold	*Titanium*	*Ceramic*	*Polymer*
Strength	Good	Good	Brittle	Good
Aesthetics	Poor	Poor	Good	Good
Cost	Higher	Cheaper	Higher	Cheaper
Biocompatibility	Good	Good	Good	Good
Battery effect	+	+	Absent	Absent
Toxicity	+	+	−	+/−

Ceramic, on the other hand offers good aesthetic value but it is relatively more expensive. Besides it has the disadvantage of being brittle and is liable to crack under high stress, leading to failure of the appliance. Metallic alloys have another disadvantage when used in the mouth—the so called "battery effect" as they are good conductors of electricity. This may cause headaches and fatigue in some patients. There is also a possibility of metal allergy caused by release of metal ions from the dental devices. Some experts are of the view that because of these disadvantages, the use of even non-allergic and non-toxic metals should be avoided in dental devices. As against these materials the use of polymers offers several advantages over metal and ceramic devices. A medical-grade polymer incorporates the desired strength, stiffness, toughness, biocompatibility as well as good aesthetics. In addition, they are cost effective.

Though aesthetically at par with ceramic as suitable material for dental appliances, polymers have the added advantage of strength. While ceramic is brittle, polymer can sustain higher compression and grinding action of teeth. Another advantage of polymer is that its modulus can be modified to suit the supporting natural tissue or natural teeth in the mouth. Their performance is also better with respect to biting and chewing sensation as well, which is close to natural feeling.

Tooth supporting structures (bone, cementum and periodontal ligament) are destroyed in chronic inflammatory disease called periodontitis. Repair of tooth supporting structure is a major goal of oral reconstructive therapy. Periodontal disease causes loss of supporting tissues. This, if left untreated leads to tooth loss. Restorative therapy of periodontal tissue is therefore the goal in preventive dentistry. However, the attempts at restoration of tooth supporting structures have often been unpredictable and disappointing. Advances in nanotechnology have recently opened a new therapeutic approach and provided a major breakthrough in this sphere. In this approach nano-fibrous polymer scaffolding (mimicking collagen architectures), modified with apatite (mimicking bone mineral), and containing microspheres for delivery of bioactive factors (mimicking development and reparative signalling cascade) is used in periodontal osseous defects. The system promotes the activities of cells at the healing sites (osteoblasts, cementoblasts and signal molecules). This induces guided cell proliferation, differentiation and tissue neogenesis in three-dimensional tissue formation. Moreover, the scaffolding delivery system allows better permeation of nutrients, metabolites and signalling molecules necessary for cell proliferation and differentiation [59].

3.5.3. Regenerative Dentistry

Tooth loss is a common feature and occurs in periodontal diseases, dental caries, trauma, and in several other disorders. Until recently, the only way out was to replace the lost tooth with implants made from non-biological materials. With advancements in tissue engineering it may be possible to grow tooth as a biological substitute to the currently available treatments. The process of tooth regeneration follows the common tissue engineering approach adopted in regenerating other tissues. It consists of selection of an appropriate dental tissue, addition of appropriate growth signalling factors and a polymer scaffold to guide the cells to grow to a particular shape and size. The polymer scaffold should then disappear, leaving the regenerated tooth in place [60]. Tooth polymer scaffolds have been prepared from polyglycolate/poly-L-lactate (PGA/PLLA) with the use of polyvinylsiloxane tooth moulds in the shape of human incisor and molar teeth. In experiments on rats, Young *et al.* [60] dissociated porcine third molar tooth buds and seeded them onto biodegradable polymer scaffold. After about 20-30 weeks, recognizable tooth structure formed. These contained dentin, odontoblasts, a well-defined pulp chamber, Hertwig's root sheath epithelia, putative cementoblasts and a fully formed and morphologically correct enamel. This was claimed as the first successful generation of tooth crown from dissociated tooth tissue that contained both dentin and enamel.

The aim of regenerative dentistry is to grow a new tooth in an individual's jaw from the individual's own cells. With advancement in regenerative dentistry it may be possible to develop a tooth of specific size and shape, depending on its position within the jaw [61,62].

3.6. Orthopedic Applications

3.6.1. Bone Grafts

Bone defects occur due to a number of reasons such as accidents, inflammatory diseases of bone, surgical resections etc. Repair of these defects are essential for maintaining the mechanical integrity of the skeleton. It is the first step in patients' rehabilitation. Traditionally a variety of techniques are available to tackle this problem, each having its own merits and demerits. Harvesting of autogenous bone grafts has been the main technique of repairing bone defects. But this is associated with prolonged morbidity. Use of allografts had provided an alternative approach. However, this approach is associated with high incidence of postoperative infection [63] and fractures [64]. Also, there is a potential risk of disease transmission [65].

Recently a variety of synthetic bone graft substitutes have become available [66]. These have further reduced the above complications. They have the advantage of being available in sufficient amount and are easy to sterilize. As a result, the overall morbidity has been gradually reduced. Reabsorbable synthetic polymers derived from the poly(α-hydroxy acid)s to form bone graft substitute have attracted attention and have a promising future for such treatment [67]. Hybrid polymer system (copolymers, complexes, hydrogels, blends etc.) have a wide range of applications. Bioabsorbable polymer mixtures with hydroxyapatite have been used in bone repair as they induce bioactive osteogenesis. A combination of a biointegrable polymeric mixture and a ceramic is a useful application as shape modification is easily made through usual plastics processing. Advances in modeling technology have enabled reproduction of a part of bone with standard devices both at low or high temperatures. Besides, it is achieved at a lower cost. Osteoconduction is optimized in biomaterial that mimics bone structure and chemistry. Such osteoconductive matrices are synthesized by cross-linking collagen with poly-L-lactide [68].

The three-dimensional scaffold produced by this process has the property to adhere both to hard and soft tissues and has sufficient wet cohesion in the body. Specially designed macrostructure provides programmable and more complete bone proliferation and vascularization [69]. A combination of biodegradable polymers (including starch, poly-L-lactide, polyhydroxyalcanoated, hyaluronan and collagen) and hydroxyapatite, Ca phosphate or bioactive glass refinements have been made to produce both bone and periosteal substitutes with enhanced biological and mechanical properties [70].

For wound healing a composite of biosynthetic membranes and skin substitutes based on modified collagen-chitosan-hyaluronan complex have been developed. This is expected to promote wound healing and achieve guided tissue regeneration [71]. For heterogeneous multifunctional biomaterials, a combination of 3-D cell osteogenic culture of bone marrow, a neurogenic bioactive factor, and other neurotrophins have been used [72].

3.6.2. Bone Stabilizing Materials

During the last more than 100 years repair of bone fractures was accomplished by use of metallic (iron and steel) screws, pins and other fixation-type devices for binding the fractures. Later on, more sophisticated metal alloys of titanium, zirconium, niobium and tantalum were used for the same purpose. However, the need to use substances more compatible to the human body remained and researchers came up with an alternative, namely, bioabsorbable polymers. Metallic devices were subject to corrosion inside the body causing inflammation, infection and even life-threatening injury. Alloys offered an improvement and could be implanted for a longer time. But being harder than the bone they support, alloys interfere with bone regrowth. Both metallic as well as alloy implants had to be removed once healing had occurred. This required another surgery involving more costs, trauma and other problems associated with the procedure.

Polymers offer an alternative substitute in orthopedic implant devices that have none of the problems associated with metallic and alloy devices. They are biocompatible and can be engineered to acquire properties close to natural bone. They have adequate strength to withstand weight-bearing applications while at the same time are elastic enough to allow bone growth. Besides, they are bioabsorbable and the rate of absorption, too, can be engineered to coincide with bone growth. Being absorbable in the body eliminates the subsequent need for second surgery required in metallic implants.

One serious problem is that bioabsorbable polymers release toxins when they are absorbed in the body, which may not pose a problem for small devices like sutures and staples but for weight-bearing devices can cause inflammation problems. Hopefully this problem will also be removed with advances in polymer chemistry.

3.7. Nerve Regeneration

One of the most disheartening aspects of medical science has been the inability of nerves to re-grow after damage by trauma or other factors leading to nerve degeneration. Recent researches in polymer science and tissue culture techniques have given great hope to patients with nerve degeneration.

Researches on nerve growth have revealed interesting facts. The nerves grow on things and do not just float around in the body. Researchers have been able to produce a gel, which will lay down a medium in which the nerves can regrow. Such a gel can be tailored to the requirements of nerves to be regenerated [73].

Nerves are also soft structures. Their growth will be hampered by any obstruction in their path. Therefore, scar tissue causes impediment in nerve growth. The gel structure that will form the backbone for nerve regeneration can be impregnated with scar-tissue-inhibiting and nerve tissue growth-stimulating molecules to favour nerve regeneration. Similarly, there are substances, which can inhibit certain types of cells and stimulate growth of other types. One such molecule is laninin, a protein in extracellular matrix that allows brain cells to adhere to connective tissue [74,75]. Attempts are also being made to find out the molecules that would attract nerves and ones that would repel nerves. Such molecules will behave like traffic signals directing the nerves to grow in a desired manner [76].

The nervous system is divided in two basic units, the central nervous system comprising of brain and the spinal cord, and the peripheral nervous system comprising peripheral nerves. It has been observed that nerves in the peripheral nervous system regenerate better than those in the central nervous system. The peripheral nervous system has special cells called the Schwann cells. These cells secrete molecules, which promote nerve regeneration. The Schwann cells are absent in the central nervous system. So if it can be found out as to what secretions of Schwann cells promote nerve regeneration, it may help to induce regeneration in central nervous system as well by appropriately applying these secretions even in the absence of Schwann cells in the central nervous system [67]. Stanford Ophthalmic Tissue Engineering Laboratory is modifying biodegradable scaffold technologies for use in retinal cell transplantation and for nerve regeneration to treat Macular Degeneration of Retina [78]. Polymer Nanofiber tubes also been found to boost nerve regeneration [79].

Electrically conducting polymers have beneficial effects on wound healing, including the repair of damaged nerves, cartilage, skin and bone. One such polymer is a derivative of polypyrrole. It contains the electrically conducting polypyrrole segments connected by a hydrolysable (degradable) ester linkage. The application of this polymer in the nervous system is in the treatment of spinal cord injuries and in plastic and reconstructive surgery in which peripheral nerve grafts are required [80].

With these developments and refinements of the techniques of nerve regeneration using appropriate biocompatible, biodegradable and electrically conducting polymers with addition of appropriate inhibiting and promoting factors and developments in bionic technology one day it may be possible for the paralyzed to walk, the blind to see, the deaf to hear and the memory-impaired to recollect [81].

3.8. Ophthalmology

The use of polymers in treating diseases of the eye has acquired significance, especially with respect to fabrication and implantation of intra-ocular lens and retinal degeneration. The human eyeball is about 2.5 cm in diameter and is made of three layers **(Fig. 10.13)**. The outermost layer is opaque and appears white in color. It has got fibrous tissue, which provides firmness to the eyeball. It is called sclera (scleros = hard). In front, this outer layer becomes transparent to allow passage of light into the eyeball. This portion of the outer layer is known as cornea. The middle layer extends up to the place where cornea and sclera meet and is known as choroid. It contains blood vessels. The innermost layer is the light sensitive layer where the image of outside object is formed. Behind the cornea there is a lens. It is made up of collagen fibers, proteoglycans and a special class of protein, which is so arranged as to make the lens transparent to light. A special property of lens is that its focal length changes involuntarily as per the requirement to focus the light sharply on retina. This property of the lens is called accommodation. The space in front of the lens is filled with a thin fluid called the aqueous humour. The pressure of acqueous humour is maintained at normal level by a balance between production and drainage of the fluid in this area. The space behind the lens is filled with a thicker, gel-like, avascular mass of fluid called vitreous humour. It contains thin collagen fibrils. The fluid itself contains a mixture of D-glucaronic acid and N-acetylglucosamine. Vitreous helps to provide shape to the eye. It is transparent to allow light to fall on the retina [77]. In normal vision light travels through cornea, lens and vitreous and falls on retina, which has light sensitive structures. Involuntary changes in focal length of the lens helps in sharp focusing of the image on the retina. From there the image is carried by the optic nerve to the visual centre in the brain [82]. Any changes either in transmission of light through the eye or in the light sensitive structure in retina leads to impairment or loss of vision. Diseases of cornea, acqueous humour, lens, vitreous

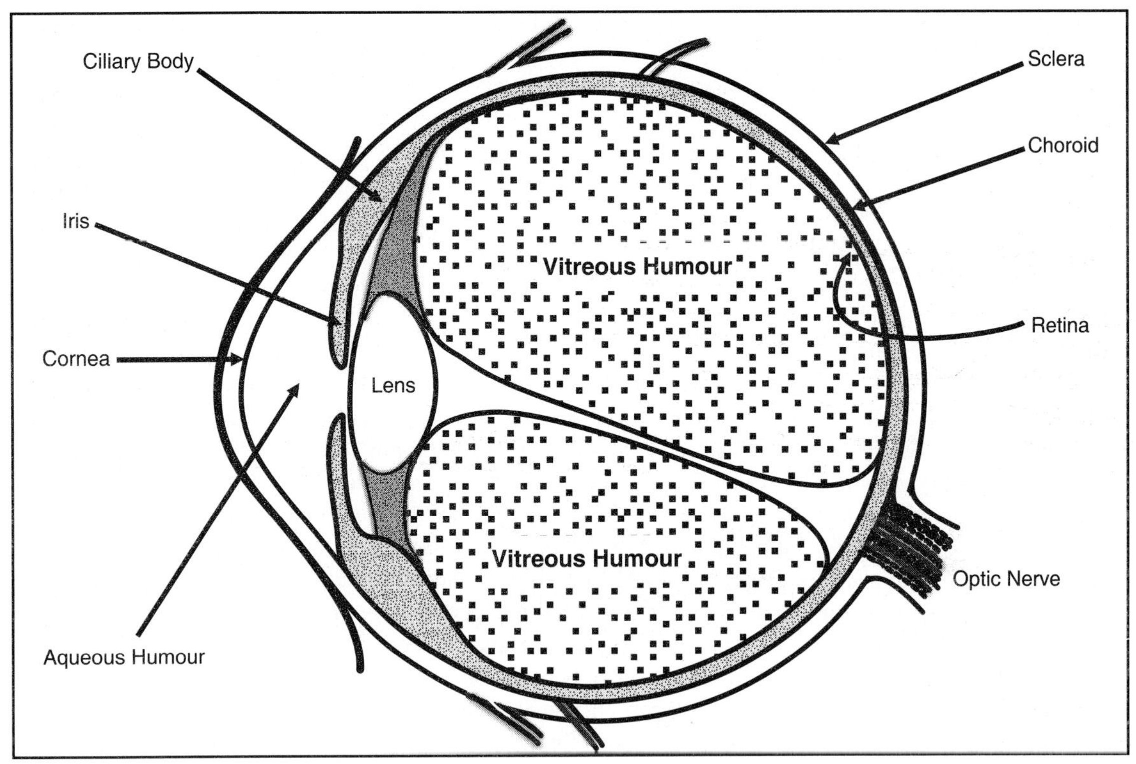

Fig. 10.13. Anatomy of normal eye.

humour and retina adversely affect the transmission and capture of the image. The diseases like glaucoma, cataract, fluidity or opacity of vitreous and retinal degeneration increase with age. Therefore, the requirement of improving and maintaining vision has gained special importance with increase in longevity.

Use of polymers has an important role especially with respect to lens, vitreous and retina. When the lens becomes opaque (a condition called cataract), it has to be removed and replaced with an artificial lens. The synthetic intraocular lens is typically made from Polymethyl methacrylate (PMMA). This material has excellent clarity and biocompatibility. Other polymers are also used as intraocular lens material. Of special significance are silicones. Lens made from this material is flexible and can be more conveniently folded for insertion into the eye, causing less damage to endothelium as compared to that after the placement of PMMA lenses. This feature is important in certain patients. Recently, claims have been made for development of a gel, which cures the lens capsule [83].

In younger people where the eye lens has not become opaque, but its power of accommodation is not adequate, adjustment in refractory power can be achieved by use of contact lens as an alternative to spectacles. Polymeric materials are also used to fabricate contact lens. Poly(methyl methacrylate) is a commonly used material for contact lens. It has the disadvantage of being rigid and has low oxygen permeability. Contact lens is placed over the cornea to add to the refractive power of the medium. Normally cornea derives oxygen from the atmosphere. Placement of contact lens hampers in transfer of oxygen to the cornea. Therefore, other materials had to be chosen for fabrication of contact lenses which have higher oxygen diffusion property [84]. However, an ideal material for contact lens is yet to be found for extended use. As of now, the contact lens has to be removed at least for the period of sleep and at other times also whenever possible. Vision Cooperative Research Centre have claimed the development of a contact lens which can be worn permanently, a permanent contact lens [85]. The placement consists of a simple operation in which a simple debridement of epithelial tissue of cornea is carried out and then a permanent contact lens made from a transparent, microporous polymer, which allows the epithelial tissue to regrow and cover the synthetic lens [83].

As already mentioned, the anterior chamber contains aqueous humour, whose pressure is maintained within normal limits by adjustment of the amount of production and drainage. Whenever it is produced in larger amount, or more commonly when there is impairment in its drainage, as happens in narrow angle glaucoma or in diabetes with formation of new vessels and in several other conditions, there occurs a rise in the pressure in the anterior chamber of the eye. This condition is called glaucoma. It is a serious condition and if untreated early, results in complete loss of vision. The main surgical intervention consists of creation of a drainage passage through sclera and cornea. Molteno developed the first successful device for this purpose in 1969 [86]. It comprised a polypropylene plate and a drainage tube made of silicone. Later, a number of improvements were made in this basic device for better drainage function. These had the silicone drainage tube with the plates fabricated from various polymers, including silicone and polymethyl methacrylate [87].

The human eye contains about 4 ml vitreous, which occupies approximately 80% of the globe of the eye. Serious eye diseases can adversely affect the vitreous. Its amount and thickness can decrease. Fibrosis in the vitreous can create a pull over the retina, which can get detached causing blindness. Such conditions require replacement of vitreous humour with artificial fluid closely mimicking the properties of the natural vitreous. Replacement of vitreous by such substances is the only method of preventing blindness. The possibilities of a polymeric substance with suitable properties lead to directing research on this aspect. Semi-crystalline poly(vinyl alcohol) (PVA) has been found to maintain the desired structural properties after injecting into the eye through a needle. Crystallinity is an important factor determining the physical properties of vitreous. This has been introduced into polyvinyl alcohol by the nucleation of crystallites through quenching an organic/aqueous PVA solution to sub-zero temperatures, followed by controlled crystal growth [88]. This polymer belongs to a group of physically cross-linked hydrogel. Some water soluble polymers (polyethyleneimine, polyvinylpyrrolidone with copolymers) have the drawback of inducing posterior detachment of vitreous because of rapid condensation.

Experiments conducted on rabbits, showed that water-soluble polymers specifically react forming complexes with components of vitreous. The vitreous was found to shrink under the effect of polyelectrolytes (vitreosynerysis) leading to posterior detachment in 1.5 hours after injection. The degree of vitreosynerysis depended on the complex-forming activity of polyelectrolytes towards the vitreous components [89]. Therefore, attempts were made to develop polymers, which exerted no toxic or traction effects on the adjacent eye structures. Production of artificial vitreous is an important step in prevention of blindness and in maintaining the integrity of the eye. Further attempts are being made to develop a polymer, which can match the physical properties of natural vitreous and permit the surgeon to manipulate the detached retina [90].

Retina is the most important structure for reception of the image as this is the only layer, which has light sensitive structures. These light sensitive structures are in the forms of rods and cones. Rods are distributed more towards the periphery of the retina and the cones are concentrated in the posterior part of the eyeball, near the place where the optic nerve enters the eye. Rods are important for night vision and cones are important for day vision. Retinal damage occurs in large number of conditions. Of special significance is a condition called retinitis pigmentosa. This is a genetic disorder characterized by degeneration of light sensitive structures of the eye. Another important cause of degeneration of retina is the ageing process itself. Two types of age related degenerations have been recognized, namely, the dry type and the wet type. The losses of light sensitive structures of retina

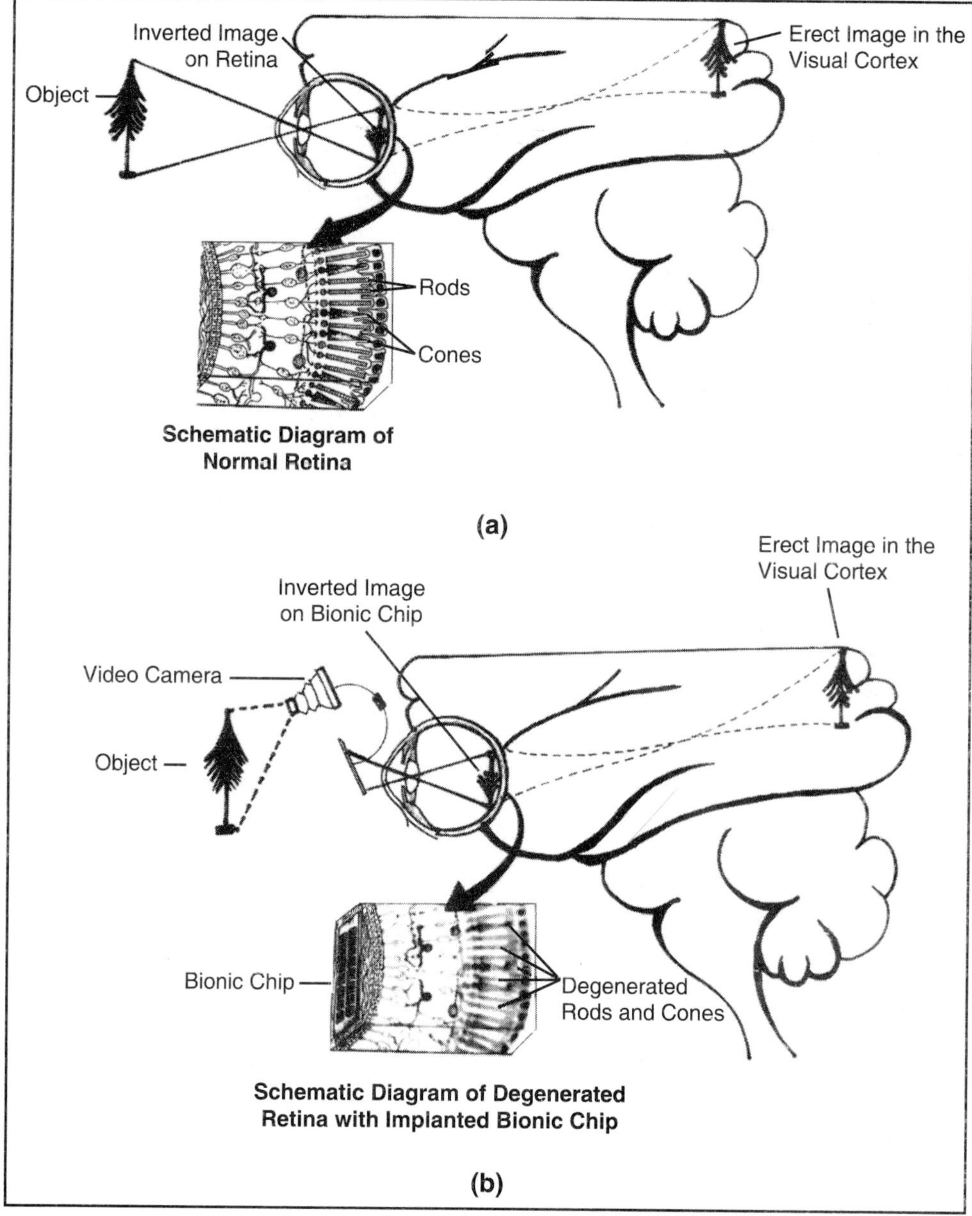

Fig. 10.14. Schematic diagrams showing the pathway of vision through (a) a normal eye and (b) through a bionic eye.

do not regenerate. The production of visual impairment or loss due to retinal degeneration is therefore a dreaded entity. Recently, an epoch making discovery, called the **'bionic eye', (Fig. 10.14)** has given hope for regaining the function of retina in patients with retinal degeneration [91]. It is a circular device of the size of a five-cent piece, which can be inserted into the eye at the site of the retina. It comprises chips that can receive the radiosignals, decode it and deliver the stimulations to retinal ganglion cells through small wires. The device comprises about 100,000 tiny ceramic detectors, each 1/20 the size of a human hair. The assemblage is too small for surgeons to handle it safely. For this reason, the arrays of detectors are attached to a polymer file one millimeter by one millimeter in size, which helps to keep them in place. After a couple of weeks of implantation, the polymer film gradually dissolves leaving only the array behind. After making an incision in the sclera (the white outer layer of the eyeball), the retina is elevated by injecting appropriate amount of fluid under it and the artificial retina is placed within that blister. The procedure is still in the experimental stage but excellent results have been claimed [92].

3.9. Ear

Human ear is divided in 3 sections, the outer, the middle and the inner ear. The external ear has two portions—the pinna- and a canal leading from it, which is called the external auditory canal. At the end of the auditory canal there is a membrane called the tympanic membrane, which separates the external ear from the middle ear. The middle ear is a box-like structure following the tympanic membrane. It has 3 small bones (called ossicles), namely, the malleus, the incus, and the stapes placed respectively from outside to inside. Mallius is attached to the tympanic membrane one side and articulates with incus on the other side. Incus, in turn articulates with stapes. The stapes sits on a window in the bone case housing the internal ear. Malleus, incus and stapes articulate with each other but are sufficiently mobile and well suited for transmission of the sound waves from external ear to the internal ear. The internal ear is also enclosed in a bony cage of the skull. It has two portions, one for appreciating and balancing the position of the body and the other for hearing. The bone housing the internal ear has a window, which opens in the middle ear. It is also covered by a membrane, which separates inner ear from the middle ear. The hearing portion of inner ear is placed in a coiled structure called cochlea (G. kochlias = a spiral) [93].

The function of the external ear is to converge the sound waves to the tympanic membrane. From there, the waves are mechanically transmitted through the three tiny bones of the middle ear to the internal ear through the window described above. The internal ear has receptors (sensitive structures), which convert the sound waves into electrical signals. These electrical signals are then carried by a nerve, called the auditory nerve, to the brain where it is finally interpreted [93].

Serious loss of hearing may occur either because of destruction of the three small middle ear bones, or because of the loss of sound sensitive structure of the inner ear. The auditory nerve may also be damaged causing loss of transmission of electrical impulses and consequent loss of hearing. But commonly it is the damage to the sound sensitive structure, which is the cause of hearing impairment in old age. Impairment or loss of hearing due to defects in conduction of sound waves through external and middle ears is called 'conduction deafness' and the loss of hearing because of damage to the sound sensitive structures or to the nerve is called 'nerve deafness' [94].

Polymers have been used to fabricate the pinna of the external ear, which may be damaged by trauma or by developmental defects. Recent advances in fabrication of small polymer appliances have made it possible to fabricate the small ear ossicles. The polymeric ossicles can be used to

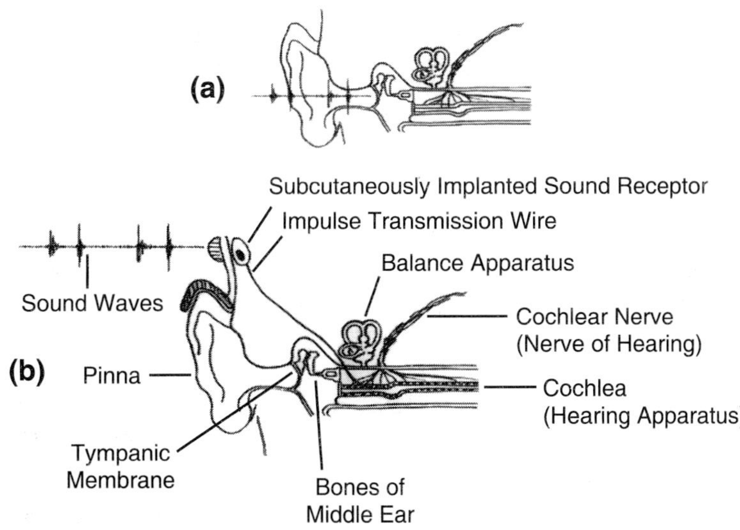

Fig. 10.15. Schematic diagram showing hearing though (a) normal and (b) bionic ear.

replace the damaged ear ossicles in conduction deafness. These have the advantage of being biocompatible and non-toxic. Further they are not corroded over a period of time. Nor do they induce any adverse local reaction [95].

However, the most interesting advance has been in the use of polymers in what is called the bionic ear or cochlear implant system or cochlear prosthesis. As described above, the hair cells in the cochlea of the inner ear act as transducers and convert the sound waves into electrical signals for transmission through the hearing nerve (the eighth cranial nerve). Loss of these receptors results in loss of the transducer function. These patients have loss of hearing in spite of the fact that they have a normal external ear and the hearing nerve. The purpose of cochlear implant system is to bypass the hair cells in the cochlea by presenting electrical stimulation directly to the auditory nerve fibers (**Fig. 10.15**). This restores the hearing function, at least partially, but that too is a great help to patients. After the insertion of chochlear implant, there is a need to treat the surrounding structures with appropriate drug or other compound in order to inhibit the growth of fibrous and the bone tissue, promote healing, prevent neuronal degeneration and/or promote neural regeneration. Polymers have a role to play not only in fabrication of the cochlear implant but also in drug delivery system to achieve these desired effects. Further, the scientists are building a new **'bionic ear'** coated in a smart plastic that boosts the growth of nerve cells in the inner ear. Professor Gordon Wallace of the Intelligent Polymer Research Institute at the University of Wollongong has used a polymer that can conduct electricity as well in addition to stimulating nerve regrowth. The polymer can be made to control the timing and site of release of drug in order to maximize interaction with nerve cells [96,97].

3.10. Breast Implants

Prosthetic breast implants are needed principally for cosmetic reasons [98] (**Fig. 10.16**). These implants have a gel filled impervious sac. Typically silicone fluid gel is used in breast implants. The

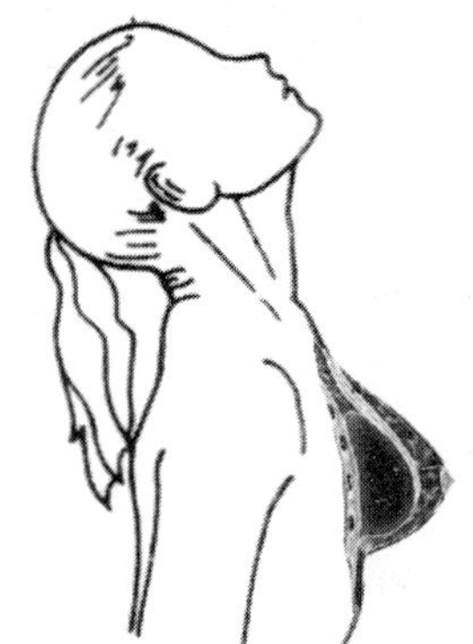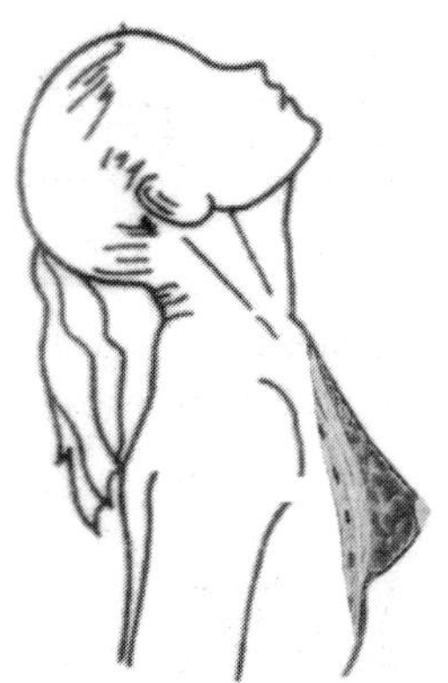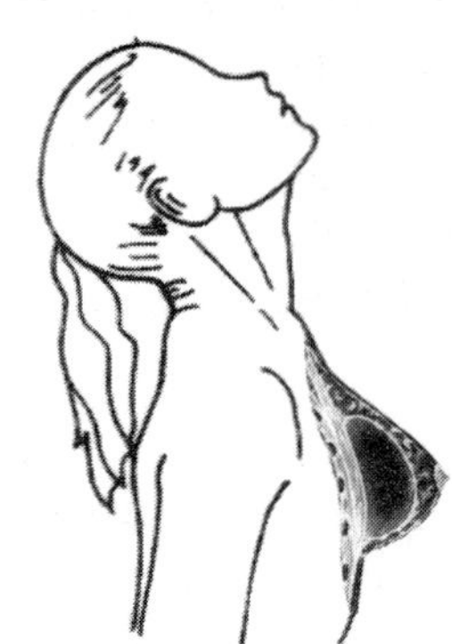

After Sub-Pectoral Implant   **Before Breast Implant**   **After Sub-Glandular Implant**

Fig. 10.16. Saline filled polymer breast implant. In pendulous breast or after mastectomy (surgical removal of breast), a saline filled polymer bag (shown dark black in the above figures) may be put to produce the desired shape to the breast for cosmetic reasons.

disadvantage is that of leakage of the gel, even in the absence of the polymeric outer sac, causing fibrosis and other adverse side effects [99]. In some cases it may be calcified requiring another surgical intervention. Further, the presence of fibrous capsule poses difficulties in detection of breast cancer [100]. Therefore, the use of silicone gel is no longer approved by United States Food and Drugs Administration [101] and saline filled breast implants are preferred. These may also leak, but presence of small amount of saline does not produce significant adverse reaction in the surrounding tissue [102]. The breast implants may become infected following piercing injury [103].

3.11. Facial Implants

Polymer material may be used in plastic surgery to fill in the gaps on the face caused by damage by injury or disease process or by surgical removal of facial structures. Earlier, such gaps were filled by tissues taken from other parts of the body (called autogenous tissues) [104,105]. Polymeric facial implants have been specially used for augmentation of malar (cheek area), paranasal (by the side of nose) and chin regions of the face [105].

The polymeric implants (called alloplastic implants) have advantage over autogenous implants with respect to availability of the material, simplification of the operative procedure, decreased exposure to anaesthesia, possibility of selecting polymeric material of desired strength, elasticity and durability for a specific procedure. It also helps in reducing the period of morbidity and the risk of infection [104,105]. Another advantage of these implants in the face is that they can be layered over other augmentation forms, such as fat, which gives it a better feel [106].

A variety of implant materials are available including, carbon based polymers, non-carbon based polymers and aliphatic polymers. Non-carbon based polymer, silicone, was first used in facial implants in 1953. Carbon based polymers have been in use for several decades. Polytetrafluoroethylene (PTFE), aliphatic polyesters, and methylmethacrylate were initially introduced in 1940s [104,105]. Later, modifications in these polymers were made to improve their quality and patient satisfaction. The commercially available improved facial implants include Gore-Tex Facial Implant, Advanta Facial Implant, Gor-Tex Strands and Multi-Strands, SoftForm, and Ultra-Soft forms, etc. These implants are usually made from Polyetrafluoroethylene or PTTE, expanded polytetrafluoroethylene, dual-porosity expanded polyterafluoroethylene and tubular expanded polytetrafluroethylene. They offer a firmer feel, a permanent solution until taken out, and are fully reversible. Some are soft with high porosity center integrated with porosity on the outer surface layer. Some are softer than others. The commercially available forms of facial include Advanta Facial Implant. These differ in physical properties depending upon the polymer characteristics [107-109].

3.12. Gynecology

Polymers have been commonly used in obstetrics and gynecology for:
1. General surgical purposes (screws, needles, clips).
2. As a drug delivery system for contraception.
3. For increasing fertility.
4. Treatment of uterine fibroid.
5. Treatment of uterine prolapse.
6. The most fascinating field of tissue engineered genital tract.

3.12.1. Contraception

Polymer carriers have been used for slow release of contraceptive drugs from intrauterine contraceptive devices. Levonorges-trel intra-uterine system (Mirena) consists of a polymer cylinder mounted on a T-shaped frame and controlling release rate membrane. The system has been found to be very effective in comparison to copper intrauterine device (**Fig. 10.17**) [110].

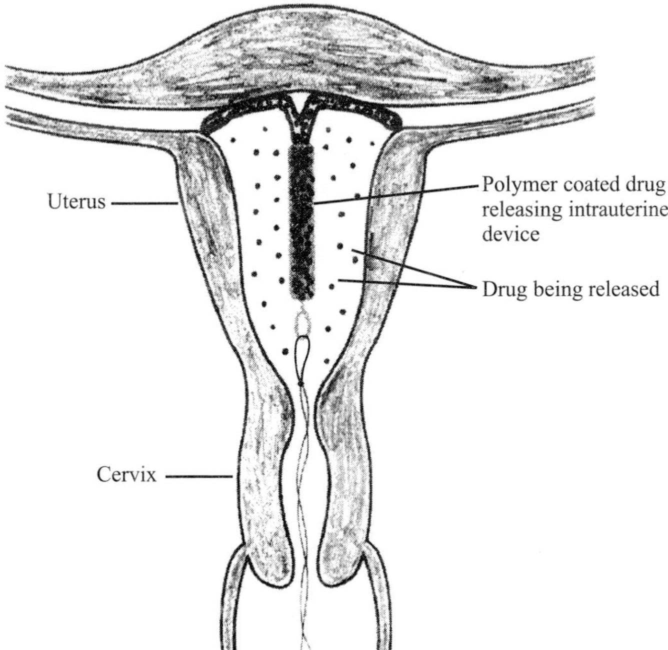

Fig. 10.17. Polymer coated drug-releasing intrauterine device for female contraception.

Experimental studies with polymer bio-implants programmed for sustained release of melatonin have reported beneficial results with respect to gonadal activity. These results were attributed to sustained release of the drug over a period of time, which was possible only with polymer bioimplants. The investigators commented that any other tested modalities of administration would not provide any appreciable results because the half-life of melatonin administered in other forms is only 2-3 hours, while the beneficial effect depends upon sustained release over a longer period of time. Experiments conducted on rats have reported the use of a polymer which when injected into the vas deferens lowers the pH sufficiently to kill the spermatozoa passing through, while the polymer itself does not degrade. This process achieves contraception in males. The polymer can be later flushed out to regain fertility. Being a non-surgical, non-occlusive and reversible process it offers an attractive method of male contraception [111].

Polymers also help in preparing a technique for transdermal fertility control system in females. The transdermal fertility control unit comprises:

1. A backing layer which is impervious to estrogen and progesterone.
2. A polymer matrix disc which is in contact with backing layer through which the hormones can be absorbed transdermally.
3. An adhesive, which anchors the dosage unit in intimate contact with the skin.

The backing layer is made from a material that is mostly impermeable and is made of such polymers as polyethylene, polypropylene, polyurethane, polyethyleneterephthalate and foils of polymer films. The trans-dermal route has following advantages over systemic medication [112]:

1. It avoids the risk of incomplete absorption and metabolisms as associated with oral therapy.
2. With trans-dermal patch it is possible to use a pharmacologically active agent with short biological half-life.
3. It is possible to program delivery of drug at the desired rate.
4. It prevents destruction of drug, which occurs in systemic circulation on oral intake.
5. It is possible to terminate delivery of the drug whenever required.

3.12.2. Improving Fertility of Sperms

A hybrid polymer delivery system has enabled scientists to enhance fertility by increasing the sperm life span and vitality in mammals. This had been achieved by using a hybrid polymer system deliver vitamin E to sperms by hybridizing it with galactose. Vitamin E, a powerful antioxidant, prevents damage to cells (sperms), thereby preserving the integrity of sperms for a longer period. This antioxidant effect causes increase in fertility in cases of shortened life span of sperms. A specific carbohydrate binding protein present on the surface of mammalian sperm cell recognizes sugar. This galactose polymer hybrid is targeted to the sperm cell. The polymer then binds to the sperm cell and is transported across the cell membrane where Vitamin E is released. It was found that the new hybrid polymer enhanced physiological properties and life span of sperms enhances fertilization rates after artificial insemination. The hybrid polymer technique could also be used to preserve the sperms in situations where storage outside the body is required, as in human infertility treatment and in the breeding of a rare endangered species [113-114].

3.12.3. Treatment of Fibroids

Fibroids are non-cancerous tumours of uterus. Until now, the surgical removal of fibroid or removal of uterus was the only available treatment. However, it is now possible to destroy the fibroids by cutting off the blood supply to the fibroid by blocking these with polymer material. In fact with the recent advances in design of polymer spheres, it seems that Uterine Fibroid Embolization will now be the preferred treatment of this common condition.

A small nick (less than ¼ of an inch) is made in the skin of the groin and a catheter is negotiated into the main artery of the thigh. The catheter is then advanced to the arteries supplying the fibroid. Small spheres or particles made of plastic or sponge material of the size of grains of sand are injected through a catheter placed in the artery supplying the fibroids. Because of decrease of blood supply there occurs death of a part of fibroid tissue resulting in shrinking of fibroid tissue [115].

Bio-Sphere Medical's microspheres have produced precisely calibrated, spherical, hydrophilic, micro-porous beads made of an acrylic co-polymer (trisacryl), which is then cross-linked with gelatin. With this product a more complete and targeted occlusion has been claimed. It has also been claimed that the hydrophilic surface and spherical shape prevent aggregation within the catheter lumen and in the vasculature, promoting ease and accuracy of delivery. The elastic properties of the microspheres

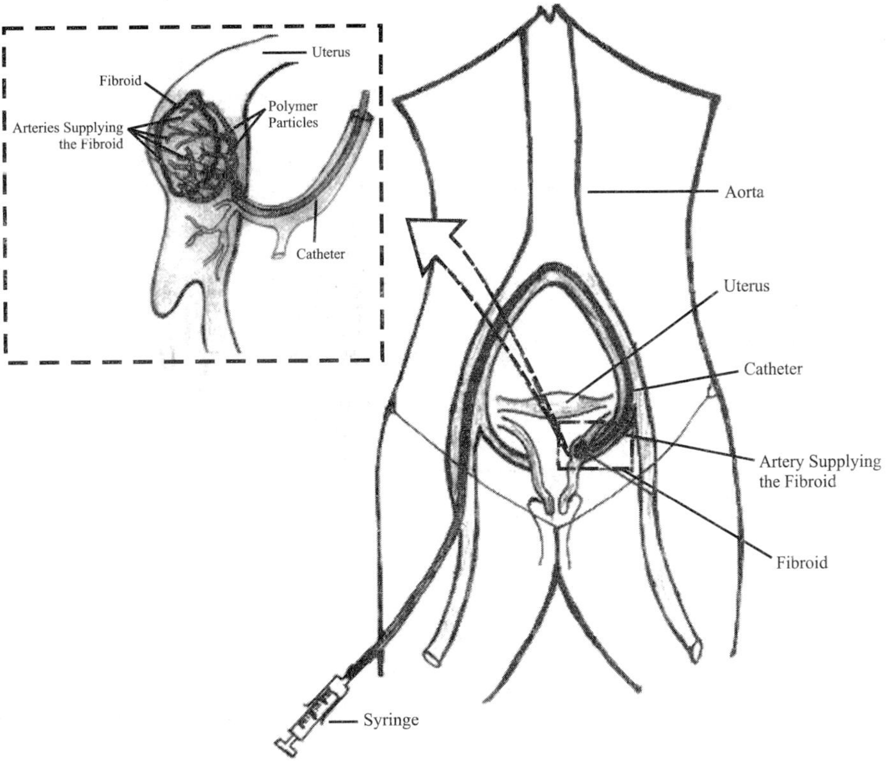

Fig. 10.18. Schematic diagram showing the technique of fibroid embolization using polymer spheres. Fibroid (a tumor of uterus) is dependent upon its rich blood supply, which can be cut off by blocking the supplying arteries by polymer particles, thus causing regression and finally disappearance of the tumor. The polymer particles are very suitable for this purpose as these can be easily injected in the feeding arteries through a simple percutaneous procedure.

allow the temporary deformation of the sphere, which facilitates the passage through small delivery systems **(Fig. 10.18)** [116].

3.12.4. Treatment of Prolapse of Uterus

Polymer scaffolds have been found suitable for the treatment of a common gynecological condition known as uterine prolapse. In this condition the support of the uterus gets weakened due to pregnancy and it descends down through the vagina. There is a 10% lifetime incidence of prolapse in general female population and the incidence increases with the number of pregnancies. A woman with three pregnancies has three-fold risk of prolapse. Currently the condition is treated surgically and often repeat surgery is required. Strengthening of the uterine support is a logical approach where removal of uterus is not desired. This uterine support is provided with a biocompatible polymer mesh. Stem cells obtained from uterus can be easily implanted on the polymer mesh. These cells differentiate into cartilage forming cells, which colonize the polymer mesh thus restoring the uterine support. The technique still in experimental stage and a lot of research is required and it may take a decade before it can be adopted in the clinic **(Fig. 10.19)** [117].

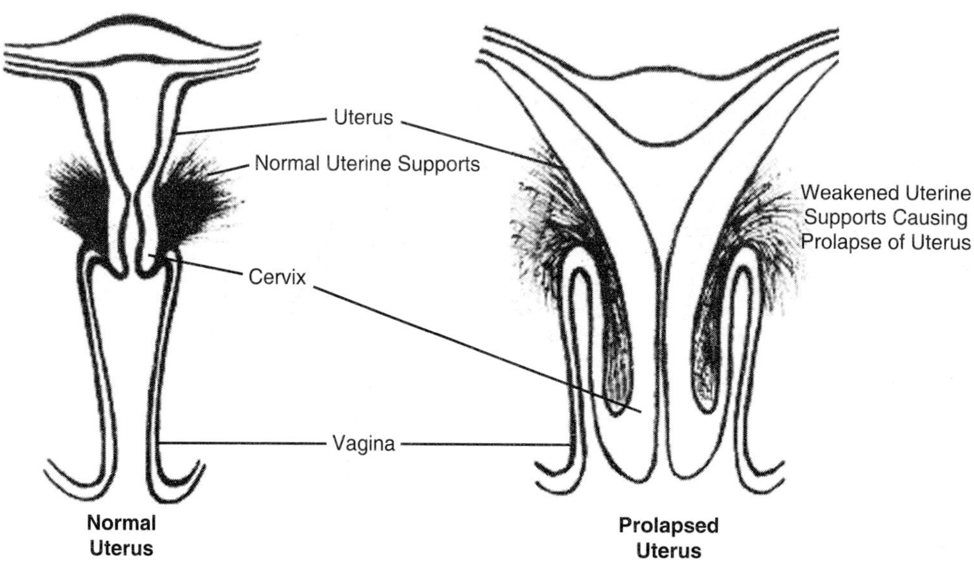

Fig. 10.19. Normal uterine support and Lax uterine support with prolapsed uterus. The lax support can be reinforced by regenerating the tissues over a polymer scaffold, thus correcting the abnormality.

3.12.5. Tissue Engineered Genital Tract

Female genital tract may be affected by several congenital malformations, including ambiguous genitalia, hypoplasia (underdeveloped) or atresia (narrowing) of vagina, cervix and uterus. These abnormalities are usually detected in adolescent females, and have profound psychological and reproductive implications [118]. The genital tract may also be damaged by tumour, trauma and severe inflammatory disease. The congenital abnormalities and acquired damage to genital tract are the commonest causes of female infertility. Till now the only modality of treatment was to reconstruct the genital tract, whenever possible, by appropriate plastic surgical methods [119]. The success of reconstruction surgery are limited. Pregnancy cannot occur if extrauterine tissues are used for reconstruction. Use of synthetic biomaterials has also failed to produce functionally desired results [120].

Developments in tissue engineering have made it possible to produce tissue engineered female reproductive organs by culturing cells from the female reproductive tract. The cultured cells are perfused on to a biocompatible matrix of appropriate polymer. With this technique the whole female reproductive tract as well as the fallopian tubes can be generated. The biocompatible matrix can be of a non-biodegradable or biodegradable three-dimensional scaffold. The artificial female reproductive organ or tissue thus developed exhibits the compliance and vasculature of natural female reproductive organs. The studies are still in experimental stage but have immense scope for treatment of congenital and acquired defects of genital tract and treatment of many cases of female infertility [121].

CONCLUSIONS AND FUTURE PROSPECTS

The use of polymers in medicine has unlimited potential in the coming years. It is dependent on the advances in polymer science in collaboration with medical and engineering scientists for the development of biomedical devices. The most prominent applications are in the field of genetic engineering, drug delivery and cell targeting, scaffold engineering, and development of biodegradable and non-biodegradable devices for medical applications. These medical applications heavily depend upon physical and chemical properties of polymers to suit in-vivo environment. Intelligent polymers with multifaceted responses to changes in biological milieu will find use in medical applications which require feedback to correct derangements in body functions, such as hypertension, cardiac arrhythmias, etc.

REFERENCES

1. C.M. Agrawal and R.B. Ray; J. Biomed.Mater.Res., 55(2), 141 (2001).
2. Barbara Breindel; [http://www.bccresearch.com/biotech/B072N.html].
3. J.C. Middleton and A.J. Tipton; Med.Plastics and Biomater., p. 31 (March/April 1998), [www.devicelink. com/mpb/archive/98/03/002.html].
4. [www.polymertech.com/press/pr_device.html].
5. Tony Blythe and David Bloor; "Electrical Properties of Polymers", 2nd Edition Cambridge University Press, Cambridge (2005).
6. B. Muir, V. Osborne, C.K. Luscombe and W.T.S. Huck; in "Encyclopedia of Nanoscience and Nanotechnology (Ed.H.S.Nalwa)", American Scientific Publishers, (2004) vol. 3, p. 497.
7. W.T.S. Huck; Int.J.Nanotechnology, 1, 119 (2003).
8. W.T.S. Huck; Chem.Commun., 4143 (2005).
9. [http://www.scienceclarified.com/everyday/Real-Life-Chemistry-Vol-3/Polymers.html].
10. N. Kashyap, S. Modi, J.P. Jain, I. Bala, S. Hariharan, R. Bharadwaj, D. Singh, R. Mahajan, N. Kumar and M.N.V.R. Kumar; CRISP, vol. 5(3), p. 7 (July-September 2004), [http://www.niper.nic.in/polymers.pdf].
11. M.S. Barrow, R.L. Jones, J.O. Park, M. Srinivasarao, P.R. Williams and C.J. Wright; Spectroscopy, 577, 18 (2004).
12. [http://www.bccresearch.com/biotech/b072N.html].

13. I. Engelberg, J. Kohn; Biomaterials, 12, 292 (1991).

14. C. Bastioli (Ed.); "Handbook of Biodegradable Polymers", Rapra Technology, Novamont SpA, Italy (2005).

15. [www.srs.ac.uk/srs/news_extras/news_polymer.html].

16. [http://web.mit.edu/newsoffice/2005/smart-plastics.html].

17. [http://www.iforum.umontreal.ca/ForumExpress/Archives/vol2no1en/article11_ang.html].

18. [http://www.secretariat.ne.jp/nanofuture/current/pdf/T2-3.pdf"].

19. [www.secretariat.ne.jp/nanofuture/current/pdf/T2-3.pdf -].

20. Allan S. Hoffman; Clin.Chem., 46, 1478 (2000), [http://www.clinchem.org/" \t "_top"].

21. G.A. Silva; Surg.Neurol., 61, 216 (2004).

22. Richard H. Smith; International Hospital Equipment & Solutions, 31(1), 6 (February/March 2005).

23. [http://www3.interscience.wiley.com/cgi-bin/abstract/109866200/ABSTRACT?CRETRY=1&SRETRY=0].

24. K.Roy; [http://www.bme.utexas.edu/faculty/shmidt/Courses/BME314/Notes/guest_lecture_BME314.pdf].

25. [http://www.tissue-engineering.net/].

26. [http://www.cbte.group.shef.ac.uk/research/mat6.html].

27. [http://www.cancerhelp.org.uk/help/default.asp?page=95].

28. R. Langer; Science, 260, 920 (1993).

29. [http://www.cems.uvm.edu/~iatridis/me208/tisseng1_intro_lee.pdf].

30. [http://www.cems.edu~iatridis/me208/tissueng_intro_lee.pdf].

31. N.R. Choudhury and H. Griesser; [http://www.unisa.edu.au/iwri/futurestudents/phdprojectsmacroporous polymer.asp].

32. G.F. Muschler, Chizunakamoto and L.G. Griffith; J. Bone and Joint Surg., 86, 1541 (2004).

33. J. Feijen; [http://pbm.tnw.utwente.nl/research%202002.htm; http://pbm.tnw.utwente.nl/body.html].

34. R.B. Mazess, Clinical Orthop.Rel.Res., 165, 239 (1982).

35. K. Naraine, C.C. Chery, E. Goetghebeur *et al.*; Eur.Surg.Res., 37(3), Page No., (2005).

36. F.J. Schoen; HST 090 (April 3, 2006).

37. S.H. Daebritz, B. Fausten, B. Hermanns, J. Schroeder, J. Groetzner, R. Autschbach, B.J. Messmer and J.S. Sachweh; Eur.J.Cardio-Thoracic Surg., 25(6), 946 (2004).

38. S.H. Daebritz, J.S. Sachweh, B. Hermanns, B. Fausten, A. Franke, J. Groetzner, B. Klosterhalfen and B.J. Messmer; Circulation, 108 Suppl. 1:II, p. 134 (Sept. 9, 2003).

39. J.A. Hyde and R.E. Phillips Jr.; J. Heart Valves Dis., 8(3), 331 (1999).

40. L. Vincent, M.D. Gott, E. Diane, D.E. Alejo and M.D. Cameron; The Annals of Thoracic Surg., 76(6), S2230 (2003).

41. B.D. Ratner, A.S. Hoffman, F.J. Schoen, J.E. Lemons (Eds.); "Biomaterials Science: An Introduction to Materials in Science, 2nd edition, Elsevier Academic Press (2004).

42. G.M. Bernacca *et al.*; J. Biomed.Mater.Res., 34, 371 (1997).

43. G.M. Bernacca *et al.*; Biomaterials, 23(1), 45 (2002).

44. [http://www.clarkson.edu/news/view.php?id=1218].

45. P. Bloomfield; Heart, 87, 583 (2002).

46. [http://www.devicelink.com/mddi/archive/02/05/0205d103a1.jpg','345','420'].

47. E. Sowton, A.D. Timmis, J.C.P. Crick, B. Griffin, A.K. Yates, P. Deverall and E. Sowton; British Heart J., 56, 115 (1986).

48. R. Mueller and T. Sanbom; Amer.Heart J., 129, 146 (1995).

49. R.A. Hill, Y. Dündar, A. Bakhai, R. Dickson and T. Walley; Eur.Heart J., 25, 902 (2004).

50. [http://www.crtonline.org/body2.cfm?id=205&ArticleID=252].

51. E. Grube, S. Silber, K.E. Hauptmann *et al.*; Circulation. 107, 38 (2003).

52. R. Fattori and T. Piva; Lancet, 361(9353), 247 (Jan.18, 2003).

53. K. Tanabe, E. Regar, C.H. Lee, A. Hoye, W.J. van der Giessen and P.W. Serruys; Curr.Pharm.Des., 10(4), 357 (2004).

54. H. Tamai, K. Igaki, E. Kyo, K. Kosuga, A. Kawashima, S. Matsui, H. Komori, T. Tsuji, S. Motohara and H. Uehata; Circulation, 102, 399 (2000).

55. P.W. Serruys, M.J.B. Kutryk A.T.L. Ong; The New England J.Med., 354(5), 483 (2006).

56. J.W. Nicholson; Int. J. Adhesion and Adhesives, 18(4), 229 (1998).

57. T.R. Pitt Ford; Restoration of Teeth, Blackwell Science Inc., (1992).

58. G.M. Brauer; J. Amer.Dental Assoc., 72(5), 1151 (1966).

59. [http://www.manufacturingcenter.com/dfx/archives/0505/0505med_feature.asp].

60. C.S. Young, S. Terada, J.P. Vacanti, M. Honda, J.D. Bartlett and P.C. Yelick; J.Dent.Res., 81(10), 695 (2002).

61. S.C. Bayne; J. Dent. Educ., 69(5), 571 (2005).

62. M.T. Duailibi, S.E. Duailibi, C.S. Young, J.D. Bartlett, J.P. Vacanti, and P.C. Yelick; J. Dent. Res., 83(7), 523 (2004).

63. W.W. Tomford, R.J. Starkweather and M.H. Goldman; J.Bone Joint Surg. (Amer.), 63, 244 (1981).

64. N.J. Phillips, A. Ali and D. Stanley; J. Bone Joint Surg. (Br.), 85B, 347 (2003).

65. S.N. Khan, F.P. Cammisa, Jr, H.S. Sandhu, A.D. Diwan, F.P. Girardi and J.M. Lane; J.Amer.Acad.Orthop.Surg,, 13(1), 77 (2005).

66. W.R. Moore, S.E. Graves and G.I. Bain; ANZ J.Surg., 71, 354 (2001).

67. A.G. Coombes and M.C. Meikle; Clin.Mater., 17(1), 35 (1994).

68. [http://www.keika.it/hybridizedpolymers.html].

69. K. Marra, P.G. Campbell, P.A. DiMilla, P. Kumta, M.P. Mooney, J. Szem and L. Weiss; Proc.Mater.Res.Soc., Fall Meeting (December, 1998).

70. D.W. Hutmacher; Tissue Engg., Suppl.l, 9, 45 (2003).

71. [http://www.keika.it/hybridizedpolymers.html].

72. S. Das; Proc.ASM Mater.Solutions Conf.Show, Columbus, OH, USA (October 18-21, 2004). [http://asm.confex.com/asm/ms2004/techprogram/paper_5305.htmASM].

73. [http://www.scicnceblog.com/community/older/2003/D/20032167.html].

74. [http://www.cwru.edu/pubs/cnews/1997/1-30/bellam.htm]

75. [http://www.scienceblog.com/community/older/2003/D/20032167.html].

76. S.K. Mallapragada and J.B. Recknor; Adv.Chem.Eng., 29, 47 (2004).

77. H. Gray, L.H. Bannister, M.M. Berry and P.L. Williams (Eds.); "Gray's Anatomy: The Anatomical Basis of Medicine & Surgery", Churchill Livingstone, 38th edition.

78. [http://visionchip.stanford.edu/polymerscaffolds.html].

79. [http://nanotechweb.org/articles/news/3/12/6/1].

80. P.C. Innis and G.G. Wallace; J.Nanoscience and Nanotechnol., 2(5), 441 (2002).

81. [http://www.usc.edu/dept/pubrel/trojan_family/summer04/imitation.html].

82. A.C. Guyton and J.E. Hall; Textbook of Medical Physiology, 10th Edition, Saunders, Philadelphia (2000).

83. [http://www.bio.csiro.au/Projects/PDFs/Opthalmic.pdf].

84. J.I. Lippman; Contact Lens Assoc. Ophthalmologists' J., 16, 287 (1990).

85. [http://www.abc.net.au/science/news/stories/s653099.htm].

86. A.C.B. Molteno; Br. J. Ophthalmol., 53, 161 (1969).

87. A.W. Lloyd, R.G.A. Faragher and S.P. Denyer; Ocular Biomater.Implants Biomater., 22, 769 (2001).

88. [http://chemistry.curtin.edu.au/research/reports/report98/polymer.html].

89. D.O. Skvorchenko, I.P. Khoroshilova-Maslova, L.D. Andreeva, I. Kh. Sharafetdinov, M.I. Shtil'man, I.A. Maklakova and D.G. Uzunian; Vestn.Oftalmol.Abstract Pubmed., 117(3), 16 (May-June 2001).

90. N. Soman and R. Banerjee;Bio-Med.Mater.Engg., 13(1), 59 (2003).

91. [http://science.nasa.gov/headlines/y2002/03jan_bioniceyes.htm].

92. [http://www.a-star.edu.sg/astar/front/studentsandscholarships/newsletter2/bionic.htm].

93. C.P. Anthony and G.A. Thibodeau; Textbook of Anatomy & Physiology, p. 338, St. Louis, Mosby (1983).

94. [http://www.emedicine.com/ped/topic931.htm].

95. [http://www.abc.net.au/science/news/stories/s1342345.htm].

96. [http://www.extenza-eps.com/WHP/doi/abs/10.1533/abib.2004.1.2.67;jsessionid=nvcsh_Os8GS7 EH9GVU?journalCode=abib].

97. [http://www.cardiff.ac.uk/biosi/staff/jacob/teaching/sensory/bio-ear.html].

98. C. Batich and D. DePalma; J. Long-Term Effects of Medical Implants, 1, 255 (1992).

99. S.L. Brown, M.S. Middleton and W.A. Berg; Alabama AJR Amer.J.Roentgenol., 175(4), 1057 (2000).

100. B.G. Silverman, S.L. Brown and R.A. Bright; Ann.Intern.Med., 124(8), 744 (1996).

101. S.L. Brown, C.M. Parmentier and E.K. Woo; Silicone gel breast implant adverse event reports to the Food and Drug Administration, 1984-1995. Public Health Rep (1998) Nov-Dec; 113(6): 535-43.

102. S.S. Strom, B.J. Baldwin, A.J. Sigurdson and M.A. Schusterman; Plastic and Reconstructive Surgery, 100, 1553 (1997).

103. M. Javaid and M. Shibu; Br. J. Plastic Surg., 52(8), 676 (1999).

104. [http://www.emedicine.com/plastic/topic53.htm].

105. [http://www.ncbi.nlm.nih.gov/entrez/query.fcgi?db=pubmed&cmd=Search&itool=pubmed_Abstract&term=%22Eppley+BL%22%5BAuthor%5D].

106. [http://www.ncbi.nlm.nih.gov/entrez/query.fcgi?db=pubmed&cmd=Search&itool=pubmed_Abstract&term=%22Sadove+AM%22%5BAuthor%5D].

107. [http://www.ncbi.nlm.nih.gov/entrez/query.fcgi?db=pubmed&cmd=Search&itool=pubmed_Abstract&term=%22Kahnberg+KE%22%5BAuthor%5D].

108. [http://www.yestheyrefake.net/facial_implants.htm].

109. [http://www.atriummed.com/facialimplantsframe.htm"\t "_blank].

110. [http://www.goremedical.com/" \t "_blank"].

111. [http://www.tissuetechnologies.com/" \t "_blank].

112. [http://www.endotext.org/female/female8/ch01s08.html].

113. M.M. Misro, S.K. Guha, SK, H.P. Singh, S. Mahajan, A.R. Ray and P. Vasudevan; Contraception, 20(5), 467(November 1979).

114. [http://www.freepatentsonline.com/5296230.html].

115. [http://www.fda.gov/cdrh/annual/fy2003/ode/part1.html].

116. [http://www.biospheremed.com/uterine_fibroid_embolization.cfm].

117. [http://www.biotechnews.com.au/index.php?id=956910805&fp=16&fpid=0].

118. D.K. Edmonds, Obstet.Gynecol.Clin.North Am., 27(1), 49 (2000).

119. "Construction of congenital uterovaginal anomalies." In: Female Reproductive Surgery" (Eds. J Rock, A Murphy, H.W. Jones Jr.) Baltimore, Williams & Wilkins, (1992).

120. M.F. Jonkman, F.M. Kauer, P. Nieuwenhuis and I.Molenaar; Artif Organs., 10(6), 475 (1986).

121. [http://www.freshpatents.com/Tissue-engineered-uterus-dt20050922ptan20050208026.php?type=description].

11

Luminescent Polymers

Yongfang Li[*] and Jianhui Hou

*CAS Key Laboratory of Organic Solids, Institute of Chemistry,
Chinese Academy of Sciences, Beijing—100080, China.*
Corresponding author's email: liyf@infoc3.icas.ac.cn

Contents

- ❏ **INTRODUCTION**
- ❏ **BASIC PROPERTIES OF LUMINESCENT POLYMERS**
- ❏ **MOLECULAR STRUCTURES OF THE LUMINESCENT POLYMERS**
- ❏ **SYNTHESIS OF LUMINESCENT POLYMERS**
- ❏ **CONCLUSIONS AND FUTURE PROSPECTS**
- ❏ **REFERENCES**

Summary: Luminescent polymers are the conjugated polymers with high photoluminescent (PL) efficiency. The most representative luminescent polymers include the soluble derivatives of poly(p-phenylene vinylene) (PPV) (emitting color from orange to green), polythiophene (PTh) (red), poly(p-phenylene) (PPP) (blue) and polyfluorene) (PF). The color of the emitted light is determined by the energy gap (E_g, in the range of 1.8~3.1 eV) of the luminescent polymers. Generally, PPP and PF emit blue light, PPV and PPE emits green to yellow light, PPyV emit green to red light, while PT emits red light.

The substituent groups cannot only improve the solubility, but also regulate the emission color and improve the EL efficiency of the luminescent polymers. In addition, the side chain substitution could also modulate the HOMO and LUMO energy levels of the conjugated polymers. Another approach to regulate the emission color is copolymerization. The copolymers of fluorene with other conjugated molecules possess special importance. The polyfluorene derivatives emit blue light with high efficiency. By copolymerization with red-emitting or green-emitting polymers, the fluorene-containing copolymers could emit colors spanning the entire visible range (red, green and blue).

Luminescent materials can be divided into two categories of fluorescent materials and phosphorescent materials. Fluorescent materials emit light when its singlet excitons decay to the ground states, while the emission of the phosphorescent materials comes from the decay of the triplet excitons to the ground states. Commonly, the luminescent polymers are the fluorescent (FL) materials.

The energy gap (E_g) of the luminescent polymer can be calculated from the longer wavelength edge (λ_{edge}, unit: nm) of the absorption spectrum, or from the difference of the onset oxidation and reduction potentials of the conjugated polymers. The cyclic voltammetry has become the most important method to measure the HOMO and LUMO energy levels of the conjugated polymers.

The luminescent polymers could be synthesized by oxidative polymerization or by chemical polymerization. Various chemical polymerization methods have been developed in the synthesis of soluble and highly efficient luminescent polymers. The chemical methods include Gilch, Wittig and McMurry route for the synthesis of PPV derivatives, Stille coupling reaction for the synthesis of PT derivatives, Yamamoto and Suzuki methods for the synthesis of PF derivatives.

1. INTRODUCTION

Since Burroughes *et al.* [1] found the electroluminescent phenomenon of poly(p-phenylene vinylene) (PPV) in 1990, luminescent polymers and polymer light-emitting diodes (PLEDs) have attracted much attentions all over the world, due to the promising applications of the PLEDs in large area plate displays [2-4]. The polymer light-emitting devices have been developed from PLEDs to polymer light-emitting electrochemical cells (LECs) [5-6], and from single-layered PLEDs to the multilayered PLEDs with hole transport layer (HTL), electroluminescent layer (ELL) and electron transport layer (ETL) [7]. The operation voltage is decreased and the electroluminescent efficiency is increased dramatically by the optimization of the device structures. In the design and synthesis of the luminescent polymers, various luminescent polymers with different structures and emitting red, green and blue light, have been prepared, and the lifetime of the PLEDs based on the luminescent polymers reached tens of thousands hours. The most representative luminescent polymers include poly(p-phenylene vinylene) (PPV) and its soluble derivatives (emitting color from orange to green), soluble polythiophene (PTh) derivatives (red), poly(p-phenylene) (PPP) and its derivatives with alkoxy side chains (blue) and soluble polyfluorene) (PF) derivatives (green to blue) (**Tab. 11.1**).

1.1. Device Structure and Operation Mechanism of PLEDs

PLED is basically composed of a luminescent polymer layer sandwiched between a transparent ITO electrode and a metal electrode, as shown in **Fig. 11.1** in which MEH-PPV is used as the luminescent polymer. The ITO electrode, which is usually modified by a conducting polymer PEDOT:PSS layer, is used as the positive electrode, the low work function metal Ca is used as the negative electrode and an Al layer is covered on Ca for the protection of the active metal. When a voltage high enough for the charge injection is applied between the positive and negative electrodes, the holes will inject from the positive electrode into the HOMO of the luminescent polymer and the electrons will inject from the negative electrode into the LUMO of the luminescent polymer. Then the injected holes and electrons move towards opposite electrodes respectively, meet and recombine to form excitons in

Tab. 11.1. Representative luminescent polymers and the colors of their emissions.

Luminescent polymer	Molecular structure	Color of emission
PTh	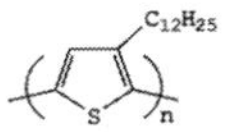	Red
CN-PPV	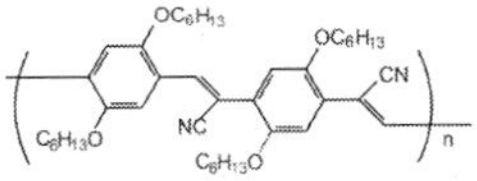	Red
MEH-CN-PPV	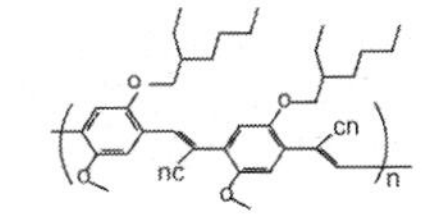	Red
MEH-PPV	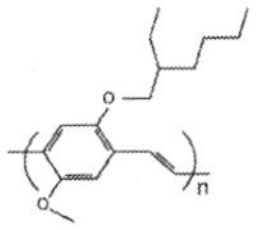	Orange
OC_1C_{10}-PPV	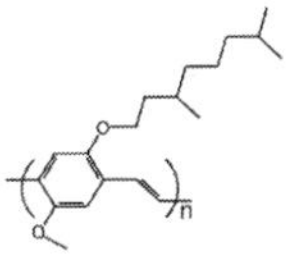	Orange
PPV	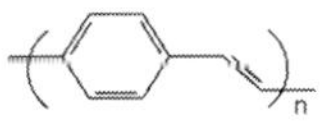	Yellow green
BP-PPV	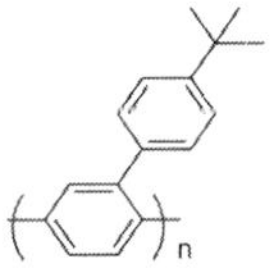	Green
DMSO-PPV	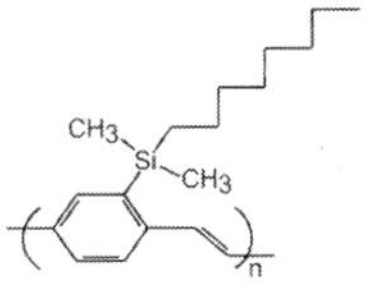	Green
BuEH-PPV	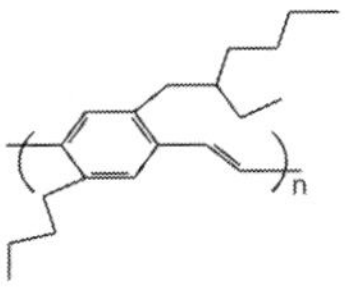	Green
BDOH-PF	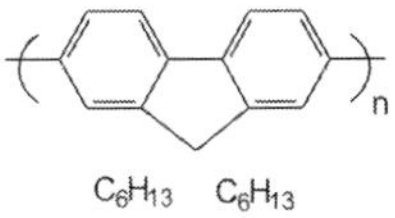	Blue
PPP		Blue
DO-PPP	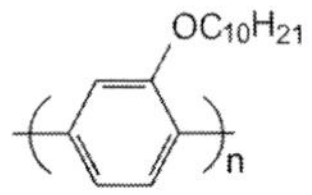	Blue

the inner part of the luminescent polymer layer. The excitons in singlet state decay to ground state and emit light.

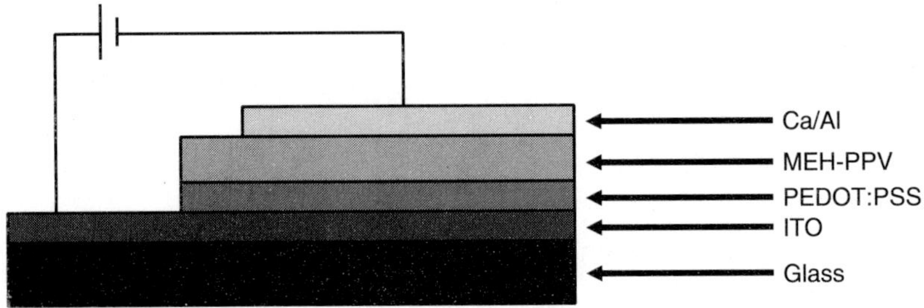

Fig. 11.1. Schematic diagram of polymer light-emitting diode based on MEH-PPV.

Generally, luminescent materials can be divided in two categories of fluorescent materials and phosphorescent materials. Fluorescent materials emit light when its singlet excitons decay to the ground states, while the emission of the phosphorescent materials comes from the decay of the triplet excitons to the ground states. Commonly, the luminescent polymers are the fluorescent (FL) materials.

For a good luminescent polymer used in PLEDs, high fluorescent quantum efficiency, higher charge carrier mobility, high thermal, electrical and optical stability and good film-forming property are the requirements.

2. BASIC PROPERTIES OF LUMINESCENT POLYMERS

2.1. Electronic Energy Levels

Luminescent polymers are the conjugated polymers with high photoluminescent (PL) efficiency. Their doped state is conductive, so, in general the luminescent polymers belong to the conducting polymers with stable intrinsic state. As mentioned above, the electroluminescence of the luminescent polymers comes from the emission decay of the excitons, which are formed by the recombination of holes and electrons injected from the positive and negative electrodes of the PLEDs respectively. The color of the emitted light is determined by the energy gap (E_g) of the luminescent polymers. The photon energies of the visible light from blue to red are 3.1~1.8 eV corresponding to the wavelength of 400~700 nm, hence, the E_g of the luminescent polymers should locate in the range of 1.8~3.1 eV.

For the charge injection in the PLEDs, the energy barrier for hole injection (ΔE_h) depends on the energy difference between the work function of the positive electrode (ITO) and the HOMO (highest occupied molecular orbital) energy level, the energy barrier for electron injection (ΔE_e) depends on the energy difference between the work function of the negative metal electrode and the LUMO (lowest unoccupied molecular orbital) energy level. The lower the energy barrier is, the easier the charge injection will be. Therefore, in order to enhance the electroluminescent performance of the PLEDs (such as reducing the operation voltage and enhancing the luminescent efficiency), we should decrease the energy barriers of the holes injection (ΔE_h) and electrons injection (ΔE_e), and make the two values close each other to reach a balanced hole and electron injection. The work function of ITO positive electrode is 4.7 eV, and that of Ca electrode (which is one of the best negative electrodes with low work functions) is 2.9 eV. Therefore, for the red luminescent polymers with E_g = 2.0 eV, the ideal luminescent polymer should possess a HOMO level of 4.8 eV and a LUMO level of 2.8 eV, while for the blue luminescent polymer with E_g = 3.0 eV, the best polymer should possess a HOMO level of 5.3 eV and a LUMO level of 2.3 eV.

Based on the above analysis, the 3 key parameters of the electronic structure of the luminescent polymers are as follows:

- Energy gap, E_g, which determines the color (wavelength) of the emission.
- HOMO level (or the ionization potential, IP), which affects the hole injection.
- LUMO level (or electron affinity, EA), which affects the electron injection.

2.2. Optical Properties

Absorption and photoluminescent (PL) spectra are of primary importance for the luminescent polymers. **Fig. 11.2** shows the absorption and PL spectra of BuEH-PPV [8], which is a green light-emitting luminescent polymer. From the longer wavelength edge (λ_{edge}, unit: nm) of the absorption spectrum, one can calculate the energy gap (E_g) of the luminescent polymer:

$$E_g = 1240/\lambda_{edge} \ (eV) \tag{11.1}$$

For BuEH-PPV, the λ_{edge} is ca. 510 nm (**Fig. 11.2**), so E_g of BuEH-PPV is ca. 2.43 eV. The E_g value calculated from **Eq. 11.1** is sometimes called the optical energy gap of the conjugated polymers. It can be seen from **Fig. 11.2** that the PL spectrum is red-shifted in comparison with its absorption spectrum and the electroluminescent (EL) spectrum is very similar to the PL spectrum. The similar PL and EL spectra indicate that the emission processes from the decay of the excitons for the EL and the PL are the same. The red shift of the PL spectrum is a general phenomenon for the conjugated polymers, and the peak wavelength difference between the PL spectrum and the absorption spectrum is called Stock's shift. The reason of the Stock's shift can be analyzed from the following aspects: firstly, different conjugation length of the polymer main chains exists in the luminescent polymers. The exciton will relax to the longest conjugated chains that possess the lowest energy, then it decays to give the PL spectrum at longer wavelength. Another possible reason for the Stock's shift is that the conformation of the ground state of the conjugated polymer chains is different from that of the excited state, and the planarity of the excited state is better than that of the ground state, so that the

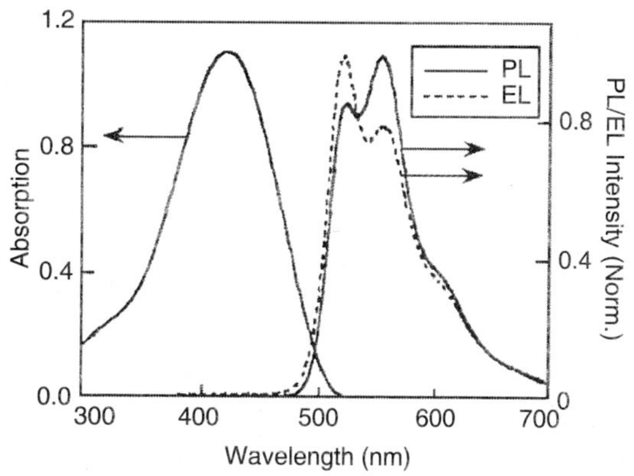

Fig. 11.2. Absorption and PL spectra of BuEH-PPV. Reproduced with permission from A.J. Heeger *et al.*; Synth.Met., 85, 1275 (1997). Copyright © 1997, Elsevier Science Ltd.

excited state possesses a lower energy and the PL spectrum red-shifted. In addition, the formation of excimers, which will be explained later in the text, also results in the red shift of the PL spectrum.

Fluorescent properties of the luminescent polymers are tightly related to the distribution of the length of the conjugated polymers. The absorption and fluorescent spectra of the luminescent polymers are quite broad, which results from the un-uniformity of the length of the conjugated main chains. Actually, the conjugated polymers can be thought as an assembly of the oligomers with different conjugation lengths. The polymer chains with different conjugation lengths show shifted absorption spectra (longer conjugation length show red-shifted absorption, while shorter conjugation length will show blue-shifted absorption), so that the broader absorption spectra appeared.

The interchain interaction of the conjugated polymers also influences their absorption and fluorescent spectra. There are three kinds of the interchain interactions, which affect the optical properties as follows.

Excimer

An excimer is a dimmer formed by an excited conjugated molecular chain and its neighbor ground state conjugated chain of the same luminescent polymer. Its stability comes from the overlap of the π-π^* orbitals of the two conjugated molecular chains. The formation of the excimer needs the two conjugated molecular chains close each other to the distance that is short enough to form the π-π^* interaction of the two molecules but long enough for avoiding the ground state interaction to form the aggregate. For the common luminescent polymers, the interchain distance of the excimers is 0.3~0.4 nm. The excimer does not affect the absorption spectra of the luminescent polymers, but it makes a long wavelength peak appear in the fluorescent spectra of the luminescent polymers, which results in broadening of the fluorescent spectra. In addition, the existence of the excimers often results in decrease of the fluorescent quantum efficiency.

Exciplex

An exciplex is a transient donor-acceptor complex between the excited state of the donor and the ground state of the acceptor [9]. The difference between the exciplex and the excimer is that the complex is formed between two different organic molecules in exciplex, while the excimer is formed between two chains of the same polymer molecules in excimer. In the exciplex, the electron in the LUMO of the excited donor will transfer to the LUMO of the acceptor and left the hole in the HOMO of the donor, then there occurs the recombination and emission decay of the transferred electron in the LUMO of the acceptor and the left hole in the HOMO of the donor. Therefore, the exciplex will show an additional PL peak, which is red-shifted in comparison with the PL peak of the donor and the acceptor. The exciplex can be utilized for producing white light emission in a bilayered or blended organic light-emitting diode [10].

Aggregate

When the distance between the conjugated chains of a luminescent polymer is short enough to make the electronic wave functions of the conjugated chains at both ground and exited state overlap over two or more conjugated chains, the wave function-overlapped conjugated chains form the aggregate. The difference between the aggregate and the excimer is that in aggregates there is strong interaction between the ground state conjugated chains, which makes a new absorption band appear at longer wavelength region in the solid film, in comparison with its dilute solution. The interchain aggregates make both absorption and fluorescent spectra broaden.

2.3. Electroluminescent Properties

As mentioned above, the electroluminescent (EL) spectrum of the luminescent polymers is usually similar to their PL spectrum and it determines the color of the light-emission. So, from the PL spectrum, we could know the color of the EL emission of the luminescent polymers.

The current-voltage (I-V) and luminance-voltage (L-V) measurements are of most importance in the studies of the electroluminescent properties of the luminescent polymers, from which the onset voltage, light-emitting efficiency, maximum luminance etc. can be obtained. **Fig. 11.3** shows the I-V and L-V curves of a typical PLED based on MEH-PPV. The onset voltage of the device is ca. 1.7 V and the luminance of the device reaches to 10^4 cd.m^{-2} at 7 V. Actually, the electroluminescent properties also relate to the device structure and fabrication processes, in addition to the nature of the luminescent polymers.

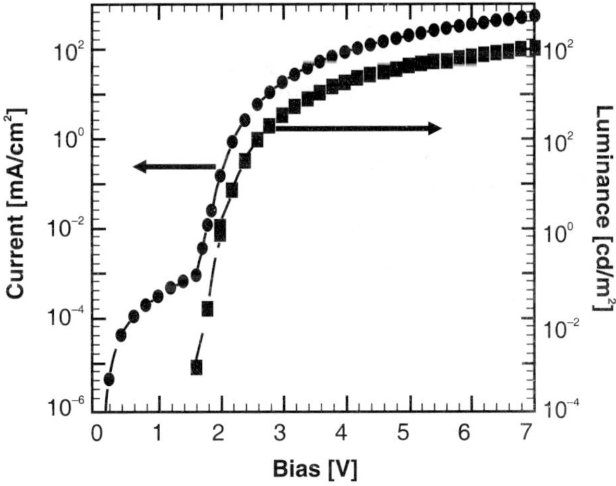

Fig. 11.3. I-V and L-V characteristics of the PLED based on MEH-PPV. Reproduced with permission from A.J. Heeger *et al.*; Chem. Phys.Lett., 217, 507 (1994). Copyright © 1994, Elsevier Science Ltd.

2.4. Electrochemical Properties

Electrochemical cyclic voltammetry of conjugated polymers have been performed intensively in recent years, since the cyclic voltammograms obtained from the measurements could reveal the electron donating or electron accepting abilities and can be used to estimate the electronic energy levels (HOMO, LUMO and E_g) of the conjugated polymers [11,12]. At real, the cyclic voltammetry has become the most important method to measure the HOMO and LUMO energy levels of the conjugated polymers.

Fig. 11.4 shows the cyclic voltammograms of MEH-PPV film on glassy carbon electrode in 0.1 M tetrabutylammonium hexafluorophosphate (Bu_4NPF_6) acetonitrile solution with Ag wire as the quasi-reference electrode [12]. The plot in the positive potential range corresponds to the p-doping/dedoping (oxidation/re-reduction) of MEH-PPV, and the plot in the negative potential range shows the n-doping/dedoping (reduction/re-oxidation) of the conjugated polymer. It can be seen that both p-doping/dedoping and n-doping/dedoping are reversible, and the onset potentials of p-doping ($\phi_{p,on}$) and n-doping ($\phi_{n,on}$) are 0.68 V and −1.49 vs. Ag wire respectively.

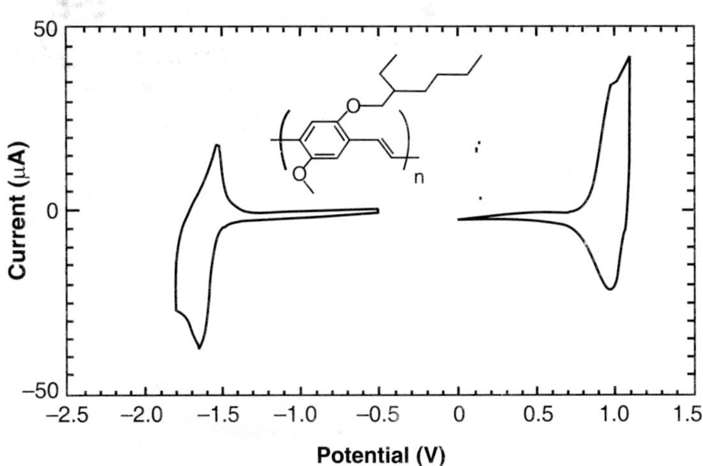

Fig. 11.4. Cyclic voltammogram of MEH-PPV film on glassy carbon electrode in Bu_4NPF_6 (0.1 M) in acetonitrile solution (silver wire was used as the reference electrode). Reproduced with permission from A.J. Heeger *et al.*; Synth.Met., 99, 243 (1999). Copyright © 1999, Elsevier Science Ltd.

From the onset potentials, we can calculate the HOMO, LUMO and E_g of the conjugated polymers according to the following equations:

$$HOMO = -e\,[\phi_{p,on}\ (V\ vs.\ SCE) + 4.4]\ or$$

$$= -e\,[\phi_{p,on}\ (V\ vs.\ Fc/Fc^+) + 4.8]\ (eV) \tag{11.2}$$

$$LUMO = -e\,[\phi_{n,on}\ (V\ vs.\ SCE) + 4.4]\ or$$

$$= -e\,[\phi_{p,on}\ (V\ vs.\ Fc/Fc^+) + 4.8]\ (eV) \tag{11.3}$$

$$E_g{}' = e\,(\phi_{p,on} - \phi_{n,on}) \tag{11.4}$$

where Fc/Fc^+ represents the ferrocene/ferrocenium reference electrode, $E_g{}'$ is the electrochemical energy gap for distinguishing with the optical energy gap E_g of the conjugated polymers (in most cases, the value of $E_g{}'$ is the same with E_g), the unit of the potentials is V, so that the unit of the HOMO, LUMO and the energy gap is eV. In order to calculate the energy level of MEH-PPV from the cyclic voltammograms shown in **Fig. 11.4**, we have to get the value of the Ag wire quasi-reference electrode referring to SCE. In the case of **Fig. 11.4**, the potential of the Ag wire is −0.01 V vs. SCE. Therefore, the HOMO, LUMO and $E_g{}'$ of MEH-PPV calculated from the electrochemical measurement were –5.07 eV, –2.90 eV and 2.17 eV respectively [12]. Generally, for the reference electrode other than SCE and Fc/Fc⁺, one needs to calibrate the reference electrode potential versus Fc/Fc⁺ by measuring the redox potential of 0.01 M ferrocene versus the reference electrode.

3. MOLECULAR STRUCTURES OF THE LUMINESCENT POLYMERS

3.1. Conjugated Main Chains

As mentioned above, the representative luminescent polymers include poly(p-phenylene vinylene) (PPV), polythiophene (PT), poly(p-phenylene) (PPP), polyfluorene (PF) and their derivatives. In addition, other conjugated polymers with their main chains containing naphthalene, pyridine and ethynylene units etc. were also utilized as the luminescent polymers. **Fig. 11.5** shows the structures of the main chains of the some representative luminescent polymers. The electronic energy structure of the luminescent polymers is mainly determined by the main chain structures, so is the light-emission color of the polymers. Generally, PPP and PF emit blue light, PPV and PPE [13] emits green to yellow light, PPyV emits green to red light [14], while PT emits red light.

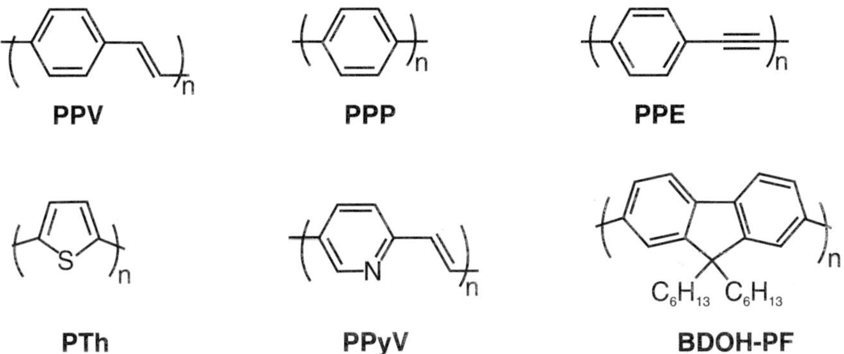

Fig. 11.5. Main chain structures (repeat units) of some representative luminescent polymers.

Besides the π-conjugated polymer main chains, polysilanes with σ-conjugation chains could be also used as luminescent polymers [15]. The bandgap of polysilanes is ca. 4 eV, and it can emit UV or near-UV light.

3.2. Side Chain Substitution

The processibility and film-forming properties of the conjugated polymers are crucial for the fabrication of high performance PLEDs. The conjugated polymers of PPV, PT and PPP etc. without substituted sidechains are insoluble and infusible, which limits their applications. The strong interchain interactions are responsible for the insolubility. To solve the problem, alkyl or alkoxy soft side chains are widely used as substituent groups on the conjugated main chains, which could reduce the interchain interactions and induce the solubility. **Fig. 11.6** shows the molecular structures of some of the side-chain substituted conjugated polymers. The substituent groups cannot only improve the solubility, but also regulate the emission color and improve the EL efficiency of the luminescent polymers. For example, MEH-PPV with alkoxy side chains and BuEH-PPV with alkyl side-chains

(Tab. 11.1) are both soluble in organic solvents such as toluene, xylene etc. The emission light color of MEH-PPV is orange-red that is a little red-shifted in comparison with that of PPV. While the emission light color of BuEH-PPV is green, a little blue-shifted in comparison with that of PPV. The PL and EL efficiencies of both polymers increased obviously in comparison with that of PPV. Similarly, the PT and PPP derivatives with the alkyl or alkoxy side chains are soluble in the organic solvents, and they are used as red and blue luminescent polymers respectively. In addition to the alkyl and alkoxy side chains, alkyl-phenylene side chains are also utilized in the PPV derivatives for getting stable and highly efficient green luminescent polymers.

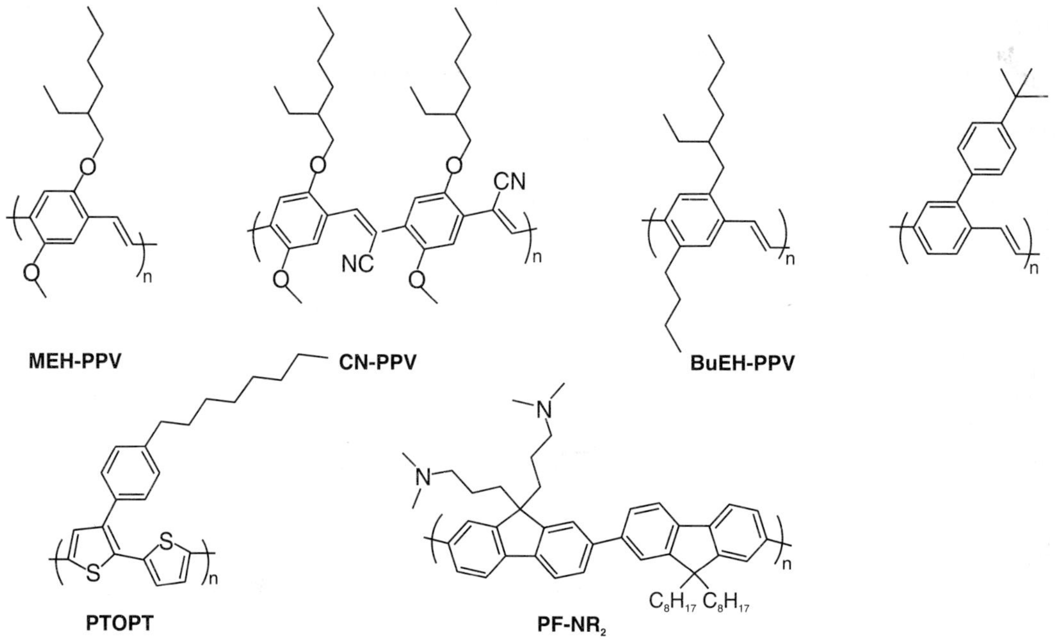

MEH-PPV

CN-PPV

BuEH-PPV

PTOPT

PF-NR$_2$

Fig. 11.6. Molecular structures of some conjugated polymer derivatives with side-chains.

Recently, Yong Cao *et al.* [16] synthesized a water-soluble PF derivative PF-NR$_2$ (**Fig. 11.6**) with a water-soluble side group. The polymer was successfully used as the cathode-modifying layer and greatly improved the electron injection of the PLEDs even with inert metal (such as Au) as the cathode [17-19]. The conjugated polymers with the sidechains of poly(ethylene oxide) (PEO) were also synthesized for the application in polymer light-emitting electrochemical cells (LECs), which will be discussed later in Section 3.5.

In addition to improving solubility and regulating emission color, the side chain substitution could also modulate the HOMO and LUMO energy levels of the conjugated polymers. For example, the substitution of the electron-withdrawing group –CN on vinylene carbon of MEH-PPV (CN-PPV) makes the HOMO and LUMO levels of MEH-PPV move downward by ca. 0.6 eV [12], so that the electron injection of CN-PPV is much easier that that of MEH-PPV.

Bredas *et al.* [20,21] studied the effect of the electron donating and accepting ability of the sustituents on the electronic energy levels of PPV, by quantum chemistry calculation (VEH method). Their calculation results for the energy gap (E_g), ionization potential (IP) and electron affinity (EA) are listed in **Tab. 11.2**. IP and EA of a conjugated polymer are the absolute values of its HOMO and

LUMO respectively. The calculated E_g values are basically the same as those obtained from the absorption spectra, which indicates that the calculated parameters of the electronic structure are quite reliable. It can be seen from **Tab. 11.2** that PPV derivatives with the electron-donating substituents of alkoxy possess lower IP and EA than those of PPV, and the decrease of IP is more obvious than EA does. The PPV derivative with the electron-withdrawing substituent of –CN makes the IP and EA increased, and the increase of EA is more obvious than IP does. Interestingly, both the substitutions of the electron-donating groups and the electron-withdrawing groups make the E_g of PPV reduced. The results mentioned above indicate that the E_g, IP and EA values of the luminescent polymers could be modulated by proper substitution of electron-donating or electron-withdrawing groups.

Tab. 11.2. Electronic parameters of PPV and its derivatives as calculated by VEH method.

PPV and derivatives	Band gap (eV)	Ionization potential (eV)	Electron affinity (eV)
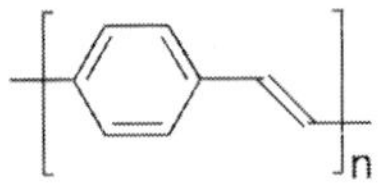 (PPV structure)	2.32	5.05	2.73
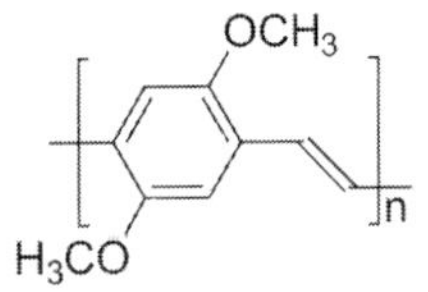 (PPV structure with OCH$_3$ / H$_3$CO substituents)	2.07	4.72	2.65
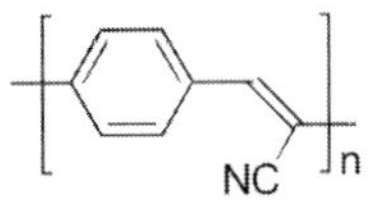 (PPV structure with NC substituent)	2.17	5.27	3.10
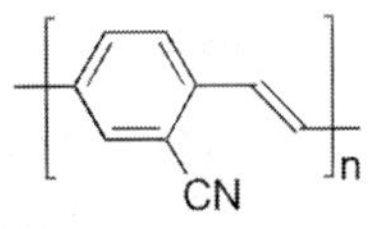 (PPV structure with CN substituent)	2.24	5.15	2.91
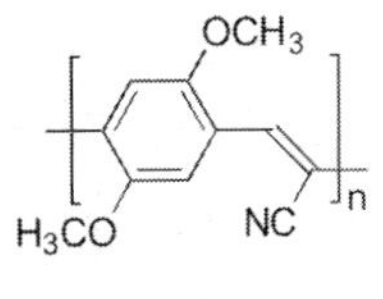 (PPV structure with OCH$_3$, H$_3$CO and NC substituents)	1.97	5.12	3.15
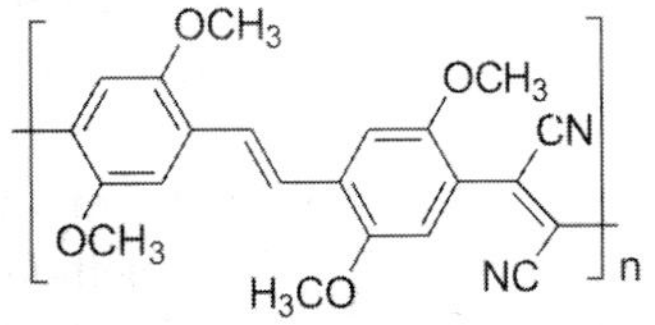 (PPV structure with OCH$_3$, H$_3$CO, CN and NC substituents)	1.74	5.08	3.34

3.3. Copolymers

A main advantage of the luminescent polymers, in comparison with inorganic luminescent semiconductors, is that the emission color could be easily regulated by the structural design of the polymers. The side chain substitution mentioned above is one approach of the structural design. Another approach to regulate the emission color is copolymerization. For example, in the copolymers of MEH-PPV and BuEH-PPV, the emission color can be regulated gradually from orange-red to green by controlling the ratio of MEH-PPV to BuEH-PPV in the copolymers. Besides the color regulation, copolymerization, sometimes, could enhance the EL efficiency [22], which may be benefited from the reduction of the exciton diffusion processes towards quenching centers.

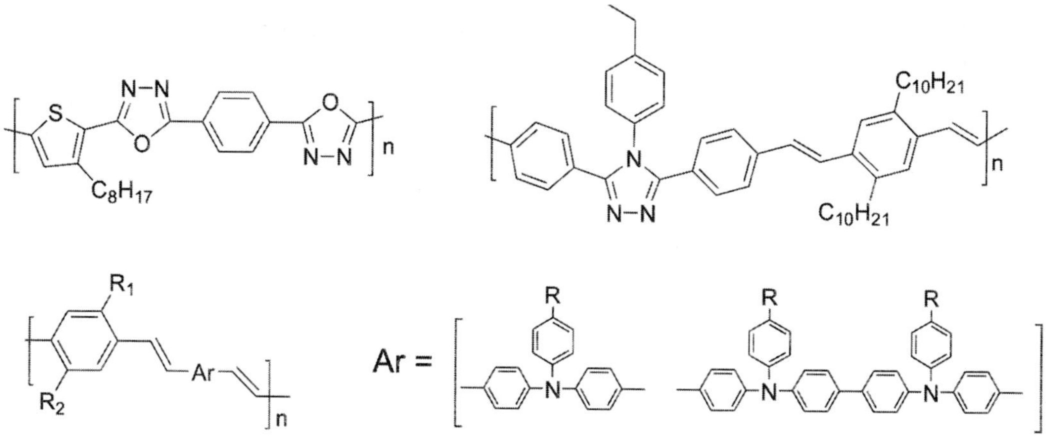

Fig. 11.7. Molecular structures of some copolymers with charge transporting segments.

Fig. 11.7 shows the molecular structures of the copolymers with charge transporting segments. The copolymerization of PT, PPP and PPV with the electron transporting molecules such as oxadiazole and triazole, could not only regulate the emission wavelength, but also improve the electron transporting property. The copolymerization of the luminescent polymers with the hole transporting molecules such as tri-phenylene-aniline (TPA) and its derivatives can improve the hole transporting property and increase the luminescent efficiency. Therefore, the copolymerization is a promising method to obtain desired luminescent polymers.

The copolymer of fluorene with other conjugated molecules possesses special importance for tuning the emission color of the luminescent polymers. The soluble polyfluorene derivatives emit blue light with high efficiency. By copolymerization with red-emitting or green-emitting polymers, the fluorene-containing copolymers could emit colors spanning the entire visible range (red, green, blue) [23]. **Fig. 11.8** shows the molecular structures of some fluorene-containing copolymers.

In addition, the emission color of the luminescent polymers is also related to the length of the conjugated chains. Reducing its conjugation length could make its emission color blue-shifted. One method to control the conjugation length is the copolymerization of the luminescent chain segment with a certain conjugation length and a non-conjugation segment (spacer). **Fig. 11.9** shows the molecular structure of this kind of copolymers [24]. By controlling the conjugation length of the luminescent chain segments of the copolymer, blue emission can be realized from the PPV copolymers with spacers.

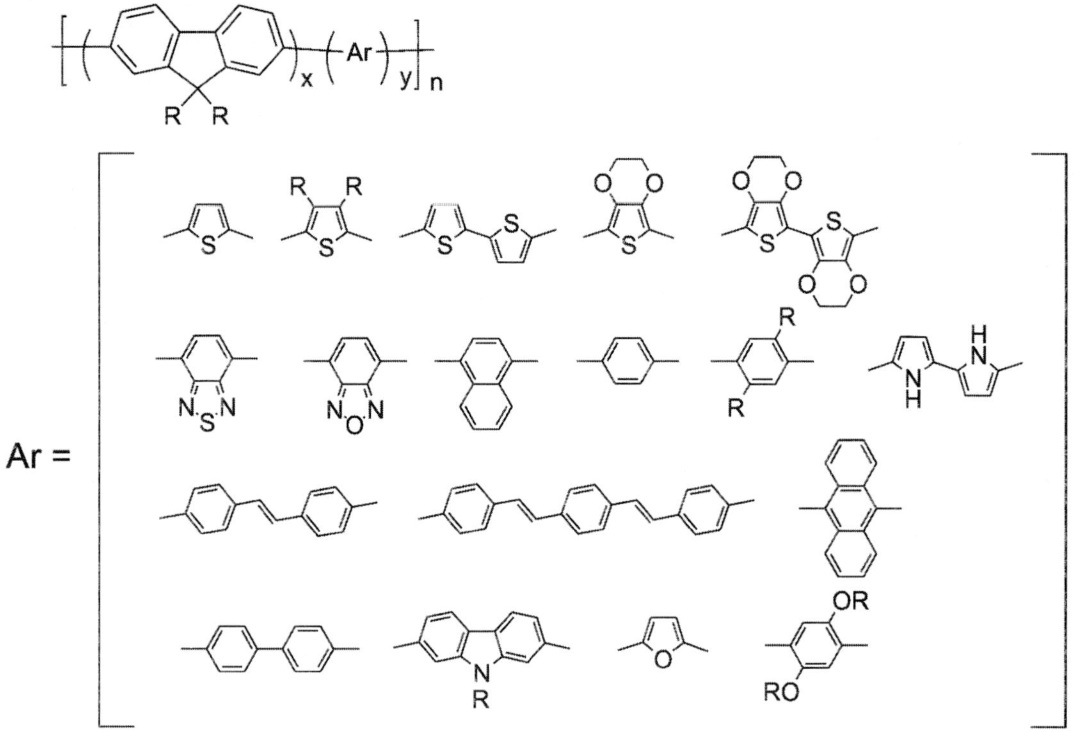

Fig. 11.8. Molecular structures of some alternating and random copolymers derived from fluorene and other aromatic monomers.

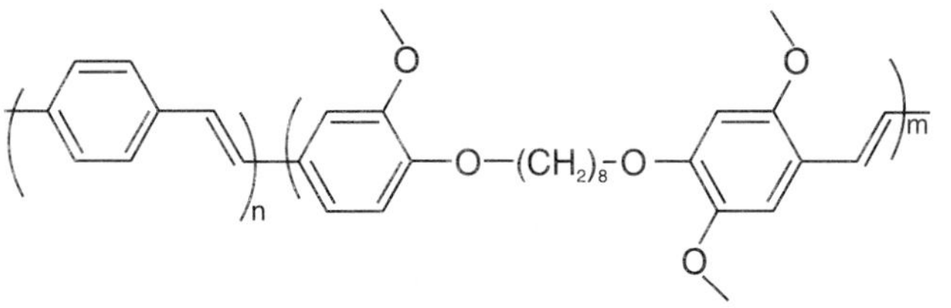

Fig. 11.9. Molecular structure of a copolymer with non-conjugated spacer.

3.4. Ladder and Comb Polymers

As mentioned above, the substitution of flexible side chains on PPP main chains could improve its solubility. But, the side-chain substitution affects little or more the coplanarity of the PPP main chains, which results in undesirable blue shift of the emission and decrease of EL efficiency. Ladder polymers of PPP as shown in **Fig. 11.10** could solve the problems [25]. The ladder polymer of PPP emits yellow color light, greatly red-shifted in comparison with original PPP. Obviously, the conjugation of the polymer main chains is greatly enhanced because of the good coplanarity of the ladder polymer, which results in the red-shifted emission. In addition, the ladder polymers also possess higher thermal stability.

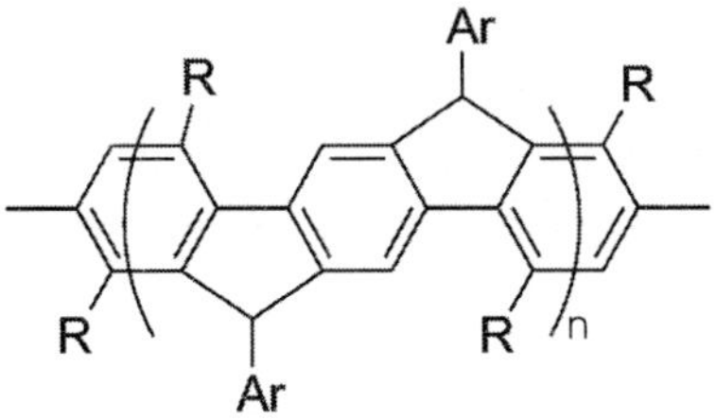

Fig. 11.10. Molecular structure of a ladder polymer of PPP.

Another effective approach to regulate the solubility and control the conjugation length (emission color) is the synthesis of comb polymer [26]. The comb polymers are composed of non-emitting flexible main chain and luminescent conjugated side chains with certain conjugation length (**Fig. 11.11**). The polymer shown in **Fig. 11.11** emits blue light.

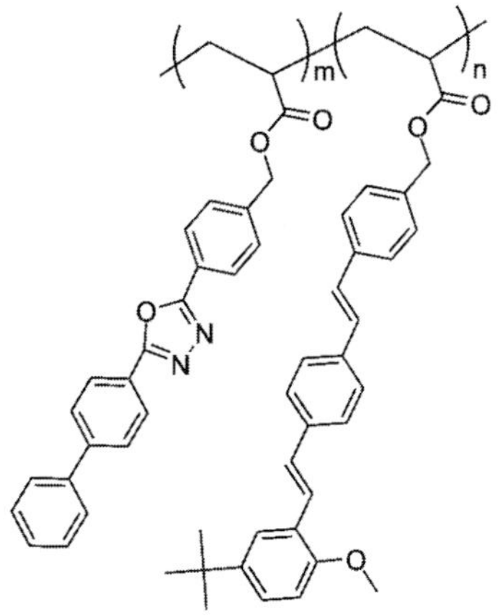

Fig. 11.11. Molecular structure of a comb polymer.

3.5. Luminescent Polymers with Ion-Conducting Segments or Side Chains

In polymer light-emitting electrochemical cells (LECs), luminescent layer is composed of the blend of luminescent polymer and polymer electrolyte (ion-conducting polymer + lithium salt) [5,6], where commonly used ion-conducting polymer is poly(ethylene oxide) (PEO). To get high performance LECs, one of the key points is the homogenous blending of the luminescent polymer and the ion-conducting polymer. However, the luminescent polymers are usually nonpolar while PEO ion-conducting polymer is very polar, so their blend is unstable thermodynamically and easy to be phase-separated. Using the bifunctional polymers with the electroluminescent and ion-conducting properties can solve the problem.

Two types of the polymers can meet the request: One is that with PEO side-chains (**Fig. 11.12a**) [27,28], another is the block copolymers with luminescent segments and PEO segments (**Fig. 11.12b**) [29,30]. The block copolymer of PPV segments with crown ether (**Fig. 11.12c**) [31] also shows the bifunctional characteristics with electroluminescent and ion-conducting properties. The copolymers shown in **Fig. 11.12b** and **Fig. 11.12c** emit blue-green light.

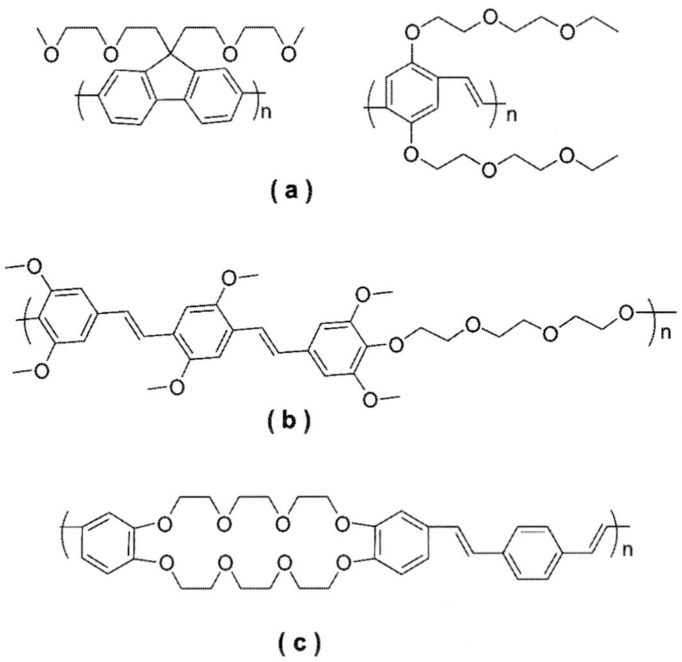

(a)

(b)

(c)

Fig. 11.12(a). Luminescent polymers with PEO side-chains, (b) bifunctional block copolymer containing PEO segments in its main chain and (c) bifunctional copolymer containing crown ether in its main chain.

3.6. White-Light Luminescent Polymers

White light-emission is very important for application of the electroluminescent devices to full color display, backlight and illumination light sources. Therefore, the studies on the white organic light-emitting diodes (WOLEDs) have attracted much attention in recent years. The WOLED is usually composed of a multilayered structure or a blended film of several components [32], which makes it complicated to fabricate the WOLED devices and difficult to get stable white light emission. Obviously, single component white-light luminescent polymers would be desirable for the fabrication of the white light-emission devices.

The single component white light emission was realized from a soluble oxadiazole-containing phenylene vinylene ether-linkage copolymer [33] (**Fig. 11.13a**) and from a copolymer of blue-emitting polyfluorene and 0.05 mol% yellow-emitting 1,8-naphthalimide derivative [34] (**Fig. 11.13b**). For the copolymer shown in **Fig. 11.13b**, the single layer PLED with a configuration of ITO/PEDOT/copolymer/Ca/Al emits white light and shows a current efficiency of 5.3 cd.A^{-1} and a power efficiency of 2.8 lm.W^{-1} at 6 V.

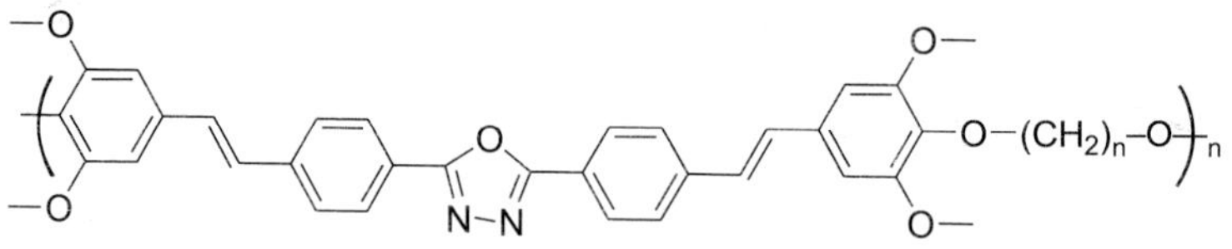

(a)

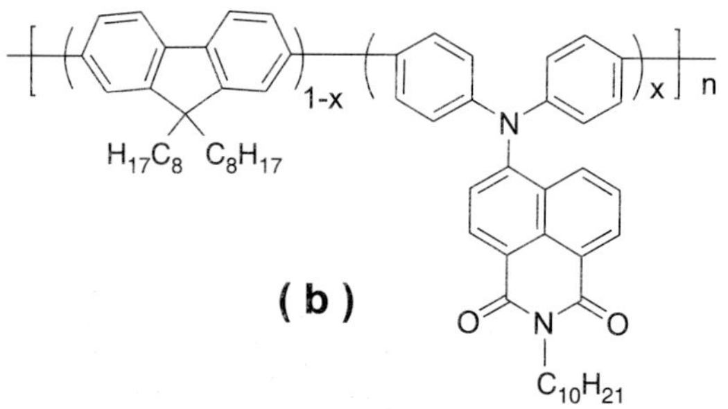

(b)

Fig. 11.13. Molecular structures of two white light emitting luminescent polymers.

3.7. Phosphorescent Polymers

All the luminescent polymers discussed above are fluorescent materials that emit light only by the decay of singlet excitons. The external quantum efficiency of the light emission of the PLEDs based on the fluorescent polymers is usually below 5% (the internal quantum efficiency is lower than 25%), because only one-fourth of the excitons formed in the polymers are in the singlet-excited states and other 75% of the excitons are in the triplet excited state which decay to their ground state without emission in the fluorescent polymers. A phosphorescent material with light emission from triplet excitons can overcome the limitation of the efficiency. An external quantum efficiency of nearly 20% (implying a 100% internal quantum efficiency) was achieved in green organic light-emitting diodes based on phosphorescent Iridium (Ir)-complex [35]. Obviously, phosphorescent polymers are desirable for increasing the EL efficiency of the PLEDs.

Recently, Tokito *et al.* [36] prepared the phosphorescent polymers involving a carbazole unit and an Ir-complex unit **(Fig. 11.14)**. The Ir-complexes were used for the phosphorescent monomer units and the carbazole unit (which possesses good hole transport ability) was bonded to the polymer backbone as the charge transport unit. The optimized ratios of the Ir-complex units to the carbazole units are 0.2 mol% for the red phosphorescent polymer (RPP), 0.6 mol% for the green phosphorescent polymer (GPP) and 1.0 mol% for the blue phosphorescent polymer (BPP) respectively, for the maximum EL efficiency. The PLEDs based on RPP, GPP or BPP exhibit red, green or blue emission with the maximum external quantum efficiency of 5.5%, 9% and 3.5% respectively [36].

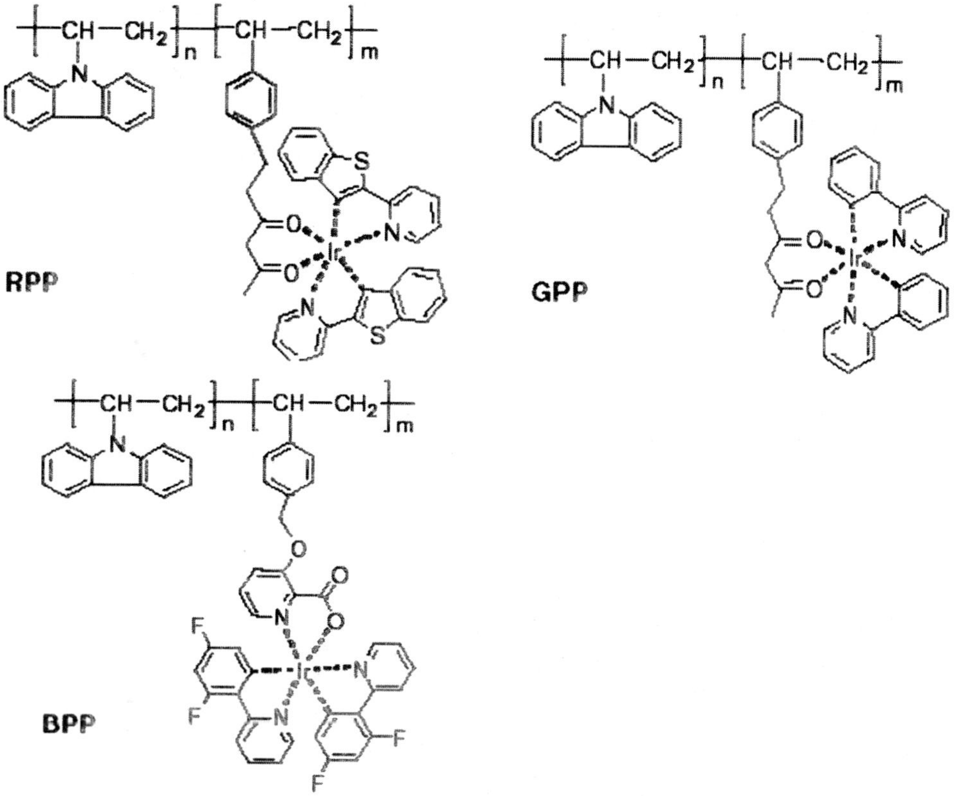

RPP

GPP

BPP

Fig. 11.14. Molecular structures of some phosphorescent polymers.

4. SYNTHESIS OF LUMINESCENT POLYMERS

4.1. Synthesis of PPVs

PPV and its derivatives can be synthesized via Wessling precursor route [37], as shown in **Fig. 11.15**. With the Wessling method, the PPV precursor is synthesized by the reaction of bis-(sulfonium halide) salts of p-xylene with base (NaOH) in water or alcohol solution. For the application in PLED, the precursor solution is spin-cast on ITO substrate, then PPV film is formed after the

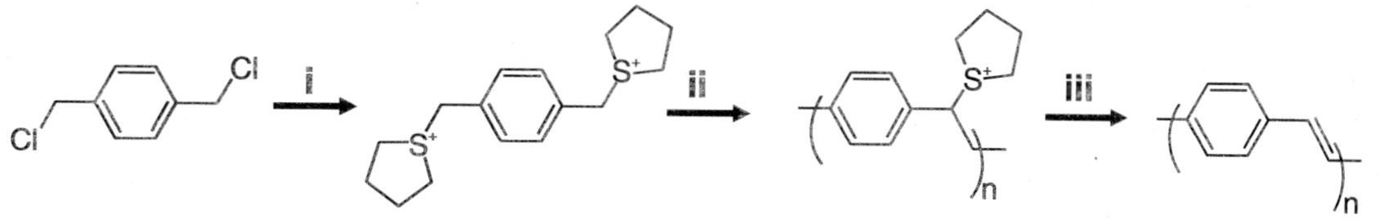

Fig. 11.15. Wessling route to prepare PPV (i) tertrahydrothiophene, MeOH, 65°C, (ii) MeOH/H$_2$O, NaOH, 0°C, then HCl, H$_2$O and (iii) 180-300°C, vacuum) [37].

precursor film is heat-treated at 180-300°C under vacuum. Vanderzande *et al.* [38] and Mullen *et al.* [39] improved the Wessling method by replacing chloride with bromide in the starting chemicals, and the temperature of the heat-treatment is decreased to 100°C for the decomposition of the precursor to PPV. The Wessling method is mainly used for the preparation of the insoluble unsubstituted PPV, although it could also be used to prepare soluble PPV derivatives, since the method is complicated for the post-treatment.

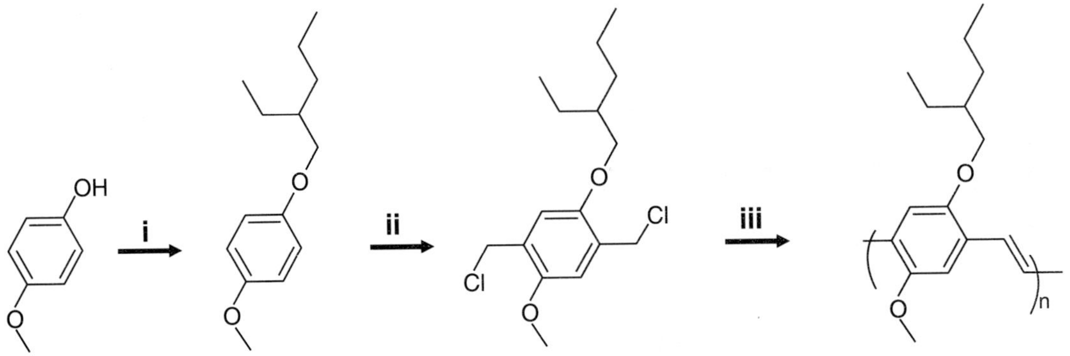

Fig. 11.16. Gilch route for the synthesis of MEH-PPV (i) CH$_3$(CH$_2$)$_3$(C$_2$H$_5$)CHCH$_2$Br, K$_2$CO$_3$, DMF, (ii) HCHO, HCl, ZnCl$_2$ and (iii) t-C$_4$H$_9$K, THF, rt or reflux) [40].

Gilch method [40] is a convenient approach to prepare soluble PPV derivatives. **Fig. 11.16** shows the synthesis route of MEH-PPV by the Gilch method. With this method, 1,4-bis(chloromethyl) (or bromomethyl) arenes is treated with about one equivalent of potassium t-butoxide in a non-hydroxylic solvent (such as tetrahydrofuran). Many soluble PPV derivatives with high molecular weight and high purity, such as MEH-PPV [41] and MDMO-PPV etc., were synthesized by this method. So the Gilch route is the most important method in the synthesis of the soluble PPV derivatives.

CN-PPV is one of the electron-accepting polymers, which can be prepared through Knoevenagel condensation [42,43] between equimolar amount of a terephthaldehyde derivative and a 1,4-diacetonnitrile-benzene derivative, as shown in **Fig. 11.17**. The condensation reaction takes place upon addition of excess potassium t-butoxide or tetrabutylammonium hydroxide in THF/t-butanol mixture at 50°C.

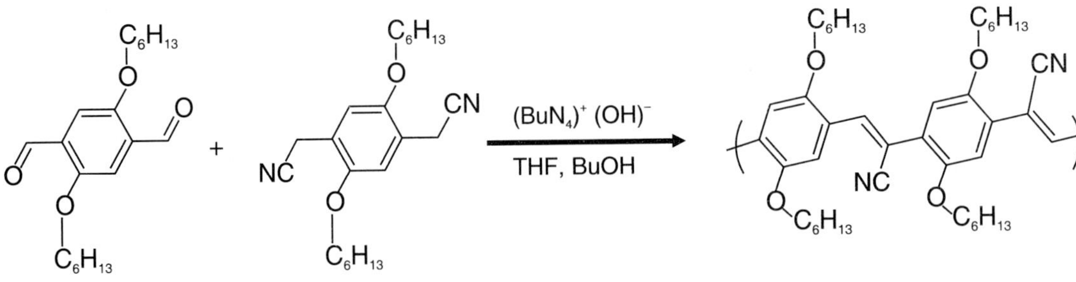

Fig. 11.17. Knoevenagel method for the preparation of CN-PPV [42,43].

Heck reaction [44], as shown in **Fig. 11.18**, is another synthetic method for the preparation of functional PPVs. In this method, the organic halides and vinylbenzene compounds are coupled for polymerization under the catalysis of Pd(0). Many functional groups, such as aldehyde, ester, nitryl, hydroxy and carboxy etc., can be attached on the side chains in the polymerization of the PPV derivatives, and a lot of solvents (such as DMF, DMA, NMP, toluene, even water) could be used in the coupling reaction. So the Heck method is superior for the synthesis of multi-functional PPV derivatives.

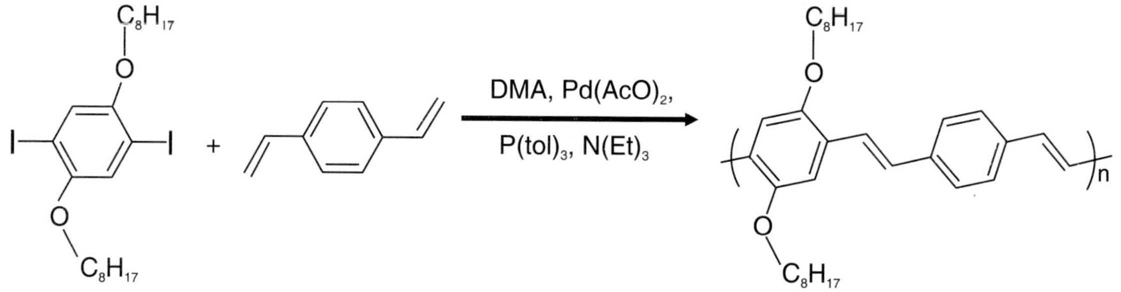

Fig. 11.18. Heck reaction for the preparation of PPV derivatives [44].

Besides the methods mentioned above, Wittig reaction route [45] and McMurry route [46] could also selected in the preparation of functional PPV derivatives.

4.2. Synthesis of Polythiophenes

Polythiophene (PT) is another important conjugated polymer, and the synthesis of PTs has been intensively studied. Among the soluble polythiophene derivatives, poly(3-alkylthiophene)s (P3ATs) are the most widely investigated. The first P3AT was prepared by a halo-Grignard coupling reaction (or Kumada cross-coupling [47,48]), as shown in **Fig. 11.19a**. In this method, a 2,5-diiodo-3-

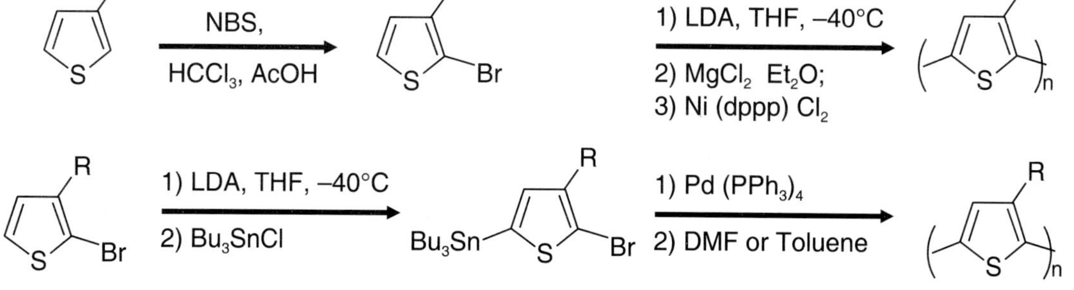

Fig. 11.19. (a) Halo-Grignard coupling reaction for the synthesis of P3AT [47,48]. (b) McCullough method to prepare regioregular P3AT [50,51] and (c) Stille coupling reaction to prepare regioregular P3AT. [58].

alkylthiophene is treated with one equivalent of Mg in THF solution, generating a mixture of Grignard species, then a catalytic amount of Ni(dppp) Cl$_2$ is added and the polymer is produced. Large quantity of P3ATs can be prepared by this method.

Because the properties of P3ATs, such as conductivity and absorption spectrum, are sensitive to the regioregularity of the polymers, so the preparation of regioregular P3ATs has drawn much attention. The first synthesis of regioregular head-to-tail (HT) coupled P3ATs was reported by McCullough and Lowe [49] in 1992. This synthetic route **(Fig. 11.19b)** regiospecifically generates 2-bromo-5-(bromomagnesio)-3-alkylthiophene, which is polymerized with the catalysis of Ni(dppp)Cl$_2$ by Kumada cross-coupling methods [50,51] to give the regioregular P3ATs with 98-100% HT-HT couplings. The regioregular P3ATs could also be synthesized by Stille coupling reaction, as shown in **Fig. 11.19c**, in which the 2-halogen-5-trialkyltin-thiophene is polymerized with the catalysis of Pd(0). The Stille method could get the regioregular P3ATs with nearly 100% HT-HT couplings.

Polythiophenes can also be prepared simply by chemical oxidation-polymerization [52] or electrochemical oxidation-polymerization [53] of thiophene monomers in organic solutions. FeCl$_3$ is usually used as the oxidant for the chemical oxidation-polymerization in a chloroform solution [52]. But the Fe impurity in the PTs prepared by the chemical oxidation-polymerization could be problem for the application in PLEDs [54].

4.3. Synthesis of PPP and PF

Soluble polyfluorene derivatives were firstly prepared by oxidative polymerization of 9-alkyl-substituted monomers [55]. However, the polymers obtained were of low molecular weight and poor purity. Pei and Yang [56] synthesized high molecular weight polyfluorenes by the nickel(0)-mediated polymerization of 9,9-disubstituted-2,7-dibromofluorene monomers (Yamamoto polymerization [57]), the synthetic route is shown in **Fig. 11.20a**. Suzuki reaction is another important synthetic route for

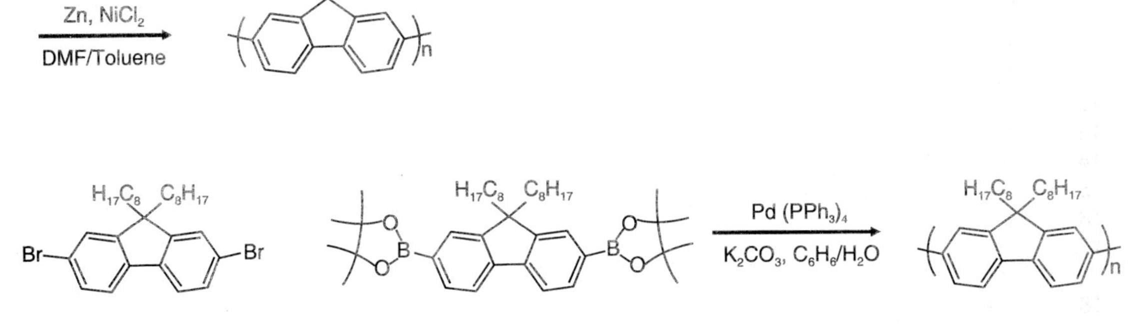

Fig. 11.20. a. Yamamoto route [52] and b. Suzuki route to synthesize soluble PFs [51].

preparing soluble PFs, as shown in **Fig. 11.20b**. In the Suzuki method, diboronic functionalized aromatic units react with diiodo (or dibromo) aromatic derivatives with a catalytic amount (2~0.5 mol%) of $Pd(PPh_3)_4$ and K_2CO_3. The synthesis of poly(p-phenylene) (PPP) and its derivatives is similar to that of PFs, Yamamoto polymerization and Suzuki polymerization can also be applied for the synthesis of soluble PPP derivatives.

5. CONCLUSIONS AND FUTURE PROSPECTS

Fifteen years have passed since the discovery of polymer light-emitting diodes (PLEDs). Luminescent polymers and the PLED devices have attracted broad attention all over the world, including academic research and display products development of companies. Hundreds of luminescent polymers have been designed and synthesized, and the device structures and performance of the PLEDs have been improved greatly. Now, the PLED displays are closing to the market applications. For the applications of PLEDs, the efficiency and stability of the luminescent polymers are of crucial importance. At present, the efficiency and stability of green polymers are high enough, but the efficiency of blue polymers and the stability of red polymers still need to be improved. In addition, the color purity and color stability of the PLED devices are other points, which need to pay attention. The problems will be overcome, and the large-scale applications of the luminescent polymers and the PLED displays will come in near future.

REFERENCES

1. J.H. Burroughes, D.D.C. Bradley, A.R. Brown, R.N. Marks, K. Mackay, R.H. Friend, P.L. Burns and A.B. Holmes; Nature, 347, 539 (1990).
2. R.H. Friend, R.W. Gymer, A.B. Holmes, J.H. Burroughes, R.N. Marks, C. Taliani, D.D.C. Bradley, D.A. Dos Santos, J.L. Bredas, M. Logdlund and W.R. Salaneck.; Nature, 397: 121 (1999).
3. W.C. Molenkamp, M. Watanabe, H. Miyata and S.H. Tolbert; J. Am. Chem. Soc., 126, 4476 (2004).
4. J. Liu, A.R. Duggal, J.J. Shiang and C.M. Heller; Appl. Phys.Lett., 85, 837 (2004).
5. Q. Pei, G. Yu, C. Zhang, Y. Yang and A.J. Heeger; Science, 268, 1086 (1995).
6. Q. Pei, Y. Yang, G. Yu, C. Zhang and A.J. Heeger; J.Am. Chem.Soc., 118, 3922 (1996).
7. M. Strukelj, F. Papadimitrakopoulos, T.M. Miller and L.J. Rothberg; Science, 267, 1969. (1995).
8. M.R. Andersson, G. Yu and A.J. Heeger; Synth. Met., 85, 1275 (1997).
9. M. Pope and C.E. Swenberg; Electronic Processes in Organic Crystals, Oxford University Press, New York, p.739 (1982).
10. (a) D.D. Gebler, Y.Z. Wang, J.W. Blatchford, S.W. Jessen, D.-K. Fu, T.M. Swager, A.G. MacDiarmid and A.J. Epstein; Appl. Phys.Lett., 70: 1644. (1997) and (b) M.Mazzeo, D.Pisignano.F.Della Sala, J. Thompson, R.I.R. Biyth, G. Gigli, G. Sotgiu and G. Barbarella; Appl. Phys. Lett., 82, 334. (2003).
11. H. Eckhardt, L.W. Shacklette, K.Y. Jen and R.L. Elsenbaumer; J. Chem. Phys., 91, 1303 (1989).
12. Y.F. Li, Y. Cao, J. Gao, D.L. Wang, G. Yu and A.J. Heeger; Synth. Met., 99, 243 (1999).
13. A. Montali, P. Smith and C. Weder; Synth. Met., 97, 123 (1998).
14. A. Kraft, A.C. Grimsdale and A.B. Holmes; Angew.Chem. Int. Ed., 37, 402 (1998).
15. L. Akcelrud; Prog. Polym. Sci., 28, 875 (2003).
16. F. Huang, H.B. Wu, D.L. Wang, W. Yang and Y. Cao; Chem. Mater., 16, 708 (2004).
17. H.B. Wu, F. Huang, Y.Q. Mo, W. Yang, D.L. Wang, J.B. Peng and Y. Cao; Adv.Mater., 16, 1826 (2004).
18. F. Huang, L.H. Hou, H.B. Wu, X.H. Wang, H.L. Shen, W. Cao, W. Yang and Y. Cao; J. Am.Chem. Soc., 126, 9845 (2004).
19. H.B. Wu, F. Huang, J.B. Peng and Y. Cao; Organic Electronics, 6, 118 (2005).

20. J.L. Bredas and A.J. Heeger; Chem. Phys. Lett., 217, 507 (1994).
21. J. Cornil, D.A. dos Santos, D. Beljonne and J.L. Bredas; J. Phys. Chem., 99, 5604 (1995).
22. P.L. Burn, A.B. Holmes, A. Kraft, D.D.C. Bradley, A.R. Brown, R.H. Friend and R.W. Gymer; Nature, 356, 47 (1992).
23. M. Leclerc; J. Polym. Sci.A: Polym. Chem., 39, 2867 (2001).
24. M. Hay and F.L. Klavetter; J. Am. Chem. Soc., 117, 7112 (1995).
25. U. Scherf and K. Mullen; Macromolecules, 25, 3546 (1992).
26. J.-K. Lee, R.R. Schrock, D.R. Baigent and R.H. Friend; Macromolecules, 28, 1966 (1995).
27. Q.B. Pei and Y. Yang; J. Am. Chem. Soc., 118, 7416 (1996).
28. G.F. He, C.H. Yang, R.Q. Wang and Y.F. Li; Display, 21, 69 (2000).
29. Q.J. Sun, H.Q. Wang, C.H. Yang and Y.F. Li; Synth.Met., 128, 161 (2002).
30. Q.J. Sun, H.Q. Wang, C.H. Yang, X.G. Wang, D.S. Liu and Y.F. Li; Thin Solid Film, 417, 14 (2002).
31. Q.J. Sun, H.Q. Wang, C.H. Yang and Y.F. Li; J. Mater. Chem., 13, 800 (2003).
32. B.W. D'Andrade and S.R. Forrest; Adv. Mater., 16, 1585 (2004).
33. Y.-Z. Lee, X.W. Chen, M.-C. Chen, S.-A. Chen, J.-H. Hsu and W.S. Fann; Appl. Phys. Lett., 79, 308 (2001).
34. G.L. Tu, Q.G. Zhou, Y.X. Cheng, L.X. Wang, D.G. Ma, X.B. Jing and F.S. Wang; Appl.Phys.Lett., 85, 2172 (2004).
35. C. Adachi, M.A. Baldo, M.E. Thompson and S.R. Forrest; J. Appl.Phys., 90, 5048 (2001).
36. S. Tokito, M. Suzuki, F. Sato, M. Kamachi and K. Shirane; Organic Electronics, 4, 105 (2003).
37. R.A. Wessling; J. Polym. Sci., Polym.Symp., 72, 55 (1985).
38. A. Beerden, D. Vanderzande and J.Gelan; Synth.Met., 52, 387 (1992).
39. R.O. Garay, U. Baier, C. Bubesk and K. Mullen; Adv.Mater., 5, 561 (1993).
40. H.G. Gilch and W.L. Wheelwright; J. Polym.Sci. Part A, Polym.Chem., 413,13 (1996).
41. D. Braun and A.J. Heeger; Appl. Phys.Lett. 1991, 58, 1982 (1990).
42. R.W. Lenz and C.V. Handlovitis; J. Org.Chem., 25, 813 (1960).
43. D.-J. Kim, S.-H. Kim, T. Zyung, J.-J. Kim, I. Cho and S.K. Choi; Macromolecules, 29(10), 3657 (1996).
44. Irina P. Beletskaya and Andrei V. Cheprakov; Chem.Rev., 100, 3009 (2000).
45. Z. Yang, B.Hu and F.E. Karasz; Macromolecules, 28(18), 6151 (1995).
46. M. Rehahn and A.D. Schluter; Makromol. Chem. Rapid Commun., 11, 375 (1990).
47. K.Y. Jen, R. Oboodi and R.L. Elsenbaumer; Polym.Mater. Sci. Eng., 53, 79 (1985).
48. G.G. Miller and R.L. Elsenbaumer; J. Chem.Soc., Chem. Commun., 1346 (1986).
49. R.D. McCullough and R.D. Lowe; J. Chem.Soc., Chem. Commun., 70 (1992).
50. K. Tamoa, K. Sumitani and M. Kumada; J. Am. Chem.Soc., 94, 4376 (1972).
51. S. Kodama, I. Naajima, M. Kumada, A. Minato and K. Suzuki; Tetrahedron, 38, 3347 (1982).
52. R. Sugimoto, S. Takeda, H.B. Gu and K. Yoshino; Chem. Express, 1, 635 (1986).
53. X.H. Li and Y.F. Li; J. Appl. Polym. Sci. 2003, 90, 940 (1990).
54. F. Chen, P.G. Mehta, L. Takiff and R.D. McCullough; J. Mater.Chem., 6, 1763 (1996).
55. M. Fukuda, K. Sawada, S. Morita and K. Yoshino; Synth.Met., 41-43, 855 (1991).
56. Q. Pei and Y. Yang; J. Am.Chem. Soc., 118, 7416 (1996).
57. T. Yamamoto, Y. Hayashi and A. Yamamoto; Bull. Chem. Soc. Jpn., 51, 209 (1978).
58. Z. Bao, W. Chan and L. Yu; Chem.Mater., 5(1), 2(1993).

12

Polymers in Electronics

V. Ramgopal Rao[*] and Saurabh Goyal

Department of Electrical Engineering, IIT Bombay, Mumbai-400076, India.
[*]Corresponding author's email: rrao@ee.iitb.ac.in

Contents

Summary: In this chapter we had a brief look at how polymers are occupying space in the field of electronics. The conjugated nature and the existence of dimerization in polymers results in their semiconducting behavior. These can be doped and their conductivity can be varied. This leads to their use in various electronic devices such as LEDs, photovoltaic cells and FETs.

1. INTRODUCTION

Twenty-five years ago, polymers were synonymous with insulators. The work at that time was mainly concentrated in studying their insulating properties and charge storage capabilities. Polymers, the name itself, would remind of the insulation of power cables, plastics and the poly-bags. Nobody would even have thought that some of the members of the known "insulator family" would actually be conducting.

The Nobel Prize for chemistry, 2000, was awarded to three people whose discovery though was the result of chanced conversations, accidental experiments and informed insights, led to a new turn in science and technology. A new field of science known as "conducting polymers" was born. Last 10 years have seen extensive research is the field of polymer electronics. The aim is to replace the current silicon technology by flexible, low cost polymer devices. In this chapter, we will see the fundamentals behind the polymer electronics and give a brief overview of the research in these 10 years and the use of polymers as electronic devices like LEDs, photovoltaic cells and field-effect transistors.

2. PHYSICS

"Conjugated Molecules" consist of long chains of alternate single and double bonds whose position can be interchanged. In this section, we will see how this conjugated nature provides semiconducting or conducting behavior to the polymers. Polyacetylene is considered the prototype of a conjugated molecule for describing molecular conduction. Polyacetylene has two equivalent conformations as shown in **Fig. 12.1**. Each carbon atom in polyacetylene possesses four valence electrons. The σ-bonds, which make up the polymer chain, consumes two of them. The hydrogen atom is bonded with the third electron. The double bond is formed of a σ-bond and a π-bond. Hence, the fourth electron occupies a p_z-orbital, which forms the π bond of the double bond. The p_z-orbital is perpendicular to the other orbitals. It overlaps with p_z-orbital on neighboring carbon atoms to form a pair of π-molecular orbitals that are delocalized over the polymer chain.

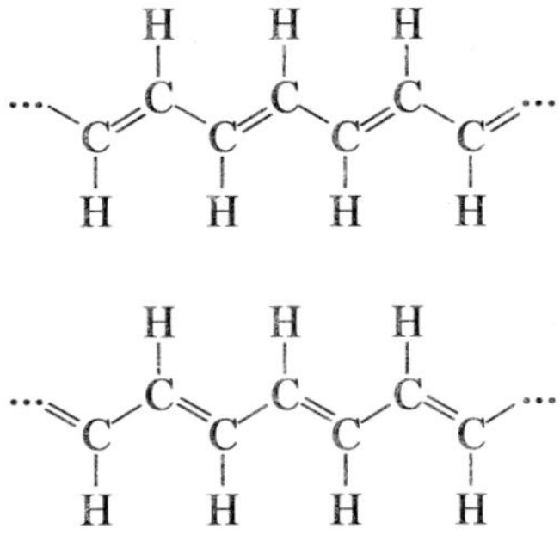

Fig. 12.1. Two equivalent conformations of polyacetylene.

Delocalization of π-electrons results in the formation of a one-dimensional band. The band is half-filled and thus the system must show metallic characteristics. This is true for a non-dimerized system. Polyacetylene and other similar conjugated polymers have two conformations **(Fig. 12.1)** and show bond dimerization. Because of this dimerization, an energy gap or a "band gap" is generated in the conjugated molecular structure, which explains the observed semiconducting behavior [1]. The lower-energy π- (or bonding) orbital is the valence band, which is also known as HOMO (Highest Occupied Molecular Orbital), while the higher-energy π^*- (or anti-bonding) orbital forms the conduction band. The conduction band is also known as LUMO or Lowest Unoccupied Molecular Orbital. The two identical structures, if joined together, lead to a discontinuity, with an electron, which does not take part in a π-bond as shown in **Fig. 12.2**. This phenomenon is called "bond-alternation" (single-double to double-single). This leads to generation of an electronic energy level known as soliton [2] in the mid-gap as shown in **Fig. 12.4**. The electron is of p_z-orbital nonbonding character. It is delocalized over approximately 14 carbon sites [1]. This delocalized electron also causes the observed magnetic susceptibility of polyacetylene. It has spin but no effective charge. The conductivity of polymers can be increased by fifteen orders of magnitude. **Fig. 12.3** compares the electrical conductivity of various polymers with the inorganic compounds. Doping, unlike conventional semiconductors, is achieved by reduction or oxidation i.e. removing or adding of electrons to the polymer. Hence, doping is also known as redox modification. Details of various

Fig. 12.2. Bond-alternation in polyacetylene.

kinds of doping including chemical, electrochemical, photo doping and charge injection doping can be found in [3] and [4]. The observed increase in conductivity can be explained by presence/appearance of energy levels in the band gap.

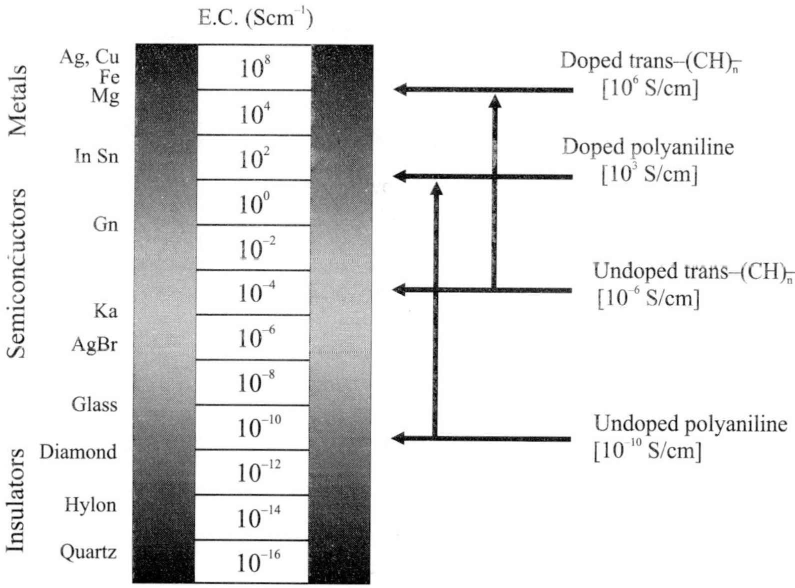

Fig. 12.3. Electrical conductivities (Scm⁻¹) of electronic polymers and related materials.

In polyacetylene, soliton is present in the band gap and hence it is easier to remove these electrons compared to the electrons in the valance band. Thus, on oxidation it is these mid-gap electrons, which are removed first, causing a decrease in spin concentration and the formation of charge carriers. Similarly, on reduction the extra electrons added go to the mid-gap states to pair off with the existing electrons. This is illustrated in **Fig. 12.4**.

When original spins are used up, i.e. solitons are finished, this process can no longer continue. Further oxidation and reduction then is forced to break π-bonds as shown in **Fig. 12.5**. This form has both spins and charge. These spin-charge pairs are known as polarons. Experiments with spin spectroscopy support this fact. Polyacetylene possesses a moderate concentration of free spins. At 1% doping level the spin spectroscopy shows almost zero spins. Spins reappeared when the doping

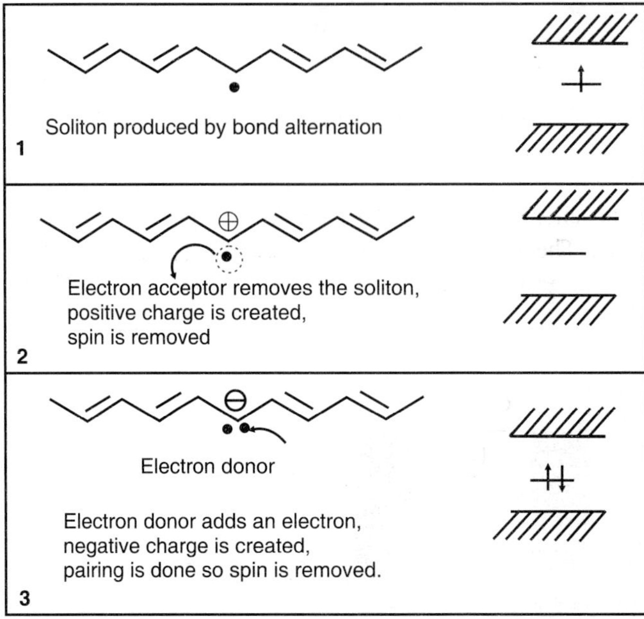

Fig. 12.4. Effect of oxidation and reduction on solitons. Reproduced with permissions from David Bott; Phys. Technol., 16, 121 (1985). Copyright © 1985, IOP Publishing Limited.

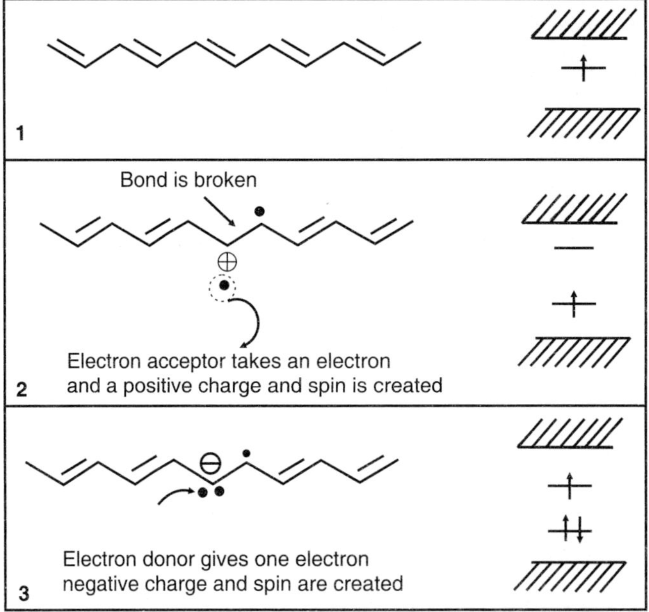

Fig. 12.5. Polyacetylene at high doping level showing broken bonds and formation of polarons. Reproduced with permissions from David Bott; Phys. Technol., 16, 121 (1985). Copyright © 1985, IOP Publishing Limited.

level was raised above 6%. Polarons result in two states being formed in the band gap as shown in **Fig. 12.5**. The formation of new states aids the conduction process, which explains the increase in conductivity on doping.

In polyacetylene, as mentioned before, the double and single bonds can be changed and both structures have same energy. In other polymers e.g. polyaromatics like polypyrrole, polythiophene and poly(propiolic anhydride), a rearrangement with equal energy is not possible. Hence, conduction takes place through polaron mechanisms. Also, polyacetylene is the only polymer, which shows charge transport by solitons. In other conjugated polymers, each bond alternation generates a double gap and conduction is by polaronic energy levels [1].

3. POLYMER LIGHT EMITTING DIODES

Electroluminescence is a phenomenon where a material emits light in response to an electric current passed through it, or to a strong electric field. Electroluminescence requires injection of electrons from one electrode and injection of holes from the other electrode. The recombination of electron and hole, followed by the radiative decay of the exciton (the excited electron-hole pair) may result in emission of light.

With the discovery of conducting polymers, and the fact that they have a band gap, aroused interest amongst the scientists and engineers to decipher their use in light emitting diodes. The potential advantages were very clear. The possibility of making low cost, robust, flexible and large area displays attracted everyone. The luminescent properties can be varied by tailoring the molecules. Hence diodes emitting different colors can be formed. For example, for a polymer (say PPV and derivatives) the color may change from yellow-greenish to orange [6] to blue [7]. The conjugated polymers, that have larger semiconductor gaps, and that can be prepared in a sufficiently pure form to control the non-radiative decay of excited states that decreases efficiency, can show high quantum yields in electroluminescent devices.

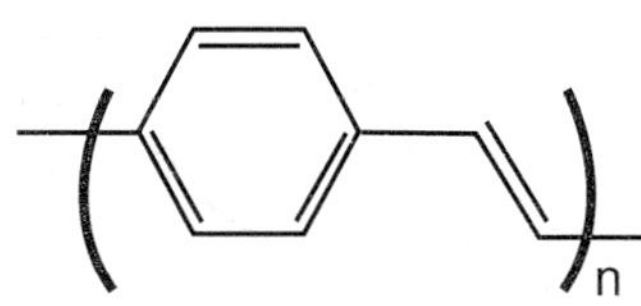

Fig. 12.6. Poly(p-phenylene-vinylene) (PPV).

Electroluminescence using conjugated polymers was first observed in 1990 [8]. Poly(p-phenylene-vinylene) (**Fig. 12.6**) was used as it could be conveniently made into high quality films and showed strong photoluminescence. PPV film was sandwiched between two metallic electrodes as shown in **Fig. 12.7**. One of these electrodes must be semi-transparent, so that light emission can take place normal to the plane of the device.

Luminescence in such a device is achieved, when the diode is biased sufficiently for the injection of positive and negative charge carriers from electrodes. Such diodes can easily be formed. Semiconducting polymer layer can be solution-processed and spin coated onto the transparent electrode ITO (Indium-Tin-Oxide). The top level electrode is formed by thermal evaporation of the metal. The device formed is as shown in **Fig. 12.7**.

Polyphenylenevinylene (PPV) has a band gap of about 2.5 eV. This produces yellow-green luminescence [9]. Working of PPV based LED can be demonstrated by energy-level-diagram as

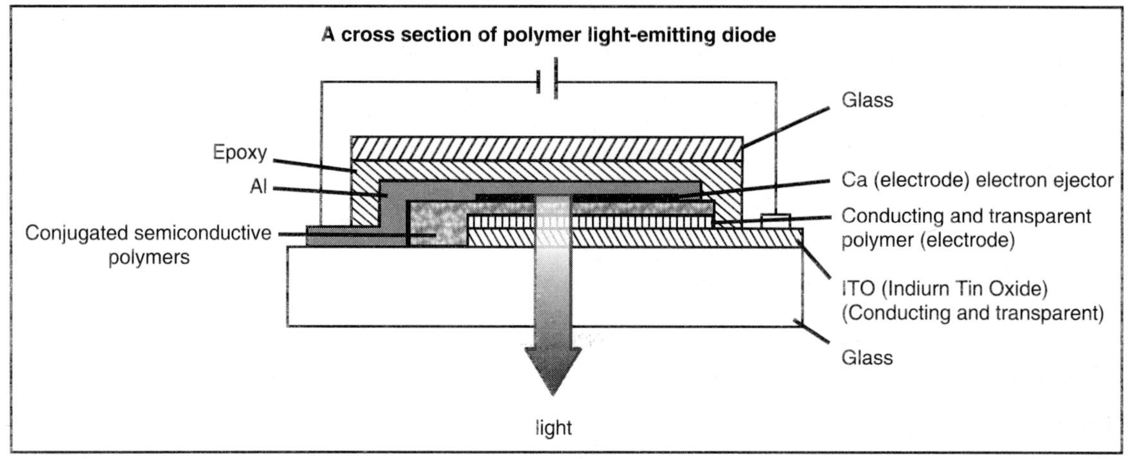

Fig. 12.7. Schematic diagram of Polymer Light Emitting Diode. Reproduced with permission from Alan J. Heeger; Semiconducting and metallic polymers: the fourth generation of polymeric materials. Nobel Lectures in Chemistry 1996-2000 (Ed. Ingmar Grenthe), World Scientific Publishing Company, Stockholm. (2003). Copyright © Nobel Foundation.

shown in **Fig. 12.8**. As said earlier, electrons and holes combine with one another within the polymer film and form neutral bound excitons. Exciton is a bound excited state that can decay by emitting a photon. In polymers, there are free states for electrons above LUMO and free states for holes below HOMO (valance band states). Electrons have to see the barrier of ΔE_e as they pass from the electrode (Aluminum) to the polymer. Similarly, holes have a barrier of ΔE_h. Hence, the negative electrode is chosen to have a low work function so that electrons can be easily injected, while the positive electrode has a high work function to facilitate the hole injection. ITO has a relatively high work function and is used as hole-injecting electrode. Al, Mg, or Ca are used as electron injecting materials. The efficiency of the first, simple LED made of PPV, ITO and Al was very low. One of the main reasons of the low efficiency was high value of ΔE_e. If both hole-injecting and electron-injecting electrodes have relatively low barriers for charge injection, high current densities and light emission is produced at low voltages. Calcium, with a relatively lower work function, can't be used as it is highly susceptible to atmospheric degradation.

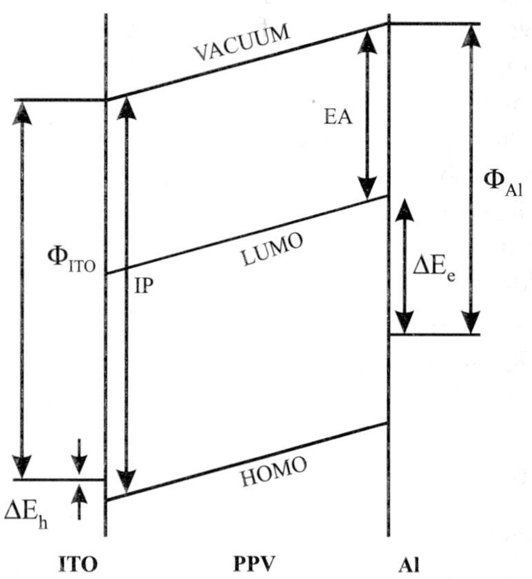

Fig. 12.8. Energy Band Diagram of an LED under forward bias. Reproduced with permission from R.H. Friend *et al.*; Nature, 397,121 (January 1999). Copyright © 1999, Nature.

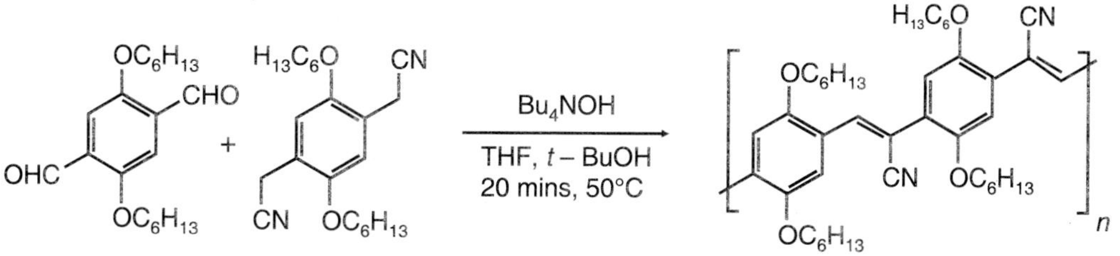

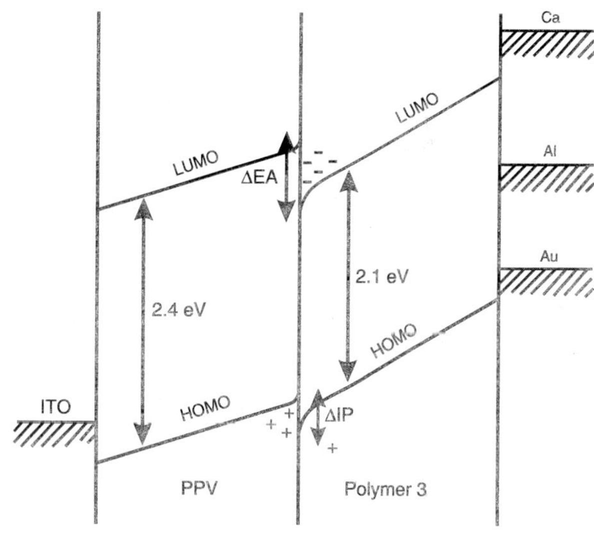

Fig. 12.9. Synthesis of a new polymer of low electron affinity. Reproduced with permission from N.C. Greenham *et al.*; Nautre, 365, 628 (October 1993). Copyright © 1993, Nature.

Heterostructure architectures have provided an interesting and feasible solution. Two polymers are used, each to transport electrons and holes. A new polymer was formed in [10] as shown in **Fig. 12.9** having low electron affinity. The schematic energy level diagram for the new diode looked as in **Fig. 12.10**. High efficiencies were reported for the same.

New polymers are also being developed to get various colors from the LEDs. Blue emission is achieved by dioctyl-substituted polyfluorene [11]. Red emitting polymers have been developed by Dow Chemical Company and others [12]. Once the success is achieved in producing various color producing polymers with sufficient lifetime and reliability, color polymer displays will be formed from them. The estimated market for polymer displays is billions of dollars. Additional revenue is expected in the areas of indicators, backlights, and general lighting. Cell phone displays and personal digital assistant consumer devices adds to the list. Hence, it becomes obvious that both commercial and non-commercial research is attracted towards the field.

Fig. 12.10. Energy level diagram of a heterostructure diode. Reproduced with permission from N.C. Greenham *et al.*; Nature, 365, 628 (October 1993). Copyright © 1993, Nature.

4. PHOTOVOLTAIC DIODES

Photovoltaics are semiconducting materials that absorb light at energies above the band gap, leading to the separation of positive and negative charge carriers. The phenomenon is the reverse of the luminescence. These charges are collected at opposite electrodes, giving rise to an open-circuit voltage that can be converted to current, if an external load is applied.

Excitons, which are formed when an electron and hole capture one another, have to be split in a photovoltaic diodes made of conjugated polymers. As we saw in earlier section, the excitons' formation and their radiative decay lead to electroluminiscence. In a photovoltaic diode, the process is reverse. So it does not seem natural that polymers would be suitable for photovoltaics as excitons formed are stable, as discussed in the last section. Excitons can be generated either by electric current as in LEDs or by energy of a photon, i.e. shining light on the polymer. Photovoltaic polymers diodes generate voltages by causing the excitons to split and redistribute their charge. Exciton dissociation was shown [13] to be efficient at interfaces between materials with different electron affinities and ionization potentials. The electron should go from lower electron affinity material to the polymer having higher electron affinity. The hole would be accepted by the material with lower ionization potential.

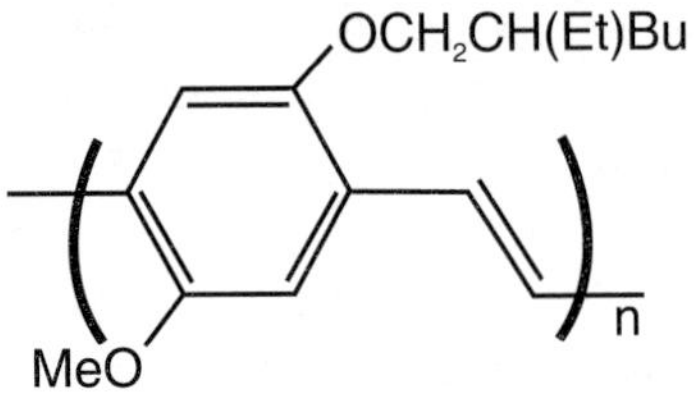

Fig. 12.11. MEH-PPV.

R = C_6H_{13}

Fig. 12.12. CN-PPV.

Heterojunctions made of donor-acceptor polymers, hence, were found suitable for photovoltaic applications. One such heterojunction [15] was made using MEH-PPV and CN-PPV as seen in **Fig. 12.13**. Cyano-PPV, or CN-PPV is used as electron acceptor (**Fig. 12.12**). It has electron-accepting cyano groups attached to the phenylene rings, which helps to lower the energy of both the valence and conduction bands. The addition of cyano groups to a dialkoxy derivative of PPV increases the ionization potential and electron

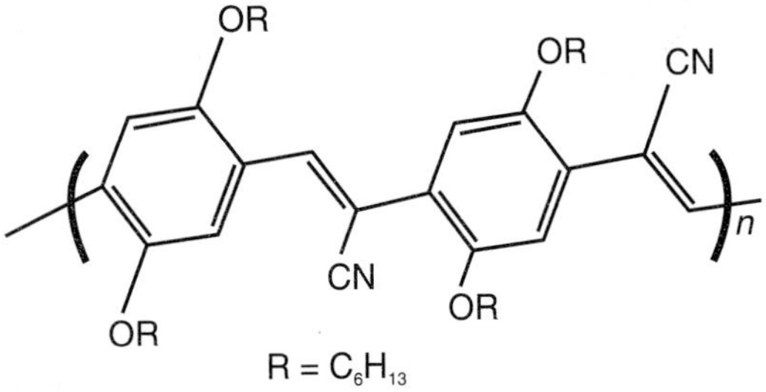

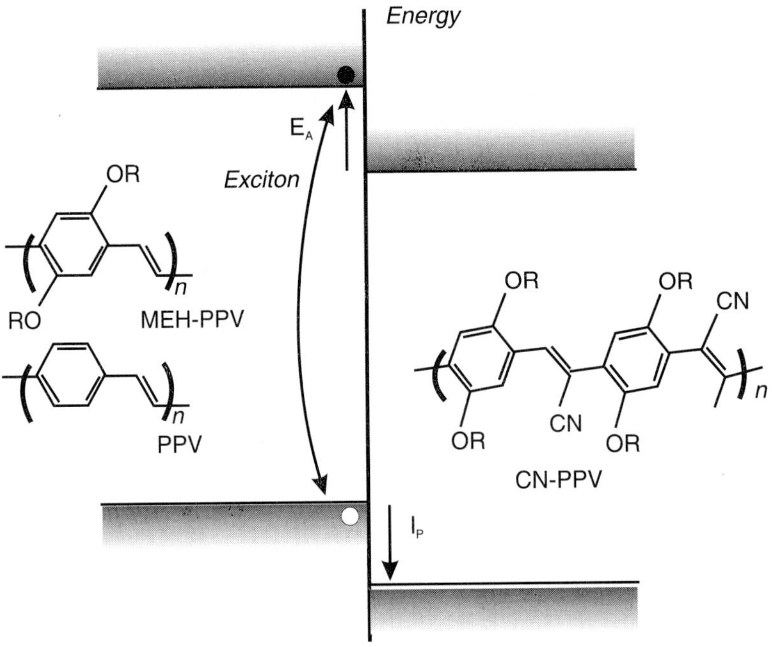

Fig. 12.13. Heterojunction made of CN-PPV and MEH-PPV/PPV. Reproduced with permission from R.H. Friend; Pure Appl. Chem., 73(3), 425 (2001). Copyright © 2001, IUPAC.

affinity by $\approx$0.5 eV [15]. MEH-PPV is a derivative of PPV in which two of the hydrogen atoms on each phenylene ring are replaced by O-CH_3 and O-$CH_2CH(C_2H_5)C_4H_9$ (**Fig. 12.11**). It is used as an electron donor. In MEH-PPV-CN-PPV heterojunction, holes move into the MEH-PPV and the electrons move into the CN-PPV. The difference in electron affinity of two polymers must be greater than the binding energy of the exciton for the dissociation to be favorable.

To increase the efficiency of the heterojunction, a novel idea was proposed [15,16]. Heterojunction was distributed throughout the film thickness by producing an inter-penetrating network of electron accepting and hole accepting polymers. Buckminsterfullerene C_{60} was used as acceptor [16] and MEH-PPV as donor in a composite heterojunction or internal heterojunction. Inter-penetrating network of donor-acceptor molecules leads to volume distribution of the photo-generation sites, which enhances exciton dissociation.

Because of low efficiencies as compared to inorganic photovoltaic cells, polymer photovoltaic cells have to cover a long way of research and experimentation before they can replace the currently used inorganic cells. The research currently going on is showing positive results and the day is not too far when solar cells would be made at very low cost using polymers.

5. POLYMER TRANSISTORS/CONDUCTING POLYMER BASED FETs

A thin film transistor is one of the most promising applications for conjugated semiconducting films. Several groups had attempted the fabrication of field-effect transistor using organic dyes and polyacetylene. The source-drain current were found to be slightly modulated by applied gate bias. First successful field effect transistor using conducting polymers was demonstrated by Koezuka and coworkers [17]. They demonstrated that it is possible to control the current flowing between source and drain through gate.

Polythiophene was used as a semiconducting material. The device that was fabricated is shown in **Fig. 12.14**. N-type silicon covered with SiO_2 was used as substrate. Gold was used for source and drain electrodes. The distance between the two gold electrodes was 10 μm. This determines the channel length of the transistor. Gold was deposited using conventional photolithography and vacuum deposition techniques.

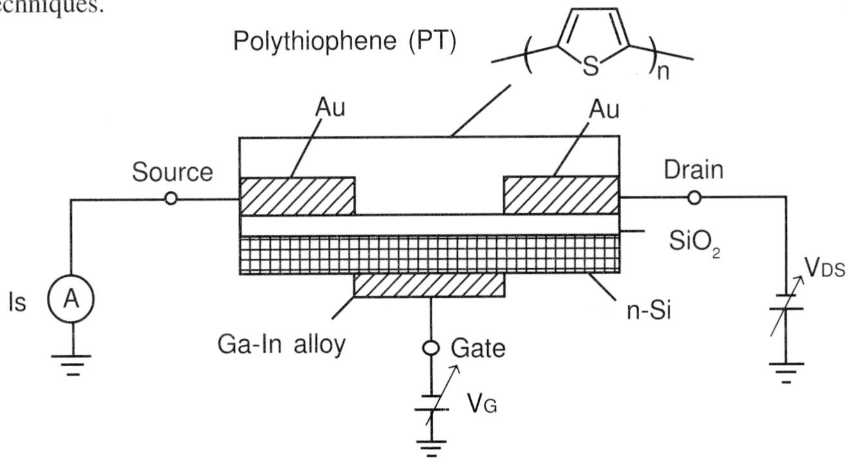

Fig. 12.14. Schematic diagram of FET. Reproduced with permission from A. Tsumara *et al.*; Appl. Phys. Lett., 49(18), 1210 (1986). Copyright © 1986, American Institute of Physics.

The polythiophene film, utilized as semiconductor, was electrochemically prepared under nitrogen atmosphere. 2-2'-bithiophene, a dimer of thiophene, was used as a starting material. The polymer film was grown between two gold electrodes using reaction constituting of 2-2'-bithiophene, tetraethylammonium perchlorate and acetonitrile. When current was passed a 1400 Å thick polythiophene layer was formed. The fabricated device had channel length of 10 µm and width of 2 mm. SiO_2 acts as gate oxide and silicon itself acts as gate electrode.

The device grown as mentioned above didn't show any field effect directly. Polythiophene formed was highly doped with ClO_4^- during its preparation. When the film was electrochemically reduced/undoped at a preselected potential vs. standard calomel electrode (SCE), the field effect began. The device with undoping done at 0 V showed the best transistor characteristics. **Fig. 12.15** shows the source current (I_S) vs. drain voltage (V_{DS}) curves for the various values of gate voltages with semiconductor undoped at 0 V (vs. SCE). The characteristics indicate

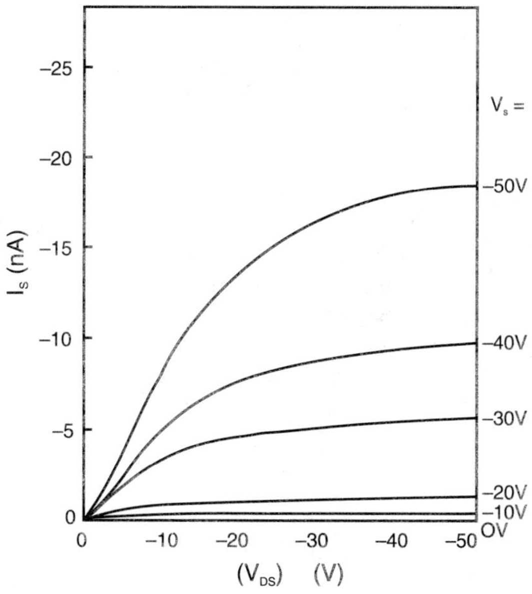

Fig. 12.15. *Is* vs V_{DS} characteristics at various gate voltages. Reproduced with permission from A. Tsumara *et al.*; Appl. Phys. Lett., 49(18), 1210 (1986). Copyright © 1986, American Institute of Physics.

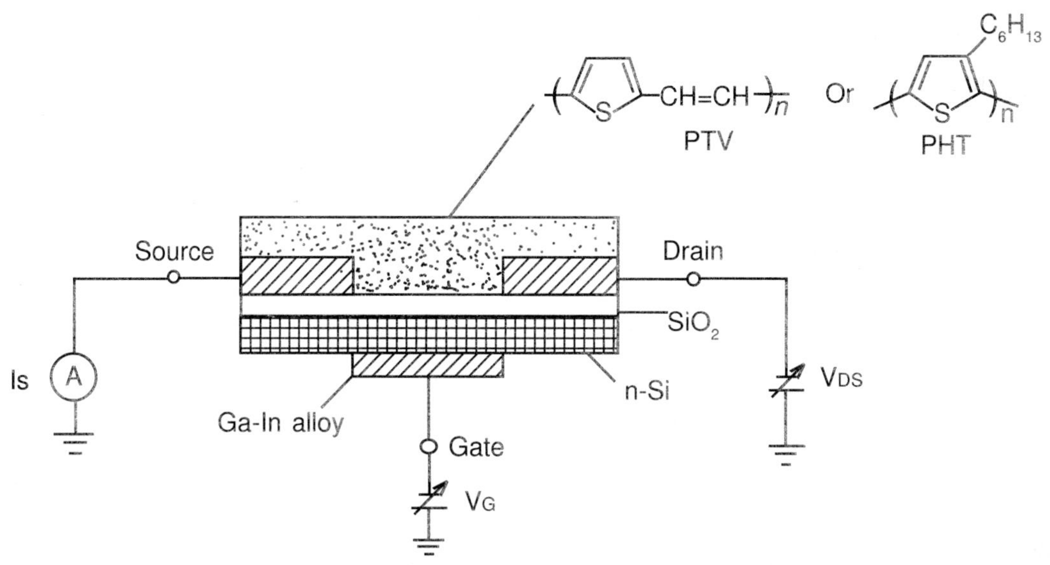

Fig. 12.16. FET structure. Reproduced with permissions from A. Tsumura *et al.*; Synth. Metals, 41-43, 1181 (1991). Copyright © 1991, Elsevier Science Ltd.

that the polythiophene is a *p*-type semiconductor. The source current was observed to be modulated by a factor of 10^2 to 10^3 by varying the gate voltages from 0 V to –50 V.

After the demonstration of field-effect transistor using polythiophene, several other polymers were tried. Spin coated poly(3-hexyl-thiophene) (PHT) and poly(2,5-thienylene-vinylene) (PTV) films [18] and electropoly-merized polyaniline [19] were the most widely studied polymers among them. The device structure remained the same (**Fig. 12.16**). The PHT was prepared according to Yoshino's method [20]. The PHT film was spin-coated on the substrate from its chloroform solution and it was doped with iodine by dissolving I_2 in the solution in advance. The device formed showed enhancement type FET characteristics (**Fig. 12.17**). The device was found to be unstable and the conductivity of the layer decreased rapidly as time passed. When the conductivity was high there was a large leakage current between source and drain, but as time passes the device characteristics changed as seen in **Fig. 12.17**. The transistors formed with non-doped PTV showed stable and excellent FET characteristics with large source current. PTV worked as a good semiconductor without any doping, hence showed stable characteristics [18].

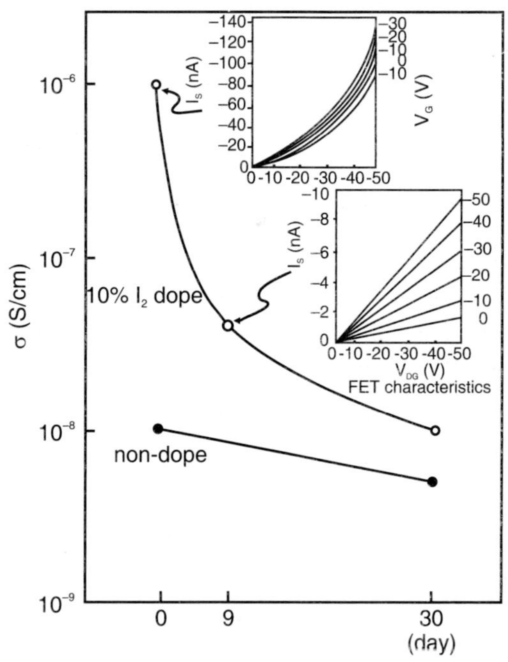

Fig. 12.17. Time dependent FET characteristics and electrical conductivity of FET with PHT as the semiconductor. The effect of I_2 doping is also shown. Reproduced with permissions from A.Tsumura *et al.*; Synth. Metals, 41-43, 1181 (1991). Copyright © 1991, Elsevier Science Ltd.

Mobility of the semiconducting layer is one of the important parameters of FETs. The first transistor developed by Koezuka *et al.* had the carrier mobility of about 10^{-5}cm^2 V^{-1}s^{-1}. Amorphous silicon transistors have mobility in the range of 0.1 to 1 cm^2 V^{-1}s^{-1}. After the advent of first conducting based polymers transistor, there was a quest to improve the carrier mobilities. The mobility is related to self-organization of the polymer chains in the solid state [21]. Highly stacked π-conjugated oligomers and polymers such as sexithiophene and regioregular P3AT were reported to give higher field-effect mobilities of 10^{-2} to 1 cm^{2}V^{-1}s^{-1} in their neutral state [22]. These values are much greater than the mobilities of amorphous π-conjugated molecules such as polyacetylene in their neutral or lightly doped states. It has also been suggested that purification of semiconducting materials to reduce the carrier scattering by impurities can lead to an increase in the mobility [23]. Garnier *et al.* [24] have obtained the same level of carrier mobility as that of amorphous silicon in the insulated-gate TFT with vacuum deposited α-sexithienyl. The semiconducting material was purified during evaporation. As early as 1993, there was an attempt to increase the mobility of PTV by H. Fuchigami *et al.* [23]. They used the concept that large π-conjugation length results in wide valence and conduction bands. This leads to higher carrier mobility. PTV was used as semiconducting layer, and was developed from a precursor using HCl as catalyst (reaction showed in **Figs. 12.18 & 12.19**). It

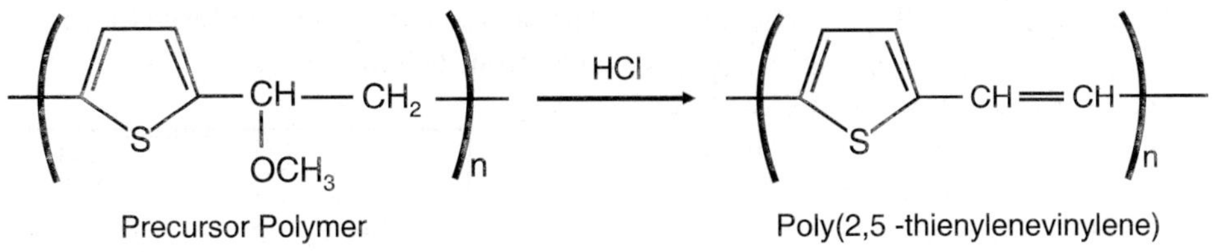

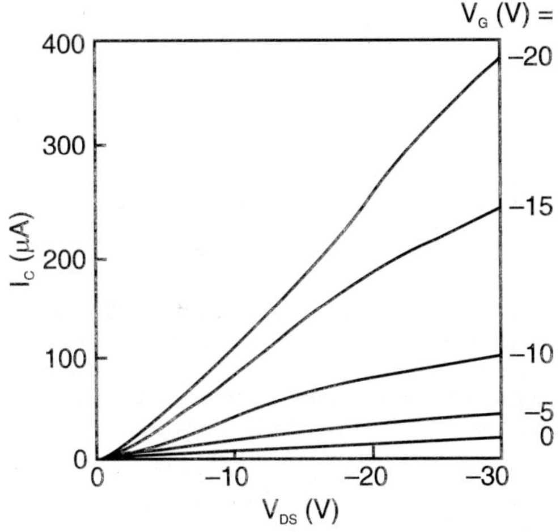

Fig. 12.18. Formation of PTV using HCl [23].

was observed (**Fig. 12.20**) that the mobility of semiconducting PTV was proportional to the amount of conversion from the precursor. The carrier mobility was calculated to be 0.22 cm^2 V^{-1}s^{-1} when conversion was 98% complete. As the reaction proceeded, the appearance and shift of visible spectra was observed. This was attributed to π - π^* transition in PTV and enlargement of π-conjugation length. Polyaniline has also been used as semiconducting layer [19,25,26] and the mobility of HCl doped PANI was reported to be as high as 5.38 cm^2 V^{-1}s^{-1}, though the same was not stable in air.

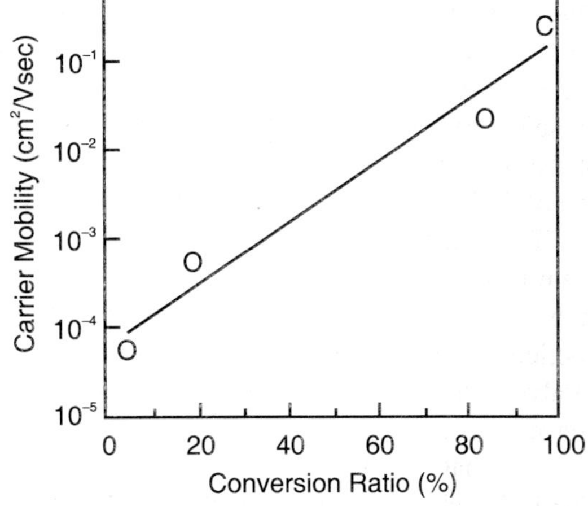

Fig. 12.19. Channel current vs. drain voltage of FET. The conversion ratio from the precursor was 98%. Reproduced with permission from H. Fuchigami et al.; Appl. Phys. Lett., 63(10), 1372 (1993). Copyright © 1993, American Institute of Physics.

Fig. 12.20. Carrier mobility vs. the conversion ratio. Reproduced with permission from H. Fuchigami et al.; Appl. Phys. Lett., 63(10), 1372 (1993). Copyright © 1993, American Institute of Physics.

As the properties of organic semiconductors materials improved, the interest in making FETs from polymers started to grow. Focus slowly shifted to all polymer transistors where all the layers of transistors are made of one kind of polymer or the other. All polymer transistors have characteristics that are difficult to achieve with materials and methods used for conventional electronics. These include mechanical flexibility, durability, lightweight, printable over large area and low cost. The manufacturing techniques do not involve high temperature processes. They are compatible with high-speed reel-to-reel techniques for fabrication. Plastic circuits have immense potential in future electronics. Electronic paper, wearable sensors, low-cost smart cards, and RF-ID tags, flexible plastic microphones are some of the potential applications of polymer transistors.

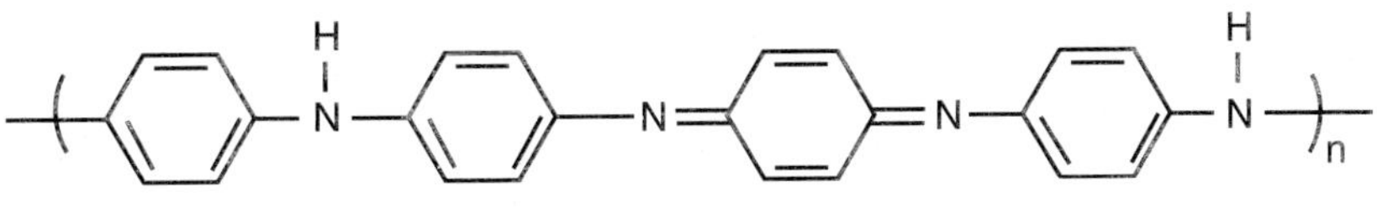

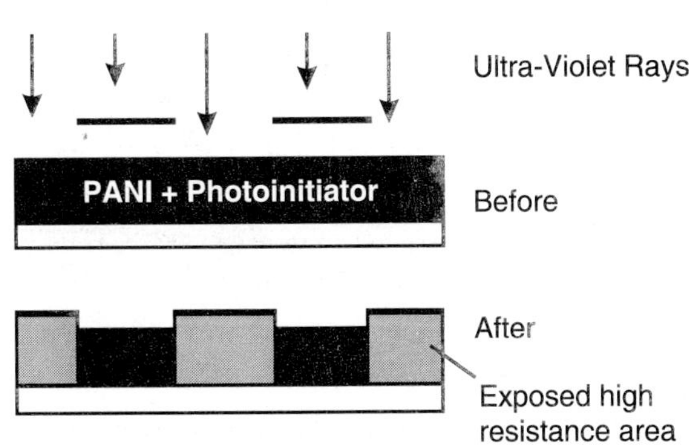

Fig. 12.21. Polyaniline.

One of the earliest successful attempts to fabricate an all polymer transistor was in 1998 at Philips Research Labs [27]. They used photopat-terning of doped polyaniline (PANI) (**Fig. 12.21**). Polyaniline doped with camphorsulfonic acid was dissolved in *m*-cresol. A photoinitiator, 1-hydroxycyclohexylphenylketon was then added. When exposed to light, this solution changed to non-conducting leucoemeraldine form as shown in **Fig. 12.22**. The sheet resistance of leucoemeraldine is 10^{14} ohm/square. Conducting tracks in nonconducting leucoemeraldine were formed by selectively exposing the spin-coated layer of the polyaniline solution on a substrate such

Fig. 12.22. Photopatterning of polyaniline [27].

as polyimide. These conducting tracks were used as interconnects and as electrodes of the transistor.

To form the transistor, a three-mask process was used. The first mask was used to form source drain using photolithography. After the source-drain layer was formed, the semiconducting layer was spin coated on to it. Polythenylenevinylene (PTV) was used as semiconducting layer. PTV was applied by conversion of a spin-coated precursor film at elevated temperature using HCl gas as a catalyst. Onto this PTV layer, polyvinylphenol (PVP) layer was spin coated. PVP is an insulator, hence used as gate dielectric. The top gate layer was formed in a similar way as drain-source layer. The schematic of the transistor is shown in **Fig. 12.23**. Channel lengths down to 2 µm were obtained. The I_D-V_D characteristics observed are shown in **Fig. 12.24**. Mobility of charge carriers was observed to be 3×10^{-4} cm^2 V^{-1}s^{-1}. The p-transistors were in accumulation mode. The transistor is on at 0 V bias. Hysteresis and current saturation were observed. The clockwise hysteresis was explained by charge trapping in PVP.

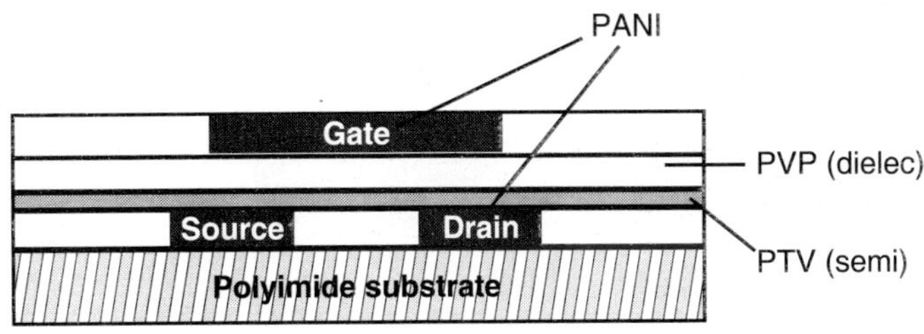

Fig. 12.23. Schematic diagram of cross section of a top-gate all polymer transistor. Reproduced with permission from C.J. Drury *et al.*; Appl. Phys. Lett., 73(1), 108 (1998). Copyright © 1998, American Institute of Physics.

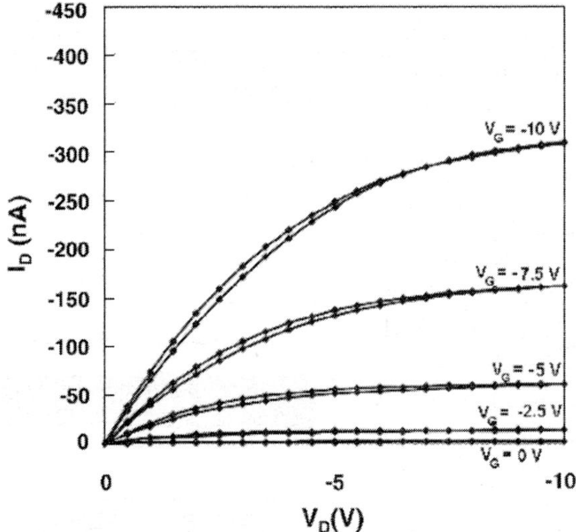

Fig. 12.24. Transistor characteristics of all polymer transistor at various gate voltages. Drain bias swept from 0 to –0 V and back. Reproduced with permission from C.J. Drury *et al.*; Appl. Phys. Lett., 73(1), 108 (1998). Copyright © 1998, American Institute of Physics.

While making polymer transistors special care should be taken for stack integrity. The previously deposited films must not dissolve in the solvents of new layer. In the above case PTV is dissolved in dichloromethane, which is a non-solvent for polyaniline [27]. PTV itself is insoluble in common organic solvents. The top gate geometry prevents degradation of semiconductor as it shields it from environment.

The geometry shown in **Fig. 12.23** is top gate geometry with gate at top. The other kind of geometry used is bottom gate geometry as shown in **Fig. 12.25**. This structure was also demonstrated at Philips research labs in 2000 [28].

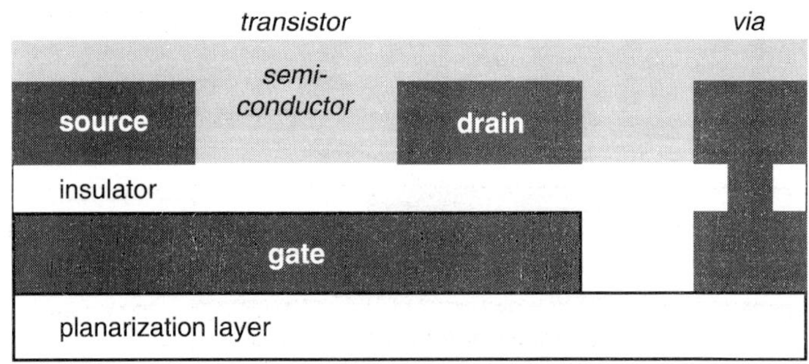

Fig. 12.25. Schematic diagram of vertical cross section of a bottom gate FET. Reproduced with permission from G.H. Gelineck *et al.*; Appl. Phys. Lett., 77(10), 1487 (2000). Copyright © 2000, American Institute of Physics.

In bottom gate transistor the gate is formed first by PANI using photopatterning. The insulator layer is then spin coated and source-drain layer is patterned. On the top, the semiconductor layer was formed. P3HT/ Pentacene was used as the semiconducting layer. High mobilities of 0.01 cm^2 V^{-1}s^{-1} were achieved in Pentacene.

Polyaniline, which was being used as conducting layer, suffers from certain drawbacks. The *m*-cresol is a toxic material. Also, the lithography required deep-UV, 254 nm light source. Hence, researchers started to look for materials that would help in scaling of technology. Poly(3,4-ethylenedioxythiophene) stabilized with polystyrenesulfonic acid (PEDOT/PSS) **(Fig. 12.26)** was used as an alternative material [29]. PEDOT/PSS can be applied with water and is stable in both acidic and basic media. PEDOT/PSS has high conductivity of about 200 ohm. cm^{-2}. Instead of normal photolithography, which was used to pattern polyaniline, a "negative" photolithography was used. The process is based on photocrosslinking.

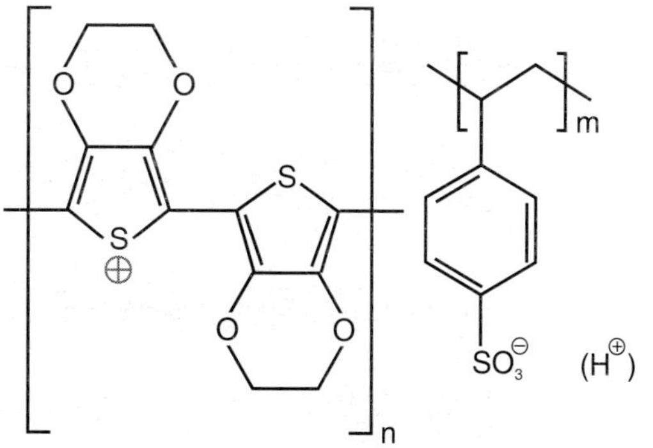

Fig. 12.26. PEDOT (left) stabilized with PSS (right).

The non-illuminated parts are removed as opposed to the illuminated ones in normal photolithography. The photo-initiators used are bisazides, polyazides or polydiazonium salts. Crosslinking of PEDOT is achieved when light falls on it.

Stability of polymers in air is another crucial point. Most of the above-mentioned semiconductors are stable in vacuum and have a shelf life in years. But in air, they degrade very fast. For example, in air, P3HT doping degrades very fast. Maximum current modulation decreases to one-tenth in few seconds [28]. PTV has a few days of shelf life in air where as for pentacene it ranges to about few months.

In top gate geometry, semiconductor layer is protected from air. In 2004, Siemens-AG reported the fabrication of an air stable transistor using top gate geometry [30]. PEDOT/PSS was used as electrodes. P3HT was chosen as semiconductor for the transistor. PMMA was used as insulator. PEDOT/PSS, because of its excellent properties, is now widely used as a material for electrodes for all polymer transistors. A new method for photolithography was proposed by M.S. Lee [31]. An oxidant film was first spin coated on PET substrate by spin-coating the oxidant solution. The oxidant solution used was prepared by dissolving ferric *p*-toluenesulfonate (FTS) as the oxidant and PVA as the matrix polymer.

Fe^{3+} in the oxidant film, when exposed to UV light through a photomask, converts to Fe^{2+} and is no longer able to oxidize. Hence, no oxidation can take place in the exposed area. This layer of patterned oxidant was then exposed to 3,4-ethylenedioxythiophene (EDOT) or pyrrole vapor, resulting in formation of PEDOT or PPy only in unexposed area. The micro-patterning can be obtained with good resolution by this method as shown in **Fig. 12.27**. Channel lengths up to 5 μm were obtained.

Using the above technique, a transparent FET was formed [31]. A rectangular gate electrode with a width of 1 mm and a thickness of 400 nm was first formed on a PET layer using photolithography. The dielectric layer such as PVCN or EPOXY/MAA was formed on top

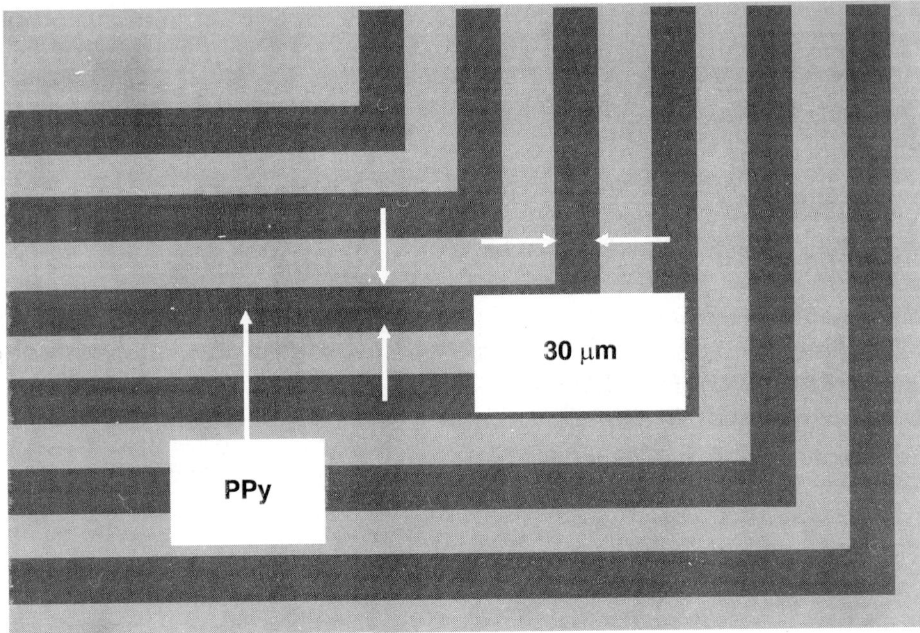

Fig. 12.27. Micropattern of conducting polymer. Reproduced with permissions from Myung Sub Lee *et al.*; Thin Film Solids, 477(1-2), 169 (2004). Copyright © 2004, Elsevier Science Ltd.

(Figs. 12.28 & 12.29). These polymers are photo-cross-linkable. These are first spin-coated and then exposed to UV with wavelength of 254 nm for about half an hour. Another layer of PEDOT/PPy was formed on top that acts as source-drain and active channel layer. The schematic of the FET formed is as shown in **Fig. 12.30**.

Till now, we have seen the fabrication of all polymer transistors using photolithography. Issues like stack integrity and solvent compatibility are important in the process flow using lithography. Masks are also needed in the process, which increases the cost of the process.

Ink-jet printing is the alternative technique used to pattern the gate, source and drain electrodes [32], requiring no lithography. The process could be made extremely cheap using such an approach. On a glass substrate, source and drain are patterned by inkjet printing. Semiconducting and insulating layer are then spun coated. Gate layer is again printed over the top. The resulting geometry is shown in **Fig. 12.31**.

However, the method above has its own advantages and disadvantages. As said above, it is cheaper as no mask needs to be made for the complete process. It gives us flexibility to choose polymers. Moreover, the method is

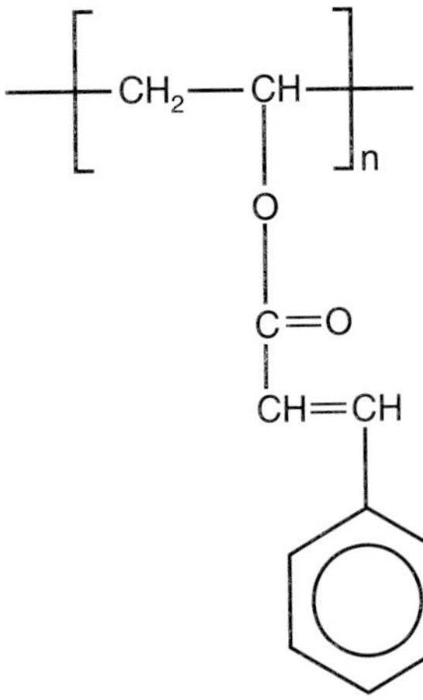

Fig. 12.28. PVCN.

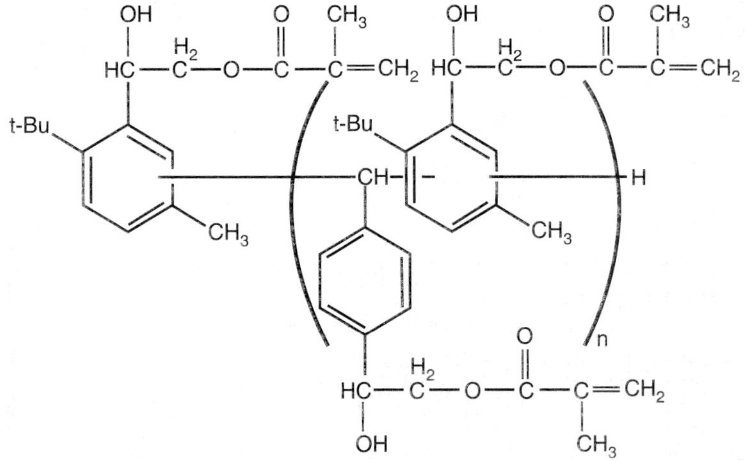

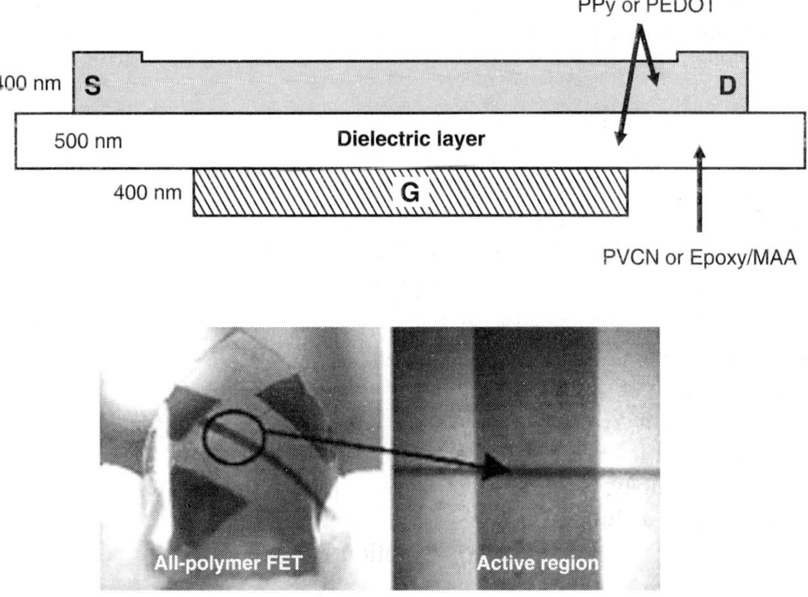

Fig. 12.29. Epoxy/MAA. Reproduced with permissions from Myung Sub Lee *et al.*; Thin Film Solids, 477(1-2), 169 (2004). Copyright © 2004, Elsevier Science Ltd.

Fig. 12.30. Structure of all polymer FET with optical transparency. Reproduced with permissions from Myung Sub Lee *et al.*; Thin Film Solids, 477(1-2), 169 (2004). Copyright © 2004, Elsevier Science Ltd.

faster than the process involving lithography. On the other hand, there may be statistical variations in flight directions and spreading of droplets on substrate. Hence, channel lengths are of the order of 20-50 µm. Because of large channel lengths the performance of transistor is limited.

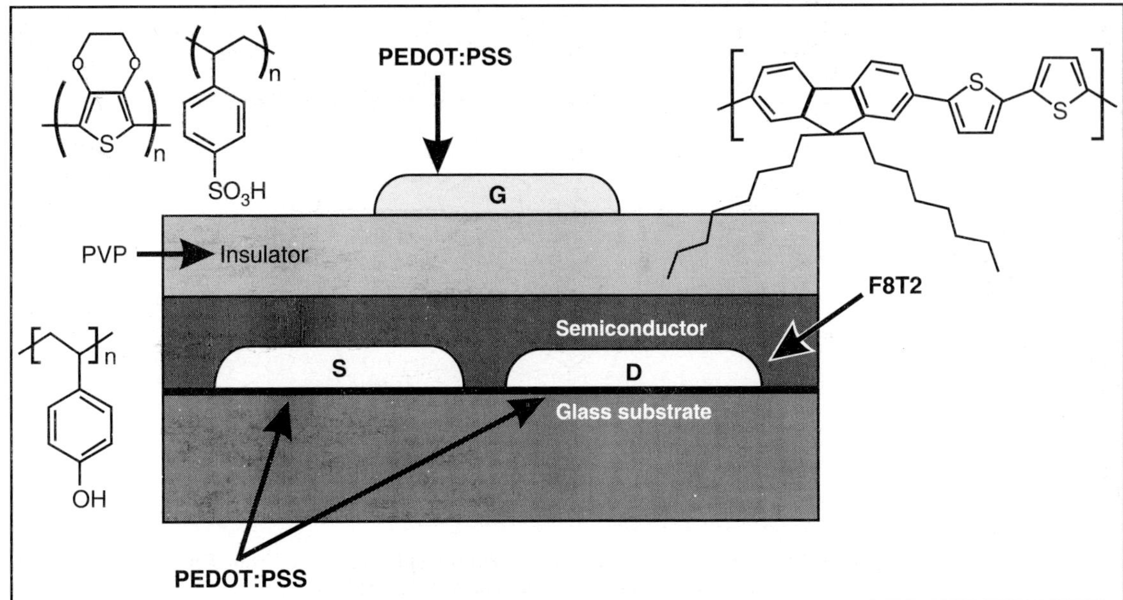

Fig. 12.31. Structure of all polymer transistor where source, drain and gate were printed by inkjet technique. Semiconducting and insulating layers were spin coated/inkjet-printed. Reproduced with permission from T. Shimoda and T. Kawase; All-polymer thin film transistor fabricated by high-resolution inkjet-printing. Proc. International Solid-State Circuits Conference (15-19 Feb., 2004). Copyright © 2004, IEEE.

In order to reduce the channel length, a substrate energy pattern is created to direct the flow of droplets. Hydrophilic glass and hydrophobic polyimide are used to restrict the placement of droplets. The spreading of water-based conducting polymer ink droplets on a hydrophilic substrate is confined with a pattern of narrow hydrophobic (water repelling) regions. The polyimide was patterned by plasma etching of 500 Å polyimide film [34]. The droplets were spread at a distance from the polyimide line that is sufficiently small, i.e. droplets would have spread and crossed over the region where the polyimide line is, but because of the hydrophobic nature of polyimide this is prevented giving rise to narrow channel lengths. This is a single mask process, as the masking step is needed only for patterning the polyimide. After the source-drain layers are printed, the semiconducting and dielectric layer can be printed or spun coated. **Fig. 12.32** shows the schematic before and after printing. Optical Micrographs are also shown. Devices up to 5 µm have been formed using this technique [34]. At IIT Bombay, our group is involved in producing an all polymer vertical transistor, which would address the scaling issues described above in addition to optimizing on the current, drives. The work is being carried out in collaboration with the Nanyang Technological University, Singapore.

In addition to photolithography and printing methods, lamination techniques [35] and micro-contact printing [36] were also used to fabricate polymer transistors. Many other polymers in addition to what are mentioned above have been used. The complete research in polymer transistors aims to produce low cost, high mobility and small channel length transistors. Looking at the present pace of research and promising results, it can be said with certainty that electronic newspaper would no longer be a dream.

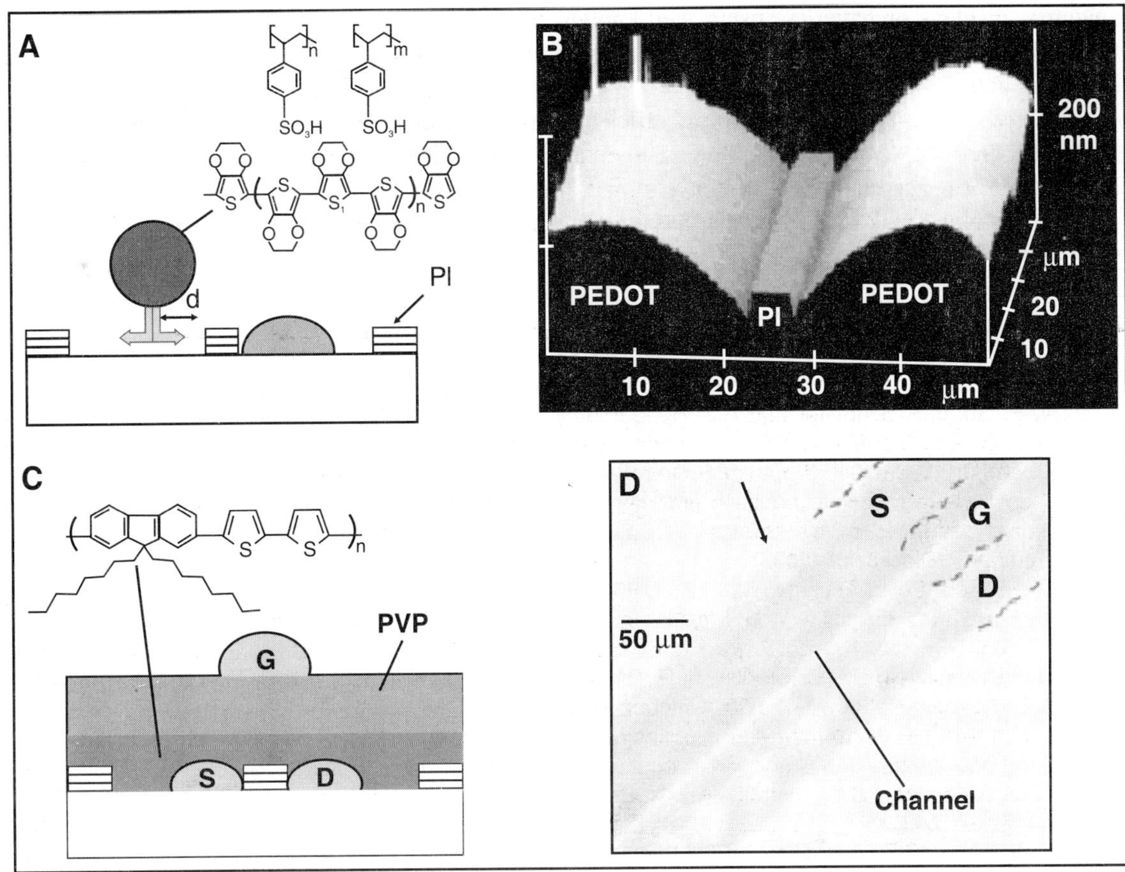

Fig. 12.32. A. Schematic of ink-jet printing (IJP) on a prepatterened surface. B. AFM image showing channel length = 5 μm. C. Complete Schematic. D. Optical Micrograph of an IJP TFT (L = 5 μm). Reproduced with permissions from H. Sirringhaus *et al.*; Science, 290, 2123 (Dec. 2000). Copyright © 2000, AAAS.

CONCLUSIONS AND FUTURE PROSPECTS

There exists a band gap in the conjugated polymers that can be easily altered by chemical modifications. Hence, LEDs of various colors can be formed using conjugated polymers. The light emits by the radioactive decay of *excitons* formed by recombination of electrons and holes. The reverse happens in photovoltaic cells where excitons are formed by external energy sources like electrical energy or light energy. Voltages are generated when excitons are split into electrons and holes. The semiconducting nature of these polymers led to their use as semiconducting layer in transistors. Various polymers have been experimented with. Mobility and stability are the major issues that determine the performance and life of the transistor formed. The conducting layers are also being replaced by high conducting polymers in "all polymer transistors".

Due to their numerous advantages, such as their low cost, flexibility, light weight, impact resistance and case of processing, polymers are the future of electronics. Polymer electronics will make it possible to produce simple, flexible system with active components (such as transistors, antennas, displays or batteries) using well-known, time-proven, and cost-effective additive printing techniques that employ high-yield roll-to-roll processes. The market size is estimated to be several billion dollars.

Presently, low performance and low efficiency are major challenges that are preventing polymer electronic products to enter the commercial markets. Scientists across the world are struggling to achieve performance equivalent to conventional silicon electronics. Promising results from research all over the globe makes one believe that the slogan of today *"plastic instead of silicon"* will be a truth in few years.

REFERENCES

1. Robero M. Faria and O.N. Oliveria Jr.; Brazilian Journal of Physics, 29(2), 360-370 (1999).
2. W.P. Su, J.R. Schrieffer and A.J. Heeger; Phys. Rev. B, 22, 2099 (1980).
3. Alan G. MacDiarmid; Synthetic metals: a novel role for organic polymers. **Nobel Lectures in Chemistry** 1996-2000 (Ed. Ingmar Grenthe), World Scientific Publishing Company, Stockholm (2003).
4. Alan J. Heeger; "Semiconducting and metallic polymers: the forth generation of polymeric materials. **Nobel Lectures in Chemistry** 1996-2000 (Ed. Ingmar Grenthe), World Scientific Publishing Company, Stockholm (2003).
5. David Bott; Phys. Technol., 16, 121 (1985).
6. M. Ferreira, O. Onitsuka, W.B. Stockton and M.F. Rubner; ACS Symposium Series, 672, 437 (1997).
7. Y. Kim, S. Kwon, D. Yoo, M.F. Rubner and M.S. Wrighton; Chem. Mat., 9, 2699 (1997).
8. J.H. Burroughes, D.D.C. Bradley, A.R. Brown, R.N. Marks, K. Mackay, R.H. Friend, P.L. Burns and A.B. Holmes; Nature, 347, 539 (October 1990).
9. R.H. Friend, R.W. Gymer, A.B. Holmes, J.H. Burroughes, R.N. Marks, C. Taliani, D.D.C. Bradley, D.A. Dos Santos, J.L. Bredas, M. Logdlund and W.R. Salaneck; Nature, 397, 121 (January 1999).
10. N.C. Greenham, S.C. Moratti, D.D.C. Bradley, R.H. Friend and A.B. Holmes; Nautre, 365, 628 (October 1993).
11. A.W. Grice, D.D.C. Bradley, M.T. Bernius, M. Inbasekaran, W.W. Wu and E.P. Woo; Appl. Phys. Lett., 73, 629 (1998).
12. Mark T. Bernius, Mike Inbasekaran, Jim O'Brien and Weishi Wu; Adv. Mater., 12(23), 1737 (2000).
13. C.W. Tang; Appl. Phys. Lett., 48(2), 183 (1986).
14. R.H. Friend; Pure Appl. Chem., 73(3), 425 (2001).
15. J.J.M. Halls, C.A. Walsh, N.C. Greenham, E.A. Marseglia, R.H. Friend, S.C. Moratti and A.B. Holmes; Nature, 376, 498 (August 1995).
16. N.S. Sariciftci, L. Smilowitz, A.J. Heeger and F. Wudl; Science, 258, 1474 (1992).
17. A. Tsumara, H. Koezuka and T. Ando; Appl. Phys. Lett., 49(18), 1210 (1986).
18. A. Tsumura, H. Fuchlgami and H. Koezuka; Synth. Metals, 41-43, 1181 (1991).
19. Y. Renkuan, Y. Shucheng, Y. Hong and J. Ruolian; Synth. Metals, 41-43, 727 (1991).
20. R. Sugimoto, S. Takeda, H.B. Gu and K. Yoshino; Chem. Express, 1, 635 (1886).
21. H. Sirringhaus, P.J. Brown, R.H. Friend, M.M. Nielsen, K. Bechgaard, B.M.W. Langeveld-Voss, A.J.H. Spiering, R.A.J. Janssen and E.W. Meijer; Synth. Metals, 111&112, 129 (2000).
22. J.G. Laquindanum, H.E. Katz, and A.J. Lovinger; J. Am. Chem. Soc., 120, 664 (1998).
23. H. Fuchigami, A. Tsumura and H. Koezuka; Appl. Phys. Lett., 63(10), 1372 (1993).
24. Francis Garnier, Gilles Horowitz, Xuezhou Peng and Denis Fichou; Adv. Mater., 2(12), 592 (1990).
25. Chin-Tsou Kuo and Wen-Hong Chiou; Synth. Metals, 88, 23 (1997).
26. Chin-Tsou Kuo, Shou-Zheng Weng and Rung-Lung Huang; Synth. Metals, 88, 101 (1997).
27. C.J. Drury, C.M.J. Mutsaers, C.M. Hart, M. Matters and D.M. de Leeuw; Appl. Phys. Lett., 73(1), 108 (1998).
28. G.H. Gelineck, T.C.T. Geuns and D.M. de Leeuw; Appl. Phys. Lett., 77(10), 1487 (2000).

29. F.J. Touwslager, N.P. Willard and D.M. de Leeuw; Appl. Phys. Lett., 81(24), 4556 (2002).
30. Henning Rost, Jurgen Ficker, J.S. Alonso, L. Leenders and I. McCulloch; Synth. Metals, 145, 83 (2004).
31. Myung Sub Lee, Han Saem Kang, Hyun Suk Kang, Jinsoo Joo, Arthur J.Epstein, Jun Young Lee; Thin Film Solids, 477(1-2), 169 (2004).
32. T. Kawase, H. Sirringhaus, R.H. Friend and T. Shimoda; All polymer thin film transistors fabricated by high-resolution ink-jet printing, Tech. Dig. of IEDM, p. 623, (2000).
33. T. Shimoda and T. Kawase; All-polymer thin film transistor fabricated by high-resolution ink-jet printing. Proc. of the International Solid-State Circuits Conference (15-19 Feb., 2004).
34. H. Sirringhaus, T. Kawase, R.H. Friend, T. Shimoda, M. Inbasekaran, W.Wu and E.P. Woo; Science, 290, 2123 (Deccember 2000).
35. Yueh-Lin Loo, Takao Someya, Kirk W. Baldwin, Zhenan Bao, Peter Ho, Ananth Dodabalapur, Howard E. Katz and John A. Rogers; Soft, conformable electrical contacts for organic semiconductors: high-resolution plastic circuits by lamination. Proc. of the National Academy of Sciences, 99(16),10252 (August 2002).
36. John A. Rogers, Zhenan Bao, Kirk Baldwin, Ananth Dodabalapur, Brian Crone, V.R. Raju, Valerie Kuck, Howard Katz, Karl Amundson, Jay Ewing and Paul Drzaic; Paper-like electronic displays: large-area rubber-stamped plastic sheets of electronics and microencapsulated electrophoretic inks. Proc. of the National Academy of Sciences, 98(9), 4835 (April 2001).

13

Ion Conducting Polymers

Ashok Kumar* and D. Saikia

Department of Physics, Tezpur University, Tezpur—784028, Assam, India.
**Corresponding author's e-mail: ask@tezu.ernet.in*

Contents

Summary: Electrochemical energy storage and conversion devices such as advanced rechargeable lithium batteries and fuel cells are in growing demand in consumer, industrial and military applications as the power source for numerous wireless telecommunication devices and portable information technologies and for the forthcoming electric and hybrid vehicles on the world's transportation scene. Electrochemical sensors are yet another emerging field where countless industrial, environmental and biomedical applications are found. The overall performance of an electrochemical device depends on the choice of cathode, anode, electrolyte and the electrode-electrolyte interfacial properties. Polymeric materials as electrolytes in electrochemical devices are of immense interest because of their flexibility and superior mechanical properties, ease of fabrication as thin films and their ability to form good contact with electrode materials. Ion-conducting polymers are solutions of salts in polar polymers in which a macroscopically solid state is achieved by entanglement or cross-linking whereas microscopically they are dynamically disordered. Ionic motion in the polymer electrolytes is strongly coupled to the constantly fluctuating polymer chain segments on a time scale of the order of atomic hopping time. The present chapter reviews the polymer electrolytes with regard to different types, models and mechanisms of ionic conduction in ion conducting polymers/polymer electrolytes. The theoretical aspects of ion transport kinetics in composite and gel polymer electrolytes have also been reviewed. The newest electrolytes are the polymer gel electrolytes, which contain aprotic organic liquid electrolyte immobilized by a cross linked structure of a suitable polymer. Polymer gel electrolytes are different from solution type Polyethylene Oxide (PEO) based electrolytes that contain polymer modified by organic liquid as plasticizer. Preparation and characterization of polymer gel electrolyte systems viz., PVDF:(PC+DEC):LiClO$_4$ and P(VDF-HFP):(PC+DEC):LiClO$_4$ have been described. The results on the two polymer gel electrolyte systems with regard to ionic conductivity, X-ray Diffraction, Fourier Transform Infrared Spectroscopy and Scanning Electron Microscopy have been discussed.

1. INTRODUCTION

Ionic conduction in solids has been known since 1830's when Faraday observed the ionic conduction in nonmetallic solids Ag$_2$S and PbF$_2$ [1]. Later Kohlrausch [2] initiated a systematic study of ionic conduction in solids. Tubandt and his group [3] established the Faraday's laws of electrolysis for solid ion conductors. Theoretical models to explain the transport of electricity through ionic solids by the flow of ions were developed by Frenkel [4], Wagner [5] and Schottky [6]. The first detailed study of the structural basis for high ionic conductivity in a crystalline solid was carried out by Ketelaar [7] in Ag$_2$HgI$_4$. Reuter and Hardel [8] synthesized a new material Ag$_3$SI, which showed exceptionally high ionic conductivity (10^{-2} S/cm) above 235°C. Bradley and Greene [9] and Owens and Argue [10] independently discovered a new family of superionic conductors MAg$_4$I$_5$ (M = K, Rb, NH$_4$, etc.). These materials exhibit conductivities of $\approx$ 0.3 S/cm at room temperature. Yao and Kummer [11] reported an unusually high ionic conductivity in the Na$_2$O-Al$_2$O$_3$ (Na-β-alumina) system. Liang [12] discovered that the dispersion of fine insulating Al$_2$O$_3$ particles in LiI enhances the Li$^+$ ion conductivity of LiI by about two orders of magnitude at room temperature from 10^{-7} to 10^{-5} S/cm. All highly conducting solid electrolytes owe their conductivities to disorder regions in their structures.

Depending on structure solid ion conductors can be classified as crystalline, amorphous (glassy), composite and polymeric ionic conductors. In crystalline ionic conductor, the carrier concentration may be defined as the density of defects with reference to the perfect lattice and each carrier is identically situated having the same mobility [13]. Alkali halides, Na-β-aluminas, Nasicon, stabilized ZrO$_2$, etc. are some examples of crystalline ionic conductors. In amorphous ionic conductors large amount of free volume assumes all the ions to be potentially conducting but in a broad distribution of states [14]. Lack of grain boundaries, isotropic and generally higher conductivities are the distinct advantages of amorphous (glassy) ionic conductor over their crystalline counterparts [15]. Alkali germanates and lithium niobate glasses are examples of amorphous ionic conductors. In composite ionic conductors, ionic conductivity is believed to occur in the thin interfacial regions surrounding the dispersed particles [16]. Several phenomenologies have been invoked to explain the enhancement in the conductivity of composite electrolytes. Jow and Wagner [17] and Shahi and Wagner [18] proposed that the dispersion of the insulating particles in the host matrix produces a space charge layer at the matrix/particle interface and thus facilitates the ionic motion. LiI (Al$_2$O$_3$) is one of the examples of composite ion conductor.

The most promising class of ionic conductors of today's world is the polymeric ionic conductors. Ionically conducting phases, free from low molecular weight solvents, based on the dissolution of salts in suitable ion-coordinating polymers are key components in new types of batteries for portable electronic devices and electric cars. Batteries for such applications require polymer electrolytes with improved properties, e.g. a high ionic conductivity, high cation transport number and improved chemical, thermal and electrochemical stability. Main reason for using polymeric electrolytes over other solid electrolytes lies in their flexibility and conformability to the electrodes. P.V. Wright carried out first measurements of ionic conductivity in polymer-salt complexes in 1973 in collaboration with David Fenton and crystallographer John Parker [19]. Armand, Chabagno and Duclot [20] investigated a wide range of PEO-alkali salts complexes and introduced the Vogel-Tammann-Fulcher relation for LiSCN and CsSCN complexes. Armand's proposal that they can be used as 'polymeric solid electrolytes' in secondary lithium batteries apparently initiated a burst of research activities amongst electrochemists worldwide. At Sheffield, in the late 1970s, Lee [21] and Payne [22] embarked upon an investigation of morphology and its relation with conductivity showing that whereas fully saturated crystalline complexes gave poor conductivities, the higher conductivity-lower activation energy regime could be extrapolated back to ambient temperature in fully amorphous PEO networks. At about the same time, Killis, Le Nest and Cheradame [23] at St. Martin d'hères had independently reached similar conclusions regarding the amorphous PEO by preparing urethane based PEO networks.

With confirmation of the predominance of amorphous phase mobility by Berthier and coworkers [24], the 1980s saw the development of a number of strategies to suppress crystallinity in polyethoxy systems. The branched or comb polyphosphazenes of Shriver, Allcock and coworkers at Pennsylvania [25] and Ward's group at Leeds [26] and the methoxy copolymers developed by Booth's group at Manchester [27] were among the more successful of these. At the same time there were significant theoretical and mechanistic developments including the dynamic percolation model proposed by Ratner and coworkers [28] at Northwestern. Angell at Arizona State University [29] and Torrell at Chalmers [30] discussed the relation between the conductivity and matrix relaxation times. The correlation between conductivity, glass transition temperature and the mechanical relaxation (the WLF equation) was investigated by Cheradame and coworkers [31] and Watanabe [32].

One of the most important steps in the understanding of polymer electrolyte is that the ionic conductivity is a property of amorphous elastomeric phases [24]. High molecular weight amorphous polymers above their glass transition temperatures may exhibit mechanical properties similar in most ways to those of a true solid, which is a result of chain entanglement and crosslinking of various types. At microscopic level, however, local relaxation processes may still provide liquid like degrees of freedom somewhat similar to that in an ordinary molecular liquid.

Since the discovery that a number of polymer-salt complexes exhibit considerable ionic conductivity, much research effort has been directed to find the optimal combination of host polymer and dopant salt for fast ionic transport. Conductivity of solvent free polymer electrolytes is increased by two different ways:

1. Suppression of crystallization of polymer chains to enhance polymer chain mobility.
2. Increase in the carrier concentration.

The suppression of crystallization of polymer chains to improve polymer chain mobility can be realized by cross-linking, co-polymerization, comb formation (side chains and dendritic polymers), polymer alloy (including inter penetrating network) and inorganic filler blend. In the combination of cross-linking and co-polymerization, Cheradame *et al.* [33] achieved an ionic conductivity of about 5×10^{-5} S/cm at 25°C by cross-linking block copolymers of EO and PO and in the comb formation, Hall *et al.* [34] obtained a ionic conductivity value of 2×10^{-4} S/cm at 25°C by adding PEO side chains to polysyloxane chains. Watanabe *et al.* [35] extended this method by synthesizing a dendritic polymer by attaching PEO chains to glycidyl ether side chains.

The increase in the carrier concentration can be realized by use of highly dissociated salts and increase in salt concentration. For the polymer electrolytes wherein anions are not fixed in the polymer, polymer-salt systems with salts having small lattice energy, large anions and plasticity to the polymer have been studied. Vallee *et al.* [36] obtained ionic conductivity value of 4×10^{-5} S/cm at 25°C with a lithium trifluorosulfonylimide:PEO system, drastically improving the ionic conductivity from 1×10^{-7} S/cm at 25°C with a NaI:PEO system. For the polymer electrolytes with anions fixed in the polymer (single ion conductor), Kobayashi *et al.* [37] got conductivity value of 1×10^{-7} S/cm at 25°C with a carboxylate system, Benrabah *et al.* [38] obtained conductivity value of 6×10^{-7} S/cm at 25°C with a sulfonate system and Fujinami *et al.* [39] achieved conductivity value of 2×10^{-5} S/cm at 25°C with a siloxy aluminate system. Angell *et al.* [40] described a new type of ionic conductors 'polymer-in-salt' materials in which lithium salts are mixed with small quantities of the polymers polypropylene oxide and polyethylene oxide and they found ionic conductivity of the order of 10^{-3} S/cm. Watanabe *et al.* [41] obtained the same order of conductivity 1×10^{-3} S/cm at ambient temperature for polypyridinium, pyridinium and aluminium chloride system.

Considerable efforts have been devoted to the development and improvement of the electrolyte's ionic conductivity for electrochemical device applications [42]. For that purpose a new system has received much attention wherein a polymer matrix is swollen in a plasticizer to get plasticized or gel polymer electrolyte. Main concepts behind the development of gel polymer electrolyte are (i) the gel polymer electrolyte by physical cross-linking and (ii) the gel polymer electrolyte by chemical cross-linking. For the physical cross-linking gel polymer electrolyte, Feuillade *et al.* [43] began research with PAN system and later Tsuchida *et al.* [44] obtained conductivity value of 1×10^{-3} S/cm at 25°C with a PVDF system, and Iijima *et al.* [45] achieved conductivity value of 1×10^{-3} S/cm at 25°C with a PMMA system. For chemical cross-linking gelled polymer electrolyte, Feuillade *et al.* [43] began research with a P(VDF-HFP) cross-linking system and Morita *et al.* [46] obtained conductivity value of 1×10^{-3} S/cm at 25°C with a PEO system. Introduction of nanoscale inorganic fillers such as TiO_2, Al_2O_3 and SiO_2 to the polymer electrolytes has proved effective to enhance the mechanical and electrochemical properties of the electrolyte by forming particle networks into the polymer bulk, inhibiting crystallization, reorganization of polymer chains and interacting with lithium ionic species [47-49].

2. CLASSIFICATION OF ION CONDUCTING POLYMERS

2.1. Polymer-Salt Complexes

The original solvent free polymer electrolytes introduced over two decades ago consisted of polymers such as polyethylene oxide (PEO, T_m ~65°C, crystallinity ~85%) and polypropylene oxide (PPO) or their blend complexed with suitable salts such as $LiClO_4$, $LiCF_3SO_3$ etc. [50-53]. Salts dissolve in only those polymers for which exothermic ion-polymer interactions compensate for lattice energy of the salt so as to reduce the free energy of the system [54]. Glass-transition temperature of the polymer normally increases by 50-100 K after dissolution of salt [55,56]. The polymer hardens, stiffens and increases in conductivity by a factor of up to 10^5. Reasonable conductivity can be achieved at 100°C (~10^{-5} S/cm) in the polymer-salt complex. At ambient temperature PEO is a poor conductor due to high crystallinity (<10^{-8} S/cm). Poor ionic transport properties of PEO type solid polymer electrolytes used in lithium batteries at ambient temperatures cause concentration polarization and limit the performance of such batteries.

2.2. Plasticized Polymer Electrolytes

In this type of electrolytes small amount of low-molar mass polar liquids are added to first type of electrolytes [57-60]. Plasticized polymer electrolytes represent a compromise between polymer and liquid electrolytes. Liquid plasticizers, however, typically lead to worsening of the mechanical properties of the electrolyte and increasing reactivity towards the metal electrodes. Plasticizers can increase ionic conductivity by a factor of 100.

2.3. "Polymer-in-Salt" or "Rubbery" Electrolytes

"Polymer-in-salt" or "rubbery" electrolytes were discovered by Angell and co-workers [40,41,61,62]. In these electrolytes high molar-mass polymers are dissolved in low-temperature molten salt mixtures. The salt typically has a low glass transition temperature such that a rubbery material is formed when the salt and the polymer are mixed together. Polypropylene oxide, polyethylene oxide [40], polypyridinium [41], polyacrylonitrile [61] etc. are used as host polymer with lithium and aluminium salts such as $LiN(CF_3SO_2)$, $LiCF_3SO_3$, $AlCl_3$ etc. Although good ambient lithium-ion conductivities and high electrochemical stability have already been realized, the salt tends to crystallize at lower temperatures, preventing their practical use.

2.4. Composite Polymer Electrolytes

In this type of electrolytes inorganic (ceramic) particles are introduced into a polymer electrolyte [63-67]. In addition to improvements in the ionic conductivity, the mechanical strength and the interfacial stability are also enhanced in the composites. Common nanometer-sized ceramic powders are Li_3N, Al_2O_3, SiO_2, TiO_2, zeolites etc. Though the conductivities of composite electrolytes are dramatically improved from those without the nanoparticles, ambient temperature conductivities still remain relatively low for real practical applications. Enhancement of conductivity in this type of electrolyte is attributed to reduction of crystallinity and polymer-ceramic grain boundaries with high defect concentration that may allow better ion transport as smaller particles give better conductivities at same T_g and crystallinity.

2.5. Gel Polymer Electrolytes

The latest development in the field of aprotic electrolytes is presented by the gel polymer electrolytes. These electrolytes are prepared by immobilizing a nonaqueous electrolyte solution within an inactive structural polymer matrix (or monomer with polymerization agent such as dibenzoyl peroxide) or by increasing the viscosity of a liquid electrolyte by adding a soluble polymer [68]. Polymer gel electrolytes combine the ionic conductivity of liquid electrolytes with the attractive mechanical and processing advantages of polymers. They are basically salt-solvent-polymer hybrid systems in which first a salt solution is prepared and then it is immobilized with the help of a suitable polymer matrix. Since the electrolyte molecules can preferentially solvate ions, coordinating polymers like PEO are no longer necessary and may be replaced by more robust polymers like polyvinylidenefluoride (PVDF) [69-72], polyacrylonitrile (PAN) [73-76], polymethyl methacrylate (PMMA) [77-80] and poly(vinylidenefluoride-hexafluoropropylene) P(VDF-HFP) [81-84] copolymers. In gel polymer electrolytes the salt generally provides free/mobile ions that take part in the conduction process and the solvent helps in solvating the salt and also acts as a conducting medium whereas the polymer is reported to provide mechanical stability by increasing the

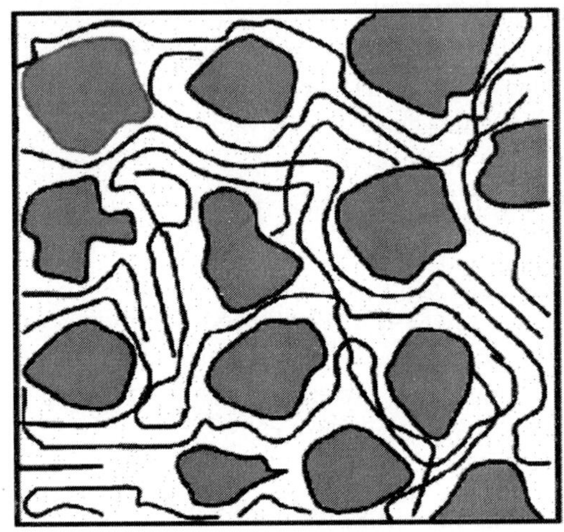

Fig. 13.1. Sketch of gel polymer electrolyte with micropores containing "free" solvent within a swollen polymer matrix.

viscosity of the electrolyte [85]. Gel polymer electrolytes are receiving much attention due to some of their unique properties like high value of conductivity at room temperature (10^{-2}-10^{-4} S/cm), ease of preparation, wide range of composition and hence wider control of properties, good adhesive properties suitable for lamination, good thermal/electrochemical stability etc., but also suffer the same disadvantages as the plasticized electrolytes, namely, release of volatiles and increased reactivity towards the metal electrodes. Common solvents used for gel polymer electrolytes are ethylene carbonate (EC) and propylene carbonate (PC). **Fig. 13.1** shows the schematic diagram of gel polymer electrolyte with micropores.

3. ION TRANSPORT IN POLYMERS (MECHANISMS AND MODELS)

Polymers are randomly bonded chain structures formed by repetition of similar units, in which the chains are of sufficient length to confer on the material some additional properties not possessed by the individual units. In the molten state or in solution, these polymer chains are randomly coiled and, even when apparently solid, at least part of the material is amorphous i.e. the randomness persists. Within the solid polymer, it is possible for the chains to align themselves in a systematic way by chain folding or by the formation of single or multiple helices, for at least part of the their length. These regions possess long-range order and are therefore crystalline. Crystallization is initiated at many locations within a polymer and the crystalline domains that develop are not coherently oriented with respect to each other, i.e. some portions of each long chain molecules cannot be aligned within an individual crystalline region and these portions remains amorphous.

For solvent free polymer electrolytes, such as heteropolymers PEO, salts dissolve because the lone pair electrons on the polymer chain oxygen atoms coordinate with the cations. The cation transport process in polymer electrolytes can be envisaged as a 'roll-on' mechanism in which a cation is initially coordinated to several oxygen (or other) atoms [86]. The linkages to one or more of the polymer segments that lie behind the direction of motion of the cation break and new linkages are formed in the forward direction. The cation motion is clearly facilitated by the flexing of the polymer chain segments, which allows old links to be broken and new attachments to be made. For high molecular weight polymer hosts' chain diffusion is small and makes little contribution to mechanisms for ion transport. Low barriers to bond rotation allow segmental motion of the polymer chain, thus providing a mechanism for ion transport (conductivity is determined largely by the local mobility (segmental motion) of the polymer segments). **Fig. 13.2** shows the schematic representation of cation hopping of polymer electrolyte in different ways [87].

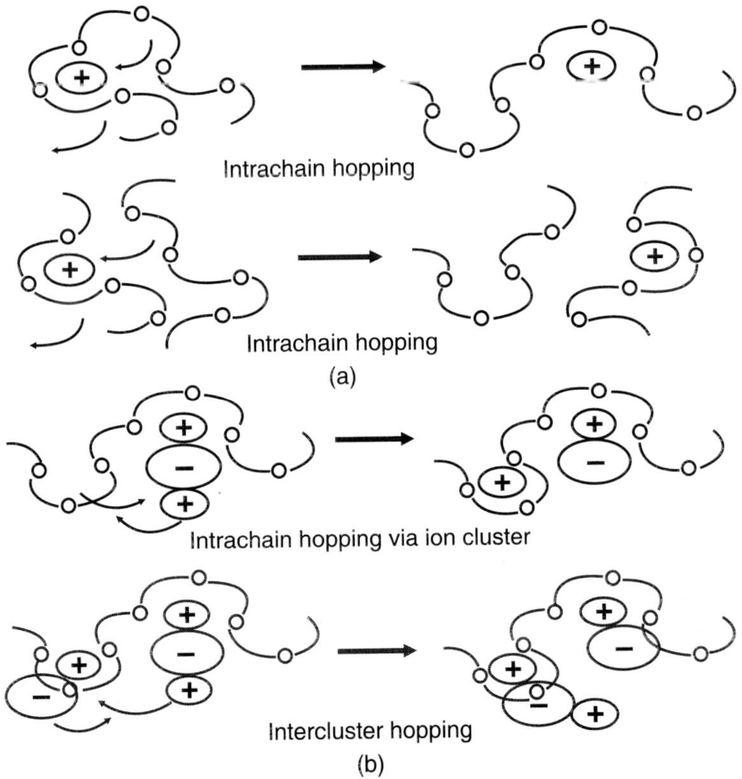

Fig. 13.2. Cation motion in polymer electrolytes [87].

For polymer gel electrolytes, polymer is an important constituent along with salt and solvent. The salt provides ions for conduction and the solvent helps in the dissolution of the salt and also provides the medium for ion conduction. The conductivity of gel polymer electrolytes can be explained by 'Breathing polymeric chain model' proposed by Chandra *et al.* [88]. According to this model, polymer gel electrolytes generally consists of free ions, ion aggregates and polymer chain dispersed in the gel matrix. The breathing of the polymer by folding/unfolding up of its chains results in density/pressure fluctuations at the microscopic level, which assist the motion of ions along with the dissociation of ion, aggregates which results in an increase in conductivity. The polymer chains act on ion pairs to effect dissociation and the dissociated ions in the solvent are further solvated by the polymer chains which results in a change in carrier concentration (n) and mobility (μ) [89].

3.1. Arrhenius Model

Polymer gel electrolyte is a semi-solid hybrid system wherein ionic conduction essentially occurs through the liquid organic phase trapped in constantly flexing polymeric molecular chains above the glass transition temperature of the polymer. Therefore, for polymer gel electrolyte systems, the Arrhenius equation for conductivity:

$$\sigma = \sigma_0 \exp\left(-E_a / kT\right) \tag{13.1}$$

where σ_0 is the pre-exponential factor, E_a is the activation energy and k is the Boltzmann's constant, often provides a good representation of the conductivity-temperature relationship. Linear graph of log σ vs $1/T$ is often called an Arrhenius plot, as for an Arrhenius process the logarithm of the relevant performance parameter depends linearly on the reciprocal of temperature. The slope of the curve gives the activation energy for the process. In fact, diffusion is the relevant activated transport process rather than conductivity and the appropriate performance parameter is the diffusion coefficient. The Nernst-Einstein equation provides the link between diffusion and conductivity, but it reveals that D is proportional to (σT) rather than σ, and it would be more appropriate to plot log (σT) instead of log σ. For polymer gel electrolytes the temperature range covered is often quite short and the variation in $1/T$ is substantially greater than the change in log T so that latter can be considered approximately constant and log (σT) can be replaced by log σ [90].

3.2. Vogel-Tamman-Fulcher (VTF) Model

For the solid electrolytes, which involve ion hopping between fixed sites, graphs of log σ versus $1/T$ are straight lines, but for the polymer electrolytes the log σ versus $1/T$ plots are often curved. Such behavior would be expected for materials such as polymer electrolytes in which conduction occurs in the amorphous region and the conduction process is better described by the Vogel-Tammann-Fulcher (VTF) equation [91-93]:

$$\sigma = AT^{-\frac{1}{2}} \exp\left(-E'_a / k(T - T_0)\right) \text{ or } \sigma = \sigma_0 \exp\left(-B/k(T - T_0)\right) \tag{13.2}$$

where B is a constant, whose dimensions are that of energy but which is not simply interpreted as an activation term. In configurational entropy terms, T_0 is the temperature at which the probability of configurational transition tends to zero and it is generally regarded as having a value around 20°C to 50°C below glass-transition temperature (T_g). The value of conductivity becoming vanishingly small as T_0 approaches T.

If the conductivity versus temperature dependence curve is linear in larger temperature region then it is said to be Arrhenius. VTF (curved) behavior can be modeled as Arrhenius (linear) behavior by dividing the entire temperature range into smaller temperature regions. The interconnection between Arrhenius and VTF behavior of σ (T) are widely reported and discussed in literature [94]. This behavior is rationalized by arguing that since VTF dependence is governed by the energy interval $k(T\text{-}T_0)$ and the Arrhenius dependence by the energy kT (where k Boltzmann constant), for $T \gg T_0$ [95] i.e. when T_0 is quite smaller than T, the curvature of conductivity versus temperature plot becomes small and VTF equation approaches Arrhenius equation.

3.3. William-Landel-Ferry (WLF) Model

In an amorphous polymer above its glass transition temperature, a single empirical function can describe the temperature dependence of all mechanical and electrical relaxation processes. The temperature dependence at a number of relaxation and transport processes can be described by WLF equation (Williams-Landel-Ferry) [96]

$$\log\left[\frac{\eta(T)}{\eta(T_s)}\right] = \log a_T = -\frac{C_1(T-T_s)}{C_2+T-T_s} \tag{13.3}$$

where a_T is the mechanical shift factor (expresses the fluidity or inverse relaxation time or relaxation rate), η is the viscosity, T_s is reference temperature (usually $T_s = T_g + 50\ K$) and C_1 and C_2 are constants.

3.4. Dynamic Bond Percolation (DBP) Model

Kohlrausch summation and Nernst-Einstein relations do not fit well in polymer ionics and the deviations become more significant as the salt concentration increases. In the former case ions move by a local liquid like process rather than by hopping from site to site in an ordered polymer host. To describe conduction mechanism in polymers adequately, it is important to model it microscopically. Dynamic bond percolation (DBP) theory [97] is a microscopic model for diffusion in dynamically disordered systems. The model assumes that mobile ions move from site to site within the polymer host and the sites lie on a regular lattice. Assuming simple first-order hopping chemical kinetics for the ion motion, the probability of finding an ion at site j can be written as:

$$\frac{dP_i}{dT} = \sum_j (P_j\, W_{ji} - P_i\, W_{ij}) \tag{13.4}$$

where W_{ji} is hopping rate from site j to site i. The difference between dynamic disordered hopping and ordinary hopping is that the hopping rates W_{ij} are themselves time dependent, and evolve on the time scale of relaxation of the polymer host. The model of **eq. 13.4** becomes specific to a percolation situation when it is assumed that the W_{ij}'s have only two values: each W_{ij} can be either 0 (probability of $1-f$) or a constant value, w (probability f). The assignment of any given bond between cells *i* and *j* as available (value *w*) or unavailable (value 0) itself evolve in time. The simplest assumption is that the value of the bonds as available or unavailable is reassigned randomly after a renewal time, τ_{ren}. This DBP model is then characterized by the two time scales τ_{ren} and the hopping time, $1/w$. The renewal time is determined by the local microviscosity of the solvent. The rate of renewal is the inverse local relaxation time that determines bond breakage near the mobile ion.

3.5. Free Volume Model

Cohen and Turnbull [98] first proposed the "free volume theory" of transport phenomena in glass forming materials. They derived the equation for diffusion coefficient, *D*, in a liquid:

$$D = ga^* u \exp\left[-\frac{\gamma V^*}{V_f}\right]$$
(13.5)

where *g* is a geometrical factor, a^* is approximately equal to the molecular diameter, *u* is the average speed of the molecules, γ is a numerical factor introduced to correct for overlap of free volume and V^* is the critical volume just large enough to permit another molecule to jump in after the displacement. Cohen and Turnbull defined the free volume as [98]:

$$V_f = \bar{V} - V_0$$
(13.6)

where $\bar{V}$ is the average volume per molecule in the liquid and V_0 is the van der Waals volume of the molecule. Using the Nernst-Einstein equation:

$$\sigma = \frac{q^2 nD}{kT}$$
(13.7)

it follows that:

$$\sigma = \frac{q^2 nga^* u}{kT} \exp(-\gamma V^* / V_f)$$
(13.8)

In these equations, *n* and *q* are the concentration and charge of the charge carriers, respectively, *k* is the Boltzmann's constant and *T* is the absolute temperature.

Free volume theory in this form has been applied to polymer electrolytes by several workers [99,100]. This theory predicts that at a constant free volume, the conductivity should decrease weakly with increasing temperature because kinetic theory requires that *u* vary as $T^{1/2}$. Since the remaining terms in **eq. 13.8** are approximately temperature independent, **eq. 13.8** can be rewritten as:

$$\sigma = \frac{C}{T^{1/2}} \exp(-\gamma V^* / V_f)$$
(13.9)

where C includes the charge carrier concentration. But the experimental data show that the electrical conductivity at constant specific volume increases strongly with temperature. The conclusion that **eq. 13.9** is not supported by the data is based on the assumption that the free volume is proportional to the macroscopic volume. Cohen and Turnbull made the assumption that:

$$V_f = V_0 \left(\exp\left[\int_{T_0}^{T} \alpha dT \right] - 1 \right)$$
(13.10)

where α is the thermal expansion coefficient [67]. A slightly different approximation often made is that the thermal expansion of the free volume is the difference in thermal expansion coefficients of the bulk material above and below T_g [100]. However, when the microscopic volume is held constant, the ionic conductivity increases strongly with temperature. Consequently, if free volume varies as the macroscopic volume, as follows from the definition given in **eq. 13.6**, free volume theory is not capable of representing the constant volume electrical conductivity for a typical polymer electrolyte. The disagreement between theory and experiment is not surprising since, as discussed by Cohen and Turnbull [98], the theory was originally developed for simple van der Waals liquids and metallic liquids. Inability of free volume theory to account for the phenomena governed by segmental motions has been widely pointed out in the literature [101,102].

One of the first attempts to modify free volume theory to account for the temperature variation of physical phenomena was made by Macedo and Litovitz [103]. They employed the reaction rate theory of Eyring [104] to arrive at the following equation for the shear viscosity:

$$\eta = \left(\frac{RT}{E_v}\right)^{1/2} \frac{(2mkT)^{1/2}}{V^{2/3}} \exp\left[\frac{\gamma V^*}{V_f} + \frac{E_v}{kT}\right] \tag{13.11}$$

where r is the gas constant, E_v is the height of the potential barrier between equilibrium positions, V is a quantity roughly equal to the volume of a molecule and m is the molecular mass. Next, the Stokes-Einstein equation:

$$\eta = \frac{kT}{6\pi Dd} \tag{13.12}$$

where d is the molecular radius, is used. It is known that the Stokes-Einstein equation breaks down for fragile glasses near T_g [105]. The Nernst-Einstein equation i.e. **eq. 13.7**, is used to obtain the following equation for the conductivity:

$$\sigma = \frac{q^2 n V^{2/3}}{6\pi d (2mkT)^{1/2}} \left(\frac{E_v}{kT}\right)^{1/2} \exp\left[-\frac{\gamma V^*}{V_f} - \frac{E_v}{kT}\right] \tag{13.13}$$

At constant volume, **eq. 13.8** can be rewritten as:

$$\sigma = \frac{B}{T} \exp\left[-\frac{E_v}{kT}\right] \exp\left[-\frac{\gamma V^*}{v_f}\right] \tag{13.14}$$

Because of the temperature dependence in the first exponential term, the **eq. 13.14** can account for a strong increase of the conductivity at constant volume and was used by Macedo and Litovitz to successfully reproduce the pressure and temperature dependence of the shear viscosity of several liquids [103]. These include several materials exhibiting VTF or WLF behavior. **Eq. 13.14** can also account for VTF or WLF behavior because that behavior is represented by the free volume factor.

4. KINETICS OF ION TRANSPORT IN POLYMER ELECTROLYTES

The conductivity (σ) of a material can be expressed by the relation:

$$\sigma = \sum_i \mu_i n_i q_i \tag{13.15}$$

where μ_i, n_i, q_i are the mobility, carrier concentration and charge of the i type of species respectively. The polymer electrolytes contain no significant conjugation within the polymer backbone and the salts on which they are based have negligible electronic conductivities. Hence, the electrons and holes do not contribute to the summation in **eq. 13.15.** Both the cation and anion contribute to the conductivity [106]. This represents a complication in the simple interpretation of the temperature dependence of conductivity.

Experimentally, one observes fairly straightforward behavior of the temperature dependence of conductivity in homogeneous electrolytes. The straight or curved plots observed when the conductivity is plotted against temperature inverse can be fit in the Arrhenius or VTF [91-93] forms respectively:

$$\sigma T = \sigma_0 \exp^{-E_A/kT} \tag{13.16}$$

$$\sigma = \sigma_0 \exp^{-B/k(T-T_0)} \tag{13.17}$$

In the Arrhenius form, E_A is the usual activation energy, whereas in the VTF form, B is a constant with dimension of energy but it is not simply interpreted as an activation term, k is Boltzmann's constant and T_0 is a reference temperature usually associated with the ideal glass transition temperature at which free volume disappears or the temperature at which the configurational entropy becomes zero. T_0 is usually lies 30–50 K below T_g.

The early investigations of the groups in Grenoble generally showed curved plots, corresponding to VTF behavior [20,23,107,108]. Cheradame and co-workers [23,33,107,108] discussed these plots in terms of chain segment mobility of the polymer host material. They used the relation of polymer chain viscosity to glass transition temperature that is summarized in the Williams-Landel-Ferry [94] relation:

$$\log \frac{\eta(T)}{\eta(T_s)} = \log a_T = \frac{-C_1(T - T_s)}{C_2 + (T - T_s)} \tag{13.18}$$

where T_s is an arbitrary reference temperature, usually $T_s = T_g + 50$ K, a_T is called the mechanical shift factor and C_1 and C_2 are universal constants. Often the WLF equation is coupled with the empirical observation known as Walden's rule [109]:

$$D\eta = Const/r_i \tag{13.19}$$

Stokes-Einstein relation can be written as:

$$D = kT/6\pi\eta r_i \tag{13.20}$$

where D is the diffusion coefficient and r_i is the radius of the ion. Again one can write the Nernst-Einstein relation as:

$$\sigma = DNq^2 / kT \tag{13.21}$$

where N is the number of carriers and q is the charge. Putting the value of D of **eq. 13.20** in **eq. 13.21**:

$$\sigma = \left(\frac{kT}{6\pi\eta r_i}\right)\left(\frac{Nq^2}{kT}\right)$$

$$\sigma = \frac{Nq^2}{6\pi\eta r_i}$$

$$\eta = \frac{Nq^2}{6\pi\sigma r_i} \tag{13.22}$$

Using **eq. 13.22** in **eq. 13.8**, one obtains the temperature dependence of conductivity in the WLF form [94]:

$$\log \frac{Nq^2/6\pi\sigma(T)r_i}{Nq^2/6\pi\sigma(T_s)r_i} = \frac{-C_1(T-T_s)}{C_2+(T-T_s)}$$

or

$$\log \frac{\sigma(T_s)}{\sigma(T)} = \frac{-C_1(T-T_s)}{C_2+(T-T_s)}$$

or

$$\log\frac{\sigma(T_s)}{\sigma(T)} = \frac{C_1(T-T_s)}{C_2 + (T-T_s)} \tag{13.23}$$

On the basis of this, Cheradame's group has argued that WLF behavior is the rule in polymer electrolytes and the fluidity of the polymer chain segments largely determines the conductivity. Writing the VTF equation in the form:

$$\log \frac{\sigma(T_s)}{\sigma(T)} = -\frac{B}{k}\left[\frac{1}{T-T_0} - \frac{1}{T_s-T_0}\right] \tag{13.24}$$

Comparing **eqs. 13.23** and **13.24** we get:

$$\frac{C_1(T-T_s)}{C_2+(T-T_s)} = \left[\frac{1}{T-T_0} - \frac{1}{T_s-T_0}\right]$$

or

$$C_1 = C_1\frac{B}{k(T_s-T_0)}\left[\frac{C_2+(T-T_s)}{(T-T_0)}\right] \tag{13.25}$$

Identifying:

$$C_2 = T_s - T_0,$$

$$C_2 = \frac{B}{k(T_s-T_0)} \tag{13.26}$$

or

$$B = kC_1C_2$$

In an attempt to understand how the conductivity mechanism works, quasi-thermodynamic theories [96,110-114], originally developed to deal with molten salts and neat polymers, have in fact been applied with some success to consideration of transport properties in polymer electrolytes. These theories are based on considerations involving the critical role of the glass transition temperature (T_g) and of the so-called equilibrium glass transition temperature T_0. Above T_g, the polymeric material becomes macroscopically rubbery rather than glassy [110-114]. The concept of equilibrium glass transition temperature T_0 is based on the kinetic feature of T_g, depending on the rate of cooling, one can observe different glass transition temperatures and T_0 is idealized as the temperature at which free volume vanishes or at which the excess configurational entropy of the material vanishes. The theoretical scheme that treats T_0 in terms of volume is called free volume theory [98,110].

The free volume model is the simplest way to understand the polymer segment mobility. It states that as temperature increases, the expansivity of the material produces local empty space i.e. free volume into which ionic carriers, solvated molecules or polymer segments themselves can move. The overall mobility of the material is determined by the amount of volume present in the material. One can obtain, for the diffusivity D, the form:

$$D = BRT \exp\left(-V^* / V_f\right) \tag{13.27}$$

When the volume is expanded in terms of the volume at the glass transition temperature plus a linear term, the free volume theory yields the form [36,107,115]:

$$D = D_0 T \exp\left(-\frac{a}{T - (T_g - C_2')}\right) \tag{13.28}$$

where the constant a and C_2' are both inversely proportional to the free volume thermal expansion factor. Rewriting **eq. 13.28** as:

$$D = D_0 T \exp\left(\frac{-a}{C_2 + T - (T_g - C')}\right) \tag{13.29}$$

where $C_2 \equiv C_2' - C'$. If $C' = 0$, the free volume argument yields the WLF relation of **eq. 13.18**.

Druger, Nitzan and Ratner [95,115-119] have developed a dynamic percolation model for description of ion transport in polymer electrolytes. Percolation theory is extremely useful in understanding transport processes in disordered media and has been recently used in the interpretation of the free volume behavior of several disordered materials like, polymers, molten salts etc. This is a microscopic model that characterizes the ionic motion in terms of jumps between neighboring positions.

The rate of jumping (i.e. of ionic motion) between any two sites (i.e. between two different positions in the material) is then represented in terms of simple first order chemical kinetics, using the master equation:

$$\frac{dP_i}{dT} = \sum_j (P_j W_{ji} - P_i W_{ij}) \tag{13.30}$$

The dynamic percolation model takes into account the dependence of ionic motion rates on the fluidity, or rate of segmental motion, of the polymer host. A characteristic rate of renewal, $\lambda = \tau_{ren}^{-1}$

is defined which characterizes the rate at which a motion pathway from one site to another becomes available for the ion to move. In static percolation theory applied, for example, to electron hopping in amorphous metals, the rates W_{ij} of **eq. 13.31** are taken to be:

$$W_{ji} = 0 \text{ (probability } 1 - f)$$

$$W_{ji} = w \text{ (probability } f) \tag{13.31}$$

with w being some average rate and f be the fraction of available bonds. For polymer electrolytes above T_g, the segmental motion changes the local coordination environment of the ion with a characteristic time t_{ren}, so that a jump which is unavailable at time t can become available at time $t + t_{ren}$ because of chains reorientation. The dynamic percolation model is characterized by the parameters f, w and t_{ren}. It has some interesting features [97,115-119] such as:

1. For observation time long compared to the renewal time, the motion is always diffusive i.e. the mean squared displacement is always proportional to time.
2. The diffusion coefficient is, in general, proportional to (λ) the average rate of renewal. This corresponds well both to the wealth of experimental data indicating that the ionic motion is modulated by the segmental motions of the polymer host and to the expectation, that structural reorganization and conductivity arise from the same motion mechanism.
3. It is possible to show in great generality that:

$$D_0(\omega + i\lambda) = D(\omega) \tag{13.32}$$

    i.e. the diffusion coefficient at frequency ω in the dynamic percolation problem may be found from the diffusion coefficient in the static percolation problem, analytically continued to frequency $\omega + i\lambda$.
4. The factor 'f' giving the number of available jumps, will be substantially different for cationic (strongly solvated) and anionic (weakly solvated) motions.

This model has the attractive feature of including the effects of segmental motion on ionic conduction, but not directly includes interionic interaction. Thus while it is the best microscopic model currently available for understanding the ionic conduction in polymer electrolytes, it is inadequate to fully expound inertial dynamics or interionic interactions.

5. KINETICS OF ION TRANSPORT IN COMPOSITE POLYMER ELECTROLYTES (EFFECTIVE MEDIUM THEORY APPROACH)

For composite polymer electrolyte it has been recognized that changes in polymer microstructure occur at the interface between matrix electrolyte and the dispersed filler [120,121] and the enhancement in conductivity is connected with the existence of a highly conductive layer at the polymer filler interface. Therefore, in composite systems there are three components with different electrical properties, viz.,

1. A highly conductive layer covering the surface of a grain or particle additive.
2. Dispersed insulating grains or particles.
3. A polymer matrix ionic conductor. We can consider a dispersed grain covered with the thin interface layer as a unit and call it a composite grain **(Fig. 13.3)** [122].

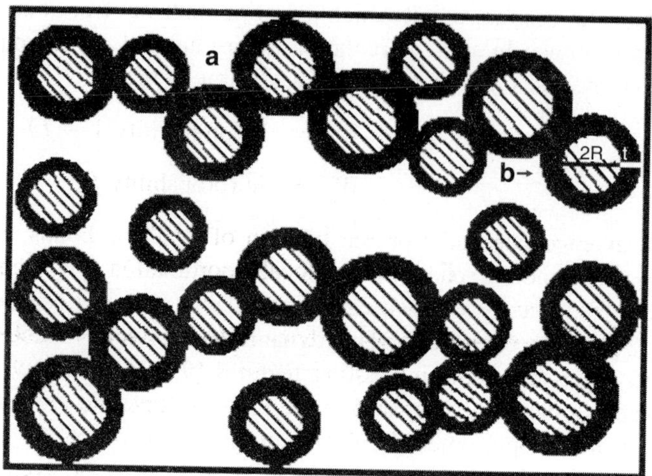

Fig. 13.3. Schematic morphology of (a) quasi-two-phase composite electrolyte and, (b) composite grain unit consisting of insulating grain and interface layer [122].

According to the Maxwell-Garnett rule [123], the equivalent conductivity σ_c of the composite unit can be expressed as:

$$\sigma_c = \sigma_1 \frac{2\sigma_1 + \sigma_2 + 2Y(\sigma_2 - \sigma_1)}{2\sigma_1 + \sigma_2 - Y(\sigma_2 - \sigma_1)} \qquad (13.33)$$

where σ_1 and σ_2 are the conductivity of interface layer and that of dispersed insulating grain respectively, Y is the volume fraction of dispersed grain in a composite unit and for spherical geometry can be calculated as:

$$Y = 1 \Big/ \left(1 + \frac{t}{R}\right)^3 \qquad (13.34)$$

where t is the thickness of interface layer and R is the radius of dispersed grain.

According to the above considerations, the composite electrolyte can be treated as quasi two-phase mixture consisting of an ionically conducting polymer matrix and composite units. The electrical properties of such a quasi two-phase mixture can simply be described by Bruggeman equation [124] or the simple effective medium equation introduced by Landauer [125]. These simple approaches ignore the local field effects and are suitable only for the description of quasi- two-phase mixture for volume fraction of composite units less than 0.1. In the quasi-two-phase mixture as the volume fraction of composite units increases, the composite grains gradually join together to form agglomerated clusters and then the local field effects cannot be ignored. Nan and Smith [126], Nan [127] and Nakamura [128] have considered the following limiting situations to incorporate this fact that (i) the matrix phase is the polymeric electrolyte and isolated composite units are scattered throughout and (ii) the matrix phase consists of overlapping or touching composite units with a small amount of dispersed polymeric electrolyte in the interstices.

Because the local field effect is a geometrical effect and has no relation with the conductivity of the two phases, and since the matrix phase screens the effect of dispersed phase in the limiting composite unit [126], the equivalent conductivity bounds σ_3^a and σ_c^a of the two limiting cases mentioned above, can be calculated according to a simplified form of Nakamura's equations [127,128]:

$$\sigma_3^a = \sigma_3[(d-1)-(d-1)V_c]/(d-1+V_c) = 2\sigma_3(1-V_c)/(2+V_c) \tag{13.35a}$$

$$\sigma_c^a = \sigma_c(d-1)V_c/(d-V_c) = 2\sigma_c V_c/(3-V_c) \tag{13.35b}$$

where $V_c = V_2/Y$, is the volume fraction of composite units, V_2 is the volume fraction of dispersed grains in the bulk electrolyte, σ_3 is the conductivity of the matrix polymeric electrolyte and 'd' is the dimensionality of the system ($d = 3$ for spherical grains). After introducing the improved conductivity parameters **(eq. 13.35)** in the self-consistent EMT equation suggested by Kirpatrick [129], the conductivity σ_m of the composite electrolyte can be expressed as:

$$\left(\frac{V_2}{Y}\right)(\sigma_c^a-\sigma_m)\Big/\left[\sigma_c^a+\left(\frac{1}{p_c}-1\right)\sigma_m\right]+\left(\frac{1-V_2}{Y}\right)(\sigma_3^a-\sigma_m)\Big/\left[\sigma_3^a+\left(\frac{1}{p_c}-1\right)\sigma_m\right]=0 \tag{13.36}$$

$$\frac{V_2/Y(\sigma_c^a-\sigma_m)}{[\sigma_m+p_c(\sigma_c^a-\sigma_m)]}+\frac{(1-V_2/Y)(\sigma_3^a-\sigma_m)}{[\sigma_m+p_c(\sigma_3^a-\sigma_m)]}=0 \tag{13.37}$$

where p_c is the continuous percolation threshold for the composite grains. Since the composite grains are allowed to overlap in the region of interface layer, p_c can be taken to be 0.28 for this random mixture on the basis of general percolation theory [127,130].

According to **eq. 13.37**, one arrives at the highest enhancement in ionic conductivity when the composite units fill the total volume of the electrolyte, i.e., when $V_c = 1$. This occurs for a characteristic value of $V_2 = V_2^*$ given by:

$$V_2^* = \left(1+\frac{t}{R}\right)^{-3} \tag{13.38}$$

For concentrations of filler exceeding V_2^*, the quasi-two-phase system consists of a mixture of composite grains and dispersed bare insulating grains and no matrix electrolyte other than that present in the composite units. For such a situation, **eq. 13.17** can be rewritten in the following form:

$$\frac{(1-V_2)(\sigma_c^b-\sigma_m)}{[\sigma_m+P_c(\sigma_c^b-\sigma_m)]}+\frac{(V_2-V_2^*)(\sigma_2^b-\sigma_m)}{[\sigma_m+P_c(\sigma_2^b-\sigma_m)]}=0 \tag{13.39}$$

where σ_c^b and σ_2^b (the modified conductivity parameters for the second limiting case) are calculated according to the method proposed by Nan [127] and Nakamura [128].

$$\sigma_c^b = 2\sigma_c(1-V_2+V_2^*)/[2(1+V_2)-V_2^*] \tag{13.40a}$$

$$\sigma_2^b = 2\sigma_2(V_2-V_2^*)/[3-V_2+V_2^*] \tag{13.40b}$$

Eqs. 13.40a and **13.40b** also consider local field effects. Here P_c is the percolation threshold of the dispersed bare insulating grains that do not overlap. Therefore P_c in **eq. 13.39** is different from the p_c in **eq. 13.37** and has been taken as the percolation threshold for a general random mixture i.e., 0.15 [126,127]. **Eq. 13.37** and **eq. 13.39** are used to calculate the conductivity σ_m of composite polymer electrolytes.

This model is valid only at temperatures lower than the melting point of the crystalline PEO phase. Above the melting point of the crystalline PEO phase, the conductivity of the matrix polymer ionic conductor is equal to the conductivity of its amorphous phase. Hence **eqs. 13.35a** and **13.37** should be rewritten in the following form:

$$\sigma_3^a = \sigma_{am}\left[(d-1)-(d-1)V_c\right]/(d-1+V_c) = 2\sigma_{am}(1-V_c)/(2+V_c) \tag{13.41}$$

$$V_2/Y(\sigma_c^a - \sigma_m)/\sigma_m + p_c(\sigma_c^a - \sigma_m)] + (1-V_2/Y)(\sigma_{am}^a - \sigma_m)/[\sigma_m + p_c(\sigma_{am}^a - \sigma_m)] = 0 \tag{13.42}$$

Here σ_{am} is the conductivity of the polymer matrix equal to the conductivity of the amorphous polymer phase in this temperature (above the melting temperature of PEO, i.e., $> 60°C$) range. It can be seen that in this temperature range two different approaches should be considered. The first is to consider the conductivity of the amorphous interface layer that is the same as for the amorphous phase of the matrix polymer ionic conductor. Generally the properties of the coating layer are different from those of the amorphous electrolyte phase, since the filler has an effect on the coating layer conductivity. The system is then treated as a mixture of two amorphous phases and its conductivity is calculated according to **eqs. 13.39** and **13.42**.

The applicability of the model presented above has been tested by Wieczorek and co-workers [120-122,131-133] for PEO based composite polymer electrolytes with dispersed inorganic and organic fillers. They found that for volume fraction of dispersed phase greater than V_2^*, the conductivity drops more rapidly than predicted by the model presented above. It has been shown that the $^1/_R$ ratio decreases for concentrations of filler exceeding V_2^* i.e. the concentration for which the maximum of conductivity is reached. Moreover, conductivity of an interface amorphous layer changes with the change of filler concentration. With the help of DSC data it was shown that T_g of the composite electrolyte varies with filler concentration. Since the structure of interface layer is highly amorphous, the temperature dependence of conductivity would follow the VTF relation:

$$\sigma = AT^{-1/2}\exp(-E_a'/k(T-T_0)) \tag{13.43}$$

where E_a' is pseudo-activation energy for conduction, A is a pre-exponential factor and T_0 is a quasi equilibrium glass transition temperature usually 20-50°C lower than T_g. It was also assumed that the pseudo-activation energy E_a' and the pre-exponential factor A were independent of the concentration of fillers. Therefore, the interface layer conductivity is only dependent on T_0 and decreases with an increase in T_g. The T_g values taken from DSC experiments were used for calculation of interface layer conductivity (σ_1) for electrolytes with different concentrations of filler. To generalize the variation in T_g on the basis of the work of Bares [134] and Schneider and diMarzio [135] have proposed following empirical equation for the T_g of immiscible blends and mixed phase system:

$$T_g = K_0 + K_1[f(V_2/S_2)] + K_2[f(C_s,V_2)] \tag{13.44}$$

where V_2 and S_2 are respectively the volume fraction and surface area of the filler, C_s is the molar concentration of the dopant salt in the complex, and K_0, K_1 and K_2 are adjustable parameters. It has been found that for a system with different concentrations of filler additive and a fixed concentration of dopant salt, the best fit to the experimental data can be obtained when T_g is approximated by the relation:

$$T_g = K_0 + K_1 V_2 + K_2 V_2^2 \tag{13.45}$$

In this equation, the effect of salt is included in K_0 that is identical to the T_g for pristine (without filler) polymer electrolyte. K_1 describes the influence of filler added on T_g of the composite system and K_2 is connected with polymer-filler-salt interaction.

On the other hand, estimation of the t/R parameter is based on the assumption that an increase in amorphous phase content for the composite electrolyte in comparison to the pristine system is due to the amorphous phase present in the surface layer. Therefore on the basis of **eq. 13.34**, the volume fraction of this component and hence the t/R parameter can be approximated [121] as:

$$1/(1+t/R)^3 = V_2/(V_1+V_2) \quad \text{or} \quad R = \{(V_2+V_1)/V_2\}^{1/3} - 1 \tag{13.46}$$

where V_2 is the volume fraction of filler and V_1 is the volume fraction of surface layer.

The inclusion of variation of T_g and t/R parameters with filler concentration in the model for the calculation of the composite electrolyte conductivity has been found to fit well with the experimental results for a series of composite polymer electrolytes [122,132,136].

6. P(VDF-HFP)/PVDF:(PC+DEC):LiClO$_4$ GEL POLYMER ELECTROLYTE SYSTEMS

6.1. Preparation

Polymer electrolyte films are synthesized by solution casting technique consisting of the following steps:

1. PVDF polymer is dissolved in dimethyl sulfoxide (DMSO) as required.
2. P(VDF-HFP) copolymer is dissolved in acetone or tetrahydrofuran.
3. Salt, LiClO$_4$ is dissolved in propylene carbonate (PC) and/or diethyl carbonate (DEC). PC has high dielectric constant ($\varepsilon = 64.6$) but has high viscosity ($\eta = 2.53$) also whereas DEC has low dielectric constant ($\varepsilon = 2.82$) but has low viscosity ($\eta = 0.748$). PC+DEC (50% by volume each) solvent is used as a compromise for high dielectric constant and low viscosity to achieve high ionic conductivity.
4. Both the polymer and salt solution are mixed together in a beaker and magnetically stirred during continuous heating for 12-14 hours at 50°C.
5. The above viscous solution is then cast on glass plates/petri dishes and allowed to dry at room temperature.

Fig. 13.4 shows the block diagram of the solution casting technique. This procedure provides mechanically stable, free standing and flexible films of thickness in the range of 30 μm^{-1} mm.

Poly vinylidenefluoride (PVDF) is a semicrystalline polymer. When copolymerized with hexafluropropylene (HFP), the degree of crystallinity is greatly reduced. This considerably enhances the flexibility of the resulting copolymer as compared to PVDF homopolymer. To compare the results with PVDF:(PC+DEC):LiClO$_4$ system, P(VDF-HFP):(PC+DEC):LiClO$_4$ electrolyte system

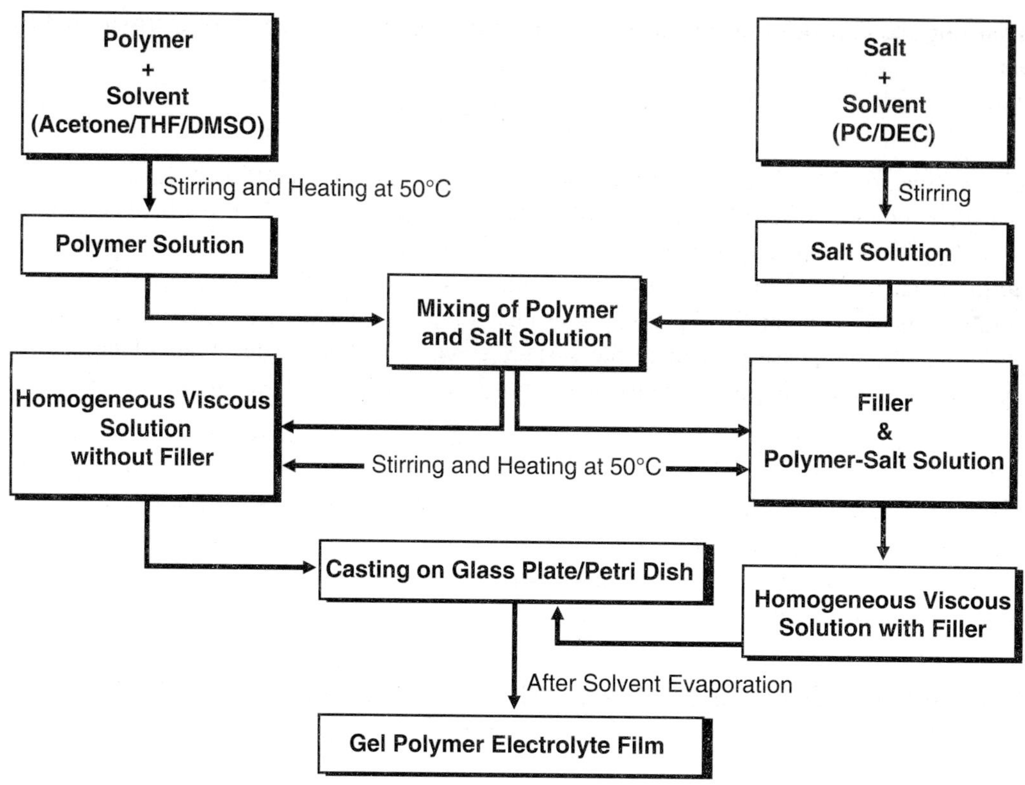

Fig. 13.4. Block diagram of solution casting technique of sample preparation.

was also studied. For this system, P(VDF-HFP) copolymer is dissolved in tetrahydrofuran (THF) and LiClO$_4$ and (PC+DEC) are added as described for previous system to get polymer gel electrolyte film. A comparative study has been carried out between PVDF polymer and P(VDF-HFP) copolymer based gel electrolytes with regard to their ionic conduction behavior.

6.2. Characterization, Results and Discussion

6.2.1. Ionic Conductivity Measurement

Hioki 3532-50 LCR HiTester has been employed for the complex impedance (modulus Z and phase angle θ) measurements. The instrument is interfaced with a computer to collect the data. It has a built in frequency synthesizer which has frequency range 42 Hz to 5 MHz. The impedance measurements are carried out at a temperature interval of 10°C from room temperature to 100°C. Sufficient time is allowed at each temperature for thermal equilibration and reproducibility of the data. The ionic conductivity is then calculated from the relation $\sigma = l\,/(R_b r^2 \pi)$ where l and r represent thickness and radius of the sample membrane discs respectively. R_b is the bulk resistance of the gel electrolyte obtained from complex impedance measurements. It is widely accepted that R_b could be obtained from the intercept on the real axis at the high frequency end of the Nyquist plot of complex impedance [137,138]. According to the theoretical analysis given by Watanabe and Ogata [32] two semicircles should appear in impedance spectrum for a symmetric cell i.e. one at higher frequencies corresponding to bulk electrolyte impedance and other at lower frequencies related to the interfacial impedance. Also it is reported [139-141] that high frequency semicircle does not appear in practical impedance plots for plasticized P(VDF-HFP) membranes. This phenomena is quite reasonable since the facile mobility in liquid and gel-type electrolyte systems, when compared with solid polymer electrolytes, indicates that ions possess small dielectric relaxation times and hence the capacitive effect of the bulk electrolyte in the spectra are inconsequential [142]. In the lower frequency region inclined spike is observed instead of second semicircle since Cu electrodes, which have been used in the present work, are perfectly blocking to lithium ions within the applied ac modulation range (0.01V). Causes responsible for ideality of the blocking electrodes commonly encountered in real measurements are still controversial, however, surface roughness of the electrodes and the faradic reaction occurring at the electrode/electrolyte interface seem to remain two preferential explanations

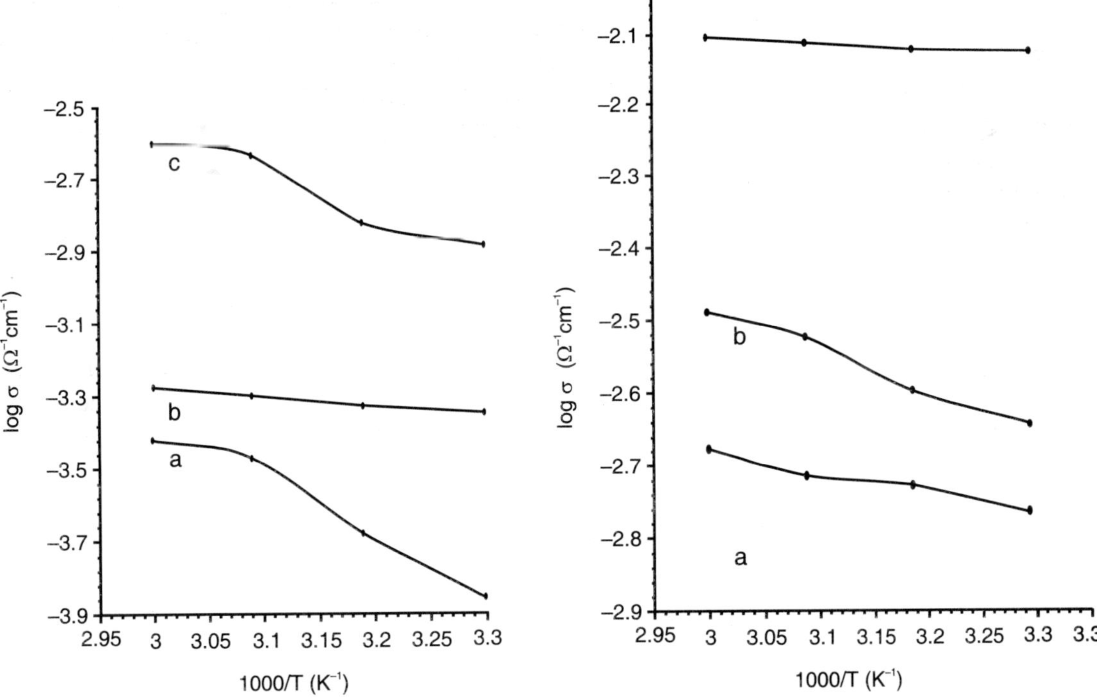

Fig. 13.5. Temperature dependence of ionic conductivity of PVDF-(PC+DEC)-LiClO$_4$ gel polymer electrolyte system with composition (wt%), (a) 25:70:5, (b) 25:65:10, (c) 25:60:15. Reproduced with permission from D. Saikia *et al.*, Electrochim Acta, 49, 2581 (2004). Copyright © 2004, Elsevier Science Ltd.

Fig. 13.6. Temperature dependence of ionic conductivity of P(VDF-HFP)-(PC+DEC)-LiClO$_4$ gel polymer electrolyte (a) 25:70:5 wt%, (b) 25:65:10 wt% and (c) 25:60:15 wt%. Reproduced with permission from D. Saikia *et al.*, Electrochim Acta, 49, 2581 (2004). Copyright © 2004, Elsevier Science Ltd.

[143]. High frequency region is independent of the electrode condition and the low frequency region behaves much like constant phase element (CPE) and is very sensitive to the roughness and electrode condition [142].

Figs. 13.5 & 13.6 show the conductivity versus temperature inverse plots of PVDF and P(VDF-HFP) based gel polymer electrolytes for different plasticizer-salt ratio with fixed amount of PVDF and P(VDF-HFP). It is evident from the figures that the ionic conductivity increases with the increase of salt concentration. The increase of the salt concentration would lead to increase in carrier ion concentration in the systems. The same trend is observed for all the temperatures studied from 30°C to 60°C. The maximum on the conductivity-salt concentration curve may be explained in terms of formation of ion pairs. Up to the concentration level, corresponding to the conductivity plateau, more and more number of ions is made available for electrolytic conduction. At higher concentration in low dielectric media, essentially two types of ion pairs and solvated ion pairs are present [144]. These may be represented by the equilibria:

$$Li + S_n + ClO_4^- = (Li+)S_nClO_4^- \text{ and } Li + S_n + ClO_4^- = (Li + ClO_4^-)S_{n-m} + S_m \quad (13.47)$$

where S represents the solvating species. Re-dissociation of such ion pairs can occur due to long range Coulombic forces giving rise to free ions, which contribute to conductance [145].

The high ionic conductivity in an electrolyte is obtained by increasing both ionic mobility and concentration of ionic charge carriers. The motion of ions in gel polymer electrolytes is liquid-like in which the movement of ions through the polymer matrix is assisted by the large amplitude segmental motion of the polymer backbone [146-148]. From the figures it is observed that the ionic conduction in both PVDF and P(VDF-HFP) based gel polymer electrolyte system obeys the VTF (Vogel-Tamman-Fulcher) relation [91-93], which describes the transport properties in a viscous matrix [149-152]. It supports the idea that the ions move through the ion conducting plasticizer rich phase.

Tab. 13.1. Conductivity of PVDF-(PC+DEC)-LiClO$_4$ gel polymer electrolyte at 30°C with varying composition (wt%). Reproduced with permission from D. Saikia *et al.*, Electrochim Acta, 49, 2581 (2004). Copyright © 2004, Elsevier Science Ltd.

PVDF	PC	DEC	LiClO$_4$	Conductivity (S/cm)
25	35	35	5	1.4×10^{-4}
25	32.5	32.5	10	4.5×10^{-4}
25	30	30	15	1.3×10^{-3}

The conductivities for gel polymer electrolytes containing different weight ratios of PVDF, (PC+DEC) and LiClO$_4$ salt are presented in **Tab. 13.1**. From the **Tab. 13.1**, it may be observed that the highest conductivity ($\approx 1.3 \times 10^{-3}$ S/cm) composition at room temperature is 25-wt% PVDF, 60 wt% (PC + DEC) and 15-wt% LiClO$_4$. The conductivity data for P(VDF-HFP) based gel polymer electrolytes containing different weight ratios of P(VDF-HFP), (PC+DEC) and LiClO$_4$ salt are presented in **Tab. 13.2**. From the **Tab. 13.2**, it may be seen that the highest conductivity ($\approx 7.5 \times 10^{-3}$ S/cm) composition at room temperature is 25 wt% P(VDF-HFP), 60 wt% (PC+DEC) and 15 wt% LiClO$_4$. Higher conductivity in P(VDF-HFP) copolymer system could be attributed to the higher amorphicity due to steric hindrance provided by CF$_3$ pendant group in HFP monomers that is randomly mixed with the VDF monomers in the polymer chain. Higher amorphicity provides mobile Li$^+$ ion more free volume giving rise to higher conductivity. On the other hand pure PVDF has higher degree of crystallinity because of single monomer throughout the polymer chain providing less free volume to the mobile Li$^+$ ion, resulting in lower conductivity.

Tab. 13.2. Conductivity of P(VDF-HFP)-(PC+DEC)-LiClO$_4$ gel polymer electrolyte at 30°C with varying composition (wt%). Reproduced with permission from D. Saikia *et al.*, Electrochim Acta, 49, 2581 (2004). Copyright © 2004, Elsevier Science Ltd.

PVDF-HFP	PC	DEC	LiClO$_4$	Conductivity (S/cm)
5	35	35	5	1.7×10^{-3}
25	32.5	32.5	10	2.2×10^{-3}
25	30	30	15	7.5×10^{-3}

The mobility of ionic species is an important parameter to consider when designing new polymer electrolytes for batteries [153]. The concentration gradients formed due to the non-unity transference number of lithium ions in polymer electrolytes, cause a decay of the current to lower steady state values. Thus, evaluation of polymer electrolytes requires consideration of transference numbers and ionic conductivities. Total ionic transference number of gel polymer electrolyte is measured by Wagner's polarization technique [154,155] which is used to determine the ionic contribution to the total charge transport by measuring the residual electronic current passing through the electrolyte. The Wagner polarization cell Ag/polymer gel electrolyte/Ag is prepared to measure the transport number. A fixed small dc potential (300 mV) is applied across the blocking electrodes and the current passing through the cells is measured as a function of time for about 5-7 hours to allow the samples to become fully polarized. Initial total current (I_T), which is the sum of ionic (I_i) and electronic (I_e) currents ($I_T = I_i + I_e$) and final current after polarization, which is only the electronic current I_e, are measured. The ionic transference number is calculated using the relation:

$$t_i = \frac{I_i}{I_{total}} = \frac{I_T - I_e}{I_T} \tag{13.48}$$

Ionic transference number of PVDF:(PC+DEC):LiClO$_4$ and P(VDF-HFP):(PC+DEC):LiClO$_4$ gel polymer electrolyte (GPE) systems with varying compositions is measured with Ag/GPE/Ag cell systems. **Tabs. 13.3 & 13.4** show the ionic transference number for the Ag/Electrolyte/Ag cell system using PVDF and P(VDF-HFP) based gel polymer electrolytes. High values of transport numbers from 0.90-0.93 in PVDF and from 0.94-0.98 in P(VDF-HFP) based gel polymer electrolytes suggest that the charge transport in these polymer electrolyte systems is predominantly ionic accompanied by mass transport and electronic contribution to the total current is negligible.

Tab. 13.3. Ionic transference numbers of PVDF-(PC+DEC)-LiClO$_4$ gel polymer electrolyte. Reproduced with permission from D. Saikia *et al.*, Electrochim Acta, 49, 2581 (2004). Copyright © 2004, Elsevier Science Ltd.

Gel polymer electrolyte system (GPE) PVDF-(PC+DEC)-LiClO$_4$	Transference numbers Ag/GPE/Ag
25:70:5 wt%	0.90
25:65:10 wt%	0.91
25:60:15 wt%	0.93

Tab. 13.4. Ionic transference numbers of P(VDF-HFP)-(PC+DEC)-LiClO$_4$ gel polymer electrolyte. Reproduced with permission from D. Saikia *et al.*, Electrochim Acta, 49,2581 (2004). Copyright © 2004, Elsevier Science Ltd.

Gel polymer electrolyte system (GPE) P(VDF-HFP)-(PC+DEC)-LiClO$_4$	*Transference numbers Ag/GPE/Ag*
25:70:5 wt%	0.94
25:65:10 wt%	0.95
25:60:15 wt%	0.98

6.2.2. X-ray Diffraction Studies and Degree of Crystallinity (K)

X-ray diffraction is one of the most important structural characterization tools used in materials science. X-ray diffraction patterns give information about different crystal parameters like crystallite size, d-spacing, diffraction planes, structure, phase and lattice constants. The intensities and the angles of diffracted X-ray beams are related to atomic arrangement of the crystal. XRD is mainly used to measure the degree of crystallinity of the polymer electrolyte samples, which is an important parameter affect with ionic conductivity. X-ray diffraction patterns for various polymer electrolyte samples have been recorded using the Philips X'pert Pro Diffractometer employing Cu K$_\alpha$ radiation. The generator is operated at 30 kV and 20 mA and scanning speed is fixed at 2°/min in 2θ. The value of 2θ is taken from 3° to 100°.

Generally a polymer is neither fully crystalline nor fully amorphous. The degree of crystallinity of a polymer can be measured by X-ray diffraction. A typical X-ray diffraction spectrograph consists of a plot of X-ray counts received by a detector versus the scattering angle of the detector as shown in **Fig. 13.7.** Depending upon its degree of crystallinity typical X-ray diffractogram of a polymer sample has sharp peaks superimposed on a broad amorphous hump as shown in **Fig. 13.8.** Total area under the diffractogram is the sum of the crystalline peaks and broad amorphous hump. For example, if a typical X-ray diffractogram has two crystalline peaks with areas A1 and A2 superimposed on a broad amorphous hump with an area of A3, then the percentage crystallinity, K, of the polymer will be:

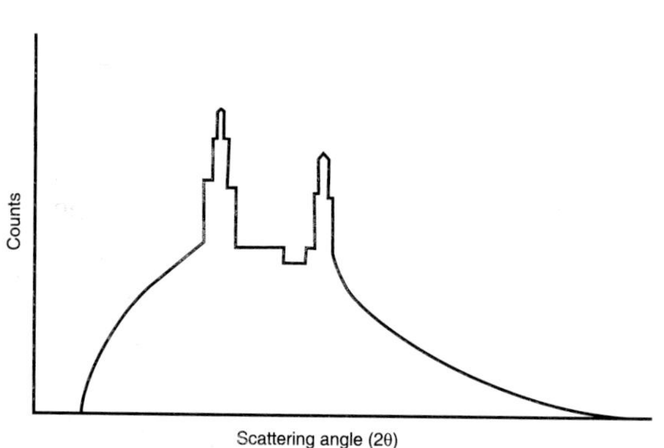

Fig. 13.7. Typical X-ray diffractogram. Reproduced with permission from D. Saikia *et al.*, Electrochim Acta, 49, 2581 (2004). Copyright © 2004, Elsevier Science Ltd.

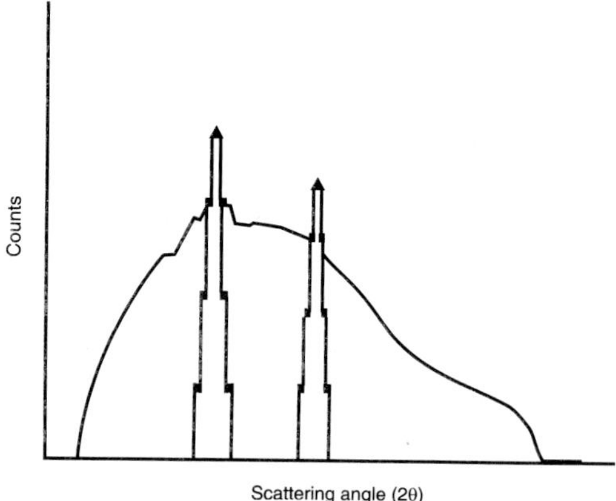

Fig. 13.8. XRD pattern with crystalline peaks and amorphous hump. Reproduced with permission from D. Saikia *et al.*, Electrochim Acta, 49, 2581 (2004). Copyright © 2004, Elsevier Science Ltd.

$$K = \frac{k_c\left(A1 + A2\right)}{k_c\left(A1 + A2\right) + k_a\left(A3\right)} \times 100 = \frac{\left(A1 + A2\right)}{\left(A1 + A2\right) + \dfrac{k_a}{k_c}\left(A3\right)} \times 100 \tag{13.49}$$

where k_a and k_c are the proportionality constants for the amorphous and crystalline phases, respectively. Usually for polymers the values of k_a and k_c are comparable, therefore assuming $k_a = k_c$, the percentage crystallinity will be:

$$K = \frac{A1 + A2}{A1 + A2 + A3} \times 100 = \frac{S}{S_0} \times 100 \tag{13.50}$$

where S is the sum of areas of all the crystalline peaks and S_0 is the sum of areas of crystalline peaks and amorphous hump i.e. total area under the diffractogram. Area has been calculated by dividing the X-ray diffractogram into minute square (0.5×0.5 mm^2) grids and counting the number of grids. The amount of crystallinity in a polymer depends on the secondary valence bonds, which can be formed, the structure of the polymer chain (degree of order), the physical treatment of the polymer, the thermal history of the polymer and the molecular weight of the polymer.

Fig. 13.9 shows the XRD patterns of PVDF, P(VDF-HFP), LiClO$_4$, P(VDF-HFP): (PC+DEC):LiClO$_4$ and PVDF:(PC+DEC):LiClO$_4$ systems respectively. PVDF is crystalline and shows its characteristic peaks at $2\theta = 18.2°$, $20°$, $26.6°$ and $38°$ (**Fig. 13.9a**). For P(VDF-HFP) samples in **Fig. 13.9b,** three peaks appear at $2\theta = 18.4$, 20 and 26.6 which correspond well with the (100)+(020),

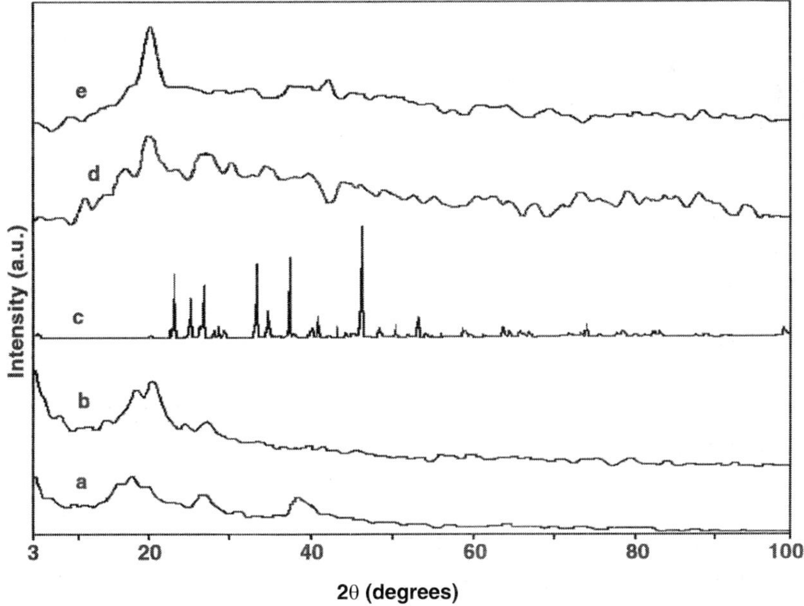

Fig. 13.9. XRD spectra of (a) PVDF, (b) P(VDF-HFP), (c) LiClO$_4$, (d) P(VDF-HFP)-(PC+DEC)-LiClO$_4$ (25:60:15 wt%) and (e) PVDF-(PC+DEC)-LiClO$_4$ (25:60:15 wt%) gel polymer electrolytes. Reproduced with permission from D. Saikia *et al.*, Electrochim Acta, 49, 2581 (2004). Copyright © 2004, Elsevier Science Ltd.

(110) and (021) reflections of crystalline PVDF [156]. This is a confirmation of partial crystallization of the PVDF units in the copolymer to give an overall semi-crystalline morphology for P(VDF-HFP). **Figs. 13.9d & 13.9e** show the X-ray diffractograms of PVDF and P(VDF-HFP) based gel polymer electrolyte systems. The degree of crystallinity (K), calculated for pure PVDF is 31.4% and for pure P(VDF-HFP) is 30%. For PVDF:(PC+DEC):LiClO$_4$ gel polymer electrolyte, K is calculated to be 20.45% and for P(VDF-HFP):(PC+DEC):LiClO$_4$ gel polymer electrolyte, K reduces to 13.64%. It is observed that crystallinity of gel polymer electrolyte is greatly reduced after the addition of PC+DEC and LiClO$_4$. Moreover, peaks corresponding to LiClO$_4$ **(Fig. 13.9c)** are not observed indicating that LiClO$_4$ is completely dissolved in (PC+DEC) solvents in the polymer matrix and does not remain a separate phase in the electrolytes. The intensity of the crystalline peaks is decreased and a noticeable broadening of the area under the peaks was observed. This is a clear indication of the reduction in the crystalline phase in the polymer complex at room temperature [157]. It is also observed that P(VDF-HFP) based electrolytes are more amorphous than PVDF based electrolytes as peaks are sharper and more in number indicating larger crystallinity in PVDF based electrolyte. Higher amorphicity results in higher conductivity in P(VDF-HFP) based electrolytes as compared to PVDF based electrolytes.

6.2.3. Fourier Transform Infra-red Spectroscopy

Fourier transform infrared (FTIR) spectroscopy is a powerful analytical tool for characterizing and identifying organic molecules and investigation of polymer structure. Using the IR spectrum, chemical bonds (functional groups) and the molecular structure of organic compounds can be identified. The wavelength of infrared radiation absorbed is characteristic of stretching/bending vibrational mode of a chemical bond. By interpreting the infrared absorption spectrum, the chemical bonds in a molecule can be determined. FTIR spectra of pure compounds are generally so unique that they are like a molecular "fingerprint". FTIR spectra of polymer electrolyte samples are studied using a Nicolet Impact 410 spectrophotometer. It is a high performance FTIR spectrophotometer with 1 cm^{-1} resolution. Spectra were scanned in the mid IR region of 4000-400 cm^{-1}. In the present work FTIR as used to investigate polymer-salt-filler interactions in the polymer electrolyte systems investigated. The IR spectra of these materials vary according to their compositions and are able to show the occurrence of complexation and interaction(s) between the various constituents [158,159].

    Fig. 13.10 shows the FTIR spectra of PVDF based electrolytes. Frequency 3023 cm^{-1} is assigned to C–H stretching vibrations of PVDF. Frequency 1640 cm^{-1} is assigned to C = O stretching vibration. Frequency 1415 cm^{-1} is assigned to C–F stretching vibration of PVDF. Frequency 1270 cm^{-1} is assigned to C–O stretching vibrations of plasticizer. Frequencies 1172–1066 cm^{-1} are assigned to– C–F– and –CF$_2$–stretching vibrations. Frequency 880 cm^{-1} is assigned to vinylidene group of PVDF. New peak observed in the polymer electrolyte system at frequency 837 cm^{-1} is assigned to C–Cl stretching vibrations. The vibrational peaks of PVDF (3021, 1400, 1183, 976, 876 cm^{-1}) are shifted to (3023, 1415, 1171, 950, 880 cm^{-1}) in the polymer electrolyte.

    Fig. 13.11 shows the FTIR spectra of P(VDF-HFP), LiClO$_4$ and P(VDF-HFP):(PC+DEC):LiClO$_4$ gel polymer electrolytes. The frequency 3000 cm^{-1} is assigned to C – H stretching vibrations of PVDF. The frequency 1791 cm^{-1} is assigned to – CF = CF$_2$ group. Frequency 1636 cm^{-1} is assigned to C = O bonding. Frequencies 1483 cm^{-1} and 1400 cm^{-1} are assigned to – CH$_3$ asymmetric bending and C – F stretching vibrations of plasticizer, propylene carbonate and diethyl carbonate and

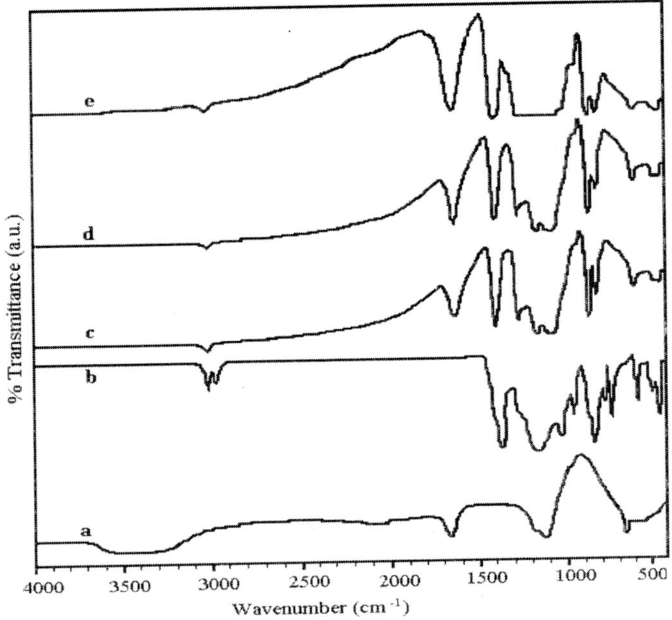

Fig. 13.10. FTIR spectra of (a) LiClO$_4$, (b) PVDF (c) PVDF-(PC+DEC)-LiClO$_4$ (25:70:5 wt%), (d) PVDF-(PC+DEC)-LiClO$_4$ (25:65:10 wt%) and (c) PVDF-(PC+DEC)-LiClO$_4$ (25:60:15 wt%). Reproduced with permission from D. Saikia *et al.*, Electrochim Acta, 49, 2581 (2004). Copyright © 2004, Elsevier Science Ltd.

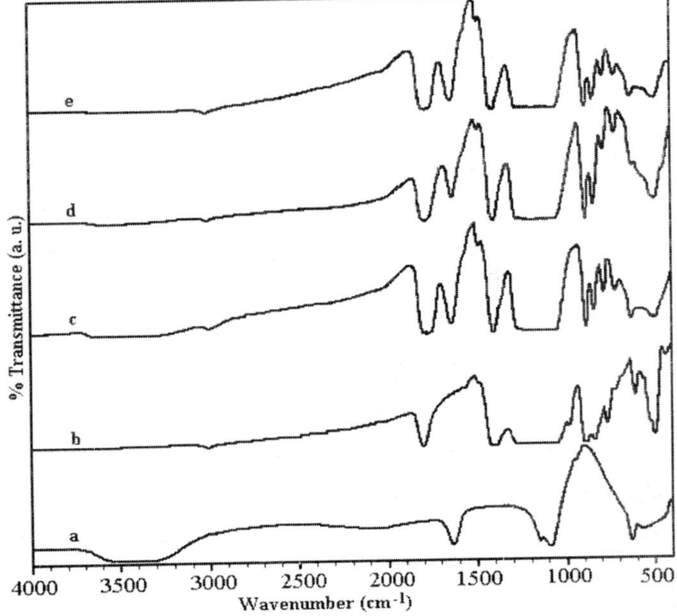

Fig. 13.11. FTIR spectra of (a) LiClO$_4$, (b) P(VDF-HFP), (c) P(VDF-HFP)-(PC+DEC)-LiClO$_4$ (25:60:15 wt%), (d) P(VDF-HFP)-(PC+DEC)-LiClO$_4$ (25:65:10 wt%) and (e) P(VDF-HFP)-(PC+DEC)-LiClO$_4$ (25:70:5 wt%). Reproduced with permission from D. Saikia *et al.*, Electrochim Acta, 49, 2581 (2004). Copyright © 2004, Elsevier Science Ltd.

P(VDF-HFP) respectively. Frequencies 1290-1060 cm^{-1} are assigned to –C–F– and –CF$_2$– stretching vibrations. Frequency 881 cm^{-1} is assigned to vinylidene group of polymer. Frequencies 840–560 cm^{-1} are assigned to C–Cl stretching vibrations. The vibrational peaks of P(VDF-HFP) (3004, 1797, 1417, 882 and 839 cm^{-1}) are shifted to (3000, 1791, 1401, 881 and 837 cm^{-1}) in the polymer electrolyte. New absorption peaks at 781 cm^{-1} and 716 cm^{-1}, which correspond to C–Cl stretching [158,160] are observed in the electrolyte.

Appearance of new peaks at frequency 837 cm^{-1} in PVDF:(PC+DEC):LiClO$_4$ and at 781 and 716 cm^{-1} in P(VDF-HFP):(PC+DEC):LiClO$_4$ which could be ascribed to C–Cl stretching vibration indicates that polymer carbon atoms interact with chlorine of ClO$_4^-$ ions. Shifting and formation of new peaks suggest the polymer-salt interaction occur in both PVDF and P(VDF-HFP) based gel polymer electrolyte systems.

6.2.4. Scanning Electron Micrograph Studies

The scanning electron microscope (SEM) is one of the most versatile instruments available for the examination and analysis of microstructural and morphological features of solid objects. The primary reason for the SEM's usefulness lies in its high depth of focus and high resolution that can be obtained when bulk objects are examined. The SEM study is carried out using JEOL JSM-35CF, scanning electron microscope with resolution 5 nm –15 nm to examine the surface morphology, porosity and distribution of filler particles in the gel polymer electrolytes. The equipment has a range of magnification from 10x to 180,000x. Each specimen is gold coated using a sputtering unit (International Scientific Instruments PS-2 coating unit) before taking the micrographs. The

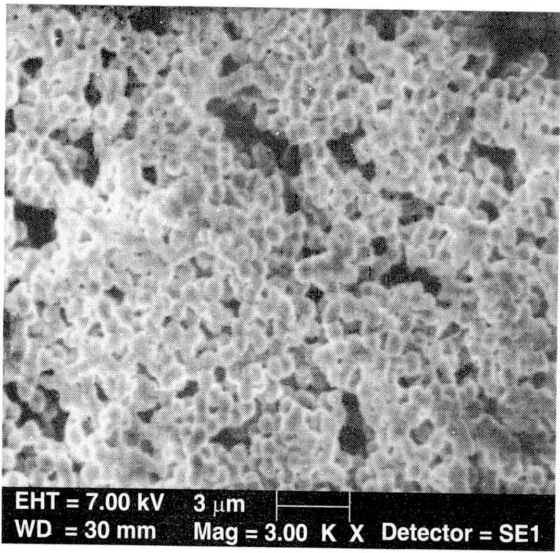

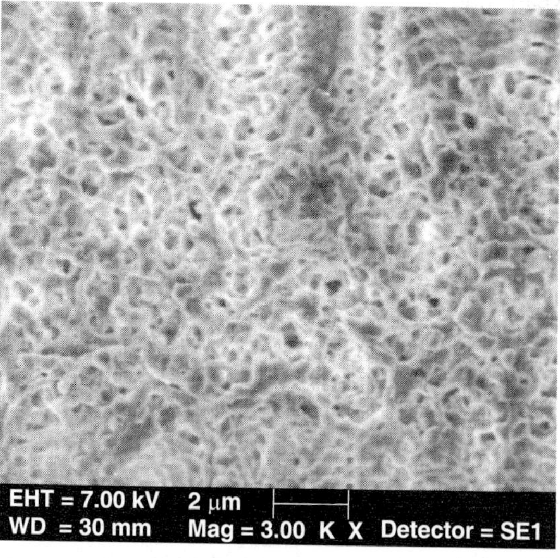

Fig. 13.12. SEM image of PVDF-(PC+DEC)-LiClO$_4$ (25:65:10 wt%) gel polymer electrolyte, Magnification 3000X. Reproduced with permission from D. Saikia *et al.*, Electrochim Acta, 49, 2581 (2004). Copyright © 2004, Elsevier Science Ltd.

Fig. 13.13. SEM image of P(VDF-HFP)-(PC+DEC)-LiClO$_4$ (25:60:15 wt%) gel polymer electrolyte, Magnification 3000X. Reproduced with permission from D. Saikia *et al.*, Electrochim Acta, 49, 2581 (2004). Copyright © 2004, Elsevier Science Ltd.

micrographs are taken at 20 kV accelerating voltage and magnification is fixed according to need from 3000× to 10000×.

Figs. 13.12 & 13.13 show the scanning electron micrograph of PVDF:(PC+DEC):LiClO$_4$ (25:65:10 wt%) and P(VDF-HFP):(PC+DEC):LiClO$_4$ (25:60:15 wt%) gel polymer electrolyte systems. A macroscopic phase separation between the dense and porous layer is noticed. Ordered and almost spherical polymer particles of about 1.5 μm and void spaces are observed in this system **(Fig. 13.12)**. The void spaces are certainly the result of solvent removal [161] resulting in larger pore size. In case of P(VDF-HFP):(PC+DEC):LiClO$_4$ gel polymer electrolyte **(Fig. 13.13)**, although the macroscopic phase separation is absent, the film is still porous with microscopic phase separation. For this system, though the surface morphology is uniform but the ordered spherical grain structure is absent. The microstructure is disordered throughout the sample and the pore size is smaller. The result suggests that P(VDF-HFP) has higher solvent retention ability since the pores in microstucture occur due to solvent removal [161,162]. SEM results are consistent with the view that higher conductivity and ionic transport number in P(VDF-HFP) based electrolytes arise from the increased amorphicity and solvent retention ability in these gel electrolytes.

7. CONCLUSIONS AND FUTURE PROSPECTS

Electrochemical materials and devices are playing an ever-increasing role in modern technology driven society. The widespread usage of portable electronic devices has initiated a growing need for larger capacity, smaller size, lightweight and lower priced rechargeable batteries. Lithium ion batteries have become the predominant battery technology for handheld electronic applications in the recent years due to their high energy, high voltage, good cycle life and excellent storage characteristics. The continuing quest to improve the electrical transport properties of polymer electrolytes is driven by their potential technological importance for electronic and energy storage devices. At the same time these materials offer excellent opportunities for theoretical studies focused on the mechanism underlying their ionic transport properties. In the present chapter theory of polymer electrolytes has been presented with regard to kinetics of ion transport in polymers. Synthesis, characterization and results of ionic conductivity, XRD, FTIR and SEM measurements on P(VDF):(PC+DEC):LiClO$_4$ and P(VDF-HFP):(PC+DEC):LiClO$_4$ polymer gel electrolyte systems have been analyzed.

Ionic conductivity enhancement in PVDF:(PC+DEC):LiClO$_4$ and P(VDF-HFP):(PC+DEC):LiClO$_4$ types of electrolytes can be attributed to the formation and redissociation of ion-pairs due to long range Coulombic forces giving rise to free ions which contribute to conductance. The motion of ions in gel polymer electrolytes occurs in liquid-like manner in which the movement of ions through the polymer matrix is assisted by the large amplitude segmental motion of the polymer backbone as indicated by FTIR analysis which suggests polymer-salt interaction for both PVDF and P(VDF-HFP) based gel polymer electrolytes. Maximum ionic conductivity for PVDF:(PC+DEC):LiClO$_4$ and P(VDF-HFP):(PC+DEC):LiClO$_4$ is found to be 1.3 × 10^{-3} S/cm and 7.5 × 10^{-3} S/cm at 303 K respectively. Higher conductivity in P(VDF-HFP) based electrolyte in comparison to PVDF based electrolyte could be attributed to higher amorphicity of the former electrolyte system due to steric hindrance provided by bulky pendant –CF$_3$ group in HFP monomer unit of the copolymer. Higher amorphicity provides higher flexibility to the polymer chains and mobile Li$^+$ ion more free volume giving rise to higher conductivity. This result is confirmed by XRD analysis which reveals that degree of crystallinity is reduced from 31.4% in PVDF to 20.45% in PVDF based gel polymer electrolyte system and 30% in P(VDF-HFP) to 13.64% in P(VDF-HFP) based gel polymer electrolyte system. Total ionic transport number was found to be higher for P(VDF-HFP) based electrolytes than that of PVDF based electrolytes. SEM results are consistent with the view that higher conductivity and ionic transport number in P(VDF-HFP) based gel polymer electrolytes arise from the increased amorphicity and higher solvent retention ability than PVDF based gel polymer electrolytes.

Though tremendous progress has been made for improving the performance of ion conducting polymers, there is a vast scope of further development of high ion conducting gel polymer electrolyte materials and fabrication of electrochemical devices like high energy density rechargeable lithium batteries, supercapacitors, sensors etc. employing them. It is generally believed that ionic conduction is a property of the amorphous phase and that ion association, ion-polymer interactions and local relaxations of the polymer strongly influence the ionic mobility. The local mobility of the ion-solvating polymer segments plays a key role for the ion transport. However, much remains unknown about the nature of the ion association processes, the ion-polymer interactions and the role that they play in ionic conductivity of the electrolytes. Dispersion of insulating and semiconducting inorganic nanoparticles into polymer electrolytes so as to increase the particle-polymer interface area and their investigations by Small Angle X-ray spectroscopy requires attention of the researchers. In polymer electrolytes both cations and anions contribute to conductivity. Movement of anions in polymer electrolytes deteriorates device the performance, therefore cationic transport number needs to be enhanced. Intensive research on cationic monoconducting (single ion conducting) polymer electrolytes is going on, but researchers are far away from the goal. Research in this field has tremendous future prospects.

REFERENCES

1. M. Faraday; in Experimental investigations in electricity, Quaritch, London, No. 1340 (1839).
2. F.W.G. Kohlrausch; Ann. Phys., 17, 642 (1882).
3. C. Tubandt and S. Eggert; Z. Anorg. Chem., 110, 196 (1920).
4. J. Frenkel; Z. Physik, 35, 652 (1926).
5. C. Wagner; Z. Phys. Chem., B22, 181 (1933).
6. W. Schottky; Z. Phys. Chem., B29, 335 (1935).
7. J.A.A. Ketelaar; Trans. Faraday Soc., 34, 874 (1938).
8. B. Reuter and K. Hardel; Nature Wissenchaften, 48, 161 (1961).
9. J.N. Bradley and P.D. Greene; Trans. Faraday Soc., 62, 2069 (1966).
10. B.B. Owens and G.R. Argue; Science, 157, 308 (1967).
11. Y.F.Y. Yao and J.T. Kummer; J. Inorg. Nucl. Chem., 29, 2453 (1967).
12. C.C. Liang; J. Electrochem. Soc., 120, 1289 (1973).
13. M.A. Ratner and A. Nitzan; Solid State Ionics, 28/30, 3 (1988).
14. H. Tuller, D.P. Button and D.R. Uhlmann; J. Non-Cryst. Solids, 40, 93 (1980).
15. J.N. Mundy and G.L. Jin; Solid State Ionics, 21, 305 (1986).
16. J. Maier; Mat. Res. Bull., 20, 383 (1985).
17. T. Jow and J.B. Wagner Jr.; J. Electrochem. Soc., 126, 1963 (1979).
18. K. Shahi and J.B. Wagner Jr.; J. Electrochem. Soc., 128, 6 (1981).
19. D.E. Fenton, J.M. Parker and P.V. Wright; Polymer ,14, 589 (1973).
20. M.B. Armand, J.M. Chabagno and M. Duclot; in Fast Ion Transport in Solids, (Ed. P. Vashisha, J.N. Mundy and G.K. Shenoy), North Holland, Amsterdam (1979).

21. C.C. Lee and P.V. Wright; Polymer, 23, 681 (1982).
22. D.R. Payne and P.V. Wright; Polymer, 23, 690 (1982).
23. A. Killis, J.F. Le Nest and H. Cheradame; Makromol. Chem. Rapid Commun., 1, 595 (1980).
24. C. Berthier, W. Gorecki, M. Minier, M.B. Armand, J.M. Chabagno and P. Rigaud; Solid State Ionics, 11, 91 (1983).
25. P.M. Blonsky, D.F. Shriver, P. Austin and H.R. Allcock; J. Am. Chem. Soc., 106, 6854 (1984).
26. D.J. Bannister, G.R. Davies, I.M. Ward and J.E. McIntyre; Polymer, 25, 1600 (1984).
27. C.V. Nicholas, D.J. Wilson, C. Booth and J.R.M. Giles; Br. Polym. J., 20, 289 (1988).
28. M.A. Ratner and A. Nitzan; Faraday Discuss. Chem Soc., 88, 19 (1989).
29. C.A. Angell; Solid State Ionics, 9/10, 3 (1983).
30. L.M. Torrel and C.A. Angell; Br. Polym. J., 20, 173 (1988).
31. H. Cheradame; in IUPAC Macromolecules (Ed. H. Benoit and P. Rempp) p. 351, Pergamon, New York (1982).
32. M. Watanabe and N. Ogata; in Polymer Electrolyte Reviews I, (Ed J.R.MacCallum and C.A.Vincent), Elsevier, London (1987).
33. A. Killis, J.F. Le Nest, A. Gandini, H. Cheradame and J.P. Cohen; Solid State Ionics, 14, 231 (1984).
34. P.G. Hall, G.R. Davies, J.E. McIntyre, I.M. Ward, D.J. Bannister and K.M.F. Le Brock; Polym. Commun., 27, 98 (1986).
35. J. Nishimoto, N. Furuya and M. Watanabe; Extended Abstracts of 62nd Meeting of Japanese Electrochemical Society, 3J12 (1995).
36. A. Vallee, S. Besner and J. Prud'homme; Electrochim. Acta, 37, 1579 (1992).
37. N. Kobayashi, M. Uchiyama and E. Tsuchida; Solid State Ionics, 17, 307 (1986).
38. D. Benrabah, S. Sylla, F. Alloin, J-Y. Sanchez and M. Armand; Electrochim. Acta, 40, 2259 (1995).
39. T. Fujinami, A. Tokimune, M.A. Mehta, G.C. Rawsky and D.F. Shriver; Chem. Mater., 9, 2236 (1997).
40. C.A. Angell, C. Liu and E. Sanchez; Nature, 362, 137 (1993).
41. M. Watanabe, S. Yamada, K. Sanui and N. Ogata; J. Chem. Soc. Chem. Commun., 11, 929 (1993).
42. H.Y. Sun, H.J. Sohn, O. Yamamoto, Y. Takeda and N. Imanishi; J. Electrochem. Soc., 146, 1672 (1999).
43. G. Feuillade and Ph. Perche; J. Appl. Electrochem., 5, 63 (1975).
44. E. Tsuchida, H. Ohro and K. Tsunemi; Electrochim. Acta, 28, 591 (1983).
45. T. Iijima, Y. Toyoguchi and N. Eda; Denki Kagaku, 63, 619 (1985).
46. M. Morita, T. Fukumasa, M. Motoda, H. Tsutsumi, Y. Matsuda, T. Takahashi and H. Ashitaka; J. Electrochem. Soc., 137, 3401 (1990).
47. B. Kumar, L.G. Scanlon and R.J. Spry; J. Power Sources, 96, 337 (2001).
48. P.S. Anantha and K. Hariharan; J. Phys. and Chem. of Solids, 64, 1131 (2003).
49. C. Capiglia, P. Mustarelli , E. Quartarone, C. Tomasi and A. Magistris; Solid State Ionics, 118, 73 (1999).
50. F. Croce, G.B. Apptecchi, L. Persi and B. Scrosati; Nature, 394, 456 (1998).
51. M.B. Armand; Adv. Mater., 2, 278 (1990).
52. Z. Gadjourova, Y.G. Andrew, D.P. Tunstall and P.G. Bruce; Nature, 412, 520 (2001).
53. M.J. Reddy and P.P. Chu; Solid State Ionics, 149, 115 (2002).
54. C.A. Vincent; Progr. Solid State Chem., 17, 145 (1987).
55. A.J. Patrick, M.D. Glasse, R.J. Latham and R.G. Linford; Solid State Ionics, 18/19, 1063 (1986).
56. J.J. Fontanella, M.C. Wintergill and J.P. Calame; J. Polym Sci. Polym. Phys. Ed., 23, 113 (1985).
57. I. Kelly, J.R. Owen and B.C.H. Steele; J. Power Sources, 14, 13 (1985).
58. L.R.A.K. Bandara, M.A.K.L. Dissanayake and B.-E. Mellander; Electrochim. Acta, 43 (10/11), 1447, (1998).
59. S. Mitra and A.R. Kulkarni; Solid State Ionics, 154/155, 37 (2002).
60. M. Forsyth, D.R. MacFarlane, A. Best, J. Adebahr, P. Jacobsson and A.J. Hill; Solid State Ionics, 147, 203 (2002).

61. Z. Wang, W. Gao, L. Chen, Y. Mob and X. Huang; Solid State Ionics, 154/155, 51, (2002).

62. W. Xu, L.M. Wang and C.A. Angell; Electrochim. Acta, 48, 2037, (2003).

63. B. Kumar and L.G. Scanlon; J.Power Sources, 52, 261, (1994).

64. P.P. Chu, M.J. Reddy and H.M. Rao; Solid State Ionics, 156, 141, (2003).

65. W.J. Lee, H.R. Jung, M.S. Lee, J.H. Kim and K.S. Yang; Solid State Ionics, 164, 65, (2003).

66. J. Zhou and P.S. Fedkiw; Electrochim. Acta, 48, 2571, (2003).

67. B. Kumar; Journal of Electroceramics, 5(2), 127, (2000).

68. J.M. Tarascon, A.S. Gozdz, C. Schmutz, F. Shokoohi and P.C. Warren; Solid State Ionics, 86, 49, (1996).

69. Y. Saito, C. Capigila, H. Yamamoto and P. Mustarelli; J. Electrochem Soc., 147(5), 1645, (2000).

70. G. Girish Kumar and S. Sampath; Solid State Ionics, 160, 289, (2003).

71. G. ̄ukowska, M. Rogowska, A. Wojda, E. Zygadlo-Monikowska, Z. Florjañczyk and W. Wieczorek; Solid State Ionics, 136/137, 1205, (2000).

72. A. Magistris, P. Mustarelli, E. Quartarone, P. Piaggio and A. Bottino; Electrochim. Acta, 46 1635, (2001).

73. H.P. Chen and J.W. Fergus; J. Mater. Sci. Lett., 21, 285, (2002).

74. K.M. Abraham, H.S. Choe and D.M. Pasquariello; Electrochim. Acta, 43(16/17), 2399, (1998).

75. Y.W. Chen-Yang, H.C. Chen, F.J. Lin, C.C. Chen; Solid State Ionics, 150, 327, (2002).

76. A. Zalewska, I. Pruszczyk, E. Sulek and W. Wieczorek; Solid State Ionics, 157, 233, (2003).

77. O. Bohnke, G. Frand, M. Rezrazzi, C. Rousselot and C. Truche; Solid State Ionics, 66, 97, (1993).

78. Y.K. Yarovoy, H.P. Wang and S.L. Wunder; Solid State Ionics, 118, 301, (1999).

79. H. Ericson, C. Svanberg, A. Brodin, A.M. Grillone, S. Panero, B. Scrosati and P. Jacobsson; Electrochim. Acta, 45, 1409, (2000).

80. M. Deepa, N. Sharma, S.A. Agnihotry, S. Singh, T. Lal and R. Chandra; Solid State Ionics, 52/153, 253, (2002).

81. Y. Wang, J. Travas-Sejdic and R. Steiner; Solid State Ionics, 148, 443, (2002).

82. A. M. Stephan and D. Teeters; Electrochim. Acta, 48, 2143, (2003).

83. Y. Saito, A. M. Stephan and H. Kataoka; Solid State Ionics, 160, 149, (2003).

84. T. Michot, A. Nishimoto and M. Watanabe; Electrochim. Acta, 45, 1347, (2000).

85. S.S. Sekhon, P. Pradeep and S.A. Agnihotry; in Solid State Ionics: Science and Technology (Eds. B.V.R. Chowdari, K. Lal, S.A. Agnihotry, N. Khare, S.S. Sekhon, P.C. Srivastava and S. Chandra), p. 217, World Scientific Publishing Co., Singapore (1998).

86. R.G. Linford; in Applications of Electroactive Polymers, (Ed. B. Scrosati), p. 1, Chapman and Hall, London (1993).

87. F.M. Gray; in Polymer Electrolytes, The Royal Society of Chemistry: Letchworth, HN, 1997

88. S. Chandra, S.S. Sekhon and N. Arora; Ionics, 6, 112 (2000).

89. S.S. Sekhon; Bull. Mater. Sci., 26(3), 321 (2003).

90. J.Y. Song, C.L. Cheng, Y.Y. Wang and C.C. Wan; J. Electrochem. Soc., 149(9), A1230 (2002).

91. H. Vogel; Z. Phys., 22, 645 (1922).

92. V.G. Tammann and H.G. Hesse; Anorg. Allg. Chem., 19, 245 (1926).

93. G.S. Fulcher; J. Am. Ceram. Soc., 8, 339 (1925).

94. J.R. MacCallum and C.A. Vincent; in Polymer Electrolyte Reviews, (Ed. J.R. MacCallum and C.A. Vincent), Elsevier, London (1987).

95. Th. Joykumar Singh, T. Mimani, K.C. Patil and S.V. Bhat; Solid State Ionics, 154/155, 21 (2002).

96. M.L. Williams, R.F. Landel and J.D. Ferry; J. Am. Chem. Soc., 77, 3701 (1955).

97. S.D. Druger, A. Nitzan and M.A. Ratner; J. Chem. Phys., 79(6), 3133 (1983).

98. M.H. Cohen and D. Turnbull; J. Chem. Phys., 31, 1164 (1959).

99. M. Duclot, F. Alloin, O. Brylev, J-Y. Sanchez and J.L. Souquet; Solid State Ionics, 136/137, 1153 (2000).

100. D. Bamford, G. Dlubek, A. Reiche, M.A. Alam, W. Meyer, P. Galvosas and F. Rittig; J. Chem. Phys., 115, 7260 (2001).

101. G. Williams; in Dielectric Spectroscopy of Polymeric Materials, Fundamentals and Applications, (Ed. J.P. Runt and J.J. Fitzgerald), American Chemical Society, Washington DC (1997).

102. M. Mierzwa, G. Floudas, P. Stepanek and G. Wegner; Phys. Rev., B 62, 14012 (2000).

103. P.B. Macedo and T.A. Litovitz; J. Chem. Phys, 42, 245 (1965).

104. S.N. Glasstone, K. Laidler and H. Eyring; in The Theory of Rate Processes, McGraw-Hill, New York (1941).

105. F.R. Blackburn, C.Y. Yang and M.D. Ediger; J. Phys. Chem., 100, 18249 (1996).

106. R. Dupon, D.H. Whitmore and D.F. Shriver; J. Electrochem. Soc., 128, 716 (1981).

107. H. Cheradame; in IUPAC Macromolecules, (Ed. H. Benoit and P. Rempp), p. 251, Pergamon, New York (1982).

108. A. Killis, J.F. LeNest, A.Gandini and H.Cheradame; J. Polym. Sci., Polym. Phys. Ed., 19, 1073 (1981).

109. S. Smedley; in The Interpretation of Ionic Conductivity in Liquids, Plenum Press, New York (1980).

110. M.H. Cohen and G.S. Grest; Phys. Rev., B 21, 4113 (1980).

111. J.H. Gibbs and E.A. DiMarzio; J. Chem. Phys., 28, 373 (1958).

112. G. Adam and J.H. Gibbs; J. Chem. Phys., 43, 139 (1965).

113. M.J. Goldstein; J. Phys. Chem., 77, 667 (1973).

114. C.A. Angell and W. Sichina; Ann. N. Y. Acad. Sci., 279, 53 (1976).

115. M.A. Ratner; in Polymer Electrolyte Reviews, (Ed. J.R. MacCallum and C.A. Vincent), Elsevier, London (1987).

116. S.D. Druger, A.Nitzan and M.A. Ratner; Solid State Ionics, 9/10, 1115 (1983).

117. C.S. Harris, A.Nitzan, M.A.Ratner and D.F.Shriver; Solid State Ionics, 18/19, 151 (1986).

118. S.D. Druger, M.A. Ratner and A. Nitzan; Phys. Rev., 331, 3939 (1985).

119. S.D. Druger, M.A. Ratner and A. Nitzan; Solid State Ionics, 18/19, 106 (1986).

120. W. Wieczorek, K. Such, S.H. Chung and J.R. Stevens; J. Phys. Chem., 98, 9047 (1994).

121. W. Wieczorek, K. Such, Z. Florjanczyk and J.R. Stevens; J. Phys. Chem., 98, 6840 (1994).

122. W. Wieczorek and M. Siekierski; J. Appl. Phys., 76, 2220 (1994).

123. J.C. Maxwell; in A Treatise on Electricity and Magnetism, Vol. 1, p. 435, 2nd Ed., Clarendon Press, Oxford (1881).

124. D.A.G. Bruggeman; Ann. Phys. (Leipzig), 24, 636 (1935).

125. R. Landuer; J. Appl. Phys., 23, 779 (1952).

126. C.W. Nan and D.M. Smith; Mater. Sci. Eng., B10, 99 (1991).

127. C.W. Nan; Prog. Mater. Sci., 37, 1 (1993).

128. M. Nakamura; Phys. Rev., B 29, 3691 (1984).

129. S. Kirpatrick; Rev. Mod. Phys., 45, 574 (1973).

130. T. DeSimone, R.M. Stratt and S. Demoulini; Phys. Rev. Lett., 56, 1140 (1986).

131. Z. Florjanczyk, W. Krawiec, D. Greszta, W. Wieczorek and M. Siekierski; Bull. Electrochem., 8, 524 (1992).

132. J. Przyluski, M. Siekierski and W. Wieczorek; Electrochim. Acta, 40, 2101 (1995).

133. J. Przyluski, W. Wieczorek and Z. Florjanczyk; in Solid State Ionics: Materials and Applications, (Ed. B.V.R. Chowdari, S. Chandra, S. Singh and P.C. Srivastava), p. 209, World Scientific Publ. Co., Singapore (1992).

134. J. Bares; Macromolecules, 8, 244 (1975).

135. H.A. Schneider and E.A. DiMarzio; Polymer, 33, 3453 (1992).

136. M.C. Wintersgill, J.J. Fontanella, J.P. Calme, M.K. Smith, T.B. Jones, S.G. Greenbaum, K.G. Adamic. A.N. Shetty and C.G. Andeen; Solid State Ionics, 18/19, 326 (1986).

137. M. Watanabe, K. Sanui, N. Ogata, T. Kobayashi and Z. Ohtaki; J. Appl. Phys., 57(1), 123 (1985).

138. K.M. Abraham, Z. Jiang and B. Carroll; Chem. Mater., 9(9), 1978 (1997).

139. J.Y. Song, Y.Y. Wang and C.C. Wan; J. Electrochem. Soc., 145, 1207 (1998).

140. K.M. Abraham, M. Alamgir and D.K. Hoffman; J. Electrochem. Soc., 142, 683 (1995).

141. G.B. Appetecchi, F. Croce, A. DePaolis and B. Scrosati; J. Electroanal. Chem., 463 248 (1999).

142. J.Y. Song, Y.Y. Wang and C.C. Wan; J. Electrochem. Soc., 147(9), 3219 (2000).

143. P.G. Bruce; in Polymer Electrolyte Review-I, (Ed. J.R. MacCallum and C.A. Vincent), p. 42, Elsevier, New York (1987).
144. M. Morita, F. Tachihara and Y. Matsuda; Electrochim. Acta, 32, 299 (1987).
145. A.M. Stephan, Y. Saito, N. Muniyandi, N.G. Renganathan, S. Kalyanasundaram and R.N. Elizabeth; Solid State Ionics, 148, 467 (2002).
146. H. Cheradame and P. Niddam-Mercier; Faraday Discuss. Chem. Soc., 88, 77 (1989).
147. M.A. Ratner and D.F. Shriver; Mater. Res. Soc. Bull., 14(9), 39 (1989).
148. K.M. Abraham; in Applications of Electroactive Polymers, (Ed. B. Scrosati), p. 75, Chapman & Hall, London (1993).
149. S. Rajendran, T. Uma and T. Mahalingam; Euro. Polym. J., 36, 2617 (2000).
150. Z. Wen, T. Itoh, M. Ikeda, N. Hirata, M. Kubo and O. Yamamoto; Journal of Power Sources, 90, 20 (2000).
151. H. Wang, H. Huang and S.L. Wunder; J. Electrochem. Soc., 147(8), 2853 (2000).
152. K. Hayamizu, Y. Aihara and W.S. Price; J. Chemical Physics, 113(11), 4785 (2000).
153. A. Ferry, M.M. Doeff and L.C. DeJonghe; Electrochim. Acta, 43, 1387 (1998).
154. J.B. Wagner and C.J. Wagner; Chem. Review, 26, 1597 (1957).
155. S. Chandra; in Superionic Solids-Principle and Application, NHPC, Amsterdam (1981).
156. S. Abbrent, J. Plestil, D. Hlavata, J. Lindgren, J. Tegenfeldt and A. Wendsjo; Polymer, 42, 1407 (2001).
157. C.J. Leo, G.V. Subba Rao and B.V.R. Chowdari; Solid State ionics, 148, 159 (2002).
158. C.S. Kim and S.M. Oh; Electrochim. Acta, 45, 2101 (2000).
159. D.L. Pavia, G.M. Lampman and G.S. Kriz; in Introduction to Spectroscopy, Harcourt college Publ., USA (2001).
160. R.M. Silverstein, G.C. Bassler and T.C. Morrill; in Spectroscopic Identification of Organic Compounds (Fifth Ed.), John Wiley & Sons. Inc, USA (1991).
161. C.S. Kim and S.M. Oh; Electrochim Acta, 46, 1323 (2001).
162. A.M. Stephan and Y. Saito; Solid State Ionics, 148, 475 (2002).

14

Specialty Coatings and Adhesives

Sharif Ahmad[*] **and Ufana Riaz**

*Materials Research Laboratory, Department of Chemistry,
Jamia Millia Islamia, New Delhi—110025, India.
Corresponding author's email: sharifahmad_jmi@yahoo.co.in

Contents

Summary: The paints and coatings industry is a mature industry that has been undergoing a rapid change in technology. Considering the finite nature of traditional petrochemical derived compounds in addition to economic and environmental considerations, the key goal of the industrialists will be the utilization of sustainable raw materials for the development of polymers so as to replace the polymers derived from petrochemicals. In this chapter, specific areas in solvent-borne coatings, water-borne coatings, UV radiation curable coatings, that are ripe for future innovations and further development are discussed with special emphasis towards the recently developed sustainable resource based coating materials. The structure and reactivity associated with the alkyd, polyester, acrylic, amino, epoxy, and polyurethane resins commonly used in coating formulations are described with preparation, performance, and cure reactions.

While adhesives play an important role in technological world as well as day-to-day life. In view of its significance this chapter covers a brief history of development of adhesives, various primitive and advanced adhesives such as epoxy, phenol formaldehyde, urea formaldehyde with special emphasis on conductive adhesives, water-borne adhesives and sustainable resource based adhesives as well as their mechanisms and characteristic properties have been discussed.

1. SPECIALTY COATINGS

1.1. Introduction

The technological advancements in the industrial sector across the world have intensified the demand for high performance coatings during the past few decades. Increasing labour cost, service life expectations as well as environmental compulsions have expedited the development of several specialty coatings based on molecular engineering, which are increasingly being popularized owing to their superior performance, cost effectiveness, and ease of application. Growing public awareness on safety, health and environment have compelled the government to adopt strict regulations for controlling the utilization of hazardous air pollutants, heavy metals as well as waste minimization. This has posed a significant challenge before the coating technologists and urged them to evolve eco-compliant high performance coating materials.

Specialty polymeric coatings can be categorized as a special class of coatings that are engineered and designed to provide aesthetic sense and high performance surface protection under specific environmental conditions. The characteristic features of these coatings are:
- Their ability to withstand stress strain exerted by different kinds of exposure in aggressive environments.
- To protect the metallic or concrete structure from corrosion and erosion.
- Aesthetic retention.

While adhesives form an integral part of a variety of products and coatings that offer challenges to create new and complex designs of materials. The significance of adhesive bonding as structure-joining technology is increasing because of its numerous advantages compared with other joining methods, being used in many industries like automotive and aircraft construction [1,2]. In general, an adhesive is defined as a material which when applied to the surfaces of materials can join them together and resist separation. Classification of adhesives is done either by their chemical nature or by their mode of applications. For a substance to be adherent it must:
- Completely wet the surface, that is, it must spread and make a contact angle approaching zero.
- Form an intimate contact between the molecules of the adhesive and the atoms and molecules in the surface.
- Must harden into a cohesively strong solid by chemical reaction, loss of solvent or water or by cooling.

The strongest adhesives are known to solidify by chemical reaction or by some physical interactions.

1.2. Glimpse of Upcoming Coating Industry

The coating manufacturing companies worldwide are passing through an era of convergence in which most of the companies have practically equalized their position in terms of price, quality and service expectations of customers. Rapid development of polymer science, synthetic pigments and additives has enabled paint and coating formulations to grow at a much faster pace [3]. Some of the emerging polymeric coating formulations that are technologically more challenging as well as significant are given in the following subsections.

1.2.1. New Generation Hybrid Coatings

Organic polymers generally suffer from certain drawbacks such as poor flexibility, poor strength, lack of temperature resistance etc. To overcome such drawbacks of the individual polymers, hybrid coatings are formed by the combination of the synergistic properties of two polymers via grafting, blending or by the formation of interpenetrating polymer network. Types of the organic as well as inorganic modifications carried out in the form of hybrid coatings are given in **Tab. 14.1**.

Tab. 14.1. Types of organic modifications in hybrid coatings.

Product	*Organic modifications*	*Inorganic*
Epoxy-polysiloxane	Hydrogenated epoxy	Siloxane
Acrylic epoxy polysiloxane	Acrylic epoxy	Amino silane
Epoxy-polysiloxane	Hydrogenated	Amino silane
Acrylic polysiloxane	Acrylic	Amino silane
Urethane-polysiloxane	Urethane acrylate	Amino silane

1.2.1.1. Modifications Based on Epoxy Resins

Epoxy resins have widened the horizon beyond expectations owing to their high performance, durable coating and ease of application. They can be chemically classified into:

    (a) Bisphenol—A epoxy resins.
    (b) Bisphenol—F epoxy resins.
    (c) Modified epoxy resins.

Major disadvantages of epoxy resins are low fracture energy, low hydrophobicity, poor weathering as well as thermal resistance and low impact resistance, which limit its commercial utility. The organic as well as inorganic modification of epoxy offers diverse and unique properties such as good processibility, flexibility, durability, toughness and high thermo-oxidative stability to the resin, along with good abrasion resistance, scratch hardness, chemical resistance, weatherability, and UV resistance.

1.2.1.2. Epoxy Acrylates

They are produced by heating together low molecular weight epoxy resins with acrylic acid/methacrylic acid at 120°C [4]. They are mainly used for fabrics or fiberglass mats etc. The epoxy-acrylate resins are cured by melamine formaldehyde. The presence of acrylate and melamine in epoxy resins enhances the physico-mechanical and corrosion resistance properties. The performance (**Tab. 14.2**) shows that on increasing the loading of melamine formaldehyde (MF), a progressive increase in the scratch hardness values is observed. The coatings of the optimum composition show scratch hardness of 4.8 kg. The performances of the coated mild steel samples in salt mist chamber and humidity environments show that no appreciable weight loss is observed except slight loss in gloss after 650 h and 280 h respectively. The samples show high UV resistance owing to the non-availability of tertiary hydrogen at the α position of the carbonyl group in epoxy-methacrylate (EMAC). These coatings also show improved resistance in aqueous and acidic environments. The presence of ester linkages in the polymer chain lends the poor performance of these coatings in alkaline environment. The superior performance of EMAC/MF coatings over epoxy-acrylate (EAC) EAC/MF coatings can

Tab. 14.2. Film properties of the EAC/MF and EMAC/MF resin systems. Reproduced with permission from S. Ahmad *et al.*; J. Appl. Polym. Sci., 95, 494 (2005). Copyright © 2005, John Wiley and Sons Ltd.

Resin code[a]	Scratch hardness (kg)	Impact resistance (lbs/in.)	Gloss at 45°	Bending mm/diameter)	Humid resistance (h)	Salt spray (h; 3.5 wt% at 35.5°C)	QUV weatherometer (h)	Chemical Resistance Test[b]		
								HCl (5 wt%, 10 days)	H_2SO_4 (2 wt%, 10 days)	NaOH (5 wt%, 10 days)
EAC/MF-10	2.7	150	60	5.2	360	180	12	A	B	C
EAC/MF-20	3.6	150	60	6.8	480	192	14	A	B	C
EAC/MF-30	4.8	150	64	7.1	600	216	14	A	A	A
EAC/MF-40	5.0	150	64	9.6	612	204	14	B	B	A
EMAC/MF-10	2.9	175	60	4.4	504	192	13	A	B	C
EMAC/MF-20	3.78	200	70	5.0	614	216	15	A	A	B
EMAC/MF-30	5.0	200	70	5.6	658	228	17	A	A	A
EMAC/MF-40	5.0	200	70	6.0	652	222	15	A	A	B

[a]The last digit of the resin code indicates the weight percentage of MF.
[b]A = unaffected; B = loss in gloss; C = loss in gloss; D = film cracked at partially removed.

be attributed to the presence of pendant methyl group, which protects the ester linkages from hydrolysis [5].

1.2.1.3. Epoxy-siloxane Hybrid Coatings

Epoxy-siloxane hybrid coatings are often known as *"Polysiloxane"* or *"Engineered Polysiloxane"* coatings. They exhibit unique blend of the properties of organic inorganic binders in an interpenetrating polymer network with outstanding corrosion resistance and weatherability. The use of inorganic siloxane polymer in the backbone and the presence of highly stable, hydrolysis resistant Si-O-Si linkage in the network makes such coatings far superior than the conventional organic polymers based on C-O-C linkages usually present in epoxies and polyurethanes in terms of corrosion resistance, heat resistance, chemical resistance and weatherability [6]. Due to the exceptionally higher dissociation energy, Si-O-Si linkages are not degraded easily which accounts for its exceptional weatherability and aesthetic retention abilities showing an outstanding performance under alkaline media; the coated panels are found to be unaffected in the presences of NaOH (10 wt%) for 325 hrs. This can be attributed to the hydrophobic nature of the silicones and methyl groups, which hinders the ingress of water molecules. Highly cross-linked network prevents the easy penetration of chemicals thereby enhancing the anticorrosive performance of these systems [7].

Their anti-dust property is attributed to the low surface tension of siloxanes and outstanding surface properties of siloxane-containing polymers. Polysiloxane coatings conform to all *VOC* and *HAP* regulations and are free from isocyanides [8]. Hence, these products can serve as an effective barrier coat against epoxy intermediate and polyurethane topcoat based systems. They are nowadays recommended for structural steel, power plants, offshore industry, concrete walls, flooring, anti-graffiti coatings etc. as they exhibit the following valuable properties:

- Self-priming coating.
- Excellent corrosion resistance.
- Excellent chemical resistance.

- Excellent color and gloss retention.
- High volume solid.
- Low temperature cure.
- Compliance to HAPS and VOC regulations and isocyanine free.
- Tolerance to humidity during drying.
- High build application by standard equipment.

In these coating systems, the siloxane behaves as chain entanglements rather than as a flexibilizer. The long, mobile siloxane chains assist in the formation of a highly cross-linked network between the two (epoxy siloxane) ES molecular units **(Scheme 14.1)** through either of the amine curatives. As discussed earlier, the free amino groups may further react with the free pendant and terminal hydroxyls as well as a terminal epoxide-ring **(Scheme 14.1)**, leading to the formation of a highly cross-linked network and resulting in a tremendous increase in the scratch hardness, impact resistance, and bending test values **(Tab. 14.2)**. The gloss values were higher for 1,6-diamino hexane modified epoxy siloxane (ES–AH) and 1,3-diamino propane modified epoxy siloxane (ES–AP) systems than for 1,6-diamino hexane modified epoxy (E–AH) and 1,3-diamino propane modified epoxy (E–AP) systems for similar reasons; the highest value was achieved for ES–AP because of a highly cross-linked and dense structure. The systems were subjected to an anticorrosive performance evaluation under the influence of various corrosive chemicals. The ES (ES–AH and ES–AP) systems coated specimens showed superior performance in acidic as well as saline environments in comparison with E–AH and E–AP [9]. The ES–AP particularly showed an outstanding performance under alkaline media; the coated panels remained unaffected in the presence of NaOH (10 wt%) for 240 hrs [9]. The siloxane-incorporated systems also remain unaffected in aqueous environment for 45 days. This can be attributed to the hydrophobic nature of the silicones and methyl groups [10]. The highly cross-linked network of ES hinders the easy penetration of chemicals into the paint system and thus enhances the anticorrosive performance of the ES–amine coatings over plain E–amine coatings, particularly the ES–AP system.

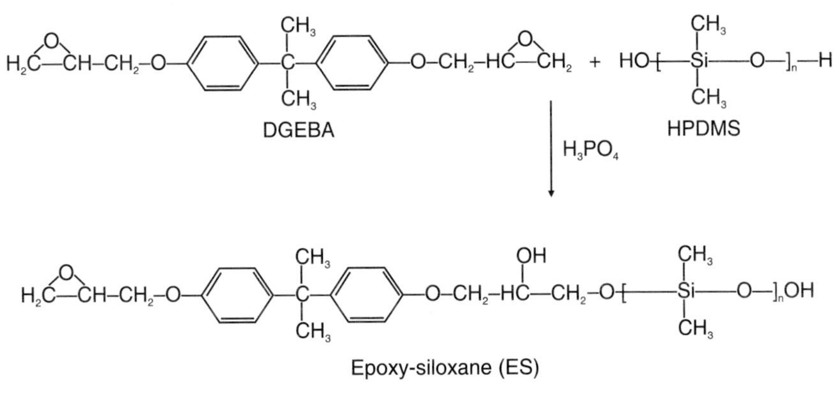

Scheme 14.1. Epoxy-siloxane hybrid coatings. Reproduced with permission from S. Ahmad *et al.*; Prog. Org. Coat., 54, 248 (2005). Copyright © 2005, Elsevier Science Ltd.

1.2.2. Antifouling Coatings

Antifouling paints and coatings have been used for centuries to prevent the attack of marine organisms. Such coatings incorporate certain toxic compounds such as cuprous oxide, organotins that are toxic to most forms of fouling. Though the use of toxins to take care of fouling growth is being practiced for over 200 years, the environmental authorities are pressurizing the marine industries to discontinue the usage of toxins under the coatings regulation act [11]. Some of the options available to the coating manufacturers which are currently being used as effective substitutes for tin and copper based coatings are discussed below:

1.2.2.1. Anti-fouling Coatings Based on Ablative Resins

Ablative resins are synthesized from polymers having labile functional groups such as esters. Hydrolysis of such molecules creates a hydrophilic site on the polymer backbone, which contributes to the antifouling performance through its poisonous properties [12-14]. When a sufficient amount accumulates on the surface, it becomes water-soluble and washes out all the diatoms and algae including their juvenile forms.

1.2.2.2. Fluoropolymer Based Coatings

Coatings based on fluoropolymers achieve their effectiveness through bond shearing between fouling and coating. This is accomplished by constructing a well organized surface of closely packed fluorinated groups driving its surface energy as low as possible thereby stabilizing the surface so that it resists rearrangement and penetration of marine adhesives [14].

1.2.2.3. Silicone Based Coatings

Coatings based on silicones have also been tested extensively for resistance to marine fouling. The most effective route to fouling free performance of silicone coatings is to encourage the release of fouling organisms from the surface by predisposing the adhesive joint between foulant and the coating to quick and easy failure through peeling [15]. All currently available products based on silicone contain fugitive oils, which have minimal environmental impact when these oils leach from the coating. The maximum service life of these coatings is approximately two years.

1.2.3. Eco-friendly Coatings

The progress of industrial development in the world has brought increased pollution of atmosphere with volatile organic substances. In context of the growing concern for environment, attempts have been made by the coating industries to reduce the consumption of solvents in coatings formulation [16]. Since the emission of Volatile Organic Compound (VOC) is a major cause of the environmental pollution, intensive studies have been going for the formulation of low VOC or zero VOC eco-friendly coating systems given in the forthcoming subsections.

1.2.3.1. Water-borne Coatings

Water-borne coatings have gained much popularity in recent years owing to their low VOC and find applications in architectural coatings, can coatings as well as industrial coatings. They can be classified as:

1.2.3.1.1. Aqueous dispersions coatings

They are characterized by large particle size (>0.3 μm) whereas water reducible coatings are characterized by smaller particle range (<0.1 μm). Water-borne polyurethanes have several advantages over solvent-borne polyurethane coatings in terms of:
- Higher toughness.
- Better abrasion resistance.
- Superior chemical resistance.
- Greater hardness.
- Light stability.
- Good elasticity.
- Polyurethane dispersions.

They also exhibit better performance as well as zero toxicity. These systems are one-component products and require no "on site mixing". They can be tailor-made to suit most applications and can be blended with alkyd, vinyl or acrylic emulsions and can also be modified with fatty acids to improve the coating performance of these systems [17].

1.2.3.1.2. Epoxy water-borne coating systems

Epoxy resins in general are not water-soluble but are modified technically with acrylic polymers to make their water-borne coatings [16,17].

1.2.3.2. Powder Coatings

Powder coatings are defined as coatings formed by the application of coating powder to a substrate and fused into continuous films by the application of heat or radiant energy [19]. They contain finely divided organic polymer, which contain pigments, fillers as well as additives. It is essential that powder-coating compositions must remain in free flowing and finely divided state for a reasonable period after they are manufactured and packaged. Resins used for the manufacturing the powder coatings are epoxy and polyester resins. Powder coatings can be categorized as:

1.2.3.2.1. Thermoplastic powder coatings

They undergo curing during stoving of the coating. The film formation is a result of melting and sintering of the powder particles using surface tension [20]. The range of the molecular weight of polymers used in powder coating lies in between 20,000-2,00,000 Dalton atomic units.

1.2.3.2.2. Thermosetting powder coatings

They are based on low molecular weight resins and undergo cross-linking as well as curing at elevated temperatures [20]. The curing agent employed in such coatings is of high functionality and low molecular weight. Advantages of thermosetting over thermoplastic powder coatings include the availability of a wide range of polymers and curing agents with:

- Excellent film performance.
- Excellent adhesion to substrate.
- Low manufacturing costs.
- Ease of application.

1.2.3.3. Radiation Cure Coatings

The growing demand of solvent-less coating for the protection of structural steel is attributed to the benefits offered by such coatings with respect to application safety, environmental control, enhanced productivity improvement and performance benefits [21]. Some of the salient features of such coatings are the:

- Capability to produce high film builds in single coat or fewer numbers of coats.
- Less possibilities of coating failure due to absence of interlayer contaminants, solvent entrapment and inter-coat adhesion problem when applied in single coat.
- Minimal volume shrinkage.
- More amenable to application in confined areas like ballast tank and interlining of tanks.
- Minimum disruption and down time curing after application.
- Absence of VOC ensures ecological and workers safety.
- Reduced application cost, labor cost and automated application.

The use of low viscosity polymers such as epoxy, PU, vinyl esters and development of solvent free hardeners and reactive diluents offer wide formulation flexibilities to the coating technologists to develop products for tailor made applications [21]. Epoxy coatings are used as high performance internal tank lining for ballast tanks, crude, solvent, acid and water storage tanks. Polyurethane is extensively used in protection of pipelines, tanks, vessels, bridges, concrete containment areas and engineered structures. The main advantage of solvent-less polyurethanes over solventless epoxies is the outstanding abrasion and impact resistance coupled with higher flexibility having pot life between 5-40 mins. The hardeners used in solvent-less polyurethane are oligomeric MDI adduct which is ecologically safe because of the absence of free isocyanate which has a nonvolatile nature [21].

1.3. Upcoming Strategy: "Plant/Crop-based Renewable Resources, 2020 Vision"

The production, modification, and processing of polymers is most important to coating industry. The world production of polymers has now by far exceeded the production of steel (by weight) and was close to 180 million tons in the year of 2000. Europe has a share of *ca.* 28 percent (50 million tons) equivalent to a production value of *ca.* 100 billion Euro [22].

Chemistry, engineering, physics, and geology are just a few of the sciences that have been applied in the petrochemical industry to impact our lives in ways that were difficult to imagine just 50 years ago. This industry has been very successful in creating a range of products: from high performance jet fuel to basic building blocks and petropolymers such as polypropylene, styrene, acrylonitrile, polyvinylidene chloride, and polycarbonate. The petrochemical industry is capital intensive and has built a considerable infrastructure to handle and process fossil fuels. The United States uses approximately 13.9 million barrels per day of hydrocarbon inputs, mostly for various types of fuel. On the other hand, the fact that fossil fuel resources are finite cannot be disputed. It may be more important to consider the potential for price sensitivity as supply peaks, rather than to debate a theoretical time point when the oil will run out. Supplementing the use of petrochemicals with renewable resources in more than minor volumes must start soon. The research has started to accomplish immediately. Irrespective of the debate on the timing of a supply-side decline in fossil fuels, demand continues as the population expands and standards of living in the emerging nations

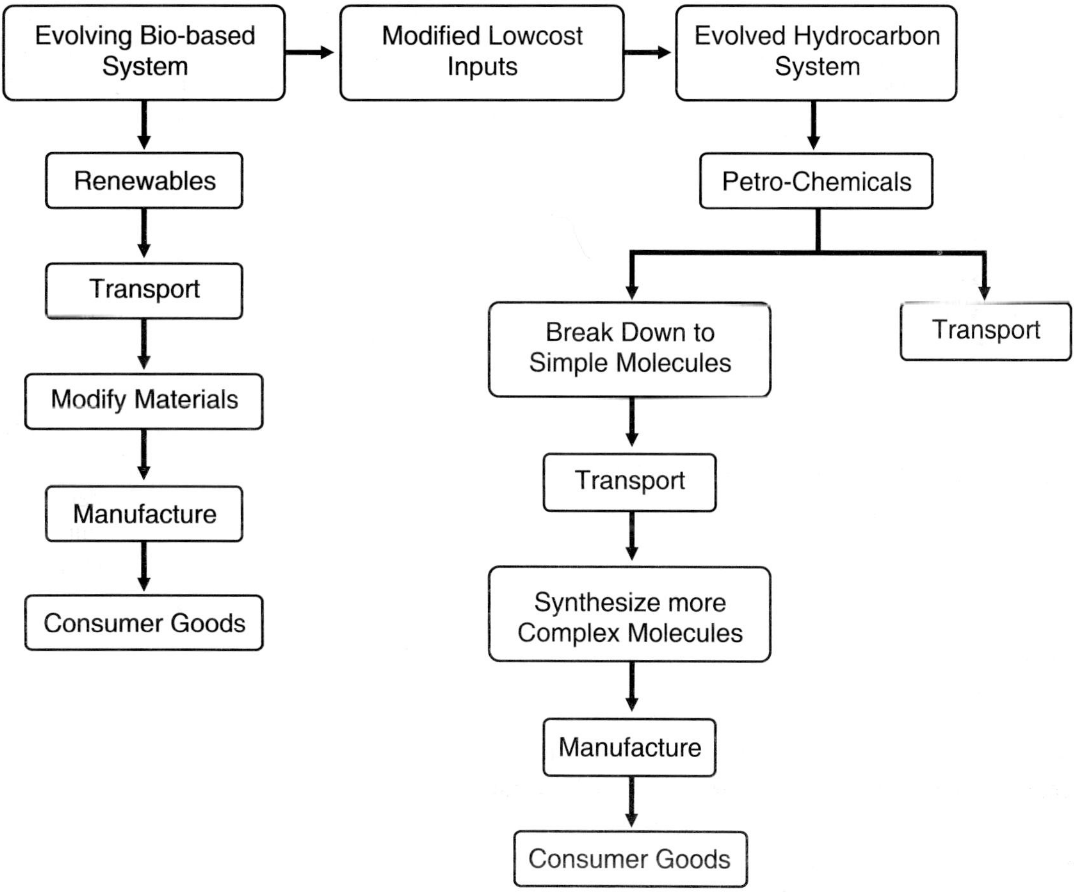

Fig. 14.1. Plant/Crop-based Renewable Resources 2020: A Vision to Enhance U.S. Economic Security Through Renewable Plant/Crop-Based Resource use can be found on the Internet at: http://www.oit.doe.gov/agriculture.

increase. It is projected that long before renewable resources become a replacement for fossil fuels, they will become necessary as a supplement. Thus, for any one of several reasons, it is important that attention should be devoted to the development of a renewable resource base for industrial raw materials particularly coatings industry.

Original *"Plant/Crop-based Renewable Resources 2020"* vision is called for simultaneous progress along a technology "front" with directed focus on limiting "hot spots" as required. Focusing on the appropriate research with correct timing and with communication to the broad but related points in the system is required with coordinated direction. Currently, plants/crops are used for biomass and a range of raw materials such as starch, protein, fatty acids, and isoprene-based compounds. Forestry is found to be a major contributor of the renewable resource based materials such as pulp and paper. Soybeans are used in printing inks and paints. Corn enters several industrial use channels via the wet mill fermentation process. However, the relative volume contribution remains low. A new opportunity has been presented by genetic engineering, which promises to allow metabolic manipulation to produce desired functional materials (**Fig. 14.1**).

1.3.1. Coatings Based on Sustainable Resource: Development of Anti-Corrosive Coatings Based on Seeds Oils

The synthesis of different polymers from oil is economically, academically, scientifically and environmentally significant because of their environment friendly nature, low cost, abundance and possible biodegradability. Vegetable oil has a number of excellent properties that could be utilized in producing a number of valuable polymeric materials [23].

Vegetable oils may be employed as a sustainable resource to replace chemicals and their products obtained from petroleum feed stock and also to expand the existing range of chemicals and raw materials, which can lead to the development of novel products like polymers [24]. A number of low molecular weight polymers such as alkyd, polyurethane, epoxies, polyetheramide and polyesteramide have been developed from vegetable seeds oils and are discussed in the proceeding section.

1.3.1.1. Polyesteramide Based Anticorrosive Coatings

Vegetable oil based polyesteramide is an amide-modified alkyd that has superior characteristics over normal alkyds in terms of hardness, ease of drying and water vapor resistance [25,26]. The resin shows satisfactory chemical resistivity and durability but it lacks alkali resistance. The high melting point and high curing temperature cause difficulties in their application as coating materials. The reaction **Scheme 14.2** for the synthesis of polyesteramide is given below:

To achieve the desirable properties and overcome the drawbacks of polyesteramide, various modifications have been carried out through the incorporation of alumina, urethane, butylated melamine formaldehyde (BMF) and poly (styrene-co-maleic anhydride) in the backbone of the polyesteramide [27]. It is reported that the incorporation of metals/metalloids and organic moieties like vinyl, ether, urethane, poly (styrene-co-maleic anhydride) to polyesteramide reduces the drying time and curing temperature as well as enhances thermal stability, scratch hardness and protective efficiency [27].

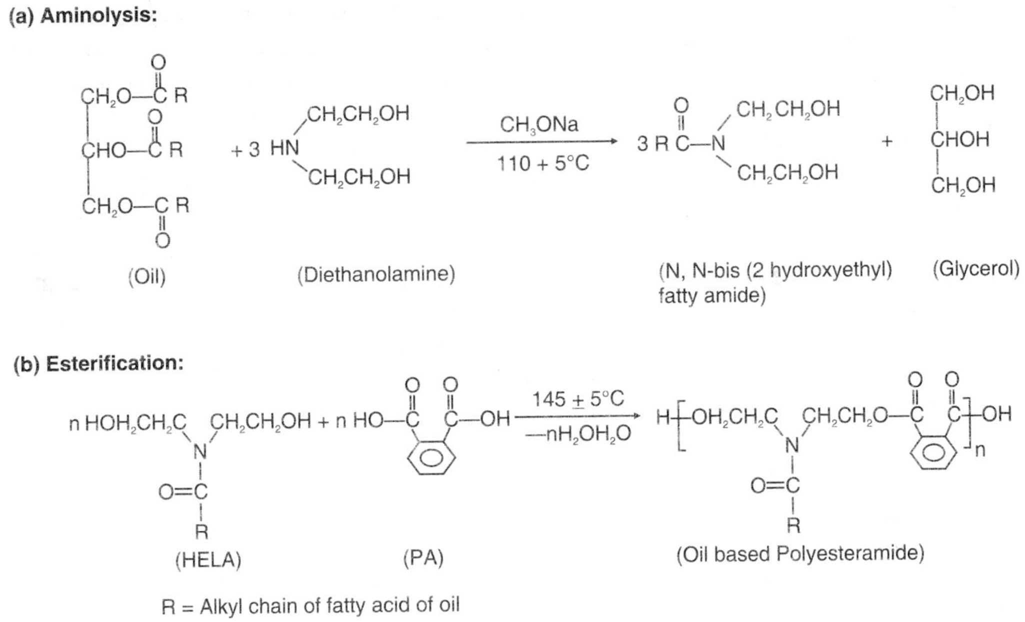

(a) Aminolysis:

$$\text{(Oil)} + 3\ \text{HN(CH}_2\text{CH}_2\text{OH)}_2 \xrightarrow[110 \pm 5°C]{CH_3ONa} 3\ R\text{C}-\text{N(CH}_2\text{CH}_2\text{OH)}_2 + \text{Glycerol}$$

(Oil) (Diethanolamine) (N, N-bis (2 hydroxyethyl) fatty amide) (Glycerol)

(b) Esterification:

$$n\ \text{HOH}_2\text{CH}_2\text{C}\cdots\text{CH}_2\text{CH}_2\text{OH} + n\ \text{HO}-\text{OC}\cdots\text{CO}-\text{OH} \xrightarrow[-nH_2O]{145 \pm 5°C} \text{Oil based Polyesteramide}$$

(HELA) (PA) (Oil based Polyesteramide)

R = Alkyl chain of fatty acid of oil

Scheme 14.2. Synthesis of polyesteramide.

1.3.1.2. Alumina-Incorporated Polyesteramide Coatings

To improve the physico-mechanical and chemical resistance properties as well as lower the curing temperature of vegetable seeds oils based polyesteramides, their respective alumina-incorporated polyesteramides resins have been synthesized [28]. The resins and their coatings have been tested for their chemical, physico-mechanical and chemical/corrosion resistance properties. Alumina incorporated polyesteramide exhibits far improved properties than plain polyesteramide. The reaction scheme for alumina incorporated polyesteramide is given in the **Scheme 14.3**.

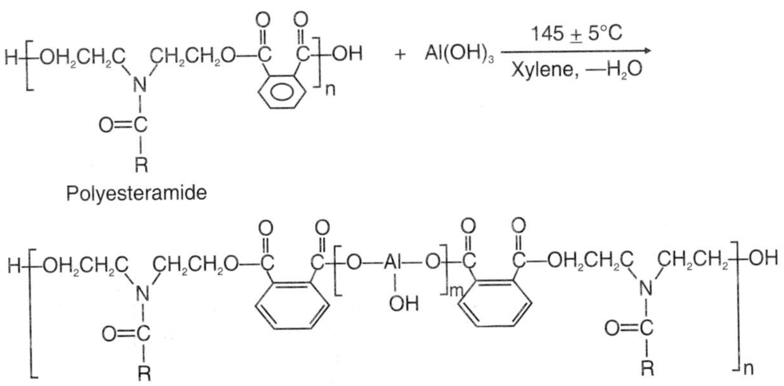

$$\text{Polyesteramide} + \text{Al(OH)}_3 \xrightarrow[\text{Xylene, } -H_2O]{145 \pm 5°C}$$

Polyesteramide

Scheme 14.3. Synthesis of alumina incorporate polyesteramide. Reproduced with premission from Ahmad *et al.;* **J. Macromal. Sci., PAC, Part A, 43(9), 1409 (2006). Copyright © 2006, Taylor and Francis Group LLC.**

The presence of alumina in polyesteramide resin enhances the corrosion protective efficiency of their coating in different corrosive environment viz. aqueous, alkaline, acidic and saline medium over those of *Annona Squamosa* oil based polyesteramide (ASPEA), *Pongamia Glabra* oil based polyesteramide (PGPEA), Linseed oil based polyesteramide (LPEA) (**Tab. 14.3**). Coatings of such resins are found to be unaffected for 240 h in aqueous, acidic as well as saline environments while in alkaline medium the Al-ASPEA and Al-LPEA show slight cracks and films are partially removed in two hours whereas Al-PGPEA exhibits only a loss in gloss during this period and the cracks in film begin after this period. These properties can also be correlated to the incorporation of alumina in the backbone of polyesteramide chains, higher flexibility and greater linearity of the chains and optimum cross-linking in the resin with higher oleic acid content. The hydroxyl group on aluminum in the polymer chains is expected to provide good adhesion between the polymeric coating and metal surface. Hydroxyl groups oriented away from the substrate surface would also provide the extra cross-link site, which can improve the anti-corrosive properties.

Tab. 14.3. Physico-chemical, mechanical and chemical/corrosion resistance properties of ASPEA-46, PGPEA-71, Al-LPEA-22, Al-ASPEA-46 and Al-PGPEA-71 resins. Reproduced with premission from S. Ahmad et al.; J. Macromal. Sci., PAC, Part A, 43(9), 1409 (2006). Copyright © 2006, Taylor and Francis Group LLC.

Resin code*	ASPEA-46[19]	PGPEA-71[17]	Al-LPEA-22[33]	Al-ASPEA-46	Al-PGPEA-71
PHYSICO-CHEMICAL ANALYSIS					
H.V. (%)	6.02	5.08	14.98	13.05	10.64
S.V. (mg KOH/g)	132.00	106.00	118.00	120.00	96.00
I.V. (g I_2/100 g)	25.00[1]	28.00	20.00	13.00	15.00
Sp. Gravity (g/ml)	0.958	1.012	0.976	1.086	1.043
Refractive index	1.507	1.524	1.543	1.543	1.560
PHYSICO-MECHANICAL PROPERTIES					
Baking temp.** (°C)	220	210	200	165	165
Baking time** (min)	30	20	10	15	15
S.H. (Kg)	3.0	3.5	2.5	4.0	4.5
I.R. (lb/inch)	100	100	150	200	250
Gloss at 60°C	90	90	135	150	175
Bending (1/8 inch)	Passes	Passes	Passes	Passes	Passes
Abrasion resistance (wt loss in mg after 500 cycle)	0.7	0.7	—	0.3	0.3
CHEMICAL/CORROSION RESISTANCE					
H_2O (10 days)	e	e	e	e	e
NaOH (5%) 2 hrs	a	a	b[2]	b	c
HCl (5%) 10 days	d[3]	d[3]	e	e	e
NaCl (3.5%) 10 days	d[4]	d[5]	e	e	e

*Last digit indicates wt% oleic acid in the oil. The coating passes adhesion test without showing any visible damage. H.V.: hydroxyl value; S.V.: saponification value; I.V.: iodine value; Sp. gravity: specific gravity; S.H.: scratch hardness; I.R.: impact resistance; a: film completely removed; b: film slightly cracked and partially removed; c: loss in gloss; d: slight loss in gloss; and e: unaffected, 1-after revised analysis, 2-after 1 h., 3-after 48 h., 4-after 2 h., and 5-after 72 h.

**Optimum baking time and temperature, which resulted in best coating properties.

1.3.1.3. Poly(styrene-co-maleic anhydride)-modified Polyesteramide-based Anticorrosive Coatings

Poly(styrene-*co*-maleic anhydride), a bifunctional acrylic copolymer, is capable of reacting with alcohols, amines, and other groups to produce many derivatives because of the anhydride groups in its backbone, is used as a modifier of latex paints to increase adhesion and gloss in the resin of low baking temperature and time [27]. The replacement of phthalic anhydride by maleic anhydride in the case of alkyds has leads to shorter stoving time and alkyds of better color and drying properties [27]. The loading of Styrene-co-maleic anhydride polyesteramide (SCPEA) copolymer on polyesteramide (PEA) backbone leads to a considerable decrease in baking temperature of the SCPEA coatings. The coatings exhibit a high degree of gloss with higher copolymer loading. The reaction scheme for the formation is given in **Scheme. 14.4.**

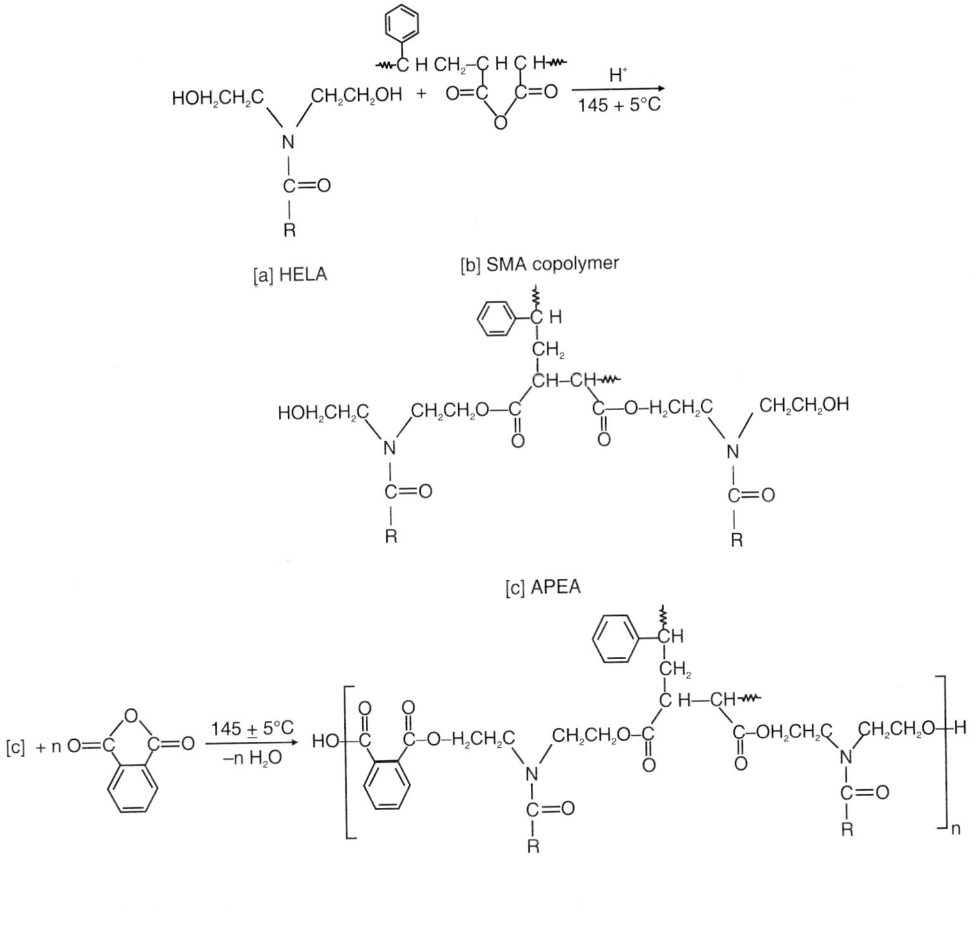

Scheme 14.4. Synthesis of air drying polyesteramide (APEAPh), R-alkyl chain of fatty acid of oil. Reproduced with permission from Zafar *et al.*; Prog. Org. Coat., 51, 250 (2004). Copyright © 2004, Elsevier Science Ltd.

The coatings of all the compositions show good adhesion to the metal substrate due to the presence of anhydride groups. The chemical resistance of SCPEA coatings is better than those of plain PEA (**Tab. 14.4**). The coatings of SCPEA are unaffected under aqueous environment. A considerable resistance to alkali (5 wt% NaOH, 3 h) is observed in the resin with optimum amount of styrene, whereas the plain PEA coatings get completely removed within 1 h. Among all compositions, resin with 15 wt% copolymer loading i.e. SCPEA-15 showed the highest resistance in acidic environment such that only a slight loss in gloss after 240 h is observed. Above 15 wt% copolymer loading, exhibit a deterioration in coating properties, in particular, the acid and alkali resistance, which can be correlated to the strain developed in polymeric chain, which adversely affects both the physico-mechanical and the anticorrosive behavior of the coatings.

Tab. 14.4. **Physico-mechanical and anticorrosive properties of APEA and APEAPh. Reproduced with permission from F. Zafar *et al.*; Prog. Org. Coat., 51 250 (2004). Copyright © 2004, Elsevier Science Ltd.**

| | | | | | | | CHEMICAL RESISTANCE | | | |
Resin code*	Drying time (min)	Scratch hardness (kg)	Impact resistance passes	Gloss at 45°C	Bending 1/8 in.	H_2O (10 days)	NaOH (5%) (2 h)	HCl (2%) (10 days)	NaCl (3.5%) (10 days)	Xylene (10 days)
APEA	20	2.5	150	65	Passes	b	f	e	c	f
APEAPh-1	20	3.0	150	67	Passes	d	a	c	d	f
APEAPh-2	16	2.8	150	67	Passes	f	a	c	d	f
APEAPh-3	15	2.7	150	69	Passes	f	c	f	f	f

a: film completely removed; b: loss in gloss; c: slight loss in gloss and adhesion; d: slight loss in gloss; c: colour change and slight loss in adhesion; and f: unaffected.

1.3.1.4. Solvent Free Polyesteramide

To overcome the use of volatile organic solvents (VOCs) used during processing and application of resins, which are ecologically harmful, a solvent free method has been developed to prepare polyesteramide [30]. The resin is found to show improved physico-mechanical and chemical resistance properties along with thermal stability as compared to other commercially synthesized vegetable oil based polyesteramides.

1.3.1.5. Oil Based Conducting Polymer Coatings

Oil based conducting blend coatings of polyaniline and poly (esteramide urethane) have also been investigated for their physico-chemical, thermal, morphological, conductivity and anti-corrosive coating characteristics. The increase in loading of polyaniline was found to show superior corrosion protective performance due to the formation of a dense, adherent, non-porous blend coating, which act as chemical barriers and restrict the diffusion of water and the corrosive media [31].

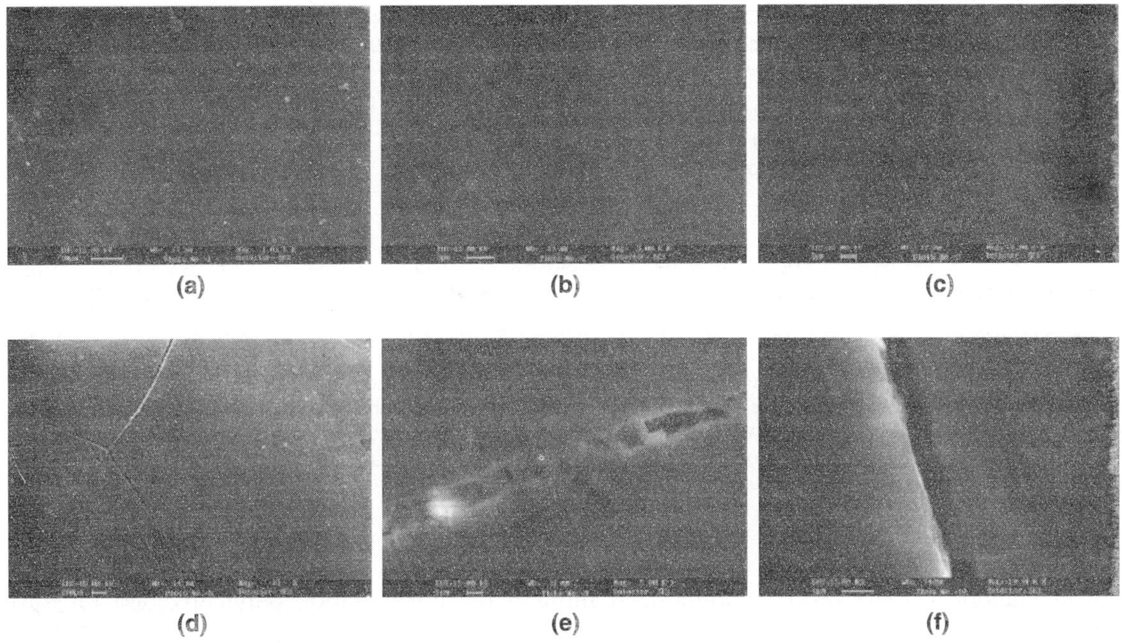

Fig. 14.2. SEM micrograph of PANI/CPEAU coating. (a) uncorroded specimen (mag. 1000×), (b) specimen immersed in 5% NaOH (mag. 3000×), (c) specimen immersed in 5% HCl (mag. 5000×), (d) specimen immersed in 5% NaOH (mag. 2000×), (e) specimen immersed in 5% NaOH (mag. 5000×), and (f) specimen immersed in 5% HCl (mmag. 10,000×). Reproduced with permission from Ahmad *et al.*; Polym. Adv. Technol., 16, 541 (2005). Copyright © 2005, John Wiley and Sons Ltd.

The micrographs of the corroded specimens of 8-PANI/CPEAU coatings immersed in 5% NaOH and 5% HCl, show no apparent change in the morphology even upto 120 and 360 hr exposure respectively in the two corrosive media (**Fig. 14.2**). However, fine cracks are noticed in the SEM micrograph of 5% NaOH (**Fig. 14.2**), corroded specimen after 240 hr exposure, which are visually invisible. Higher magnification of this micrograph exhibits slight separation and formation of blisters, leading to the partial dissolution of the coating after 240 hr. Likewise, the SEM micrograph of 8-PANI/CPEAU blend coating after 480 h exposure in 5% HCl, shows blistering of some portions of the coating and formation of a kind of tunnel causing deep localized removal of coating material while the rest of the coated area remains intact showing strong adhesion between film and metal surface. The coating thus prevents the spread of the corrosive ions and further development of pits occurring on the substrate. It can be concluded that the films of 8-PANI/CPEAU blend coatings remain completely intact as well as show good adherence to the metal surface upto 120 hr in case of 5% NaOH medium and upto 360 hr incase of 5% HCl medium (**Tab. 14.5**). However, the formation of localized linear cracks is observed after 360 h in case of 5% NaOH and 480 hr in case of 5% HCl leading to the separation of the film.

Tab. 14.5. Corrosion Studies of CPEAU and PANI/CPEAU. coatings. Reproduced with permission from Ahmad *et al.*; Polym. Adv. Technol., 16, 541 (2005). Copyright © 2005, John Wiley and Sons Ltd.

Sample	H_2O (15 days)		5% NaOH (15 days)		5% HCl (15 days)		5% NaCl (15 days)	
	Before aging	After aging	Before aging	After aging	Before aging	After aging	Before aging	After aging
CPEAU	a	—	f	—	e	—	f	—
2-PANI/CPEAU	a	b	c	c	b	c	d	e
4-PANI/CPEAU	a	a	c	c	a	c	c	d
8-PANI/CPEAU	a	a	b	c	a	b	b	c

a: the film remains intact and unaffected (slight loss in gloss observed after 10 days); b: the film remains unaffected till 8 days (shows loss in gloss after 8 days); c: the film remains intact till 5 days (shows slight dissolution and discoloration after 5 days); d: the film remains unaffected till 3 days (shows discoloration after 3 days and dissolution after 8 days); e: the film remains intact and unaffected till 2 days (shows dissolution and loss in gloss after 2 days); and f: film completely removed within 2 hours.

1.3.2. Polyetheramide

Polyetheramide (PEtA) resin is synthesized through the condensation polymerization of *N,N*-bis(2-hydroxy ethyl) linseed oil fatty amide with bisphenol-A **(Scheme. 14.5)**. The systems attain a good combination of the properties of ether and fatty amide [32]. Studies also confirm that the factors that govern the physico-mechanical properties also influence the anticorrosive, thermal as well as curing behavior of the resin.

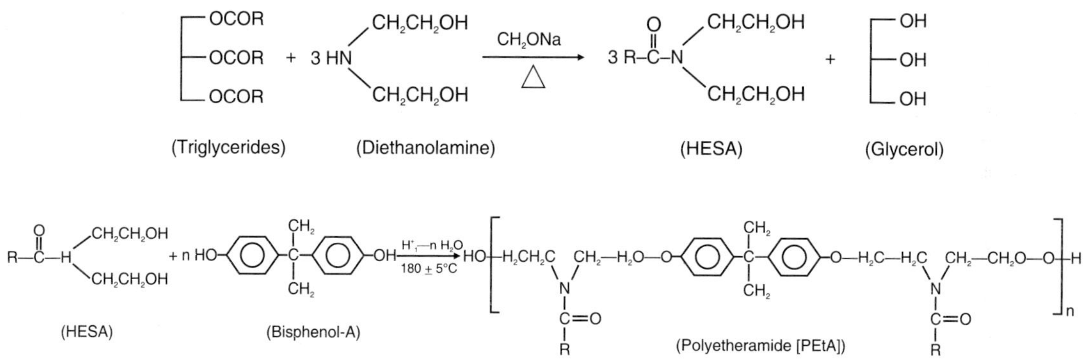

Scheme 14.5. Synthesis of polyetheramide. Reproduced with permission from Alam *et al.*; Prog. Org. Coat., 50, 224 (2004). Copyright © 2004, Elsevier Science Ltd.

1.3.2.1. Urethane Modified Polyetheramide Based Anticorrosive Coatings

Modification of polyetheramide has been carried out by urethane linkages formed during the addition reaction of free hydroxyl groups of fatty amide and isocyanate moiety of toluylene-2,4-diisocyanate (TDI) **(Scheme. 14.6)**. The chemical resistance test exhibits that these coatings show no damage upto 240 h [32]. The physico-mechanical and chemical resistance performance evaluation shows better results with the increased loading of TDI in PEtA. 20-30-wt% **(Tab. 14.6)**. The urethane modified polyesteramide coatings have a superior physico mechanical corrosion resistance performance-as compared to polyetheramide [32]. Among different compositions, the urethane-modified polyetheramide with 30% loading of TDI in the coatings gives the best performance.

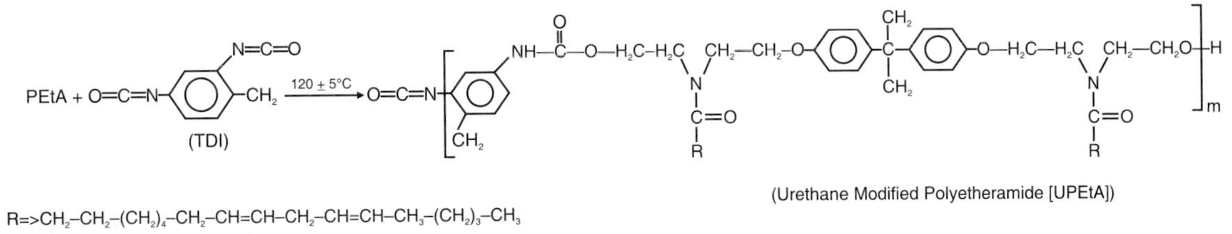
PEtA + O=C=N
N=C=O
CH$_2$
(TDI)
120 ± 5°C
O=C=N
NH—C—O—H$_2$C–H$_2$C
CH$_2$—CH$_2$-O
CH$_3$
O—H$_2$C—H$_2$C
CH$_2$—CH$_2$O
H
N
C=O
R
m
(Urethane Modified Polyetheramide [UPEtA])
R=>CH$_2$–CH$_2$–(CH$_2$)$_4$–CH$_2$–CH=CH–CH$_2$–CH=CH–CH$_3$–(CH$_2$)$_3$–CH$_3$

Scheme 14.6. Synthesis of urethane modified polyetheramide. Reproduced with permission from Alam *et al.*; Prog. Org. Coat., 50, 224 (2004). Copyright © 2004, Elsevier Science Ltd.

Tab. 14.6. Physico-mechanical and chemical resistance of urethane modified polyetheramide. Reproduced with permission from Alam *et al.*; Prog. Org. Coat., 50, 224 (2004). Copyright © 2004, Elsevier Science Ltd.

Resin code	*Scratch hardness (1.2 kg)*	*Impact resistance 150 lb/ in. (passes)*	*Bending test 1/8 in.*	*Gloss at 45°*	*H_2O (10 days)*	*5% NaOH (2 h)*	*Xylene (10 days)*	*3.5% NaCl (5 days)*	*5% HCl (6 days)*
UPEtA-15	Fails	Fails	Fails	—	—	—	—	—	—
UPEtA-20	Passes	Passes	Passes	40	d	b	d	b	c
UPEtA-25	Passes	Passes	Passes	41	d	c	e	c	d
UPEtA-30	Passes	Passes	Passes	41	d	c	e	c	d

a: film completely removed; b: film partially removed; c: loss in gloss; d: slightly loss in gloss; e: unaffected.

1.3.2.2. Polyetheramide-butylated Melamine Formaldehyde Based Anticorrosive Coatings

The polyetheramide-butylated melamine formaldehyde resins exhibit satisfactory physico-mechanical performance when subjected to scratch hardness, impact resistance and bend tests. This may be attributed to the presence of flexible ether linkages and amide groups in the resin, which are responsible for good adhesion. The presence of bisphenol-A and melamine moieties imparts toughness to the coating systems [33]. The additional cross-linking brought about via thermal polymerization of unsaturation along with the hard nature of melamine and bisphenol-A introduces brittleness in the polymer. This results in the deterioration of the physico-mechanical properties of the coatings. The proposed structure of PEtA and the scheme for curing reaction of PEtA with BMF shows that the length of polyether linkages is presumably shorter. Closely spaced aromatic rings of bisphenol-A and melamine moieties separate these short polyether linkages. Such an alignment causes steric hindrance to the free rotation of shorter polyether links, introducing rigidity to the system [33]. The chemical resistance tests performed in acid, alkaline as well as saline environments show superior corrosion resistance performance up to 40 wt% loading of BMF. As the BMF content and curing temperature exceed the above-mentioned values, the chemical resistance deteriorates.

1.3.3. Vegetable Oil Based Epoxy

Aliphatic epoxies obtained from vegetable seeds oils are marked by good adhesion, flexibility and corrosion resistant properties. These attributes of oil epoxies render them suitable for versatile applications as plasticizers, diluents and corrosion protective coatings [34,35]. However, the main drawback of oil epoxies, which limits their extensive application, is their poor load bearing ability and toughness along with sluggish cure rate can be attributed to the presence of long aliphatic triester chains with internally embedded epoxide groups. At ambient temperatures, curing times of the order of a week are generally observed with aliphatic amines. Trans, in situ hydroxylation through epoxidation of oil is an effective way to overcome some of these drawbacks of the polymer chain [36]. The modified oil epoxy has substantial hydroxyl groups in the backbone. The hydroxyl groups of the modified epoxy may be successfully cured, by acid/anhydride, isocyanate and melamine type. Common acid/anhydride curing agents may reduce the alkali resistance of the resin and urethane bonds formed by isocyanate curing of hydroxyl groups have low thermal stability. Melamine resins are characterized by greater hardness, water, alkali and solvent resistance with an adequate thermal stability provided by the s-triazine ring. In case of epoxies, properties of the final product are greatly affected by the selection of curing agent as well as time and temperature of cure [36].

Seeds of *Pongamia glabra* (common name: karanj, family: Leguminaceae) yield pongam/honge oil-a bitter, red brown, non-drying, non-edible oil, 27-36% by weight of seeds, used for tanning leather, soap, as an illuminating oil and as a folk medicine. Ayurvedic medicine prescribes the oil for biliousness, eye ailments, itch, leucoderma, rheumatism, skin diseases, worms and wounds. Yunani system advocates its use as styptic and vermifuge, for fever, hepatalgia, leprosy, lumbago, piles, scabies and ulcers. Pongamia oil has a high content of triglycerides having a good amount of unsaturation. Only a small portion of oil is reported to be utilized while a major share still remains unutilized [38-40]. Pongamia oil may be put to use for subsequent polymer production through its epoxidation at unsaturation. Aliphatic epoxies derived from seeds oil have been used as reactive diluents, plasticizers and as cationic UV curable coatings. However, their poor load bearing capacity and sluggish cure rate due to long aliphatic triester chains with internally embedded epoxide groups limits their extensive application. Trans, in situ hydroxylation through epoxidation of oil may prove to be an effective alternative to overcome the above-mentioned drawbacks.

Unsaturation of seeds oil can be effectively utilized through epoxidation for the development of an anti-corrosive coating material. *Annona squamosa* oil epoxy (AOE) (**Scheme. 14.7**) based coatings, exhibit good physico-mechanical as well as anticorrosive performance [37]. Satisfactory results have been achieved in alkaline environment.

Scratch hardness values reveal that AOE-2, AOE-4 and AOE-5 systems show a higher scratch hardness compared to AOE-1 and AOE-3. This can be attributed to the aliphatic nature of adipic acid, which is known to confer flexibility to the systems while rigid aromatic PA moieties impart hardness. However, as compared to AOE-2 and AOE-4, AOE-5 has lower scratch hardness. AOE-5 has a greater number of aromatic moieties per unit molecular chain length of the resin compared to other systems [37]. An increased number of aromatic moieties, after an optimum amount of loading, reportedly produce steric hindrance within the polymeric film network, which hinders free movement

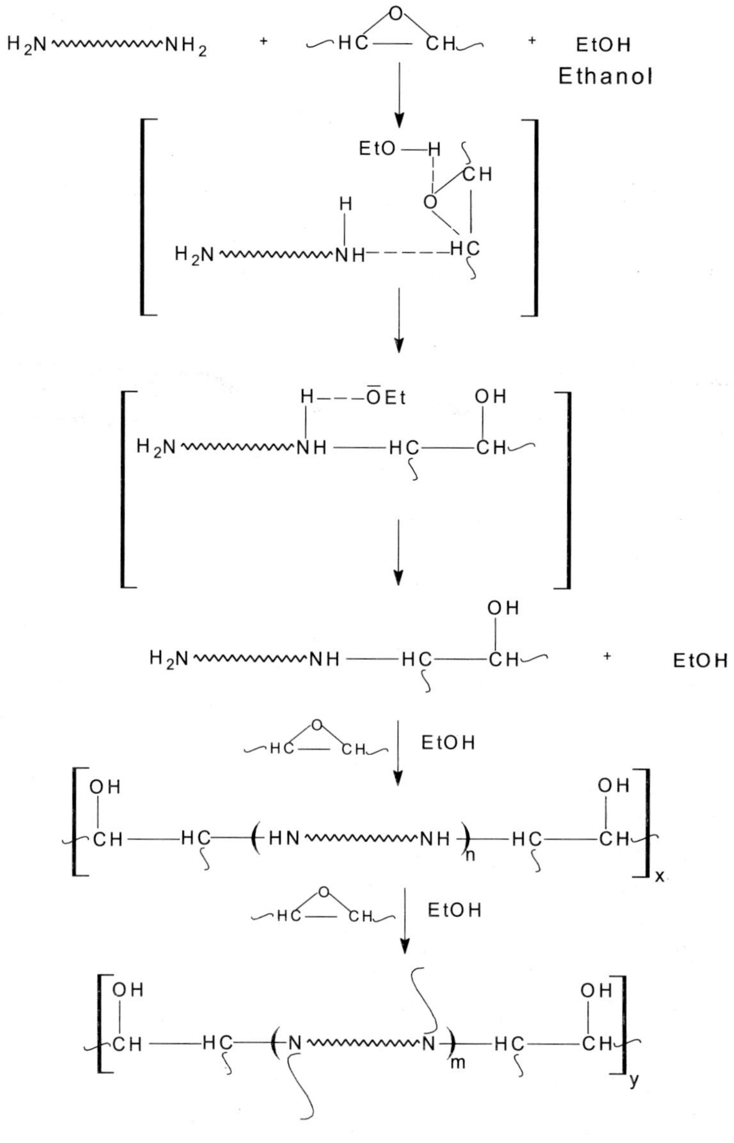

Scheme 14.7. Synthesis of AOE.

of polymeric chains and deteriorates the scratch hardness along with impact resistance of coatings. All these systems pass 1/8-inch conical mandrel bend test as a typical characteristic of oil epoxy based coatings. Adhesion of these AOE-curing agents systems is primarily attributed to polar nature of hydroxyl groups formed during curing reaction. Number of polar hydroxyl and amide groups is same in all systems, consequently all systems show similar results (no loss) for adhesion tests.

Abrasion resistance tests show better results for AOE-2 and AOE-4 than AOE-1 and AOE-3 while AOE-5 shows still inferior results. Abrasion resistance of films is partly governed by their adhesion to substrate. Since adhesion tests show same results for all epoxy-curing agents systems, it is suggested that the role of adhesion seems to be insignificant in determining the abrasion resistance of these coating systems.

The uncoated panels underwent minimum weight loss in water while coated panels (Al-24345 and mild steel) remained unaffected in aqueous medium (distilled water) supporting that these coating materials render the coated substrate impermeable to water and thus they can successfully protect both the substrates (Al and mild steel) from aqueous corrosion. On immersion in 10 wt% HCl, 10 wt% NH_4OH and 10 wt% NaOH for 288 hrs, the films remained intact with only a slight loss in gloss.

1.3.4. Curing Agents for Oil Epoxy

Butylated melamine formaldehyde (BMF) is generally selected as a curing agent in coating formulations owing to its compatibility with hydrocarbon solvents and with other film forming materials [41]. The resulting cured films of BMF and hydroxylated linseed oil epoxy (HLOE) exhibit a combination of superior features of aliphatic epoxy and melamine resin such as higher values of gloss, adhesion, flexibility, chemical and thermal resistance. These systems may safely be used upto 200°C.

HLOE-BMF systems, with varying content (10-50 wt%) of BMF, were baked at different temperatures for different time periods. The systems cured at 110-120°C for 20 minutes and post-cured at 170-190°C for 10 minutes were found to show better performance. Scratch hardness values increased from 0.9 kg (HLOE-10) to 1.9 kg (HLOE-30) with increased content of BMF, upto 30 wt% loading, beyond which deterioration in values was observed. With successive loading of BMF, upto 30 wt%, all the systems passed 150 lb/inch impact resistance test and 1/8-inch conical mandrel bend test. Beyond 30 wt% loading of BMF, the above values of impact resistance and bend test deteriorated in a regular fashion as in the case of scratch hardness. HLOE undergoes etherification reaction on treatment with BMF and the resulting ether linkages provide flexibility and adhesion properties to the resin [42]. The structure and physico-chemical characterization of HLOE indicate that a substantial number of hydroxyl moieties are present in HLOE backbone. Thermal polymerization through unsaturation of modified oil epoxy, renders a highly cross-linked network. An increase in BMF content in HLOE is subsequently accompanied by an increase in ether linkages and presumably a greater number of cross-links per unit chain length. A higher number of ether linkages render better adhesion property as well as flexibility to polymeric coatings. Thus, with successive loading of BMF, a highly cross-linked structure is attained with improved adhesion and flexibility, which facilitates an increase in scratch hardness values along with impact resistance and bending ability upto HLOE-30. With increased content of BMF, that is beyond 30 wt% loading, a higher cross link density achieved by curing reaction (first stage cure) as well as by thermal polymerization (second stage cure), presumably exceeds its optimum required limit hard nature of melamine, beyond 30 wt% addition of BMF, may further hinder the free rotation (bending ability) of fatty triester chains and produce brittleness in the polymeric films, causing deterioration in scratch hardness, impact resistance and bend test values. HLOE-BMF systems exhibit good gloss (77 to 80) at 45°. HLOE-30 showed better physico-mechanical performance compared to its other counterparts. HLOE-30 was also tested for its chemical resistance performance. Coated panels underwent a loss in gloss and weight when immersed in 3 wt% HCl solution after 24 hrs. Similar performance was noticeable in 3 wt% NaCl in 7½ hrs. The coated panels do not withstand 3 wt% NaOH solution for five hrs and undergo loss in gloss, adhesion and weight of the films within this specified period of time.

1.3.5. Blends of Oil Epoxy with Commercial Polymers

Blends of Linseed oil epoxy (LOE) with commercial polymers such as poly(methylmethacrylate) (PMMA) and polystyrene (PS) have been found to be synergistic in nature [43]. Coatings developed from these blends exhibit improved performance in acid and alkaline environments. Linseed oil epoxy (LOE) and poly(methylmethacrylate) (PMMA) blend coating systems cured with melamine formaldehyde (MF) LOE-PMMA-MF show high scratch hardness than LOE:PS:MF systems which is attributed to high polarity of the PMMA causing enhanced molecular interaction with LOE thus imparting higher rigidity to the paint system. LOE-PMMA-MF coating systems show higher UV resistance as well as acid and alkali resistance owing to the better adhesion to the metal substrate. The higher UV resistance of LOE-PMMA-MF coatings may be attributed to the absence of the aromatic ring as well as tertiary hydrogen at α position to carbonyl group [43].

1.3.6. Blends of Oil Epoxy with Commercial Epoxy

Long chain aliphatic epoxy resins and oils are largely used for increasing the fracture energy of epoxy resins. Blending of DGEBA epoxy resin with linseed oil epoxy shows far improved physico-mechanical and anti-corrosive properties [44]. It is observed that corrosion performance of DGEBA-LOE over plain DGEBA.DGEBA-LOE coatings form flexible films with good bending ability and reduced corrosion at edges especially in acid and aqueous environments [44]. However, on increasing the LOE concentration, resistance to alkali decreases, which may be presumably due to the saponificable nature of ester linkages in LOE.

1.4. Next Generation Coatings

1.4.1. Coatings Based on Nanotechnology

In the field of coatings benefits of nanotechnology can be achieved by enhancing the physico-mechanical properties of polymeric coatings such as scratch hardness, dirt resistance etc. Nanotechnology has become an important tool for upgrading the performance of paints and coatings and various nanoparticles such as nano silica, nano zinc, nano rutile are being extensively used in coatings formulation by incorporating them as a binder or as a filler [45]. Use of nano particles for making creamer, anti-friction, anti-stain as well as ultra-scratch coatings has become the focus of all paint companies worldwide. Major polymer producers like BAYER, BASF and DSM are already using nanomaterials as fillers for producing nanocomposites [46].

1.4.2. Smart Coatings

Smart coatings refer to coatings synthesized with intelligent polymers or materials that exhibit unusual functionality in an innovative way. A few examples of such class include sensor-based coatings, Thermo-sensitive coatings, nano particle based coatings, photocatalytic coatings etc.

Conductive polymers within a paint system can act as artificial polymeric neurons capable of giving response like our neural network—a property which is generally utilized for making sensor based coatings [47]. The use of smart materials for corrosion sensing purposes relies on the interaction of such polymers with the corrosive environment which can be manifested by color change, pH change or change in conductivity with progress in corrosion and can be used for indicating corrosion damage. Several initiatives have been launched for the use of smart coatings as corrosion sensors but are futuristic in nature till date.

Thermochromic polymers are used for rapid visual assessment of temperature and find application as temperature indicating coatings [48]. Incorporation of photoactive pigments helps in application defects such as pinholes; scratches etc with the help of UV lamp and such optically active coatings find application for marine tanks where a defect free coating is required.

The incorporation of nanomaterials in coatings also finds application in designing easy clean coating, ultra scratch coatings as well as antistatic coating materials.

1.4.3. Biomimetic Coatings

Biomimetic coatings are created by imitating some natural or biological phenomena such as:
- Physical structures.
- Natural products.
- Physiological processes.

They represent the next generation of advance coating like the sol gel hybrid coating systems, nanocomposite based coatings and smart coatings to achieve novel properties, which are not possible by using conventional technologies [49]. Biomimetic coatings are an example of the intelligence revolution that can be achieved by coating chemists by working in coordination with biologists thus demonstrating how man made systems can be benefited by adapting with natural mechanisms [50-51]. Their advantages are:
- Energy efficiency.
- Minimizing VOC.
- Better adaptation to environment conditions.

The paradigm shift in coating material design discussed above offers new market opportunities for biomass/agricultural polymer materials.

1.4.4. Hyper-branched Polymer Based Coatings

Hyper-branched polymers are a new class of polymeric materials belonging to a class of macromolecules with highly branched structure and a large number of end groups [52]. The structure of these polymers has a great impact on their physical and chemical properties. Hyper-branched polymers such as polyesters, alkyds, epoxies and polyacrylates are applicable especially for coatings.

1.4.4.1. Hyper-Branched Polyesters

Hyper-branched polyesters involves first reacting a starter polyol with chain extender to form gencration branched core [53]. Solvent-less hyper-branched polyol is obtained by polyesterification of trimethylolpropane with propionic acid. Hyper-branched polyesters are used in a variety of coating applications such as coil coating, powder coating, and coating for automobiles.

1.4.4.2. Hyper-Branched Alkyds

Hyper-branched alkyds provide the same solution as polyesters by meeting the stringent requirement of volatile organic content. They are developed by modifying hyper-branched polyesters with tall oil fatty acids [54]. They are suitable for producing decorative paints.

1.4.4.3. Hyper-Branched Epoxies

They are reported based on the copolymerization of glycerol with a range of mono-epoxies of different hydrophilicity by using boron trifluoride as catalyst and find application in automotives, as additives to improve mechanical properties [55].

1.4.4.4. Hyper-Branched Polyurethane

Hyper-branched polyurethanes are produced by reacting multifunctional poly(isocyanates) with multifunctional amines. They show good solubility in common organic solvent. High thermal stability, mechanical strength and toughness are achieved and find application as UV cured coatings [56-57].

2. SPECIALTY ADHESIVES

2.1. Brief History

The history of adhesives dates back to the early Egyptian civilization. The ancient adhesives were derived from casein curd, which was used to disguise the treasures of *Tutankhamun* [58]. A detailed account of the formulation of these ancient adhesives is however available in medieval European history [59]. In the medieval period aqueous mixtures of casein with lime were used for gluing furniture, paint pigment binding etc. Some other formulations of primitive adhesives included animal blood, soybean etc. Many archaeological surveys have discovered the use of animal blood as an adhesive in lime mortar, cement and wood as it contains serum albumin and serum globulin that are 100% adhesive functional proteins [60]. During World War II and twenty years thereafter, blood adhesives were used extensively in the manufacture of interior grades of structural plywood, which were then replaced with the Soybean adhesives that were developed in later half of the 19th century. Polymer-based adhesives that have replaced most of the conventional adhesives, however, have invaded the 20th century.

2.2. Adhesion and the Theories of Adhesion

The good adhesion is developed when strong interfacial forces with a very close contact hold two surfaces together where one of the materials may be a polymer and the other an inorganic material such as glass or metal substrate. Much of the adhesion discussed here will be focused on polymer-based adhesives.

2.3.1. Adhesion of Polymers to the Substrates

Polymer adhesion is based on coupling mechanisms operating at molecular level. The nature of polymers and type of forces involved in adhesion play an important role in polymeric adhesion. In case of polymer adhesion, it is important to take into account the bulk material properties of the polymer as well as the type of forces involved to develop adhesion. Most of the cross-linked polymer materials gain their strength from entanglements between the chains. To form a strong polymeric adhesion, it is necessary for the polymer network to be continuous across the interface and this can be achieved by:

Chain interdiffusion: Coupling of the polymeric chains by interdiffusion occurs when polymers are miscible with each other and miscible to form a uniform interface. Welding is the most common form of interdiffusion coupling.

Interface coupling of chains: This involves the utilization of a specific coupling of chains such as that of diblock copolymers, which are commercially used for coupling between phases of polymer blends. A diblock copolymer consists of linear chains each of which contains chemically distinct repeating units.

Chemical reaction to form coupled chains: This involves the insertion of a small percentage of chemically modified chains between bulk polymers that can react with the polymers to form coupling chains at interface. A classical example is the introduction of maleic anhydride in polypropylene to induce coupling with Nylon 6 where maleic anhydride interacts with the hydroxyl of polyamide to form a grafted copolymer of polypropylene and Nylon 6 at the interface. Nature of polymerization is also one of the most important factors of chemical reactions in polymer adhesives as most of the polymer-based adhesives are hardened by addition polymerization or by condensation polymerization.

2.3. Synthetic Adhesives

A number of perplexing adhesives are commercially available which can be simplified by classifying them according to their mode of application. Some major classifications are described in the upcoming sections.

2.3.1. Epoxies

Epoxies dominate the field of structural adhesives due to their better wetting ability, excellent mechanical properties and high chemical and thermal resistance. However, fully cured epoxy resins exhibit high brittleness and high capacity of water uptake, which limits their applications in humid environments. Numerous researches have been carried out to determine the most important factors governing the durability of epoxy resin bonds to aluminum [61,62]. It has been found that the distilled water is more damaging to the environment than saline water. The increase in temperature causes a growth in the rate of strength loss, and the presence of stable oxides on aluminum substrate surface provides that the damages which are too severe in nature [63]. However, all researchers are in agreement that the behavior of an adhesive joint depends on many other factors and therefore every adhesive–adherend system has to be specifically studied. The cross-linking of epoxy resins can be achieved through two different reaction mechanisms, such as polymerization by addition and by steps [64,65]. The most common curing process is based on the addition reaction of a hardener. Depending on the cross-linking agent, the final properties of epoxy networks are different [64]. When the curing agent is an aliphatic amine, the curing process frequently occurs at room temperature, but it is slow and incomplete. Therefore, the obtained networks presents low glass-transition temperature

(T_g) and a high ability to carbonate and to absorb water. On the other hand, epoxy resins cured with aromatic amines generally present good thermal and chemical resistance. The advantages of anhydride hardeners are their low contraction, viscosity and excellent thermal resistance. But epoxy-anhydride networks present high water uptake capacity. The homopolymerisation is another epoxy curing procedure, which consists of polymerization steps initiated through tertiary amines [65-68]. Although the reaction mechanism is still a controversy, it is believed that the initiator forms adducts with oxirane groups (initiation). The adducts react with themselves and other epoxy rings (propagation).

Adhesive properties of different epoxy resins are evaluated in terms of the lap shear strength using aluminum adherends. It has durability in various humid environments. The micro-cohesive mechanisms are directly correlated with the adhesive strength (**Fig. 14.3**). The durability of studied epoxy adhesives under different humid environments show that humidity causes higher damage when the temperature is high (**Fig. 14.4**). Nevertheless at room temperature humidity, the temperature affects scarcely the adhesive strength. Comparing the ageing at 23°C and 80% RH with the reference wedge test (23°C and 30% RH), a slight decrease in the adhesive strength is observed as the temperature increases. The epoxy resins show little changes in the tensile strength with the temperature, which has been associated with the different thermal dilatation behavior between the aluminum substrates and epoxy resin, whose linear expansion coefficients are 24.3×10^{-6} and around 40×10^{-6} °C^{-1} respectively. In all cases, the humidity causes a decrease in the adhesive resistance but the magnitude of damage depends on the epoxy adhesive employed. It is observed that if higher amount of water is absorbed by the adhesive, the produced damage is higher. It is known that the

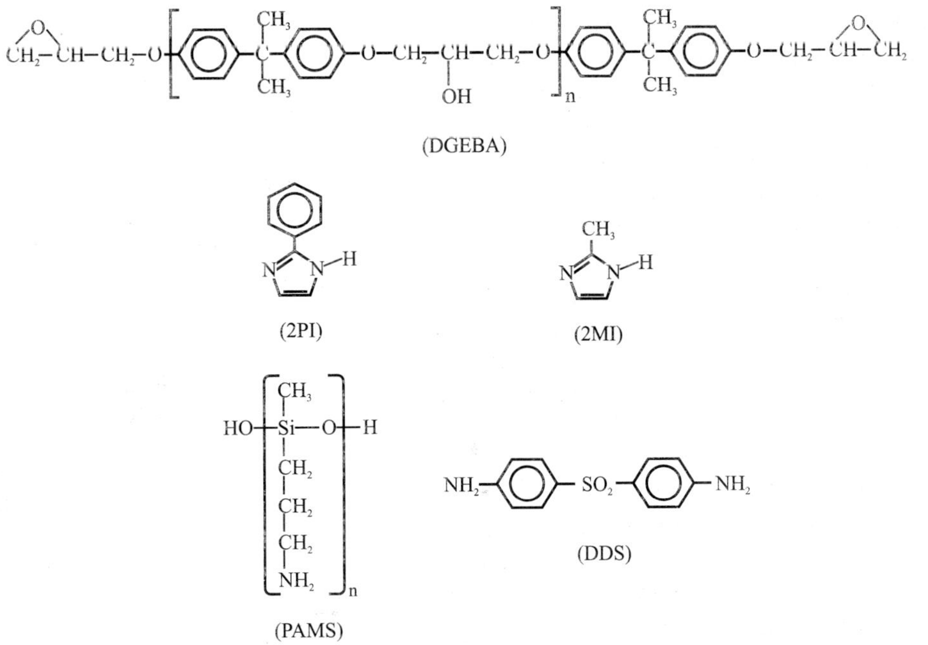

(DGEBA)

(2PI)

(2MI)

(PAMS)

(DDS)

Fig. 14.3. Chemical structures of epoxy monomer (DGEBA), imidazole initiators (2 PI and 2 MI) and used amine hardeners (DDS and PAMS). Reproduced with permission from Prolongo *et al.*; Int. J. Adhesion and Adhesives, 26, 125 (2006). Copyright © 2006, Elsevier Science Ltd.

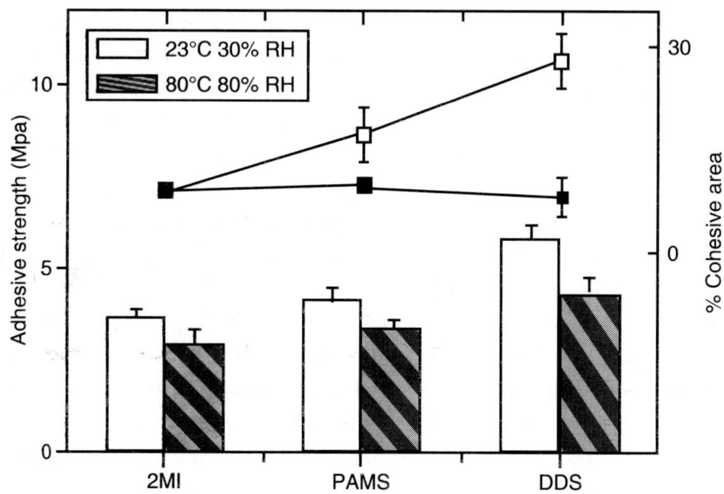

Fig. 14.4. Adhesive strength of different epoxy resins. Homopolymerized epoxy resin (2MI), DGEBA/PAMS (PAMS) and DGEBA/DDS resin (DDS). Cohesive area percentages measured at reference conditions (23°C and 30% RH) ( ) and after wedge test (80°C and 80% RH) (•). (The abscise origin of the right axis is displaced to avoid overlaps). Reproduced with permission from prolongoa *et al.*; Int. J. Adhesion Adhesives, 26, 125 (2006). Copyright © 2006, Elsevier Science Ltd.

water diffusion along the interface can be many times faster than water permeation through the adhesive [61]. A clear relation exists between the decreasing adhesive strength of the lap shear probes and the amount of absorbed water by the wetness. The decrease in the adhesive strength due to the hydrothermal ageing is lower for the homopolymerised epoxy resin due to its low tendency to absorb water. This behavior can be explained by the structure of the epoxy network, whose hydroxyl concentration is negligible. The presence of OH groups is an important factor governing the water uptake and the subsequent damage due to van der Walls interactions. Comparing the studied epoxy/amine systems, DGEBA/PAMS resins present low water uptake ability. In fact, the measured percentage of absorbed water by this system is very low in comparison to other epoxy resins. This behavior can be explained by the hydrophobic nature of the polyaminosiloxane [69]. Several different water uptake effects can cause the decrease in the adhesive resistance. One of them is the adhesive plasticisation. The water penetration inside the epoxy network produces the rupture of interchain interactions, which causes the stresses relaxation. This relaxation gives rise to the decrease in T_g of the network named commonly politicization effect [70-72] and can be analyzed through the T_g measurements before and after wedge test. It is observed that the decrease in T_g is not linearly proportional to the absorbed water amount. The T_g decreasing seems to depend on the absorbed water amount, the nature of the epoxy resin and on the ageing temperature. It is observed that the water entering into the epoxy network causes a higher decrease in the glass transition temperature (T_g) as the ageing temperature increases. On the other hand, the decrease in T_g caused by the water entrance is different depending on the epoxy network taken into account. The reason can be the different interactions between the absorbed water molecules and epoxy network. In the literature, it has been demonstrated that the absorbed water molecules found in the network has two different states namely: bound and free water [71,73]. Epoxy adhesives also constitute of a hardener, which come in a variety of formulations since there are many epoxy resins and many hardeners available.

These adhesives have good strength, do not produce volatiles during curing and have a very low shrinkage value. Owing to the flexibility and low peel strength, epoxy-based adhesives can be brittle, but they produce extremely strong durable bonds with many materials.

2.3.2. Polyimides

The bulk of the work dealing with polyimides as high performance adhesives has been investigated by the workers at NASA [74-76]. The "lap shear test" is carried out to evaluate their performance in which the following variables are studied:
- Effect of conditions during tape making.
- Effect of bonding time, bonding temperature and bonding pressure.
- Effect of aging time and aging temperature on room temperature lap-shear strengths.
- Effect of high testing temperature on lap-shear strengths.
- Effect of some of the solvents and boiling water on lapshear strengths.
- Effect of surface treatments.

All the studies to date dealing with polyimide adhesion systems involve the preparation of adhesive tapes and their subsequent bonding mechanism. While many polyimide systems are thermosets in origin, the use of this process for thermoplastic polyimides is due to unavailability of melt processible resins and the lack of thermal stability of the polymers above their high melting points. Secondly, the crystallization ability of the past polyimides rapidly deteriorates once taken to melt temperatures. Hence in order to preserve crystallinity (which often occurs in the thermoplastic systems during the imidization process), the solvent route is preferred.

2.3.3. Polyurethanes

These are highly reactive chemical formulations and cure very fast providing strong adhesion, impact resistance and low temperature strength. They are generally used with primers. Polyurethane chemistry has added valuable contributions to the adhesive bonding technology making available many varied raw materials for the production of adhesives with a wide range of performance characteristics [77,78]. Polyurethane adhesives vary widely in composition and are used in many different applications in various market segments.

Polyurethane adhesives are normally defined as those adhesives that contain a number of urethane groups in the molecular backbone, which are formed during use, regardless of the chemical composition of the rest of the chain. Thus a typical urethane adhesive may contain, in addition to urethane linkages, aliphatic and aromatic hydrocarbons, esters, ethers, amides, urea and allophanate groups [79,80]. An isocyanate group reacts with the hydroxyl groups of a polyol to form the repeating urethane linkage. Isocyanate groups will react with water to form a urea linkage and carbon dioxide as a by-product. Linear thermoplastic polyurethanes may be obtained by using compounds with two reactive groups such as diisocyanate and diols. When polyols with three or more hydroxyl groups (i.e. a functionality of 3 or more) are reacted with an isocyanate, or when isocyanates with three or more isocyanate groups are reacted with a polyol, the resulting polymer is crosslinked. The amount of crosslinking affects the stiffness of the polymer. Contrary to linear polymers, crosslinked polymers do not flow when heated. All structural adhesives are crosslinked because this eliminates creep (deformation under constant load). In reaction systems where there is an excess of isocyanate, crosslinking reactions may occur. These reactions form linkages of allophanate and biuret. Urethane based systems make good adhesives for a number of reasons:

- They effectively wet the surface of most substrates.
- They can interact with the substrate through polar interactions (e.g. hydrogen bonding).
- Their relatively low molecular weight/ small molecular size allows them to permeate porous substrates (for reactive adhesives).
- They can form covalent bonds with substrates that have active hydrogen atoms (for reactive adhesives).
- Through molecular composition the adhesive stiffness, elasticity and crosslinking can be tailored to suit specific needs.

The most commonly used isocyanates in polyurethane adhesives are MDI (methylene diphenyl diisocyanate) and TDI (toluene diisocyanate), both being aromatic isocyanates in nature. Aliphatic isocyanates are also used but less often. Different types of polyols are used for the production of adhesives. The most commonly used polyols are polyether (polypropylene glycol) and polyester (adipate based) polyols. Polyurethane adhesives can be classified as the following.

2.3.3.1. Non-reactive Polyurethane Adhesives

These are solvent-borne adhesives. Polyurethane solvent adhesives consist of a high molecular weight hydroxyl terminated polyurethane (MW approximately 100,000 Daltons) dissolved in a solvent. The polyurethanes are obtained by reacting a high molecular weight polyester diol with a diisocyanate and differ in solution viscosity and crystallization tendency. The polymer solutions are applied to both surfaces to be bonded. The solvents are allowed to evaporate and the surfaces are then pressed together, where interdiffusion of the polymer chains occur.

2.3.3.2. Water Based Polyurethane Adhesives

These adhesives are high molecular weight polyurethanes dispersed in water (PU dispersions, or PUD's) [81]. The water carrier is eliminated during use, leaving the coalesced polymer to form the adhesive bond.

2.3.3.3. Reactive Polyurethane Adhesives

One-component adhesives: They are liquid isocyanate-terminated polyurethanes with a relatively high molecular weight (prepolymer) and rather have a low remaining isocyanate content. The prepolymers are prepared by reacting an excess of isocyanate with high molecular weight polyester or polyether polyols. The free isocyanate groups react with moisture from the environment to form urea linkages. If the functionality of the prepolymer is larger than two (i.e. contains more than two isocyanate groups per molecule), the film will be cured by chemical crosslinking.

Two-component adhesives: Such adhesives consist of two relatively low molecular weight components: the polyol and isocyanate. When the two components are mixed, they form urethane groups in the adhesive films. The polyols are usually of the ether or ester type. The isocyanates and polyols employed may have a functionality of two or higher. In the latter, a crosslinked polymer film is formed.

2.3.4. Woodbinders

Aromatic isocyanates (predominantly MDI) are used as binders to manufacture oriented strand board (OSB), medium density fiberboard (MDF) and particleboard. For these boards, the isocyanate is blended with wood strands, fibers and chips, respectively [81]. The cure takes place in a press at about 200°C. The curing reaction is predominantly via the reaction with water and thus the formation of urea groups. Additionally, a wide range of reactions with wood components occur depending on the temperature, moisture content and the specific location within the wood matrix.

2.3.5. Silicone Based Adhesives

Silicone based adhesives are known for their flexibility and high temperature resistance. They are available in single or two part forms. They are used as shower sealants and on curing liberate alcohol or acetic acid. Their adhesion to the surface is fair but durability is good. Two-form version requires an additional hardener to be mixed with the resin and provides better adhesion and curing than the single form version [81].

2.3.6. Urea Formaldehyde

Arguably the most important development in this century, for the furniture industry, was the development and introduction of urea formaldehyde (UF) adhesives. This was followed a few years later by the introduction of chipboard, with UF resins as binding agents. Chipboard, together with the more recently developed medium density fiberboard or MDF (which also has UF resins as binding agents), now form the basis of most of today's cabinet furniture. New bonding techniques using resistance and radio-frequency heating of glue lines to accelerate the cure of these new synthetic adhesives and shorten assembly times were also developed in the early 1940s, principally to meet the military needs of the World War (II) [81]. In the immediate post-war years, the experience gained using the new adhesives and techniques by furniture companies transferred to aircraft production during the war, was quickly applied to cabinet furniture production. One disadvantage of UF adhesives is their relatively long cure time at room temperature, but the introduction of polyvinyl acetate adhesives (PVAc) in the 1950s, with their reasonably short setting times at normal ambient temperature, offered an alternative to animal glue for jointing. Later, cross-linked versions of urea formaldehyde, a thermosetting material, resulting from the chemical reaction between urea and formaldehyde under slightly acid conditions, joined the group. The rate of this reaction, and cure time, is accelerated by heat. When curing takes place, further heating (within limits) has no effect on the material, which shows good heat resistance. It also has reasonable moisture resistance, and these two properties, combined with good adhesion to wood and wood-based materials and low cost, has ensured its extensive use in the furniture industry over more than half a century. UF adhesives are two-part systems, with the resin normally supplied as a liquid, to which must be added a powder hardener, which converts the liquid adhesive into its solid state [81]. Curing of the adhesive starts when resin and hardener are mixed, and to obtain a reasonable pot life for the adhesive, hardeners contain additives, which slow down the rate of reaction and extend pot life. Unfortunately, pot life and cure time are related. Extending pot life will lengthen cure time and so a compromise is necessary between the two. Heat will increase the speed of reaction, however, so that a UF adhesive which takes several hours to cure at normal room temperature will require only a few minutes at 80°C and less than 1 min at 125°C. UF adhesives are therefore ideal for veneering when hot pressing conditions can be used for fast production rates. Pot life is also affected by heat, and if considerable variations in temperatures occur throughout the year, different hardeners may be required for winter and summer working. Some UF adhesives are available in powder form, containing both resin and hardener, and in this case addition of water triggers the reaction. UFs in this form are useful when only small quantities are required, and so are popular in small workshops. UF adhesives develop excellent adhesion to wood and wood-based materials, and produce strong, durable bonds, which have good creep, temperature and moisture resistance. They have some gap-filling properties and, quite importantly, are of low cost. The only disadvantages of this family of adhesives is that they are two-part systems and have a long cure time at normal room temperatures They are used extensively for hot pressing and also for edge-jointing solid wood panels and applying substantial wood clippings to the edges of chipboard based panels. The last two operations require radio frequency heating to accelerate cure if high production rates are needed.

2.3.7. Hot Melt Based Adhesives

They are solid at normal ambient temperatures and need to be heated to a liquid state before application. They must remain sufficiently fluid to wet out the two surfaces during bonding. As the adhesive cools, it will revert to its solid state, completing the bond. The rate of cooling is initially high, so that bond strength develops rapidly. Care is needed, however, to ensure that the applied adhesive does not over-chill before the two surfaces are brought together, otherwise good adhesion will not be achieved and the bond will be weak, the rapid cooling rates are a disadvantage in this instance. With hot melt adhesives, the change from solid to liquid is reversible, and controlled by temperature. In one respect this is an advantage as it enables a surface to be pre-coated with adhesive, and later reactivated by heat before bonding. However, it also means that a hot melt bond is sensitive to heat, and will lose strength at elevated temperatures. EVA (ethylene vinyl acetate) hot melts are extensively and successfully used on continuous edge banding machines for applying thin decorative edgings to chipboard panels. Application temperatures of around 200°C are normally used. When service temperatures are expected to be higher than normal, other types of hot melt adhesives (e.g. moisture curable polyurethanes, polyolefines) are employed. They are based on modern plastics that undergo physical changes and are designed to be used for light loaded materials.

Reactive hot melt adhesives are highly viscous or solid prepolymers with a low melting point. The prepolymers are produced from solid (i.e. crystalline) polyester polyols and isocyanate have a low number of free isocyanate groups. The adhesive is applied above its melting point and bond through the physical process of cooling (like a conventional hot melt adhesive) as well as through reaction with ambient moisture (as a one component adhesive).

2.3.8. Plastisols

Plastiols are modified PVC dispersions, which require heat to harden and provide tough and resilient joints. Rubber plastisol adhesives based on solutions latexes are generally suited for light loading joints.

2.3.9. *Polyvinylacetate (PVAc) Based Adhesives*

Polyvinyl acetate based adhesives are dispersions of small droplets of PVAC in water. They set by loss of moisture, leaving the droplets to coalesce to form a continuous solid. Setting times depend on the moisture levels in the components being bonded, and the thickness of the adhesive. With optimum bonding conditions (i.e. low moisture contents and thin gluelines) a modern PVAC adhesive develops reasonable bond strength in about 15 min at normal ambient temperatures [81]. Damp conditions and thick adhesives extend setting times, as can be seen from the results of bond strength development tests shown in **Fig. 14.5**. They are obtained using single lap shear test pieces, prepared using standard beech slips and a PVAC adhesive. Area of bond is 25/25 mm. The beech slips are pre-conditioned to the moisture content of either 12 or 15%. The thin (0.1 mm) and thick (approximately 0.2 mm) gluelines are achieved by subjecting the bonds to a pressure of 0.7 and 0.07 MN/sq.m, respectively, whilst the adhesive cures. Wood failure in the form of fiber pullout occurs at bond strengths of around 1300 Newtons, and the degree of fiber pullout increases with increasing bond strength. Below this strength level, failure takes place totally within the adhesive. The PVAC adhesion depends on the effect of wood moisture content that influences the rate of development of PVAC bonds. They are suited for bonding paper or wood and are generally used for packaging purposes [81].

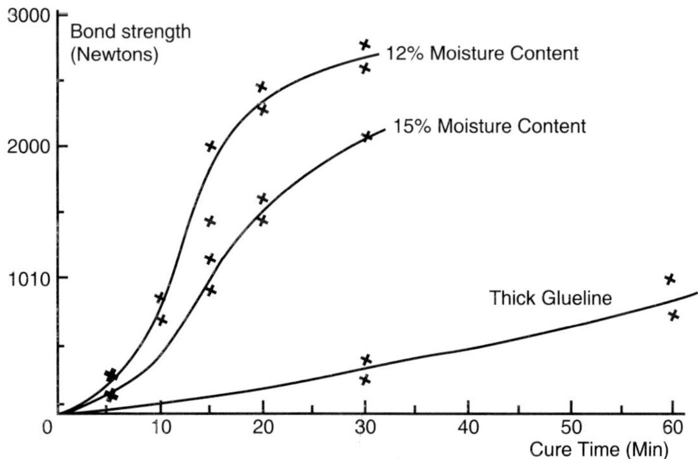

Fig. 14.5. Effect of wood moisture content and glueline thickness on the rate of development of PVAC bonds. Reproduced with permission from Tout; Int. J. Adhesion Adhesives, 20, 269 (2000). Copyright © 2000, Elsevier Science Ltd.

2.3.10. *Pressure Sensitive Adhesives*

They are used in tapes or labels and do not solidify but often withstand adverse environments. The field of radiation-curable pressure-sensitive adhesives (PSAs) has grown considerably in recent years with environmental factors demanding reduced solvent emissions and energy requirements [81]. Electron beam (eb)-curable PSAs are suitable for medical applications. These materials are in the class "urethane methacrylates" that are oligomeric polyurethanes prepared from aliphatic diisocyanates, ethanediol, and difunctional polyglycols. The isocyanate functional oligomers are end-capped with hydroxyethyl methacrylate (HEMA) and act as a tackifier to produce viscous liquids, which form PSAs upon Electron beam eb irradiation. They offer the possibility of wide performance as well as substitution of hydrophobic poly(propylene glycol) blocks with more hydrophilic polyols in the adhesive prepolymer in order to improve the moisture vapor transmission rates (MVTRs).

2.3.11. Bismaleide Based Adhesives

One of the major problems faced by air carried weapons is that they reach very high temperatures during high-speed aircraft maneuvers, which cause aerodynamic heating. Thus the use of epoxy-based adhesives becomes difficult and high temperature resistant adhesives are required which are bismaleimide-based adhesives. These materials are easy to process as compared to polyamides and the incorporation of a second phase such as carboxyl terminated butadiene acrylonitrile rubber further toughens these materials.

2.3.12. Dental Adhesives

Adhesive materials are used in dentistry because they are conservative and generally aesthetic, i.e. tooth colored. Two main classes of material are involved, the glass-ionomer cements and the composite resins. The use of adhesive materials in dentistry is growing rapidly at the present time [82]. The main dental treatments that employ adhesives are orthodontics (i.e. correction of misaligned teeth) and restorative dentistry (i.e. repair of teeth damaged by caries). In the former branch of dentistry, teeth are subjected to forces in order to bring about the desired realignment, and this is done using special wires that are attached to the teeth via brackets. These brackets have been traditionally retained by mechanical clamping, but are now increasingly retained with the aid of adhesives [83]. In restorative dentistry, adhesive materials are being employed because not only are they aesthetic, matching the tooth in color and translucency, but also they are conservative: less of the structure has to be cut away in preparing the cavity for these materials than has conventionally been necessary for amalgams. This change towards the use of adhesive materials has also affected the way in which amalgams are being used, since there are now special bonding agents designed for use with amalgam restorations. These bonding agents are designed to eliminate leakage at the margins of amalgam restorations, though at the price of an increase in the difficulty of the placement technique.

2.3.13. Orthodontic Adhesives

There were early attempts to bond orthodontic brackets using epoxy adhesives [84], but these never attracted much support within the dental profession due to the difficulties associated with using the materials and the uncertain clinical outcomes. More recently, adhesive materials similar in kind to those used in conservative dentistry have been employed which are (i) composite resins and (ii) the glass-ionomer cement. The latter has been less widely used, though its use is growing [85]. Direct bonding of orthodontic brackets using composite resins employs the acid-etch technique [87]. This technique was pioneered over 40 years ago by Buonocore [86] and involves treatment of the tooth surface with phosphoric acid [88]. Lightly filled or unfilled resins are then applied to this surface, and they form their attachments mechanically rather than through the secondary chemical bonds that are usually associated with the phenomenon of adhesion [82]. Orthodontic brackets bonded in this way are susceptible to demineralization around their edges [89,90].

2.3.14. Composite Resins

Composite resins are the most widely used of the aesthetic repair materials. They consist of bulky difunctional monomers, generally based on a substance known as bis-glycidylmethacrylate (bisGMA). BisGMA is derived from bisphenol A and methyl methacrylate, and possesses two terminal double bonds, which are capable of undergoing addition polymerization [91]. This monomer is a very viscous liquid at room temperature so to achieve more workable viscosity and to allow incorporation of sufficient powdered filler, diluents are used. These are typically also difunctional monomers, but of lower molar mass than BisGMA, such as di- or triethyleneglycol dimethacrylate.

Bonding of these materials varies according to the substrate. For enamel, the acid-etch technique is used, as described previously. It results in bonding via micromechanical attachment, as the resin flows into the vastly increased surface volume created by the etching process. Bonding is required to develop under much wetter conditions than those involving just the enamel of the tooth. Certain bonding agents are claimed to be able to make use of this source of calcium, and hence are recommended to be used over the top of the smear layer [92]. Against this, there is considerable experimental evidence that bond strengths to dentine are stronger where the smear layer has been removed [93]. There are currently a wide variety of bonding systems available to the clinician [82]. They share the character of being formulated from molecules having both polar and non-polar functional groups. One substance used extensively in these formulations is 2-hydroxyethyl methacrylate, HEMA, but other molecules have been used, including a range of organophosphorus compounds [81]. The most successful systems have been those based on multiple components and multiple applications [82], so that the dentine surface has been etched, primed and then sealed prior to bonding. Now, as a result of the demand from clinicians for faster treatments, the so-called single-bottle" bonding agents are becoming available designed to combine all the functions of the individual components in a single formulation [83], one preferably applied only once. The danger with this approach is that by simplifying the procedure, the bond to dentine may be compromised [84], either in strength or durability. Clinical studies of durability of composite resins indicate that these procedures are satisfactory and that there are few if any adhesion failures. Surface cleanliness of the tooth is vital, and inadvertent contamination of the freshly cut surface with saliva can radically reduce the strength of the bond that is subsequently formed with that surface [85]. This is illustrated by the data in **Tab.** 14.7, where surfaces of dentine and enamel were prepared, treated with three layers on bonding agent, and dried before a commercial composite resin was applied. One set was also deliberately contaminated by exposure to saliva for 15s, followed by drying for 5s. Glass-ionomer cements are materials fabricated from a water-soluble polymeric acid and a basic glass, typically a calcium fluoro-aluminosilicate [86]. They belong to the class of material known as acid-base cements, and their setting involves neutralization of the acid groups on the polymer by the powdered, solid base [87]. There are also later reactions, which contribute to the overall setting and strength development, and these involve ion-depleted species from the glass [88,89]. There are other forms of modified glass-ionomer cement available to the clinician, namely, metal-reinforced types [93]. In these materials, conventional glass ionomers have metal powders incorporated into them, these powders typically being made of the silver-tin alloy used in dental amalgams [93,94]. Alternatively, this metal can be fused to the glass to form cermets [95].

Tab. 14.7. Adhesive strength, T_g and the water uptake percentage ($\Delta W/W_0$) of the aged samples at different temperatures and relative humidity. Reproduced with permission from Prolongoa *et al.*; Int. J. Adhesion Adhesives, 26, 125 (2006). Copyright © 2006, Elsevier Science Ltd.

	$T(°C)$	RH (%)	2MI	PAMS	DDS
Adhesive strength (MPa)	23	30	3.7 + 0.2	4.1 + 0.2	5.8 + 0.3
	23	80	3.5 + 0.2	4.4 + 0.1	5.6 + 0.3
	80	30	3.3 + 0.3	3.6 + 0.2	4.8 + 0.2
	80	80	3.0 + 0.4	3.4 + 0.2	4.4 + 0.5
T_g (°C)	23	30	140	101	207
	23	80	141	91	200
	80	30	141	100	204
	80	80	133	72	115
$\Delta W/W_0$ (%)	23	80	0.9	1.2	1.5
	80	80	1.2	1.4	1.9

The main materials being bonded adhesively are composite resins, where special bonding agents are required, and glass ionomers, including resin modified glass ionomers, which are adhesives in their own right. There has also been some progress recently with bonding amalgams in place adhesively, again using special bonding agents. Assessing the success of these materials as dental treatments is difficult, though what evidence there is suggests that they rarely, if ever, suffer from failure of the adhesive. Under the demanding service conditions of the mouth, the restoration tends to fail by other means, including wear and fracture. A significant problem is that secondary caries may develop around the restoration, under which circumstances the filling needs to be removed, the additional caries cut away, and a larger filling put back in place.

2.3.15. Conducting Adhesives

Most commercial conductive adhesives are based on metals, doped metal oxides or carbon black [96]. In many cases the alkali nature of the matrix or the curing mechanism in these adhesives adversely affects the conductivity of the inherently conducting polymers (ICP). ICP-based conducting adhesives could therefore be more suitable in possible applications such as joining organic field-effect transistors [97], organic light emitting diodes [98,99] to e.g. ICP-based circuitry [100]. Such adhesives provide transparency and flexibility needed in some applications where conductive adhesives are used. Conductivity of PANI/DBSA adhesive is found to be 10^{-1} S/cm and can further increase to 1 S/cm with a PANI/DBSA content of 15-20 wt% [100].

2.3.16. Phenol Formaldehyde Based Adhesives

Phenolic resins are polycondensation products of phenol with formaldehyde in the presence of acid or alkali. These conventional resins find applications as laminate, adhesives, components of particles board, bonding agent etc. The resins have a high degree of regularity as phenols are linked through methylene groups and a single phenol is employed in most phenolic resin manufacture. Lignin has also been used to prepare resins as it forms a polymer with properties similar to phenol formaldehyde [101]. The advantages of using lignin instead of phenol have been investigated with the aim to lower the costs of production of phenol formaldehyde. Substitution of phenol by lignin also reduces the curing time and improves strength as well as water resistance [101].

2.4. Adhesives Based on Natural Resources

Uncertainty in the availability of petrochemicals and their unusual price hike posed an adverse effect on synthetic adhesive technology [102]. In such a restrictive situation, the replacement of the petrochemical-based adhesive materials by adhesives based on renewable resources becomes a necessity. Adhesives derived from natural sources and their industrial utilization has been a focus of considerable attention in recent years. Recent advances in the research and development program on adhesive systems based on natural resources demonstrate the breadth of available possibilities for using nature's storehouse for durable adhesive formulation [103]. Substantial progress in adhesives based on natural products has been made in the last two decades to prepare adhesive formulations that come close to meeting today's exacting industrial standards.

Although high priority is given to the research based on natural gums and adhesives, transfer of know-how from the laboratory bench to industrial practice has not been possible. The price differential between the synthetic adhesives and formulations based on renewable resources and inadequate knowledge about the natural polymers prohibit industrialization [103].

A number adhesives and gums based on renewable resources are available in nature, most are used in other fields (i.e., textile printing, food products, cosmetics, mining industries, etc.) [104-106]. Due to the variability in the chemical composition and physical characteristics (that depends on the location and extraction process), the use of natural gums and extracts in adhesive technology is restricted. The complex chemical structure of renewable resource–based polymers prohibits their increasing use in the adhesive field.

2.4.1. Xanthum Gum

Xantham gum is a microbial polysaccharide produced in pure culture fermentation by the bacterium *Xanthomonas campestris,* originally obtained from the rutabaga plant [107]. A number of improvements in polysaccharide manufacture have been made by several research groups [108-111]. Xanthan gum is one of the most important commercial polysaccharides with exceptional rheological properties. It is a dry, cream-color powder with 11% moisture content, 9% ash content, specific gravity of 1.5, and bulk density of 836 kg/m [112] (**Fig. 14.6**).

Xantham gum prevents separation of insoluble particles and therefore finds use in the agricultural and pharmaceutical fields. Industrial application of xanthan gum includes textile printing, pigment and dye suspension, and cleaners, adhesives, and polishes [113]. Oil field applications of xanthan gum [114,115] include drilling mud, workover and completion fluids, fracturing, and enhanced oil recovery due to its well-known remarkable rheological properties.

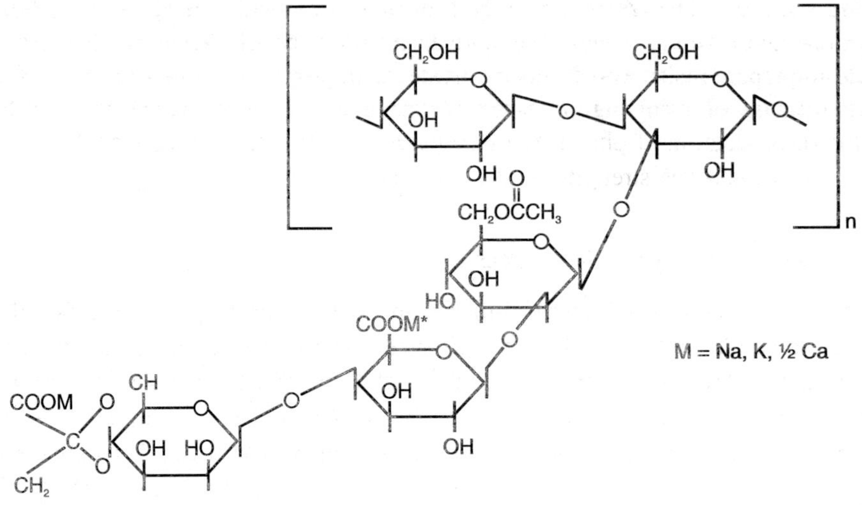

M = Na, K, ½ Ca

Fig. 14.6. Chemical structure of xanthum gum.

2.4.2. *Guar Gum*

Guar gum, a galactomanan from the seed endosperm of the legume *Cyamposis tetragonolobus*, is an important food stabilizer used in a variety of food products, ranging from sauces to ice cream. It is a dirty white-to-yellow powder, the particles of which do not have a characteristic shape [98]. It is a unique carbohydrate polymer because it gives the highest viscosity at equivalent concentrations compared to other carbohydrate polymers [116]. Guar gum has a typical analysis of 2.5% crude fiber, 10%-15% moisture, 5%-6% protein, and 0.5%-0.8% ash [116] (**Fig. 14.7**).

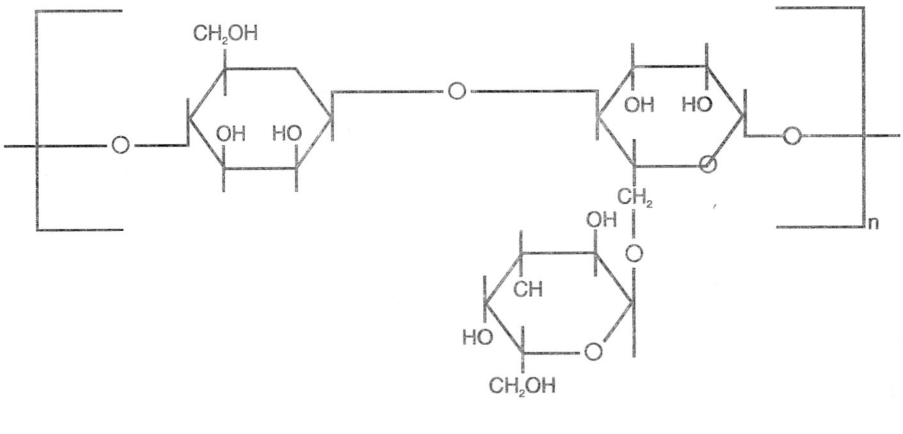

Fig. 14.7. Chemical structure of guar gum.

Guar gum is the most extensively used gum in both food and industrial applications due to its low cost and the above-mentioned properties. It is used as a flocculent or flotation agent, foam stabilizer, filtration aid, and water-treating agent. In the textile industry, it is used as a sizing agent and thickener for dyestuffs. It facilitates wet end processing and improves the properties of paper. It is also used as a gelling agent [117] and thickening agent for slurry explosives.

2.4.3. Locust Bean Gum

Locust bean gum is obtained from the carob tree or *Ceratonia siliqua,* which is grown throughout the Mediterranean region. Pods produced by the carob tree consist of husk, embryo, and endosperm [118], the source of the gum, which is separated from the tough outer husk and yellow embryo tissue. Typical analyses of locust bean gum revealed that the gum powder contains 10%-12% moisture, 1.2% ash, 5% acid insoluble matter, 6% protein, 10-ppm lead, and 3-ppm arsenic [119]. Locust bean gum and derivatives are excellent film formers and can be used as alkali-resistant paints in textile printing and as fiber-bonding and beater additives in the papermaking industries. The effects of the agar content and the use of locust bean gum on the gel properties of ceramic suspensions were investigated by Zhang *et al.* [120]. In many applications, it is used in combination with xanthan gum, agar, and carrageenan, at considerable cost savings. It imparts a smooth, chewy texture or the consistency of ice cream. It is also used for carpet dyeing, paper products, and wet end addition **(Fig. 14.8)**.

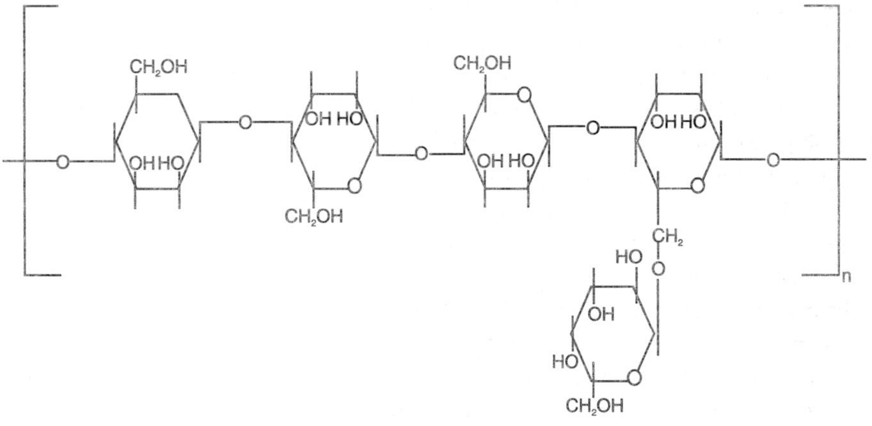

Fig. 14.8. Chemical structure of locust bean gum.

2.4.4. Gum Arabic

Gum Arabic is a dried exudate from a species of the acacia tree *(Acacia senegal)* found in various tropical and semitropical areas of the world [121]. The acacia trees produce gum only under adverse conditions (i.e., lack of moisture, poor nutrition, and high temperatures). Gums of good quality are usually pale and fairly easily soluble in water, but are ionic and give a weakly acidic reaction. The gum contains 15% moisture and 5% ash. The aqueous viscosity depends on the origin of the products, the electrolyte content, the pH level, and pretreatment. The addition of gum Arabic to water results in lowering of the surface tension. The freezing point of gum Arabic solutions decreases as the concentration of gum Arabic increases. Gum Arabic has wide and varied industrial uses, but is mainly used as a stabilizer and thickener in foods [122]. The largest use of gum Arabic is as a crystallization inhibitor in sugar syrup and confections [123]. Other uses are in the preparation of flavor materials. Gum Arabic is widely used to stabilize beverage emulsions and for emulsifying the flavor oils used in these beverages [124]. It has limited application in the cosmetic and pharmaceutical industries as a binder and emulsion stabilizer and in lithography, for which gum Arabic is used widely in plating and etching solutions. It is also used as a blend with polyvinyl alcohol in sustained release of antimicrobial drugs [125].

2.4.5. Kendu Fruit Adhesive

Kendu fruit adhesive (KFA) is a fruit gum obtained from kendu fruit. Kendu is botanically known as *Dispyros cardifolia roxb* [126] belonging to the family Ebenceae and genus *Diospyros*. It is one of the common trees grown in tropical countries like India, Malay, and tropical Australia [127]. It is a hardy tree, attaining a height of 4 to 5 m. Some of the basic aspects of kendu were reported by Sharma *et al.* in 1986 [128]. Certain investigations on basic chemistry and some physicochemical properties of KFA have been reported by Parija *et al.* [129]. The KFA is light brown and semiliquid with a turbidity of 12,000 NTU, a pH of 5.3, and with a fruity odor. It contains about 70% moisture; 6.6% tannin and lignin; 15% total sugar, including 13.7% reducing sugar; and a number of metallic and nonmetallic constituents. As the KFA is rich in lignin and tannin, it can be used as a substitute for phenol in resol formulations to achieve cost-effectiveness. Its bond strength is comparable to that of synthetic adhesives for certain plastic-plastic adhesion.

Natural gums and adhesives are increasingly used in different technical fields (i.e., food, cosmetic, pharmaceutical applications, and heavy ion separations) because of their abilities to form gels and stabilizing properties. Lack of knowledge of the exact qualitative and quantitative distribution of adhesive components in the natural adhesives prohibits their use in the adhesive field. However, these adhesives can be used for binding wood-wood and other conventional interfaces. The bond strength of interfaces can be improved by optimization of the formulations of the adhesives through chemical modification, prevulcanization, and addition of suitable extenders. Natural gums like alginic acid and KGD can be used as novel ecofriendly complexing agents and adsorbents for the separation of different heavy metal ions from the industrial effluents. Adequate information and systematic investigations are to be carried for the maximum utilization of these nature-made adhesives as a compliment and substitute for the nonrenewable petroleum-based synthetic adhesives.

3. CONCLUSIONS AND FUTURE PROSPECTS

As discussed in this chapter, the environmental attributes of being annually renewable and biodegradable, in contrast to the current petroleum based plastics will be a major drive for the entry of sustainable resource based polymers or those derived from agricultural feedstock into the market place. However, cost and performance requirements will dictate whether or to what extent these new coating materials will be successful in replacing the current products.

While recent developments taking place in the industry sector are driven by factors such as costs, new application areas, health and safety as well as environmental issues. The adhesives industry has not escaped the myriad of environmental regulations written over the last decade and is developing technologies that meet the government regulations. Solvent-borne products which release volatile organic compounds (VOCs) on curing are being phased out and replaced by systems that are gaining in popularity such as water-borne adhesives, hot melt adhesives and radiation cured adhesives. The prime focus is on the development of new methodologies and products that allow sufficient adjustments to meet the future environmental demands.

ACKNOWLEDGEMENTS

The authors wish to thank the Materials Research Group of the Department of Chemistry, Jamia Millia Islamia (Central University), New Delhi, India for providing significant information on oil based polymeric coatings.

Appendix I: Full forms of abbreviations used in the chapter.

Full form	Abbreviation	Full form	Abbreviation
Annona squamosa oil epoxy	AOE	Linseed oil epoxy	LOE
Annona squamosa oil based polyesteramide	ASPEA	Melamine formaldehyde	MF
Butylated melamine formaldehyde	BMF	Polyaniline/ coconut oil polyesteramide urethane	PANI/CPEAU
1,6-diamino hexane modified epoxy siloxane	ES–AH	Polyesteramide	PEA
3-diamino propane modified epoxy	E–AP	Polyetheramide	PEtA
1,3-diamino propane modified epoxy siloxane	ES–AP	Polymethylmethacrylate	PMMA
Diglycidyl ether of bisphenol-A	DGEBA	Polystyrene	PS
Epoxy- acrylate	EAC	Pongamia glabra oil based polyesteramide	PGPEA
Epoxy-methacrylate	EMAC	Pongamia oil epoxy	POE
Hydroxylated linseed oil epoxy	HLOE	Styrene-co-maleic anhydride polyesteramide	SCPEA
Linseed oil based polyesteramide	LPEA	Toluylene-2,4-diisocyanate	TDI

REFERENCES

1. A. Higgins; Int. J. Adhesion Adhesives, 20, 367 (2000).
2. P. Pearce, B.C. Ennis, I. Grabovac and C.J. Morris; Adhesion, 47, 123 (1994).
3. H. Baumann, M. Buhler, H. Fochem, F. Hirsinger, H. Zoebelein and J. Falbe; Angewandte Chemie Int. Ed. (Eng.), 27, 41 (2003).
4. S. Ahmad, S.M. Ashraf, S.N. Hassan and Abul Hasnat; J. Appl. Polym. Sci., 95, 954 (2005).
5. K.J. Saunders; in Organic Polymer Chemistry, Chapman and Hall, London (1988).
6. J.M. Keijman; Surf. Coat. Int. A, 04, 142 (2003).

7. A.A. Kumar, M. Alagar and R.M.V.G.K. Rao; J. Appl. Polym. Sci., 81, 38 (2001).
8. C.S. Wu, Y.L. Liu, Y.S. Chiu, Polymer, 43, 4277, (2002).
9. Sharif Ahmad, S.M. Ashraf and Eram Sharmin, Ash Mohammad and Manawwer Alam; J. Appl. Polym. Sci., 100(6), 4981 (2006).
10. T.V.T. Velan and I.M. Bilal; Bull.Mater.Sci., 23, 425 (2000).
11. M. Callow; Chem. Ind., 123, 51, (1990).
12. K. Kroyer; Great Britain Patent No. 1, 307, 001 (1973).
13. http://dsm.iea.org
14. R.F. Bradey Jr.; Polym. Paint Colour J., p. 18 (March 2000).
15. Lori Hauffmann and Harold Hower; J. Protective Coat. Linings, p. 5 (August 2003).
16. R.J. Parmar, N.V. Patel and J.S. Parmar; Paint India, 53(2), 31 (2003).
17. www.dic.co.jp/eng/
18. M. Angelovici, H. Dodiuk and S. Kenig; Int. J. Adhesion and Adhesives, 11(1), 43 (1991).
19. D. Pai; Paint India, 46(5), 33 (2001).
20. P. Venkateshu and P.A. Mahanwar; L(III)9, 29 (2003).
21. S.C. Jain; Paint India, 54(3), 37 (2004).
22. http://www.oit.doe.gov/agriculture/ Chemical-Technical Utilisation of Vegetable Oils Final Conference Proceedings Bonn, Germany, (June 20-21, 2000).
23. K. Hill; Pure Appl. Chem., 72(7), 1255 (2000).
24. L.E. Gast, W.J. Schneider and J.C. Cowan; J. Amer. Oil Chem. Soc., 43, 418 (1966).
25. J. Economy; J. Macromol. Sci. Chem. A, 21, 1705 (1984).
26. S. Ahmad, S.M. Ashraf, A. Hasnat, S. Yadav and A. Jamal; J. Appl.Polym. Sci., 82, 1855 (2001).
27. Fahmina Zafar, Eram Sharmin, S.M. Ashraf and Sharif Ahmad; J. Appl.Polym. Sci., 92, 2530 (2004).
28. T.K. Roy, V.M. Mannari and D.A. Raval; J. Sci.Ind.Res., 56, 159 (1997).
29. F. Zafar, S.M. Ashraf and S. Ahmad; Prog. Org. Coat., 51, 250 (2004).
30. Sharif Ahmad, S.M. Ashraf and Ufana Riaz; Polym. Adv. Technol., 16, 541 (2005).
31. Manawwer Alam, Eram Sharmin, S.M. Ashraf and Sharif Ahmad; Prog. Org. Coat., 50, 224 (2004).
32. Sharif Ahmad, S.M. Ashraf, Eram Sharmin, Mubina Nazir and Manawwer Alam; Prog. Org. Coat., 52, 85 (2005).
33. J. Chen, M.D. Soucek, W.J. Simonsick and R.C. Celikay; Polymer, 43, 5379 (2002).
34. J.V. Crivello and R. Narayan; Chem. Mater., 4, 692 (1992).
35. S. Ahmad, S.M. Ashraf, E. Sharmin, F. Zafar and Abul Hasnat; Prog.Cryst.Growth Ch., 45,83 (2002).
36. Sharif Ahmad, F. Naqvi, Eram Sharmin and Kanak Lata Verma; Prog. Org. Coat, 55(3), 268 (2006).
37. S. Ahmad, S.M. Ashraf, F. Naqvi, S. Yadav and A. Hasnat; Prog. Org. Coat., 47, 95 (2003).
38. S.K. Karmee, P. Mahesh, R. Ravi and A. Chaddha; J. Am. Oil Chem. Soc., 81, 425 (2004).
39. A.H. Bandivdekar and S.B. Moodbidri; Archives of Andrology, 48, 9 (2002).
40. Sharif Ahmad, Eram Sharmin, L. Imo and S.M. Ashraf; Prog. Org. Coat., 50, 47 (2004).
41. M. Maciejewski, M. Kedzierski, E. Bednarek and E. Rudnik; Polym. Bull., 48, 251 (2002).
42. Sharif Ahmad, S.M. Ashraf, Sanjay Kumar, Manawwer Alam and Abul Hasnat; Ind. J. Chem. Tech., 12, 193 (2005).
43. Sharif Ahmad, S.M. Ashraf and Abul Hasnat; Paint India, 47 (2002).
44. Chemical Engineering and News (C-E & N), Cover Story on Nanomaterials (Sept. 2003).
45. Klaus Rose; Proc. Conf. on Org. Inorg. Hybrids II, Paper 3, UK (May 28-29, 2002).
46. D. Zarate and D. Gaughen; J. Protective Coat. Linings, p. 17 (May 2003).
47. K. Pianoforte; Coatings World, (Nov 2002).
48. A. Sellinger; Nature, 394, 256 (1998).
49. R. Quack; Paint India, LIII(7), 39 (2001).
50. M.E. Callow, J.A. Callow, Biofouling; 15, 49 (2000).
51. B. Klajnert and B. Maria; Acta Biochimica Polonica, 48(1), 198 (2001).
52. B. Branko and T&B. Branslav; European Coat. J., 6, 36 (2004).
53. C. Green; Polym. Paint. Colour J., V189 (4420), 36 (1999).

54. A.T. Royappa, N. Dalal and M.W. Giese; J. Appl. Polym. Sci., 2, 9290 (2001).
55. A.V. Ambade and A.J. Kumar; J. Appl. Polum. Sci., Polym. Chem., 39(9), 1295 (2001).
56. B. Bruchmann; Polymer Research, BASF (2001).
57. J. Scheirs and E.T. Long; Modern Polyesters: Chemistry and Technology of Polyesters and Copolyesters, Wiley Interscience, New York (2003).
58. F.P.L. Report, Animal Glues: Their manufacture, Testing, and Preparation (June, 1955).
59. F.P.L. Report, Blood Albumin Glues: Their Manufacture, Preparation and Application, No. 492, p. 13 (May, 1959).
60. F.A. Keimel, Historical Development of Adhesives and Adhesive Bonding in Handbook of Adhesive Technology (Eds. A. Pizzi and K.L. Mittal), Marcel Dekker, Inc., New York, p. 432 (1994).
61. K.B. Armstrong; Int. J. Adhesion Adhesives, 17, 89 (1997).
62. A. Rider and P. Chalkley; Int. J. Adhesion Adhesives, 24, 95 (2004).
63. A.J. Kinloch; Adhesion and Adhesives Science and Technology, Chapman and Hall, New York (1990).
64. J.P. Pascault, H. Sautereau, J. Verdu and R.J.J. Williams; Thermosetting Polymers, Marcel Dekker, New York (2002).
65. R.F.T. Stepto (Ed.); Polymer Networks: Principles of their Formation, Structures and Properties, Blackie Academic & Professional, London (1998).
66. S.K. Ooi, W.D. Cook, G.P. Simon and C.H. Such; Polymer, 41, 3639 (2000).
67. S.G. Prolongo, J.C. Cabanelas, T. Fine and J.P. Pascault; J. Appl. Polym. Sci., 3(6), 2678 (2004).
68. S.G. Prolongo and J.C. Cabanelas; Baselga J. Macromol. Symp., 198, 283 (2003).
69. J.C. Cabanelas, S.G. Prolongo, B. Serrano, J. Bravo and J. Baselga; J. Mater.Process. Technol., 6863, 1 (2003).
70. A. Higgins; Int. J. Adhesion Adhesives, 20, 367 (2000).
71. C.H. Shen and G.S. Springer. J. Composite Mater., 10, 2 (1976).
72. H.G. Carter, K.G. Kibler, J. Composite Mater., 12, 118 (1978).
73. W.W. Wright; Composites, 201 (1991).
74. D.J. Progar and J.F. Dezern; J. Adhesion Sci. Tech., 4, 305 (1989).
75. D.J. Progar and T.L. St.Clair; J. Adhesion. Sci. Tech., 8, 67 (1994).
76. D.J. Progar and T.L. St. Clair; J. Adhesion. Sci. Tech., 5, 711 (1991).
77. K. Frisch and S.L. Reegen; Advances in Polyurethane Science and Technology, Vol. 1, Technomic, Westport (1984).
78. Z. Wirpsza; Polyurethanes Chemistry, Technology and Applications, PTR Prentice Hall, London (1993).
79. L. Nielsen; Mechanical Properties of Polymers and Composites, Marcel Dekker, New York (1994).
80. J. Brandrup and E.H. Immergut; Polymer Handbook, Wiley, New York (1989).
81. R. Tout; Int. J. Adhesion Adhesives, 20, 269, (2000).
82. J.W. Nicholson; Int. J. Adhesion Adhesives, 18, 229 (1998).
83. K.J. Anusavice; Skinner's Science of Dental Materials, Saunders, Philadelphia (1996).
84. G.V. Newman, Amer. J. Orthodon., 51, 901 (1965).
85. C. Charles; Biomaterials, 19, 589 (1998).
86. M.G. Buonocore; J. Dental Res., 34, 849 (1955).
87. J.P. Fricker; Amer. J. Orthod. Dentofacial Orthop., 113, 384 (1998).
88. R. Van Noort; Introduction to Dental Materials, Mosby, London (1994).
89. L. Gorelick, A.M. Geiger and A.J. Gwinnett; Amer. J. Orthodon., 81, 93 (1982).
90. B. Ogaad, G. Rolla and J. Arends; Amer. J. Orthodon., 94, 68 (1988).
91. D.C. Watts; in Materials Science: a Comprehensive Treatise (Eds. R.W. Cahn, P. Haasan and E.J. Kramer), Vol. 14 (Williams DF, editor), VCH, Weinheim, Germany, (1991).
92. W.H. Douglas; Br. Dent. J., 17, 209 (1989).
93. A. Sam and D.F. Williams; Clin. Mater., 14, 243 (1993).
94. J.W. Nicholson and G. Singh; Biomaterials, 17, 2023 (1997).

95. H.A. Alhadainy and A.I. Abdalla; Amer. J. Dent., 9, 77 (1996).

96. P. Kathirgamanathan; Polymer, 34, 2907 (1993).

97. H. Sirringhaus, N. Tessler and R.H. Friend; Synth. Met., 102, 857 (1999).

98. Y. Yang; MRS Bulletin, 22, 44 (1997).

99. J.R. Sheats, Y.L. Chang, D.B. Roitman and A. Stocking; Accounts of Chemical Research, 32, 193 (1999).

100. T. Makela, S. Jussila, H. Isotalo and S. Pienimaa; Synth. Met., 101, 705 (1999).

101. R.C. Patrick; in Treatise on Adhesion and Adhesives, Vol. 2, Marcel Dekker, NewYork (1969).

102. A.A. Lambuth; Adhesives from Renewable Resource: Historical Perspective and Wood Industry Needs. ACS Symp. Ser. (1989).

103. R.K. Goel and R.K. Gupta; Industrial Adhesives and Gums, 1st Ed. Small Business Pub., New Delhi, India (1980).

104. R.J. Gettens and G.L. Stout; Painting Materials, Van Nostrand Reinhold, New York (1942).

105. K.C. Dartey and R.G. Sanderson; Inform Int. News Fats, Oils Rel. Mater., 7(6), (1996).

106. K.T.T. Wang; J. Forest Sci., 13, 1 (1998).

107. I.W. Cottrell, K.S. Kang and P. Kovacs; In Handbook of Water Soluble Gums and Resins (Ed. R.L. Davidson), McGraw-Hill Book Company, NewYork, Ch. 24 (1980).

108. X. Zhao, H. Ma, X. Hongwu and R. Ban; J. Chem. Ind. Eng., 48(2), 2 (1997).

109. A. Amanullah, S. Satti and A.W. Nienow; Biotechnol. Prog., 14(2), 265 (1998).

110. S.-T. Yang, Y.M. Lo and D. Chattopadhyay; Biotechnol. Prog., 14(2), 259 (1998).

111. F. Garcia-Ochoa, V.E. Santos and A. Alcon; Enzyme Microbiol. Technol., 23(1-2), 75 (1998).

112. W. Meer; in Handbook of Water Soluble Gums and Resins (Ed. R.L. Davidson), McGraw Hill Book Co., New York, Chap. 7 (1980).

113. W.H. McNeely and K.S. Kang; Xantham and Some Other Biosynthetic Gums in Industrial Gums, 2nd Ed., Academic Press, New York (1973).

114. W.M. Kulicke, R. Oertel, M. Otto, W. Kleinitz and W. Littmann; Hydrocarbon Technol., 43(12), 471 (1990).

115. Y.R. Xie and N.S. Hettiarachchy; J. Amer. Oil Chem. Soc., 75 (6), 729 (1998).

116. J.K. Seaman; In Handbook of Water Soluble Gums and Resins (Ed. R.L. Davidson), McGraw-Hill, New York, Ch. 6 (1980).

117. E.L. Hirst; J. Chem. Soc., 70, 1 (1942).

118. W.S. Arbuckle; In Ice Cream, AVI Publishing Co., Westport CT, p. 84 (1986).

119. C. Griffith; Manuf. Chem., 20, 321 (1949).

120. T. Zhang, S. Blackburn and J. Bridgwater; J. Br. Ceram. Trans., 93(6), 229 (1994).

121. W. Meer; In Handbook of Water Soluble Gums and Resins (Ed. R.L. Davidson), McGraw Hill, New York, Ch. 8 (1980).

122. A. Goto and K. Ishiwatar, Japan Patent No. 2001017114 (2001).

123. The Printing of Textile Fabrics; ICI: (1959).

124. F. Thevenet; In Flavor Encapsulation (Eds. S.J. Risch and G.A. Reineccius), Amer. Chem. Soc., Washington DC, 37, (1988).

125. V. Kushwaha, A. Bhowmick, B.K. Behera and A.K. Ray; Artif.Cells, Blood Substit. Immobil. Biotechnol., 26(2), 1073 (1998).

126. K.K. Kirtikar and B.P. Basu; Ind. Med. Plants, 2, 1500 (1935).

127. J.F. Dastur; Medicinal Plants of India and Pakistan, D.B. Taraporevala Sons and Co. (1962).

128. S. Sharma, S.S. Sharma and A.K. Machroo, Sci. Rep., 452 (July 1986).

129. S. Parija, M. Misra and A.K. Mohanty; J. Polym. Mater., 14, 297 (1997).

15

Recombinant Polypeptides in Therapeutics

M.S. Nayeem[a] and Rizwan H. Khan[b]*

[a]Department of Chemistry, Aligarh Muslim University, Aligarh—202 002, India.
[b]Interdisciplinary Biotechnology Unit, Aligarh Muslim University, Aligarh—202 002, India.
*Corresponding author's email: rizwanhkhan@hotmail.com

Contents

- **INTRODUCTION**
- **THERAPEUTIC BLOOD RELATED PRODUCT**
- **THERAPEUTIC ENZYMES**
- **THERAPEUTIC ANTIBODIES**
- **THERAPEUTIC HORMONES**
- **THERAPEUTIC INTERFERONS AND INTERLEUKINS**
- **REFERENCES**

Summary: Technological advances in drug development and biological sciences are allowing for the rapid development of new diagnostic methods and drugs based on biological molecules, including proteins and nucleic acids. Currently, several categories of rDNA products, viz. hormones of therapeutic interest, haemopoietic growth factors, blood coagulation products, thrombolytic agents, anticoagulants, interferons, interleukins and therapeutic enzymes are being produced using rDNA technology for human benefit. In view of increasing importance of rDNA products for human healthcare, this chapter provides general/recent information on different aspects, of therapeutic applications of selected rDNA products, especially hormones of therapeutic interest, haemopoietic growth factors, human blood coagulation products, thrombolytic agents, anticoagulants, human interferons, human interleukins and therapeutic enzymes.

1. INTRODUCTION

Before the advent of biotechnology era, therapeutically useful proteins were primarily isolated from human plasma fractions. Products marketed by the pioneers in this industry, such as Cutter Laboratory (now Bayer Corp.), Baxter Healthcare Corp. and Alpha Therapeutics Corp. constituted only a limited number of proteins, e.g., human serum albumin, clotting factor VIII and IX, immunoglobulins, fibrins and antithrombins. Recent advances in various discipline of biotechnology, including progress in genomics and proteomics, have accelerated the discovery and market introduction of an increased number of protein products.

As a class, protein therapeutics include:

1. Naturally occurring human proteins, primarily plasma proteins.
2. Recombinant copies of naturally occurring proteins.
3. Mutated or modified version of a naturally occurring protein having higher efficiency, lower toxicity or altered functionality.
4. Monoclonal antibodies.

While some protein drugs are still being isolated from plasma pool, most of the proteins generated for therapeutic use currently, are either recombinant proteins or monoclonal antibodies. This chapter summarizes the advances that have taken place in recent years in the field of proteins, enzymes, hormones, antibodies and interferons and interleukins as therapeutic products. In 1973, the first gene technology experiment was published, which laid the basis for recombinant therapeutic protein. Less than 10 years later in 1982, the first recombinant protein (human insulin, Genentech) came on to the market. From 1985, onwards, recombinant proteins have gained increasing importance as therapeutic agent. Majority of therapeutic proteins currently available are naturally occurring proteins (insulin, growth hormone, interferon, enzymes, antibodies, hormones, erythropoietin, factor VIII, tissue-plasminogen activator, etc.). The advantages of recombinant products are:

- The provision of a drug that could not be produced by conventional methods e.g. erythropoietin.
- The provision of a safe product free from human pathogenic viruses e.g. factor VIII or hepatitis vaccine.

These proteins are chimeras and fusion proteins or mutated or deleted proteins (muteins), most of which are still at an early stage of development. Methods such as localized mutagenesis, deletion mutagenesis and the manufacture of humanized or single chain antibodies have provided major technical impetus in this area. These are expressed in *E. coli*, immortalized Chinese Hamster Ovary (CHO) cells, transgenic animals and plants, cultivated mammalian cells and BHK (baby hamster kidney) cells etc.

Many biopharmaceutical substances lack stability and/or are not absorbable in a medically useful form through the gastrointestinal tract, the lungs, or the skin. In the gastrointestinal tract, for example, digestive chemicals normally break down protein products. Even injection may not ensure effective delivery to the target cells. To be effective, many injected drugs need to survive transport through the liver and encounters with enzymes. Therefore, how biopharmaceuticals are delivered is very critical. Drug-delivery innovations relevant to biopharmaceuticals include those described below.

Liposomes

A liposome is a microscopic fatty droplet designed to carry a therapeutic substance, especially to specific bodily tissues. The liposomal outer membrane and the outer membrane of the target cell can fuse, whereupon the liposome empties into the cell. Liposomal encapsulation of a therapeutic substance enables increasing the accumulation of the active ingredient in target tissues and controlling the spread of the active ingredient to nontarget tissues, where it might do harm.

Immunotoxins

An immunotoxin is a combination of a monoclonal antibody and a toxic (e.g., anticancer) substance. Because it responds only to specific antigens, the MoAb component limits the toxic effects of the immunotoxin to target (e.g., tumor) cells.

Prodrugs

A prodrug is any medical compound designed to work only after the body or a specific type of tissue in the body has activated it. Prodrugs are useful when the "active" drug is too toxic for nonspecific or general distribution to bodily tissues, when absorption of the "active" drug is poor, or when the body breaks down the "active" drug prematurely. For example, a prodrug that can be activated by only one type of enzyme will work only in tissues that produce that enzyme. Such a prodrug can thus spare nontarget tissues toxic effects. The introduction into tumor cells of genes for enzymes that can activate anticancer prodrugs, a prodrug-activating gene therapy, has been well studied.

Polyethylene Glycol

Frequent injections of a therapeutic protein can result in harmful immune responses. Adding polyethylene glycol (PEG) to therapeutic proteins increases their stability in the body and lengthens the time they stay in the bloodstream, thus decreasing the number of injections needed. PEG can contribute to the treatment of severe combined immunodeficiency disease (SCID). SCID, a hereditary disorder, renders even ordinarily trivial infections so deadly to children that institutionalization or isolation is necessary for their survival. Neither bone marrow transplants nor daily infusions of leukocytes, the conventional treatments, are always effective against SCID. Deficiency of the enzyme adenosine deaminase (ADA) causes about one-third of all cases. Adding PEG to recombinant ADA enables effective weekly infusions, as PEG slows the breakdown of ADA in the body. PEG likewise slows the breakdown of another enzyme, L-asparaginase, which the body produces naturally. Pegaspargase, a combination of PEG and recombinant L-asparaginase, can improve the condition of children with lymphoblastic leukemia.

Broadly the currently used recombinant therapeutic proteins could be classified according to the natural source of these proteins in the following categories:

- Blood and related products.
- Enzymes.
- Antibodies.
- Hormones and growth factors.
- Interferons and interleukins.

The general schemes for the production of these recombinant proteins as well as the construction of hybrid genes are illustrated in **Figs. 15.1 & 15.2** respectively.

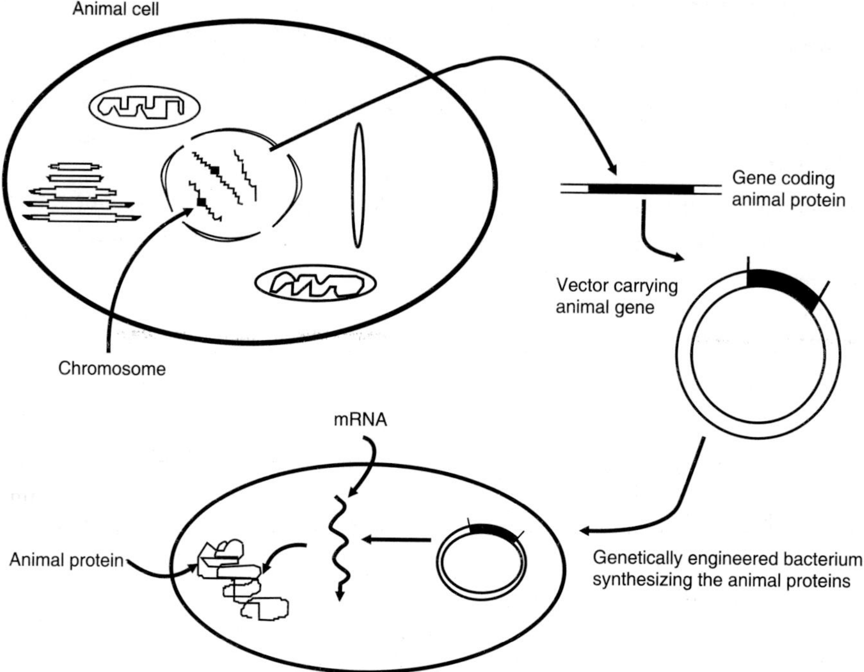

Fig. 15.1. General scheme for the production of animal protein by a bacterium.

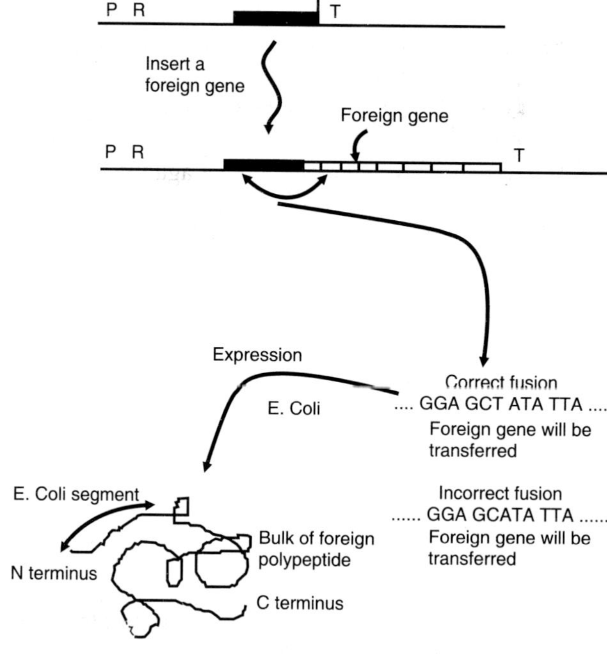

Fig. 15.2. The construction of a hybrid gene and the synthesis of a fusion protein.

2. THERAPEUTIC BLOOD RELATED PRODUCT

2.1. Blood Clotting Factors

A variety of plasma proteins are an integral part of the blood clotting process (Factor I to XIII). The vast majority of hereditary diseases characterized by poor coagulation response results from deficiency of blood factors VIII and IX. Genetic defects, altering the amino acid sequence of blood factor VIII and IX include haemophilia A, von Willebrands diseases (vWD) and haemophilias.

Factor VIII complex composed of two separate gene products. The smaller (170 k Da) polypeptide exhibits coagulant activity and is often designated VIII:C. This polypeptide is coded for by the factor VIII gene. The larger polypeptide, designated von Willebrand factor (VIII:vWB) is predominantly associated with platelet adhesion. This factor is coded for by HvWB gene. Upon synthesis individual von Willebrand factor polypeptide, polymerize forming large multimeric structures. The product of the factor VIII genes, (VIII:C polypeptide) then associates with the multimeric VIII:vWB, which may be co-purified from plasma). The overall structure displays a molecular mass in excess of 1 million Da approximately 15% of which is carbohydrate.

Failure to synthesize VIII:C results in classic haemophilia (Haemophilia A), while failure to synthesize VIII:vWB results on von Willebrand's disease. In the case of haemophilia A, the VIII:vWB gene product is synthesized as usual; however, von Willebrand's disease is characterized by absence of both factor VIII: C and VIII:vWB. Patients suffering from von Willebrand's diseases actually synthesize normal factor VIII:C; however, this polypeptide is rapidly degraded as stabilization of this factor requires its associates with the VIII:vWB polypeptide. Haemophilia B, also known as Christmas disease, results from a deficiency of factor IX. Its clinical consequences are identical to those of classic haemophilia but it does not occur as frequently as the latter disease. The nature and severity of the clinical features of haemophilia depend upon the level of the factor in the plasma. Patients with very low level (e.g. < 1 percent of normal quantity) of factor VIII:C or factor IX are likely to experience frequent and spontaneous bots of bleeding. Person with higher level of active factor (3-5 percent or above) experience less and never clinical symptoms.

Normally the management of these bleeding disorders is achieved by administering concentrates of the relevant deficient factor, which in turn can be obtained by suitable fractionation techniques from plasma obtained from healthy human donors. Recently plasma derived factors VIII preparations have been purified to a greater extent using high-resolution chromatographic technique, including immunoaffinity chromatography.

Recombinant DNA techniques facilitate the production of recombinant blood factors (**Fig. 15.3**). To date several such products gained regulatory approval for medical use. As most block factors display post-translational modifications, (Glycosidation, proteolytic processing and in some cases γ-carboxylation) eukaryotic expression systems are used. Products thus far approved or produced in engineered CHO cells (e.g. Benefix and Re Factor) or BHK cells (e.g. Novoseven). Recombinant blood factors can also be produced in the milk of transgenic animals and the first cloned sheep (Dolly) was also harboring the gene for human factor IX.

After initial recovery from the productive system, recombinant blood factors are subject to a number of downstream processing/chromatographic purification steps in order to yield a purified product, which is usually marketed in lyophilized format.

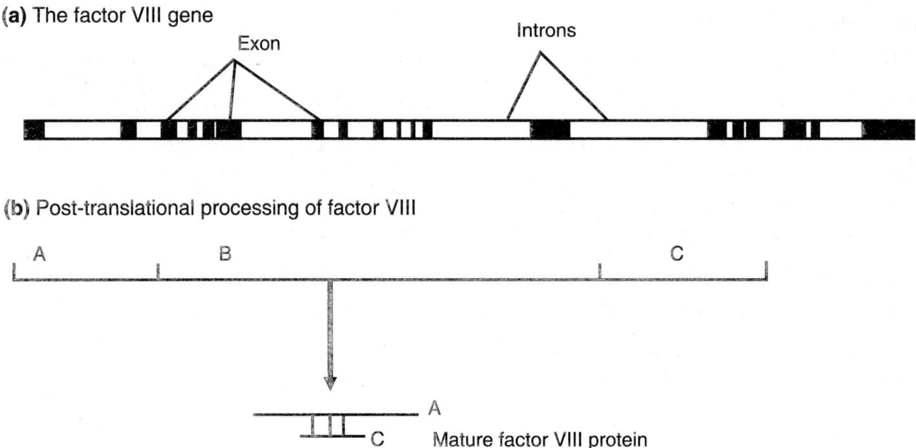

Fig. 15.3. The factor VIII gene and its translation product.

2.2. Anticoagulants

Among the traditional anticoagulants Hirudin is present in the saliva of European Leech (Hirudo medicinalis). Hirudin was reported in the 1880s, though its characterization was not undertaken until the 1950s. The Hirudin gene was cloned in the mid 1980s and has subsequently been expressed in a number of host systems. The polypeptide consists of 65 amino acids and has a molecule weight of 7 k Da. The polypeptide contains a sulfated tyrosine residue at position 63 and is also characterized by a high content of acidic amino acids towards the c-terminal end. The overall conformation of the molecule consists of a globular domain, which is stabilized by three sites of intermolecular disulfide bridges and an elongated c-terminal region.

The anticoagulant activity of Hirudin stems from its ability to bind thrombin i.e. factor IIa tightly. This results in inactivation of the thrombin molecule. Thrombin not only catalyses the proteolytic cleavage of fibrinogen thus forming fibrin and hence promoting clot formation, but also plays a role in activation of factors V, VIII and XIII. Binding of hirudin to thrombin masks both thrombin's fibrinogen binding site and its catalytic site. Hirudin appears to have a number of therapeutic advantages over some other anticoagulants:

- Hirudin acts directly on thrombin.
- It does not require a cofactor to exert its inhibiting effect.
- High doses are less likely to promote hemorrhage.
- It is a particularly weak immunogen.

Over the past number of years recombinant hirudin has become available and two hirudin-based-products (trade names Revasc and Refludon) are now approved for general medical use. Both products are produced in Saccharomyces cerevisiae strain transformed with an expression plasmid housing a synthetic nucleotide sequence coding for hirudin. The products are secreted by the yeast into the fermentation medium from where they are recovered and purified. Both products display identical biological activity to that of native hirudin. Their structure differs from the native molecule only by the absence of a sulfate group on Tyrosine 63.

The saliva of same species of leech contains polypeptide other than hirudin, which exhibit anticoagulant activity. One such anticoagulant is the protein antistasin. Antistasin has molecular mass of approximately 15 k Da and exerts its anticoagulant effect by inhibiting factor X a. This protein is currently the subject of study as it also displays anti-tumour activity. A number of other proteins obtained from various species of leech may also be of potential clinical significance. Specific examples include destabilase, an enzyme which catalyses the depolymerization of fibrin clots and decorsin which seems to inhibit platelet aggregation by interacting with a platelet surface glycoproteins.

Ancord is a serine protease with anticoagulant activity. This enzyme has a molecular mass of approximately 35 k Da and is highly glycosylated. It is purified from the venom of Malaysian pit viper. Ancord catalyses the proteolytic cleavage of micro particulate fibrin molecules prior to clot formation. However, the enzyme has no effect on blood clot once they are formed. It continues to be evaluated in clinical trial.

2.3. Thrombolytic Agents

Tissue plasminogen activator, urokinase and streptokinase are proteins utilized therapeutically as thrombolytic (clot degrading) agents. As such they are used in a variety of situation including treatment of myocardial infarction, embolism, strokes and deep vein thrombosis. Many such products are administered over a relatively short time periods subsequent to thrombus formation. Such treatment is then followed by administration of a suitable anticoagulant for larger period of time.

Human t-PA preparations produced by recombinant DNA methods have been available commercially since the late 1980s. Activase was the first recombinant tPA product to be approved (in 1987) for therapeutic use. It is produced in engineered CHO cells, which harbor a cDNA sequence coding for natural human t-PA. The additional products listed in the table are all engineered forms of t-PA in which three of the five natural t-PA domains are removed. The resultant smaller (355 amino acid) molecule consists as of t-PA's catalytic domain and Kringle 2 domain (the latter is involved in fibrin binding selectively). These engineered forms of t-PA are produced is recombinant *E. coli* and accumulate intra-cellularly as inclusion bodies.

Downstream processing thus includes denaturation and refolding steps, subsequently to cellular homogenization. As *E. coli* cannot undertake post-translational modifications, the products are unglycosylated. Despite the molecular differences between the native and engineered forms of t-PA all exhibit the same clinical efficacy as thrombolytic agents.

Several different t-PA preparations are currently in development or undergoing clinical trial. Recombinant t-PA has most recently been produced in the milk of transgenic animals. t-PA produced in transgenic animal, is predominantly the two chain form of the enzyme. The protein may be purified by a combination of techniques such as acid fractionation, hydrophobic interaction chromatography and immunoaffinity chromatography. As expected, the glycosylation pattern observed on the purified product differs somewhat from the patterns obtained when the protein is produced by the more conventional method of CHO cell culture (Tab. 15.1).

Tab. 15.1. Approved therapeutic blood related products [1,2].

Year of approval	Product	Company	Expression system	Indications
1987	Activase (Alteplase, rh-t-PA)	Genetics	CHO cells	Acute myocardial infarction
1992	Recombinate (rh Factor VIII	Baxter Health Care/Genetics Institute	Animal cell line	Haemophilia A
1993	Kogenate (rh Factor VIII)	Bayer	BHK Cell	Haemophilia A
1993	Bioclate (rh Factor VIII)	Centeon	CHO Cell	Haemophilia A
1995	Novoseven (rh Factor VIIa	Novo-Nordisk	BHK Cell	Some form of haemophilia
1996	Ecokinase (Reteplase, rtPA; differs from human t-PA in that 3 of its 5 domain have been deleted.	Galenus Mannheim	E. coli	Acute myocardial infarction
1996	Retavase (Reteplase, rt-PA	Bochringer Manheim	E. coli	Acute myocardial infarction
1996	Rapilysin (Reteplase, rt-PA)	Bochringer Manheim	E. coli	Acute myocardial infarction
1997	Benefix (rh Factor IX)	Genetic Institute	CHO Cell	Haemophilia B
1999	Re Facto (Maroctocog-alta i.e. B-domain deleted rh Factor VIII)	Genetics Institute	CHO Cell	Haemophilia A

3. THERAPEUTIC ENZYMES

Enzymes as drug have two important features that distinguish them from all other types of drug. First, enzymes often can bind and act as their target with great affinity and specificity. Second, enzymes are catalytic and convert multiple target molecules to the desired products. These two features make enzymes specific and potent drug that can accomplish therapeutic biochemistry in the body that small molecules cannot. These characteristics have resulted in the development of many enzyme drugs for a wide range of disorders.

In 1987, the Food and Drug Administration (FDA) approved the first recombinant enzyme drug, Activase (altiplase, recombinant human t-PA). This clot buster enzyme is used for the treatment of heart attack caused by the blockage of coronary artery by a clot. This was the second recombinant protein drug to be marketed (the first genetically engineered drug was insulin in 1982). The enzymes (the first two have been discussed in the previous section, therefore, rest of the enzymes will be discussed in the forthcoming paragraphs) currently being used in therapy are:

- Ancord → anticoagulant.
- Tissue plasminogen activator → thrombolytic agent.
- Clotting factor VII and IX → treatment of clotting disorder.
- Asparaginase → treatment of some type of cancer.
- Urokinase → thrombolytic agent.
- DNase.
- β-glucocerebrosidase.
- Super oxide dismutase.

Asparaginase

The enzyme is a tetramer with a molecular mass in the region of 130 k Da. As such it may be purified from yeast, fungi and bacteria such as *E. coli*. Asparaginase catalyses the hydrolysis of the amino acid, Asparagine, yielding Aspartic acid and ammonia as evident from **scheme 15.1**.

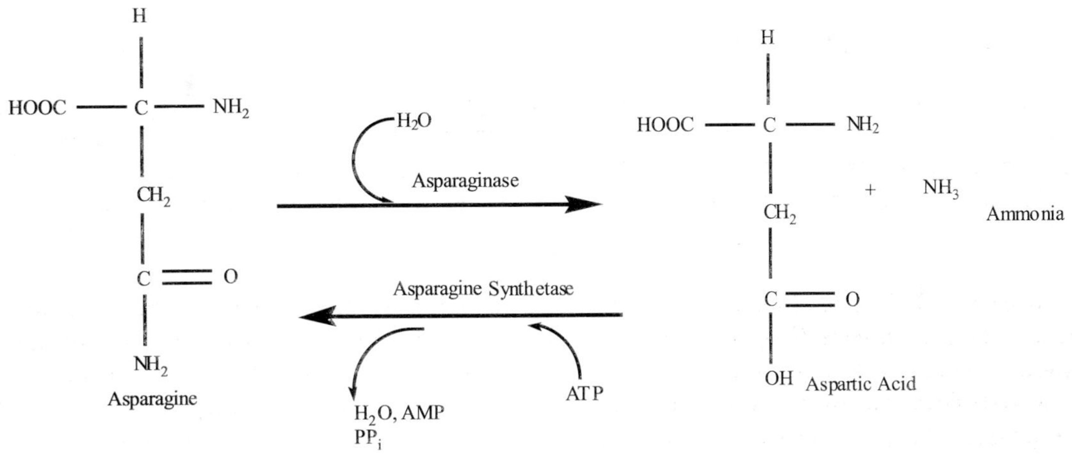

Scheme 15.1. Hydrolysis of asparagine.

All cells require this enzyme to sustain normal metabolic activity. Although most human cells are themselves capable of synthesizing this amino acid, certain malignant cells lack this ability. Intravenous administration of asparaginase results in the rapid depletion of serum asparagine level which normally ranges between 0.5 and 1.5 mg/100 ml. Protein synthesis in malignant cells incapable of synthesizing asparagines is thus severally compromised. In contrast, untransformed body cells begin to synthesize their own asparagines.

This enzyme is mostly used in the treatment of certain forms of childhood leukemia. Asparaginase preparations used clinically are normally purified from *E. coli* or from Erwinia Chrysanthemi but recombinant DNA technology now facilitates the synthesis of asparaginase.

DNase I

This enzyme is a glycoprotein consisting of 260 amino acids. It shows homology with bovine DNase. This enzyme hydrolyses, DNA into small fragment. Recombinant DNase is produced in CHO cell line. It is marketed under the trade name pulmozyme and is used to reduce mucous viscosity and enable the clearance of airway secretion in patients with cystic fibrosis. This therapeutic enzyme was approved is 1993 and sponsored by Genentech Inc.

Glucocerebrosidase

Glucocerebrosidase a glycolipid normally found in small quantities in certain cell type. Deficiency of glucocercbrosidase leads to intra-cellular accumulation of this lipid resulting in swollen organs due to affected macrophage anemia, bone pain and some time neuronal damage. Recombinant enzyme is produced in CHO cell line and is marketed under trade name ceredase and cerezyme. Ceredase approved in 1991 is used for enzyme replacement therapy is patients with Gaucher's disease type I and Cerezyme approved in 1994 is used for patients with type I, II and III Gaucher's disease.

Superoxide Dismutase

Superoxide dismutase catalyses the conversion of highly reactive superoxide (O_2^-) to hydrogen peroxide H_2O_2. This enzyme plays an important role in all aerobic organism, as the superoxide is highly reactive and can cause serious cellular damage by reacting with unsaturated fatty acids in membranes. During normal metabolic activity oxygen molecule is directly reduced to water.

$$O_2 + 4e^- + 4H^+ \rightarrow 2H_2O \tag{15.1}$$

Incomplete oxidation of the oxygen molecular invariably generates molecules such as superoxide (O_2^-), hydrogen peroxide (H_2O_2) and hydroxyl (OH) ion, all of which are extremely reactive. Two specific enzymes protect cell from the harmful effect of superoxide and hydrogen peroxide. These are superoxide dismutase and catalase. The enzyme superoxide dismutase catalyses the conversion of superoxide to hydrogen peroxide and catalase. Catalyses the conversion of hydrogen peroxide into oxygen and water (Superoxide dismutase isolated from bovine liver or erythrocytes has been available for a number of years and has been used as an anti-inflammatory agent).

Tab. 15.2. Approved therapeutic enzymes [3,4].

Year of approval	Product	Company	Indications
1990	Adagen adenosine deaminase	Enzon Inc.	Enzyme replacement therapy for Adenosine deaminase in severe immunodeficiency disease
1991	Ceredase cerezyme/ glucocerebrosidase	Genzyme	Gaucher disease
1993	Pulmozyme/DNase	Genentech	Cystic fibrosis
1994	Oncaspar/PEGylated- L-Asparaginase	Enzon	Acute lymphoblastic leukemia
1996	Sucraid/ Sacrosidase	Orphan Medical Inc	Congenital sucrase isomaltase deficiency
2002	Elritek/Rasburicase	Sanoli Synthelabo Research	Hyperuricemia associated with malignancy or chemotherapy
2003	Fabrazyme	Genzyme corporation	Fabry disease
2003	Aldurazyme/Laronidase	BioMann Pharmaceuticals Inc.	MPS-1
2003	Rectagar/ Galactosidase-A	Trans Karyotic Therapy	Fabry disease

$$O_2^- + O_2^- + 2\ H^+ + \text{Superoxide dismutase} \rightarrow H_2O_2 + O_2 \qquad (15.2)$$

$$H_2O_2 + H_2O_2 + \text{Catalase} \rightarrow 2\ H_2O + O_2 \qquad (15.3)$$

These enzymes have been of interest to the pharmaceutical industry for many years but they have never fulfilled their promise even in the PEGylated form. Surprisingly these enzymes have been shown to prolong the life of (the nematode) caenorhobditis elegans and this effect may translate to mammal. Future version of these enzymes may help to reduce organ injury in hemorrhagic shock.

Besides these enzymes, many enzymes are used since long in different clinical indications. e.g. trypsis, papain and collagenase have often been used as debriding agents.

Chymopapain injection called chemonucleolysis is used for the treatment of sciatica, a medical condition characterized by back and leg pain caused by degradation of an intervetebral disc **(Tab. 15.2)**.

4. THERAPEUTIC ANTIBODIES

Throughout the 1990s, innovative recombinant DNA technology, including chimerization and humanization, enhanced the clinical efficiency of antibodies and led to the recent wave of approvals by the FDA for therapeutic, immunoglobulin (Ig) and Fab molecules (monovalent antibody fragment) produced by proteolysis. Advances in immunoconjugate and chimeric technology have revitalized the "magic bullet" concept of antibody based immunotherapeutics for the treatment of cancer. A multitude of strategies are utilized. Conjugated antibodies are common strategy to obtain both specific targeting and an improved therapeutic strength. The conjugation of antibodies is mainly made with different toxins or radioisotopes. There are several differences between naked and conjugated antibodies in regards to R&D, manufacturing and the market experience. Effect on bulk disease, curative possibilities, target spectrum and technical expertise needed are low for naked antibodies while these are high for conjugated antibodies.

The focus of targeted therapies is to develop Monoclonal Antibody (MOAB) that binds to antigens present preferentially or exclusively on tumor cells. One such focus is epidermal growth factor receptors (EGFRs). Because EGFRs are expressed in greater number on tumor cells and are involved in the signaling pathways that regulate cell division, repair and survival, inhibiting these receptors may inhibit growth or progression of EGFR-expressing tumors. EGFRs are associated with hormone-sensitive tumor such as breast and prostate cancer as well as nonhormone specific cancers such as colon, lung, head and neck and ovarian. Another focus of targeted therapy is vascular endothelial growth factor (VEGF). VEGF is a necessary part of angiogenesis, the development of new blood vessels needed for tumor growth. In B-cell disorders, MOABs target cell surface molecules found on certain cells. These molecules are called clusters of differentiation (CD). The number that appears after CD identifies the antigen such as CD15, CD20, CD33 and CD55. MOAB therapy can be a good choice for treating breast cancer because it can be directed towards a specific antigen, such as the HER-2 growth factor receptor. 20-30% of breast cancers are HER-2 positive, which is a cell surface receptor that sends a growth signal to the nucleus through the tyrosine kinase intracellular signaling pathway. All of these strategies have been exploited and MOABs developed against these receptors are FDA approved for various indications.

The discovery of hybridoma technology by Kohler and Milstein in 1975 heralded a new era in antibody research and clinical development. Mouse hybridomas were the first reliable source of monoclonal antibodies and were developed for a number of *in vivo* therapeutic applications. Innovative recombinant DNA technology, including chimerization and humanization, enhanced the clinical efficiency of mouse antibodies and led to the recent wave of approvals by the FDA for therapeutic immunoglobulin (Ig) and Fab molecules. These developments have continued, and in 2002 the FDA approved the first radiolabeled antibody for cancer immunotherapy (Zevalin).

The list of approved antibody therapeutics against cancer and against viral and inflammatory diseases is growing rapidly, with more than 30 antibodies in late-phase clinical trials.

Intact antibodies provide high-specificity, high-affinity targeting reagents and are usually multivalent. In addition, intact antibodies comprise Fc domains, which can be important for cancer immunotherapy through their abilities both to recruit cytotoxic effector functions and to extend the serum half-life, mediated by the neonatal Fc receptor.

The Fc-induced effector functions are often unwanted and can be simply removed by proteolysis of infected antibodies to yield monovalent Fab fragments. Proteolysis, however, does not easily yield molecules smaller than a Fab fragment, and microbial expression of single-chain Fv (scFv) is currently the favored method of production. In scFvs, the variable (V_B and V_L) domains are stably tethered together with a flexible polypeptide linker. In comparison with whole antibodies, small antibody fragments such as Fab or scFv exhibit better pharmacokinetics for tissue penetration and also provide full binding specificity because the antigen-binding surface is unaltered. However, Fab and scFv are monovalent and often exhibit fast off-rates and poor retention time on the target. Therefore, Fab and scFv fragments have been engineered into dimeric, trimeric or tetrameric conjugates to increase functional affinity through the use of either chemical or genetic cross-links.

Bispecific antibodies contain two different binding specificities fused together and, in the simplest example, bind to two adjacent epitopes on a single target antigen, thereby increasing the avidity. Alternatively, bispecific antibodies can cross-link two different antigens and are powerful therapeutic reagents, particularly for recruitment of cytotoxic T cells for cancer treatment. Bispecific antibodies can be produced by fusion of two hybridoma cell lines into a single 'quadroma' cell line; however, this technique is complex and time-consuming, and it produces unwanted pairing of the heavy and light chains. Far more effective methods to couple two different Fab modules incorporate either chemical or genetic conjugation or fusion to adhesive heterodimeric domains, including designed CH3 domains. Bispecific antibodies provide an innovative alternative therapeutic.

The efficiency of antibodies *in vivo*, for example in cancer therapy, lies in their capacity to discriminate among tumor-associated antigens at low levels. Immunotherapy has been more successful against circulating cancer cells than solid tumors because of better cell accessibility. This is illustrated by the FDA approval of intact antibodies: Rituxan for the treatment of non-Hodgkin lymphoma and Campath and Mylotarg for the treatments of leukemia **(Tab. 15.3)**. Only two monoclonal antibodies have been approved for the treatment of solid tumors: Herceptin for the treatment of breast carcinoma and PanoRex for colon cancer (in Germany). Although the mechanisms of action are still under investigation, Herceptin appears to utilize Fc receptors and angiogenesis, whereas Rituxan activates apoptosis through receptor dimerization.

Tab. 15.3. Approved therapeutic antibodies [5,6].

Year	Product name	Product type	Specificity	Company	Indication
1986	Orthoclone OKT3 muromomab	Mouse	CD3	Ortho Biotech	Transplant rejection
1994	ReoPro abciximab	Chimeric Fab	Gpllb/gplla	Centocor	Cardiovascular disease
1997	Rituxan retuximab	Chimeric	CD20	Idec	Non-Hodgkin lymphoma
1997	Zenapax daclizumab	Humanized	CD25	Protein Design Lab	Transplant rejection
1998	Remicade inflixamab	Chimeric	TNF-α	Centocor	Crohn disease, rheumatoid arthritis
1998	Simulect basiliximab	Chimeric	CD25	Novartis	Transplant rejection
1998	Synagis palivizumab	Humanized	RSV	MedImmune	Respiratory syncytial virus
1998	Herceptin trastuzumab	Humanized	Her-2	Genentech	Metastatic breast cancer
2000	Mylotarg Gemtuzumab Ozogamicin	Humanized	CD33	Wyeth Ayerst	Acute myeloid leukemia
2000	CroFab	Ovine Fab	Snake venom	Protherics	Rattlesnake antidote
2001	DigiFab	Ovine Fab	Digoxin	Protherics	Digoxin overdose
2001	Campath Alemtuzumab	Humanized	CD52	Millenium and Ilex Partner L.P.	Chronic lymphocytic leukemia
2002	Zevalin Ibritumomab Tiuxetan	Mouse	CD20	Idec Pharmaceutical Corp.	Non-Hodgkin lymphoma
2003	Bexxar Tositumomab	antibody conjugated with I-13	CD-20	Corixa Corp.	Non-Hodgkin lymphoma
2004	Avastin Bevacizumab	Humanized monoclonal antibody in combination with intravenous 5-Fluorouracil	VEGF	Genentech Inc	Metastatic cancer
2004	Erbitux Cetuximab	Anti-EGFR antibody	EGFR	Imclone System Inc. NY	Colorectal cancer

Radiolabeled antibodies are important clinical reagents for both tumor imaging and therapy and also provide an effective evaluation of pharmacokinetics. The choice of radionuclide dictates the application. For example, Zevalin is approved for lymphoma therapy as a rapidly cleared, intact mouse antibody to match the clearance rates of yttrium-90. Therapeutic administration requires a balance between long dissociation rates at the target site and slow blood clearance, which can lead to accumulation in the liver and high radiation exposure of other tissues.

Production of antibodies for preclinical and clinical trials has been evaluated in numerous expression systems, including bacteria, yeast, plant, insect and mammalian cells. Mammalian or plant cells are favored for intact antibodies and, occasionally, also for expression of scFvs, diabodies and minibodies.

5. THERAPEUTIC HORMONES

Humulin or recombinant human insulin is the first genetically engineered human recombinant therapeutic. Elily & Co. introduced this to the market in 1982 for the treatment of diabetes. Insulin is a polypeptide hormone that is produced by the β-cells of the pancreatic islets of Langerhans. The mature insulin consists of two polypeptide chains, the A and B chain joined by two disulphides cross links and has a molecular mass of 5.8 k Da. The insulin A chain normally contains 21 amino acid residue whereas the b chain consists of 30 residues. The A chain contains one intrachain disulphide linkage. Self-association of native form of insulin because of hydrogen bonds between amino acids residue towards the C terminal of insulin B chain becomes a rate-limiting factor for insulin absorption. That is why a diabetic must take insulin 30-40 minutes prior to meal.

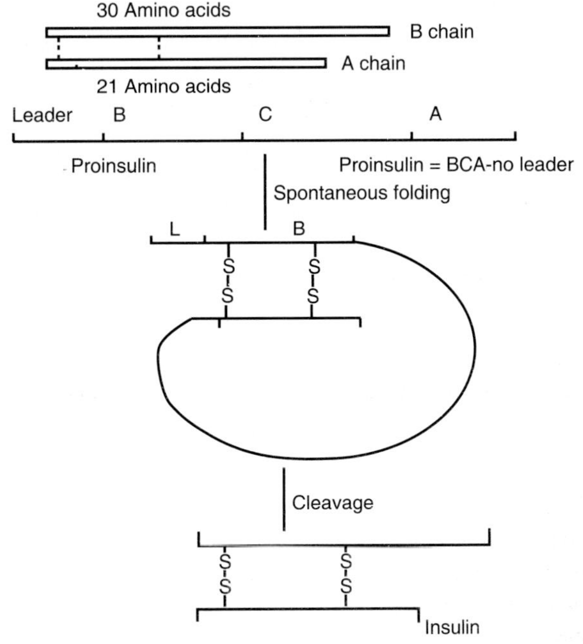

Fig. 15.4. The structure of insulin molecule and its synthesis by processing from proinsulin.

However this sometime becomes inconvenient and results in hypoglycaemia if the person skips the meal for any reason. Altering the sequence of amino acids found towards the C terminus of insulin B chain can reduce the propensity of these molecules to self associate. Insulin lispro differs from native insulin in that the position of proline and lysine residue at position B28 and B29 respectively of native insulin are reversed in this engineered insulin. Insulin aspart differs from native insulin in that the proline residue at position B28 is replaced by aspartic acid.

Glucagon is a polypeptide hormone synthesized in the pancreas by the α cells of the islets of Langerhans. Glucagon consists of 29 amino acids and has a molecular mass of approximately 3.5 k Da. Glucagon is a hyperglycaemic hormone and is occasionally used clinically in order to reverse insulin-induced hypoglycaemia in diabetic patients.

Human growth hormone is a polypeptide consisting of 191 amino acid residues. It consists of two intra chain disulphide links and has a molecular mass of 21 k Da. The expression of the cDNA for HGH in *E. coli* facilitated the large-scale production of ample quantities of growth hormone. The recombinant product is identical in sequence to the native molecule with the exception of an additional methionine residue, due to the presence of an AUG start codon at the 5′ end of the cDNA. The recombinant production of different hormones is schematically exemplified in **Figs. 15.4 to 15.6.**

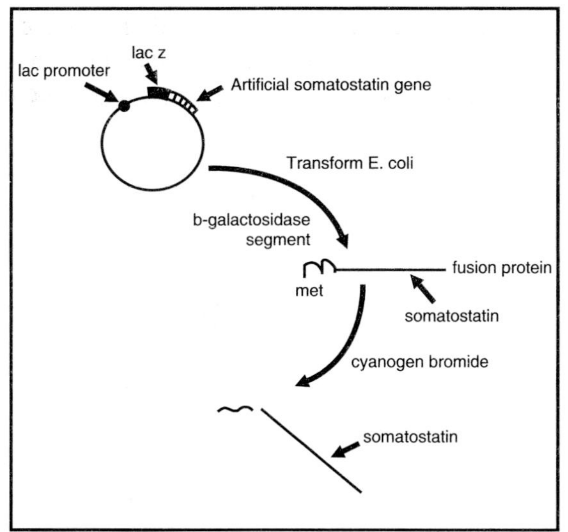

Fig. 15.5. Recombinant somatostatin production.

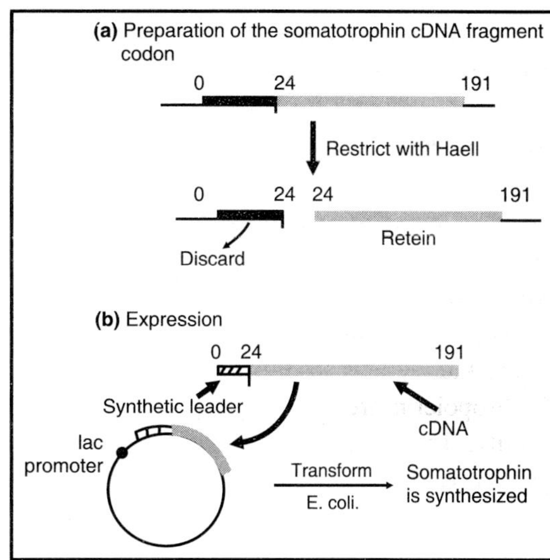

Fig. 15.6. Recombinant somatotrophin production.

Tab. 15.4. Approved therapeutic hormones [7].

Year	Product	Company	Indication
1982	Humulin/Human insulin	Eli Lily	Diabetes
1982	Novolin/Human insulin	Novonortis	Diabetes
1985	Protropin/HGH	Genentech	HGH deficiency
1989	Epogen/Erythrpoetin	Amgen	Anaemia
1990	Procrit/ Erythrpoetin	Johnson & Johnson	Anaemia
1993	Neutropin/HGH	Genentech	HGH deficiency
1995	Norditropin/HGH	Novo Nordisk	HGH deficiency
1995	Genotropin/HGH	Pharmecia & Upjohn	HGH deficiency
1995	Biotrpin/HGH	Bio Technology General	HGH deficiency
1996	Humalog/Insulin Lispro	Eli Lily	Diabetes
1996	Serostim/HGH	Serono Laboratories	AIDS-associated wasting (Cachexia)
1996	Humatrope/HGH	Eli Lily	HGH deficiency
1996	Saizen/HGH	Serono Laboratories	HGH deficiency
1996	Fertinex/FSH	Serono Laboratories	Stimulation of ovulation during assisted reproduction
1996	Puregon/FSH	N V Organon	Anovulation and superovulation
1997	Liprolog/Insulin Lyspro	Eli Lily	Diabetes
1997	Follistim/FSH	Organon	Stimulation of ovulation during assisted reproduction
1997	Neorecorman/Erythrpoetin	Roche	Anaemia
1998	Gonal-F/FSH	Serono Laboratories	Stimulation of ovulation during assisted reproduction
1998	Glucagen	Novo Nordisk	Hypoglycaemia
1998	Glucagon for injection	Novo Nordisk	Hypoglycaemia
1999	NovoRapid/Insulin aspart	Novo Nordisk	Diabetes

The most commonly used gonadotropins include follicle-stimulating hormone (FSH) Leutinizing hormone(LH) and human chorionic gonadotropin (HCG). In the male FSH stimulates the production of spermatocytes, LH stimulates production of testosterone and in the female these hormones play critical role in regulating the reproductive cycle. In the pregnant female, the corpus luteum is maintained by human chorionic gonadotropin, which is secreted by the placenta. HCG is also found in the urine and this forms the basis of pregnancy detection kits. Recombinant FSH, LH, and HCG have all been produced and several of them have got regulatory approval for medical use **(Tab. 15.4)**.

Erythropoietin is a glycoprotein of 36 k Da, 60 per cent of which is carbohydrate. It is produced by the kidney and stimulates the production of R.B.C. from their precursor stem cell. Recombinant Erythropoietin are indistinguishable from native form in terms of primary sequence and biological activity. They are produced in eukaryotic cell lines mainly CHO cell lines in order to facilitate glycosylation.

6. THERAPEUTIC INTERFERONS AND INTERLEUKINS

Besides hormones additional regulatory factors promoting the production, activation/regulation of cells constituting the immune system and synthesized mainly by leukocytes are called cytokines. The first cytokine to be identified and studied was interferon, which was discovered initially in 1957. Since then different types of interferon such as alpha, beta, gamma, omega and tau were isolated and characterized along with their major biological source and activity. Because of its central role in promoting immunological and inflammatory response interferon attracts clinical interest. Recombinant interferon–alphas approved for general medical applications include 'Roferon A', 'Alferon A' and 'Infergen'. Approved indications are summarized in **Tab. 15.5**. Among these, Infergen is different in the sense that it is not a naturally occurring molecule. This synthetic interferon alpha produced by recombinant means in *E. coli*, contains the most frequently observed amino acids in each corresponding position of several naturally occurring interferon alpha subtypes. It displays better antiviral and antiproliferative property as compared to other approved recombinant interferon alphas. Interferon beta produced in engineered CHO cell line, Avonex and Rebif, are approved for general medical use. They are glycosylated like native form. However Betasteron and Betaferon are recombinant interferon beta produced in *E. coli* and they lack glycosylation. But the lack of glycosylation has no significant effect on medical efficacy. Actimmune is the recombinant human interferon gamma produced in *E. coli* and is approved for chronic granulomatous disease. Chronic granulomatous disease is a genetic condition, characterized by the occurrence of recurrent often life threatening infections. The genetic defect inhibits a reduced nicotinamide adenine dinucleotide phosphate (NADPH) oxidase system active in phagocytes, thus inhibiting the production of various oxidative substances, which phagocytes use to destroy microorganism.

The interleukins constitutes another class of cytokines. At least nineteen different interleukins are known and their biological activity and source of production are well characterized. All of these are polypeptide regulatory factors, which initiate their biological activity by binding on cell surface receptor on sensitive cells. Some interleukins such as IL-2 and IL-11 are already employed in the treatment of a number of medical applications while others are currently being assessed in clinical

Tab. 15.5. Approved threpeutic interferon and interleukins [9,10].

Year	Product	Company	Indication
1986	Roferon-A/Interferon alpha-2a	Hoffmann-La Roche	Hairy cell leukemia, AIDS related Kaposi's sarcoma (1988), Chronic myelogenous leukemia (1995), Hepatitis C (1996)
1986	Intron A/Interferon alpha-2b	Alpha Therapeutics	Hairy cell leukemia, genital warts, AIDS related Kaposi's sarcoma (1988), Hepatitis C (1991), Hepatitis B (1992), Malignant melanoma (1995), Follicular lymphoma in conjunction with chemotherapy (1997)
1989	Alferon N/Interferon alpha-N3	Interferon Science	Genital warts
1993	Betasteron/Interferon-b-1b	Berlex/Chiron	Multiple Sclerosis
1996	Avonex/Interferon beta-1a	Biogen	Multiple Sclerosis
1999	Wellferon/Interferon alpha-N1	Glaxo Wellcome	Hepatitis C
1997	Infergen/Interferon alpha	Amgen	Hepatitis C
1995	Betaferon/Interferon beta-1b	Scherring AG	Multiple Sclerosis
1998	Rebif/Interferon beta-1a	Ares Serono Europe	Multiple Sclerosis
1990	Actimmune/Interferon g-1b	Genentech	Chronic granulomatous disease
1992	Proleukin aldesleukin/ Interleukin-2	Chiron	Renal Carcinoma, Metastatic melanoma (1998)
1997	Neumega oprelvekin/ interleukin-11	Genetics Institute	Chemotherapy induced thrombocytopenia

trials. IL-2 is a single chain glycosylated polypeptide of molecular mass 15-20 k Da and consists of 133 amino acid residues. It plays an important role in regulating both innate and acquired immunity. It promotes growth and differentiation of T and B lymphocytes, NK cells, monocytes and macrophages. Native IL-2 is secreted in small amount by activated T helper cells. However recombinant DNA technology now facilitates large-scale production using both native human IL-2 gene and cDNA. The resultant proteins produced in *E. coli* though, not glycosylated as the native form, exhibits biological activity similar to the native form. Lack of glycosylation seems to have little, if any, adverse effect on the biological activity of the product. IL-11 is a 23 k Da polypeptide predominantly produced by fibroblast and bone marrow stroma cells. It is haemopoietic growth factor, which stimulates thrombopoisis. Recombinant human IL-11 produced in engineered *E. coli* strain has recently been approved for the treatment and prevention of chemotherapy induced thrombocytopenia, a condition characterized by a low blood platelets.

REFERENCES

1. Gary Walsh; Proteins Biochemistry and Biotechnology, John Wiley & Sons Ltd. (2002).
2. J. Lusher; Biodrugs, 13(4), 289 (2000).

3. Mvellard; Curr. Opin. Biotech., 14, 1 (2003).

4. J.M. Mates, C. Perez-Gomez and I. Nunez de Castro; Clin. Biochem., 32(8), 595 (1999).

5. P.A. Gregory and M.W. Louis; Nat. Biotech., 23, 1147 (2005).

6. P. Hudson; Exp. Opin. Invest. Drugs. 9(6), 1231 (2000).

7. D. Oxender; Novel Therapeutics from Modern Biotechnology, Springer-Verlag, Godalming (1999).

8. R. Prevost and D. Pharm; Pharmacotherapy 18(5), 1001 (1998).

9. G. Kovecker; International J. Clin. Pract., 54(9), 590 (2000).

10. E.C. Borden, D. Lindner, R. Dreicer, M. Hussein and D. Pereeboom; Semin. Cancer. Biol., 10(2), 125 (2000).

16

Electrically Conducting Polymer Composites

Inamuddin[a] and Amir-Al-Ahmed[b*]

[a]*Analytical and Polymer Research Laboratory, Department of Applied Chemistry, Faculty of Engineering and Technology, Aligarh Muslim University, Aligarh—202 002, India.*
[b]*Sensor Research Laboratory, Department of Chemistry, University of the Western Cape, Private Bag X17, Bellville—7535, Republic of South Africa.*
[*]*Corresponding author's email: amiral@rediffmail.com*

Contents

- **INTRODUCTION**
- **STABILITY AND DEGRADATION OF CONDUCTING COMPOSITES**
- **APPLICATIONS**
- **CONCLUSIONS AND FUTURE PROSPECTS**
- **ACKNOWLEDGEMENTS**
- **REFERENCES**

Summary: Although, there has been substantial research involving electrically conducting polymers over the last two decades, however, practical uses of these polymers in real world applications are still limited by two major obstacles, one is lack of processibility and other instability to electrical and environmental decay. In this connection, research has motivated to the investigators to study of the composite materials based on electrically conducting polymers, while processibility and environmental stability, mechanical properties, the thermostablity of the conductivity, will be induced in these materials by using functionalized doping agents and traditional thermoplastic resins as well as inorganic materials. As results in this chapter, several methods to produce composite materials and their electrical conductivity measurement studies have been reviewed with a variety of applications.

1. INTRODUCTION

1.1. Composite Materials

It is not clear that, how man understood the fact that, mud bricks made with lined straws inside make building sturdier and also last long. The Pharaohs made their slaves to use straw to prepare bricks to enhance the structural integrity of their buildings, which reflects the wisdom of the dead civilization. So mankind has been aware of composite materials since several thousand years before Christ and has applied innovations to improve the quality of life. Though, composites were known to mankind since prehistoric times, the concept and technology went through a sea change with better understanding of the basics facts like the bonding mechanism between the matrix and dispersoids, dispersoid size and distribution, morphological features etc. [1].

In Longman's Dictionary the "composite materials" is simply defined as "something combining the typical or essential characteristics of individuals making up a group". This short statement grasps the very essence of this new material, which has recently emerged with numerous applications in the automotive, aerospace, electronic and wears industry [1].

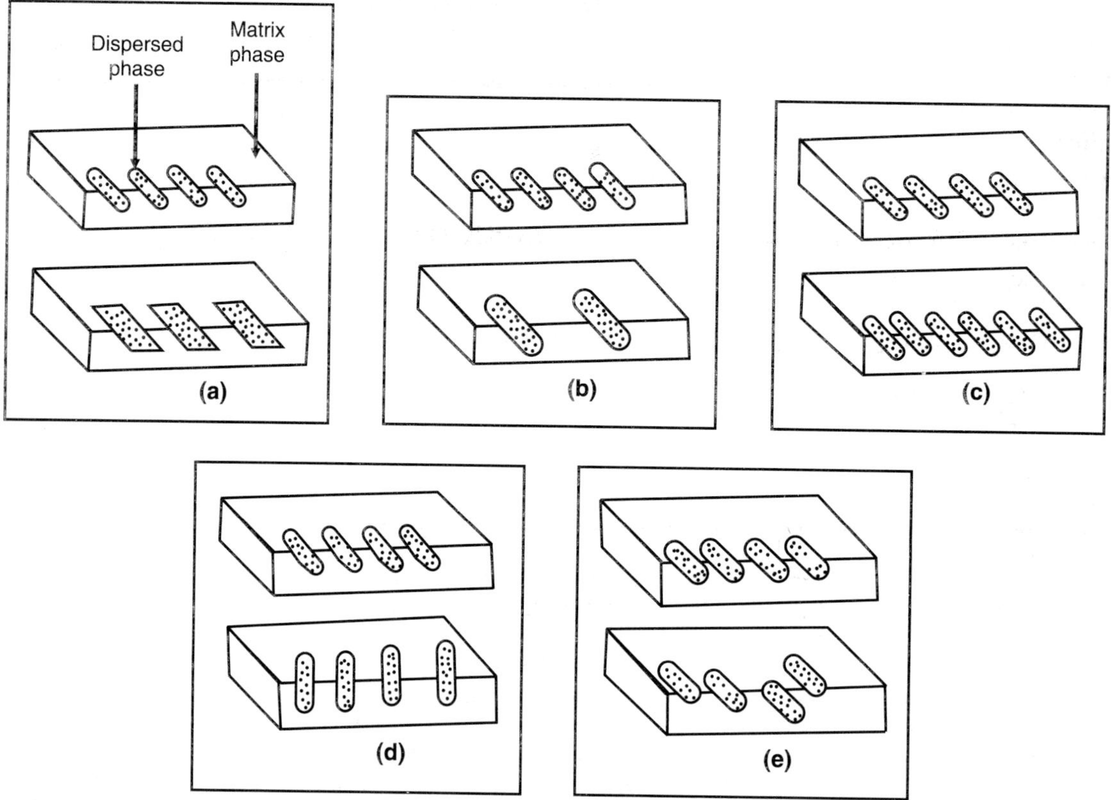

Fig. 16.1. Schematic representations of the various geometrical and spatial characteristics of particles of the dispersed phase (a) shape, (b) size, (c) concentration, (d) orientation and (e) distribution that may influence the properties of composites.

1.2. Classification of Composite Materials

Many composite materials are composed of just two phases; one termed as the matrix, which is continuous and surrounds the other phase, which normally called the dispersed phase. The properties of composites are a function of the properties of the constituent phases, their relative amounts and the geometry of the dispersed phase (i.e. the shape of the particles and particle size, distribution and orientation); these characteristics are schematically represented in **Fig. 16.1**.

One simple scheme for the classification of composite materials [1,2] is shown in **Fig. 16.2**, which consists of three main divisions:

    (a) Particle-reinforced (dispersed phase is equiaxed, i.e., particle dimensions are approximately the same in all directions).

    (b) Fiber-reinforced (dispersed phase has the geometry of a fiber).

    (c) Structural composites (combinations of composites and homogeneous materials).

As noted in **Fig. 16.2**, large particle and dispersion-strengthened composites are the two sub-classifications of particle-reinforced composites. The distinction between these, is based upon reinforcement or strengthening mechanism. The term 'large' is used to indicate that particle-matrix interactions cannot be treated on the atomic or molecular level. Some polymeric materials to which fillers have been added are really large-particle composites. Again, the fillers modify or improve the properties of the material and/or replace some of the polymer volume with a less expensive material—the filler. In dispersion-strengthened composites, materials may be strengthened and hardened by the uniform dispersion of several volume percent of fine particles (that may be metallic or nonmetallic; oxide materials are also often used) of a very hard and inert material. Again, the strengthening mechanism involves interactions between the particles and dislocations within the matrix, as with precipitation hardening; and the strengthening is retained at elevated temperatures and for extended time periods because the dispersed particles are chosen to be unreactive with the matrix phase.

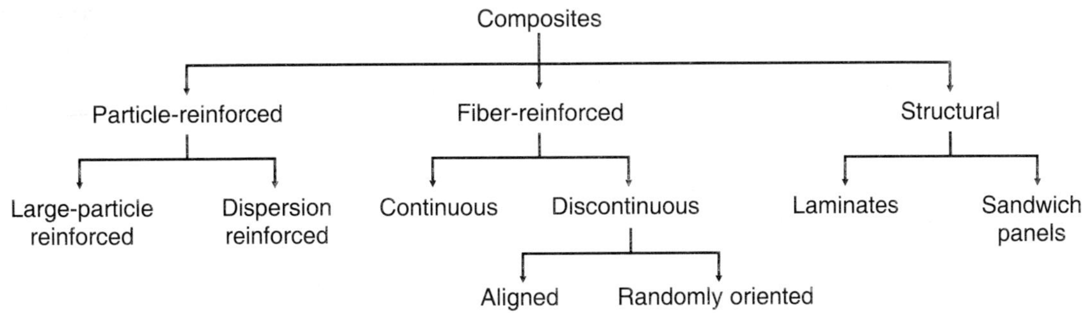

Fig. 16.2. Classification scheme for the various composite materials.

Technologically, the most important composites are those in which the dispersed phase is in the form of a fiber. Here the potential for reinforcement efficiency is maximum and design goals of these composites often include high strength and/or stiffness on a weight basis. These composites with exceptionally high specific strengths and moduli have been produced with low-density fiber and matrix materials. As noted in **Fig. 16.2**, fiber-reinforced composites are sub-classified by fiber length. For short fiber composites, the fibers are too short to produce a significant improvement in strength. On the basis of diameter, fiber reinforcements are classified as whiskers, fibers, or wires. These types of composites are sometimes classified according to matrix type, for example, polymer-matrix, metal-matrix and ceramic-matrix. Polymer-matrices are the most common, which may be reinforced with glass, carbon and aramid fibers. Service temperatures are higher for metal-matrix composites, which also utilize a variety of fiber and whisker types. The objective of many polymer-matrix and metal-matrix composites is a high strength and/or specific modulus, which require matrix materials having low densities. With ceramic-matrix composites, the design goal is to increased fracture toughness. Other more advanced composites and promising engineering materials are carbon-carbon (carbon fibers embedded in a pyrolyzed carbon matrix) and the hybrids (containing at least two different fiber types), laminar composites (composed of two-dimensional sheets or panels that have a preferred high strength direction) and sandwich panels (consist of two strong and stiff sheets, or faces, separated by a layer of less dense material, or structure, or core, which has lower stiffness and lower strength).

Structural composite is normally composed of both homogeneous and well designed composite materials, the properties of which depend not only on the properties of the constituent materials but also on the geometrical design other structural elements. Mainly composites consists of a bulk material i.e. matrix and a reinforcement of some material. So in another way, today's most artificial composites can be divided in four main groups depending on the matrix material as [1,2]:

1. Polymer matrix composites (PMC).
2. Metal matrix composites (MMC).
3. Ceramic matrix composites (CMC).
4. Carbon and graphite matrix composites.

1.3. Polymer Matrix Composites

Polymer matrixes composites are obtained by the incorporation of some basic structural material in the form of particles, whiskers, fibers or a mesh into the polymer matrix. Main characteristics sought in an additive that in turn get conferred on the composite, are the elastic rigidity, tensile and fatigue strength, hardness and appropriate electrical and magnetic properties. So a polymer matrix composite can be considered the combination of one or more materials with a polymer matrix to produce the desired material with properties from individual components [1,2]. Chemically, the matrix materials may be classified in two main categories:

1. Thermosettings.
2. Thermoplastics.

Polymers make ideal matrices as they can be processed easily, possess lightweight and offer desirable mechanical properties. The matrix which provides the continuous phase to the composite should have some distinct properties, like, it should provides a uniform distribution of the structural and environmental load to the reinforcing material through a good adhesion and a strong interface with the reinforcement; it should protect the surface of the composite against abrasion, wear-tear and corrosion (all of these factors initiate fracture); it should absorbs the impact of load and minimizes stress concentrations by enhancing the fracture toughness and it should resist high temperature and withstand repeated cycling of operations, especially under hygroscopic conditions and thus prevents or delays the onset of micro-cracking in the composite.

To meet the demand of materials of improved performance, commercial polymers are always mixed together with various additives of monomeric or polymeric in nature. It is aimed that the additives will act synergistically with the polymer and will meet the combined requirements of a particular application.

1.4. Conducting Polymeric Composites

The simplest conducting polymeric composite consists of a fine metal powder dispersed uniformly throughout an insulating polymeric matrix. Composites based on silver powder can be made with electrical conductivity as high as 10^6 Sm^{-1} at a loading of 85% by weight, where the insulating matrix serves essentially as an adhesive to hold the metal powder in position without disrupting the metal-metal contact. The metal powder composites are unsatisfactory for many applications because of poor mechanical strength, high weight to volume ratio and high level of electrical conductivity. The art of making a good conducting composite is to use the minimum quantity of conductive component to achieve the required degree of electrical performance. There are two main factors related to it:

1. Quality of inter-particle contacts.
2. Shape and size of conductive particles [3,4].

Tab. 16.1. Electrical conductivity of conducting polymer coating on polyester fabric. Reproduced with permission from Dhawan *et al.*; Synth.Met.,125, 389 (2002). Copyright © 2002, Elsevier Science Ltd.

Conducting polymer composite fabric	Conductivity (Scm^{-1})
Insulating polyester fabrics	> 10^{-10}
Polyaniline on polyester surface (doped with Cl$^-$)[a]	6.6×10^{-4}
Polyaniline on coated polyester surface (doped with Cl$^-$)[b]	$6.25 - 1.42 \times 10^{-3}$
Polyaniline on coated polyester surface (doped with Cl$^-$)[c]	$1 \times 10^{-1} - 1.66 \times 10^{-2}$
Polyaniline on high silica fabric	$5 \times 10^{-2} - 3.57 \times 10^{-2}$

[a]Single coating of conducting polymer (polyaniline) on insulating polyester fabrics.
[b]Double coating of polyaniline on coated polyester fabrics.
[c]Triple coating of polyaniline on coated polyester fabrics.

Conducting polymers can also be used to prepare a conducting composite with an insulating polymer matrix like that of poly(vinyl chloride) (PVC), poly(methylmethacrylate) (PMMA) and poly(styrene) (PS) etc. Where the conducting polymer provides the electronic conduction and the non-conducting polymer matrix acts as a solid adhesive to keep the conducting polymer particles together and render mechanical strength without any contribution to the electrical conduction. The conducting polymer composite was prepared by coating conducting polyaniline (PANI) on insulating fabrics like high silica cloth and polyester cloth wherein the coated fabrics retains the flexibility of the conventional fabrics and the electrical conductivity of the composite. The electrical conductivity of polyaniline coated on the fabrics measured by two-probe method [5] is given in **Tab. 16.1**. The perusal of **Tab. 16.1** indicates that whereas the insulated polyester fabrics show a conductivity of 10^{-10} Scm^{-1}, a single coating of polyaniline on polyester fabric shows a conductivity of 6.6×10^{-4} Scm^{-1}. On again coating the polymer on the coated polyester fabrics decreases the electrical conductivity to 6.25×10^{-3}-1.42×10^{-3} Scm^{-1}, whereas the third coating of the polymer on coated surface decreases the electrical conductivity to 1×10^{-1}-1.66×10^{-2} Scm^{-1}. Similarly, coating of

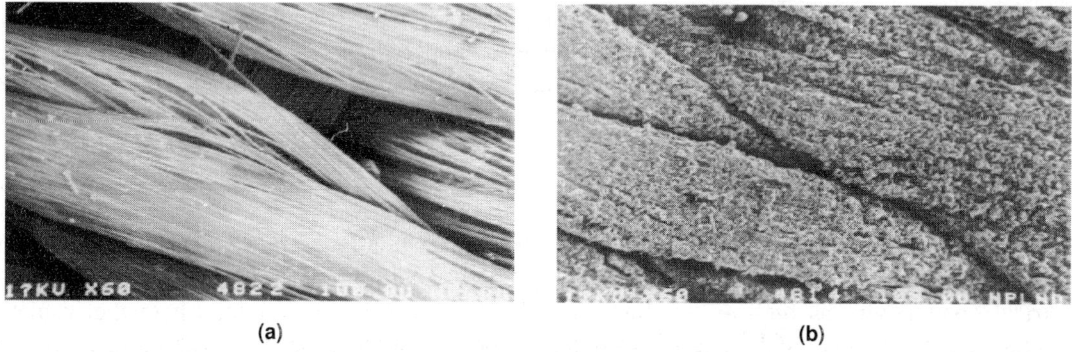

Fig. 16.3. Scanning electron micrograph at 60 magnification of (a) blank silica cloth and (b) polyaniline coated silica cloth. Reproduced with permission from Dhawan *et al.*; Synth.Met. 125, 389 (2002). Copyright © 2002, Elsevier Science Ltd.

conducting polyaniline on high silica cloth shows a conductivity of 5×10^{-2}-3.57×10^{-2} Scm^{-1}. This shows that three coatings of the polymer polyaniline are required on the insulating polyester fabric or high silica cloth in order to achieve electrical conductivity of 1×10^{-1}-3.57×10^{-2} Scm^{-1}. The conducting polymer coated fabrics can then be used for the shielding of electromagnetic interference or for the dissipation of electrostatic charge. The study of scanning electron micrographs of the blank high silica cloth (**Fig. 16.3(a)**) and polyaniline coated silica cloth (**Fig. 16.3(b)**) show a uniform deposition of conducting polyaniline on the fabrics [5]. The dc electrical conductivity measurements of polyaniline:maleic anhydride (PANI:MAc) composites prepared by P.C. Deb *et al.* [6] were carried out by using four-probe method at ambient temperature. **Tab. 16.2** lists the conductivity of PANI:MAc composites. It may be seen that the conductivity is very low (~10^{-6} Scm^{-1}) at MAc concentration of 0.12 M. On increasing the MAc concentration, the conductivity increases to 10^{-1} Scm^{-1}. The lower conductivity of the composites at low MAc concentrations may be attributed to low doping resulting from low concentration of dopant in the reaction mixture. As the concentration of dopant is increased, the conductivity increases rapidly. It may be seen that in spite of low doping level of PANI:MAc composites as compared to PANI:oxalic acid composites [7], the conductivity of the former, synthesized by higher concentration of MAc, is higher. This may be attributed to the presence of double bond and hydrogen bonding by free –COOH groups in MAc, if any, in the dopant structure.

Tab. 16.2. Electrical conductivity of polyaniline salts with maleic anhydride. Reproduced with permission from Samui *et al.*; Synth.Met., 125, 423 (2002). Copyright © 2002, Elsevier Science Ltd.

Salt	*Counter anion value (n)*	*Conductivity (Scm^{-1})*
PANI-MAc I	0.33	1.55×10^{-6}
PANI-MAc II	0.40	1.02×10^{-4}
PANI-MAc III	0.45	8.05×10^{-3}
PANI-MAc IV	0.48	6.55×10^{-2}
PANI-MAc V	0.50	1.05×10^{-1}
PANI-MAc VI	0.52	6.82×10^{-2}
PANI-MAc VII	0.51	2.25×10^{-1}

1.4.1. Organic:Organic Conducting Composites

During the last decades, extensive research efforts have followed the initial discovery of intrinsically conducting polymers (ICP's) containing repeating units of oxidized or reduced monomers, first reported by MacDiarmid and co-workers [8]. After that, the interest in ICP's has developed through three stages:

1. An initial interest motivated by their unique properties and practical possibilities.
2. A decline in interest owing to difficulties in processing and poor mechanical properties.
3. Renewed interest following the discovery of solution and melt processibility of PANI in the early 1990s [9-14].

In recent years, there has been some optimism that striking advances in understanding the chemistry and physics of ICP's [15] will support the development of large scale applications, witnessed by the award of the Nobel Prize in Chemistry to Heeger, MacDiarmid and Shirakawa in 2000. Manufacturers and researchers, who are actively involved in research and development in this field, have also recognized this trend. For example, some leading companies have discussed their strategy and advances in applications of ICP's at two European events: 'Commercializing Conductive Polymers' in February 2002 and 2003 in Brussels and Barcelona, respectively owing to their interesting electrical properties and their potential applications in various fields like electronics, microelectronics and medicine, corrosion protection [16], light emitting diodes (LEDs) [17], electromagnetic interference (EMI) shielding [18] and drug delivery [19], rechargeable batteries, gas separation membranes, antistatic agents and electroluminescent diodes (ELDs) [20-23], modified electrodes, chemical and bio-sensors etc. Because of their important advantages most of the works with electrically conducting polymers have been focused on three main classes of polymeric materials viz. polyacetylene (PA) and its derivatives, polyphenylenes (PPh) and its derivatives, polyheterocyclics such as polypyrrole (PPy) and polythiophene (PTh), polyaniline (PANI) etc. These polymers are relatively stable under ambient conditions and can be conveniently synthesized via oxidative or electro-chemical polymerization [24-27]. Oxidizing agents, such as iodine and $FeCl_3$, are convenient for oxidatively polymerizing monomers and to produce conductive polymers [26].

However, their poor mechanical properties and processibility constitute major obstacles to their extensive applications [28]. To improve the mechanical properties and processibility many kinds of method have been used, including the introduction of long alkyl groups into the main chain [29,30], the synthesis of soluble precursors [31,32], the preparation of conducting polymer composites [33] and so on [29,34]. Among these methods, the preparation of composites is the easier and more effective. In this consequent many kinds of conducting polymer composites, which combine the electrical conductivity of PPy, PANI, PTh and their derivatives with good mechanical properties of insulating polymers viz. polyethylene (PE) [34,35], PVC [36-39], poly(ethylene oxide) (PEO) [40,41], poly(vinyl alcohol) (PVAL) [42-46], poly(etherketone) (PEK) [47] poly(tetrahydrofuran) [48], polystyrene [49,50], polyurethane [51,52], ethylene vinyl acetate copolymer [53,54], polyamides and polyimide [55-58], rubber [59,60], polycarbonate [61] and poly(acrylonitrile) [62] have been reported.

This trend has been driven by the need to replace traditional inorganic conducting fillers and to improve the processibility of conducting polymers, along with their mechanical properties and stability. These composite materials have introduced conducting polymers to practical applications in different fields, including electromagnetic shielding and microwave absorption [63-66], static electricity dissipation [67-69], heating elements (clothing, wall papers, etc.) [70,71], conducting glues [72], conducting membrane materials [73-74], paint coatings for anticorrosion protection [75] and sensor materials [76,77].

As a consequence, conducting composites are very close to the industrial applications mentioned above [63-77]. Nevertheless, the choice of the best method to produce composites with specified characteristics remains an unresolved problem. The problem arises because the processing method may significantly determine the properties of the manufactured composite materials. Thus, our present discussion will be confined to the preparation and electrical properties of the composites of polyheterocyclics because of their attractive properties like relative ease of synthesis, good environmental stability and electrical conductivity.

1.4.2. Organic:Inorganic Composite Materials

It is well-accepted fact that the progress of mankind today is directly or indirectly dependent on advanced technological materials (high performance materials) that perform better and open new dimensions in research and development. Among the major developments in materials in recent years are composite materials. In fact, composites are now one of the most important classes of engineered materials, as they offer several outstanding properties as compared to conventional materials. These materials have found increasingly wider utilities in the general areas of chemical sensors, chromatography, fabrication of selective materials and electrical and optical applications.

Composite materials formed by the combination of inorganic materials and organic polymers are attractive for the purpose of creating high-performance or high functional polymeric materials. Of particular interest is the molecular level combination of two different components that may lead to new composite materials that are expected to provide different possibilities, termed "organic-inorganic hybrid" materials. These hybrid materials usually show properties intermediate between those of plastics and glasses (ceramics). Accordingly, hybrids can be used to modify organic polymer materials or to modify inorganic glassy materials. In addition to these characteristics, the hybrid materials can be considered new composite materials that exhibit very different properties from their original components (organic polymers and inorganic materials), especially in the case of molecular level hybrids. Therefore, organic-inorganic composite materials are of intensive interest in the field of contemporary materials chemistry as these materials can exhibit synergetic extraordinary properties such as electrical, magnetic and optical properties [78], which arise from the synergism between the properties of the organic and inorganic components. There are several routes for the preparation of these materials, but probably the most prominent one is the incorporation of inorganic building blocks in organic polymers by sol-gel process. These materials have gained much interest due to their remarkable change in properties such as mechanical [79], thermal [80-83], electrical [84] and magnetic [85] compared to pure organic polymers. Additionally, the properties of the composite materials depend on the morphology of the phases viz. organic and/or inorganic network, which has to be controlled over several length scales. Therefore, the development of such materials is a 'land of multidisciplinarity' [86], where chemists, physicists, material scientists and engineers have to work closely together to fully exploit this technical opportunity for creating materials and device with benefits of the best of the two worlds, namely, inorganic and organic. Commencing a chemical point of view, one can distinguish between several ways to incorporate inorganic systems in organic polymers depending on the interactions between the moieties: materials with strong (covalent, coordination, ionic): weak (Van-der Waals, hydrogen-bonds, hydrophilic–hydrophobic balance) or without chemical interactions between the two components **(Fig. 16.4)** [86,87]. Based on the structural distinction, Sanchez and Ribot classified the organic-inorganic hybrid materials into two classes [88].

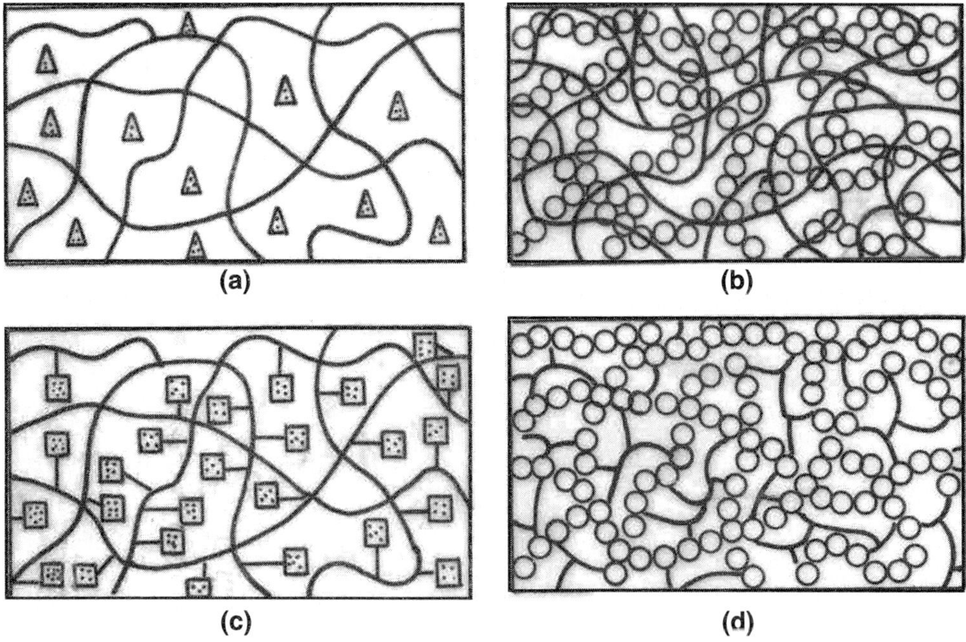

Fig. 16.4. Different kinds of inorganic-organic composite materials. (a) embedding of the inorganic moiety into the organic polymer, (b) interpenetrating networks (IPNs) with chemical bonds, (c) incorporation of inorganic groups by bonding to the polymer backbone and (d) dual inorganic-organic hybrid polymer.

Class I (network modifiers) in which the organic polymer or organic molecules are simply embedded into the inorganic matrix. No covalent bonds exist between both phases only weak interaction such as Van-der Walls forces, electrostatic or H-bonds. As an example of this class I are the organic dyes or biomolecules incorporated in the porous inorganic gel via the H bond between organic and inorganic parts the transparency and no phase separation could be achieved.

Class II (network formers) in which the organic and inorganic parts establish covalent or iono-covalent bonds. To this class correspond the hybrid materials that incorporate functionalized alkoxysilanes such as poly(dimethylsiloxane). This sol-gel method is extensively used to obtain amorphous materials where the homogeneity is directly designed on a molecular scale, ranging from single-phase to multi-phase systems. The combination of several precursors for the synthesis of nanostructured materials hybrid polymers are an open way in the chemical route and can be used to form nanoparticles, coating, fibers or bulk solids with the corresponding technological applications.

Hybrid materials made in this way are termed creamers (ceramic polymers), ormosils (organically modified silicates) or ormocers (organically modified ceramics) [89]. Through the combinations of different inorganic and organic components in conjunction with appropriate processing methods, various types of primary and secondary bonding can be developed leading to materials with unique combination of properties which cannot be achieved by other materials. The concept of 'molecular level mixing' between two different materials can also be considered. Since the late 1980s, molecular level combination between organic polymers and inorganic materials has been of interest. Review articles concerning hybrid materials have been reported by Saegusa and Chujo [90], Sanchez and co-workers [86,88], Schubert *et al.* [91], Novak [92], Loy and Shea [93], Mark [94], Johan N. Hay and Steve J. Shaw [95] and others [96]. During this period, the study of organic-inorganic hybrid materials focused on the following points: firstly, exploration of new preparative methodology for hybrid materials; secondly, new combinations between different materials; thirdly, fictionalization of hybrid materials; and fourthly, modifications of hybrids for industrial applications. It is necessary to point out some important composite materials that are believed to have evolved and will result in interesting molecular hybrid materials in the near future. But due to numerous papers published on polymer organic–inorganic composite materials, it is impossible to completely review this field. Therefore, this review will focus on the most common principal concepts used to incorporate organic systems into an inorganic polymeric matrix and the resulting properties of such materials. The reader is referred to the literature cited for a more detailed description.

1.4.3. Electrically Conducting 'Organic:Inorganic' Composites

The conjugated backbone of electrically conducting organic polymers is responsible for their electroactive character and, therefore, possesses good tunable electrical conductivity as well as electrochromic properties [97-100]. But they are chemically sensitive and have poor mechanical properties and pose processibility problems. The inherent instability is also due to highly unsaturated backbone of conjugated polymers. Stability problems have, therefore, caused many research groups to search for conjugated polymers of high stability. To meet the demand of materials of improved performance, explosive research is going on to synthesize the composites (combinations of desirable properties of each component) of 'organic-organic' and 'organic-inorganic' nature. In the view of the above-mentioned facts, researchers have shown much interest in the study of electrically conducting behavior of 'organic-inorganic' composite materials [91,101-105]. Special interest today is focused on composite system having high conductivity at ambient and sub-ambient temperatures, since they find unique applications, such as separators in high power and rechargeable lithium batteries. Moreover, composite materials composed of oxides or polyvalent metal acid salts and conducting polymers have brought out more fields of applications, such as smart windows, toners in photocopying, conducting paints etc. [106-108].

These organic-inorganic electrically conducting composites may be prepared by mixing a conducting material in an insulating or a conducting polymer matrix. The insulating polymer matrix acts as a solid adhesive, which keeps the conducting components together and provides mechanical strength without any contribution in electrical conduction. Thus, one or more materials (for e.g. insulating polymers, organic molecules, metal powder, inorganic compounds etc.) can be combined with electrically conducting polymer matrix to produce a new conducting material with different physical, mechanical, thermal and electrical properties. Thus, the synthesis of polymeric-inorganic composites has received a great deal of attention because it provided new materials with special mechanical, chemical, electrochemical and optical as well as magnetic properties [109-143]. Various research groups have also successfully applied conventional dispersion polymerization techniques to the preparation of sterically stabilized particles of electrically conducting polymers such as PPy and PANI. The preparation by dispersions solves some problem of their limited processibility. A wide range of steric stabilizers based on various water-soluble polymers, e.g. poly(2-vinylpyridine) [144, 145], poly(vinylalcohol) [146,147], poly(vinylalcohol-co-vinylacetate) [148-155], poly(oxyethylene) [156-158], poly(N-vinylpyrrolidone) [150,151,159,160], poly(vinyl methyl ether) [161-163], proteins [164] and cellulose ethers [165] have been reported by various research groups. Armes *et al.* have recently shown that polyaniline and polpyrrole colloids can be prepared by using colloidal silica [166-170] and, incase of poypyrrole, also tin oxide sols [171] instead of polymeric stabilizer. The electrical conductivity measurements of polyaniline-silica composite [171] prepared by dispersion polymerization are illustrated **Fig. 16.5**. Its conductivity at 25°C is 6.1×10^{-2} Scm^{-1}. As expected with semi-conductors, the conductivity increases with increasing temperature (**Fig. 16.5**). The electrical conductivity of the polyaniline-silica composite is virtually the same as that of PANI:PVAL or polyaniline:poly(N-vinylpyrrolidone) (PVP) composites [172-173] of comparable polyaniline content (**Fig. 16.5**).

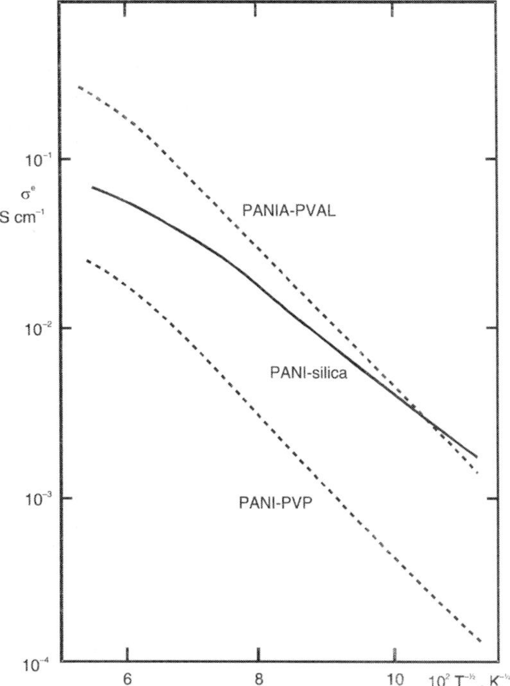

Fig. 16.5. Temperature dependence of electrical conductivity, σ_e for a PANI (37.8 wt%)—silica composite (full line) and its comparison with PANI (48.3 wt%)—PVA and PANI (39.4 wt%)—PVP composites (broken lines). Reproduced with permission from Stejskal *et al.*; Macromolecules, 29, 6814 (1996). Copyright © 1996, American Chemical Society.

Consequently, the electrical properties are independent of the type of the stabilizer, particulate or polymer, used for the preparation of polyaniline. Recently, several groups have also combined conductive polymers with transition metal oxides to generate hybrid organic-inorganic composites [174-177], which possess higher reversible capacity [174], redox cyclability [176] and structure stability [177]. Among the transition metal oxides, several researchers studied [178-180], the V_2O_5:PANI and V_2O_5:PPy nanocomposites. E. Ruckenstein *et al.* [181] synthesized V_2O_5:PANI nanocomposite, which showed relatively high conductivity (10^{-2} Scm^{-1}). The steps of the intercalative polymerization in the mesostructured V_2O_5 are presented schematically in **Fig. 16.6**. Consequently, new direction towards, electronic and optoelectronic devices based on conjugated conducting polymers:inorganic composites comprising semiconducting materials as the active component include light emitting diodes, field effect transistors and photovoltaic devices. Nevertheless, electronic devices based on inorganic materials still dominate the market due to their well defined and the wide spectrum of electronic properties. But the approach of organic-inorganic composite materials could quickly lead to a new generation of inexpensive computer displays or solar cells that are either flexible or embedded within curved plastic, glass or other materials. Because, wholly inorganic semiconductors require processing at high temperatures, making it impossible to embed circuits in plastic or other

Step 1

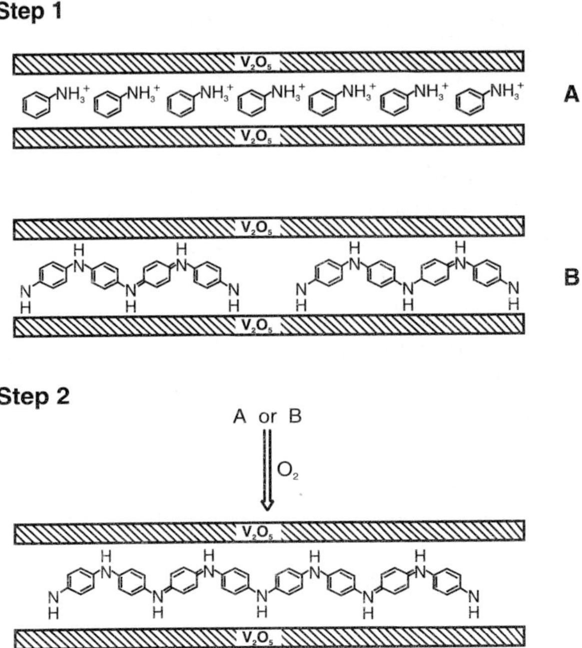

Step 2

Fig. 16.6. Synthesis of a mesostructured V₂O₅:Polyaniline nanocomposite. Reproduced with permission from Li and Ruckenstein; Langmuir, 18, 6956 (2002) Copyright © 2002, American Chemical Society.

heat-sensitive materials. On the other hand, the new hybrid materials "self-assemble," crystallizing from a liquid solution at low temperatures to form alternating organic and inorganic layers of material with the correct semiconducting properties.

1.5. Synthesis Strategies

Although chemists from the sol-gel scientific community initially have worked out organic:organic and or organic-inorganic composites and have attracted a great deal of attention in material science, but at the time of advancement in materials composites are elaborated by researchers coming from a variety of disciplines, polymers chemists, solid-state chemists, catalysis and material researchers etc. Each of these communities elaborate composites using their own tools, specific disciplinary methods and more important their own raw materials. It is not seldom to see that a polymer chemist will work out a composite system having an emphasis on the polymer side of the composite, using even pre-formed polymers, capped oligomers etc. Sol-gel and inorganic chemists will preferably use as precursors, silicon or metal alkoxides or even inorganic building units such as clusters or nanoparticles. They can also use lamellar inorganic compounds as host for organic components. Many names have been given to these materials: ceramers, polycerams, ormosils or ormocers. These names were labeled by materials science researchers coming from different sides (polymer scientists, glass or ceramic scientists, organometallic scientists). Nanomers from materials were the colloidal side of the hybrids chemistry is preponderant and so on. However it is now commonly accepted that a molecular approach for the synthesis of hybrids reflects better the wide opportunities offered by this compounded chemistry. Thus, a critical challenge in the design of these hybrid organic-inorganic and/or inorganic-organic systems is a control of mixing between the two dissimilar phases.

1.6. Percolation in Conducting Composites

The word percolation has been used in the study of flow of fluids in a porous media. Basically, the percolation theory studies the formation and structure of clusters in a large lattice whose sites or bonds are present with a certain probability. The percolation theory is applied in different sectors of our society, from oil fields to forest fires to social science and science. Thus percolation theory is a statistical and geometrical approach to explain the shape of the electrical conductivity curve in composite materials. Percolation theory proceeds from a statistical distribution of the conducting particles, which corresponds to a maximum of entropy. It thus denies any interaction between matrix and conducting particles. As the concentration of conducting particles increases, they become closer to each other and at the critical concentration (Φ_c), they are sufficiently close together or even touching, leading to the conduction of charges as evident from schematic diagram given in **Fig. 16.7.** In carbon black composites, the increase in electrical conductivity is not linear; instead a slow increase in electrical conductivity is followed by a sudden jump at a certain concentration of carbon black in the composites, which is then followed by a slow increase. This sequence of events is called percolation. The carbon black concentration at which the conductivity jumps is known as critical concentration (Φ_c) [182].

2. STABILITY AND DEGRADATION OF CONDUCTING COMPOSITES

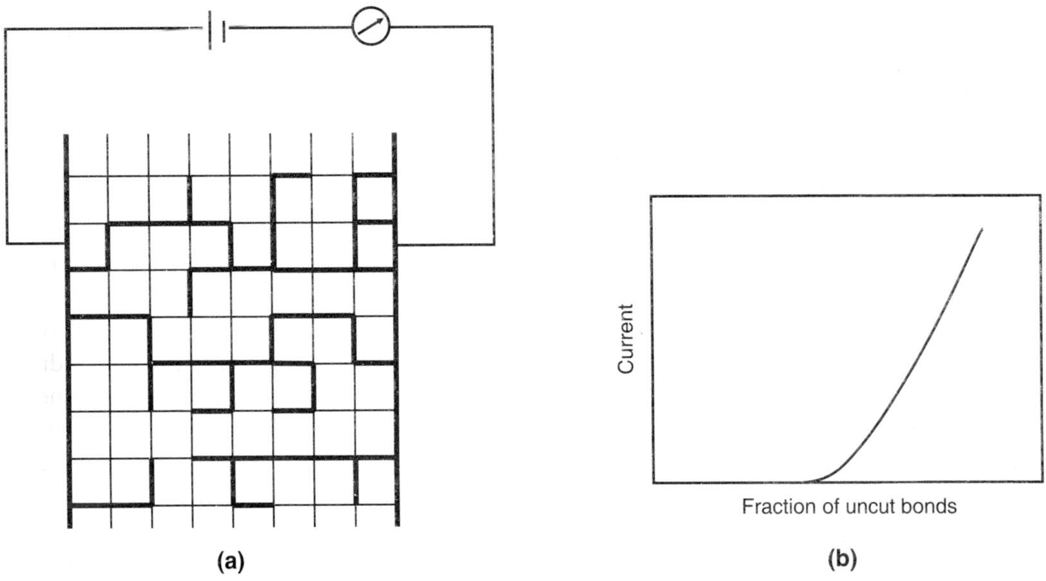

Fig. 16.7. (a) Bond percolation model of a square lattice and (b) Variation in electrical conductivity vs. the increase in the number of connection between two electrodes.

2.1. Stability and Stabilization

The ability of a polymer to retain its useful properties, i.e., structural integrity defined as the stability and the preventive measures undertaken to retain its usefulness, are collectively known as "Polymer Stabilization". There are a variety of methods of stabilizing such systems some of them are as follows:

1. By incorporating antioxidants such as benzoquinone and hindered phenols or by using radical traps such as azo-bis-isobutylronitrile in the system.
2. By ion implantation or by predoping the material with a strong electron acceptor prior to oxygen exposure.
3. By encasing the polymer in a system with reduced oxygen and moisture permeability.
4. By synthesizing new polymers with less susceptibility to intrinsic degradation to oxygen and to moisture, even at elevated temperature.
5. By preparing environmentally stable composites [183-185]. A schematic diagram showing the important factors in the study of stability and stabilization of these polymeric materials is presented in **Fig. 16.8**.

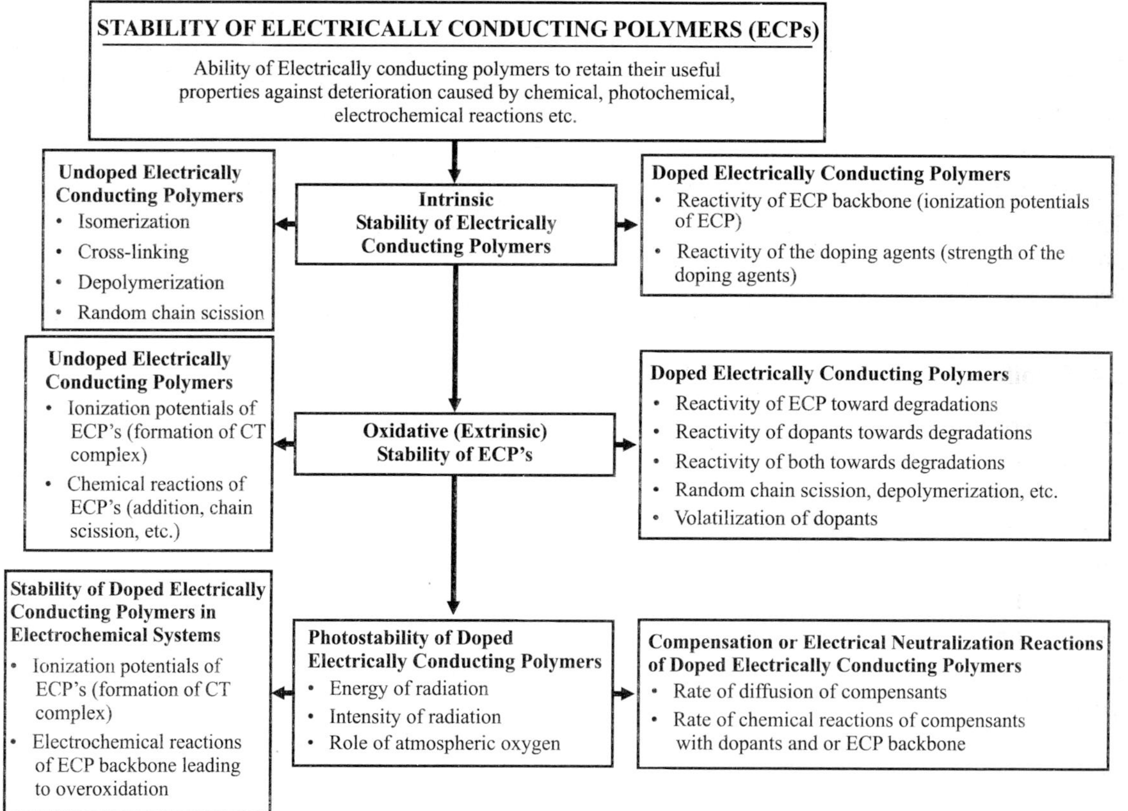

Fig. 16.8. Important factors in the study of stability and stabilization of electrically conducting polymers and composites.

2.2. Thermal Stability and Degradation

Polymer undergoes chemical reactions just like any typical low molecular weight organic compound, leading to the deterioration of useful polymeric properties. The degradation of polymer is characterized by uncontrolled change in molecular weight or constituents of the polymer. The process of deterioration of the useful polymeric properties involving chemical reactions is called as "degradation". There are many external causes of degradation of polymeric materials such as heat, light, mechanical stress, oxygen, ozone, moisture, atmospheric pollutants etc. along with the factors effective at the time of processing. Also, the presence of reactive sites in the polymer (e.g. superoxides, defects, chemically reactive groups etc.) may degrade the polymer properties with or without combination of external factors.

The pristine conjugated polymers have been reported to contain electronic spins leading to the inter-and intra-chain reactions between these reactive sites which can alter the chemical structure even when they are pure, affecting their dopability and, hence, the electroactivity. There are two main factors, which affect the intrinsic degradation of conjugated polymers viz. reactivity of polymer backbone and the reactivity of dopant. The oxidative degradation of most polymers proceeds via the chemical reaction of peroxy radicals.

2.3. Photostability and Degradation

Sunlight consists of IR and visible radiations, apart from high-energy UV radiations (200-380 nm) of the electromagnetic spectrum. The conjugated bonds, present in conducting polymers, undergo n-π^*, π-π^* and σ-σ^* transitions very easily, leading to formation of free radicals on exposure to sunlight. The UV radiations contain enough energy to cause C-C, C-N and C-O homolytic bond fission. Thus produced free radicals can react with atmospheric oxygen leading to oxidation accompanied by depletion of chain length of the polymer.

2.4. Stability and Degradation in Chemical Systems

Of late, conductive polymers are being used as electrode materials in electrical storage devices, i.e., in non-rechargeable (primary) and rechargeable (secondary) batteries. Besides their high electrical conductivity, they also have high selectivity to electrode reactions, low catalytic activity towards side reactions, sufficient mechanical strength, fabricability, low cost etc. The electrode materials must also possess high stability towards degradation reactions during the passage of current or the storage. The degradation of electrode materials leads to instability in electrode potential with time that takes place due to diffusional processes occurring at the electrode and other side reactions such as cross-linking, chemical reactions of dopant ions with polymer etc. For a commercial battery, a good self-life (capability to retain its charged state) as well as ability to repeat charging and discharging many hundred times is a pre-qualification. It has been observed that electroactive polymers are distinct from traditional inorganic electrode materials as electrically conducting polymers neither deteriorate nor redeposit during charging and discharging processes and, hence, they may be expected to give long life storage systems. The durability studies of polymeric electrodes may be done galvanostatically or potentiostatically to evaluate their life in battery application [186,187].

2.5. Overview of Stability and Degradation of Conducting Composites

2.5.1. Polyaniline Based Composites

Sifullina *et al.* [188] reported the electrochemical synthesis of polyaniline composites in porous polyethyleneterepthalate (PET) and polyethylene (PE) matrices prepared by solvent crazing. Electrochemical polymerization of aniline in porous matrices produced highly dispersed polyaniline of morphology different from that of polyaniline prepared in absence of matrices. Distribution of polyaniline in the porous matrices was observed to be dependent on the mode of solvent crazing and the extent of drawing of the solvent crazed polymer.

Kalhori *et al.* [189] reported the surface grafting of polyaniline on silica by chemical polymerization. Whereas the change in electrical conductivity of composite of polyaniline and poly(ethylene-co-vinylacetate) was investigated by Tsanov *et al.* [190]. It was observed that the fraction of polyaniline at which a conductivity jump occurs is not constant during aging as well as during storage. The electrical conductivity of the films of low polyaniline content (up to 2.5 wt%) increased by several orders of magnitude over a period of eight months storage. It was proposed that the polyaniline phase undergoes flocculation, which subsequently forms a continuous conductive network. The electrical conductivity jump was attributed to the dynamic interfacial interactions between the constituents of the composites.

Conductive blends prepared from plasticized cellulose acetate and polyaniline protonated with sulfonic acid, phosphonic acid and phosphoric acid were observed to be highly transparent. Films casted from m-cresol solution showed a percolation threshold below 0.5 wt% and excellent mechanical properties [191]. The significance of the dispersion of the PANI phase for conductivity properties of its blends and their dependence on a matrix polymer was also demonstrated by Zilberman *et al.* [192,193]. They investigated VersiconeTM melt-mixed blends with thermoplastic polymers such as PS, PS plasticized with DOP, PCL, CoPA, LLDPE and LDPE. The blending temperature was chosen depending on the matrix polymer. Thus, blend temperatures were given by PS or LLDPE at 180°C, plasticized PS at 150°C, CoPA at 165°C, LDPE at 130°C and PCL at 70°C. The results showed that the blend morphology and the level of interaction between components of the blends strongly affected the electrical conductivity of the blend, as interpreted from the graphical data for the dependence of the electrical conductivity as a function of PANI:TSA content [192,193]. These data showed that percolation began in the range of ~5–10 wt% PANI:TSA for hetero-chain polar polymers (PCL, CoPA) and plasticized PS-DOP blends. By contrast, PANI:TSA blends with non-polar carbochain polymers (LLDPE, LDPE, PS) were conductive only at PANI:TSA loadings higher than ~30 wt%. SEM and TEM studies of the blend morphology displayed large agglomerates (5–50 μm) of the PANI particles within LLDPE and PS, indicating the absence of a continuous network of PANI even at 20 wt% PANI:TSA. In contrast, the blends of CoPA or PS:DOP exhibited dispersed small PANI particles (0.1–0.5 μm). Moreover, the PANI:TSA particles in the plasticized PS:DOP matrix were smaller than those in the CoPA matrix and even more; in the PCL-based blend with the lowest percolation threshold (~5 wt% PANI:TSA) only a few very small particles were registered. As a consequence, the higher conductivity at relatively low PANI:TSA content in PCL and CoPA was assigned to the higher levels of dispersability and structuring of the PANI-TSA particles within the matrix polymer. Thus, the PCL matrix in the PANI:TSA (20 wt%)/PCL and PANI:TSA (5 wt%):PCL blends showed spherulitic crystallization, in which the spherulites in the 5/95 blend were similar to those of the neat PCL and larger than those obtained for the 20/80 blend. The included PANI:TSA particles were located around the spherulites in the amorphous regions [192,193], as usual for additives in semicrystalline polymers. Hence, the doped PANI network located within these regions, leading to reduction of the percolation concentration. This accounts the significantly lower percolation threshold for the PCL-based blends (5 wt%) in comparison with that for the amorphous-matrix-based blends, PANI:TSA:CoPA (10-15 wt%) and PANI:TSA:PS-DOP (10 wt%) [192,193]. These interpretations accord with those of Shacklette *et al.* [194] discussed above on the effects of mismatches in surface energy and in solubility parameter on the distribution and reaggregation of PANI particles in a matrix polymer.

A green colored porous vycor glass composite with polyaniline was prepared by oxidative polymerization of aniline on aniline impregnated porous vycor glass. Exposure of composite to NH_4OH solution leads to a reversible color change from dark green to dark blue [195]. Jan *et al.* [196] prepared composites of polyaniline with vinylidene chloride-co-methylacrylate and -co-butyl acrylate copolymers by chemical method. Appropriate polymerization conditions made it possible to obtain composites containing ~5 to 50% of polyaniline in the composite and the electrical conductivity was observed to be 1.5 Scm^{-1}.

Charge transport studies were reported on camphor sulfonic (CSA) acid doped composite of polyaniline with nylon-12 (CSA:PANI:Nylon-12) and toluene sulfonic acid (TSA) doped composite of polyaniline with carbon black (CB) (TSA:PANI:CB). The CSA:PANI:Nylon-12 showed a temperature dependent insulator-metal transition and anomalous high microwave adsorption whereas the TSA:PANI:CB showed an unusual insulator to metal transition from the individual materials to the composites. A common mechanism was suggested for the inhomogeneous charge transport due to the formation of ordered regions during the polymerization of conducting polyaniline in both the composites [197].

Electrical conductivity of polyaniline and polycarbonate composite could be increased more than 15 orders of magnitude by a small variation in amount of polyaniline in composite as explained by percolation model. The percolation threshold and critical exponent of electrical conductivity suggests the anisotropy in electrical conductivity while highly conducting composite is found to be stable up to 160°C in terms of electrical conductivity [198].

Percolation threshold of electrical conductivity of emulsion polymerized polyaniline-chlorosulfonatedpolyethylene copolymer and polyaniline-styrene-butadiene terpolymer composites were observed to be 21% and 70% respectively. However, it increased two orders of magnitude and exhibited a new percolation threshold 3.0% for both the composites after secondary doping by m-cresol. The secondary doping noticeably increased permanent set change with polyaniline content as well as changed the ductile fracture into the brittle fracture. Secondary doping did not show any marked effect on tensile strength and elongation [199].

Young *et al.* [200] reported the preparation of electrically conductive polyaniline:polystyrene composites by in-situ polymerization and blending. The electrical conductivity of polyaniline-polystyrene composites was improved with increasing amount of polyaniline and reached to a high value of 0.1 Scm^{-1} at a content of 12 wt% as calculated by elemental analysis. The composites were found very soluble in organic solvents such as chloroform, xylene and N-methylpyrrolidone (NMP), while electrical conductivity of these composites, doped with dodecylbenzenesulfonic (DBSA) acid and thermally treated at 180°C for 3 hours showed good stability.

Polyaniline:polystyrene composite has been prepared by E. Ruckenstein *et al.* [201] in a two-step method. In the first step, an emulsion with the appearance of a gel was prepared starting from a solution of polystyrene and aniline in benzene as the dispersed phase and a solution of sodium dodecylsulphate (SDS) in water as the continuous phase. In second step an oxidant dissolved in an aqueous solution of hydrochloric acid was added drop-wise to the emulsion with vigorous mechanical stirring in order to polymerize aniline and to dope the PANI formed. Depending upon the PANI content in the composites, a green to dark blue solid material was obtained. For electrically conductive polymer composites, there exists a sharp change in conductivity when the volume fraction of the conductive component reaches a percolation threshold. The critical volume percent depends on the shape and distribution of the conductive particles. The critical volume concentration was in the range of 2-10 vol%, which is much lower than the 20-30 vol% observed for polymer composites [202]. The lower critical volume indicated that the PAN1 was well dispersed within the composite. Prepared composites were not brittle and had electrical conductivities as high as 3 to 5 Scm^{-1}.

PANI:PMMA composites [203] have been prepared in the similar way via an emulsion method. The conductivity of the samples exhibits a percolation threshold at a low PANI content of approximately 4 to 10 wt%. The materials have been processed either by cold or hot pressing. Those prepared by hot pressing can have electrical conductivity as high as 2 Scm^{-1} and good mechanical properties, while those prepared by cold pressing have somewhat higher conductivities but somewhat less good mechanical properties. A conductive nylon-6:PANI composite for rechargeable battery electrode and for liquid crystal display with improved mechanical strength and electrical stability was also reported [204].

Phase separation and conductive pathways formation was observed in polyaniline:polyethylene-co-vinylacetate composite films resulting in the formation of polyaniline enriched lower side layers during storage. A distinct difference in the electrical conductivity of the two sides of the films was detected. A mathematical interpretation of the evolution in percolation behavior of polyaniline: polyethylene-co-vinylacetate composites versus time is presented. The unstable conducting properties (both the electrical conductivity jump and phase separation phenomenon) of polyaniline:polyethylene-co-vinylacetate composites over time revealed their dissipative nature supporting the dynamic interfacial model of conductive polymer composites [205]. Zheng *et al.* [206] used electrochemical technique to synthesize highly conducting polyaniline:polyvinylacetate composite film using perchloric acid as a dopant as well as oxidant. Electrical conductivity was observed to be 0.173 Scm^{-1}.

Synthesis and characterization of highly conducting polyaniline:clay nanocomposite with extended chain conformation was reported by Qiuju *et al.* [207]. The conductive emeraldine salt form of polyaniline was inserted into the galleries of montmorillonite to produce the hybrid with highly conducting polyaniline. The product obtained was a nanocomposite and over 90% of polyaniline chains were inserted between the layers. It was also found that the polymerization is a diffusion-limited process [207]. A composite of polyaniline encapsulating titanium oxide (TiO$_2$) with nanometer size has been synthesized by in-situ emulsion polymerization. Particle dimensions have been measured and the nature of the association between the components have been studied using scanning electron microscope (SEM) and transmission electron microscope (TEM) techniques. The interaction between PANI and TiO$_2$ and the nature of chain growth have been investigated and explained with the help of FTIR spectroscopic studies. The improvement in thermal stability and crystallinity of nanocomposites of polyaniline and TiO$_2$ has been evaluated by using TGA and XRD. The mechanism of charge transport and photo induced charge transfer in composites has also been reported [208].

Polyaniline composites with vinyl chloride, vinylidine chloride, methyl methacrylate and butyl acrylate using Haloflex dispersion were prepared by Beadle *et al.* [209]. The loading of PANI component in such composite may be determined from their nitrogen content and can be readily controlled over a very wide range (0.9 to 88%) by simply varying the initial copolymer latex concentrations. The conductivity of the composite samples indicates typical percolation threshold behavior with a percolation concentration of approximately 5-10 wt/wt%. The temperature dependence of conductivity is approximately linear over the temperature rang 132-287 K. This is consistent with either one-dimensional variable range hopping model or alternatively a fluctuation induced tunuelling model.

Polyaniline thermally blended with butadiene-styrene rubber at different weight composition and then was capillary extruded to fibers. Microscopic analysis on the extrudates revealed that polyaniline was deformed during the process to produce elongated structures, i.e., ellipsoids or even short fibers in the blends. Electrical measurements were performed and it was found that blends with more than 20 weight percentage polyaniline could produce an electrical conductive composite with a good level of conduction. The relationship between the volume conductivity and content of polyaniline in the blends showed characteristics of a percolation system, with a threshold as low as 5-weight% of polyaniline [210].

Oyama *et al.* [211] reported a composite of 2,5-dimercapto-1,2,4-thiodiazole and polyaniline on a copper current collector that provides high charge density exceeding 225 Ah/Kg cathode with average discharge voltage at 3.4 V. The composite cathode showed excellent rate capability and cyclability of >500 cycles. While large increase in the charge density to 550 Ah/Kg cathode is achieved by adding elemental sulfur (S_8) to the composite cathode.

A conducting elastomeric foam composite from an elastomer foam and polyaniline was reported by Bessette *et al.* [212]. Only ~5% of conductive polymer is required for insulator to conductor transition and the electrical conductivity of the composite foam could be effectively controlled between 10^{-7} and 10^{-1} Scm^{-1} by varying the amount of oxidant used and/or by variation in the copolymer composition. J.Y. Bergeron [213] has been chemically synthesized poly(aniline-co-N-butyl aniline) by 0.1 M perchloric acid and ammonium peroxidisulphate as oxidant. The effect of the actual composition (in mole fractions of aniline units F_1) of chemically synthesized matrix on their electrical and physical properties depends on their reactivity ratios. The reactivity ratios to chemical oxidation of the monomer pair were found to be controlled by their oxidation potentials Epa. The aniline co-monomer has a lower reactivity than the N-butyl aniline co-monomer, their reactivity ratios r_1 and r_2 being respectively 0.4 and 8.9. Consequently, a very large discrepancy between feed composition f_1 and co-polymer composition F_1 was found which suggested poly(aniline-co-N-butylaniline) have quasi-block structure with well-defined polyaniline and poly(N-butylaniline) sequences [213]. The co-polymers showed a percolation transition in the conductivity for aniline contents higher than 15%. They are probably the first known example of a 'quasi composite' where both conducting and insulating segments are located within the same macromolecule.

Several protonated poly(o-anisidine):polyvinylalcohol composites were prepared using different types of acids such as sulfuric, p-toluenesulfonic, camphorsulfonic and p-dodecylbenzenesulfonic acids. The linear dependence of the log of electrical conductivity on the variation of humidity was observed for all the composites, which was caused by the salt-base transition of the conducting polymer, i.e., by the movement of free acid between the active sites of the conducting polymer and the strongly bound water existing in polyvinylalcohol, which in turn depended directly on the environmental humidity. The response time of the composites to humidity was shortened with a decrease in the size of the dopant anions [214]. The composite of polyaniline and its derivatives with polyvinylalcohol were observed to be useful materials as humidity sensors as reported by Shiigi *et al.* [215]. Certain composites like poly(o-phenylene diamine) and poly(o-aminophenol) with polyvinylalcohol exhibited linear dependence of electrical conductivity versus atmospheric humidity whereas those of poly(m-phenylenediamine) and poly(o-toluidine) with polyvinylalcohol showed non-linear dependence on humidity without hysteresis. H. Shiigi *et al.* [216,217] also prepared composites of polyaniline and its derivatives with polyvinylalcohol. They developed sensors based on this composite to detect carbon dioxide and humidity.

A conducting composite of polyaniline and polycarbonate (PC) was prepared by a blending using chloroform as a solvent by Lee *et al.* [218]. Polycarbonate containing sulfo group enhance the Coulombic interaction between two phases of the composite. The effect of ionic groups in sulfonated polycarbonate was monitored using measurements of both the mechanical and thermal properties. The electrical conductivity was increased to 7.5 Scm^{-1} on doping with camphorsulfonic acid or dodecylbenzenesulfonic acid. The effect of interactions of PANI with its host polymer for composites of PANI:DBSA:PC prepared by an inverted emulsion polymerization method developed in accord with Ruckenstein *et al.* [219-222] was confirmed by Jeon *et al.* [223]. Investigating the effect of DBSA concentration in the emulsion reaction mixture on the final composite conductivity, they found that the electrical conductivity of the composite increased by about three-fold from a value of 4.5×10^{-3} Scm^{-1} (at 16.7% PANI) as the mole ratio of DBSA:aniline was increased from 0.75 to 3. FTIR spectroscopy on the composite showed the existence of hydrogen bonding between PANI and PC, which increased the glass transition temperature with increasing PANI content. Moreover, comparison of differential scanning calorimetry (DSC) and conductivity data showed that the electrical conductivity increased around the glass transition temperature. The authors explained this by the fact that the PANI chains contacted more frequently and facilitated electron transfer through the hydrogen bonding between PANI and PC. In addition, the tensile strength of the composite decreased with PANI content below the percolation threshold (13 wt%) of PANI (**Fig. 16.9(a)**). This suggested that PANI functioned as a defect in the PC matrix in accord with scanning electron microscopy data, which showed an inhomogeneous distribution of PANI in the PC matrix below 13-wt% of PANI [223]. In contrast, the continuous increase of the tensile moduli of the composites (**Fig. 16.9(b)**)

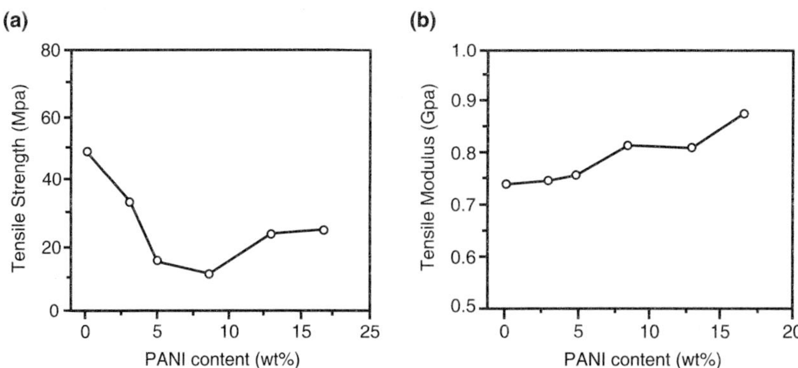

Fig. 16.9. (a) Tensile strength and (b) tensile modulus of the PC and PANI:PC composites as a function of PANI content in composite. Reproduced with permission from Jeon *et al.*; Synth.Met., 104, 95 (1999). Copyright © 1999, Elsevier Science Ltd.

is attributed to the higher rigidity of PANI molecules [223]. It is difficult to demonstrate discrete PANI and PC components by SEM above the percolation threshold (13 wt%), suggesting a fine distribution of PANI in the matrix. Together with the mechanical behavior [223], this suggests that the structure of the PANI:PC composite is changed at high content of PANI due to a physical— chemical interaction (e.g. hydrogen bonding) of the component. This interaction may also be displayed by improved thermal stability of PANI:PC blends [224] at the time of investigation of electronic transport properties of PANI:TSA:PMMA and PANI:TSA:PVC blends. Kaiser *et al.* [225] found that blending PANI with PMMA and PVC increased the conductivity [226], especially at lower temperatures **(Fig. 16.10)**. This increase was ascribed to lessening of insulating barriers around PANI particles in these blends. The temperature dependence of the conductivity in PANI blends was well described by a series combination of quasi-1D metallic resistivity and tunneling (between small metallic islands). However, it should be stressed again that unlike these PMMA and PVC cases and than that of the unblended PANI:TSA.

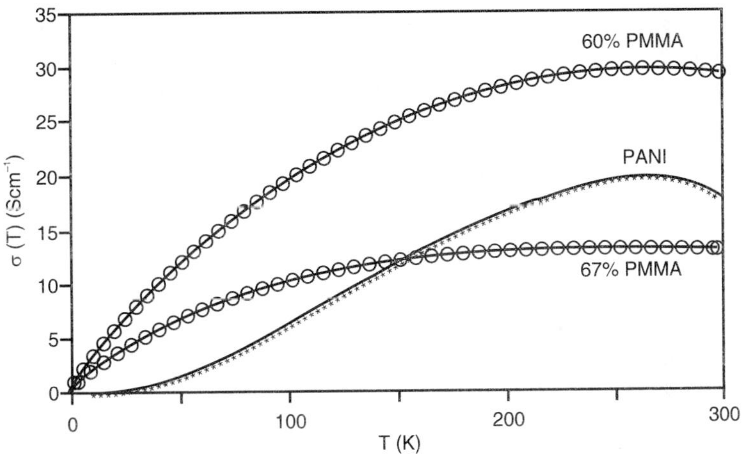

Fig. 16.10. Temperature dependence of conductivity of PANI:TSA:PMMA blends (with 60 & 67% PMMA) and unblended PANI:TSA. Reproduced with permission from Subramaniam *et al.*; Solid State Commun., 97, 235 (1996). Copyright © 1996, Elsevier Science Ltd.

Iroh and Rajagopalan [227] electrochemically codeposited polyaniline and polypyrrole on carbon fibers under potentiostatic conditions. Electropolymerization was carried out by varying the applied potential and the feed ratio of monomers. Weight gain plots indicated that polyaniline is preferentially formed when the monomer ratio is 90% aniline:10% pyrrole, however, more polypyrrole was formed as pyrrole concentration increased. The thermal stability of the composites was observed to be between the stability of homopolymers of the components. The morphology of the composite was affected by the polymerization potential and monomer feed ratio.

An improvement in the electrical conductivity of polyaniline:nylon-6 composite fabrics was observed on surface modification of nylon-6 fabrics by various plasma treatments by Oh *et al.* [228]. Electrical conductivities of polyaniline:nylon-6 composite fabrics were highly increased by ultrasonic treatment, which assisted the diffusion of aniline into the inside of nylon fabrics by cavitation and vibration. The fabric conductivity was also observed to increase with increase in monomer concentration and the polyaniline deposition cycles.

Nylon-6 based PANI composites [229] were also prepared by diffusion-controlled system. The conductivity of PANI:Nylon-6 composite films is 1.12×10^{-2} Scm^{-1}. Thermal stability of PANI:Nylon-6 composite films doped with various protonic acids such as HCl, benzenesulfonic acid (BSA), sulfosalicylic acid (SSA) and TSA was investigated by Byun *et al.* [230]. The conductivities of all the doped PANI:Nylon-6 composite films decayed at elevated temperatures in air with the loss of conductivity being most pronounced at 150°C or higher, which is above both the glass transition temperature of the matrix polymer and the dissociation temperature of the dopants. By applying Arrhenious equation:

$$\log k = \log A - (E_a/2.303R)\,(1/T) \tag{16.1}$$

All the plots of lnk vs 1/T did not give linear results. The conductivity degradation kinetic behavior of the composite films doped with SSA and TSA was first order but those of HCl and BSA doped films were not. The non-linear behavior suggests that multiple mechanisms of degradation are operating in case of HCl and BSA doped composite films such as degradation of the dopants and subsequent reaction of the products of the degradation with polyaniline or a reaction between the dopants and polyaniline backbone. Park *et al.* [231] prepared polyaniline:polyacrylonitrile (PANI:PAN) composite by electrochemical polymerizing aniline on the polyacrylontrile (PAN) coated platinum electrode and they compared the properties on the composite with that of the polyaniline.

2.5.2. *Other Composites*

Conducting composites consisting of polybithiophene and based on porous crosslinked polystyrene as host polymer were synthesized in the vicinity and above the percolation threshold by chemical oxidative polymerization by M. Morsil *et al.* [232]. Measurements of electrical properties versus the temperature for different degrees of doping with Cu(ClO$_4$)$_2$ were carried out on this composite and it was observed that the room temperature conductivity depends strongly on the doping level and decreases from 0.4 Scm^{-1} for the highest doped sample (y = 2.08) to 0.13 Scm^{-1} for the lowest doped sample (y = 0.83) as evident from **Fig. 16.11**. The experimental variations are weakly thermally activated for all compounds. The activation energy is found to be nearly equal to 30, 54 and 27 meV for y = 2.08, 1.25 and 0.83, respectively.

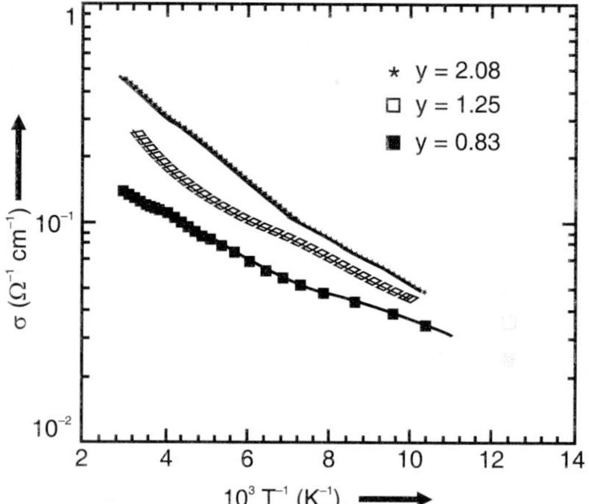

Fig. 16.11. Electrical conductivity vs. I0³ T⁻¹(K) on conducting Cu(ClO$_4$)$_2$:Polybithiophene:polystyrene composites with different dopant concentrations. Reproduced with permission from Morsil *et al.*; Synth.Met., 76, 273 (1996). Copyright © 1996, Elsevier Science Ltd.

Thermoelectric power (TEP) of this composite (**Fig. 16.12**) shows positive low values and exhibits a quasi linear dependence on temperature. At room temperature, the TEP is found to be nearly equal to 45 μVK^{-1} for the highest doped sample and decreases down to 20 μVK^{-1} for the least doped sample. The experimental curves have been fitted on the basis of a linear quasi-metallic model for which thermoelectric power S is written as AT+B: The predominant AT term can be related to the 'metallic' conduction within the clusters. A is found to be nearly equal to 0.13, 0.09 and 0.08 PV μVK^{-2} for y = 2.08, 1.25 and 0.83, respectively.

The B term is associated with hopping mechanisms between conducting clusters. For the two samples of lower concentration, B is equal to 2 μVK^{-1}, whereas it takes the value 6 μVK^{-1} for the highly doped sample. There is a strong contrast between the behavior of the TEP, which is quasi metallic and the thermally activated variations of the electrical conductivity. This apparent conflict has also been seen in highly doped polyacetylene [233] and in poly(vinylalcohol):polypyrrole composites [234], with resistivity being dominated by insulating barriers and thermopower by metallic regions between the barriers.

The method of chemically initiated oxidative modification of polypropylene (PP) particles in suspension by pyrrole was used for preparation of conductive polypropylene:polypyrrole (PP:PPy) composites [235]. Their properties were compared with either PP:PPy composites prepared by melt mixing of virgin polypropylene with chemically synthesized polypyrrole or with polypropylene:carbon black composites prepared also by melt mixing. As a result, the influence of the amount of polypyrrole and carbon black in the composites on their electrical conductivity was investigated. It is well known that at low loading of the conductive filler the modest conductivity increase is observed with rising tiller content. Sudden conductivity increase occurs in relatively narrow concentration range around so-called percolation threshold. At this concentration a conductive network is formed within the insulating phase and a dramatic increase in conductivity by several orders of magnitude is observed. This phenomenon was broadly studied and well described in several reviews [236,237]. The conductivity comparison of PP:PPy and PP:carbon black (CB) composites is shown in **Fig. 16.13**. The PP:PPy composites with the concentration of pyrrole varying from 1 to 10 wt% were prepared by chemical modification at the same reaction conditions. Bulk conductivity of virgin

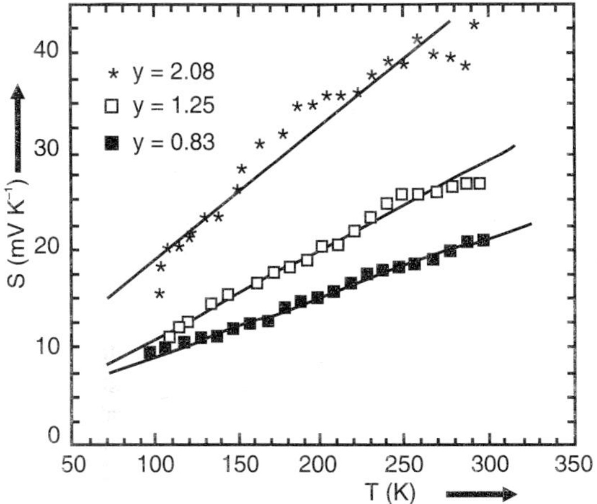

Fig. 16.12. Thermoelectric power(S) vs T (°C) of conducting Cu(ClO$_4$)$_2$:Polybithiophene:polystyrene composites with different dopant concentrations. Reproduced with permission from Morsil *et al.*; Synth.Met.,76, 273 (1996). Copyright © 1996, Elsevier Science Ltd.

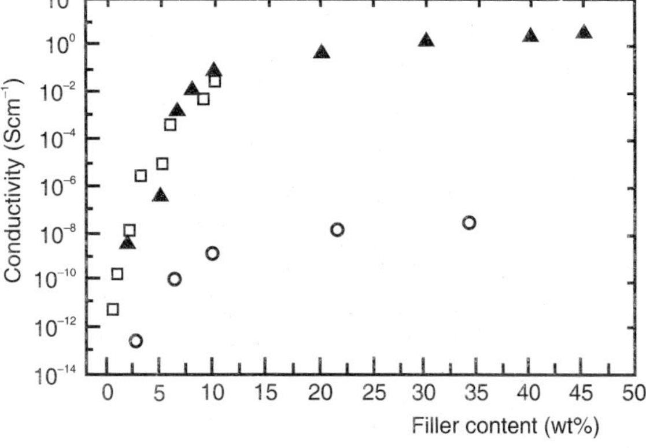

Fig. 16.13. Conductivity dependence on filler content of (□) PP:Ppy composites prepared by Method I, (O) PP:PPy composites prepared by Method II and (Δ) PP:CB composites prepared by Method III. Reproduced with permission from Omastova *et al.*; Synth.Met., 102, 1251 (1999). Copyright © 1999, Elsevier Science Ltd.

PP was found to be about 10^{-16} Scm^{-1}. Even a very small PPy amount present in composites results in a significant conductivity increase. The increase of the content of PPy from 1.1 wt% to 8.9 wt% in PP:PPy composites results in an enormous enhancement in conductivity by eight orders of magnitude. The melt mixed PP:PPy composites show considerably lower conductivity than the composites prepared by method I. The difference in conductivity is about 7 orders of magnitude for the same PPy concentration and the conductivity values do not exceed 10^{-1} Scm^{-1}, even when the PPy content is 34.2 wt%. The conductivity of PP:CB composites showed similar tendency to that of PP:PPy composites prepared by chemical modification. The estimated percolation threshold of CB content was found to be between 5 and 8 wt%. Tieke and Gabriel [238] have reported polypyrrole:polyimide (PPY:PI) composite films prepared by electrochemical and chemical oxidation of pyrrole on a polyimide-coated electrode (I) or by exposing polyimide films containing ferric chloride to pyrrole vapor (II). Type (I) films consist of a sequence of three layers with the polyimide sandwiched between two layers of PPy. The maximum conductivity was found to be ~10 Scm^{-1} for the type I and 5×10^{-4} Scm^{-1} for the type (II) materials. The composite films are thermally stable up to 350°C.

A surface absorption method was also used to obtain conductive polyimide films with poly(3–methylthiophene) and polypyrrole [239]. The materials were prepared by soaking the polyimide films in neat pyrrole or 3-methylthiophene and then polymerizing by the oxidative method. These methods use the imidised polymer (viz. the polyimide) as the substrate and therefore will have to contend with limited diffusion of the conducting pieces or its precursor into the rigid host matrix. Furthermore, these techniques specifically deal with absorbing the monomer and subsequently polymerizing it on the polyimide surface. While Tieke and Gabriel [238] do indicate that their composites are stable at elevated temperatures, in general, the nature of the interaction between the guest and the host, the extent of polymerization and the stability of the composite to the thermal and solvent environments are not apparent from the above methods.

A layer of poly(3-dodecylthiophene)(p3ddt) is embedded at the surface of polyimide by permeation of the polythiophene (in solution of tetrahydrofuran (THF)) into a solution of polyamic acid in n-methyl pyrrolidinone (NMP) or dimethylacetamide (DMAc) [240]. The resulting composite could be imidised chemically and the conducting polymer was stable in the composite even after solvent extraction. The sheet resistance of the composite decreased rapidly upon exposure to iodine; this effect was similar to, though slower than that observed for pure dodedcylthiophene. Also, the composite recovered conductivity after a few hours of exposure to the dopant subsequent to de-doping. PPY:PI composites were prepared by the electrochemical polymerization of pyrrole onto a polyimide-precoated electrode by F. Selampinar [241] and the conductivity of the composite films depends on the pyrrole content. Almost equal conductivities on both sides of the films suggest that the homogeneity was achieved, at least, in terms of conductivity.

J. Yang *et al.* [242] prepared polypyrrole:polypropylene (PPy:PP) conducting polymer composite films with good mechanical properties and high electrical conductivity using a two-direction permeation/diffusion polymerization method. The maximum conductivity of these composite films, 6 Scm^{-1} is of the same order of magnitude as that of pure polypyrrole, 4.5 Scm^{-1}, which was synthesized by chemical oxidation of pyrrole monomer in aqueous solution. The composite films exhibit quite uniform mechanical properties in the film plane. Young's modulus is in the range 0.4-0.8 GPa, tensile strength 40-85 MPa and elongation at break 90-200%, maintaining the mechanical properties of polypropylene in the composite. The mechanical properties of these highly conductive materials are summarized in **Tab. 16.3**.

Tab. 16.3. Mechanical properties of polypyrrole:polypropylene composite films. Reproduced with permission from Yang *et al.*; Polymer, 37,793 (1996). Copyright © 1996, Elsevier Science Ltd.

Mechanical properties	PP microporous films	Ppy/PP composite films[a]	Pure PPy films
Young's modulus (Gpa)	0.4-1.2	0.4-0.8	—
Tensile strength (MPa)	60-120	40-85	23.6
Elongation at break (%)	50-200	90-200	3

[a]Reaction time, 5 min.; concentration of pyrrole, 0.2 M; concentration of oxidant, 0.5 M; temperature, 10°C.

M. Tang *et al.* [243] also reported the electrical conductivity for a polymer composite consisting of PPy and an insulating host, polystyrene (PS). The host polymer was blended with pyrrole monomer using either supercritical carbon dioxide (SC CO_2) or high-pressure liquid carbon dioxide (HPL CO_2) as the carrying solvent. After the blending process, the blended host polymer was doped with an oxidizing agent. **Fig. 16.14** shows the electrical conductivity of the polymer composite as a function of the doping time where a 2.25 M $FeCl_3$ aqueous solution was used as an oxidizing agent. A rapidly increasing curve is observed at the beginning of doping period. It is observed that, for doping times longer than 20 min, the conductivity curve shown in **Fig. 16.14**. Hence a doping time of 20 min was used in this experiment

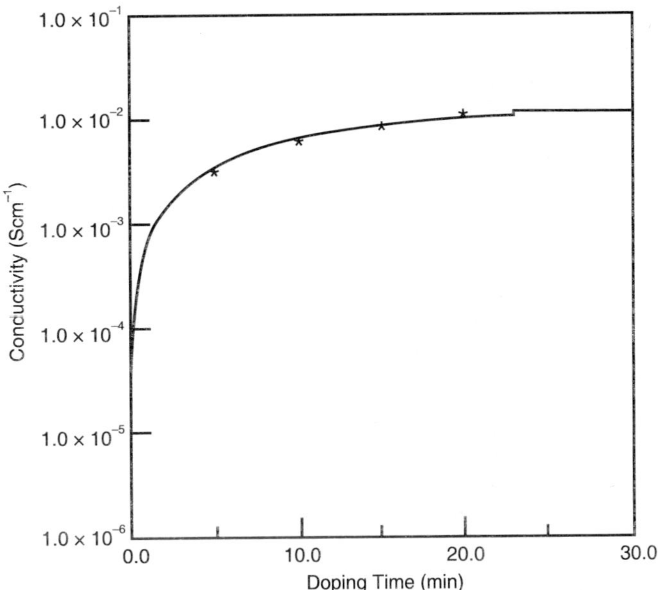

Fig. 16.14. Plot of the electrical conductivity of the PPy:PS composite against the doping time. Reproduced with permission from Tang *et al.*; Europ. Polym.J., 39, 143 (2003). Copyright © 2003, Elsevier Science Ltd.

in order to obtain maximum electrical conductivity for the composites. The doping time for other concentrations of iron chloride was also taken as 20 min. Comparisons of the conductivity of composites were made at the same doping time and on oxidant concentration. After the doping process, the sample was dried under vacuum for 72 h. The blending process was operated at three temperature and pressure conditions with the same density of CO_2. Various concentrations of the oxidizing agent were used in the doping process. The conductivities of the composites formed at direct operating conditions were measured and the results are shown in **Fig. 16.15**. It is observed that the electrical conductivity increases significantly, when CO_2 changes from the high-pressure liquid state to the supercritical fluid state. Electrochemical synthesis of conducting polymer composites of polythiophene with natural rubbers was achieved by S. Yigit [244].

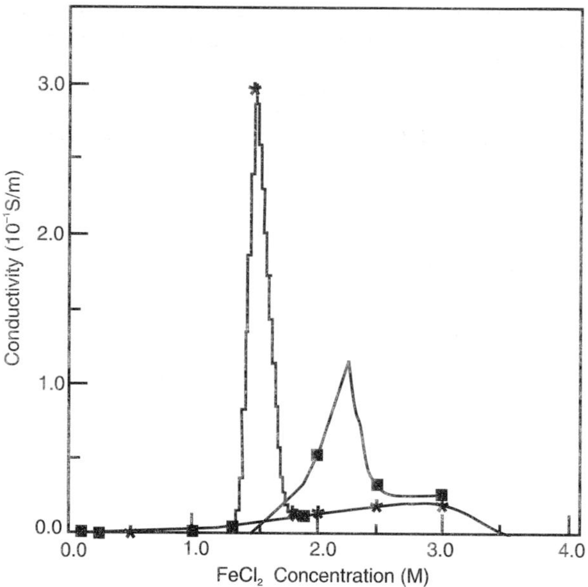

Fig. 16.15. Plots of the electrical conductivity against the concentration of $FeCl_3$. The blending conditions in CO_2 are (a) (▲), at 30°C and 7.95 MPa, (b) (■), at 40°C and 10.5 Mpa and (c) (•) at 50°C and 13.14 MPa. Reproduced with permission from Tang *et al.*; Europ.Polym.J., 39, 143 (2003). Copyright © 2003, Elsevier Science Ltd.

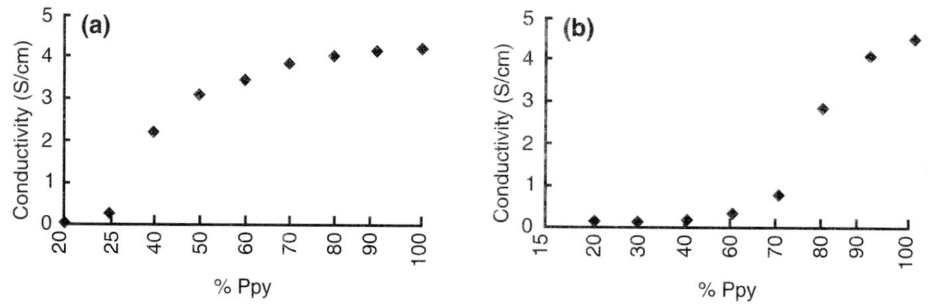

Fig. 16.16. Electrical conductivity vs. weight percent of PPy. (a) Pin:PPy composite films in the first system and (b) Pin:PPy composite films in the second system. Reproduced with permission from Bozkurt *et al.*; Synth.Met., 82, 41 (1996). Copyright © 1996, Elsevier Science Ltd.

Conducting polymer composite of polypyrrole:polyindene (PPY:Pin) was prepared via electrochemical methods by A. Bozkurt *et al.* [245]. In this study, two different approaches were utilized. In the first, the electro-initiated polymerization of indene on a platinum electrode was achieved at 2.0 V versus Ag/Ag^+ in acetonitrile (AN). Then the polyindene-coated electrode was used for the electrochemical polymerization of pyrrole at 1.0 V versus Ag/Ag^+. In the second case, electrochemical coating of platinum electrode with polypyrrole at 1.0 V versus Ag/Ag^+ was carried out and indene was polymerized on the conducting polymer at 2.0 V versus Ag/Ag^+ in acetonitrile medium. The results of electrical conductivity measurements revealed that the solution sides and the

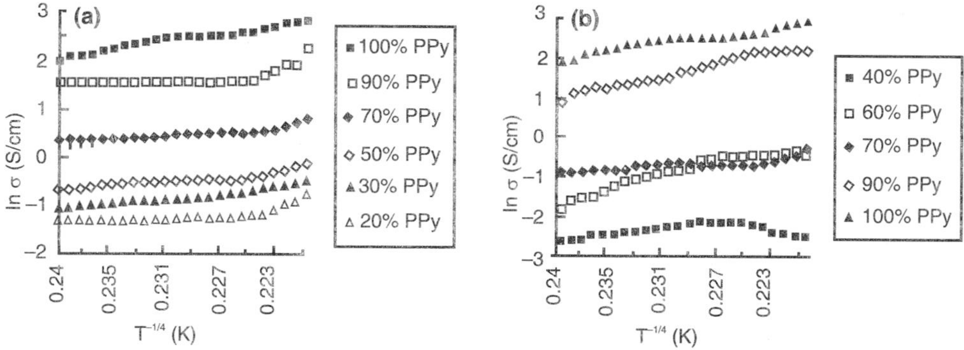

Fig. 16.17. Temperature dependence of electrical conductivity as a function of percent composition of PPy. (a) first system and (b) second system. Reproduced with permission from Bozkurt *et al.*; Synth.Met., 82, 41 (1996). Copyright © 1996, Elsevier Science Ltd.

electrode sides of the films have the same conductivities. This gives an idea about the homogeneity of the films, at least in terms of conductivity. It is interesting to note that in the first system (Pin first, PPy later) the conductivities are not greatly affected by the percolation composition, after 70% PPy content (**Fig. 16.16(a)**). However, in the second system (PPy first, Pin second), the conductivities of the films are low until 50% PPy content has been reached (**Fig. 16.16(b)**). The temperature dependence of electrical conductivity of the polypyrrole:polyindene composite films were also reported and represented in **Fig. 16.17(a) & (b)**. **Fig. 16.17(a)** shows Pin:PPy copolymer, which has a T$^{-1/4}$ dependence regardless of the PPy content, whereas for PPy:Pin composite with compositions below 50% PPy contents the conductivity first increases up to around 370 K, then the reverse behavior is observed, i.e., it decreases with increasing temperature. Conducting PPy:PI composite prepared by J.O. Iroh *et al.* [246] was studied as electrochemical impedance spectroscopy at different applied dc polarization potentials. The properties of the coatings were found to differ significantly between pure PI (polyindene) and the PPy:PI composite, depending on the applied potential and on the amount of the conducting polymer. An additional time constant related to the PPy:PI interface appears when PPy is added to PI. The pseudocapacitance of the PPy:PI composite increased significantly with increased cathodic potential **Fig. 16.18(a)**. This was explained by additional doping of PPy by the PI matrix. While, **Fig. 16.18(b)** shows the behavior of the capacitance of the PPy:PI composite films at negative working electrode potentials. The capacitance increases exponentially with the increase of the absolute value of the potential. This process is stopped by delamination of the composite film from the electrode surface that occurs at high potentials.

Polycarbonate based PPy composite [247] have been obtained by polymerizing pyrrole with FeCl$_3$, in the presence of polycarbonate dissolving in chloroform and then precipitating in a non-solvent for both the polymers. For composites containing less than 15% pyrrole, its conductivity ranges from 10^{-2} to 2 Scm^{-1}. Pyrrole:rubber composites [248] were prepared in two steps using the inverted emulsion pathway. In the first step, inverted emulsion was generated by dispersing an aqueous solution of ferric chloride in an organic solution of a host polymer (styrene-butadiene-styrene) (SBS), styrene-ethylene-butylene-styrene (SES) or styrene-isoprene-styrene and nonionic surfactant with stirring. In the second step, a solution of pyrrole in toluene was introduced drop wise in the emulsion generated in the first step and PPy was thus formed rapidly (within 10-15 min).

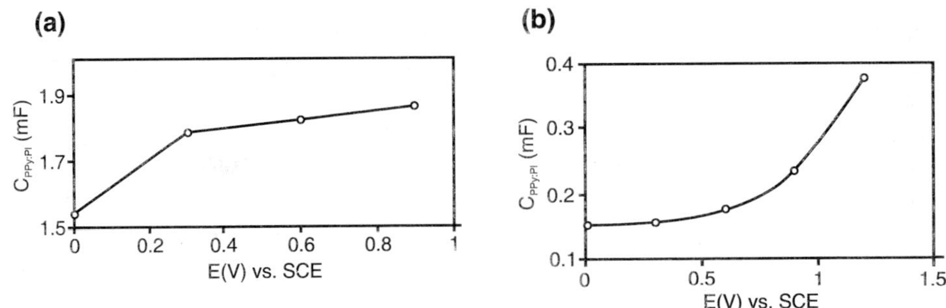

Fig. 16.18. Capacitance of PPy:PI composite at anode potentials. (a) cathode potential and (b) 0.01 M KPF$_6$ in AN. Reproduced with permission from Iroh and Levine; J. Power Sources, 117, 267 (2003). Copyright © 2003, Elsevier Science Ltd.

Methanol was introduced to prepare the composite after polymerization was completed. The ratio of pyrrole:host polymer, the surfactant and the organic solvent employed in the inverted emulsion, the ratio of aqueous phase/organic phase in the inverted emulsion, the nature of the host polymer and even the organic solvent used to wash the composites were important factors that affected the conductivities and mechanical properties of the composites.

Composites containing polyvinyl chloride and PPy (PVC:PPy) prepared electrochemically were reported by many researchers [249-251]. Uosaki *et al.* [249] prepared it by casting PVC from dimethylformamide (DMF) solution on glassy carbon electrode followed by depositing PPy over the wet PVC coated glassy carbon electrode. De Paoli *et al.* [250] reported preparation of PVC:PPy film, which consists of coating of PVC on Pt electrode from THF solution, followed by electropolymerisation of PPy. PPy content in these films can be varied by varying polymerization time. Conductivity was found in the range of 5-50 Scm^{-1}. The mechanical properties of electrochemically prepared PVC:PPy are similar to those of pure PVC and can be improved further by addition of polychloroprene rubber as a plasticizer. PVC:PPy and brominated polyvinyl carbazole based PPy composite were also prepared by Niwa *et al.* [251]. The surface resistance and the optical transmission of the alloy film can be controlled by the polymerization conditions and semitransparent conducting films with the transmittance of 60% and the conductivity 2 Scm^{-1} can be obtained. In case of PVA:PPy composite [252], initially PPy forms at higher current density at the solution/PVA interface whereas at lower current density the polymerization takes place initially at the electrode/PVA interface. This suggests that the diffusion of pyrrole in the PVA was slow and limiting the polymerization at higher current densities. The rate of polymerization must depend on the rate of diffusion of monomer to the electrode, which in turn must be dependent on the extent of the swelling of the matrix polymer and perhaps the existence of channels within the matrix. The conductivity of the composite was found to be in the range 0.1-10 Scm^{-1}.

Polyacrylonitrile:polypyrrole based composite [253] was prepared electrochemically on the polyacrylonitrile coated electrode. The composite films are much easier to peel off and to handle than pure PPy. Its conductivity is in the range of 10^{-3} to 10^{-2} Scm^{-1}, which are somewhat lower than those of PPy homogeneous film prepared with corresponding anion (ranging between 10 to 100 Scm^{-1}. U. Geissler *et al.* [254,255] reported PPy composite on poly(N-vinylcarbazole) matrix synthesized by different methods. In the first method poly(N-vinylcarbazole) matrix was first deposited on the electrode followed by electrodeposition of PPy, the reverse way also have been used, i.e., electrodeposition of PPy followed by deposition of N-vinyl carbazole and the third method consists of simultaneous polymerization of both the monomer and the polymers on the electrode surface. The conductivity of the composite depends on the reaction scheme, i.e., the way in which they are formed. The conductivity value of 0.05 Scm^{-1} can be achieved with a small amount of pyrrole.

Styrene based PPy composites [256] was prepared by copolymerizing styrene with 4-chloromethyl styrene and refluxed overnight with added a solution of Potasio-pyrrole in THF medium. Here pyrrole moiety was grafted on the co-polymer matrix. The conductivity of the composite is ~10 Scm^{-1}.

3. APPLICATIONS

The possible applications of the composite materials are diverse and the future of these materials appears bright. But in this review only a limited number of applications can be exemplary introduced.

3.1. Optical Devices

Optics was certainly one of the first applications of hybrid materials [257]. Thus, organic:inorganic/organic:organic hybrid materials with high transparency are expected to be new optical materials such as optical fiber, wave-guide and optical lens. Materials with high refractive index, low density and good transparency in visible region are in demand as optical lenses especially for glasses. In order to implement optoelectronic or photonic properties into devices, the materials have to fulfill high optical quality demands and have to be stable over a long period. Furthermore, the control of the refractive index as well as the thickness and therefore, the processibility of the materials are important. Due to its low optical losses (high optical quality), silica glass and composites made from it seem to be very useful for these devices [258,259]. For example, photoconducting composites were prepared by the incorporation of cadmium sulfide particles in a polyvinylcarbazole matrix. The resulting materials allowed the tuning of the band gap of the sensitizing nanocrystals so that their spectral response was adjusted to suit a particular wavelength of operation [260,261].

Thus, the optical properties of organic-organic/organic-inorganic composites have attracted much interest, particularly their transparency and 'active' optical properties. Transparent products can be obtained from layered systems under certain conditions. Hybrids of clay and nematic liquid crystals prepared by Kawasumi *et al.* [262], exhibit interesting electro-optical properties. These materials can be transparent **(Fig. 16.19(a))** or opaque **(Fig. 16.19(c))** depending on the frequency of the applied electric field. The effect is reversible and can be repeated many times. The transparent and opaque states exhibit a memory effect after switching off the field **(Fig. 16.19(b) and Fig. 16.19(d))**. Such materials have potential applications in optical storage devices, displays and light-controlling glass. Another interesting example of optoelectronic 'organic-inorganic' hybrid is TiO_2:polyacrylate, which shows good visible light transparency as compared to TiO_2. The development of optically activated hybrid organic:inorganic mesostructured silica composites has led to a number of advanced optical applications, such as optical switches and sensors and low threshold wave-guide microlasers. However, the low refractive index of silica composites (n = 1.43) requires that the optically active layer be supported by an ultra low refractive index layer for wave guiding to occur. This obstacle could be overcome by using a higher refractive index inorganic component. Recently, D. Stucky *et al.* [263] demonstrated a new synthetic approach for high refractive index, dye-activated,

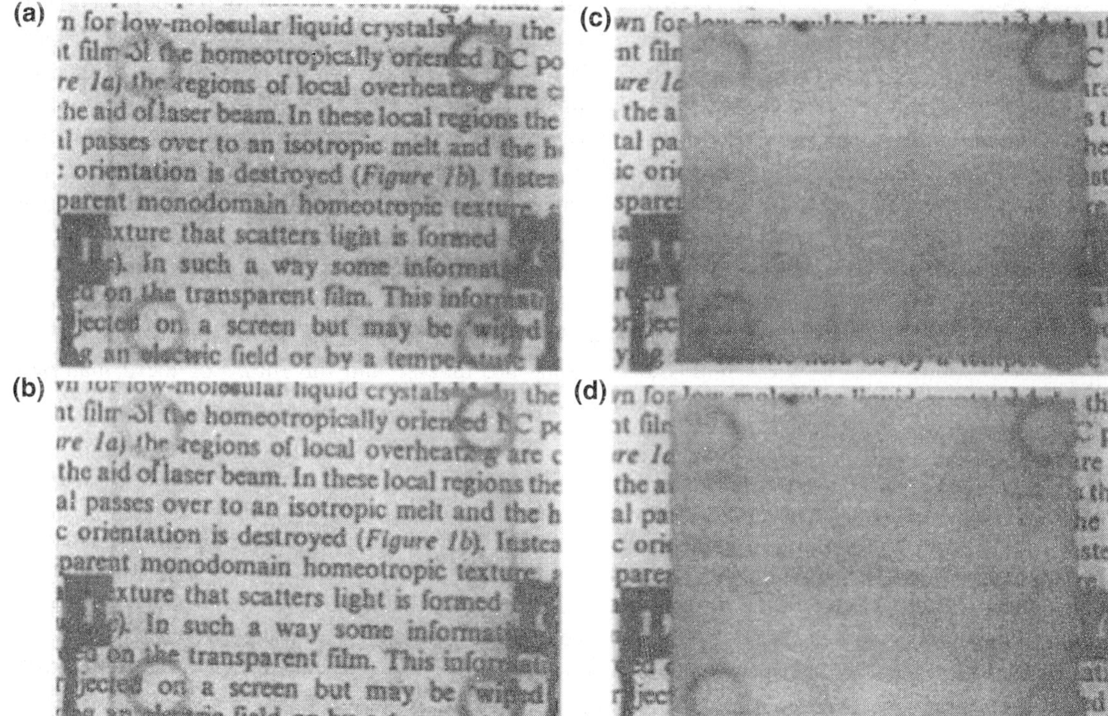

Fig. 16.19. Electro-optical properties of clay:nematic liquid crystal hybrids. (a) low frequency, on, (b) low frequency, off, (c) high frequency, on and (d) high frequency, off. Reproduced with permission from Kawasumi *et al.*; Mater. Sci. Eng., C6, 135 (1998). Copyright © 1998, Elsevier Science Ltd.

hybrid mesostructured materials utilizing a fluorocarbon stabilized titania precursor. These titania composites can readily be processed into solid, optically transparent, crack free fibers and planar wave-guides.

Several research groups [264-270] have studied light emitting diodes using hybrid composites. Compared to classical organic or polymer based diodes, those fabricated with composites as an emitting layers exhibited low turn-on voltage [264,265,267,268], improved stability [269]. Depending on the nature of material as well as the adopted structure, the obtained emitted light could be that of the organic [269,271] or the inorganic [266,267] or both of them as a function of the applied voltage [263,264]. Color conversion of light from short wavelength to longer one could also be achieved using particular arrangement of the layers [271].

Among conjugated polymers poly(p-phenylene vinylene)s have been studied extensively for their electroluminescence and potential applications in light emitting diodes (LEDs) because of its easy processibility and relatively good optical and electrical performance. On the other hand, porous silicon was found to have high conductivity and to exhibit interesting electroluminescence properties. Optical study of composites made of porous silicon and a non-conjugated polymer has been performed by Guha *et al.* [268,272] using mixed structures but no LEDs result was reported. The synthesis of intercalated hybrids of smectic clay and a substituted poly(p-phenylene vinylene) has also been reported. These hybrids exhibit electroluminescence and the luminescence appears to be color tunable depending on the degree of intercalation. Organically modified silicates (ormosil) and organically modified ceramics (ormocer) synthesized with sol-gel route have also been intensively studied in the last ten years due to their wide spectrum of integrated optics applications such as the fabrication of passive and active wave guides, sensors, electro-optical modulators and solid-state microlasers [273].

The desirable electrical and optical properties exhibited by the hybrid perovskites, along with the potential for simple, low-cost processing techniques, make it interesting to consider building devices with these materials. These useful optical characteristics make the perovskites attractive as potential emissive materials in electroluminescent (EL) devices. The first attempts to induce EL in these materials involved attaching silver paint contacts to single crystals of $(C_6H_5C_2H_4NH_3)_2$ $(CH_3NH_3)Pb_2I_7$ [274]. More recently, EL devices were prepared, with a structure analogous to that of traditional organic light emitting diodes (OLEDs), but with hybrid $(R-NH_3)_2PbI_4$, perovskite light-emitting layers [275,276]. In addition to their potential applications in 'organic-inorganic light emitting diodes systems, semiconducting hybrid perovskites are also attractive as a new class of channel materials for thin-film field-effect transistors (TFTs) [277]. The 'organic:inorganic' hybrid materials are interesting for this apaplication because they can combine the higher carrier mobilities of ionic and covalently bonded inorganic semiconductors with the simple, low-cost and low-temperature thin-film techniques that make organic semiconductors exciting as alternative channel materials. Recently, C.R. Kagan *et al.* [277] demonstrated the first 'organic:inorganic' hybrid material as the semiconducting channel in a TFT, using the hybrid perovskite $(C_6H_5C_2H_4NH_3)_2SnI_4$.

3.2. Conductors

Electrically conducting composite materials show various applications in electronic and photonic systems. Most of them show the electrical conduction behavior in the semiconductor region and hence they can be used as semiconducting materials. Nowadays, these materials have been used in the preparation of potentiometric sensors, i.e., gas sensors, chemical sensors, biosensors, ion-sensors as well as ion-selective electrodes, which are of vital analytical and environmental interest. The integration of chemically sensitive membranes with solid state electronics has led to the evolution of miniaturized, mass produced potentiometric probes known as ion-selective field effect transistors (ISFETs). Thus, the electrical and electronic properties exhibited by solid-state materials are crucial in a large number of inorganic as well as organo-inorganic materials applications [278-282]. These unique electronic properties result from their extended structures, where strong interactions between the atoms, ions or molecules occur throughout the lattice. In terms of conductivity, the behavior ranges from insulating through semiconducting to metallic and superconducting. Many types of electrically conducting composite materials classified as electrolytes or polymer ionics have been developed and characterized in recent years [283], for applications such as solid-state lithium batteries or supercapacitors have been produced using 'organic:inorganic' polymeric systems formed by the mixture of organic polymers and inorganic moieties prepared by the sol–gel techniques. In these systems at least one of the network forming species should contain components that allow an interaction to the conducting ions. This is often realized using organic polymers, which allow an interaction with the ions such as poly(ethylene oxide) (PEO) [284-286]. Solid electrolytes, having conductivities of 10^{-2} to 10^{-1} Scm^{-1}, are required in several systems operating either with high current densities (electrolyzers, batteries, etc.) [287], or at very low current levels (gauges, electrochemical memories, coulometers, etc.) [288], in order to avoid excessive Joule-heat losses or excessive cell impedance. Furthermore, high conductivities are required for materials employed in the preparation of charged membranes or in thermoelectric generators [289]. Only a few solid electrolytes are presently known to exhibit such a favorable conductance and most of them only at high temperature. The combination of conducting polymers with inorganic species is another example for nanocomposites with a potentially important application. A variety of combinations have already been tested and especially polyaniline and polypyrrole show an interesting potential in combination with iron oxides, barium titanate, platinum and tungsten oxide for magnetical, energy storage, catalytic and electrochromic devices, respectively [290]. Metallo and metal free phthallocyanines and their polymers are also known for almost a century and have become one of the most intensively studied macrocycles due to their resemblance to biologically occurring porphyrins [291]. They possess useful optical, electronic, electrical, photoelectric and electrochemical properties and hence, have become ideal candidates for use in the fabrication of novel electronic devices [292-303], molecular electronic devices [304-308], gas sensors and detectors [309-312]. They can be reversibly doped to achieve metallic conductivity by exposure or co-crystallization with a variety of electron acceptors. Thus, the electrical conductivity can also be improved by effecting refinements in these polymers via thermal and reactive treatments. Proton conducting composites were also obtained by the formation of interpenetrating networks (IPNs) from silicon alkoxide end-capped PEO, phenyltriethoxysilane using monododecylphosphate or phosphotungstic acid to induce the proton conductivity [313]. High temperature protonic conducting polymers' membranes provide new technological applications in the electrochemical devices including electrochromic displays, chemical sensors, fuel cells and others. Organic/inorganic composite membranes, consisting of SiO_2:PEO hybrids, are remarkable family of isotropic, amorphous polymer material, which shows good protonic conductivities at high temperatures above 100°C. At the same time hybrid materials with ionic conductivity have also been described. Lithium salt first dissolved in a suitable phase, an organically modified silica (amines, sulphonic acid, sulfonamides [314] or silica polymer networks such as poly(ethylene glycol) (PEG), poly(propylene glycol) (PPG) etc. [315]. Transport phenomena of cations in polymeric structures have been widely studied and demonstrate that reasonable values of Li^+ mobility are reached if the surrounding solvating media also experiences a high degree of mobility [316], such as those measured in polymers above T_g. In PEG:silica structures, the dependence of ionic conductivity (σ_{Li+}) on temperature demonstrate polymer-like behavior, in agreement with thermal analysis and the NMR experiments which present a value of T_g in the range –60°C to + 20°C, depending upon the exact material composition. Ionic conduction in organic-inorganic composite based on poly(propylene glycol) has been described by R.F. Bianchi *et al.* [317]. At last we can say that more and more compounds presenting electronic properties are also under study. Redox targets have been entrapped in hybrid matrices, leading to photo- or electroactive materials [318]. These properties have also been used to study the change of structure during the sol-gel transition via the determination of self-diffusion coefficients [319]. These properties are also expected for redox sensors and biosensors. Because of the flexibility of the chemistry, the redox properties of materials can be tuned. Very promising results arise from electronic conductors such as silica:polypyrrole, silica:polyaniline interpenetrated networks and V_2O_5:polypyrrole layered structures. The search of anisotropic conductivity properties is a clear evidence by this last example. Recently, semiconductor research has also found interesting isolating materials that can be used in organic transistors and chips [320].

3.3. Sensors

Organic or inorganic semiconductors have been reported to change their conductivities when exposed to variety of organic and inorganic vapors. Thus, these materials can be expected to behave as sensors. Composite materials of tin oxide and derivatives of polypyrrole [311], gave reversible changes in electrical resistance at room temperature when exposed to a variety of organic vapors. Composite materials containing 2.5% polymer by mass were fabricated and exposed to low concentrations of ethanol, methanol, acetone, methyl acetate and ethyl acetate vapors [321-324]. The composite materials were found to give more significant and reversible decrease in electrical resistance in comparison with sensors constructed solely of tin dioxide or polypyrrole. These materials could be used in the quality control of foodstuff, especially in the early detection of soft rot in potato cubers. Some researchers incorporated preformed polypyrrole and poythiophene into clay (montmorillonite) by the interaction of colloidal nanoparticles of the polymers with the colloidal layered host [325]. This method using a colloid-colloid reaction [326] might provide a general route to incorporation of intractable polymers within layered host structures that can be exfoliated, such as smeccice clays [327], metal disulfides and some metal oxides. These composite materials have potential to be used as hybrid sensors.

Chemical sensing properties of the electrochemically prepared polypyrrole:poly(vinyl alcohol) (PPy:PVA) films were studied by exposing them to NH_3 gas [328]. The results shown in **Fig. 16.20** reveals that the sample has a very good sensitivity towards NH_3. A remarkable increase in resistance of the sample is observed within 5 min from purging and this change is reversible within 5% of the initial resistance. Increasing the concentration of NH_3 more obvious changes are observed but an NH_3 concentration above 10% results in almost reversible change of resistance.

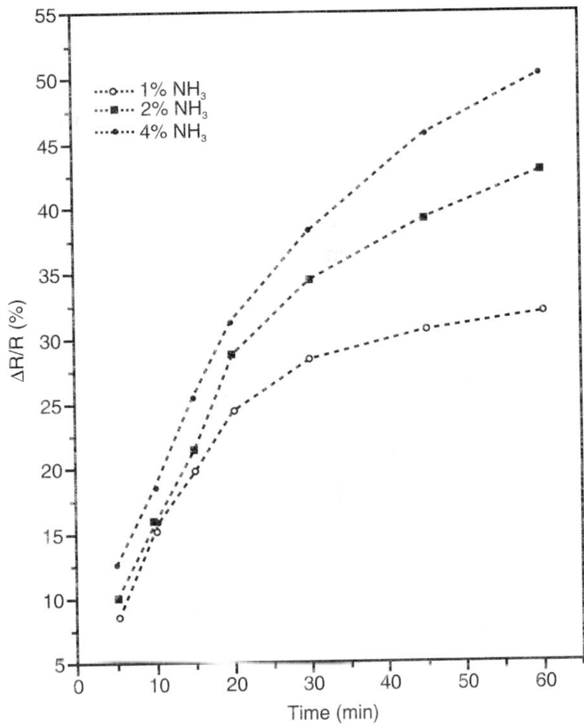

Fig. 16.20. Resistance changes recorded for PPy:PVA:IPNs on exposure to NH_3-argon mixtures of different NH_3 (%). Reproduced with permission from Gangopadhyay and De; Sensors and Actuators, B 77, 326 (2001). Copyright © 2001, Elsevier Science Ltd.

3.4. Ion-Exchangers

Ion-exchange technique is one of the most common and effective treatment methods for liquid waste for a long time. It is a well-developed technique that has been employed for many years in both the nuclear industry and other industries. Wastewaters containing toxic metals are generated as a byproduct in various industries. Various techniques have been developed for treatment of such waste streams. Ion-exchange process is one the important techniques for the treatment and identification of industrial waste. For this purpose, first inorganic ion-exchangers and later on organic resins have been used. Organic polymers as ion exchangers are well known for their uniformity, chemical stability and control of their ion exchange properties through synthetic methods [329]. The inorganic ion exchange materials besides other advantages are important in being more stable to high temperature and radiation field than the organic ones [330]. However, the inorganic adsorbents have their own limitations. For instance, these materials in general are reported to be not very much reproducible in behavior, less stable in concentrated acidic and basic media and high cost. Fabrication of the inorganic adsorbents into rigid beads type media suitable for column operation is also quite difficult and cannot be used in a convenient way in case when the impurities from a large volume of effluent are to be removed. Further they have generally worse mechanical and chemical strength than the organic counterpart because of inorganic nature [331].

In order to obtain a combination of these advantages associated with polymeric and inorganic materials as ion exchangers, attempt have been made to develop ion exchangers by combination of multivalent metal acid salts and organic conducting polymers (polyaniline, polypyrrole and polythiophene etc.), providing a new class of 'organic-inorganic' composite ion exchangers with better chemical, mechanical, thermal and magnetic properties as compared to pure inorganic compound and organic polymers. With respect to the ion exchange properties composite materials show enhancement in granulometric properties, good ion exchange capacity, reproducibility and possessing good selectivity for heavy metals indicating its useful environmental applications. Few such excellent ion exchange materials have been developed in our laboratory and successfully being used in environmental analysis [332-336].

The electrical conductivities of these organic-inorganic composite cation-exchange materials (as prepared and 0.5 M HCl treated) were determined from the measurement of conductivity of the samples using the four-probe method of conductivity measurement for semiconductors. The current-voltage data so generated by a 4-in-line probe dc electrical conductivity-measuring instrument was processed for calculation of resistivity (ρ_o) using the following equation:

$$\rho_o = (V/I) \times 2\,\pi S \tag{16.2}$$

where V is the voltage (V) and I is the current (A). Since the thickness of the sample is small compared to the probe distance a correction factor for it has to be applied and the corrected resistivity may be calculated as:

$$\rho = \rho_i / G_7(W/S) \tag{16.3}$$

where ρ is the corrected resistivity in ohm cm, $G_7(W/S)$ is the correction factor used in the case of non-conducting bottom surface and it is a function of W, thickness of the sample under test (cm) and S, probe spacing (cm); i.e.,

$$G_7(W/S) = (2S/W)\log_e 2 \tag{16.4}$$

Thus, the electrical conductivity (σ) was calculated using the following equation:

$$\sigma = 1/\rho \tag{16.5}$$

where σ is the electrical conductivity in Scm^{-1}.

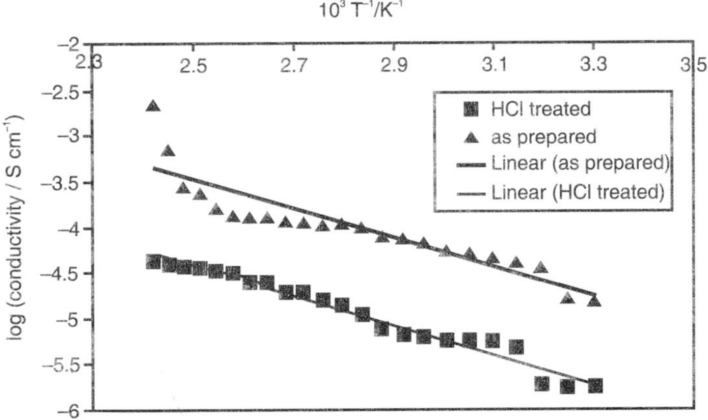

Fig. 16.21. Arrhenius plots for polypyrrole:Th(IV) phosphate composite material. Reproduced with permission from Khan *et al.*; Mater. Res.Bull., 40, 289 (2005). Copyright © 2005, Elsevier Science Ltd.

The variations of electrical conductivity (σ) of the composites by raising temperatures (between 35°C to 200°C) are carried out on polypyrrole:Th(IV) phosphate and polyaniline:Sn(IV) phosphate composites [337,338]. On examination, it was observed that the electrical conductivity of the composites increase with the increase in temperature and the values lie in the order of 10^{-6} to 10^{-3} Scm^{-1}, i.e., in the semiconductor region. To determine the nature of dependence of electrical conductivity on temperature plots of log σ versus $10^{3}T^{-1}/K^{-1}$ were drawn (**Figs. 16.21 & 16.22**) and they followed Arrhenius equation similar to other semiconductors [183]. On examination, it was observed that the composite material show enhanced electrical conductivity on exposure to HCl, due to the charge-transfer reaction between polypyrrole and polyaniline component of the composites and doping agents, HCl:

$$[\text{Polyaniline Sn(IV) phosphate}] + n\,HCl \rightarrow [\text{Polyaniline}(nH^+)(nCl^-)\,Sn(IV)\text{phosphate}] \qquad (16.6)$$

$$[\text{Polypyrrole Th(IV) phosphate}] + n\,HCl \rightarrow [\text{Polpyrrole}(nH^+)(nCl^-)\,Th(IV)\text{phosphate}] \qquad (16.7)$$

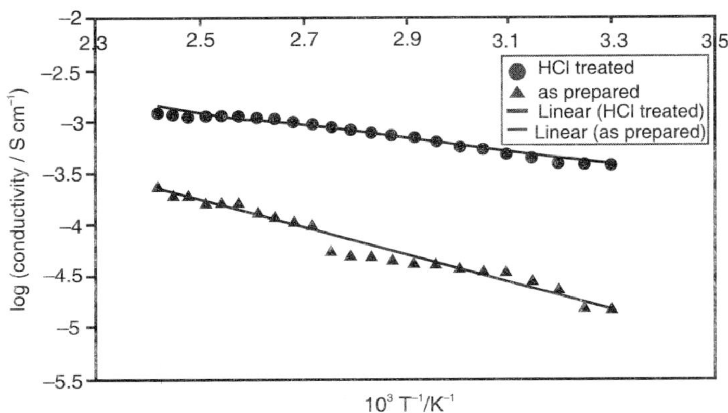

Fig. 16.22. Arrhenius plots of polyaniline:Sn(IV) phosphate composite material. Reproduced with permission from Khan and Inamuddin, React. Funct. Polym., 66, 1649 (2006). Copyright © 2006, Elsevier Science Ltd.

The dependence of the electrical conductivity through the bi-phasic systems (polypyrrole:Th(IV) phosphate, polypyrrole:polyantimonic acid and polyaniline:Sn(IV) phosphate composites; prepared with different concentrations of pyrrole and aniline monomers) on the concentration of conducting phase (i.e. polypyrrole and polyaniline) was examined (**Tab. 16.4**). A slight increase in electrical conductivity for these composites is followed at a certain pyrrole and aniline concentration by a sudden jump, which is again followed by moderate increase. At about 23.33% and 27% pyrrole for polypyrrole:Th(IV) phosphate [339] and polypyrrole:polyantimonic acid [340] respectively and 8% aniline concentration for polyaniline:Sn(IV) phosphate [338] (critical concentration of conducting phase), the sharp rise in electrical conductivity is observed that could possibly be explained on the basis of percolation theory [341]. Thus, we can say that main factor that made the composites electrically conductive is the presence of polypyrrole or polyaniline in sufficient amount. Thus major part of electrical conductivity of the composite is due to the incorporation of polypyrrole or polyaniline in the composites. It is observed that the main constituent that made the composite electrically conductive is polypyrrole or polyaniline when in sufficient amount. Thus, the electrical conductivity of the composite is due to oxidized polypyrrole and polyaniline respectively; the conducting properties will depend on the percolation behavior of the conducting phase. Thus major part of electrical conductivity of the composites is due to the incorporation of polypyrrole and polyaniline.

The weight loss vs. temperature characteristics of the organic:inorganic composite polyaniline:Sn(IV) phosphate can vary significantly as shown in **Fig. 16.23**, which also indicates the utility of TGA for providing relative thermal stability of this composite. The thermal degradation

Tab. 16.4. Values of electrical conductivity at ambient temperature for the various composites with different concentrations of monomer. A portion of this table is reproduced with permission from Khan *et al.*; J.Electroanal.Chem., 572, 67 (2004). Copyright © 2004, Elsevier Science Ltd.

Polypyrrole:Th(IV) Phosphate Composites

Pyrrole (%)	16.66	20.00	23.33	26.66	30.00	33.33	36.66
Electrical conductivity (Scm^{-1})	5.92×10^{-8}	6.15×10^{-8}	2.53×10^{-5}	3.47×10^{-5}	3.54×10^{-5}	3.82×10^{-4}	4.90×10^{-4}

Polypyrrole:Polyantimonic Acid Composites

Pyrrole (%)	23.33	26.67	30.00	33.33	36.67	40.00
Electrical conductivity (Scm^{-1})	3.94×10^{-9}	1.83×10^{-5}	4.51×10^{-5}	7.40×10^{-5}	8.90×10^{-5}	2.37×10^{-4}

Polyaniline:Sn(IV) Phosphate Composites

Aniline (%)	6	8	10	12	14	16
Electrical conductivity (Scm^{-1})	9.46×10^{-5}	2.54×10^{-4}	5.23×10^{-3}	6.48×10^{-3}	6.93×10^{-3}	8.01×10^{-3}

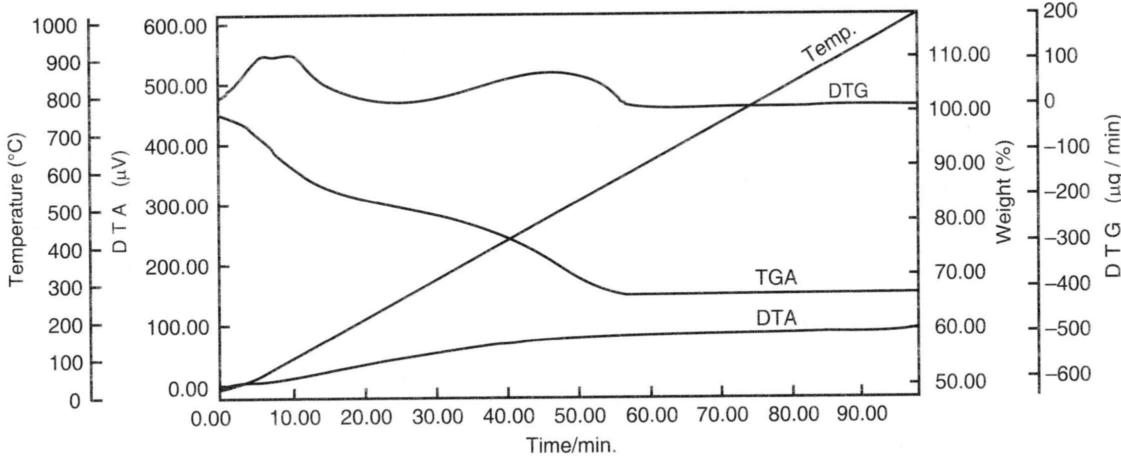

Fig. 16.23. Simultaneous TGA-DTA curves of as-prepared polyaniline:Sn(IV) phosphate. Reproduced with permission from Khan and Inamuddin; React. Funct. Polym., 66, 1649 (2006). Copyright © 2006, Elsevier Science Ltd.

curve for this organic:inorganic composite was fairly constant as the results indicate a continuous weight loss of mass (about 12.42%) up to 135°C, which may be due to the removal of external water molecule [342]. A slow weight loss (~9.02%) observed between 135°C and 398°C (may be due to the condensation of phosphate group) to pyrophospahte groups. Further weight loss (~12.03%) between 398 and 601°C may be due to the complete decomposition of the organic part of the material. At 601°C onwards, a smooth horizontal section represents the complete formation of the oxide form of the material. The scanning electron microscope (SEM) photographs of these composite confirm the formation organic:inorganic composite. SEM photographs of polypyrrole, Th(IV) phosphate and polypyrrole:Th(IV) phosphate [338] and polypyrrole, polyantimonic acid polypyrrole:polyantimonic acid [341] obtained at different magnifications (**Figs. 16.24 & 16.25**) indicating the binding of inorganic ion-exchange material with organic polymer, i.e., polypyrrole. The SEM pictures showed the difference in surface morphology of organic polymer, inorganic precipitate and composite material. It has been revealed that after binding of polypyrrole with Th(IV) phosphate and polyantimonic acid the morphology has been changed. Because polypyrrole, Th(IV) phosphate and polyantimonic acid showed granular, fibrous and plate like structure respectively.

Fig. 16.24. Scanning electron microphotographs (SEM) of (a) chemically prepared polypyrrole at the magnification of 3500×, (b) Th(IV) phosphate at the magnification of 3000× and (c) polypyrrole:Th(IV) phosphate composite system at the magnification of 2500×. Reproduced with permission from Khan *et al.*; Mater.Res.Bull., 40, 289 (2005). Copyright © 2005, Elsevier Science Ltd.

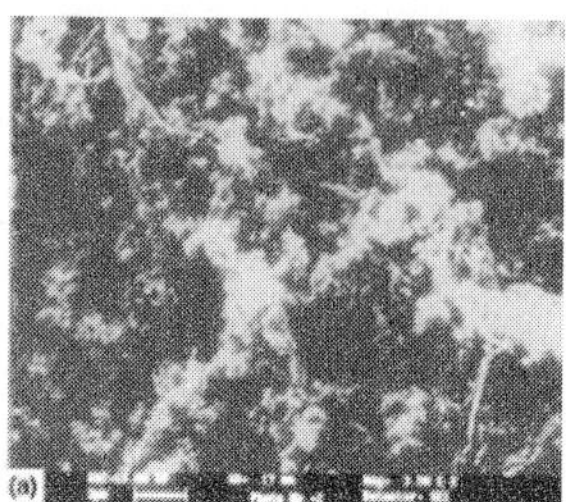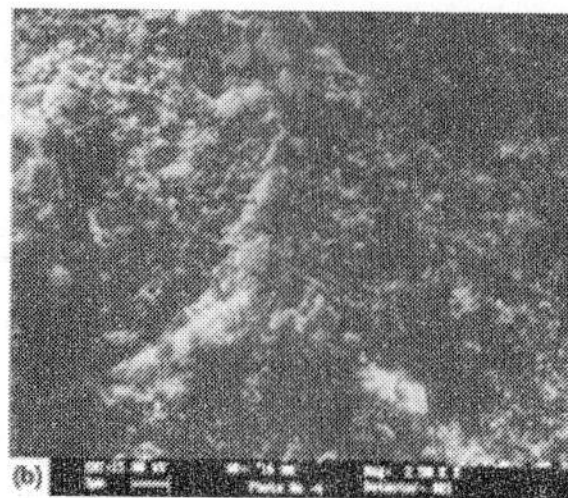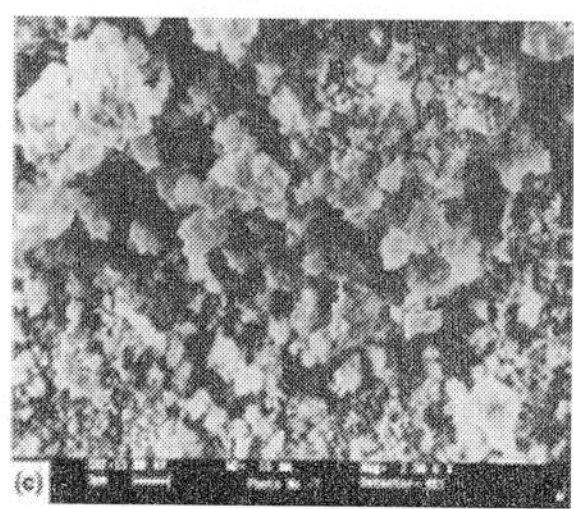

Fig. 16.25. Scanning electron microphotographs (SEM) of (a) chemically prepared polypyrrole at a magnification 3500×, (b) polyantimonic acid at a magnification of 2000× and (c) polypyrrole:polyantimonic acid composite system at a magnification of 2500×. Reproduced with permission from Khan *et al.*; J.Electroanal.Chem., 572, 67 (2004). Copyright © 2004, Elsevier Science Ltd.

CONCLUSIONS AND FUTURE PROSPECTS

We have concluded that the electrically conducting polymers have established their place as important constituents of various electronic and photonic devices due to their characteristics properties. Electrically conducting polymers have recently emerged as a leading contender for numerous applications in the industry such as automotive, aerospace, electronics, electrical etc. Therefore, these materials are finding applications from coating to lubricants to solid-state technology to biotechnology. To meet the demand of the materials of improved performance, commercial polymers are always mixed together with various additives of monomeric or polymeric in nature. It is aimed that the additives will act synergistically with the polymer and will meet the combined requirement of a particular application. While besides the electrical conductivity like metals, electrically conducting composites have fascinated research and development groups worldwide due to a wide range of other desirable properties such as architectural flexibility, environmental stability, ease of fabrication, light weight, mechanical properties, environmental stability, reasonable and well defined high electrical conductivities, ease of fabrication and processiblity etc. Some possible applications of electrically conducting composites may include—rechargeable batteries, electrochromic devices, optoelectronics, photovoltaics, field effect transistors, display devices, printed circuit board, chemical indicators, biosensors, gas sensors, solar cells, electroplating, conducting textiles, electrochemically switchable devices, ion-exchangers, transparent coatings, adhesives, radiation detectors, schottky diodes, fillers, solid lubricants, corrosion inhibitors, photocatalysis and so on. Thus, we can say that conducting composites are enjoying a fair good demand from industry. Therefore, challenging research on electrically conducting composite materials is in progress, which is of tremendous future prospects. In this regard, when a decision is made for manufacturing electrically conducting composites, one has to plan some research work to regulate and adopt known data in the field to fit the particular conditions and materials relevant to future use, or even to find newer and newer results to accommodate the properties needed in the desired materials.

ACKNOWLEDGEMENTS

One of the authors of this chapter (Inamuddin) is highly thankful to his Ph.D. Supervisor Dr. Asif Ali Khan and the Chairman, Dr. Ali Mohammad, Department of Applied Chemistry, Aligarh

Muslim University, Aligarh, for his invaluable guidance and constant encouragement. I especially appreciate the freedom he has provided me to write this chapter. Special thanks are also due to Mr. Fareed Mahdi, Department of Civil Engineering, Aligarh Muslim University, Aligarh, who provided the computer facility for the completion of this chapter. Amir-Al-Ahmed is thankful to Prof. Emmanuel Iuwoha, Department of Chemistry, University of the Western Cape and Prof. M. Ali Hossain, School of Environmental Science and Management, Independent University of Bangladesh. Authors are also thankful to Dr. Faiz Mohammad (editor) for his kind invitation to write this chapter.

Appendix 1. Full forms of abbreviations used in this chapter.

Term used	*Full form*
AN	Acetonitrile
BSA	Benzenesulfonic acid
CB	Carbon black
CSA	Camphor sulfonic acid
Ceramers	Ceramic polymers
CMC	Ceramic matrix composites
CoPA	Copolyamide 6/6.9, a random copolymer of 51% -[-HN-$(CH_2)_5$-CO-]- and 49% -[-HN-$(CH_2)_6$-NH-CO-$(CH_2)_7$-CO-]-
DOP	Dioctyl phthalate
DBSA	Dodecylbenzene sulfonic acid
DSC	Differential scanning calorimetry
DMAc	Dimethyl acetamide
DMF	Dimethyl formamide
ECPs	Electrically conducting polymers
EO	Ethylene oxide
EL	Electroluminescent
EMI	Electromagnetic interference shielding
ELDs	Electroluminescent diodes
FTIR	Fourier transform infrared spectroscopy
ICP's	Intrinsically conducting polymers
IPNs	Interpenetrating networks
ISFETs	Ion-selective field effect transistors
LEDs	Light emitting diodes
LLDPE	Linear low density polyethylene
LDPE	Low density polyethylene

MMC	Metal matrix composites
MAc	Maleic acid
NMP	N-methylpyrrolidone
Ormocers	Organically modified ceramics
Ormosils	Organically modified silicates
OLEDs	Organic light emitting diodes
PEK	poly(ether ketone)
PMMA	Poly(methylmethacrylate)
PVC	Poly(vinyl chloride)
PANI-TSA	Polyaniline-p-toluenesulfonic acid
PS	Poly(styrene)
PANI	Polyaniline
PANI-MAc	Polyaniline:maleic anhydride
PPy	Polypyrrole
PTh	Polythiophene
PVP	Poly(N-vinylpyrrolidone)
PVAL	Poly(vinyl alcohol)
PS	Polystyrene
PA	Polyacetylene
PCL	Poly-ε-caprolactone
PPV	Poly(p-phenylenevinylene)
PET	Polyethyleneterepthalate
PE	Polyethylene
PPh	Polyphenylenes
P3ddt	Poly(3-dodecyl thiophene)
PP	Polypropylene
PI	Polyimide
PEO	Poly(ethylene oxide)
PEG	Polyethylene glycol
PPG	Poly(propylene glycol)
PPy-PVA	Polypyrrole:poly(vinyl alcohol)
Pin	Polyindene
PMC	Polymer matrix composites
PC	Polycarbonate
PANI-PAN	Polyaniline:polyacrylonitrile
PAN	Polyacrylontrile

PP-PPy	Polypropylene:polypyrrole
PPy-PI	Polypyrrole:polyimide
PPy-PP	Polypyrrole:polypropylene
PPy-PS	Polypyrrole:polystyrene
PPY-Pin	Polypyrrole:polyindene
SEM	Scanning electron microscope
SDS	Sodium dodecylsulphate
SSA	Sulfosalicylic acid
SES	Styrene-ethylene-butylene-styrene
SBS	Styrene-butadiene-styrene
TEP	Thermoelectric power
THF	Tetrahydrofuran
TFTs	Thin-film field-effect transistors
TSA	p-toluenesulfonic acid
TEM	Transmission electron microscope
TGA	Thermogravimetric analysis

REFERENCES

1. S.C. Sharma; "Composite Materials", Narosa Pub. House, New Delhi, India (2000).
2. W.D. Callister, Jr.; in "Material Science and Engineering, An Introduction", 520, John Wiley and Sons, Chichester (2001).
3. A.R. Blythe; in "Electrical Properties of Polymers", 123, Cambridge University Press, Cambridge (1979).
4. D.C. Bott; in "Handbook of Conducting Polymers" (T.A. Skotheim, Ed.), 2, 1191, Marcel Dekker, New York (1986).
5. S.K. Dhawan, N. Singh and S. Venkatachalam; Synth.Met., 125, 389 (2002).
6. A.B. Samui, A.S. Patankar, R.S. Satpute and P.C. Deb; Synth.Met., 125, 423 (2002).
7. S. Palaniappan; Macromolecules, Current Trends, in proceedings of the International Symposium on Macromolecules, Vol. 1 Allied Publisher, India, 1995, p. 315.
8. C.K. Chiang, C.R. Fincher, Jr., Y.W. Park, A.J. Heeger, H. Shirakawa, E.J. Louis, S.C. Gau and MacDiarmid; Phy.Rev.Lett., 39, 1098 (1977).
9. Y. Cao, P. Smith and A.J. Heeger; Synth.Met., 48, 91 (1992).
10. A.J. Heeger; Synth.Met., 57, 3471 (1993).
11. A. Pron, J.E. Osterholm, P. Smith, A.J. Heeger, J. Laska and M. Zagorska; Synth.Met., 57, 3520 (1993).
12. Y. Cao, P.Smith and A.J. Heeger; Synth.Met., 57, 3514 (1993).
13. L.W. Shacklette, C.C. Han and M.H. Luly; Synth.Met., 57, 3532 (1993).
14. O.T. Ikkala, J. Laakso, K. Vakiparta, E. Virtanen, H. Ruohonen, H. Jarvinen, T. Taka, P. Passiniemi, J.E. Osterholm, Y. Cao, A. Andreatta, P. Smith and A.J. Heeger; Synth.Met., 69, 97 (1995).
15. A.G. MacDiarmid; Synth.Met., 84, 27 (1997).
16. G.S. Akundi and J.O. Iroh; Polymer, 42, 9645 (2001).
17. C. Zhang and A. Heeger; J.Appl.Phys., 84, 1579 (1998).
18. S. Koul, R. Chandra and S.K. Dhawan, Polymer, 41, 9305 (2000).

19. K. Kontturi, P. Pentti and G. Sundholm; J.Electroanal.Chem., 453, 231 (1998).

20. D. Jr. MacInnes, M.A. Druy, P.J. Nigrey, D.P. Naires, A.G. MacDiarmid and A.J. Heeger; J.Chem. Soc., Chem. Commun., 317 (1989).

21. M.R. Anderson, B.R. Mattes, H. Reiss and R.B. Kaner; Synth Met., 1151, 41 (1991).

22. T.A. Skotheim (Ed.); 'Handbook of Conducting Polymers', Marcel Dekker, New York (1986) Vol. I.

23. J.H. Burroughes, D.D.C. Bradley, A.R. Brown, R.N. Marks, K. Mackay, R.H. Friend, P.L. Burns and A.B. Holmes; Nature, 347(6293), 539 (1990).

24. A.F. Diaz and K.K. Kanazawa, J.Chem. Soc., Chem. Commun., 635 (1979).

25. F. Jonas, L. Schrader; Synth.Met., 41, 831 (1991).

26. M. Nakata, M. Taga and K. Hideo; J.Polymer, 24, 437 (1992).

27. J. Laska, K. Zak and A. Pron; Synth.Met., 84, 117 (1997).

28. J.M. Machado, F.E. Karasz and R.W. Lenz; Polymer, 29,1412 (1988).

29. M. Sato, S. Tanaka and K.J. Kaeriyama; J.Chem. Soc., Chem. Commun., 873 (1986).

30. S. Hotta, S. Rughooputh, A.J. Heeger and F. Wudl; Macromolecules, 20, 212 (1987).

31. I. Murase, T. Chinishi, T. Naguchi and M. Hirooka; Polym. Commun., 25, 327 (1984).

32. J.D. Capistran, D.R. Gagnon, S. Antoun, R.W. Lenz and F.E. Karasz; Polym. Prepr., 25, 282 (1984).

33. A.A. Ahmad; Ph.D. Thesis, Aligarh Muslim University (2004).

34. H.V. Dijk, Aagaard and R. Schellekens; Synth.Met., 55, 1085 (1993).

35. K. Yoshino, X.H. Yin, S. Morita, Y. Nakanishi, S. Nakagawa, H. Yamamoto, T. Watanuki and I. Isa; Jpn J.Appl.Phys., 32, 979 (1993).

36. M. Nakata and H. Kise; J.Polym, 25, 91 (1993).

37. Y.A. Dubitsky and B.A. Zhubanov; Synth.Met., 53, 303 (1993).

38. M. DePaoli, S. Panero, P. Prosperi and B. Scrosati: Electrochim. Acta, 35, 145 (1990).

39. O. Niwa, M. Hikita and T. Tamamura; Macromol. Chem., Rapid. Commun., 6, 375 (1985).

40. J.F. Rabek, J. Lucki, H. Kereszti, B. Krische, B.J. Qu and W.F. Shi; Synth.Met., 45, 335 (1991).

41. S. Radhakrishnan and D.R. Saini; Synth.Met., 58, 243 (1993).

42. T. Oho and S. Miyata; J.Polym., 18, 95 (1986).

43. A. Pron, M. Zagorska, W. Fabianowski, J.B. Raynor and S. Lefrant; Polym. Commun., 28, 193 (1987).

44. M. Makhlouki, J.C. Bernede, M. Morsli, A. Bonnet, A. Conan and S. Lefrant; Synth.Met., 62, 101 (1994).

45. S.E. Lindsey and G.B. Street; Synth.Met., 10, 67 (1985).

46. G.B. Street, S.E. Lindsey, A.I. Nazzal and K.J. Wyne; Mol. Cryst. Liq. Cryst., 118, 137 (1985).

47. F. Selampinar, U. Akbulut, E. Yildiz, A. Gungor and L. Toppare; Synth.Met. 89, 111, (1997).

48. Y. Kang, M.H. Lee and S.B. Rhee; Synth.Met., 47, 157 (1992).

49. S.Hotta, S.D.D.V.Rughooputh and A.J.Heeger; Synth.Met., 22, 79.(1987)

50. E. Ruckenstein and J.S. Park; J.Appl. Polym. Sci., 42, 925 (1991).

51. X.T. Bi and Q.B. Pei,; Synth.Met., 22, 145 (1987).

52. H.T. Chiu, J.S. Lin and C.M. Huang; J.Appl. Electrochem., 222, 358 (1992).

53. J.E. Osterholm, J. Laakso and P. Nyholm; Synth.Met., 28, 435 (1989).

54. K.S. Ho, K. Levon, J. Mao and W.Y. Zheng; Synth.Met., 3591, 55 (1993).

55. L.H. Dao, X.F. Zhong, A. Menikh, R. Paynter and F. Martim; Annual Tech. Conf. SPE, 49, 783 (1991).

56. A.A. Ahmed and F. Mohammad; J.Solid State Phenomena, 111, 95 (2006).

57. M.B. Meador, D.H. Green, J.V. Auping, J.R. Gaier, L.A. Ferrara, D.S. Paradopoulos, J.W. Smith and D.J. Keller; J.Appl. Polym. Sci., 63, 821 (1997).

58. F. Selampinar, U. Akbulur, T. Yalchin, S. Suzer and L. Toppare; Synth.Met. 62, 201 (1994).

59. Y. Sun and E. Ruckenstein; Synth.Met., 74, 145 (1995).

60. M. Bardet, M. Guinaudeu, C. Bourgeoisat and H. Cherin; Synth.Met., 359, 41 (1991).

61. H.L. Wang, L. Toppare and J.E. Fernandez; Macromolecules, 23, 1053 (1990).

62. Y.H. Park and M.H. Man; J.Appl. Polym. Sci., 45, 1992, (1973).

63. A. Bhattacharya and A. De; Prog. Solid State Chem., 24, 141 (1996).
64. T. Taka; Synth.Met., 41, 1177 (1991).
65. D. Cottevieille, A. Le Mehaute, C. Challioui, P. Mirebeau, J.N. Demay; Synth.Met., 101, 703 (1999).
66. T. Makela, J. Sten, A. Hujanen and H. Isotalo; Synth.Met., 101, 707 (1999).
67. F. Jonas and L. Schader; Synth.Met., 41, 831 (1991).
68. F. Jonas and G. Heywang; Electrochim Acta, 39, 1345 (1994).
69. J.D. Stenger-Smith; Prog. Polym. Sci., 23, 57 (1998).
70. C. Han and L.W. Shacklette; US Patent 5, 378, 404 (1995).
71. A.K. Bakhshi; Bull. Mater. Sci., 18, 469 (1995).
72. K. Hanhi, V. Lonnberg, K. Pyorala, V. Loennberg, K. Pyoeraelae; WO Patent 9, 706, 213 (1997).
73. V. Misoska, J. Ding, J.M. Davey, W.E. Price, S.F. Ralph and G.G. Wallace; Polymer, 42, 8571 (2001).
74. J.M. Davey, S.F. Ralph, C.O. Too, G.G. Wallace and A.C. Partridge; React. Funct. Polym., 49, 87 (2001).
75. B. Wessling and J. Posdorfer; Electrochim.Acta; 44, 2139 (1999).
76. R. Gangopadhyay and A. De; Sens. Actuator B, 77, 326 (2001).
77. S. Koul, R. Chandra and S.K. Dhawan; Sens. Actuator B, 75, 151 (2001).
78. C.R. Kagan, D.B. Mitzi and C.D. Dimitrakopoulos; Science, 286, 945 (1999).
79. A. Okada and A. Usuki; Mater. Sci. Engg., C3:109 (1995).
80. J.W. Gilman; Appl. Clay Sci., 15, 31 (1999).
81. J.W. Gilman, C.L. Jackson, A.B. Morgan, J.R. Harris, E. Manias, E.P. Giannelis, M. Wuthenow, D. Hilton and S.H. Phillips; Chem. Mater., 12, 1866 (2000).
82. D. Porter, E. Metcalfe and M.J.K. Thomas; Fire Mater., 24, 45 (2000).
83. M. Zanetti, S. Lomakin and G. Camino; Macromol. Mater. Engg., 279, 1 (2000).
84. S.P. Armes; Polym. News, 20, 233 (1995).
85. D.Y. Godovski; Adv. Polym. Sci., 119, 79 (1995).
86. P. Judeinstein and C. Sanchez; J.Mater.Chem., 6, 511 (1996).
87. U. Schubert and N. Husing; Synthesis of Inorganic Materials, Weinheim, Wiley-VCH, 2000.
88. C. Sanchez and F. Ribot; New J.Chem., 18, 1007 (1994).
89. E. Bescher and J.D. Mackenzie; Materials Sciences and Engineering, Chap. 6, 145 (1998).
90. Y. Chujo and T. Saegusa; Ad. Polym. Sci., 100, 11 (1992).
91. U. Schubert, N. Husing and A. Lorenz; Chem. Mater., 7, 2010 (1995).
92. B.M. Novak; Adv. Mater., 6, 422 (1993).
93. C.A. Loy and K.J. Shea; Chem. Rev., 5, 1431 (1995).
94. J.E. Mark; J.Appl. Polym.Sci., Appl. Polym.Symp., 60, 273 (1992).
95. J.N. Hay and S.J. Shaw; Europhysics News, 34(3) (2003).
96. J.E. Mark, C.Y. Lee and P.A. Bianconi (Eds.); Hybrid Organic-Inorganic Composites, American Chemical Society Symposium Series, Vol. 565, Washington, American Chemical Society (1995).
97. K.F. Webb and A.S. Teja; in Proc. of the Eight Int. Conf. on Properties and Phase equilibria for Product and Process Design, The Netherlands, Noordwijkerhout (1998).
98. A.G. MacDiarmid and A.J. Epstein; Faraday Discuss., Chem. Soc., 88, 333 (1989).
99. A.G. MacDiarmid and A.J. Epstein; Conducting Polymers, Science and Technology, Second Brazilian Polymer Conference, Plenum Publishing Corp., Brazil (1993).
100. S. Roth and W. Graupner; Synth.Met., 57, 3623 (1993).
101. A.A. Khan, M.M. Alam and F. Mohammad; Electrochim. Acta, 48, 2463 (2003).
102. K. Gurunathan, D.P. Amalnerkar, D.C. Trivedi; Mater. Lett., 57, 1642 (2003).
103. N.K. Raman, M.T. Anderson and C.J. Brinker; Chem. Mater., 8, 1682 (1996).
104. J. Wen and G.L. Wilkens; Chem. Mater.8, 1667 (1996).
105. I. Honma, S. Nomura and H. Nakajima; J.Memb. Sci., 185, 83 (2001).
106. M.D. Butterworth, R. Corradi, J. Johal, S.F. Lascelles, S. Maeda and S.P. Armes; J. Colloid Interface Sci., 174, 510 (1995).

107. S. Maeda and S.P. Armes; Mater. Chem., 4, 935 (1994).

108. S. Maeda, M. Gill and S.P. Armes; Polym. Mater. Sci. Engg., 4, 935 (1994).

109. R. Schollhorn; Chem. Mater., 8, 1747 (1996).

110. P.G. Romero; Adv. Mater., 13, 163 (2001).

111. Y. Wang and N. Herron; Science, 273, 632 (1996).

112. M.G. Kanatzidis, R. Bissessur, D.C. DeGroot, J.L. Schindler and C.R. Kannewurf; Chem. Mater., 5, 595 (1993).

113. S. Higashika, K. Kimura, Y. Matsuo and Y. Sugie; Carbon, 37, 354 (1999).

114. L. Wang, M.R. Lane, P. Brazis, C.R. Kannewurf, Y. Kim, W. Lee, J. Choy and M.G. Kanatzidis; J. Am. Chem. Soc., 122, 6629 (2000).

115. R.C. Patil, S. Radhakrishnan, S. Pethkar and K. Vizaymahanan; J.Mater. Res., 16, 1982 (2001).

116. S.L. Roberts, C.A. Koval and R.D. Noble; Ind. Eng. Chem. Res., 39, 1673 (2000).

117. A.D. Pomogailo; Russ. Chem. Rev., 69, 53 (2000).

118. C.L. Beaudry and L.C. Klein; ACS Symp. Ser., (Nanotechnology) 622, 382 (1996).

119. M. Inagaki, S. Katahira and T. Nakamura; Adv. Sci. Technol., 13, 277 (1999).

120. K. Checkiewicz, G. Zukowska and W. Wieczorek; Chem. Mater., 13, 379 (2001).

121. H.L. Frisch, B. Xi, Y. Qin, M. Rafailovich, N.L. Yang and X. Yan; High Perform. Polym., 12, 543 (2000).

122. A.J.G. Zarbin, O.L. Alves and M.A. DePaoli; Synth.Met., 84, 107 (1997).

123. W. Feng, E. Sun, A. Fujii, H. Wu, K. Niihara and K. Yoshino; Bull. Chem. Soc. Jpn., 73, 2627 (2000).

124. D. Gang, S.K. Banerji and T.E. Clevenger; Proc. Conf. Hazard. Waste Res., Kansas, U.S.A., (1999) 64.

125. G.T. Gomez, M.C. Lira and P.R. Gomez; Bol. Soc. Esp. Ceram. Vidrio, 39(3), 391 (2000).

126. C. Perruchot, M.M. Chehimi, M. Delamas, J.A. Eccles, T.T. Steele and C.D. Mair; Synth.Met., 113, 53 (2000).

127. H.J. Chai, J.W. Kim, S.G. Kim and M.S. Jhan; Polym Prepr., 40, 813 (1999).

128. M.C. Lira and P.R. Gomez; J. Solid State Chem., 147, 601 (1999).

129. W. Quiju, X. Zhijian, Q. Zorgneng and W. Fosong; Gaofenzi Xuebau (Chinese), 53, 551 (1999).

130. S. Sinha Ray and M. Biswas; Mater. Res.Bull., 34, 1187 (1999).

131. Y. Pan, G.W. Zhang and X.Y. Su; J.Mater. Sci., 29, 5757 (1994).

132. J.M. Bae, I. Honma and S. Hirakawa; J. Korean Phys. Soc., (Suppl. 4th Int. Conf. on Electronic. Materials'98), 35, 315 (1995).

133. I. Honma, S. Nomura and H. Nakajima; Mater. Res. Soc. Symp. Proc. (Nanophase and Nanocomposite Materials III), 581, 381 (2000).

134. G.J.F. Demets, F.J. Anaissi and H.E. Toma; Electrochim. Acta, 46, 547 (2000).

135. B.P. Grady and W.B. Genetti; 58th Annu. Tech. Conf. Soc. Plast. Eng., 2, 143 (2000).

136. T. Furukawa; Shaho-Daiichikogyo Seiyaku Kabushiki Kaisha, 514, 2 (2000).

137. D.D.L. Chung; Polym. Polym. Composites, 8, 219 (2000).

138. E. Shouji and A.D. Buttry; Proc. Electrochem. Soc., 99, 208 (2000).

139. L.P. Cheng, D.J. Lin and K.C. Yang; J. Membr. Sci., 172, 157 (2000).

140. R. Gong, J. Guan and R. Yuan; Mat. Sci. Ed., 15, 62 (2000).

141. V.D. Noto, M. Fauri, M. Vittadello, S. Lavina and S. Biscazzo; Electrochim. Acta, 46, 1587 (2001).

142. T. Uma, T. Mahalingam and U. Stimming; Mater. Chem.Phys., 82, 478 (2003).

143. Y. Yang-Yen and C. Wen-Chang; Mater. Chem.Phys., 82, 388 (2003).

144. S.P. Armes and M. Aldissi; J.Chem.Soc., Chem.Commun., 88 (1989).

145. S.P. Armes, M. Aldissi, S.F. Agnew and S. Gottesfeld; Langmuir, 6, 1745 (1990).

146. H. Eisazadeh, K.J. Gilmore, A.J. Hodgson, G. Spinks and G.G. Wallace; Colliods Surf. A, 103, 281 (1995).

147. T. Nagaoka, H. Nakao and K. Ogura; Anal. Sci., 12, 119 (1996).

148. S.P. Armes, M. Aldissi, S.F. Agnew and S. Gottesfeld; Mol. Cryst. Liq. Cryst., 190, 63 (1990).

149. N. Gospodinova, P. Mokreva and L. Terlemezyan; J.Chem.Soc., Chem. Commun., 923 (1992).

150. J. Stejskal, P. Kratochvil, N. Gospodinova, L. Terlemezyan and P. Mokreva; Polymer, 33, 4857 (1992).

151. N. Gospodinova, L. Terlemezyan, P. Mokreva, J. Stejskal and P. Kratochvil; Eur. Polym. J., 29, 1305 (1993).

152. J. Stejskal, P. Kratochvil, N. Gospodinova, L. Terlemezyan and P. Mokreva; Polym. Int., 32, 401 (1993).

153. C.L. Huang, R.E. Partch and E. Matijevic; J. Colloid Interface Sci., 170, 275 (1995).

154. H. Eisazadeh, K.J. Gilmore, A.J. Hodgson, G. Spinks and G.G. Wallace; Colloids Surf, A 103, 281 (1995).

155. J. Stejskal, P. Kratochvil and M. Spirkova; Polymer, 36, 4135 (1995).

156. B. Vincent and J.W. Waterson; J.Chem.Soc., Chem., Commun., 683 (1990).

157. P. Tadros, S.P. Armes and S.Y. Luk; J.Mater.Chem., 2, 125 (1992).

158. H. Eisazadeh, G. Spinks and G.G. Wallace; Polym. Int, 37, 87 (1995).

159. J. Stejskal, P. Kratochvil and M. Helmstedt; Langmuir, 12, 3389 (1996).

160. C. DeAnnitt and S.P. Armes; J. Colloid Interf. Sci., 150, 134 (1992).

161. P. Banerjee, M.L. Digar and S.N. Bhattacharyy and B. Mandal; Eur. Polym. J., 30, 499 (1994).

162. P. Banerjee, S.N. Bhattacharyya and B.M. Mandal; 11, 2414 (1995).

163. P. Banerjee and B.M. Mandal; Macromolecules, 28, 3940 (1995).

164. M. Gill, J. Mykytiuk, S.P. Armes, J.L. Edwards, T. Yeates, P.J. Moreland and C.J. Mollett; J.Chem. Soc., Chem. Commun, 108 (1992).

165. D. Chattopadhyay and B.M. Mandal: Langmuir, 12, 1585 (1996).

166. M. Gill, S.P. Armes, D. Fairhurst, S. Emmett, T. Pigott, G. Idzorek; Langmuir, 8, 2178 (1992).

167. M. Gill, F.L. Baines and S.P. Armes; Synth.Met.,1029, 57 (1993).

168. S. Maeda and S.P. Armes; J. Colloid Interf. Sci., 159, 257 (1993).

169. S. Maeda and S.P. Armes; J. Mater.Chem., 4, 935 (1994).

170. R. Flitton, J. Johal, S. Maeda and S.P. Armes; J. Colloid. Interf. Sci., 173, 135 (1995).

171. J. Stejskal, P. Kratochvil, S P Armes, S.F. Lascelles A. Riede, M. Helmstedt, J. Prokos and I. Krivka, Macromolecules, 29, 6814 (1996).

172. I. Krivka, J. Prokes, R. Kuzel, J. Stejskal, S. Nespurek and P. Kratochvil; Proc. 10th Eur. Microelectronics Conf., Copenhagen, Denmark, (1995) 228.

173. I. Krivka, R. Kuzel, J. Prokes, J. Stejskal, S. Nespurek and P. Kratochvil; Proc. 23rd Int. Conf. Microelectronics, Terme Catez, Slovenia (1995) 257.

174. F. Leroux, B.E. Koene and L.F. Nazar; J. Electrochem. Soc., 143, 181 (1996).

175. E. Shouji and D.A. Buttry; Langmuir, 15, 669 (1999).

176. M. Lira-Cantu and P. Gomez-Romero; J.Electrochem. Soc., 146, 2029 (1999).

177. F. Huguenin, M.T.D. Gambardella, R.M. Torresi, S.I. deTorresi and D.A. Buttry; J. Electrochem. Soc., 147, 2437 (2000).

178. C.G. Wu, D.C. DeGroot, H.O. Marcy, J.L. Schindler, C.R. Kannewurf, Y.J. Liu, W. Hirpo and M.G. Kantzidis; Chem. Mater., 1996, 8, (1992).

179. H.P. Oliveira, C.F.O. Graeff, C.A. Brunello and E.M.J. Guerra; Non-Cryst.Solids, 273, 193 (2000).

180. G.J.F. Demets, F.J. Anaissi and H.E. Toma; Electrochim. Acta, 46, 547 (2000).

181. Z.F. Li and E. Ruckenstein; Langmuir, 18, 6956 (2002).

182. E.J. Samuelsen and J. Mardalen; in "Handbook of Organic Conductive Molecules and Polymers", (H.S. Nalwa Ed.), 3, 522, John Wiley and Sons, Chichester (1997); http://ist-socrates.bereley.edu/~iqwu/ paper1/paper.html

183. F. Mohammad; in "Handbook of Advance Electronic and Photonic Materials and Devices" (H.S. Nalwa, Ed.) vol. 8, p. 322, Academic Press, CA, USA (2001).

184. F. Mohammad, D. Phil. Thesis, University of Sussex (1987).

185. F. Mohammad, in "HandBook of Advance Electronic and Photonic Materials and Devices" (H.S.

Nalwa, Ed.), vol. 8, p. 321, Academic Press, CA, USA (2001).

186. F. Mohammad, in "Handbook of Organic Conductive Molecules and Polymers" (H.S. Nalwa, Ed.), vol. 3, p. 797, John Wiley, Chichester (1997).

187. F. Mohammad, D. Phil.Thesis, University of Sussex (1987) p. 80.

188. S.A. Sifullina, L.M. Yarysheva, A.V. Volkov, A.L. Volynskii and N.F. Bakeev; Vysokomol. Soed in., Ser. A, Ser. B (russ), 38, 1172 (1996).

189. H. Kalhori, S. Zehzad, S.A. Fakhri and A.A. Entezami; J.Polym. Sci. Technol., 1, 29 (1996).

190. T. Tsanov and L. Terlemezyan; Polym. Composites, 5, 438 (1997).

191. A. Pron, M. Zagorska, Y. Nicolan, F. Genond and M. Nechtschein; Synth.Met., 84, 89 (1997).

192. M. Zilberman, A. Siegmann and M. Narkis; J.Macromol. Sci. Phys., B 37, 301 (1998).

193. M. Zilberman, A. Siegmann and M. Narkis; J. Macromol. Sci. Phys., 339, 333 (2000).

194. L.W. Shacklette, C.C. Han and M.H. Luly; Synth.Met., 57, 3532 (1993).

195. A.J.G. Zarbin, M.A. De Paoli and L.O. Alves; Synth.Met., 84, 107 (1997).

196. J. Przyluski and G. Zukowska; Polymer, 42, 229 (1997).

197. G. Du, J. Avlyanov, C.Y. Wu, K.G. Reimer, A. Benatar, A.G. MacDiarmid and A.J. Epstein; Synth.Met., 85, 1339 (1997).

198. X.H. Yin, K. Yoshino, K. Hashizume and I. Isa; Jpn. J.Appl.Phys., Part-I, 36, 3557 (1997).

199. Y.M. Ma, X.Y. Wang, J.S. Guo and H.Q. Xie; Gaodenh Xuexiao Huaxue Xuebao (Chinese), 18, 1560 (1997).

200. S.Y. Oh, H.C. Koh, J.W. Choi, H.W. Rhee and H.S. Kim; Polym. J., 29, 404 (1997).

201. E. Ruckenstein and S. Yang; Synth.Met., 5, 283 (1993).

202. E.C. Cooper and B. Vincent, J.Phys.D, Appl.Phys., 22, 1580 (1989).

203. S. Yang and E. Ruckenstein; Synth.Met., 59, 1 (1993).

204. T.H. Ahu and Y.H. Park; Kolon. Ind. Inc., S. Korea, Cl. Co. 8J7/12, 706 (1997).

205. T. Tsanov and L. Terlemezyan; Polym. Polym. Composites, 6, 39 (1998).

206. J.B. Zheng, H. Wang, W. Feng and H.C. Wu; Semicond.Photonics Technol., 5, 216 (1999).

207. Q. Wu, Z. Xue, Z. Qi and F. Wang; Gaofenzi Xuebao, 5, 551 (1999).

208. W. Feng, E. Sun, A. Fujii, H. Wu, K. Niihara and K. Yoshino; Bull. Chem. Soc. Jpn., 73, 2627 (2000).

209. P. Beadles, S.P. Armes, S. Gottesfeld, C. Mombourquette, R. Houlto W.D. Andrews and A.P.Agnew; Macromolecules, 25, 2526 (1992).

210. R.H. Cruz-Eatrada and M.J. Flokes; J.Mater. Sci., 35, 5065 (2000).

211. N. Oyama; Macromol. Symp., 159 (Polymer Science and Industrial Research in the Fast-Changing Age), Wiley-VCH Verlag. GmbH., 221 (2000).

212. M.D. Bessete, R.A. Wwiss, P.P. Gan, C. Erkey and Y. Fu; World Properties, Inc., USA, A 5, 11 (2000).

213. J.Y. Bergeron and L.H. Dao; Macromolecules, 25, 3332 (1992).

214. K. Ogura, R.C. Patil, H. Shiigi, T. Tonosaki and M. Nakayama; J.Polym. Sci., Part A: Polym. Chem., 38, 4343 (2000).

215. H. Shiigi, R.C. Patil, M. Nakayama, T. Tonosaki and K. Ogura; Recent Res. Dev. Electrochem., 2, 217 (1999).

216. H. Shiigi, T. Oho, T. Tonosaki and K. Ogura; J.Electrochem.Soc., 69, 997 (2001).

217. H. Shiigi, R.C. Patil, M. Nakayama, T. Tonosaki and K. Ogura; Electrochem., 2, 217 (1999).

218. W.J. Lee, Y.J. Kim and S. Kaang; Synth.Met., 113, 237 (2000).

219. S. Yang and E. Ruckenstein; Synth.Met., 59, 1 (1993).

220. E. Ruckenstein and S. Yang; Synth.Met., 53, 283 (1993).

221. E. Ruckenstein and Y. Sun; Synth.Met., 74, 107 (1995).

222. Y. Sun and E. Ruckenstein; Synth.Met., 74, 145 (1995).

223. B.H. Jeon, S. Kim, M.H. Choi and I.J. Chung; Synth.Met., 104, 95 (1999).

224. T. Jeevananda, Siddaramaiah, V. Annadurai and R. Somashekar; J.Appl. Polym. Sci., 82, 383 (2001).

225. A.B. Kaiser, C.K. Subramaniam, P.W. Gilberd and B. Wessling; Synth.Met., 69, 197 (1995).
226. C.K. Subramaniam, A.B. Kaiser, P.W. Gilberd, C.J. Liu and B. Wessling; Solid State Commun., 97, 235 (1996).
227. J.O. Iroh and R. Rajagopalan; Surf. Eng., 16, 481 (2000).
228. K.W. Oh, J.H. Seong and S.H. Kim; Polymer (Kor.), 24, 673 (2000).
229. S.W. Byun and S.S. Im; Synth.Met., 57, 3501 (1993).
230. S.W. Byun and S.S. Im; Synth.Met., 69, 219 (1995).
231. Y.H. Park, Y.II. Kim, D.H. Baik and C.R. Park; Mole. Cryst. Liq. Cryst., 370, 327 (2001).
232. M. Morsil, A. Bonnet, F. Samir, V. Jousseaume and S. Lefrant; Synth.Met., 76, 273 (1996).
233. A.B. Kaiser; Synth.Met., 45, 183 (1991).
234. M. Makhlouki, M. Morsli, A. Bonnet, A. Conan and S. Lefrant; J.Appl. Polym. Sci., 99, 443 (1992).
235. M. Omastova, I. Chodak and J. Pionteck; Synth.Met., 102, 1251 (1999).
236. A.I. Medalia; Rubber Chem. Technol., 59, 432 (1986).
237. F. Carmona; Physica, A 157, 461 (1989).
238. B. Tieke and W. Gabriel; Polymer, 31, 20 (1990).
239. M.B. Meador, D.H. Green, J.V. Auping, J.R. Gaier, L.A. Ferrara, D.S. Paradopoulos, J.W. Smith and D.J. Keller; J.Appl. Polym. Sci., 63, 821 (1997).
240. W. Jinwei and M.P. Srinivasan; Synth.Met., 105, 1 (1999).
241. F. Selampinar, U. Akbulat and L. Toppare; Synth.Met., 84, 185 (1997).
242. J. Yang, Y. Yang, J.Hou, X. Zhang, W. Zhu and M. Xu; Polymer, 37, 793 (1996).
243. M. Tang, T.Y. Wen, T.B. Du and Y.P. Chen; Europ. Polym. J., 39, 143 (2003).
244. S. Yigit, J. Hacaloglu, U. Akbulut and L. Toppare; Synt. Met., 79, 11 (1996).
245. A. Bozkurt, U. Akbulut and L. Toppare; Synth.Met., 82, 41 (1996).
246. J.O. Iroh and K. Levine; J. Power Sources, 117, 267 (2003).
247. S. Pouzet, N.L. Bolay, A. Ricard and F. Josse; Synth.Met. 55, 1079 (1993).
248. Y. Sun and E. Ruckensen; Synth.Met., 72, 1 (1995).
249. K. Uosaki, K. Ohazaki and H. Kita; J. Polym. Sci., Part A, Polym. Chem., 28, 399 (1990).
250. M.A. DePaoli, R.J. Waltman, A.F. Diaz and J. Bargon; J.Polym. Sci., Part A, Polym. Chem., 23, 1687 (1985).
251. Niwa, M. Hikita and T. Tamamura; Appl.Phys. Lett., 46, 444 (1985).
252. G.B. Street, S.E. Lindsey, A.I. Nazzal and K.J. Wyne; Mol.Cryst. Liq.Cryst., 118, 137 (1985).
253. Y.H. Park and M.H. Han; J.Appl. Polym. Sci., 45, 1973 (1992).
254. U. Geissler, M.L. Hallensleben and L. Toppare; Synth.Met., 40, 239 (1991).
255. U. Geißler, M.L. Hallensleben and L. Toppare; Synth.Met., 55, 1483,(1993).
256. G.B. Street, S.E. Lindsey, A.I. Nazzal and K.J. Wyne; Mol. Cryst. and Liq.Cryst., 118, 137 (1985).
257. J.D. MacKenzie and D.R.U Irich (Eds.); Sol-Gel Optics, Proc. SPIE (1990) p. 1328, (1992) p. 1758, (1994) p. 2288.
258. P.N. Prasad, F.V. Bright, U. Narang, R. Wang and R.A. Dunbar; J.D. Jordan and R. Gvishi; ACS Symp. Ser. 585, 317 (1995).
259. R. Gvishi, U. Narang, G. Ruland, D.N. Kumar and P.N. Prasad; Appl. Organomet. Chem., 11, 107 (1997).
260. J.G. Winiarz, L. Zhang, M. Lal, C.S. Friend and P.N. Prasad; Chem.Phys., 245, 417 (1999).
261. J.G. Winiarz, L. Zhang, M. Lal, C.S. Friend and P.N. Prasad; J.Am. Chem. Soc., 121, 5287 (1999).
262. M. Kawasumi, N. Hasegawa, A. Usuki and A. Okada; Mater. Sci. Eng., C6, 135 (1998).
263. H.B. Michael, W. Shannon. Boettcher, L. Evelyn. Hu and G.D. Stucky; J.Am. Chem. Soc., 126, 10826 (2004).
264. V.I. Colvin, M.C. Lamp and A.P. Alivatos; Nature, 370, 354 (1994).
265. M.C. Schlamp, X. Peng and A.P. Alivatos; J.Appl.Phys., 82, 5837 (1997).
266. H. Mattoussi, L.H. Radzilowski, B.O. Dabbousi, M.G. Bawendi and F.M. Rubner; J.Appl.Phys., 83, 7965 (1998).
267. K. Tada, M. Hamaguchi, A. Hosano, S. Yura, H. Haradaand and K. Yoshini; Jpn. J.Appl.Phys., 36,

418 (1997).

268. S. Guha, G. Hendershot, D. Peebles, P. Stiner, F. Kolowskiand and W. Lang; Appl.Phys. Lett., 64, 613 (1994).

269. S.A. Carter, J.C. Scott and P.J. Brock; Appl.Phys. Lett., 71, 1145 (1997).

270. Y. Yang, J. Huang, S. Liu and J. Shen; J.Mater. Chem., 7, 131 (1997).

271. S. Guha, R.A. Haight, N.A. Bojarczuk and D.W. Kisker; J.Appl.Phys., 82, 4126 (1997).

272. S. Guha, P. Steiner, F. Kolowskiand and W. Lang; Thin Solid Films, 225, 119 (1995).

273. J. Wen and G.L. Wilkes; Chem. Mater., 8, 1667 (1996).

274. X. Hong, T. Ishihara and A.V. Nurmikko; Solid State Commun., 84, 657 (1992).

275. M. Era, S. Morimoto, T. Tsutsui and S. Saito; Appl.Phys. Lett., 65, 676 (1994).

276. T. Hattori, T. Taira, M. Era, T. Tsutsui and S. Saito; Chem.Phys. Lett., 254, 103 (1996).

277. C.R. Kagan, D.B. Mitzi and C. Dimitrakopoulos; Science, 286, 945 (1999).

278. N. Lakshmi and S. Chandra; J.Mater. Sci., 37, 249 (2002).

279. A.L. Laskar and S. Chandra; Superionic Solids and Solid Electrolytes: Recent Trends, Academic Press, London (1989).

280. T. Takahashi; High Conductivity Solid-State Conductors-Recent Trends and Applications, World Scientific, Singapore (1989).

281. S. Chandra, N. Singh and B. Singh; Solid State Commun., 57, 519 (1986).

282. S. Chandra; Superionic Solids: Principles and Applications, North Holland, Amsterdam (1981).

283. F. Corce, G. Gerace, G. Dautzenberg, G.P. Passerini, Appetecchi and B. Scrosati; Electrochim. Acta, 39, 2187 (1994).

284. K. Dahmouche, M. Atik, N.C. Mello, T.J. Bonagamba, H. Panepucci, M. Aegerter and P. Judeinstein; Mater. Res. Soc. Symp. Proc., 435, 363 (1996).

285. L.M. Bronstein, C. Joo, R. Karlinsey, A. Ryder and J.W. Zwanziger; Chem. Mater., 13, 3678 (2001).

286. P.H. DeSouza, R.F. Bianchi, K. Dahmouche, P. Judeinstein, R.M. Faria and T.J. Bonagamba; Chem. Mater.,13, 3685 (2001).

287. S. Pizzini and G. Bianchi; Chim. Ind. (Milan), 55, 966 (1973).

288. G. Alberti; Inorganic Ion Exchange Membranes, R. Passino, (Ed.), Pontificiae Acdemiae Scientiarum Sciripta Varia, Rome, pp. 629 (1976).

289. J.T. Kummer and N. Weber; U.S. Patent 3, 455 (1968) 233.

290. R. Gangopadhyay and A. De; Chem. Mater., 12, 608 (2000).

291. J. Simon and J.J. Andre; Molecular Semiconductors, Springer, Berlin, Chap. 3 (1985).

292. M. Kaneko and D. Wohrle; Adv. Polym. Sci., 84, 141 (1988).

293. Phthalocyanines Properties and Applications, in Leznoff, C.C., Lever, A.B.P.VCH, NY (Eds.), 1989, All chapters and references therein.

294. S. Venkatachalam and V.N. Krishnamurty; Indian J.Chem., Sect. A, 33, 683 (1994).

295. T.J. Marks; Angew. Chem. Int. Ed. Eng., 29, 857 (1990).

296. P. Petit, P. Turek and J.J. Andre; Synth.Met., 18, 59 (1978).

297. J.J. Andre, J. Simon, R. Even, B. Boudjema, G. Guiland and M. Maitrot; Synth.Met., 18, 683 (1987).

298. H. Li and T.F. Guarr; J.Chem.Soc., Chem. Commun., 832 (1989).

299. H. Li and T.F. Guarr; J.Electroanal. Chem., 317, 189 (1991).

300. H. Li and T.F. Guarr; Synth.Met., 38, 243 (1990).

301. G.G. Roberts, M.C. Petty, S. Baker, M.T. Fowler and N.J. Thomas; Thin Solid Films, 132, 113 (1985).

302. S. Venkatachalam, V.N. Veena Vijayanathan; in: J.C. Solomone (Ed.) Encyclopedia of Polymer Materials 6, CRC, Boca Raton, p. 4221 (1995).

303. S. Venkatachalam; in H.S. Nalwa (Ed.) Handbook of Organic Conducting Molecule and Polymers 2, Wiley, New York, Ch. 17 (1997).

304. E. Orti, J.L. Bredas and C. Clarisse; J.Chem.Phys., 92, 1228 (1990).

305. E. Orti and J.L. Bredas, J.Chem.Phys., 89, 1009 (1988).

306. A.B.P. Lever, M.R. Hempstead, C.C. Leznoff, W. Lin, M. Melnik, W.A. Nevin and P. Seymour; Pure

Applied Chem., 58, 467 (1986).

307. R.A. Collins and K.A. Mohammad; J.Phys., D2, 154 (1988).
308. Y. Sadaoka, T.A. Jones and W. Gopel, Sens. Actuators, B1, 148 (1990).
309. M.S. Nieuwenhuizen, A.J. Noderlof and A.W. Barendsz; Anal.Chem., 60, 230 (1988).
310. M.S. Nieuwenhuizen, A.J. Noderlof and A.W. Barendsz; Anal.Chem., 60, 235 (1988).
311. J.J. Miasic, A. Hooper and B.C. Tofield; J.Chem.Soc., Faraday Trans., I, 82, 1117 (1986).
312. A. Belgachi and R.A. Collins; J, Phys. D: Appl.Phys., 23, 223 (1990).
313. I. Honma, Y. Takeda and J.M. Bae; Mater Res Soc.Proc., 576, 257 (1999).
314. J.Y. Sanchez, A. Denoyelle and C. Poinsingnon; Poly.Adv. Technol., 4, 99 (1993).
315. P. Judeinstein, J. Titman, M. Stamm and H. Schmidt; Chem.Mater., 6, 127 (1994).
316. C.A. Vincent; in Electrochemical Sconce and Technology of Polymers-2, (Ed. R.G. Linford), Elsevier, Amsterdam, p. 47 (1990).
317. R.F. Bianchi, P.H. Souza, T.J. Bonagamba, H.C. Panepucci and R.M. Faria; Synth.Met., 102, 1186 (1999).
318. C. Sanchez, B. Alonso, F. Chapusot, F. Ribot and P. Audebert; J.Sol-Gel Sci. Technol., 2, 161 (1994).
319. P. Audebert, P. Griesmar, P. Hapiot and C. Sanchez; J.Mater. Chem., 2, 1293 (1992).
320. D. Vuillaume and F. Rondelez; La Recherché, 275, 461 (1995).
321. T. Maekawa, J. Tamaki, N. Miura, N. Yamazoe and S. Matsushima; Sensors Actuators B Chem., 9, 63 (1992).
322. J. Gardner, E. Hines and H.C. Tang; Sensors and Actuators B Chem., 9, 9 (1992).
323. K. Persaud and G. Dodd; Nature, 299, 352 (1962).
324. T. Nenov and S. Yordanov; Sensors and Actuators B. Chem., 8, 117 (1992).
325. C.O. Oriakhi and M.M. Lemer; Mater. Res.Bull., 30, 723 (1995).
326. J.P. Lemmon and M.M. Lemer; Chem. Mater., 6, 207 (1994).
327. Vaiar, H. Ishii and E. Giannelis; Chem. Mater., 5, 1694 (1993).
328. R. Gangopadhyay and A. De; Sensors and Actuators, B 77, 326 (2001).
329. A. Clearfield; Solv. Extrn. Ion Exch., 18, 655 (2000).
330. M. Qureshi and K.G. Varshney; Inorganic Ion Exchangers in Chemical Analysis, CRC Press Inc, Boca Raton, FL, (1991).
331. F. Sebesta; J. Radioanal. Nucl.Chem., 220, 77 (1997).
332. A.A. Khan and M.M. Alam; Analytica Chimica Acta, 504, 253 (2004).
333. A.A. Khan and M.M. Alam; React.Funct.Polym., 55, 277 (2003).
334. A.A. Khan, Inamuddin and M.M. Alam; React.Funct.Polym., 63, 119 (2005).
335. R. Niwas, A.A. Khan and K.G. Varshney; Coll.Surf. A: Physicochem. Engg.Asp., 150, 7 (1999).
336. R. Niwas, A.A. Khan and K.G. Varshney; Indian J.Chem., 37A, 469 (1998).
337. A.A. Khan, Inamuddin and M. Mezbaul Alam; Mater. Res.Bull., 40, 289 (2005).
338. A.A. Khan and Inamuddin; React. Funct. Polym., 66, 1649 (2006).
339. Inamuddin; M. Phil. Dissertation, Aligarh Muslim University, Aligarh, India (2004) p. 161.
340. A.A. Khan, M.M. Alam, Inamuddin and F. Mohammad; J.Electroanal. Chem., 572, 67 (2004).
341. D. Stauffer; Introduction to Percolation Theory, Taylor and Francis, London (1985).
342. C. Duval; Inorganic Thermogravimetric Analysis, Elsevier, Amsterdam, p. 315 (1963).

Index